Blood Cell Biochemistry

Volume 7

Hematopoietic Cell Growth Factors and Their Receptors

Blood Cell Biochemistry

Series Editor

J. Robin Harris, *Institute of Zoology, University of Mainz, Mainz, Germany*

Blood Cell Biochemistry

Volume 7
Hematopoietic Cell Growth Factors and Their Receptors

Edited by
Anthony D. Whetton
Leukaemia Research Fund Cellular Development Unit
UMIST
Manchester, England

and

John Gordon
The Medical School
University of Birmingham
Birmingham, England

Springer Science+Business Media, LLC

ISSN 1078-0491

ISBN 978-1-4757-7052-0 ISBN 978-0-585-31728-1 (eBook)
DOI 10.1007/978-0-585-31728-1

Originally published by Plenum Press, New York in 1996
Softcover reprint of the hardcover 1st edition 1996

10 9 8 7 6 5 4 3 2 1

Contributors

Arne N. Akbar Department of Clinical Immunology, The Royal Free Hospital, London NW3 2PF, United Kingdom

Jacques Banchereau Schering-Plough, Laboratory for Immunological Research, 69571 Dardilly, France

Robert A. Briddell Department of Developmental Hematology, Amgen Inc., Thousand Oaks, California 91320

Hal E. Broxmeyer Departments of Medicine and Microbiology/Immunology and Walther Oncology Center, Indiana University School of Medicine, Indianapolis, Indiana 46202-5121

Christophe Caux Schering-Plough, Laboratory for Immunological Research, 69571 Dardilly, France

Rachel S. Chapman Molecular Pharmacology Group, School of Biological Sciences, University of Manchester, Manchester M13 9PT, United Kingdom.

David Cosman Department of Molecular Biology, Immunex Research and Development Corporation, Seattle, Washington 98101

Ana Cumano Unité de Biologie Moléculaire du Gène, Institut Pasteur, Paris 75724, France

Rene Devos Roche Research Gent, 9000 Gent, Belgium

Françoise Dieterlen-Lièvre Institut d'Embryologie du CNRS et du Collège de France, Nogent-sur-Marne, France

Caroline Dive Molecular Pharmacology Group, School of Biological Sciences, University of Manchester, Manchester M13 9PT, United Kingdom

Fan Dong Department of Hematology, Dr. Daniel den Hoed Cancer Center and Erasmus University, Rotterdam, The Netherlands

Caroline A. Evans Leukaemia Research Fund Cellular Development Unit, Department of Biochemistry and Applied Molecular Biology, UMIST, Manchester M60 1QD, United Kingdom

Pierre Garrone Schering-Plough, Laboratory for Immunological Research, 69571 Dardilly, France

Isabelle Godin Institut d'Embryologie du CNRS et du Collège de France, Nogent-sur-Marne, France

Christopher D. Gregory Department of Immunology, School of Medicine, University of Birmingham, Birmingham B15 2TT, United Kingdom

Yves Guisez Roche Research Gent, 9000 Gent, Belgium

Margaret M. Harnett Division of Biochemistry and Molecular Biology, University of Glasgow, Glasgow G12 8QQ, United Kingdom

Andrew W. Heath Department of Medical Microbiology, University of Sheffield Medical School, Sheffield S10 2RX, United Kingdom

Barbara L. Kee The Wellesley Hospital Research Institute, Toronto, Ontario M4Y 1J3, Canada

Taisei Kinoshita DNAX Research Institute of Molecular and Cellular Biology, Palo Alto, California 94304; *present address*: Institute of Molecular and Cellular Biosciences, University of Tokyo, Bunkyo-ku, Tokyo 113, Japan.

Johan Kips Department of Respiratory Diseases, University Hospital, Depintelaan, 9000 Gent, Belgium

Yong-Jun Liu Schering-Plough, Laboratory for Immunological Research, 69571 Dardilly, France

Clair Mappin Department of Rheumatology, The University of Birmingham, Birmingham B15 2TT, United Kingdom

Andrew N. J. McKenzie The MRC Laboratory of Molecular Biology, Cambridge CB2 2QH, United Kingdom

Ian K. McNiece Department of Developmental Hematology, Amgen Inc., Thousand Oaks, California 91320

Atsushi Miyajima DNAX Research Institute of Molecular and Cellular Biology, Palo Alto, California 94304; *present address*: Institute of Molecular and Cellular Biosciences, University of Tokyo, Bunkyo-ku, Tokyo 113, Japan.

Anne-Marie O'Farrell DNAX Research Institute of Molecular and Cellular Biology, Palo Alto, California 94304

C. J. Paige The Wellesley Hospital Research Institute, Toronto, Ontario M4Y 1J3, Canada

Renaat Peleman Department of Respiratory Diseases, University Hospital, Depintelaan, 9000 Gent, Belgium

Andrew Pierce Leukaemia Research Fund Cellular Development Unit, Department of Biochemistry and Applied Molecular Biology, UMIST, Manchester M60 1QD, United Kingdom

Darrell Pilling Department of Rheumatology, The University of Birmingham, Birmingham B15 2TT, United Kingdom

Geert Plaetinck Roche Research Gent, 9000 Gent, Belgium

Mike Salmon Department of Rheumatology, The University of Birmingham, Birmingham B15 2TT, United Kingdom

Jan Tavernier Roche Research Gent, 9000 Gent, Belgium

Nydia G. Testa Cancer Research Campaign Departments of Experimental Haematology and Medical Oncology, Christie Hospital NHS Trust, Manchester M20 9BX, United Kingdom

Ivo P. Touw Department of Hematology, Dr. Daniel den Hoed Cancer Center and Erasmus University, Rotterdam, The Netherlands

Jose van der Heyden Roche Research Gent, 9000 Gent, Belgium

Andrew Weaver Cancer Research Campaign Departments of Experimental Haematology and Medical Oncology, Christie Hospital NHS Trust, Manchester M20 9BX, United Kingdom

Preface

Historically, the field of hematopoietic growth factor research began with the work of Carnot and Deflandre—in 1906 they suggested that the rate of erythropoiesis is regulated by a humoral factor found in the blood, namely, erythropoietin. From this comparatively early start, accelerating progress has been made in erythropoietin research, which demonstrates the general trends in this field of study. Erythropoietin was purified to homogeneity by 1977 (from enormous quantities of urine from aplastic anemia patients). Subsequently, the gene for erythropoietin has been cloned (1985), and massive quantities of this growth factor have been produced for clinical trials (late 1980s onward). Erythropoietin has become established as a pharmaceutical product of great value in the treatment of a number of diseases, most notably chronic renal failure. Once the ligand had been cloned, interest turned to the erythropoietin receptor, which was cloned in 1989. Since then, structure/function studies have been performed on receptor mutants, cellular signaling events downstream from the occupied receptor have been identified, and the specific producer cell types and molecular stimuli for erythropoietin production have been thoroughly investigated, as has the regulation of erythropoietin gene transcription. This schedule of events since the 1970s typifies that seen for a number of hematopoietic growth factors. Along the way, the hematopoietic growth factors have been recognized as members of the cytokine family of signaling molecules that are important in a number of different physiological and pathological situations (see below).

Cytokines are a group of proteins that are produced by a variety of different cell types in a number of different organs. They can act in a paracrine or autocrine manner to potentiate survival (or death), proliferation, and development. Generally there has been a realization that these proteins play a pivotal role in the immune and inflammatory responses as well as in embryogenesis, homeostasis, growth, and development. The corollary of this is that the cytokines or their antagonists (which are sometimes naturally occurring) have immense clinical potential. This in part explains the massive growth in cytokine literature over the past few years. The decrease in the reports that can be found under the subject heading "lymphokines" is in part a result of the realization that the lymphopoietic growth and development factors are members of the larger family, the cytokines. As for the massive increase in publications on such cytokines as G-CSF, this is not just a manifestation of the fact that the cloning of the cytokines made the substances available to far more experimenters; it is linked to the realization that there are clinical benefits to be had from the use of cytokines. Success with G-CSF and erythropoietin in the clinic has stimulated a hunt for other cytokines that stimulate myelopoiesis and lymphopoiesis. This led to the realization that many cytokines are pleiotropic, stimulating stem, myeloid progenitor, and lymphoid cell proliferation and development (e.g., interleukin 11 or IL-11, IL-4). Furthermore, other cytokines such as IL-6 or IL-1 act on not only hematopoietic cells but also on other cell types (e.g., hepatocytes). The complexities of the field have also multiplied with the discovery of growth inhibitors, membrane-bound growth factors, soluble receptors in the plasma, and a role for integrins and extracellular matrix in hematopoietic cell development.

In recent years a common approach to research in lymphopoiesis and experimental hematology has developed. There is the realization that the differentiation of myeloid cells can be defined in the same way as has been employed in immunology: immunologists recognized relatively early that the process of differentiation is a complex progression involving the loss and gain of distinct cell surface antigens recognized by antibodies and given cluster designations (CD) for major cell surface determinants. This approach, and the use of flow cytometry to sort cells on the basis of cell surface determinant expression, is now used widely to study stem and myeloid progenitor cells as well as lymphoid cells. Not too surprisingly, some of the CDs are for cell surface receptors that recognize cytokines, and if there is one particular area where great strides forward have been made in the past few years, it is the definition of specific cell surface determinants as important signaling molecules, as reflected by a number of chapters in this book.

From the tone of this introduction the exhaustive coverage of hematopoietic growth factor research running from basic studies to the clinical use of cytokines would add up to a much larger volume than this one. We have concentrated on the current aspects of hematopoietic growth factor research, namely, receptors, signal transduction, and clinical applications. For example, David Cosman reviews aspects of a novel receptor family that is important in lymphopoiesis and programmed cell death. The area of apoptosis or programmed cell death research now looms so large in immunology and cancer research that the topic is accordingly given a whole chapter, written by Rachel Chapman, Chris Gregory, and Caroline Dive. Jan Tavernier discusses the role of IL-5 *in vivo* and describes the rationale behind treating the IL-5 receptor as a drug target for diseases unrelated to hematopoiesis and the role of IL-5 in eosinophilopoiesis and disease. Although Tavernier and colleagues consider the ligand binding site of the IL-5 receptor as a drug target there are undoubtedly pharmaceutical agents which are targeted at intracellular signaling molecules. Many groups are now working on the downstream consequences of hematopoietic growth factor receptor occupation. These are reviewed by Maggie Harnett, by Atsushi Miyajima and co-workers, and also by Caroline Evans and Andrew Pierce. Evans and Pierce, Christophe Caux and co-authors, Jaques Banchereau *et al.* in their chapter, Mike Salmon, and Ana Cumano with her co-authors consider the effects and role of cytokines in the survival, proliferation, and development of appropriate cell types. Finally, the current and future applications of myeloid colony-stimulating factors in the clinic is addressed by Andrew Weaver and Nydia Testa, and Ian McNiece provides a detailed insight into the procedures required to take a cytokine to the clinic, using stem cell factor as an example. The role of cytokines in leukemias is also covered in several chapters in this book, including that by Fan Dong and Ivo Touw.

We have endeavored to bring together a collection of reviews that highlights current research in this exciting area of cytokines, their receptors, and the application of the growing knowledge in these subjects to the clinic.

John Gordon
Tony Whetton

Contents

Chapter 1
The Hematopoietic Cytokine Receptors
Anne-Marie O'Farrell, Taisei Kinoshita, and Atsushi Miyajima

Chapter 2
Interleukin 13 and Related Cytokines
Andrew N. J. McKenzie and Andrew W. Heath

Chapter 3
The Tumor-Necrosis-Factor-Related Superfamily of Ligands and Receptors
David Cosman

Chapter 4
Cellular Signaling Events in B Lymphocytes
Margaret M. Harnett

Chapter 5

Cellular and Molecular Aspects of Myeloid Cell Proliferation and Development

Caroline A. Evans and Andrew Pierce

Chapter 6

Myelosuppressive Cytokines and Peptides

Hal E. Broxmeyer

Chapter 11
***In Vitro* Regulation of Dendritic Cell Development and Function**
Christophe Caux and Jacques Banchereau

Blood Cell Biochemistry

Volume 7
Hematopoietic Cell Growth Factors and Their Receptors

Chapter 1

The Hematopoietic Cytokine Receptors

Anne-Marie O'Farrell, Taisei Kinoshita, and Atsushi Miyajima

1. INTRODUCTION

The growth and differentiation of hematopoietic stem cells to form the vast repertoire of mature blood cells that exists *in vivo* is orchestrated by an array of intercellular signals, mediated by cytokines in association with a complex stromal microenvironment. Cytokines are a diverse group of glycoproteins, expressed constitutively or inducibly by a wide variety of cell types, in membrane-bound or secreted forms (reviewed in Nicola, 1989; Arai *et al.*, 1990; Howard *et al.*, 1993). In addition to controlling hematopoietic development, cytokines mediate many physiological responses, such as immunity, inflammation, and antiviral activity. A single cytokine can exhibit multiple functions depending on its target cell type, and different cytokines often show similar biological functions on the same target cell population (Metcalf, 1986). Combinations of cytokines can interact synergistically (Metcalf and Nicola, 1991; Heyworth *et al.*, 1988, 1992) or antagonistically (reviewed in Graham and Pragnell, 1990; Ruscetti *et al.*, 1991) to give novel responses. In addition, many cytokines trigger the release of other cytokines (Dinarello *et al.*, 1986; Fibbe *et al.*, 1986; Yang *et al.*, 1988). Thus, a complex network is formed among various types of cells through cytokines.

Cytokines bind to specific transmembrane receptor proteins expressed on target cells. Binding of a cytokine to its receptor triggers intracellular signal transduction processes, ultimately leading to altered gene expression and other cellular changes (Ihle *et al.*, 1994a; Kan *et al.*, 1992). Molecular cloning of cytokine receptor (CR) genes in the last decade has revealed that CRs can be grouped into several novel receptor families. Members of one

Anne-Marie O'Farrell, Taisei Kinoshita, and Atsushi Miyajima DNAX Research Institute of Molecular and Cellular Biology, Palo Alto, California 94304; *present address for T. K. and A. M.*: Institute of Molecular and Cellular Biosciences, University of Tokyo, Bunkyo-ku, Tokyo 113, Japan.

Blood Cell Biochemistry, Volume 7: Hematopoietic Cell Growth Factors and Their Receptors, edited by A. D. Whetton and J. Gordon. Plenum Press, New York, 1996.

family, the CR superfamily, are defined by extracellular domain sequence and structural homology (Cosman, 1993; Miyajima *et al.*, 1992). These new receptor families are distinct from classical growth factor receptors with intrinsic tyrosine kinase activity or hormonal receptors with seven transmembrane domains. One important conclusion from the cloning of many CR genes is that the multiple functions of a given cytokine are mediated, in most cases, by the same receptor and therefore must be explained by differential activation of distinct intracellular pathways. Interestingly, many members of the CR superfamilies are heteromeric. In this chapter we describe the heteromeric structure of hematopoietic cytokine receptors, which provides a basis for overlapping biological functions among different cytokines.

2. CHARACTERISTICS OF THE CYTOKINE RECEPTOR SUPERFAMILY

2.1. Classification of Cytokine Receptors

The majority of receptors for hematopoietic cytokines, including many interleukins (IL) and colony-stimulating factors (CSF), belong to the CR superfamily and are more specifically referred to as class I CRs. A second group within the CR superfamily, designated as class II receptors, are structurally related to class I receptors and include the receptors for interferon (IFN) and interleukin-10 (IL-10).

All receptors of the CR superfamily exhibit a common tertiary structure consisting of approximately 200 amino acid residues arranged to form 14 antiparallel β strands (Bazan, 1989, 1990) (Figure 1). These β strands form two barrel-like structures of about 90 amino acids each, which are characteristic of adhesive proteins, with significant evolutionary resemblance to the fibronectin type III domain (Bazan, 1990; Patthy, 1990). Class I receptors have two pairs of periodically spaced cysteine residues with one neighboring tryptophan residue, predicted to form the core structure of the N-terminal half of the extracellular region. They also contain a highly conserved Trp-Ser-X-Trp-Ser stretch (WSxWS box, where x is any amino acid) at the C-terminal end (Figure 1) (D'Andrea *et al.*, 1989a; Idzerda *et al.*, 1990; Miyajima *et al.*, 1992). The WSxWS box is predicted to lie in the base of the ligand-binding crevice (Bazan, 1990), although the precise role of this motif is unclear (DeVos *et al.*, 1992; Yoshimura *et al.*, 1992). In class II receptors, the WSxWS box is replaced by a more degenerate sequence, and one of the two pairs of conserved cysteines is now found in the C-terminal half of the extracellular region (Bazan, 1990; Ho *et al.*, 1993). The class I and II receptors are also functionally related by the use of similar signaling pathways, as described below (see Section 6.2).

A number of CRs belong to other receptor superfamilies: members of the tumor necrosis factor receptor (TNFR) family exhibit characteristic extracellular domain cysteine-rich motifs but do not possess the domain structure common to class I and II receptors (Bazan, 1993). The receptors for a subset of hematopoietic cytokines such as M-CSF, SCF and *flt3/flk2* belong to the classical tyrosine kinase (TK) growth factor receptor family (Ullrich and Schlessinger, 1990). The receptors for TGF-β and its relatives form a unique receptor family that possess serine/threonine kinase domains (Massague *et al.*, 1994). Chemotactic cytokines such as IL-8 (Holmes *et al.*, 1991), also known as chemokines, have receptors with seven transmembrane regions. The interleukin-1 receptor belongs to the immunoglobulin receptor superfamily (Sims *et al.*, 1988).

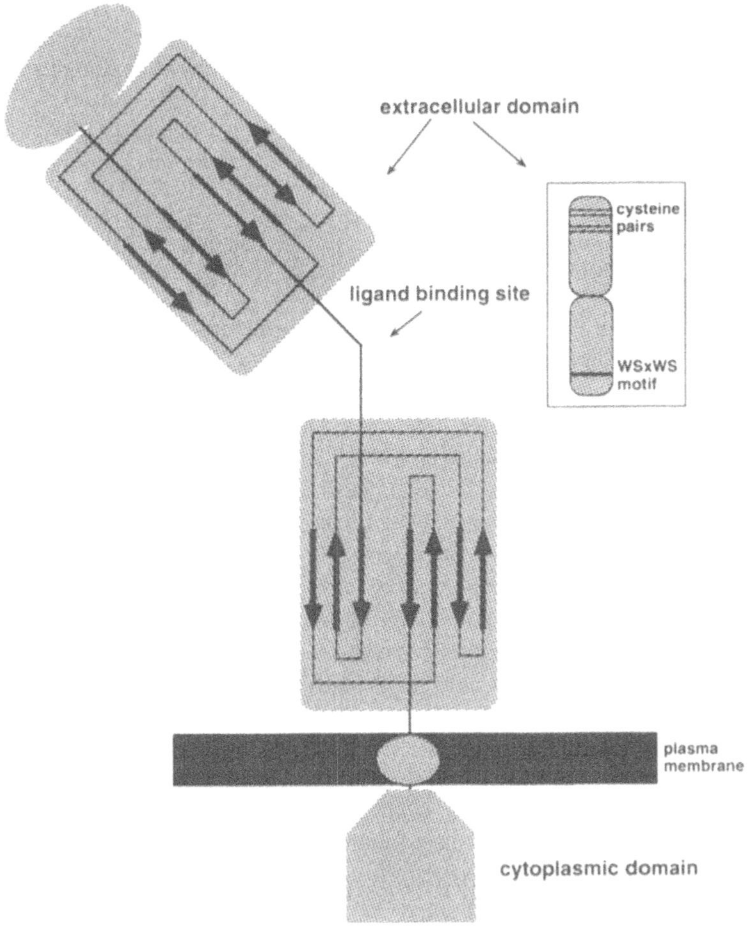

FIGURE 1. Domain structure of CRs. The structure of IL-3Rα is depicted as a representative example. Black arrows denote β strands, which form two barrel-like structures analogous to fibronectin (FN) type III modules. The figure insert indicates the extracellular domain motifs, two conserved cysteine pairs and a WSxWS sequence.

2.2. Multimeric Structure

A fundamental characteristic that pertains to many members of the CR superfamily is that functional high-affinity receptors are composed of multiple subunits. These heteromeric receptors are typically composed of a unique subunit, responsible for ligand binding in the receptor complex, and a common subunit, which is involved in signal transduction (reviewed in Cosman, 1993). This multimeric nature in part explains ambiguities that arose during initial characterization and cross-linking studies of cytokine receptors, i.e., ligand cross-linking of multiple proteins, the existence of both low- and high-affinity ligand-binding sites, and cross competition for receptor binding among a subset of cytokines. More importantly, the sharing of receptor subunits that mediate signal transduction provides a molecular basis for the functional redundancy of cytokines. The existence of a common receptor component was first demonstrated for the IL-3, GM-CSF, and IL-5 receptor complexes (Miyajima, 1992). Subsequently, two further molecules were identified as shared receptor subunits, glycoprotein 130 (gp130) and the interleukin 2 receptor γ (IL-2Rγ) subunit (reviewed in Kitamura *et al.*, 1994).

In this review, members of the class I CR superfamily have been grouped according to their common receptor subunits and the conservation of additional functional domains in ligand-specific subunits.

2.3. Soluble Receptors

Many cytokine receptors exist not only as transmembrane cell surface proteins but also as soluble isoforms. Soluble receptors are generated either by alternative mRNA splicing or by protease cleavage of membrane-bound receptors (Heaney and Golde, 1994; J. S. Rose and Heinrich, 1994). Soluble receptors specifically bind ligand with an affinity equivalent to that of their membrane-bound counterparts and have been detected in urine, sera, and cell culture supernatants. Furthermore, in certain pathological conditions, increased concentrations of soluble receptors have been detected, such as the soluble interleukin-2 receptor α (IL-2Rα) subunit in serum of ovarian cancer patients (Barton *et al.*, 1993).

What are the functions of these molecules? Several possibilities exist. Soluble receptors may act as receptor antagonists. For example, the soluble interleukin 4 receptor (sIL-4R) sequesters ligand and competes with membrane-bound receptors (Maliszewski *et al.*, 1990). In fact, administration of sIL-4R inhibits IL-4 function *in vivo* (Fanslow *et al.*, 1991). Alternatively, soluble receptors may act as receptor agonists, as in the case of soluble growth hormone receptor (sGHR), which binds GH, preventing its degradation, and releases it to bind the transmembrane GHR (Herington *et al.*, 1986a,b; Leung *et al.*, 1987). A third type of soluble receptor can transduce signals in association with heterologous transmembrane receptor subunits and thereby confers cytokine responsiveness to target cells, for example, the soluble interleukin 6 receptor α (sIL-6Rα) (Taga *et al.*, 1989).

Interestingly, many different viruses carry genes that encode soluble forms of CRs or receptor-like molecules. For example, the shope fibroma virus genome contains a gene that encodes a soluble form of tumor necrosis factor receptor (TNFR) (Howard *et al.*, 1990; Smith *et al.*, 1990).

3. CLASS I CYTOKINE RECEPTORS

3.1. The βc Family

3.1.1. βc; A Common Receptor β Subunit

Interleukin 3, also known as multicolony-stimulating factor (multi-CSF), has a broad spectrum of activity on hematopoietic cells. IL-3 acts on multipotential progenitors, is unique in its ability to stimulate proliferation of all committed myeloid progenitors (Metcalf, 1986; Schrader, 1986), and also enhances functional activity of a range of mature hematopoietic cells (Lopez *et al.*, 1987; Cannistra *et al.*, 1988). A second cytokine, GM-CSF, originally defined as a colony-stimulating factor for granulocytes and macrophages, is also a multilineage growth regulator and exhibits a subset of IL-3 activities (Metcalf, 1985;

Koike *et al.*, 1987). In contrast, a third cytokine, IL-5, has a narrower range of target cells, primarily eosinophilic progenitors, eosinophils, and basophils, but exerts similar effects to IL-3 and GM-CSF on these populations (Sanderson, 1992; Goodall *et al.*, 1993). Yet the intracellular signals induced by these cytokines in common target cells are virtually indistinguishable (Kanakura *et al.*, 1990: Murata *et al.*, 1990). In human hematopoietic cells, IL-3, GM-CSF, and IL-5 cross-compete for receptor binding (Park *et al.*, 1989; Lopez *et al.*, 1990). The cloning and characterization of the receptors for IL-3, GM-CSF, and IL-5 have provided an explanation for the observed functional redundancy and cross-competition for receptor binding among these cytokines.

The high-affinity receptors for murine and human IL-3, GM-CSF, and IL-5 are each composed of two subunits, α and β (Miyajima *et al.*, 1993). The α subunits are ligand specific and bind their ligands with low affinity (see Section 3.1.2). The β subunit (βc), originally named AIC2B, is a class I CR with a long cytoplasmic domain common to the IL-3, GM-CSF, and IL-5 receptors (Figure 2). This βc subunit by itself does not detectably bind any cytokine but is required to form high-affinity receptors for IL-3, GM-CSF, or IL-5 (Hayashida *et al.*, 1990; Kitamura *et al.*, 1991a,b; Takaki *et al.*, 1991; Tavernier *et al.*, 1991) (Figure 3). The requirement of βc for high-affinity binding to each cytokine explains the cross-competition observed in human hematopoietic cells among IL-3, GM-CSF, and IL-5. Furthermore, βc plays an essential role in signal transduction (Sakamaki *et al.*, 1992). The use of a common signal-transducing subunit provides a basis for the common activities elicited by IL-3, IL-5, and GM-CSF.

Interestingly, the mouse receptors are more complex because of the presence of an

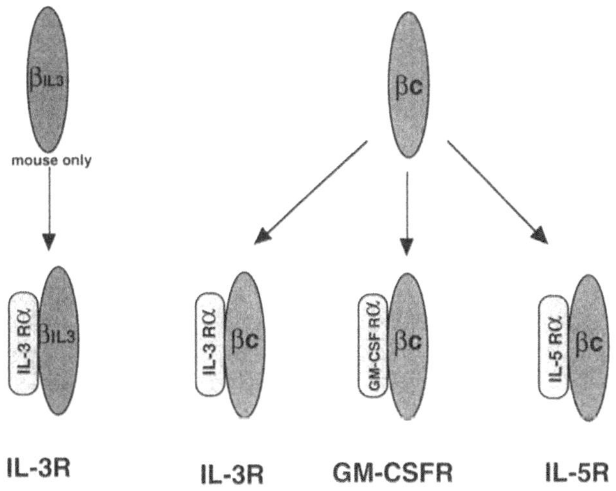

FIGURE 2. βc, a common subunit for IL-3, GM-CSF, and IL-5 receptors. Functional receptors for murine and human IL-3, GM-CSF, and IL-5 are heterodimers consisting of a ligand-specific α subunit that binds ligand with affinity and a common β subunit, βc, that does not by itself bind ligand. In addition, unique to IL-3 and in mice, a further β subunit, designated here as βIL3 exists. Ligand binding to the Rα subunits is thought to trigger heterodimerization of α and β subunits to form a high-affinity receptor complex.

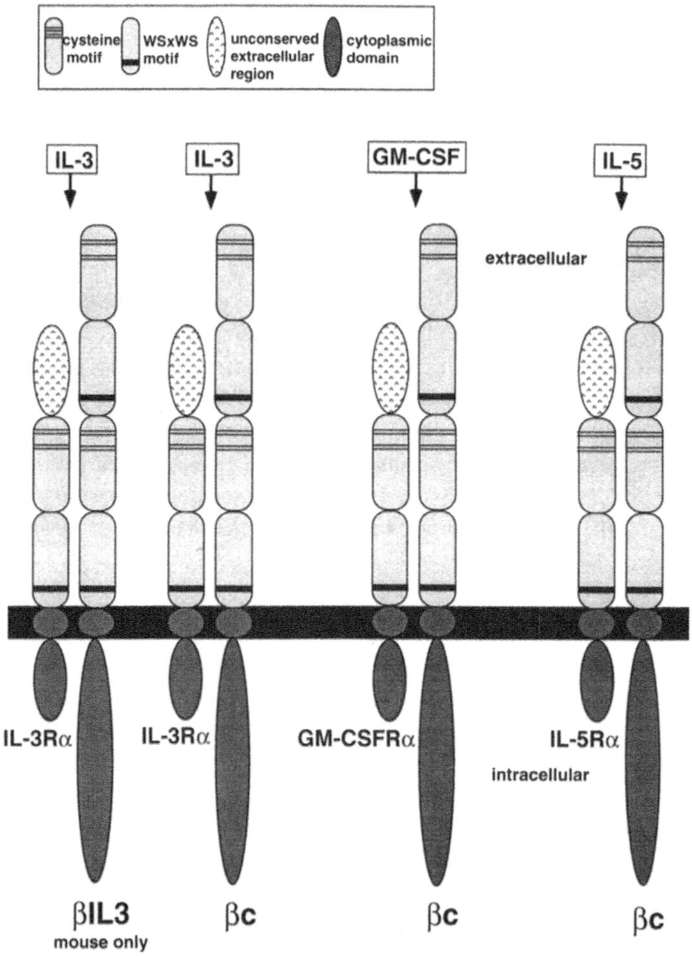

FIGURE 3. Schematic structures of βc family of class I CRs. The extracellular domains of the α and β receptor subunits of the IL-3, GM-CSF, and IL-5 receptors contain one and two CR domains, respectively, each of about 200 amino acids, containing the four conserved cysteine residues, and a WSxWS box at N- and C-terminal ends, respectively. βc and βIL-3 contain large cytoplasmic domains, each of about 430 amino acids.

additional β subunit (βIL3, originally termed AIC2A) specific to the IL-3R (Itoh *et al.*, 1990) (Figures 2 and 3). The βIL3 and βc subunits are 91% identical (Gorman *et al.*, 1992). The genes for βc and βIL3 are both localized on chromosome 15 and exhibit identical exon–intron structures, suggesting that these genes were generated relatively recently by gene duplication. Interestingly, whereas βc alone does not bind any cytokine, βIL3 binds IL-3 with low affinity and forms a high-affinity IL-3 receptor only with the IL-3Rα subunit. The two β subunits are coexpressed in various hematopoietic cells, and high-affinity murine receptors formed with either βc or βIL3 bind IL-3 with almost the same affinity and transduce similar signals (Hara and Miyajima, 1992). Hence, the physiological role of the βIL3 subunit is unclear.

To address the roles of βc and βIL3, mutant mice lacking either one of these β subunits were generated by gene targeting. As predicted, bone marrow cells from mice lacking βc did not respond to either GM-CSF or IL-5 but responded normally to IL-3 in *in vitro* colony-forming assays. Because of the lack of GM-CSF responsiveness, these mice developed pulmonary proteinosis (Nishinakamura *et al.*, 1995) similar to GM-CSF-deficient mice (Stanley *et al.*, 1994). Eosinophil number was significantly lowered in βc-deficient mice, which may be attributed to lack of IL-5 responsiveness. In contrast, no obvious defect was observed in the βIL3-deficient mice. Mice lacking both β subunits are required to evaluate the role of IL-3, IL-5, and GM-CSF in hematopoiesis.

3.1.2. The α Subunits for IL-3, GM-CSF, and IL-5 Receptors

The α subunits of the IL-3, GM-CSF, and IL-5 receptors are unique, and each binds its cognate ligand with low affinity (*Kd* values of 100 nM, 5 nM, and 30 nM, respectively). The IL-3Rα (Hara and Miyajima, 1992), GM-CSFRα (Gearing *et al.*, 1989; Park *et al.*, 1992), and IL-5Rα (Mita *et al.*, 1989; Takaki *et al.*, 1991) share a common structural organization, with an extracellular N-terminal region of approximately 100 amino acids, a CR family domain, and a cytoplasmic domain of approximately 50 amino acids, in both human and mouse cells (Miyajima *et al.*, 1991) (Figure 3). These receptor components are essential for ligand binding in the high-affinity (100 pM, 50 pM, 150 pM, for IL-3, GM-CSF, and IL-5 receptors, respectively) receptor complexes. The α subunits are also necessary for signal transduction, as truncation of their cytoplasmic domains abrogates signal transduction (Takaki *et al.*, 1994; T. Kitamura and T. Kinoshita, unpublished results). Although the precise role of α-subunit cytoplasmic domains in signal transduction has not yet been defined, two conserved cytoplasmic sequences, box 1 and box 2, have been identified in the IL-3, GM-CSF, and IL-5Rα subunits (reviewed in Ihle *et al.*, 1994a; Kitamura *et al.*, 1991b; Takaki *et al.*, 1994) (see Section 6.3).

Although IL-3, IL-5, and GM-CSF exhibit common biological functions, they also exhibit clearly unique functions on different target cell populations (Arai *et al.*, 1990; Heyworth *et al.*, 1991). Such functional differences may be mediated by unique signaling functions of the IL-3Rα, GM-CSFRα, and IL-5Rα. Alternatively, specific functions may simply correlate with different patterns of expression of each cytokine Rα subunit. For example, IL-3Rα is expressed by a wide range of hematopoietic lineages and some endothelial cells (Colotta *et al.*, 1991; Hara and Miyajima, 1992; Korpelainen *et al.*, 1993; Sato *et al.*, 1993a), whereas expression of the IL-5Rα appears to be restricted to eosinophils, basophils, and some mouse B cells. Because of limitations of sensitivity, expression studies using antibodies or ligand binding have not conclusively resolved this issue. To address this question, Takagi *et al.* (1995) generated transgenic mice that constitutively express the IL-5Rα subunit in various hematopoietic lineages. Interestingly, IL-5 exhibits multi-CSF activity on bone marrow cells from IL-5Rα transgenic mice indistinguishable from that of IL-3. This experiment clearly indicates that the normally restricted activity of IL-5 results from restricted expression of the IL-5Rα subunit. Once the IL-5Rα is expressed in a wide range of hematopoietic lineages, it has a potential similar to that normally observed for the IL-3Rα.

A naturally occuring mutation in the IL-3Rα gene has been identified in several mouse strains, including AJ and AKR. The splicing of IL-3Rα mRNA is defective in these mice as a result of a five-base-pair deletion at a splice branch point within intron 7 (Ichihara *et al.*, 1995). It had previously been shown that bone marrow cells from these mice do not form colonies in response to IL-3 (Hapel *et al.*, 1992; Breen *et al.*, 1993), and it is now apparent that this IL-3 nonresponsiveness correlates with impaired expression of IL-3Rα (Ichihara *et al.*, 1995). These IL-3 nonresponder mice show apparently normal hematopoiesis, raising questions about the absolute requirement for IL-3 during hematopoiesis.

3.1.3. Cytokine and Receptor Interactions

Although no significant homology is apparent among the primary structures of IL-3, IL-5, and GM-CSF, these cytokines have similar four-α-helix tertiary structures (Sprang and Bazan, 1993). Site-directed mutagenesis has shown that the amino termini of IL-3, IL-5, and GM-CSF are crucial for receptor binding (Shanafelt *et al.*, 1991). This N-terminal region encodes the first α-helix of the conserved helical core structure, and within this α-helix, residue 21 (glutamic acid) is critical for high-affinity binding of hGM-CSF to its receptor (Lopez *et al.*, 1992; Shanafelt and Kastelein, 1992). In IL-3 and IL-5, this Glu residue is conserved (Glu 22 and Glu 13, respectively), and its substitution impairs high-affinity binding of IL-3 or IL-5. Therefore, this common Glu residue in the first helix may interact with the βc subunit. Recently, a model has been proposed whereby Glu 21 of GM-CSF interacts with a short region of βc: Tyr 365, His 367, and Ile 368 (Woodcock *et al.*, 1994). These amino acids in βc are essential for GM-CSF or IL-5 binding but not for IL-3 binding. Therefore, additional regions of βc are implicated in IL-3 binding, which may account for the larger βc-mediated affinity conversion observed for the IL-3Rα (500- to 1000-fold) than for the GM-CSFRα or IL-5Rα.

3.2. The gp130 Family of Receptors

3.2.1. The gp130 Protein, A Common Receptor Subunit

Interleukin 6 (IL-6), interleukin 11 (IL-11), leukemia inhibitory factor (LIF), oncostatin M (OSM), ciliary neurotrophic factor (CNTF) and cardiotropin 1 (CT-1) are structurally and functionally related cytokines. Although each of these cytokines has unique biological activities, groups of these cytokines exhibit similar functions on common target cells such as hematopoietic cells, osteoclasts and hepatocytes (Boulton *et al.*, 1991; T. M. Rose and Bruce, 1992). For example, IL-6, LIF, OSM, and CNTF promote macrophage development of the mouse M1 cell line (Bruce *et al.*, 1994).

This functional overlap is now explained by the use of a common receptor subunit, gp130. gp130 is a glycoprotein of molecular mass 130 kDa, first identified as the high-affinity converter or signal-transducing component of the human IL-6R (Hibi *et al.*, 1990). Subsequently it has become apparent that gp130 is an essential component of many functional cytokine receptors, including the receptors for IL-11, LIF, OSM, CNTF, and

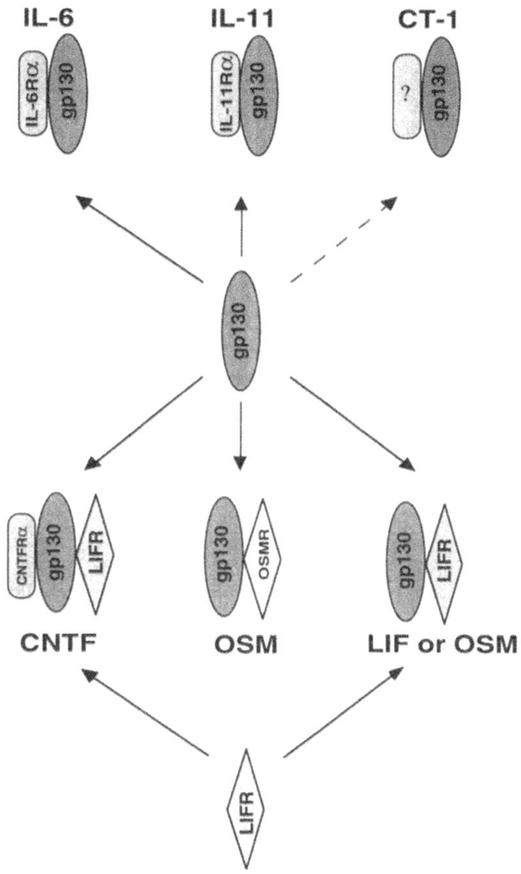

FIGURE 4. gp130, a common subunit for IL-6, IL-11, LIF, OSM, and CNTF receptors. Functional receptors for IL-6 and IL-11 consist of specific ligand-binding α subunits and the common signaling subunit gp130. The combination of gp130 and LIFR constitutes functional receptors for both LIF and OSM. OSM also has a type II receptor, consisting of gp130 and a unique OSMR subunit. The CNTFR consists of a specific α subunit, LIFR, and gp130. gp130 binds OSM with low affinity, and LIFR binds LIF with low affinity. The cloned IL-12 receptor chain (not illustrated) contains a gp130-like domain and exists in dimeric form.

possibly CT-1 (Figure 4). The role of gp130 in receptor complexes is twofold: to mediate high-affinity cytokine binding and as a signal transducer. Intriguingly, gp130 also exhibits low-affinity binding properties for OSM (see Section 3.2.3b). gp130 is 918 amino acids long, with a single transmembrane domain. The extracellular region of gp130 contains a CR family domain, three additional units of a fibronectin type III domain (Patthy, 1990), and a region that is homologous to the chicken protein contactin (Taga *et al.*, 1989; Hibi *et al.*, 1990) (Figure 5). A soluble form of gp130, that antagonizes the function of membrane-bound gp130 has been identified (Davis *et al.*, 1993; Narazaki *et al.*, 1994a).

3.2.2. The IL-6 and IL-11 Receptors

3.2.2a. The IL-6 Receptor. Interleukin 6 has multiple functions. For example, it is a B-cell differentiation factor, a hepatocyte-stimulating factor, a growth factor for human myeloma and mouse plasmacytoma cells, a neural growth factor, and has a number of roles in myelopoiesis (reviewed in Kishimoto *et al.*, 1992b). The IL-6R consists of two polypeptide chains, an 80-kDa IL-6Rα and gp130 (Figure 5). The IL-6Rα binds IL-6 with low

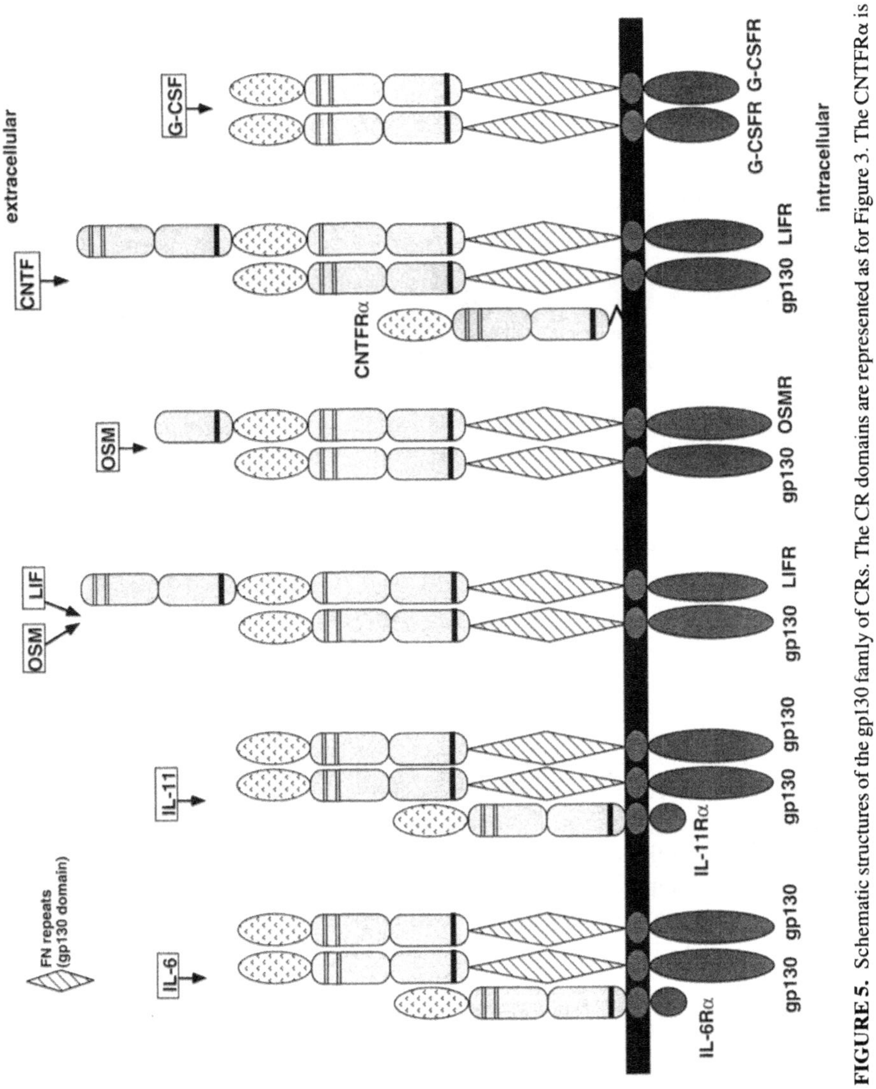

FIGURE 5. Schematic structures of the gp130 famly of CRs. The CR domains are represented as for Figure 3. The CNTFRα is linked to the membrane by a novel glycosyl-phosphatidyl linkage. Binding of IL-6 to IL-6Rα triggers its association with a homodimer of gp130 (a similar mechanism may be employed by the IL-11R). Binding of LIF, OSM, or CNTF to LIFR, gp130, or CNTFRα, respectively, induces heterodimerization of gp130 with LIFR.

affinity, and its cytoplasmic domain is dispensable for signaling (Kishimoto *et al.*, 1992a). In contrast, gp130 does not bind IL-6 but is required for high-affinity IL-6 binding and mediates signal transduction (Hibi *et al.*, 1990). The IL-6Rα has two distinctive extracellular domains (Yamasaki *et al.*, 1988). The first 100 amino acids belong to the immunoglobulin superfamily, and the second domain (approximately 250 amino acids) contains the class I CR family motifs (Figure 5). The IL-6Rα is expressed at high levels during the pathogenesis of several diseases, including multiple myeloma (Suematsu *et al.*, 1990), and an abnormal IL-6Rα is expressed on plasmacytoma cells (Sugita *et al.*, 1993).

The model for transmission of IL-6 signals is that IL-6 binding to IL-6Rα causes interaction with gp130 and triggers homodimerization of gp130 (Murakami *et al.*, 1990, 1993). Subsequently, gp130 is rapidly tyrosine phosphorylated, which ultimately leads to gene activation (Lord *et al.*, 1991; Nakajima and Wall, 1993). Activation of the IL-6R complex can be induced even when the IL-6Rα subunit is in soluble form (Taga *et al.*, 1989; Tamura *et al.*, 1991).

3.2.2b. The IL-11 Receptor. Interleukin 11 plays an important role in early hematopoiesis (Musashi *et al.*, 1991a,b) and exhibits similar biological activities to IL-6 on myeloid cells and mature B lymphocytes (Anderson *et al.*, 1992). Interleukin 11 also plays a role in erythropoiesis (Quesniaux *et al.*, 1991). An IL-11 receptor chain, IL-11Rα, that binds IL-11 with weak affinity (10 nM) has recently been isolated (Hilton *et al.*, 1994). This receptor shows highest homology to the IL-6Rα and CNTFRα (24% and 22% amino acid identity, respectively). Previous characterization of IL-11R suggested that gp130 is a component of the IL-11R because neutralizing antibodies to gp130 inhibited IL-11-dependent proliferation of the human TF-1 cell line (Fourcin *et al.*, 1994; Yin *et al.*, 1992). Consistent with this model, gp130 in combination with IL-11Rα constitutes high-affinity binding sites for IL-11, and mediates functional activities of IL-11 (Figure 5). Hence, the mechanism of IL-11R activation is likely to be similar to that of the IL-6R.

3.2.3. The LIF, OSM, CNTF, and CT-1 Receptors

3.2.3a. The LIF Receptor. Like IL-6, LIF has pleiotropic effects (Hilton and Gough, 1991), and these cytokines exhibit overlapping activities, for example, in the acute-phase hepatic response (Baumann *et al.*, 1992; Murray *et al.*, 1993). A 190-kDa LIF binding protein was first identified by Gearing *et al.* (1991) in human cells. This molecule, designated here as LIFR, is similar to βc or βIL3 in that it possesses two CR domains. The LIFR has a long cytoplasmic domain with closest homology to gp130 (Figure 5). However, LIFR binds LIF with only low affinity, insufficient to mediate the biological effects of LIF. Subsequently, Gearing and his colleagues isolated a molecule that reconstituted high-affinity LIF receptors when expressed in combination with LIFR. Interestingly, this molecule was identical to gp130 (Gearing *et al.*, 1992). The current model for receptor complex formation is that LIF binding triggers heterodimerization of LIFR and gp130 to form a high-affinity LIFR (Baumann *et al.*, 1994a). In the murine system, a potentially soluble LIFR species that is 70% homologous to the human receptor has been isolated (Gearing *et al.*, 1991).

3.2.3b. The OSM Receptor. OSM is a cytokine that is structurally and functionally related to LIF (T. M. Rose and Bruce, 1992). gp130 and LIFR together constitute a high-affinity receptor for OSM. Interestingly, the roles of gp130 and LIFR are reversed from those in the high-affinity LIFR, as gp130 shows weak specific binding for OSM, and LIFR acts as its affinity converter (Gearing *et al.*, 1992). Although OSM and LIF employ two common receptor subunits, cell lines exist that respond to OSM but not to LIF, raising the possibility that an additional OSM receptor exists (Gearing *et al.*, 1994). This issue has been clarified by the recent cloning of a novel receptor subunit, designated here as OSMR, by Mosely *et al.* (1994). The OSMR, when expressed with gp130, confers OSM responsiveness to transfected cells, and this receptor complex has been described as the OSM type II receptor (Mosley *et al.*, 1994) (Figure 5). The OSMR is highly homologous to the gp130 family of cytokine receptors, with 30% identity to LIFR, although it lacks the N-terminal region of the CR domain. It will be interesting to compare the roles of the type I and II OSM receptor complexes.

3.2.3c. The CNTF Receptor. CNTF is a neural growth factor with considerable overlapping biological activity with LIF (Hall and Rao, 1992; Ip *et al.*, 1992; Schooltink *et al.*, 1992). CNTF binds to a specific CNTFRα subunit, which shows considerable homology to the IL-6Rα (Davis *et al.*, 1991). The CNTFRα is unique in that it is covalently anchored to the cell membrane by a glycosylphosphatidylinositol linkage (which can be cleaved by a phospholipase C) and does not possess a cytoplasmic domain. Interestingly, it has been established that signaling pathways activated by CNTF involve signal transduction through gp130 and LIFR and are virtually indistinguishable from those initiated by LIF (Ip *et al.*, 1992; Stahl *et al.*, 1993). Thus, in the CNTFR system, CNTF first binds CNTFRα; then this ligand–receptor complex induces heterodimerization of gp130 and LIFR (Davis *et al.*, 1993).

3.2.3d. The CT-1 Receptor. Pennica and colleagues have recently identified a novel cytokine, cardiotropin 1 (CT-1), implicated in regulation of heart hypertrophy (Pennica *et al.*, 1994). Interestingly, CT-1 shows significant homology to LIF and CNTF, and, moreover, exhibits overlapping biological effects with these cytokines. For example, like CNTF, CT-1 promotes neuron survival. Both LIF and CT-1 stimulate the development of adrenergic neurons into cholinergic neurons and promote macrophage development of the M1 cell line. Intriguingly, CT-1 competes with LIF for binding to M1 cells, implying that these cytokines share a receptor component, possibly LIFR or gp130, or both (Pennica *et al.*, 1994). Moreover, the gp130 knockout (KO) mouse exhibits dramatic cardiac problems (T. Kishimoto, personal communication). It will thus be interesting to elucidate the subunit composition of the CT-1R and the role of cytokines that signal via gp130 in cardiac hypertrophy and heart development.

3.2.4. The IL-12 Receptor

Interleukin 12, originally known as natural killer cell-stimulating factor, is a disulfide-linked heterodimeric cytokine that stimulates NK cells and early T cells (Kobayashi *et al.*, 1989; Gately *et al.*, 1991). Interestingly, one subunit of IL-12, p35, is structurally related to IL-6, and the other subunit, p40, is structurally related to IL-6Rα, and appears analogous to a soluble receptor (Gearing and Cosman, 1991; Merberg *et al.*, 1990). A 100-kDa component of the IL-12R has recently been cloned, and its structure is similar to that of gp130 (Chua *et al.*, 1994) (Figure 5). This IL-12R subunit exists in dimeric/oligomeric form even in the absence of ligand and creates a low-affinity (3 nM) IL-12 binding site but is not functional. Thus, the complex of heteromeric IL-12 and IL-12R is reminiscent of a complex of IL-6, sIL-6Rα, and gp130. It should be noted, however, that additional receptor components must be required to constitute a functional high-affinity IL-12R.

3.2.5. The Granulocyte Colony-Stimulating Factor Receptor

Granulocyte CSF (G-CSF) was originally identified as a factor that promotes granulocytic development of hematopoietic progenitors (Nicola *et al.*, 1979, 1983; Ikebuchi *et al.*, 1988a) and also is a potent activator of mature blood cell functional activity (Vadas *et al.*, 1983; Weiser *et al.*, 1987; Molineux *et al.*, 1990). The structure of G-CSF is similar to that of IL-6, and these cytokines exhibit some similar activities; for example, both act synergistically with SCF on early hematopoietic progenitors (Ikebuchi *et al.*, 1988b; Leary *et al.*, 1992). The G-CSFR chain has a molecular mass of approximately 130 kDa (Fukunaga *et al.*, 1990a,b; Nagata and Fukunaga, 1991), and its extracellular domain exhibits three distinct regions of homology to other surface proteins. The first of these is a CR domain (200 amino acids), and the second, an N-terminal Ig-like domain (90 amino acids). Third, and perhaps most significant, the G-CSFR has a 300-amino-acid region composed of 3 Fibronectin III-like modules in tandem repeat with 46% amino acid homology to gp130 (Larsen *et al.*, 1990; Nagata and Fukunaga, 1991) (Figure 5). The cloned G-CSFR binds G-CSF with low affinity, and receptor homodimerization is necessary for high-affinity binding to occur (Fukunaga *et al.*, 1991). Thus, although the G-CSFR does not require gp130 to constitute functional receptors, the basic mechanisms of receptor activation parallels that of the IL-6R. In fact, chimeric receptors composed of the extracellular domain of the G-CSFR and the cytoplasmic domain of either gp130 or LIFR function similarly to the native G-CSFR (Baumann *et al.*, 1994b). In summary, it appears that a common mechanism is employed by IL-6-related cytokines to activate their cognate receptors and initiate signaling processes, whereby ligand binding leads to receptor dimerization, either homodimerization of gp130 (or G-CSFR), or heterodimerization of gp130 and LIFR.

3.3. The γ_c Family of Receptors

3.3.1. The IL-2 Receptor

Interleukin 2 plays a central role in the activation and maintenance of T-cell, B-cell, and NK cell responses (Howard *et al.*, 1993). The functional high-affinity IL-2R is composed of at least three receptor chains; IL-2Rα, IL-2Rβ, and IL-2Rγ (Figure 6). The IL-2Rβ and IL-2Rγ subunits are members of the class I CR family, but the IL-2Rα subunit is not (Leonard, 1992).

The IL-2Rα was the first component of the IL-2R complex to be isolated, by purification with the monoclonal anti-Tac antibody (Leonard *et al.*, 1984, 1985). The IL-2Rα binds IL-2 with low affinity (10 nM), insufficient for receptor internalization or signal transduction, which implicated additional lymphoid surface proteins in the IL 2R complex (Hatakeyama *et al.*, 1985; Kondo *et al.*, 1986; Robb and Rusk, 1986). Hatakeyama *et al.* (1989a,b) subsequently isolated the second IL-2R chain, IL-2Rβ, which has a large cytoplasmic domain and binds IL-2 with very low affinity (>100 nM). The IL-2Rβ reconstitutes high-affinity (100 pM) binding for IL-2 in combination with the α chain and is essential for receptor internalization and signal transduction in lymphoid cells (Hatakeyama *et al.*,

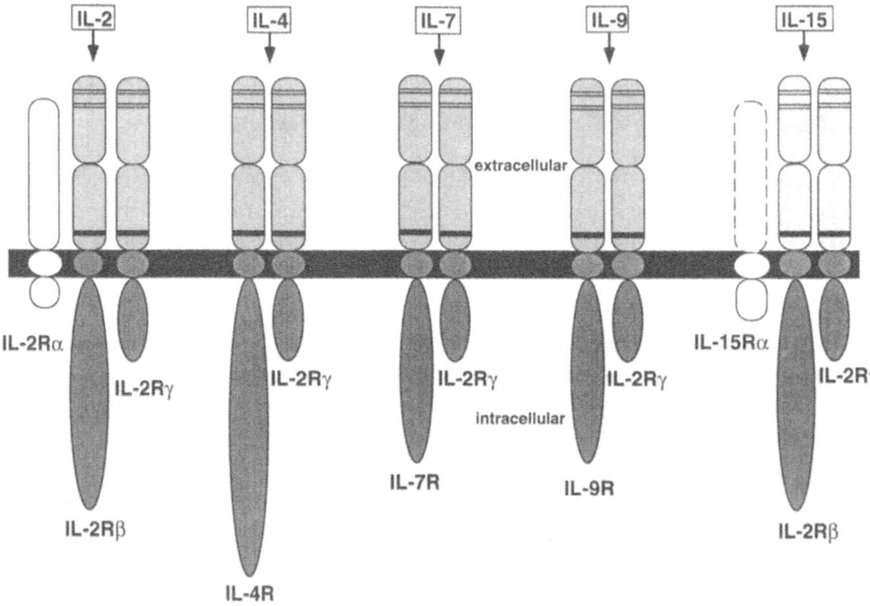

FIGURE 6. Schematic structures of the γc family of class I CRs. The CR domains are represented as for Figure 3. For the IL-4, IL-7, and IL-9 receptors, ligand binding to the IL-4R, IL-7R, or IL-9R, respectively, triggers heterodimerization with γc. For the IL-2 and IL-15 receptors, heterodimerization of the IL-2Rβ and γc subunits is likely to be important for signal transduction.

1989a,b). However, intermediate-affinity (10 nM) IL-2 binding sites had been observed, unaccounted for by these receptor components (Hatakeyama *et al.*, 1989b; Takeshita *et al.*, 1992a). This consideration, together with the observation that the combination of IL-2Rα and IL-2Rβ subunits was unable to transduce IL-2-signaling events in fibroblasts led to speculation that the functional IL-2 receptor contained an additional chain (tentatively referred to as the γ subunit). Finally, in 1993 the IL-2Rγ was purified, and its cDNA cloned, to reveal a 64-kDa protein expressed in various cell types including lymphoid cells (Takeshita *et al.*, 1992b). Characterization of IL-2Rγ resolved the above discrepancies and demonstrated that high-affinity (100 pM) binding sites for IL-2 are actually comprised of three subunits, α, β, and γ, whereas intermediate-affinity binding sites are composed of β and γ heterodimers (Takeshita *et al.*, 1992; Voss *et al.*, 1992, 1993). A heterodimer of β and γ subunits mediates IL-2 signal transduction (Nakamura *et al.*, 1994; Taniguchi and Minami, 1993), while the α subunit enhances ligand binding affinity.

X-linked severe combined immunodeficiency (XSCID) syndrome is caused by inactivating point mutations in the gene encoding IL-2Rγ (Leonard *et al.*, 1991; Noguchi *et al.*, 1993a). Mutations identified to date give rise to a premature stop codon in the IL-2Rγ coding region, resulting in the truncation of its C-terminal portion. XSCID patients exhibit severe and persistent infections as a consequence of impaired cellular and humoral immune functions. T-cell development in XSCID patients is greatly impaired, with a complete lack of T cells in some instances, reflecting a critical role for IL-2Rγ in T-cell proliferation and differentiation. However, T-cell development in immunodeficient patients who lack IL-2 gene expression (Weinberg and Parkman, 1990) or in IL-2 knockout mice is apparently normal (Schorle *et al.*, 1991). These facts implicate a role for IL-2Rγ in other cytokine receptor systems that regulate T-cell development. Consistent with this hypothesis, it has emerged that IL-2Rγ is a component of several cytokine receptor complexes, including the receptors for interleukin 4 (IL-4), interleukin 7 (IL-7), interleukin 9 (IL-9), interleukin 15 (IL-15), and possibly interleukin 13 (IL-13), cytokines that are known to affect T-cell proliferation and/or differentiation (Figure 7) (Leonard *et al.*, 1994). The IL-2Rγ has therefore been named γc.

3.3.2. The IL-4 and IL-13 Receptors

Like IL-2, IL-4 regulates T-cell development; IL-4 also has numerous effects on both resting and activated B cells (Howard *et al.*, 1993), is a mast cell growth factor, and acts synergistically on certain populations of myeloid cells (Rennick *et al.*, 1992). Initial characterization of IL-4 binding proteins revealed that IL-4 cross-links multiple proteins, one of molecular mass approximately 140 kDa and a lower-molecular-weight species of 60–75 kDa. The 140-kDa IL-4R is expressed on a variety of cell types, including lymphoid cells, myeloid cells, and nonhematopoietic cells such as fibroblasts and neuroblasts (Lowenthal *et al.*, 1988; Urdal and Park, 1988; Mosley *et al.*, 1989; Galizzi *et al.*, 1990; Idzerda *et al.*,

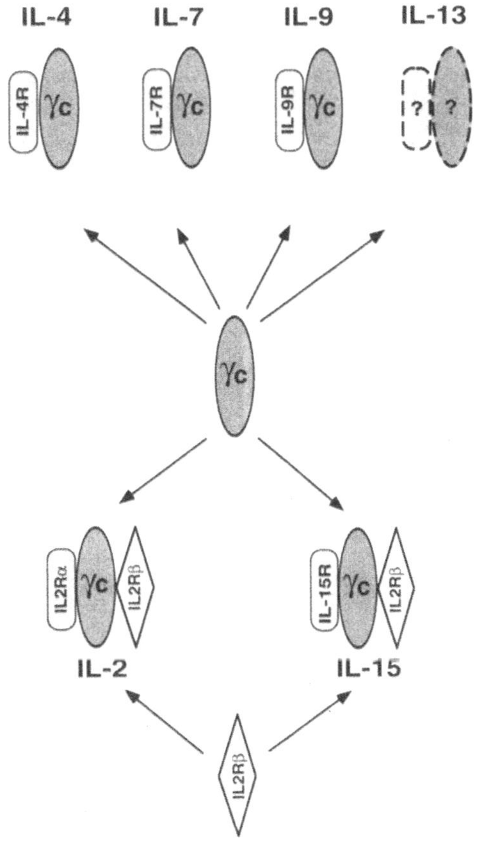

FIGURE 7. γc, a common subunit for IL-2, IL-4, IL-7, IL-9, and IL-15 receptors. The functional receptors for IL-4, IL-7, and IL-9 are heterodimers composed of a specific ligand binding subunit and the common γc subunit. The functional receptors for IL-2 and IL-15 are each composed of a specific ligand-binding subunit, the common γc subunit, and IL-2Rβ. The subunit composition of the IL-13R is not yet clear. Another cytokine, IL-16, not represented, employs the IL-7R as its signal transducer and has a unique ligand-binding subunit.

1990). The cloned human and mouse IL-4Rs have approximately 50% amino acid identity (Galizzi *et al.*, 1990; Idzerda *et al.*, 1990; Harada *et al.*, 1990; Mosely *et al.*, 1989). A soluble form of the mouse receptor, capable of high affinity IL-4 binding, has also been isolated (Mosley *et al.*, 1989).

Although the 140-kDa IL-4R binds IL-4 with high affinity (100 pM), several lines of evidence have suggested that the functional IL-4R contains an additional subunit that enhances affinity and plays a role in signal transduction (Noguchi *et al.*, 1993b; Zurawski *et al.*, 1993). Accordingly, it has recently been demonstrated that IL-4 cross-links IL-2Rγ, and the combination of IL-4R plus IL-2Rγ significantly increases IL-4 binding affinity relative to that observed with IL-4R alone. Furthermore, IL-2Rγ is required for IL-4-mediated signal transduction (Figures 6 and 7) (Kondo *et al.*, 1993; Russell *et al.*, 1993).

Interleukin 13 is a cytokine that exhibits similar functions to IL-4, for example, the modulation of B cell and monocyte activities (Minty *et al.*, 1993; Punnonen *et al.*, 1993). On the basis of competitive binding studies between IL-4 and IL-13, it has been proposed that although the cloned IL-4R does not bind IL-13, IL-4 and IL-13 cross-compete for binding to TF-1 cells, suggesting that the functional IL-4R and IL-13R share a receptor component (Zurawski *et al.*, 1993) (see Chapter 2). It is as yet unclear whether the IL-2Rγ is involved in the IL-13 receptor.

3.3.3. The IL-7 Receptor

Interleukin 7 plays an important role in lymphopoiesis, specifically in proliferation of pre-B cells and thymocytes (Namen *et al.*, 1988; Conlon *et al.*, 1989; Murray *et al.*, 1989). Goodwin *et al.* (1990) cloned the ligand-binding chain of the human and murine IL-7 receptors (64% identity) and established that this receptor chain, IL-7R, is a member of the CR superfamily, unique in that only two of the four positionally spaced extracellular domain cysteines are present (Figure 6). Expression of the IL-7R cDNA in COS cells was sufficient to confer high-affinity IL-7 binding. A soluble form of the human receptor was similarly isolated (Goodwin *et al.*, 1990).

Because IL-7 is a T-cell growth factor that, if inactivated, could account for the XSCID phenotype, Noguchi *et al.* (1993b) considered the possibility that IL-2Rγ is a component of the IL-7R. They demonstrated that IL-2Rγ is a functional component of the IL-7R (Figure 6). IL-2Rγ augments IL-7 binding affinity and the efficiency of IL-7 internalization. Furthermore, IL-2Rγ is essential for formation of functional high-affinity receptor complexes that mediate IL-7-stimulated lymphocyte development (Kondo *et al.*, 1994).

3.3.4. The IL-9 Receptor

The target populations of IL-9 include T helper cells, mast cells, erythroid progenitors, and megakaryoblastic leukemia cells (Uyttenhove *et al.*, 1988; Renauld *et al.*, 1995). Initial characterization of IL-9 receptors on murine T cells revealed an IL-9 binding protein of 54–64 kDa that binds IL-9 with relatively high affinity (100 pM) (Druez *et al.*, 1990). Subsequently, human and murine IL-9 receptors were cloned and identified as members of the class I CR family (Renauld *et al.*, 1992; Chang *et al.*, 1994). More recently, Russell *et al.* (1994) have demonstrated that an antibody to γc inhibits IL-9-dependent proliferation and, furthermore, that IL-9 cross-links γc in affinity-labeling assays, implicating γc as a component of the functional IL-9 receptor complex (Figures 6 and 7).

3.3.5. The IL-15 Receptor

The biological activities of the recently identified cytokine IL-15 show extensive overlap with those of IL-2. For example, both cytokines stimulate T cells, natural killer cells, and B cells. Consistent with overlapping biological effects, IL-15 shares two receptor subunits with IL-2, the IL-2Rβ and IL-2Rγ (Figures 6 and 7), both of which are required for signaling (Giri *et al.*, 1994a). The IL-2Rβ can bind IL-15 (affinity, 1 nM), whereas IL-2Rγ can not, although IL-2Rγ augments IL-2 binding affinity (100 pM) in combination with IL-2Rβ. Hence, IL-15 and IL-2 utilize common receptor components responsible for signaling in the IL-2 receptor complex. However IL-15 and IL-2 do exhibit distinct biological functions; certain cell populations such as early murine pre-T cells ($CD3^-$, $CD4^-$, $CD8^-$) or the 32D cell line are responsive to IL-2, but not to IL-15. Therefore, the existence of an IL-15-specific receptor subunit(s), not shared by IL-2, was considered a possibility. Recently, Giri *et al.* (1994b) have isolated a cDNA clone that encodes a unique 58-kDa protein that binds IL-15 with relatively high affinity and hence is designated IL-15Rα. The IL-15Rα is expressed on many cell types including nonlymphoid cells, exhibits an extracellular "sushi" domain (similar to the IL-2Rα, which has two extracellular "sushi" domains or a factor XIII domain), and a short 36-amino-acid cytoplasmic domain.

Thus, IL-2Rγ is involved in multiple receptor systems. However, a difference between βc and γc should be noted. The use of the common signaling receptor subunit βc by IL-3, IL-5, and GM-CSF correlates with virtually identical functions exhibited by these cytokines in the same target cells. In contrast, although IL-2, IL-4, IL-7, and IL-15 share γc, these cytokines exhibit distinct effects in common target cells. For example, IL-2 stimulates activation of Ras and supports long-term proliferation of T cells (Satoh *et al.*, 1992a; Izquierdo and Cantrell, 1993; Ravichandran and Burakoff, 1994), whereas IL-4 does not activate Ras and stimulates only transient T-cell proliferation.

3.4. Single-Chain Receptors

Several members of the CR superfamily encode proteins that appear to function as single-chain receptors (Figure 8). These include receptors for erythropoietin (EPO) (D'Andrea *et al.*, 1989b), prolactin (PRL) (Boutin *et al.*, 1988), and growth hormone (GH) (Leung *et al.*, 1987) (Figure 8).

Both GH and PRL are structurally and functionally related and cross-react with each other's receptors (Y. K. Fu *et al.*, 1992). The GHR is an unusual member of the class I CR family in that it has a degenerate WSxWS motif. Growth hormone and its receptor have been cocrystallized (DeVos *et al.*, 1992). A single GH molecule can bind two receptor molecules, at the same site on each receptor, by two different sites on the GH molecule. It has been proposed that receptor binding occurs sequentially, first preferentially by ligand site 1 and then by site 2. At low concentrations of GH, receptor binding occurs with high affinity, leading to receptor homodimerization. However, at very high concentrations of GH, each receptor binds a different molecule of GH, thereby abrogating the potential for receptor dimerization. The GH system provides a model of receptor binding that probably extends to other single-chain class I receptors (Fuh *et al.*, 1992).

Erythropoietin (EPO) is an essential cytokine for erythroid development. The EPOR is a prototypical member of the class I CR family, expressed mainly in erythroid cells (D'Andrea *et al.*, 1989b; Jones *et al.*, 1990; Winkelmann *et al.*, 1990). Reconstitution

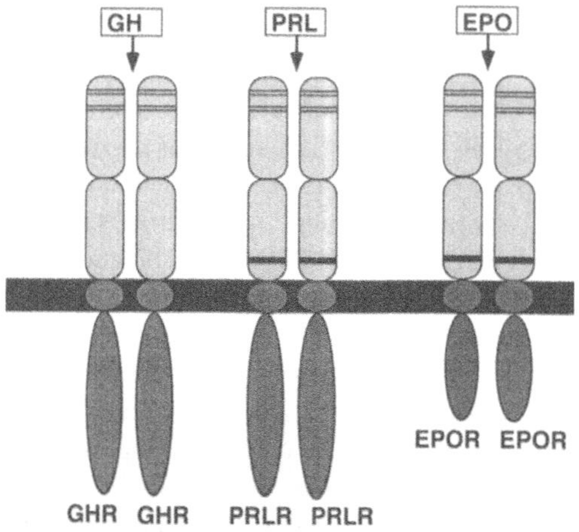

FIGURE 8. Schematic structures of single-chain class I CRs. The CR domains are represented as for Figure 3. For this group of receptors, ligand binding is thought to trigger receptor homodimerization, which constitutes high-affinity functional receptor complexes.

experiments in hematopoietic cells have indicated that the cloned EPOR chain is sufficient to mediate functional high-affinity (100 pM) binding of EPO. It should be noted, however, that there is evidence for the existence of an additional EPOR protein subunit (Landschulz *et al.*, 1989). Embryonic erythroid cells express high levels of an EPOR with a truncated cytoplasmic domain, but the full-length receptor is present on erythroblasts and erythroleukemia cells (Landschulz *et al.*, 1989).

Several activating mutations have been identified in the EPOR. A mutation that leads to receptor dimerization confers EPO-independent growth on normally EPO-dependent cells, suggesting that receptor dimerization is a critical step for signal transduction (Watowich *et al.*, 1992). Proerythroblast cell lines expressing this mutant EPOR induce erythroleukemia in mice. Also, the membrane glycoprotein gp55, encoded by Friend spleen focus-forming virus, associates with the EPOR and stimulates EPO-independent growth, although gp55 is not significantly homologous to EPO (Li *et al.*, 1990). Although the precise mechanisms of EPOR activation by gp55 is unclear, the transmembrane domain of gp55 seems to be important.

4. CLASS II CYTOKINE RECEPTORS

4.1. The Interferon Receptors

Interferons (IFNs), a group of cytokines with antiviral activity, are classified into two groups, type I, which consists of IFNs α and β (produced by virus-infected cells), and type II, IFNγ (produced by activated T cells) (Weissman and Weber, 1986). Analysis of IFN activities suggested that IFNα/β bind to a common receptor (Branca and Baglioni, 1981) and the isolation of receptor components has verified this theory.

A cDNA that encodes a human receptor for IFNα and IFNβ, the class II IFNα/βR, was isolated by genetic transfer into mouse cells (Uze *et al.*, 1990) (Figure 9). However, cross-linking experiments show at least two distinct IFNα receptor proteins, and the cloned IFNα/βR chain by itself does not mediate the full complement of activities of IFNα/β, suggesting that additional receptor components exist (Fischer *et al.*, 1990; Uze *et al.*, 1990; Vanden and Pfeffer, 1988).

The high-affinity functional IFNγR is composed of at least two chains, IFNγRα and IFNγRβ (Figure 9). The IFNγRα is a 90-kDa protein that binds IFNγ with high affinity (Aguet *et al.*, 1988) and dimerizes on ligand binding (Greenlund *et al.*, 1993). This class II receptor has two novel cytoplasmic domains, a membrane proximal 48-amino-acid domain, and a distal YDH sequence that are required for functional activity (Farrar *et al.*, 1992). The IFNγRα, however, is not sufficient to confer responsiveness to all types of IFNγ (Uze *et al.*, 1990), and IFNγRβ (localized at human chromosome 21 or mouse chromosome 16) is required (Hemmi *et al.*, 1994; Soh *et al.*, 1994). The IFNγRβ subunit does not appear to bind ligand and, together with IFNγRα, constitutes a functional receptor (Hemmi *et al.*, 1994; Soh *et al.*, 1994). Whether the IFNγRβ subunit is a component of other receptor complexes remains unknown. The myxoma virus encodes a soluble homologue of the IFNγRα, the M-T7 protein (Upton *et al.*, 1992). M-T7 is secreted from myxoma-virus-infected cells, specifically binds rabbit IFNγ, and inhibits its antiviral activity.

4.2. The Interleukin-10 Receptor

Interleukin 10 was identified as a factor that inhibits cytokine production from T helper type 1 cells. Other functions of IL-10 include macrophage deactivation, B-cell development,

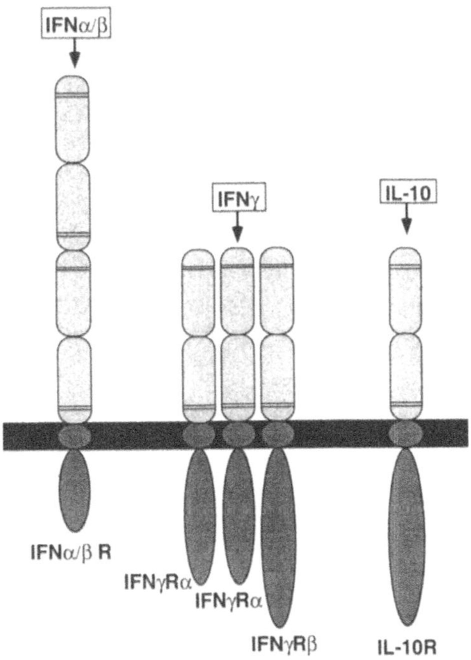

FIGURE 9. Schematic structures of class II CRs. The CR domains are represented as for Figure 3. Class II receptors contain four conserved cysteine residues, one pair at each end of the CR domain, as indicated. The WSxWS motif of class I receptors is absent, but several characteristic amino acids (P, W, W, Y) are conservatively positioned in the extracellular domain. The oligomerization mechanism for the IFNγ receptor is unclear.

and regulation of mast cell growth (reviewed in Ho and Moore, 1994). Interleukin 10 and IFNγ possess both similar and antagonistic biological activities (Moore *et al.*, 1993), raising the concept that cross-talk occurs either between signaling pathways elicited by these cytokines or by use of common receptor subunits. Epstein–Barr virus and equine herpes virus type II encode viral (v) homologues of IL-10 (Moore *et al.*, 1990; Rode *et al.*, 1993, respectively), which are likely to contribute significantly to viral infection of the host immune system. Interestingly, vIL-10 exhibits a subset of the activities of human IL-10 (Ho and Moore, 1994).

In 1993, Moore and his colleagues isolated a murine IL-10R cDNA from mast and macrophage cell lines and subsequently cloned the human counterpart from a Burkitt lymphoma cell line (Ho *et al.*, 1993; Liu *et al.*, 1994). The 90- to 100-kDa murine and human IL-10Rs are 60% identical and are class II CRs. Both mIL-10R and hIL-10R confer high-affinity IL-10 binding and mediate IL-10-stimulated proliferation when transfected into normally nonresponsive BAF3 cells (Ho *et al.*, 1993; Liu *et al.*, 1994). BAF3 transfectants also bind vIL-10 and proliferate in response to vIL-10. However, neither COS cells transfected with mIL-10R or hIL-10R nor the murine mast cell line MC/9, which proliferates transiently in response to mIL-10, can bind vIL-10. These results suggest either that distinct IL-10Rs exist, one of which confers vIL-10 responsiveness, or that human, mouse, and viral IL-10Rs are multimeric, with a shared receptor subunit.

5. OTHER CYTOKINE RECEPTOR SUPERFAMILIES

5.1. The Tumor Necrosis Factor Receptor Superfamily

The TNF superfamily comprises a rapidly emerging group of cytokines that are structurally and functionally related. These cytokines are produced as aminoterminally membrane anchored (type II) molecules [with one exception, lymphtoxin-α (LT-α)] and exhibit a distinctive β-strand formation known as a β-jellyroll fold (Bazan, 1993). The receptors for these ligands show extracellular domain homology and have recently been grouped as the TNFR family. Members of the TNFR family are defined by the presence of four blocks of approximately 40 amino acids in the extracellular domain, each containing spatially conserved cysteine repeats (Bazan, 1993; Mallett and Barclay, 1991). Crystallographic analysis of LT-α binding to its receptor suggests that cytokines of the TNF family bind as trimeric ligands to three receptor molecules (Banner *et al.*, 1993). The reader is referred to recent reviews for further details (Banchereau *et al.*, 1994; Bazan, 1993).

5.2. The Tyrosine Kinase Receptor Superfamily

Tyrosine kinase (TK) receptors contain a large glycosylated extracellular ligand-binding domain, a single membrane-spanning hydrophobic region, and a cytoplasmic domain that possesses intrinsic tyrosine kinase activity which mediates signaling (Hunter and Cooper, 1985).

On the basis of their structural characteristics, TK-Rs can be classified into four families (Ullrich and Schlessinger, 1990; Wilks, 1990). Type III TK receptors possess five extracellular immunoglobin-like (Ig) repeats and a kinase insert (KI) region that structurally divides the catalytic TK domain. The type III TK-R family includes, among others, the platelet-derived growth factor receptor (PDGFR), and *c-fms*, *c-kit*, and *flk2/flt3*, the receptors for macrophage colony-stimulating factor (M-CSF), stem cell factor (SCF), and the *flk2/flt3* ligand, respectively (Chabot *et al.*, 1988; Sherr, 1990; Matthews *et al.*, 1991). These cytokines are dimeric and trigger receptor homodimerization on binding (Ullrich and Schlessinger, 1990).

The viral homologue of the *c-fms* protooncogene product, *v-fms*, induces fibrosarcomas in cats and is constitutively active (Wheeler *et al.*, 1986). Single point mutations in the extracellular domain of the M-CSFR can constitutively induce tyrosine kinase activity, leading to ligand-independent cellular transformation, perhaps by inducing receptor aggregation (Roussel *et al.*, 1987, 1988).

5.3. The Transforming Growth Factor β Receptor Superfamily

The transforming growth factor β (TGF-β) cytokine family is a group of structurally and functionally related growth factors with a diverse array of biological activities (Massague, 1987). These cytokines are synthesized as inactive precursors and cleaved to yield biologically active disulfide-linked homo- or heterodimeric ligands. The receptors for this family are also related and form the TGF-β receptor superfamily (Massague *et al.*, 1994). As representatives of this family, TGF-β receptors are described below.

Three distinct classes of TGF-β receptor chains have been identified, receptor types I, II, and III. Receptor types I and II are transmembrane proteins with intrinsic serine/threonine kinase domains, and both receptor types display *in vitro* autophosphorylation activity (Lin *et al.*, 1992: Massague *et al.*, 1994). Functional TGF-β receptors are heteromeric, consisting of a receptor type I and a receptor type II. In this receptor complex, only type II receptors can bind TGF-β, but the kinase activities of both receptor types I and II are essential for signal transduction to occur (Bassing *et al.*, 1994a,b; Wrana *et al.*, 1994). Type I TGF-βRs can interact with multiple ligands and also with multiple type II receptors. A functional receptor consisting of two kinases presents interesting implications for signal transduction, especially when the pleiotropic effects of TGF-β are considered. Recent evidence suggests that different TGF-β activities are mediated by specific receptor types (Chen *et al.*, 1994).

The type III TGF-β receptor is a proteoglycan, betaglycan, found in both membrane-anchored and soluble forms (Andres *et al.*, 1989; Lopez-Casillas *et al.*, 1991). Betaglycan is not a signaling receptor but binds TGF-β, and may act either as an antagonist and sequester TGF-β, or present TGF-β to receptor types I and II (Lopez-Casillas *et al.*, 1993, 1994).

6. MECHANISMS OF RECEPTOR ACTIVATION AND SIGNAL TRANSDUCTION

6.1. Cytokine Receptor Signaling

The initial activation steps for receptors with intrinsic tyrosine kinase (TK) activity are ligand-induced dimerization and autotransphosphorylation of receptor monomers. When activated, TK-Rs recruit and interact with a number of SH2-domain-containing signaling molecules, by means of phosphorylated tyrosine residues (Kashishian *et al.*, 1992; Lev *et al.*, 1992; Reedijk *et al.*, 1992). As described above, ligand-induced homo- or hetero-dimerization of receptor components is also the initial step of activation of receptors of the CR superfamily. Key questions currently under investigation are: how cytokine receptors without intrinsic kinase activity transduce signals, how receptors induce signals common to various cytokines as well as specific to each cytokine, how a given receptor can elicit different biological responses, and whether distinct regions of receptor cytoplasmic domains are required to mediate such effects.

Although receptors of the class I and II CR families do not contain intrinsic kinase domains, most cytokines stimulate rapid tyrosine phosphorylation of a number of cellular substrates, including components of their receptors (Isfort and Ihle, 1990; Sorensen *et al.*, 1989; Welham *et al.*, 1992; Welham and Schrader, 1992; Izuhara and Harada, 1993; Yin and Yang, 1994). Furthermore, cytokine-stimulated tyrosine phosphorylation is essential to mediate the biological effects of cytokines (Kanakura *et al.*, 1990; Satoh *et al.*, 1992b). A number of mitogenic cytokines such as IL-2, IL-3, IL-5, GM-CSF, IL-6, G-CSF, and EPO are known to induce activation of Ras as well as its downstream cascade including Raf and mitogen-activated protein kinase (MAPK) (Duronio *et al.*, 1992; Kan *et al.*, 1992; Welham *et al.*, 1992). Certain cytokines also activate Piml kinase (Sato *et al.*, 1993b) and protein kinase C (PKC) (Heyworth *et al.*, 1993; Shearman *et al.*, 1993) and induce expression of nuclear protooncogenes, including *c-myc*, *c-fos*, and *c-jun*, and cell cycle regulators (Sherr, 1993; Ihle *et al.*, 1994a; Matsushime *et al.*, 1991, 1994). Many of these signaling molecules are also activated by TK-Rs such as EGFR, *c-kit*, or *c-fms* (Miyzawa *et al.*, 1991; Sherr, 1990; Welham and Schrader, 1992), where the receptor can directly phosphorylate signaling substrates. An important issue, therefore, is the identification of kinases that associate with members of the CR superfamily and elucidation of the mechanisms(s) by which receptor–kinase complexes couple to downstream signaling events.

Several tyrosine kinases that can associate with CRs have been identified. The p56 Lck tyrosine kinase, a member of the Src family, directly interacts with IL-2Rβ (Hatakeyama *et al.*, 1991). Another Src-like kinase, Lyn, has been implicated in IL-3 signaling, and the Fes kinase has been shown to associate with βc in response to IL-3 or GM-CSF (Hanazono *et al.*, 1993a), with IL-4R in response to IL-4 (Izuhara *et al.*, 1994), and is also activated in response to EPO (Hanazono *et al.*, 1993b). However, the physiological significance of those tyrosine kinases remains unclear, and kinase specificity varies according to the cell type.

6.2. Jaks and STATs

An abundance of evidence now indicates that a recently identified family of non-receptor tyrosine kinases, the Janus kinase (Jak) family, plays a crucial role in signaling of class I and II CRs. To date, four members of the Jak family have been identified (Jakl, Jak2, Jak3, and Tyk2) (Ihle *et al.*, 1994b). Jak kinases lack SH2 or SH3 domains but contain an active kinase domain and a kinase-like domain. The importance of Jaks in cytokine signaling was originally discovered by elegant somatic cell genetic experiments using mutant cells unresponsive to interferon (IFN), which revealed that Jakl and Tyk2, and Jakl and Jak2 molecules play essential roles in IFNα/β- and IFNγ-induced gene expression respectively (Velazquez *et al.*, 1992; Muller *et al.*, 1993; Watling *et al.*, 1993). These kinases associate directly with many cytokine receptors in the absence of cytokines and are

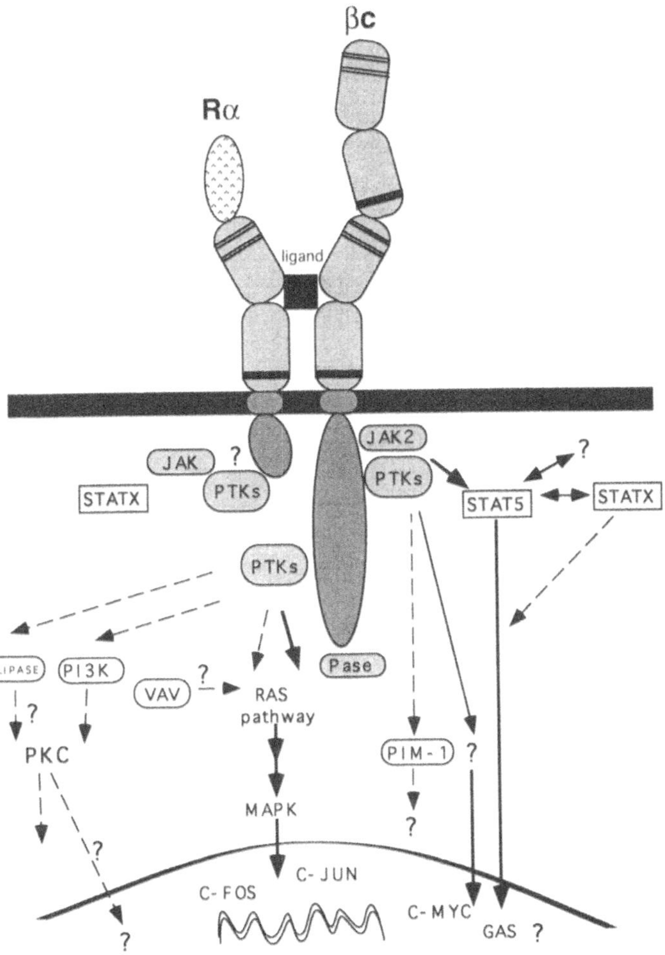

FIGURE 10. Cytokine receptors and signal transduction. The GM-CSF/IL-3 receptor is depicted as a representative receptor, and signaling pathways that may couple to these receptors are also depicted. Broken lines denote hypothetical pathways, and molecules that associate with receptor cytoplasmic domains are shaded in gray. PTK denotes protein tyrosine kinases, whose identities are not yet clear, Pase denotes phosphatase. The cytoplasmic domain of βc can be divided into functional domains, the membrane proximal region including box 1 and box 2, which mediates the JAK/STAT pathway and induction of *c-myc*, and the distal region that activates the Ras pathway.

activated and phosphorylated upon cytokine binding to the receptors (Ihle *et al.*, 1994b). For example, Jak2 binds to gp130, βc, and the receptors for EPO, GH, PRL, and G-CSF and becomes tyrosine phosphorylated and activated in response to ligand binding (Artgetsinger *et al.*, 1993; Dusanter *et al.*, 1994; Quelle *et al.*, 1994; Witthuhn *et al.*, 1993; Yin *et al.*, 1994). Interleukin 6, CNTF, and LIF activate Jak1, Jak2 and Tyk2 (Lutticken *et al.*, 1994; Narazaki *et al.*, 1994b; Stahl *et al.*, 1994), and IL-2, IL-4, IL-7, and IL-9 activate Jak1 and Jak3 (Johnston *et al.*, 1994; Miyazaki *et al.*, 1994; Russell *et al.*, 1994; Stahl *et al.*, 1994; Witthuhn *et al.*, 1994). For the IL-2, IL-4, IL-7, and IL-9 receptors, Jak3 associates with γc, and Jakl associates with a different component of the receptor complex, for example, IL-2Rβ (Y. Y. Fu *et al.*, 1992; Schindler *et al.*, 1992a, b.) Cytokine binding induces heterodimerization of receptor subunits and activation of Jakl and Jak3. Mutations in γc that disrupt its interaction with Jak3 cause XSCID, implying that Jak activation is crucial for the function of this receptor subunit.

What are the substrates of Jaks? Extensive studies on transcriptional activation mediated by the IFNRs have unveiled a novel signal transduction pathway that provides a direct link between receptor activation and gene transcription (Y. Y. Fu *et al.*, 1992; Schindler *et al.*, 1991a,b; Shuai *et al.*, 1993a, 1994). This is mediated by a recently identified novel family of transcription factors known as STATs (signal transducers and activators of transcription), which are substrates for Jaks. STATs contain a well-conserved SH2 domain and normally reside in the cytoplasm in latent forms. STATs are tyrosine phosphorylated by Jaks on cytokine stimulation, which induces dimerization of STATs, either homodimerization or heterodimerization with a different STAT protein. STAT dimers translocate to the nucleus, where they bind specific DNA sequences in the promoters of target genes (Darnell *et al.*, 1994; Shuai *et al.*, 1994).

STATs act downstream of many receptors of the class I and class II cytokine receptors (Rothman *et al.*, 1994), and to date at least six STATs have been isolated. Each STAT protein acts in association with a restricted subset of cytokine receptors. STAT1 and STAT2 were originally identified as IFNα/β-activated STATs (Y. Y., Fu *et al.*, 1992; Schindler *et al.*, 1992a). STAT1 is also activated by a number of other cytokines, including IFNγ (Shuai *et al.*, 1992), IL-6 (Feldman *et al.*, 1994), IL-10 (Lehmann *et al.*, 1994), and EGF (Y. Y. Fu and Zhang, 1993). STAT3 is identical to the acute-phase responsive factor (APRF) activated by IL-6 (Zhong *et al.*, 1994) and is also activated by EGF, IL-10, and G-CSF (Tian *et al.*, 1994). STAT4 was identified by homology to STAT1 and is activated by IL-12 (Jacobson *et al.*, 1995). Mammary gland factor, initially isolated as a prolactin- responsive protein, was found to be a member of the STAT family, now known as STAT5 (Wakao *et al.*, 1992, 1994). Interestingly, two highly homologous murine STAT5 molecules have been isolated, STAT5A and STAT5B, and both are activated by a number of cytokines including IL-2, IL-3, IL-5, GM-CSF, and EPO (Mui *et al.*, 1995; H. Wakao *et al.*, 1995). STAT6 (IL-4 STAT) was cloned as an IL-4-induced STAT in monocytes (Hou *et al.*, 1994).

Thus, class I and II CRs employ common signaling pathways to mediate their effects. The association of different receptors with particular Jaks and STATs appears to be complex. For example IL-2 activates Jakl and Jak3, but not Jak2, whereas IL-3, IL-5, and GM-CSF primarily activate Jak2 (Ihle *et al.*, 1994b). Nevertheless, STAT5 can be activated by all these cytokines. In contrast, although IL-2 and IL-4 activate the same Jaks, they appear to activate different STATs. However, different combinations of receptors, Jaks, and STATs may provide the potential for divergence of signals. Several experiments indicate that receptors must be involved in recruitment and selection of STATs. For example, STAT1 is recruited to the IFNγ receptor by means of this SH2 domain and a single phosphorylated tyrosine residue in the receptor (Shuai *et al.*, 1993b). Likewise, STAT6 appears to bind to the IL-4R via its SH2 domain (Hou *et al.*, 1994). N. Stahl *et al.* (1995) have demonstrated that a single tyrosine-containing amino acid motif at the carboxy terminal of gp130 can mediate association with STAT3. A chimeric EPOR containing this motif activates STAT3 in response to EPO, suggesting that receptors contain distinct cytoplasmic motifs that mediate association with particular STAT proteins. In contrast, a βc mutant that lacks all tyrosine residues still activates STAT5 (Mui *et al.*, 1995), suggesting that motifs other than phosphorylated tyrosines exist to select and activate STATs. Differential expression of particular STATs at particular stages of hematopoietic development and the presence of additional cellular components that affect the activation of STATs may also influence the outcome of receptor activation. Because the Jak–STAT pathway can provide rapid transport of ligand-specific signals to the nucleus, this pathway may be important to mediate the specific and diverse functions of cytokines.

6.3. Signals and Functions

6.3.1. Functional Cytoplasmic Domains

Analysis of cytokine receptors bearing sequential carboxy-terminal deletions has revealed that receptor cytoplasmic domains can be divided into several functional domains. Two relatively conserved motifs, box 1 and box 2, are generally found at the membrane-proximal cytoplasmic region (reviewed in Ihle *et al.*, 1994a). In βc, gp130, IL-2Rβ, EPOR, G-CSFR, GHR, PRLR, box 1 and box 2 are required for the association and activation of Jaks and essential for mitogenesis. However, the relationship between Jak activation and mitogenic signaling is still unclear. For example, IL-3 dependent activation of Jak2 and STAT5 is not sufficient to drive proliferation of a murine mast cell line (O'Farrell *et al.*, 1996). This membrane-proximal region is also responsible for induction of *c-myc* and *Pim-1* in the βc subunit (Sato *et al.*, 1996b). Regions of βc downstream of box 1 and box 2 are required for activation of Ras, Raf, and MAPK and induction of *c-fos/c-jun* in βc and IL-2Rβ. In the IL-2Rβ, this region is also known as the acidic region and is essential for association with the Lck tyrosine kinase (Hatakeyama *et al.*, 1989a, 1991). The C-terminal domain of the EPOR mediates association with a protein tyrosine phosphatase, PTP1C (also known as HCP), encoded by the *me* gene. This phosphatase appears to regulate EPO signaling negatively, because a naturally occuring mutation in the murine *me* locus confers hypersensitivity to EPO, and mutant EPORs that lack the C terminus are hypersensitive to EPO (de la Chapelle *et al.*, 1993). Likewise, in βc, truncation of the C terminus enhances signals such as tyrosine phosphorylation (Sakamaki *et al.*, 1992; Sato *et al.*, 1993b), and the β chain has been found to associate with a tyrosine phosphatase (Yi *et al.*, 1993).

6.3.2. Mitogenesis, Apoptosis, and Differentiation

Most cytokines have multiple effects on cells, such as regulation of cell viability, cell proliferation, differentiation, or functional activity. Are such distinct biological responses mediated by distinct signaling events? A series of cytoplasmic deletion mutants is useful to define the function of each signaling pathway. The membrane-proximal region of βc (required for induction of *c-myc* and *pim-1* as well as Jak2 and STAT5 activation) is essential for mitogenesis, more specifically the induction of DNA synthesis and cell cycle progression (Kinoshita *et al.*, 1995; Sato *et al.*, 1993b). In contrast, the distal cytoplasmic domain of βc (responsible for activation of Ras, Raf, and MAPK) is required for prevention of apoptosis. Thus, mutant βc receptors that lack the distal region induce DNA synthesis and show transient mitogenesis but can not prevent apoptosis (Kinoshita *et al.*, 1996). Similar functional cytoplasmic regions may exist in other receptors such as IL-2Rβ that have a similar domain structure to βc.

A number of cytokine receptors can not only mediate mitogenic signals, but also regulate hematopoietic differentiation. For example, G-CSF promotes development of myeloid progenitor cells into neutrophils. In addition to the morphological changes associated with neutrophil development, cells express neutrophil-specific enzymes such as myeloperoxidase and elastase, which are important for functional activity. Analysis of deletion and chimeric G-CSF receptor demonstrated that growth and differentiation signals are mediated by distinct cytoplasmic domains of the G-CSF receptor (Fukunaga *et al.*, 1993; Ishizaka *et al.*, 1993). While growth signals are delivered via the membrane-proximal region, differentiation signals are associated with a C-terminal short stretch. Supporting these molecular analyses, this latter cytoplasmic region is truncated in patients with Kostmann's neutropenia, who exhibit decreased neutrophil numbers resulting from insufficient progenitor differentiation (Guba *et al.*, 1994). Further analysis of signal transduction through this "differentiation domain" should provide critical insight into the molecular basis of hematopoietic differentiation (see Chapter 12).

7. CONCLUDING REMARKS

The majority of cytokine receptors belong to a single cytokine receptor superfamily, defined on the basis of sequence and structural homology. The ligands for these receptors are also related, implying that these two groups of molecules are descended from a primordial cytokine–receptor pair of genes. From the mass of information recently available regarding cytokine receptors, several common concepts are apparent.

Most cytokine receptors are multimeric, composed of a ligand-specific subunit and one (or more) of a number of common signal-transducing subunits. The existence of common signaling subunits therefore provides a molecular basis for the functional redundancy and cross-competition exhibited by groups of cytokines.

The signals initiated by ligand-bound receptor complexes couple to cytoplasmic signaling pathways involving tyrosine phosphorylation and ultimately converge in the nucleus. Several such signaling pathways have been identified, for example, the Ras–MAPK pathway and the Jak–STAT pathway. Members of the CR superfamily do not have intrinsic tyrosine kinase activity but in most cases can signal through an associated Jak kinase. Because Jak kinases appear to be ubiquitously expressed, cytokine receptors may simply be regarded as tyrosine receptors in disguise. Recent results point to the involvement of distinct cytoplasmic domains of receptors in distinct signaling events, which may mediate specific biological effects, providing some explanation for the pleiotropic actions of cytokines.

A major unresolved question is how cytokine receptors that share a common signaling subunit can mediate specific gene expression to effect distinct responses. Current attention is focused on the Jak–STAT pathway through the recruitment and interaction of specific STAT molecules by common and possibly by unique receptor subunits. Combinatorial effects of transcription factors are likely to regulate cell responses, and future studies will undoubtedly reveal the precise events that lead to differential gene regulation and to the dysregulation that occurs in diseased states.

ACKNOWLEDGMENTS. We would like to thank members of our lab for helpful suggestions, and Gerard Manning for review of this manuscript. DNAX Research Institute is supported by Schering-Plough Corporation.

8. REFERENCES

Aguet, M., and Mogensen, K. E., 1983, Interferon receptors, *Interferon* **5**:1–22.

Aguet, M., Dembic, Z., and Merlin G., 1988, Molecular cloning and expression of the human interferon-gamma receptor, *Cell* **55**:273–280.

Anderson, K. C., Morimoto, C., Paul, S. R., Chauhan, D., Williams, D., Cochran, M., and Barut, B. A., 1992, Interleukin-11 promotes accessory cell-dependent B-cell differentiation in humans, *Res. Immunol.* **143**: 385–400.

Andres, J. L., Stanley, K., Cheifetz, S., and Massague, J., 1989, Membrane-anchored and soluble forms of betaglycan, a polymorphic proteoglycan that binds transforming growth factor-β, *J. Cell. Biol.* **109**:3137–3145.

Arai, K., Lee, F., Miyajima, A., Miyatake, S., Arai, N., and Yokota, T., 1990, Cytokines: Coordinators of immune and inflammatory responses, *Annu. Rev. Biochem.* **59**:783–836.

Artgetsinger, L. S., Campbell, G. S., Yang, X., Witthuhn, B. A., Silvennoinen, O., Ihle, J. N., and Carter Su, C., 1993, Identification of Jak2 as a growth hormone receptor-associated tyrosine kinase, *Cell* **74**:237–244.

Banchereau, J., Bazan, F., Blanchard, D., Briere, F., Galizzi, J. P., Van, K. C., Liu, Y. J., Rousett, F., and Saeland, S., 1994, The CD40 antigen and its ligand. *Annu. Rev. Immunol.* **12**:881–922.

Banner, D. W., D'Arcy, A., Janes, W., Gentz, R., Schoenfeld, H. J., Broger, C., Loetscher, H., and Lesslauer, W., 1993, Crystal structure of the soluble human 55kd TNF receptor–human TNF-beta complex: Implications for TNF receptor activation. *Cell* **73**:431–445.

Barton, D. P., Blanchard, D. K., Michelini, N. B., Nicosia, S. V., Cavanagh, D., and Djeu, J. Y., 1993, High serum and ascitic soluble interleukin-2 receptor alpha levels in advanced epithelial ovarian cancer, *Blood* **81**: 424–429.

Bassing, C. H., Howe, D. J., Segarini, P. R., Donahoe, P. K., and Wang, X. F., 1994a, A single heteromeric receptor complex is sufficient to mediate biological effects of transforming growth factor-beta ligands, *J. Biol. Chem.* **269**:14861–14864.

Bassing, C. H., Yingling, J. M., Howe, D. J., Wang, T., He, W. W., Gustafson, M. L., Shah, P., Donahoe, P. K., and Wang, X. F., 1994b, A transforming growth factor beta type I receptor that signals to activate gene expression, *Science* **263**:87–89.

Baumann, H., Marinkovic-Pajovic, S., Won, K.-A., Jones, E. V., Campos, S. P., Jahreis, G. P., and Morella, K. K., 1992, The action of interleukin-6 and leukaemia inhibitory factor on liver cells, *in Polyfunctional Cytokines: IL-6 and LIF* (G. R. Bock, J. Marsh, and K. Widdows, eds.), pp. 100–124, Wiley, Chicester.

Baumann, H., Symes, A. J., Comeau, M. R., Morella, K. K., Wang, Y., Friend, D., Zielger, S. F., Fink, J. S., and Gearing, D. P., 1994a, Multiple regions within the cytoplasmic domains of the leukemia inhibitory factor receptor and gp130 cooperate in signal transduction in hepatic and neuronal cells, *Mol. Cell. Biol.* **14**:138–146.

Baumann, H., Gearing, D., and Zielger, S. F., 1994b, Signaling by the cytoplasmic domain of hematopoietin receptors involves two distinguishable mechanisms in hepatic cells, *J. Biol. Chem.* **269**:16297–16304.

Bazan, J.F., 1989, A novel family of growth factor receptors: A common binding domain in the growth hormone, prolactin, erythropoietin, and IL-6 receptors, and the p75 IL-2 receptor β chain, *Biochem. Biophys. Res. Commun.* **164**:788–795.

Bazan, J. F., 1990, Structural design and molecular evolution of a cytokine receptor superfamily. *Proc. Natl. Acad. Sci. U.S.A.* **87**:6934–6938.

Bazan, F., 1993, Emerging families of cytokine and receptors. *Curr. Biol.* **3**:603–606.

Boulton, T. G., Stahl, N., and Yancopoulos, G. D., 1991, Ciliary neurotrophic factor/leukemia inhibitory factor/interleukin 6/oncostatin M family of cytokines induces tyrosine phosphorylation of a common set of proteins overlapping those induced by other cytokines and growth factors, *Science* **253**:59–63.

Boutin, J.-M., Jolicoeur, C., Okamura, H., Gagnon, J., Edery, M., Shiroota, M., Banville, D., Dusanter-Fourt, I., Djiane, J., and Kelly, P.A., 1988, Cloning and expression of the rat prolactin receptor, a member of the growth factor hormone/prolactin receptor gene family, *Cell* **53**:69–77.

Branca, A. A., and Baglioni, C., 1981, Evidence that types I and II interferons have different receptors, *Nature* **294:**768–770.

Breen, F. N., Hume, D. A., and Weidemann, M. J., 1993, The effects of interleukin 3 (IL-3) on cells responsive to macrophage colony-stimulating factor (CSF-1) in liquid murine bone marrow culture, *Blood* **82:**752–761.

Bruce, A. G., Hoggatt, I. H., and Rose, T. M., 1994, Oncostatin M is a differentiation factor for myeloid leukemia cells, *Cytokine* **6:**48–54.

Cannistra, S. A., Vellenga, E., Groshek, P., Rambaldi, A., and Griffin, J. D., 1988, Human granulocyte-macrophage colony-stimulating factor and interleukin-3 stimulate monocyte cytotoxicity through a tumour necrosis factor-dependent mechanism, *Blood* **71:**672–676.

Chabot, B., Stephenson, D. A., Chapman, V. M., Besme, R. P., and Berstein, A., 1988, the proto-oncogene *ci-kit* encoding a transmembrane tyrosine kinase receptor maps to the mouse *W* locus, *Nature* **335:**88–89.

Chang, M. S., Engel, G., Benedict, C., Basu, R., and McNinch, J., 1994, Isolation and characterization of the human interleukin-9 receptor gene, *Blood* **83:**3199–3205.

Chen, R. H., Ebner, R., and Derynck, R., 1994, Inactivation of the type II receptor receals two receptor pathways for the diverse TFG-beta activities, *Science* **260:**1335–1338.

Chua, A. O., Chizzonite, R., Desai, R., Truitt, T. P., Nunes, P., Minetti, L. J., Warrier, R. R., Presky, D. H., Levine, J. F., Gately, M. K., and Gubler, U., 1994, Expression cloning of a human IL-12 receptor component, *J. Immunol.* **153:**128–136.

Colotta, F., Bussolino, F., Polentarutti, N., Guglielmetti, A., Sironi, M., Bocchietto, E., De, R.M., and Mantovani, A., 1991, Differential expression of the common beta and specific alpha chains of the receptors for GM-CSF, IL-3, and IL-5 in endothelial cells. *Virology* **180:**633–647.

Conlon, P. J., Morrissey, P. J., Nordan, R. P., Grabstein, K. H., Prickett, K. S., Reed, S. G., Goodwin, R., Cosman, D., and Namen, A. E., 1989, Murine thymocytes proliferate in direct response to interleukin 7, *Blood* **74:**1368–1373.

Cosman, D., 1993, The hematopoietin receptor superfamily, *Cytokine* **5:**95–106.

D'Andrea, A. D., Fasman, G. D., and Lodish, H. F., 1989a, Erythropoietin receptor and interleukin-2 receptor beta chain: a new receptor family, *Cell* **57:**277–285.

D'Andrea, A. D., Lodish, H. F., and Wong, G. G., 1989b, Expression cloning of the murine erythropoietin receptor, *Cell* **57:**277–285.

Darnell, J. J., Kerr, I. M., and Stark, G. R., 1994, Jak-STAT pathways and transcriptional activation in response to IFNs and other extracellular signaling proteins, *Science* **264:**1415–1421.

Davis, S., Aldrich, T. H., Valenzuela, D. M., Furth, M. E., Squinto, S. P., and Yancopoulos, G. D., 1991, The receptor for ciliary neurotrophilic factor, *Science* **253:**59–63.

Davis, S., Aldrich, T. H., Stahl, N., Pan, L., Taga, T., Kishimoto, T., Ip, N. Y., and Yancopoulos, G. D., 1993, LIFR beta and gp130 as heterodimerizing signal transducers of the tripartite CNTF receptor, *Blood* **82:**1120–1126.

de la Chapelle, A., Traskelin, A. L., and Juvonen, E., 1993, Truncated erythropoietin receptor causes dominantly inherited benign human erythrocytosis, *Proc. Natl. Acad. Sci. U.S.A.* **90:**4495–4499.

De Vos, A. M., Ultsch, M., and Kossiakoff, A. A., 1992, Human growth hormone and the extracellular domain of its receptor: Crystal structure of the complex, *Science* **225:**306–312.

Dinarello, C. A., Canin, A. J. C., Wolff, S. M., Bernheim, H. A., Beutler, B., Cerami, A., Palladino, M. A., Jr., and O'Connor, J. V., 1986, Tumour necrosis factor (cachectin) is an endogenous pyrogen and induces production of interleukin-1 (IL-1), *J. Exp. Med.* **163:**1433–1450.

Druez, C., Coulie, P., Uyttenhove, C., and Van Snick, J., 1990, Functional and biochemical characterization of mouse p40/IL-9 receptors, *J. Immunol.* **145:**2494–2499.

Duronio, V., Welham, M. J., Abraham, S., Dryden, P., and Schrader, J. W., 1992, p21ras activation via hemopoietin receptors and *c-kit* requires tyrosine kinase activity but not tyrosine phosphorylation of p21ras GTPase-activating protein, *Proc. Natl. Acad. Sci. U.S.A.* **89:**1587–1591.

Dusanter, F. I., Muller, O., Ziemiecki, A., Mayeux, P., Drucker, B., Djiane, J., Wilks, A., Harpur, A. G., Fischer, S., and Gisselbrecht, S., 1994, Identification of JAK protein tyrosine kinases as signaling molecules for prolactin. Functional analysis of prolactin receptor and prolactin-erythropoietin receptor chimera expressed in lympoid cells, *EMBO J.* **13:**2583–2591.

Fanslow, W. C., Clifford, K. N., Park, L. S., Rubin, A. S., Voice, R. F., Beckmann, M. P., and Widmer, M. B., 1991, Regulation of alloreactivity *in vivo* by IL-4 and the soluble IL-4 receptor, *J. Immunol.* **147:**535–540.

Farrar, M. A., Campbell, J. D., and Schreiber, R. D., 1992, Identification of a functionally important sequence in the C terminus of the interferon-gamma receptor, *Proc. Natl. Acad. Sci. U.S.A.* **89:**11706–11710.

Feldman, G. M., Petricoin, E. F. 3rd, David, M., Larner, A. C., and Finbloom, D. S., 1994, Cytokines that associate with the signal transducer gp130 activate the interferon-induced transcription factor p91 by tyrosine phosphorylation, *J. Biol. Chem.* **269:**10747–10752.

Fibbe, W. E., van, D. J., Billiau, A., Voogt, P. J., Duinkerken, N., Kluck, P. M., and Falkenburg, J. H., 1986, Interleukin-1 (22-K factor) induces release of granulocyte-macrophage colony-stimulating activity from human mononuclear phagocytes, *Blood* **68:**1316–1321.

Fischer, T., Rehm, A., Aguet, M., and Pfizenmaier, K., 1990, Human chromosome 21 is necessary and sufficient to confer human IFN gamma responsiveness to somatic cell hybrids expressing the cloned human IFN gamma receptor gene, *Cytokine* **2:**157–161.

Fourcin, M., Chevalier, S., Lebrun, J. J., Kelly, P., Pouplard, A., Wijdenes, J., and Gascan, H., 1994, Involvement of gp130/interleukin-6 receptor transducing component in interleukin-11 receptor, *Eur. J. Immunol.* **24:**277–280.

Fu, X. Y., and Zhang, J. J., 1993, Transcription factor p91 interacts with the epidermal growth factor receptor and mediates activation of the *c-fos* gene promoter, *Cell* **74:**1135–1145.

Fu, X. Y., Schindler, C., Improta, T., Aebersold, R., and Darnell, J. J., 1992, The proteins of ISGF-3, the interferon alpha-induced transcriptional activator, define a gene family involved in signal transduction, *Proc. Natl. Acad. Sci. U.S.A.* **89:**7840–7843.

Fu, Y. K., Arkins, S., Fuh, G., Cunningham, B. C., Wells, J. A., Fong, S., Cronin, M. J., Dantzer, R., and Kelley, K. W., 1992, Growth hormone augments superoxide anion secretion of human neutrophils by binding to the prolactin receptor, *J. Clin. Invest.* **89:**451–457.

Fuh, G., Cunningham, B. C., Fukunaga, R., Nagata, S., Goeddel, D. V., and Wells, J. A., 1992, Rational design of potent antagonists to the human growth hormone receptor, *Science* **256:**1677–1680.

Fukunaga, R., Ishizaka-Ikeda, E., and Nagata, S., 1990a, Purification and characterisation of the receptor for murine granulocyte colony-stimulating factor, *J. Biol. Chem.* **265:**14008–14015.

Fukunaga, R., Ishizaka-Ikeda, E., Seto, Y., and Nagata, S., 1990b, Expression cloning of a receptor for murine granulocyte colony stimulating factor, *Cell* **61:**341–350.

Fukunaga, R., Ishizaka-Ileda, E., Pan, C.-X., Seto, Y., and Nagata, S., 1991, Functional domains of the granulocyte colony-stimulating factor receptor, *EMBO J.* **10:**2855–2865.

Fukunaga, R., Ishizaka, I. E., and Nagata, S., 1993, Growth and differentiation signals mediated by different regions in the cytoplasmic domain of granulocyte colony-stimulating factor receptor, *Cell* **74:**1079–1087.

Galizzi, J. P., Zuber, C. E., Harada, N., Gorman, D. M., Djossou, O., Kastelein, R., Banchereau, J., Howard, M., and Miyajima, A., 1990, Molecular cloning of a cDNA encoding the human interleukin 4 receptor, *Int. Immunol.* **2:**669–675.

Gately, M. K., Desai, B. B., Wolitky, A. G., Quinn, P. M., Dwer, C. M., Podlaski, F. J., Familletti, P. C., Sinigaglia, F., Chizonnite, R., Guler, U., and Stern, A.S., 1991, Regulation of human lymphocyte proliferation by a heterodimeric cytokine, IL-12 (cytotoxic lymphocyte maturation factor), *J. Immunol.* **147:**874–882.

Gearing, D. P., and Cosman, D., 1991, Homology of the p40 subunit of natural killer cell stimulatory factor (NKSF) with the extracellular domain of the interleukin-6 receptor [letter], *EMBO J.* **10:**2839–2848.

Gearing, D. P., King, J. A., Gough, N. M., and Nicola, N. A., 1989, Expression cloning of a receptor for human granulocyte-macrophage colony-stimulating factor, *EMBO J.* **8:**3667–3676.

Gearing, D. P., Thut, C. J., Vanden Bos, T., Gimpel, S. D., Delaney, P. B., King, J., Price, V., Cosman, D., and Becmann, M. P., 1991, Leukemia inhibitory factor receptor is structurally related to the IL-6 signal transducer, gp130, *EMBO J.* **10:**2839–2848.

Gearing, D. P., Comeau, M.R., Friend, D. J., Gimpel, S. D., Thut, C. J., McCourty, J., King, J. A., Gillis, S., Mosley, B., Ziegler, S. F., and Cosman, D., 1992, The IL-6 signal transducer gp130: An oncostatin M receptor and affinity converter for the LIF receptor, *Science* **255:**1434–1437.

Gearing, D. P., Ziegler, S. F., Comeau, M. R., Friend, D., Thoma, B., Cosman, D., Park, L., and Mosley, B., 1994, Proliferative responses and binding properties of hematopoietic cells transfected with low-affinity receptors for leukemia inhibitory factor, oncostatin M, and ciliary neurotrophic factor, *J. Biol. Chem.* **269:**6215–6222.

Giri, J. G., Ahdieh, M., Eisenman, J., Shanebeck, K., Grabstein, K., Kumaki, S., Namen, A., Park, L. S., Cosman, D., and Anderson, D., 1994a, Utilization of the beta and gamma chains of the IL-2 receptor by the novel cytokine IL-15, *EMBO J.* **13:**2822–2830.

Giri, J. G., Grabstein, K., Abdieh, M., Eisenman, J., Paxton, R., Shanebeck, K., Park, L., Satoru, K., Cosman, D., and Anderson, D., 1994b, A novel T cell factor, IL-15 shares both the beta and gamma subunits of the IL-2 receptor, *Cytokine* **5:**553.

Goodall, G. J., Bagley, C. J., Vadas, M. A., and Lopez, A. F., 1993, A model for the interaction of the GM-CSF, IL-3 and IL-5 receptors with their ligands, *Growth Factors* **8:**87–97.

Goodwin, R. G., Friend, D., Ziegler, S. F., Jerzy, R., Falk, B. A., Gimpel, S., Cosman, D., Dower, S. K., March, C. J., Namen, A. E. and Park, L. S., 1990, Cloning of the human and murine interleukin-7 receptors: Demonstration of a soluble form and homology to a new receptor superfamily, *Cell* **60:**941–951.

Gorman, D. M., Itoh, N., Jenkins, N. A., Gilbert, D. J., Copeland, N. G., and Miyajima, A., 1992. Chromosomal localization and organization of the murine genes encoding the β subunits (AIC2A and AIC2B) of the interleukin 3 granulocyte/macrophage colony-stimulating factor, and interleukin 5 receptors, *J. Biol. Chem.* **267:**15842–15848.

Graham, G. J., and Pragnell, I. B., 1990, Negative regulators of haemopoiesis— current advances, *Prog. Growth Factor Res.* **2:**181–192.

Greenlund, A. C., Schreiber, R. D., Goeddel, D. V., and Pennica, D., 1993, Interferon-gamma induces receptor dimerization in solution and on cells, *J. Biol. Chem.* **268:**18103–18110.

Guba, S. C., Sartor, C. A., Hutchinson, R., Boxer, L. A., and Emerson, S. G., 1994, Granulocyte colony-stimulating factor (G-CSF) production and G-CSF receptor structure in patients with congenital neutropenia, *Blood* **83:**1486–1492.

Hall, A. K., and Rao, M. S., 1992, Cytokines and neurokines: Related ligands and related receptors, *Prog. Growth Factor Res.* **4:**139–155.

Hanazono, Y., Chiba, S., Sasaki, K., Mano, H., Miyajima, A., Arai, K., Yazaki, Y., and Hirai, H., 1993a, c-fps/fes protein-tyrosine kinase is implicated in a signaling pathway triggered by granulocyte–macrophage colony-stimulating factor and interleukin-3, *EMBO J.* **12:**1641–1646.

Hanazono, Y., Chiba, S., Sasaki, K., Mano, H., Yazaki, Y., and Hirai, H., 1993b, Erythropoietin induces tyrosine phosphorylation and kinase activity of the *c-fps/fes* protooncogene product in human erythropoietin-responsive cells, *Blood* **81:**3193–3196.

Hapel, A. J., Fung, M. C., Mak, N. K., Morris, C., Metcalf, D., and Nicola, N., 1992, Bone marrow cells from *A/J* mice do not proliterate in interleukin-3 but express normal numbers of interleukin-3 receptors, *Br. J. Haematol.* **82:**488–493.

Hara, T., and Miyajima, A., 1992, Two distinct functional high affinity receptors for mouse interleukin-3 (IL-3) *EMBO J* **11:**1875–1884.

Harada, N., Castle, B. E., Gorman, D. M., Itoh, N., Schreurs, J., Barrett, R. L., Howard, M., and Miyajima, A., 1990, Expression cloning of a cDNA encoding the murine interleukin 4 receptor based on ligand binding, *Proc. Natl. Acad. Sci. U.S.A.* **87:**857–861.

Hatakeyama, M., Minamoto, S., Uchiyama, T., Hardy, R. R., Yamada, G., and Taniguchi, T., 1985, Reconstitution of functional receptor for human interleukin-2 in mouse cells, *Nature* **59:**837–845.

Hatakeyama, M., Mori, H., Doi, T., and Taniguchi, T., 1989a, A restricted cytoplasmic region of IL-2 receptor beta chain is essential for growth signal transduction but not for ligand binding and internalization, *Cell* **318:** 467–470.

Hatakeyama, M., Tsudo, M., Minamoto, S., Kono, T., Doi. T., Miyata, T., Miyasaka, M. and Taniguchi, T., 1989b, Interleukin-2 receptor beta chain gene: Generation of three receptor forms by cloned human alpha and beta chain cDNA's, *Science* **244:**551–556.

Hatakeyama, M., Kono, T., Kobayashi, N., Kawahara, A., Levin, S. D., Perlmutter, R. M., and Taniguchi, T., 1991, Interaction of the IL-2 receptor with the *src-* family kinase p56lck: Identification of novel intermolecular association, *Science* **252:**1523–1528.

Hayashida, K., Kitamura, T., Gorman, D. M., Arai, K., Yokota, T., and Miyajima, A., 1990, Molecular cloning of a second subunit of the receptor for human granulocyte–macrophage colony stimulating factor (GM-CSF): Reconstitution of a high-affinity GM-CSF receptor, *Proc. Natl. Acad. Sci. U.S.A.* **87:**9655–9659.

Heaney, M. L., and Golde, D. W., 1994, Soluble hormone receptors, *Blood* **82:**1945–1948.

Hemmi, S., Bohni, R., Stark, G., Di, M. F., and Aguet, M., 1994, A novel member of the interferon receptor family complements functionality of the murine interferon gamma receptor in human cells, *Cell* **76:**803–810.

Herington, A. C., Ymer, S., Roupas, P., and Stevenson, J., 1986a, Growth hormone-binding proteins in high-speed cytosols of multiple tissues of the rabbit, *Biochim. Biophys. Acta* **881:**236–240.

Herington, A. C., Stevenson, J., Coulson, L., and Ymer, S., 1986b, Characterization of a soluble prolactin-binding activity in rat liver cytosol, *J. Endocrinol.* **109:**61–66.

Heyworth, C. M., Ponting, I. L. O., and Dexter, T. M., 1988, The response of haemopoietic cells to growth factors: Developmental implications of synergistic interactions, *J. Cell. Sci.* **91:**239–247.

Heyworth, C. M., Hampson, J., Dexter, T. M., Walker, F., Burgess, A. W., Kan, O., Cook, N., Vallance, S. J., and Whetton, A. D., 1991, Development of multipotential haemopoietic stem cells to neutrophils is associated with increased expression of receptor for granulocyte macrophage colony stimulating factor and altered biological responses to GM-CSF during development, *Growth Factors* **5**:87–98.

Heyworth, C. M., Whetton, A. D., Nicholls, S., Zsebo, K., and Dexter, T. M., 1992, Stem cell factor directly stimulates the development of enriched granulocyte–macrophage colony-forming cells and promotes the effects of other colony-stimulating factors, *Blood* **80**:2230–2236.

Heyworth, C. M., Dexter, T. M., Nicholls, S. E., and Whetton, A. D., 1993, Protein kinase C activators can interact synergistically with granulocyte colony-stimulating factor or interleukin-6 to stimulate colony formation from enriched granulocyte–macrophage colony-forming cells, *Blood* **81**:894-900.

Hibi, M., Murakami, M., Saito, M., Hirano, T. Taga, T., and Kishimoto, T., 1990, Molecular cloning and expression of an IL-6 signal transducer, gp 130, *Cell* **63**:1149–1157.

Hilton, D. J., and Gough, N. M., 1991, Leukemia inhibitory factor: A biological perspective, *J. Cell. Biochem.* **46**:21–16.

Hilton, D. J., Hilton, A. A., Raicevic, A., Rakar, S., Harrison-Smith, M., Gough, N. M., Begley, C. G., Metcalf, D., Nicola, N. A., and Willson, T. A., 1994, Cloning of a murine IL-11 receptor α-chain; requirement for gp130 for high affinity binding and signal transduction, *EMBO J* **13**:4765–4775.

Ho, A. S.-Y., and Moore, K. W., 1994, Interleukin-10 and its receptor, *Ther. Immunol.* **1**:173–185.

Ho, A. S., Liu, Y., Khan, T. A., Hsu, D. H., Bazan, J. F., and Moore, K. W., 1993, A receptor for interleukin 10 is related to interferon receptors, *Proc. Natl. Acad. Sci. U.S.A.* **90**:11267–11271.

Holmes, W. E., Lee, J. Kuang, W-J., Rice, G. C., and Wood, W. I., 1991, Structure and functional expression of a human interleukin-8 receptor, *Science* **253**:1278–1282.

Hou, J., Schindler, U., Henzel, W. J., Ho, T. C., Brasseur, M., and McKnight, S.L., 1994, An interleukin-4-induced transcription factor: IL-4 stat, *Science* **265**:1701–1706.

Howard, M. C., Miyajima, A., and Coffman, R., 1993, T-cell derived cytokines and their receptors, in *Fundamental Immunology,* (W.E. Paul, ed.), pp. 763–800. Raven Press, New York.

Howard, S. T., Chan, Y. S., and Smith, G. L., 1990, Vaccinia virus homologues of the Shope fibroma virus inverted terminal repeat proteins and a discontinuous ORF related to the tumor necrosis factor receptor family, *Science* **248**:1019–1023.

Hunter, T., and Cooper, T. H., 1985, Protein-tyrosine kinases, *Annu. Rev. Biochem.* **54**:897–930.

Ichihara, M., Hara, T., Takagi, M., Cho, L. C., Gorman, D. M., and Miyajima, A., 1995, Impaired interleukin-3 (IL-3) response of the *A/J* mouse is caused by a branch point deletion in the IL-3 receptor α subunit gene, *EMBO J.* **14**:939–950.

Idzerda, R. L., March, C. J., Mosely, B., Lyman, S. D., Vanden Bos, T., Gimpel, S. D., Din, W. S., Grabstein, K., Widmer, M. B., Park, L. S., Cosman, D., and Backmann, M. P., 1990, Human interleukin 4 receptor confers biolgoical responsiveness and defines a novel receptor superfamily, *J. Exp. Med* **171**:861–873.

Ihle, J. N., Witthuhn, B., Tang, B., Yi, T., and Quelle, F. W., 1994a, Cytokine receptors and signal transduction in *Bailliere's Clinical Haematology*, pp. 17–48, Bailliere Tindall, London.

Ihle, J. N., Witthuhn, B. A., Quelle, F. W., Yamamoto, K., Thierfelder, W. E., Kreider, B., and Silvennoinen, O., 1994b, Signaling by the cytokine receptor superfamily: JAKs and STATs, *Trends Biochem. Sci.* **19**:222–227.

Ikebuchi, K., Clark, S. C., Ihle, J. N., Souza, L. M., and Ogawa, M., 1988a, Granulocyte colony-stimulating factor enhances interleukin 3-dependent proliferation of multipotential hemopoietic progenitors, *Proc. Natl. Acad. Sci.* **85**:3445–3449.

Ikebuchi, K., Ihle, J. N., Hirai, Y., Wong, G. G., Clark, S. C., and Ogawa, M., 1988b, Synergistic factors for stem cell proliferation: Further studies of the interleukin-6, and granulocyte colony-stimulating factor, *Blood* **72**:2007–2014.

Ip, N.Y., Nye, S. H., Boulton, T. G., Davis, S., Taga, T., Li, Y., Birren, S. J., Yasukawa, K., Kishimoto, T., Anderson, D. J., 1992. CNTF and LIF act on neuronal cells via shared sigtnaling pathways that involve the IL-6 signal transducing receptor component gp130, *Cell* **69**:1121–1132.

Isfort, R. J., and Ihle, J. N., 1990, Multiple hematopoietic growth factors signal through tyrosine phosphorylation, *Growth Factors* **2**:213–220.

Ishizaka, I. E., Fukunaga, R., Wood, W. I., Goeddel, D. V., and Nagata, S., 1993. Signal transduction mediated by growth hormone receptor and its chimeric molecules with the granulocyte colony-stimulating factor receptor, *Proc. Natl. Acad. Sci. U.S.A.* **90**:123–127.

Itoh, N., Yonehara, S., Schreurs, J., Gorman, D. M., Maruyama, K., Ishii, A., Yahara, I., Arai, K., and Miyajima, A., 1990, Cloning of an interleukin-3 receptor: A member of a distinct receptor gene family, *Science* **247**: 324–327.

Izquierdo, M., and Cantrell, D. A., 1993, Protein tyrosine kinases couple the interleukin-2 receptor to p21ras, *Eur J. Immunol.* **23:**131–135.

Izuhara, K., and Harada, N., 1993, Interleukin-4 (IL-4) induces protein tyrosine phosphorylation of the IL-4 receptor and association of phosphatidylinositol 3-kinase to the IL-4 receptor in a mouse T cell line, HT2, *J. Biol. Chem.* **268:**13097–13102.

Izuhara, K., Feldman, R. A., Greer, P. and Harada, N., 1994, Interaction of the *c-fes* proto-oncogene product with the interleukin-4 receptor, *J. Biol. Chem.* **269:**18623-18629.

Jacobson, N. G., Szabo, S. J., Weber, Nordt, R. M., Zhong, Z., Schreiber, R. D., Darnell, J. Jr., and Murphy, K. M., 1995, Interleukin-12 signaling in T helper type 1 (Thl) cells involves tyrosine phosphorylation of signal trandsducer and activator of transcription (STAT) 3 and STAT4, *J. Exp. Med.* **181:**1755–1762.

Johnston, J. A., Kawamura, M., Kirken, R. A., Chen, Y. Q., Blake, T. B., Shibuya, K., Ortaldo, J. R., McVicar, D. W., and O'Shea, J. J., 1994, Phosphorylation and activation of the Jak-3 Janus kinase in response to interleukin-2, *Nature* **370:**151–153.

Jones, S. S., D'Andrea, A. D., Haines, L. L., and Wong, G. G., 1990, Human erythropoietin receptor: Cloning, expression, and biologic characterization, *Blood* **76:**31–35.

Kan, O., Evans, C. A., and Whetton, A. D., 1992, Cellular events stimulated by myeloid haemopoietic growth factors in *Bailliere's Clinical Haematology* (T. M. Dexter and B. I. Lord, eds.), pp. 653–679, Bailliere Tindall, London.

Kanakura, Y., Druker, B., Cannistra, S. A., Furukawa, Y., Torimoto, Y. and Griffin, J. D., 1990, Signal transduction of the human granulocyte–macrophage colony-stimulating factor and interleukin-3 receptors involves tyrosine phosphorylation of a common set of cytoplasmic proteins, *Blood* **76:**706–715.

Kashishian, A., Kazlauskas, A., and Cooper, J. A., 1992, Phosphorylation sites in the PDGF receptor with different specificities for binding GAP and PI3 kinase *in vivo*, *EMBO J.* **11:**1373–1382.

Kinoshita, T., Yokota, T., Arai, K.-I., and Miyajima, A., 1995, Suppression of apoptotic death in hematopoietic cells by signalling through the IL-3/GM-CSF receptors, *EMBO J.* **14:**266–275.

Kishimoto, T., Akira, S., and Taga, T., 1992a, Interleukin-6 and its receptor: A paradigm for cytokines, *Science* **258:**593–597.

Kishimoto, T., Hibi, M., Murakami, M., Narazaki, M., Saito, M. and Taga, T., 1992b, The molecular biology of interleukin 6 and its receptor in *Polyfunctional Cytokines: IL-6 and LIF* (G. R. Bock, J. Marsh, and K. Widdows, eds.), pp. 6–23, John Wiley & Sons, West Sussex.

Kitamura, T., Hayashida, K., Sakamaki, K., Yokota, T., Arai, K., and Miyajima, A., 1991a, Reconstitution of functional human granulocyte/macrophage colony-stimulating factor (GM-CSF): Evidence that AIC2B is a subunit of murine GM-CSF receptor, *Proc. Natl. Acad. Sci. U.S.A.* **88:**5082–5086.

Kitamura, T., Sato, N., Arai, K., and Miyajima, A., 1991b, Expression cloning of the human IL-3 receptor cDNA reveals a shared β subunit for human IL-3 and GM-CSF receptors, *Cell* **66:**1165–1174.

Kitamura, T., Ogorochi, T., and Miyajima, A., 1994, Multimeric cytokine receptors, *Trends Endocrinol. Metab.* **5:**8–14.

Kobayashi, M. L., Fitz, M., Ryan, M., Hewick, R. M., Clark, S. C., Chan, S., Loudon, R., Sherman, F., Perussia, B., and Trinchieri, G., 1989, Identification and purification of natural killer cell stimulating factor (NKSF), a cytokine with multiple biologic effects on human lymphocytes, *J. Exp. Med.* **170:**827–845.

Koike, K., Ogawa, M., Ihle, J. N., Miyake, T., Shimizu, T., Miyajima, A., Yokota, T. and Arai, K., 1987, Recombinant murine granulocyte-macrophage (GM) colony stimulating factor supports formation of GM and multipotential blast cell colonies in culture: Comparison with the effects of interleukin-3, *J. Cell. Physiol.* **131:**458–464.

Kondo, S., Shimizu, A., Saito, Y., Kinoshita, M., and Honjo, T., 1986, Molecular basis for two different affinity states of the interleukin 2 receptor: Affinity conversion model, *Proc. Natl. Acad. Sci. U.S.A.* **83:**9026–9029.

Kondo, M., Takeshita, T., Ishii, N., Nakamura, M., Watanabe, S., Arai, K., and Sugamura, K., 1993. Sharing of the interleukin-2 (IL-2) receptor gamma chain between receptors for IL-2 and IL-4, *Science* **262:**1874–1877.

Kondo, M., Takeshita, T., Higuchi, M., Nakamura, M., Sudo, T., Nishikawa, S., and Sugamura, K., 1994, Functional participation of the IL-2 receptor gamma chain in IL-7 receptor complexes, *Science* **263:**1453–1454.

Korpelainen, E.I., Gamble, J.R., Smith, W.B., Goodall, G.J., Qiyu, S., Woodcock, J.M., Dottore, M., Vadas, M.A., and Lopez, A.F., 1993, The receptor for interleukin 3 is selectively induced in human endothelial cells by tumor necrosis factor alpha and potentiates interleukin 8 secretion and neutrophil transmigration, *Exp. Cell. Res.* **206:**311–317.

Landschulz, K. T., Noyes, A.N., Rogers, O., and Boyer, S. H., 1989, Erythropoietin receptors on murine erythroid colony-forming units: Natural history, *Blood* **73:**1476–1486.

Larsen, A., Davis, T., Curits, B., Gimpel, S., Sims, J. E., Cosman, D., Park, L., Sorensen, E., March, C. J., and Smith, C. A., 1990, Expression cloning of a human granulocyte colony-stimulating factor receptor: A structural mosaic of hematopoietin receptor, immunoglobulin, and fibronectin domains, *J. Exp. Med.* **172:**1559-1570.

Leary, A. G., Zeng, H. Q., Clark, S. C., and Ogawa, M., 1992, Growth factor requirements for survival on G_0 and entry into the cell cycle of primitive human hemopoietic progenitors, *Proc. Natl. Acad. Sci. U.S.A.* **89:**4013–4017.

Lehmann, J., Seegert, D., Strehlow, I., Schindler, C., Lohmann, M. M., and Decker, T., 1994, IL-10-induced factors belonging to the p91 family of proteins bind to IFN-gamma-responsive promoter elements, *J. Immunol.* **153:**165–172.

Leonard, W., 1992, The interleukin-2 receptor; structure, functional, intracellular messengers and molecular regulation in *Interleukin-2* (J. Waxman and F. Balkwill, eds.), pp. 29–46, Blackwell Scientific Publications, London.

Leonard, W. J., Depper, J. M., Crabtree, G. R., Rudikoff, S., Pumphrey, J., Robb, R. J., Kronke, M., Svetlik, P. B., Peffer, N. J., Waldmann, T. A., and Greene, W. C., 1984, Molecular cloning and expression of cDNAs for the human interleukin-2 receptor, *Nature* **311:**626–631.

Leonard, W. J., Depper, J. M., Kanehisa, M., Kronke, M., Peffer, N. J., Svetlik, P. B., Sullivan, M., and Greene, W. C., 1985, Structure of the human interleukin-2 receptor gene, *Science* **230:**633–639.

Leonard, W. J., Noguchi, M., Russell, S. M., and McBride, O. W., 1991, The molecular basis of X-linked severe combined immunodeficiency: The role of the interleukin-2 receptor gamma chain as a common gamma chain, gamma c, *Cell* **66:**9–10.

Leonard, W. J., Noguchi, M., Russell, S. M., and McBride, O. W., 1994, The molecular basis of X-linked severe combined immunodeficiency—the role of the interleukin-2 receptor-gamma chain as a common gamma-chain, gamma(c), *Immunol. Rev.* **138:**61–86.

Leung, D. W., Spencer, S. A., Cachianes, G., Hammonds, R. G., Collins, C., Henzel, W. J., Barnard, R., Waters, M. J., and Wood, W. I., 1987, Growth hormone receptor and serum binding protein: Purification, cloning and expression, *Nature* **330:**537–543.

Lev, S., Givol, D., and Yarden, Y., 1992, Interkinase domain of kit contains the binding sites for phosphatidylinositol 3′ kinase, *Proc. Natl. Acad. Sci. U.S.A.* **89:**678–682.

Li, J. P., D'Andrea, A. D., Lodish, H.F., and Baltimore, D., 1990, Activation of cell growth by binding of Friend spleen focus-forming virus gp55 glycoprotein to the erythropoietin receptor, *Nature* **343:**762–764.

Lin, H. Y., Wang, X. F., Ng, E. E., Weinberg, R. A., and Lodish, H. F., 1992, Expression cloning of the TGF-beta type II receptor, a functional transmembrane serine/threonine kinase [published erratum appears in *Cell*, 1992, **70**(6): following 1068], *Cell* **68:**775–785.

Liu, Y., Wei, S. H., Ho, A. S., deWaal Malefyt, R., and Moore, K. W., 1994, Expression cloning and characterization of a human IL-10 receptor, *J. Immunol.* **152:**1821–1829.

Lopez, A. F., To, L. B., Yang, Y. C., Gamble, J. R., Shannon, M. F., Burns, G. F., Dyson, P. G., Juttner, C. A., Clark, S., and Vadas, M. A., 1987, Stimulation of proliferation, differentiation, and function of human cells by primate interleukin 3. *Proc. Natl. Acad. Sci. U.S.A.* **84:**2761–2765.

Lopez, A. F., Eglinton, J.M., Lyons, A. B., Tapley, P. M., To, L. B., Park, L. S., Clark, S. C., and Vadas, M. A., 1990. Human interleukin-3 inhibits the binding of granulocyte–macrophage colony-stimulating factor and interleukin-5 to basophils and strongly enhances their functional activity, *J. Cell. Physiol.* **145:**69–77.

Lopez, A. F., Shannon, M. F., Hercus, T., Nicola, N. A., Cambareri, B., Dottore, M., Layton, M. J., Eglinton, L., and Vadas, M. A., 1992, Residue 21 of human granulocyte–macrophage colony-stimulating factor is critical for biological activity and for high but not low affinity binding, *EMBO J.* **11:**909–916.

Lopez-Casillas, F., Cheifetz, S., Doody, J., Andres, J. L., Lane, W. S., and Massague, J., 1991. Structure and expression of the membrane proteoglycan betaglycan, a component of the TGF-β receptor system, *Cell* **67:**785–795.

Lopez-Casillas, F., Wrana, J. L., and Massague, J., 1993, Betaglycan presents ligand to the TGF beta signaling receptor, *Cell* **73:**1435–1444.

Lopez-Casillas, F., Payne, H. M., Andres, J. L., and Massague, J., 1994, Betaglycan can act as a dual modulator of TGF-beta access to signaling receptors: Mapping of ligand binding and GAG attachment sites, *J. Cell Biol.* **124:**557–568.

Lord, K. A., Abdollahi, A., Thomas, S. M., De Marco, M., Brugge, J. S., Hoffman-Liebermann, B., and Liebermann, D. A., 1991. Leukemia inhibitory factor and interleukin-6 trigger the same immediate early response, including tyrosine phosphorylation, upon induction of myeloid leukemia differentiation, *Mol. Cell. Biol.* **11:**4371–4379.

Lowenthal, J. W., Castle, B. E., Christiansen, J., Schreurs, J., Rennick, D., Arai, N., Hoy, P., Takebe, Y., and Howard, M., 1988, Expression of high affinity receptors for murine interleukin 4 (BSF-1) on hemopoietic and nonhemopoietic cells, *J. Immunol.* **140:**456–464.

Lutticken, C., Wegenka, U. M., Yuan, J., Buschmann, J., Schindler, C., Ziemiecki, A., Harpur, A. G., Wilks, A. F., Yasukawa, K., Taga, T., and Kishimoto, T., 1994, Association of transcription factor APRF and protein kinase Jak1 with the interleukin-6 signal transducer gp130, *Science* **263:**89–92.

Maliszewski, C. R., Sato, T. A., Vanden, B. T., Waugh, S., Dower, S. K., Slack, J., Beckmann, M. P., and Grabstein, K. H., 1990, Cytokine receptors and B cell functions. I. Recombinant soluble receptors specifically inhibit IL-1- and IL-4-induced B cell activities *in vitro, J. Immunol.* **144:**3028–3033.

Mallett, S., and Barclay, A. N., 1991, A new superfamily of cell surface proteins related to the nerve growth factor receptor, *Immunol. Today* **12:**220–223.

Massague, J., 1987, The TGF-β family of growth and differentiation factors, *Cell* **49:**437–438.

Massague, J., Attisano, L., and Wrana, J. L., 1994, The TGF-beta family and its composite receptors, *Trends Cell. Biol.* **4:**172–178.

Matsushime, H., Roussel, M. F., Ashmun, R. A., and Sherr, C. J., 1991, Colony-stimulating factor 1 regulates novel cyclins during the G_1 phase of the cell cycle, *Cell* **65:**701–713.

Matsushime, H., Quelle, D. E., Shurtleff, S. A., Shibuya, M., Sherr, C. J., and Kato, J. Y., 1994, D-type cyclin-dependent kinase activity in mammalian cells, *Mol. Cell. Biol.* **14:**2066–2076.

Matthews, W., Jordan, C. T., Wiegand, G. W., Pardoll, D., and Lemischka, I. R., 1991, A receptor tyrosine kinase specific to hematopoietic stem and progenitor cell enriched populations, *Cell* **65:**1143–1152.

Merberg, D. M., Wolf, S. F., and Clark, S. C., 1990, Sequence similarity between NKSF and the IL-6/G-CSF family [letter], *Curr. Top. Microbiol. Immunol.* **166:**13–22.

Merlin, G., Falcoff, E., and Aguet, M., 1985, ^{125}I-Labelled human interferons alpha, beta and gamma: Comparative receptor-binding data, *J Gen. Virol.* **66:**1149–1152.

Metcalf, D., 1985, The granulocyte/macrophage colony-stimulating factors, *Science* **229:**16–22.

Metcalf, D., 1986, The molecular biology and functions of the granulocyte–macrophage colony-stimulating factors, *Blood* **67:**257–267.

Metcalf, D., and Nicola, N. A., 1991, Direct proliferative actions of stem cell factor on murine bone marrow cells *in vitro.* Effects of combination with colony-stimulating factors, *Proc. Natl. Acad. Sci. U.S.A.* **88:**6239–6243.

Minty, A., Chalon, P., Derocq, J. M., Dumont, X., Guillemot, J. C., Kaghad, M., Labit, C., Leplatois, P., Liauzun, P., Miloux, B., and Casellas, P., Loison, G., Lupker, J., Shire, D. Ferrara, P., and Caput, D., 1993, Interleukin-13 is a new human lymphokine regulating inflammatory and immune responses, *Nature* **362:**248–250.

Mita, S., Tominaga, A., Hitoshi, Y., Honjo, T., Sakamoto, K., Akagi, M., Kikuchi, Y., and Takatsu, K., 1989, Characterization of high-affinity receptors for interleukin5 (IL-5) on IL-5-dependent cell lines, *Proc. Natl. Acad. Sci. U.S.A.* **86:**2311–2315.

Miyajima, A., 1992, Molecular structure of the IL-3, GM-CSF, and IL-5 receptors, *Int. J. Cell Clon.* **10:** 126–134.

Miyajima, A., Kitamura, T., Harada, N., Yokota, T., and Arai, K., 1992, Cytokine receptors and signal transduction, *Annu. Rev. Immunol.* **10:**295–331.

Miyajima, A., Mui, A.L., Ogorochi, T., and Sakamaki, K., 1993. Receptors for granulocyte-macrophage colony-stimulating factor, interleukin-3, and interleukin-5, *Blood* **82:**1960–1974.

Miyazaki, T., Kawahara, A., Fujii, H., Nakagawa, Y., Minami, Y., Liu, Z-J., Oishi, I., Silvennoinen, O., Witthuhn, B. A., Ihle, J. N., and Taniguchi, T., 1994, Functional activation of Jak1 and Jak3 by selective association with IL-2 receptor subunits, *Science* **266:**1045–1047.

Miyazawa, K. Hendrie, P. C., Mantel, C., Wood, K., Ashman, L. K., and Broxmeyer, H. E., 1991, Comparative analysis of signaling pathways between mast cell growth factor (c-kit ligand) and granulocyte–macrophage colony-stimulating factor in a human factor-dependent myeloid cell line involves phosphorylation of Raf-1, GTPase-activating protein and mitogen-activated protein kinase, *Exp. Hematol.* **19:**1110-1123.

Molineux, G., Pojda, Z., and Dexter, T. M., 1990, A comparison of hematopoiesis in normal and splenectomised mice treated with granuloycyte-colony stimulating factor, *Blood* **75:**563–569.

Moore, K. W., Vieira, P., Fiorentino, D. F., Trounstine, M. L., Khan, T. A., and Mosmann, T. R., 1990, Homology of cytokine synthesis inhbitory factor (IL-10) to the Epstein–Barr virus gene BCRFI [published erratum appears in *Science* 1990, **250**(4980):494], *Science* **248:**1230–1234.

Moore, K. W., O'Garra, A., deWaal Malefyt, R., Vieira, P., and Mosmann, T. R., 1993, Interleukin-10, *Annu. Rev. Immunol.* **11:**165–190.

Mosley, B., Beckmann, M. P., March, C. J., Idzerda, R. L., Gimpel, S. D., VandenBos, T., Friend, D., Alpert, A., Anderson, D., Jackson, J., Wignall, J. M., Smith, C., Gallis, B., Sims, J. E. Urdal, D., Widmer, M. B., Cosman, D., and Park, L. S., 1989, The murine interleukin-4 receptor: Molecular cloning and characterization of secreted and membrane bound forms, *Cell* **59:**335-348.

Mosely, B., Delmus, C., Friend, D., Thoma, B., and Cosman, D., 1994, The oncostatin-M specific receptor: Cloning of a novel subunit related to the LIF receptor, *Cytokine* **6:**554.

Mui, A. L.-F., Wakao, H., O'Farrell, A.-M., Harada, N., and Miyajima, A., 1995, Interleukin-3, granulocyte macrophage colony stimulating factor and interleukin-5 transduce signals through two STAT5 homologs, *EMBO J.* **14:**1166–1175.

Muller, M., Laxton, C., Briscoe, J., Schindler, C., Improta, T., Darnell, J. J., Stark, G. R., and Kerr, I. M., 1993, Complementation of a mutant cell line: Central role of the 91 kDa polypeptide of ISGF3 in the interferon-alpha and -gamma signal transduction pathways, *EMBO J.* **12:**4221–4228.

Murakami, M., Hibi, M., Nakagawa, N., Nakagawa, T., Yasukawa, K. Yamanishi, K., Taga, T., and Kishimoto, T., 1990, IL-6-induced homodimerization of gp130 and associated activation of a tyrosine kinase, *J. Exp. Med.* **171:**2001–2009.

Murakami, M., Narazaki, M., Hibi, M., Yawata, H., Yasukawa, K., Hamaguchi, M., Taga, T., and Kishimoto, T., 1993, Critical cytoplasmic region of the interleukin 6 signal transducer gp130 is conserved in the cytokine receptor family, *Science* **260:**1808–1810.

Murata, Y., Yamaguchi, N., Hitoshi, Y., Tominaga, A., and Takatsu, K., 1990, Interleukin 5 and interleukin 3 induce serine and tyrosine phosphorylations of several cellular proteins in an interleukin 5-dependent cell line, *Biochem. Biophys. Res. Commun.* **173:**1102–1108.

Murray, R. Suda, T., Wrighton, N., Lee, F., and Zlotnik, A., 1989, IL-7 is a growth and maintenance factor for mature and immature thymocyte subsets, *Int. Immunol.* **1:**526–531.

Murray, R., Lee, F., and Chiu, C. P., 1993, The genes for leukemia inhibitory factor and interleukin-6 are expressed in mouse blastocysts prior to the onset of hemopoiesis, *J. Immunol.* **151:**2555–2561.

Musashi, M., Clark, S. C., Sudo, T., Urdal, D. L., and Ogawa, M., 1991a, Synergistic interactions between interleukin-11 and interlukin-4 in support of proliferation of primitive hematopoietic progenitors of mice, *Mol. Cell. Biol.* **11:**4371–4379.

Musashi, M., Yang, Y. C., Paul, S. R., Clark, S. C., Sudo, T., and Ogawa, M., 1991b, Direct and synergistic effects of interleukin 11 on murine hemopoiesis in culture, *Blood* **78:**1448–1451.

Nagata, S., and Fukunaga, R., 1991, Granulocyte colony-stimulating factor and its receptor, *Prog. Growth Factor Res.* **3:**131–141.

Nakajima, K., and Wall, R., 1993, Interleukin-6 signals activating *junB* and *TIS11* gene transciption in a B-cell hybridoma, *Proc. Natl. Acad. Sci. U.S.A.* **90:**11924–11928.

Nakamura, Y., Russell, S. M., Mess. S. A., Friedmann, M., Erdos, M., Francois, C., Jacques, Y., Adelstein, S., and Leonard, W. J., 1994, Heterodimerization of the IL-2 receptor beta- and gamma-chain cytoplasmic domains is required for signalling, *Nature* **369:**330–333.

Namen, A. E., Lupton, S., Hjerrild, K., Wignall, J., Mochizuki, D. Y., Schmierer, A., Mosley, B., March, C. J., Urdal, D., Gillis, S., Cosman, D., and Goodwin, R. G., 1988, Stimulation of B-cell progenitors by cloned murine interleukin-7, *Nature* **333:**571–573.

Narazaki, M., Yasukawa, K., Saito, T., Ohsugi, Y., Fukui, H., Koishihara, Y., Yancopoulos, G. D., Taga, T., and Kishimoto, T., 1994a, Soluble forms of the interleukin-6 signal-transducing receptor component gp130 in human serum possessing a potential to inhibit signals through membrane-anchored gp130, *J. Exp. Med.* **179:**1337–1342.

Narazaki, M., Witthuhn, B. A., Yoshida, K., Silvennoinen, O., Yasukawa, K., Ihle, J. N., Kishimoto, T., and Taga, T., 1994b, Activation of JAK2 kinase mediated by the inteuleukin 6 signal transducer gp130, *Proc. Natl. Acad. Sci. U.S.A.* **91:**2285–2289.

Nicola, N. A., 1989, Hemopoietic cell growth factors and their receptors, *Annu. Rev. Biochem.* **58:**45–77.

Nicola, N. A., Metcalf, D., Johnson, G. R., and Burgess, A. W., 1979. Separation of functionally distinct human granulocyte–macrophage colony-stimulating factors, *Blood* **54:**614–627.

Nicola, N. A., Metcalf, D., Matsumoto, M., and Johnson, G. R., 1983, Purification of a factor inducing differentiation in murine myelomonocytic leukemia cells, *J. Biol. Chem.* **258:**9017–9023.

Nishinakamura, R., Miyajima, A., and Murray, R., 1995, Mice deficient for the IL-3/GM-CSF/IL-5 receptor exhibit lung pathology and impaired immune resonse while beta IL-3 receptor-deficient mice are normal, *Immunity* **2:**211–222.

Noguchi, M., Yi, H., Rosenblatt, H. M., Filipovich, A. H., Adelstein, S., Modi, W. S., McBride, A. W., and Leonard, W. J., 1993a, Interleukin-2 receptor γ chain mutation results in X-linked severe combined immunodeficiency in humans, *Cell* **73:**147–157.

Noguchi, M., Nakamura, Y., Russell, S. M., Ziegler, S. F., Tsang, M., Cao, X., and Leonard, W. J., 1993b, Interleukin-2 receptor gamma chain: A functional component of the interleukin-7 receptor, *Science* **262:** 1877–1880.

O'Farrell, A.-M., Ichihara, M., Mui, A., and Miyajima, A., 1996, Signaling pathways activated in a unique mast cell line where IL-3 supports survival and SCF is required for a proliferative response, *Blood* **87:**3655–3668.

Park, L. S., Friend, D., Price, V., Anderson, D., Singer, J., Prickett, K. S., and Urdal, D. L., 1989, Heterogeneity in human interleukin-3 receptors. A subclass that binds human granulocyte/macrophage colony stimulating factor, *J. Biol. Chem.* **264:**5420–5427.

Park, L. S., Martin, U., Sorensen, R., Luhr, S., Morrissey, P. J., Cosman, D., and Larsen, A., 1992, Cloning of the low-affinity murine granulocyte–macrophage colony-stimulating factor receptor and reconstitution of a high-affinity receptor complex, *Proc. Natl. Acad. Sci. U.S.A.* **89:**4295–4299.

Patthy, L., 1990. Letter to the editor: Homology of a domain of the growth hormone/prolactin receptor family with type III modules of fibronectin, *Cell* **61:**13–14.

Pennica, D., King, K. L., Chien, K. R., Baker, J. B., and Wood, W. I., 1994, Cardiotrophin-1, a novel cytokine that induces cardiac myocyte hypertrophy, *Cytokine* **6:**577.

Punnonen, J., Aversa, G., Cocks, B. G., McKenzie, A. N., Menon, S., Zurawski, G., de Waal Malefyt, R., and de Vries, J., 1993, Interleukin 13 induces interleukin 4-independent IgG4 and IgE syntehsis and CD23 expression by human B cells, *Proc. Natl. Acad. Sci. U.S.A.* **90:**3730-3734.

Quelle, F. W., Sato, N., Witthuhn, B. A., Inhorn, R. C., Eder, M., Miyajima, A., Griffin, J. D., and Ihle, J. N., 1994, JAK2 associates with the beta c chain of the receptor for granulocyte–macrophage colony-stimulating factors, and its activation requires the membrane-proximal region, *Mol. Cell. Biol.* **14:**4335–4341.

Quesniaux, V. F., Clark, S. C., Turner, K., and Fagg, B., 1991, Interleukin-11 stimulates multiple phases of erythropoiesis *in vitro, Proc. Natl. Acad. Sci. U.S.A.* **88:**765–769.

Ravichandran, K. S., and Burakoff, S. J., 1994, The adapter protein Shc interacts with the interleukin-2 (IL-2) receptor upon IL-2 stimulation, *J. Biol. Chem.* **269:**1599-1602.

Reedijk, M., Liu, X., Van der Geer, P., Letwin, K., Waterfield, M. D., Hunter, T., and Pawson, T., 1992, Tyr721 regulates specific binding of the CSF-1 receptor kinase insert to PI3′-kinase SH2 domains: A model for SH2-mediated receptor-target interactions, *EMBO J.* **11:**1365–1372.

Renauld, J. C., Kermouni, A., Vink, A., Louahed, J., and Van Snick, J., 1995, Interleukin-9 and its receptor: involvement in mast cell differentiation and T cell oncogenesis, *J. Leuk. Biol.* **57:**353–360.

Renauld, J. C., Druez, C., Kermouni, A., Houssiau, F., Uyttenhove, C., Van Roost, E., and Van Snick, J., 1992, Expression cloning of the murine and human interleukin 9 receptor cDNAs, *Proc. Natl. Acad. Sci. U.S.A.* **89:** 5690–5694.

Rennick, D., Moore, R. G., and Thompson-Snipes, L., 1992, IL-4 and hematopoiesis, in *IL-4: Structure and Function* (H. spits, ed.), pp. 151–168, CRC Press, Boca Raton.

Robb, R. J., and Rusk, C.M., 1986, High and low affinity receptors for interleukin 2: Implications of pronase, phorbol ester, and cell membrane studies upon the basis for differential ligand affinities, *J. Immunol.* **137:** 142–159.

Rode, H.-J., Janssen, W., and Rosen-Wolff, A., 1993, The genome of equine herpesvirus type 2 harbors an interleukin-10 (IL-10)-like gene, *Virus Genes* **7:**111–116.

Rose, J. S., and Heinrich, P. C., 1994, Soluble receptors for cytokines and growth factors: Generation and biological function, *Biochem. J.* **300:**281–290.

Rose, T. M., and Bruce, A. G., 1992, Oncostatin M is a member of a cytokine family that includes leukemia-inhibitory factor, granulocyte colony-stimulating factor, and interleukin 6, *J. Immunol.* **149:**1271–1275.

Rothman, P., Kreider, B., Azam, M., Levy, D., Wegenka, U., Eilers, A., Decker, T., Horn, F., Kashleva, H., Ihle, J., and Schindler, C., 1994, Cytokine and growth factors signal through tyrosine phosphorylation of a family of related transcription factors, *Immunity* **1:**1–20.

Roussel, M. F., Dull, J. F., Rettenmier, C. W., Ralph, P., Ullrich, A., and Sherr, C.J., 1987, Transforming potential of the *c-fms* proto-oncogene product, *Nature* **325:**549–550.

Roussel, M. F., Downing, J. R., Rettenmier, C. W., and Sher, C. J., 1988, A point mutation in the extracellular domain of the human CSF-1 receptor (*c-fms*) proto-oncogene product activates its transforming potential, *Cell* **55:**979–988.

Ruscetti, T. W., Dubois, C., Falk, L. A., Jacobsen, S. E., Sing, G., Longo, D. L., Wiltrout, R. H., and Keller, J. R., 1991, Clinical applications of TFG-β. *In vivo* and *in vitro* effects of TGF-β1 on normal and neoplastic haemopoiesis, in *Ciba Foundation Symposium 157* (R. Gregory Bock and J. Marsh, eds.), pp. 212–231, John Wiley & Son, New York.

Russell, S. M., Keegan, A. D., Harada, N., Nakamura, Y., Noguchi, M., Leland, P., Friedmann, M. C., Miyajima, A., Puri, R. K., Paul, W. E., and Leonard, W. J., 1993, Interleukin-2 receptor gamma chain: A functional component of the interleukin-4 receptor, *Science* **262:**1880–1883.

Russell, S. M., Johnston, J. A., Noguchi, M., Kawamura, M., Bacon, C. M., Friedmann, M. Berg, M., McVicar, W., Witthuhn, B.A., Silvennoinen, O., Goldman, A. S., Schmalsteig, F. C., Ihle, J., O'Shea, J., and Leonard, W. J., 1994, Interaction of IL-2R beta and gamma-c chains with JAK1 and JAK3: Implications for XSCID and XCID, *Science* **266:**1042–1045.

Sakamaki, K., Miyajima, I., Kitamura, T., and Miyajima, A., 1992, Critical cytoplasmic domains of the common beta subunit of the human GM-CSF, Il-3 and IL-5 receptors for growth signal transduction and tyrosine phosphorylation, *EMBO J.* **11:**3541–3549.

Sanderson, C. J., 1992, Interleukin-5, eosinophils, and disease, *Blood* **79:**3101–3109.

Sato, N., Caux, C., Kitamura, T., Watanabe, Y., Arai, K., Banchereau, J., and Miyajima, A., 1993a, Expression and factor-dependent modulation of the interleukin-3 receptor subunits on human hematopoietic cells, *Proc. Natl. Acad. Sci. U.S.A.* **90:**11137–11141.

Sato, N., Sakamaki, K., Terada, N., Arai, K., and Miyajima, A., 1993b, Signal transduction by the high-affinity GM-CSF receptor: Two distinct cytoplasmic regions of the common beta subunit responsible for different signaling, *EMBO J.* **12:**4181–4189.

Satoh, T., Minami, Y., Kono, T., Yamada, K., Kawahara, A., Taniguchi, T., and Kaziro, Y., 1992a, Interleukin 2-induced activation of Ras requires two domains of interleukin 2 receptor beta subunit, the essential region for growth stimulation and Lck-binding domain, *J. Biol. Chem.* **267:**25423–25427.

Satoh, T., Uehara, Y., and Kaziro, Y., 1992b, Inhibition of interleukin 3 and granulocyte–macrophage colony-stimulating factor stimulated increase of active ras·GTP by herbimycin A, a specific inhibitor of tyrosine kinases, *J. Biol. Chem.* **267:**2537–2541.

Schindler, C., Fu, X. Y., Improta, T., Aebersold, R., and Darnell, J. J., 1992a, Proteins of transcription factor ISGF-3: One gene encodes the 91- and 84-kDa ISGF-3 proteins that are activated by interferon alpha, *Proc. Natl. Acad. Sci. U.S.A.* **89:**7836–7839.

Schindler, C., Shuai, K., Prezioso, V. R., and Darnell, J.J., 1992b, Interferon-dependent tyrosine phosphorylation of a latent cytoplasmic transcription factor [see comments], *Science* **257:**809–813.

Schooltink, H., Stoyan, T., Roeb, E., Heinrich, P. C., and Rose, J. S., 1992, Ciliary neurotrophic factor induces acute-phase protein expression in hepatocytes, *New Biol.* **4:**61–65.

Schorle, H., Holtschke, T., Hunig, T., Schimpl, A., and Horak, I., 1991, Development and function of T cells in mice rendered interleukin-2 deficient by gene targeting, *Nature* **352:**621–624.

Schrader, J.W., 1986, The panspecific hemopoietin of activated T lymphocytes (interleukin-3), *Annu. Rev. Immunol.* **4:**205–230.

Shanafelt, A. B., and Kastelein, R. A., 1992, High affinity ligand binding is not essential for granulocyte–macrophage colony-stimulating factor receptor activation, *J. Biol. Chem.* **267:**25466–25472.

Shanafelt, A. B., Miyajima, A., Kitamura, T., and Kastelein, R. A., 1991, The amino-terminal helix of GM-CSF and IL-5 governs high affinity binding to their receptors, *EMBO J.* **10:**4105–4112.

Shearman, M. S., Heyworth, C. M., Dexter, T. M., Haefner, B., Owen, P. J., and Whetton, A. D., 1993, Haemopoietic stem cell development to neutrophils is associated with subcellular redistribution and differential expression of protein kinase C subspecies, *J. Cell Sci.* **104:**173–180.

Sherr, C. J., 1990, Review, colony-stimulating factor-1 receptor, *Blood* **75:**1–12.

Sherr, C. J., 1993, Mammalian G1 cyclins, *Cell* **73:**1059–1065.

Shuai, K., Schindler, C., Prezioso, V. R., and Darnell, J. J., 1992, Activation of transcription by IFN-gamma: Tyrosine phosphorylation of a 91-kD DNA binding protein, *Science* **258:**1808–1812.

Shuai, K., Ziemiecki, A., Wilks, A. F., Harpur, A. G., Sadowski, H. B., Gilman, M. Z., and Darnell, J. E., 1993a,

Polypeptide signalling to the nucleus through tyrosine phosphorylation of Jak and Stat proteins, *Nature* **366:**580–583.

Shuai, K., Stark, G. R., Kerr, I. M., and Darnell, J. J., 1993b, A single phosphotyrosine residue of Stat91 required for gene activation by interferon-gamma, *Science* **261:**1744–1746.

Shuai, K., Horvath, C. M., Huang, L. H., Qureshi, S. A., Cowburn, D., and Darnell, J. J., 1994, Interferon activation of the transcription factor Stat91 involves dimerization through SH2-phosphotyrosyl peptide interactions, *Cell* **76:**821–828.

Sims, J. E., March, C. J., Cosman, D., Wimder, M. B., MacDonald, H. R., McMahan, C. J., Grubin, C. E., Wignall, J. M., Jackson, J. L., Call, S. M., Friend, D., Albert, A. L., Gillis, S., Urdal, D., and Dower, S. K., 1988, cDNA expression cloning of the IL-1 receptor, a member of the immunoglobulin superfamily, *Science* **241:**585–589.

Smith, C. A., Davis, T., Wignall, J. M., Din, W. S., Farrah, T., Upton, C., McFadden, G., and Goodwin, R. G., 1990, T2 open reading frame from the Shope fibroma virus encodes a soluble form of the TNF receptor, *Science* **248:**1019–1023.

Soh, J., Donnelly, R. J., Kotenko, S., Mariano, T. M., Cook, J. R., Wang, N., Emanuel, S., Schwartz, B., Miki, T., and Pestka, S., 1994, Identification and sequence of an accessory factor required for activation of the human interferon gamma receptor, *Cell* **76:**793–802.

Sorensen, P., Mui, A. L., and Krystal, G., 1989, Interleukin-3 stimulates the tyrosine phosphorylation of the 140-kilodalton interleukin-3 receptor, *J. Biol. Chem.* **264:**19253–19258.

Sprang, S. R., and Bazan, J. F., 1993, Cytokine structural taxonomy and mechanisms of receptor engagement, *Curr. Opin. Struct. Biol.* **3:**815–827.

Stahl, N., Davis, S., Wong, V., Taga, T. Kishimoto, T., Ip, N. Y., and Yancopoulos, G. D., 1993, Cross-linking identifies leukemia inhibitory factor-binding protein as a ciliary neurotrophic factor receptor component, *Science* **260:**1805–1808.

Stahl, N. Boulton, T. G., Farruggella, T., Ip, N. Y., Davis, S., Witthuhn, B. A., Quelle, F. W., Silvennoinen, O., Barbieri, G., Pellegrini, S., Ihle, J. N., and Yancopoulos, G. D., 1994, Association and activation of Jak–Tyk kinases by CNTF–LIF–OSM–IL-6 beta receptor components, *Science* **253:**92–95.

Stahl, N., Farruggella, T. J., Boulton, T. G., Zhong, Z., Darnell, J. Jr., and Yancopoulos, G. D., 1995, Choice of STATs and other substrates specified by modular tyrosine-based motifs in cytokine receptors, *Science* **267:**1349–1353.

Stanley, E., Lieschke, G. J., Grail, D., Metcalf, D. Hodgson, G., Gall, J. A., Maher, D. W., Cebon, J., Sinickas, V., and Dunn, A. R., 1994, Granulocyte/macrophage colony-stimulating factor-deficient mice show no major perturbation of hematopoiesis but develop a characteristic pulmonary pathology, *Proc. Natl. Acad. Sci. U.S.A.* **91:**5592–5596.

Suematsu, S., Hibi, M., Sugita, T., Saito, M., Murakami, M., Matsusaka, T., Matsuda, T., Hirano, T., Taga, T., and Kishimoto, T., 1990, Interleukin 6 (IL-6) and its receptor (IL-6R) in myeloma/plasmacytoma, *J. Exp. Med.* **171:**2001–2009.

Sugita, T., Totsuka, T., Saito, M., Yamasaki, K., Taga, T., Hirano, T., and Kishimoto, T., 1993, Functional murine interleukin 6 receptor with the intracisternal A particle gene product at its cytoplasmic domain. Its possible role in plasmacytomagenesis, *EMBO J.* **12:**1705–1712.

Taga, R., Hibi, M., Hirata, Y., Yamasaki, K., Yasukawa, K., Matsuda, T., Hirano, T., and Kishimoto, T., 1989, Interleukin-6 triggers the association of its receptor with a possible signal transducer, gp130, *Cell* **58:**573–581.

Takagi, M., Hara, T., Ichihara, M., Takatsu, K., and Miyajima, A., 1995, Multi-colony stimulating activity of interleukin-5 (IL-5) on hematopoietic progenitors from trangenic mice that express IL-5 receptor α subunit constitutively, *J. Exp. Med.* **181:**889–899.

Takaki, S., Mita, S., Kitamura, T., Yonehara, S., Yamaguchi, N., Tominaga, A., Miyajima, A., and Takatsu, K., 1991, Identification of the second subunit of the murine interleukin-5 receptor: Interleukin-3 receptor-like protein, AIC2B is a component of the high affinity interleukin-5 receptor, *EMBO J.* **10:**2833–2838.

Takaki, S., Kanazawa, H., Shiiba, M., and Takatsu, K., 1994, A critical domain of the interleukin-5 (IL-5) receptor α chain and its function in IL-5-mediated growth signal transduction, *Mol. Cell. Biol.* **14:**7404–7413.

Takeshita, T., Ohtani, K., Asao, H., Kumaki, S., Nakamura, M., and Sugamura, K., 1992a, An associated molecule, p64, with Il-2 receptor beta chain. Its possible involvement in the formation of the functional intermediate-affinity IL-2 receptor complex, *J. Immunol.* **148:**2154-2158.

Takeshita, T., Asao, H. Ohtani, K., Ishii, N., Kumaki, S., Tanaka, N., Munakata, H., Nakamura, M., and Sugamura, K., 1992b, Cloning of the gamma chain of the human IL-2 receptor, *Science* **257:**379–382.

Tamura, T., Udagawa, N., Takahashi, N., Miyaura, C., Tanaka, S., Yamada, Y., Koishihara, Y., Ohsugi, Y., Kumaki,

K., Taga, T., and Kishimoto, T., 1991, Soluble interleukin-6 receptor triggers osteoclast formation by interleukin 6, *Proc. Natl. Acad. Sci. U.S.A.* **88:**11349–11353.

Taniguchi, T., and Miniami, Y., 1993, The IL-2/IL-2R system: A current overview, *Cell* **73:**5–8.

Tavernier, J., Devos, R., Cornelis, S., Tuypena, T., Van der Heyden, J., Fiers, W., and Plaetinck, G., 1991, A human high affinity interleukin-5 receptor (IL5R) is composed of an IL5 specific α chain and a β chain shared with the receptor for GM-CSF, *Cell* **66:**1175–1184.

Tian, S. S., Lamb, P., Seidel, H. M., Stein, R. B., and Rosen, J., 1994, Rapid activation of the STAT3 transcription factor by granulocyte colony-stimulating factor, *Blood* **84:**1760–1764.

Ullrich, A., and Schlessinger, J., 1990, Signal transduction by receptors with tyrosine kinase activity, *Cell* **61:** 203–212.

Upton, C., Mossman, K., and McFadden, G., 1992, Encoding of a homolog of the IFN-gamma receptor by myxoma virus, *Science* **258:**1369–1372.

Urdal, D. L., and Park, L. S., 1988, Studies on hematopoietic growth factor receptors using human recombinant IL-3, GM-CSF, G-CSF, M-CSF, IL-1, and IL-4, *Behring Inst. Mitt.* **83:**27–39.

Uyttenhove, C., Simpson, R. J., and Van Snick, J., 1988, Functional and structural characterization of P40, a mouse glycoprotein with T-cell growth factor activity, *Proc. Natl. Acad. Sci. U.S.A.* **85:**6934–6938.

Uze, G., Lutfalla, G., and Gresser, I., 1990, Genetic transfer of a functional human interferon alpha receptor into mouse cells: cloning and expression of its cDNA, *Cell* **60:**225–234.

Vadas, M. A., Nicola, N. A., and Metcalf, D., 1983, Activation of antibody dependent cell-mediated cytotoxicity of human neutrophils and eosinophils by separate colony stimulating factors, *J. Immunol.* **130:**795–799.

Vanden, B. C., and Pfeffer, L. M., 1988, Characterization of interferon-alpha binding sites on human cell lines, *J. Interferon Res.* **8:**803–811.

Velazquez, L., Fellous, M., Stark, G. R., and Pellegrini, S., 1992, A protein tyrosine kinase in the interferon alpha/beta signaling pathway, *Cell* **70:**313–322.

Voss, S. D., Sondel, P. M., and Robb, R. J., 1992, Characterization of the interleukin 2 receptors (IL-2R) expressed on human natural killer cells activated *in vivo* by IL-2: Association of the p64 IL-2R gamma chain with the IL-2R beta chain in functional intermediate-affinity IL-2R, *J. Exp. Med.* **176:**531–541.

Voss, S. D., Leary, T. P., Sondel, P. M., and Robb, R. J., 1993, Identification of a direct interaction between interleukin 2 and the p64 interleukin 2 receptor gamma chain, *Proc. Natl. Acad. Sci. U.S.A.* **90:**2428–2432.

Wakao, H., Schmitt-N. M., and Groner, B., 1992, Mammary-gland specific factor is present in lactating rodent and bovine mammary tissue and composed of a single polypeptide of 89kDa, *J. Biol. Chem.* **267:**16365–16370.

Wakao, H., Gouilleux, F., and Groner, B., 1994, Mammary gland factor (MGF) is a novel family of the cytokine regulated transcription factor gene family and confers the prolactin response, *EMBO J.* **13:**2182–2191.

Wakao, H., Harada, N., Kitamura, T., Mui, A. L., and Miyajima, A., 1995, Interleukin-2 and erythropoietin activate STAT5/MGF via distinct pathways, *EMBO J.* **14:**2527–2535.

Watling, D., Guschin, D., Muller, M., Silvennoinen, O., Witthuhn, B. A., Quelle, F. W., Rogers, N. C., Schindler, C., Stark, G. R., Ihle, J. N., and Kerr, I. M., 1993, Complementation by the protein tyrosine kinase JAK2 of a mutant cell line defective in the interferon-gamma signal transduction pathway, *Nature* **366:**166–170.

Watowich, S. S., Yoshimura, A., Longmore, G. D., Hilton, D. J., Yoshimura, Y., and Lodish, H. F., 1992, Homodimerization and constitutive activation of the erythropoietin receptor, *Proc. Natl. Acad. Sci. U.S.A.* **89:** 2140–2144.

Weinberg, K., and Parkman, R., 1990, Severe combined immunodeficiency due to a specific defect in the production of interleukin-2, *N. Engl. J. Med.* **322:**1718–1723.

Weiser, W. Y., Van Niel, A., Clark, S. C., Dard, J. R., and Remold, H. J., 1987, Recombinant human granulocyte/macrophage colony-stimulating factor activates intracellular killing of *Leishmania donovani* by human monocyte-derived macrophages, *J. Exp. Med.* **166:**1436–1446.

Weissman, C., and Weber, H., 1986, The interferon genes, *Prog. Nucleic Acid Res.* **33:**251–300.

Welham, M. J., and Schrader, J. W., 1992, Steel factor-induced tyrosine phosphorylation in murine mast cells. Common elements with IL-3-induced signal transduction pathways, *J. Immunol.* **149:**2772–2783.

Welham, M. J., Duronio, V., Sanghera, J. S., Pelech, S. L., and Schrader, J. W., 1992, Multiple hemopoietic growth factors stimulate activation of mitogen-activated protein kinase family members, *J. Immunol.* **149:**1683–1693.

Wheeler, E. F., Rettenmier, C. W., Look, A. T., and Sherr, C. J., 1986, The *v-fms* oncogene induces factor independence and tumorigenicity in CSF-1-dependent macrophage cell line, *Nature* **324:**377–380.

Wilks, A. F., 1990, Structure and function of the protein tyrosine kinases, *Prog. Growth Factor Res.* **2:**97–111.

Winkelmann, J. C., Penny, L. A., Deaven, L. L., Forget, B. G., and Jenkins, R. B., 1990, The gene for the human erythropoietin receptor: Analysis of the coding sequence and assignment to chromosome 19p, *Blood* **76:** 24–30.

Witthuhn, B., Quelle, F. W., Silvennoinen, O., Yi, T., Tang, B., Miura, O., and Ihle, J. N., 1993, JAK2 associates with the erythropoietin receptor and is tyrosine phosphorylated and activated following EPO stimulation, *Cell* **74:**227–236.

Witthuhn, B. A., Silvennoinen, O., Miura, O., Lai, K. S., Cwik, C., Liu, E. T., and Ihle, J. N., 1994, Involvement of the Jak-3 Janus kinase in signalling by interleukins 2 and 4 in lymphoid and myeloid cells, *Nature* **370:** 153–157.

Woodcock, J. M., Zacharakis, B., Plaetinck, G., Bagley, C. J., Qiyu, S., Hersuc, T. R., Tavernie, R. J., and Lopez, A. F., 1994, Three residues in the common β chain of the human GM-CSF, IL-3 and IL-5 receptors are essential for GM-CSF and IL-5 high affinity binding but not IL-3 high affinity binding and interact with Glu21 of GM-CSF, *EMBO J.* **13:**5176–5185.

Wrana, J. L., Attisano, L., Wieser, R., Ventura, F., and Massague, J., 1994, Mechanism of activation of the TGF-beta receptor, *Nature* **370:**341–347.

Yamasaki, K., Taga, T., Hirata, Y., Yawata, H., Hawanishi, Y., Seed, B., Taniguchi, T., Hirano, T., and Kishmoto, T., 1988, Cloning and expression of the human interleukin-6 (BSF-2/IFNB2) receptor, *Science* **241:**825–828.

Yang, Y. C., Tsai, S., Wong, G. G., and Clark, S. C., 1988, Interleukin-1 regulation of hematopoietic growth factor production by human stromal fibroblasts, *J. Cell Physiol.* **134:**292–296.

Yi, T., Mui, A. L., Krystal, G., and Ihle, J. N., 1993, Hematopoietic cell phosphatase associates with the interleukin-3 (IL-3) receptor beta chain and down-regulates IL-3-induced tyrosine phosphorylation and mitogenesis, *Mol. Cell. Biol.* **13:**7577–7586.

Yin, T., and Yang, Y. C., 1994, Mitogen-activated protein kinases and ribosomal S6 protein kinases are involved in signaling pathways shared by interleukin-11, interleukin-6, leukemia inhibitory factor, and oncostatin M in mouse 3T3-L1 cells, *J. Biol. Chem.* **269:**3731–3738.

Yin, T., Taga, T., Tsang, M.L., Yasukawa, K., Kishimoto, T., and Yang, Y. C., 1992, Involvement of IL-6 signal transducer gp130 in IL-11-mediated signal transduction, *J. Biol. Chem.* **267:**10238–10247.

Yin, T., Yasukawa, K., Taga, T., Kishimoto, T., and Yang, Y. C., 1994, Identification of a 130-kilodalton tyrosine-phosphorylated protein induced by interleukin-11 as JAK2 tyrosine kinase, which associates with gp130 signal transducer, *Exp. Hematol.* **22:**467–472.

Yoshimura, A., Immers, T., and Neumann, D., Longmore, G., Yoshimura, Y., and Lodish, H. F., 1992, Mutations in the Trp-Ser-X-Trp-Ser motif of the erythropoietin receptor abolish processing, ligand binding, and activation of the receptor, *J. Biol. Chem.* **267:**11619–11625.

Zhong, A., Wen, Z., and Darnell, J. J., 1994, Stat3: A STAT family member activated by tyrosine phosphorylation in response to epidermal growth factor and interleukin-6, *Science* **264:**95–98.

Zurawski, S. M., Vega, F. J., Huyghe, B., and Zurawski, G., 1993, Receptors for interleukin-13 and interleukin-4 are complex and share a novel component that functions in signal transduction, *EMBO J.* **12:**2663–2670.

Chapter 2

Interleukin 13 and Related Cytokines

Andrew N. J. McKenzie and Andrew W. Heath

1. INTRODUCTION

The function of CD4-positive, T helper cells is pivotal to the workings of the specific immune system. These cells direct antibody responses to protein antigens and influence cellular immunity by activating macrophages and instigating delayed-type hypersensitivity reactions. T helper cells perform these functions partially through altered expression of cell surface antigens, such as the CD40 ligand, gp39; and partially by the secretion of a large range of cytokines that have effects on various immune and hematopoetic cells.

In recent years $CD4^+$ T cells have been further divided into T helper 1 (Thl) and T helper 2 (Th2) subsets on the basis of the cytokines they secrete, which lead to different functional properties of these two subsets (Mossman and Coffman, 1989). The Thl cells are characterized by the secretion of IFNγ and IL-2 and are important in enhancing cellular immune responses. The Th2 cells are characterized by secretion of IL-4 and IL-5 and are of importance in enhancing antibody responses. The two subsets of cells appear also to have potent effects in cross-regulating each other. With the continuing discovery of novel cytokines, molecules that share many of the functions of previously described cytokines are being discovered. Examples of these include interleukin 15 (IL-15), which shares many functions with IL-2, and IL-13, which mimics some of the effects of IL-4. This chapter discusses results obtained to date with the novel cytokine IL-13 and attempts to compare its effects on cells of the immune and hematopoetic systems with those of IL-4.

Andrew N. J. McKenzie The MRC Laboratory of Molecular Biology, Cambridge CB2 2QH, United Kingdom. **Andrew W. Heath** Department of Medical Microbiology, University of Sheffield Medical School, Sheffield S10 2RX, United Kingdom.

Blood Cell Biochemistry, Volume 7: Hematopoietic Cell Growth Factors and Their Receptors, edited by A. D. Whetton and J. Gordon. Plenum Press, New York, 1996.

2. THE IL-13 GENE AND PROTEIN STRUCTURE

2.1. Cloning of Interleukin 13 cDNA

It was apparent, following the identification of Th1 and Th2 cell clones, that there was a degree of differentiation between these cell types on the basis of cytokine secretion. An induction-specific mRNA expressed relatively abundantly by a murine Th2 cell line known as C1.Ly-$1^{+}2^{-}$/9 (Brown *et al.*, 1989) corresponded to a cDNA known initially as P600. By a differential screening protocol, the P600 cDNA was identified as an mRNA induced on mitogen activation of mouse T cells. The human homologue of mIL-13 was later cloned by three separate groups using three strategies. McKenzie *et al.* (1993a) used a probe derived from the coding region of mouse IL-13 cDNA to screen several cDNA libraries derived from mRNAs of activated human T cells. The full-length human IL-13 cDNA was isolated from a library made from the $CD8^{+}$ T-cell line A10 although partial cDNAs were identified in a library from a $CD4^{+}$ clone (B21). Minty *et al.* (1993) used a differential screening approach to isolate a cDNA from an activated human lymphocyte library, and partial cDNA clones were also isolated by hybridization of mRNA from activated peripheral blood lymphocytes with a genomic DNA containing the genes for IL-4 and IL-5 (Morgan *et al.*, 1992) (see next section).

2.2. The IL-13 cDNAs and Proteins

The hIL-13 coding region has 66% nucleotide homology with that of mIL-13 and 58% amino-acid sequence identity. Rat IL-13 has also now been cloned, and the cDNA has 74% and 87% nucleotide identity with human and mouse, respectively (Lakkis and Cruet, 1993); homology is 63% and 79% at the amino acid level. The mIL-13 cDNA encodes a protein of 131 amino-acids (Brown *et al.*, 1989), and the hIL-13 cDNA encodes a protein of 131 or 132 amino acids depending on the presence or absence of a glutamine residue at position 98. A 21-amino-acid leader peptide in human IL-13 results in the N-terminal amino acid of the mature protein being Gly 21 (Minty *et al.*, 1993).

There are five cysteine residues that are conserved between the two proteins, and hIL-13 has four potential N-linked glycosylation sites, three of which are conserved in the mouse (McKenzie *et al.*, 1993a; Minty *et al.*, 1993); although COS cell transfection experiments appeared to indicate that most secreted hIL-13 was nonglycosylated, with an apparent molecular weight of around 10,000 (McKenzie *et al.*, 1993; Minty *et al.*, 1993).

The IL-4 and IL-13 amino acid sequences are approximately 30% homologous, which is low but above "background" (Zurawski *et al.*, 1993), and the proteins share a common four-α-helical bundle tertiary structure (Bazan, 1990). Residues forming the core of IL-4 are either conserved in IL-13 or have undergone conservative, hydrophobic replacement. The largest areas of change between IL-4 and IL-13 seem to be in the loop regions of the proteins (Walter *et al.*, 1992).

2.3. Genomic Structure and Location

The human and mouse IL-13 genes span regions of DNA of about 4.5 kb and are both comprised of four exons and three introns (McKenzie *et al.*, 1993b). The human IL-13 gene maps to chromosome 5q, and the mouse IL-13 gene maps to the syntenic region of chromosome 11 (McKenzie *et al.*, 1993b). This region of chromosome 5 also contains the genes for other cytokines including IL-3, IL-4, IL-5 and GM-CSF. The intron–exon structures of the genes encoding IL-4, IL-5 and GM-CSF are also shared with the IL-13 gene. Chromosomal mapping studies indicate that the hIL-13 gene is within 20 Kb upstream of the hIL-4 gene (Morgan *et al.*, 1992), and the mouse IL-13 gene appears to be in a similar position relative to the mouse IL-4 gene (A. N. J. McKenzie, unpublished observations).

3. RECEPTORS FOR IL-13

It is now apparent that cytokines often utilize multicomponent receptors in which subunits may be shared with other cytokine receptors (See Chapters 1 and 3). Several lines of evidence indicate that the primary binding subunit of the IL-4 receptor (IL-4R) can also function as a component of the IL-13R. (1) An IL-4 analogue with a single amino acid mutation in its fourth-α-helix has almost no biological activity but binds the IL-4R with high affinity and acts as an IL-4 antagonist. This analogue is also able to antagonize the biological response of IL-13 on the TF-1 cell line (Zurawski *et al.*, 1993) and block responses of IL-4- and IL-3-stimulated human B cells (Aversa *et al.*, 1993). (2) Blocking antibodies raised against the IL-4R inhibit the action of both IL-4 and of IL-13 (Zurawski *et al.*, 1995; Obiri *et al.*, 1995). (3) Cross-linking and binding studies indicate that IL-4 and IL-13 can cross-compete for binding. However, the inability of IL-13, contrasted with the ability of IL-4, to stimulate activated T cells implies that there is an independent IL-13 binding receptor chain that is not present on these cells. This is supported by data showing that IL-13 cannot bind to Il-4Rα in transfected cell lines.

The cloned IL-4 binding protein (IL-4Rα) has been characterized as a 140-kDA protein. However, the different binding characteristics of the IL-4 analogue to cells transfected with the IL-4Rα cDNA and to TF-1 cells indicated that at least one other receptor subunit is involved in IL-4 binding. On lymphoid cell types, the shared γ-chain of the IL-2R has been shown to complex with IL-4Rα and the cloned IL-7 receptor (Kondo *et al.*, 1994; Noguchi *et al.*, 1993). However, on other cell lineages that lack the γ chain, IL-4Rα still associated with a distinct 60- to 70-kDa protein. This may be the IL-13 receptor, because cross-linking studies indicate that the IL-13R is approximately 65 kDa in size (Vita *et al.*, 1995; Zurawski *et al.*, 1995). The IL-13Rs have been detected on B lymphocytes, monocytes, and nonhematopoietic cells including COS cells and renal carcinoma cell lines (Obiri *et al.*, 1995). In fact, the latter cell lines display 10- to 50-fold more IL-13R than cells of the hematopoietic lineages. The relevance of this expression is unknown. Both IL-13 and IL-4 appear to stimulate similar signal transducers and activators of transcription (STAT) proteins (Lin *et al.*, 1995). Molecular cloning of the IL-13 receptor should clarify the roles of the specific receptor subunits in different cell lineages.

4. BIOLOGICAL SOURCES OF IL-13

Northern blotting of various human tissues indicates the presence of a 1.3 kb IL-13 mRNA in activated and resting T-cell lines and clones but not in tissues from heart, brain, lung, placenta, skeletal muscle, and liver (McKenzie *et al.*, 1993a). The hIL-13 protein is produced by T cells and T-cell clones of both the $CD4^+$ and $CD8^+$ lineages following stimulation with various mitogens or with antigen (Brown *et al.*, 1989; McKenzie *et al.*, 1993a; Minty *et al.*, 1993). Following stimulation of T-cell clones, IL-13 mRNA is detectable within one hour, and protein production continues from 2 to 72 hr, whereas IL-4 production appears to occur later and to be more short lived. Despite initial results that suggested restriction to T cells, IL-13 gene transcription has been identified in a number of B-cell malignancies (Fior *et al.*, 1994).

Studies of mice infected with the parasites *Leishmania major* and *Schistosoma mansoni* indicate that IL-13 production appears to correlate with Th2-like responses (Reiner *et al.*, 1994; Wang *et al.*, 1994; Wynn *et al.*, 1994). Furthermore, IL-13 appears to be associated with Th2 type T-cell clones (Cherwinski *et al.*, 1987).

5. BIOLOGICAL ACTIVITIES OF IL-13

The two common themes to the following information are that although the activities of IL-13 and IL-4 overlap, they also show distinct function, and that the activities of human IL-13 and mouse IL-13 also overlap but are not identical. In general the activities of mouse IL-13 appear more restricted (see Figure 1).

5.1. Hematopoietic System

Interleukin 13 enhances the stem cell factor (SCF)-induced proliferation of Lin-Sca-1^+ bone marrow progenitor cells to a greater extent than found with IL-4. In contrast to IL-4, IL-13 has synergistic effects with GM-CSF on these Lin-Sca-1^+ precursors, although both cytokines have similar effects on G-CSF-enhanced colony formation. Indeed, IL-4 and IL-13 synergize with SCF and G-CSF to drive differentiation toward the monocyte/macrophage lineage, whereas SCF and GCSF alone normally lead to 90% granulocyte production (Jacobsen *et al.*, 1994). In the later stages of hematopoiesis, IL-13 has been shown to have inhibitory effects on the growth of human $CD19^+sIg^-$ B-cell precursors as well as leukemic B-cell precursors (B ALL cells) (Renard *et al.*, 1994).

5.2. Myeloid Cells

5.2.1. Monocytes and Macrophages

A large number of reports on the biological effects of IL-13 pertain to its effects on the cells of the monocyte/macrophage lineages, and it is in this area that the functions of human and mouse IL-13 and IL-13 and IL-4, respectively, appear most similar.

One of the first activities of IL-13 to be identified was its effect on proliferation of the human premyeloid cell line TF-1, a property shared by both the mouse and the human cytokines as well as by IL-4 (McKenzie *et al.*, 1993a). Interleukin 13 affects both the morphology and function of human monocytes. Addition of IL-13 to human monocyte cultures causes increased adherence and homotypic aggregation (McKenzie *et al.*, 1993a) and prolongs the life span of these cells. These effects are similar to those seen with IL-4 (te Velde *et al.*, 1988).

Cell surface phenotype changes induced by coculture with IL-13 include increased

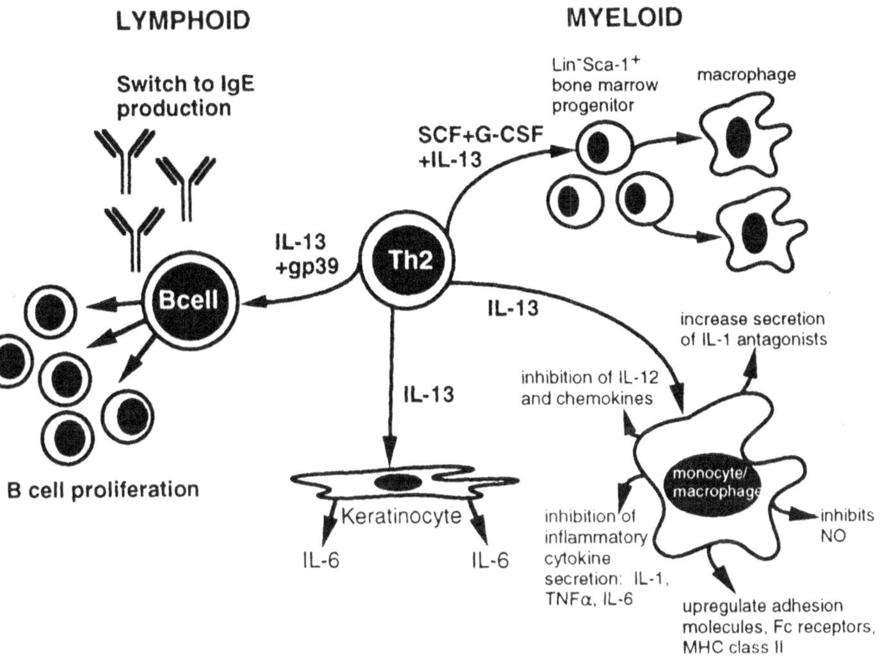

FIGURE 1. Summary of the biological effects of IL-13.

expression of a number of integrin family members. CD11b/CD18, CD11c/CD18, CD29, and CD49e expression are all enhanced by both IL-13 and IL-4, whereas CD11a/CD18, CD49b,d,f, and CD61 expression are unaffected. MHC class II, CD23, and CD13 expression are also increased, while expression of the three Fcγ receptors (CD16, CD32, CD64) and CD14 [lipopolysaccharide (LPS) receptor] is down-regulated (Vercelli *et al.*, 1988; de Waal Malefyt *et al.*, 1993).

In general, the effects of both IL-4 and IL-13 on cells of the monocyte/macrophage compartment could be described as "antiinflammatory." The down-regulation of both Fcγ receptor expression and LPS receptor expression results in a reduction of antibody-dependent cellular cytotoxicity (ADCC) and in decreased responsiveness to LPS stimulation as characterized by a diminution in chemokine and growth factor secretion, while IL1R antagonist secretion is increased (Minty *et al.*, 1993; de Waal Malefyt *et al.*, 1993). In addition, IL-13 inhibits the transcription of IFNα and both chains of IL-12 (de Waal Malefyt *et al.*, 1993). These effects of IL-13 in depressing inflammatory cytokine secretion are shared by IL-4 and IL-10, but IL-4 and IL-13 are able to mediate these effects in the presence of anti-IL-10 antibodies; thus, the antiinflammatory effects of IL-4 and IL-13 are not IL-10 dependent. The reduction of IL-12 production may be of great importance in favoring the generation of Th2-like T-cell responses, as IL-12 would appear to play a pivotal role in skewing responses toward the Th1-like profile (Afonso *et al.*, 1994; Heinzel *et al.*, 1993; Hsieh *et al.*, 1993). Although we have described several down-regulatory effects of IL-13 on monocyte function, IL-13 does not simply result in a blanket "deactivation" of the cells.

Indeed, in keeping with enhanced MHC class II expression and enhanced expression of several integrins, the antigen-presenting capacity of monocytes is increased on exposure to IL-13 (R. de Waal Malefyt, unpublished data). In addition, IL-13 is chemotactic for macrophages (Magazin *et al.*, 1994).

The antiinflammatory effects of IL-4, which are similar to those of IL-13, lead not only to a reduction in ADCC but also to decreased killing of phagocytosed bacteria such as *Salmonella* (Denich *et al.*, 1993) and parasites such as *Leishmania* (Lehn *et al.*, 1989). Although there is currently little information available, these effects will probably be mimicked by IL-13. Indeed, in experiments using mouse macrophages cultured from the bone marrow in the presence of GM-CSF, IL-13 and IL-4 both reduce the production of nitric oxide following LPS activation. This reduction in release of NO, one of the most important cytotoxic mechanisms of phagocytes, correlates with a lowering of parasiticidal activity against *Leishmania major* (Doherty *et al.*, 1993).

Interestingly, both IL-4 and IL-13 have been shown to suppress the replication of HIV in human peripheral blood monocytes (Montanier *et al.*, 1993) and bronchoalveolar macrophages (Denis and Ghadrian, 1994), although there is no effect on HIV replication in T cells (Montanier *et al.*, 1993). The mode of action of IL-13 in lowering HIV replication remains to be determined but is not related to reduced CD4 expression, and proviral DNA remains in these cells.

In summary, the effects of IL-13 and IL-4 on cells of the monocyte/macrophage lineage appear to be very similar and also appear to differ little between mouse and human.

5.2.2. Granulocytes

There is currently very little information on the effects of IL-13 on cells of the granulocyte lineage. In contrast to its effects on macrophages, IL-13 is not a chemotactic factor for neutrophils (Magazin *et al.*, 1994), although in common with its effects on macrophages, IL-13 enhances IL-1RII (decoy receptor) expression by human polymorphonuclear cells (Colotta *et al.*, 1994).

5.3. B Cells

There appears to be a large divergence in the effects of IL-13 on B cells of mouse and human. To date, IL-13 has not been shown to have any effect on mouse B cells, although IL-4 has quite potent effects in activation, costimulation for proliferation, and isotype switching. The effects of IL-13 referred to below all apply to human B cells.

The effects of IL-13 on human B cells largely overlap those of IL-4: IL-13, like IL-4, induces the up-regulation of a number of activation markers on human B cells including sIgM, CD23, CD71, CD72, and MHC class II (DeFrance *et al.*, 1994; Punnonen *et al.*, 1993), but there is no significant effect on other markers such as CD40, LFA-1, LFA-3, and B7 [although B7 expression was measured using an antibody to B7.1; thus, it is possible that expression of B70 (B7.2), the second ligand for CTLA-4, is enhanced by IL-13 (Azuma *et al.*, 1993; Freeman *et al.*, 1993)].

The activating effects of IL-13 and IL-4 allow for proliferation of human cells in the presence of a further stimulus such as that mediated through surface IgM binding or through anti-CD40 (Briere *et al.*, 1993; Cocks *et al.*, 1993; McKenzie *et al.*, 1993a). The natural signals mimicked by these *in vitro* effects are mediated through binding specific antigen and through binding of the CD40 ligand, gp39, on activated T cells. Again, although IL-4 has very similar effects on mouse B cells, IL-13 is apparently lacking in activity on B cells from the mouse (our own unpublished observations).

Although cytokines have powerful roles in isotype switching, costimulation through CD40 also appears to be necessary (Armitage *et al.*, 1992). In the mouse, IL-4 is associated with switching to the IgG_1 and IgE isotypes; indeed, IL-4-deficient mice can not be induced to produce an IgE response (Kuhn *et al.*, 1991) in common with CD40-deficient mice (Kawabe *et al.*, 1994). In the human, IL-4 is likewise associated with isotype switching to IgE and to the human equivalent of mouse IgG_1, which is IgG_4. Like IL-4, IL-13 induces isotype switching to IgE in human B cells. Because IgE-secreting cells can be produced from a naive IgD-negative population, the effect of IL-13, like that of IL-4, is in switching isotypes of individual cells rather than inducing selective outgrowth of IgE-positive cells. This hypothesis is supported by the finding that IL-13 can induce germ line ε mRNA synthesis (Punnonen *et al.*, 1993; Cocks *et al.*, 1993). Again, IL-13 has not been shown to have an effect on mouse B-cell isotype switching, but the effects of IL-4 on mouse B cells, described briefly above, are similar to those on human cells.

It should be mentioned at this point that although IL-13 has not been shown to have effects on mouse B cells *in vitro*, and IL-4 "knockout" mice in most circumstances have not produced IgE, there is a striking exception to these findings. When IL-4-deficient mice are infected with the rodent malaria parasite *Plasmodium chabaudi*, IgE is detectable in the serum, albeit at lower levels than in wild-type mice (van der Weld *et al.*, 1994). These authors suggest that IL-13 might be responsible for this isotype switching in the IL-4 "knockouts." Of course, it is also possible that an as yet unidentified cytokine is responsible. Further studies on IL-4, IL-13, and double-"knockout" mice may provide an answer.

In summary, in the natural course of a developing humoral immune response, B cells take up antigen through surface IgM receptors, and process and present the peptides to T helper cells; B cells proliferate to produce a larger number of effector (plasma) or memory cells and switch immunoglobulin isotypes from IgM to various IgG isotypes, IgE, or IgA. Both IL-4 and IL-13 (at least in the human) play a role in all of these aspects of the humoral response. Both cytokines can augment B-cell antigen presentation through increasing expression of molecules such as MHC class II, enhance B-cell proliferation in response to both antigen and T-cell membrane derived signals, and have a clear role to play in immunoglobulin isotype switching.

5.4. T Cells

Interleukin 13, unlike IL-4, has no effects on enhancing proliferation of activated T cells or T-cell clones. It appears from this lack of response, and from binding data (Obiri *et al.*, 1995; Vita *et al.*, 1995), that T cells, including human T cells, lack IL-13 receptors. This is intriguing because IL-4, as well as mediating many of the obvious effects of a Th2-like response (such as IgE production), also plays a leading role in the induction of a Th2 response both *in vitro* (Le Gros *et al.*, 1990; Swain *et al.*, 1990) and *in vivo* (Chatelein *et al.*, 1992; Coffman *et al.*, 1991). Interleukin 13 appears to mediate some of the effects of a Th2-like response in humans without the potential for positive feedback in skewing the response further toward Th2. It will be of interest to discover whether the controversial differences between highly polarized Thl/Th2 responses in the mouse and the apparent lack of such strong polarization in the human relate to differing roles for IL-13 and IL-4 in the two species.

ACKNOWLEDGMENTS. The authors would like to thank Sarah Bell for critical appraisal of the manuscript and Pauline Whitaker and Gillian Griffiths for its preparation.

6. REFERENCES

Afonso, L. C. C., Scharton, T. M., Viera, L. Q., Wysocka, M., Trinchieri, G., and Scott, P., 1994, The adjuvant effect of interleukin-12 in a vaccine against *Leishmania major*, *Science* **263:**235–237.

Armitage, R. J., Fanslow, W. C., Strockbine L., Sato, T. A., Clifford, K. N., Macduff, B. M., Anderson, D. M., Gimpel, S. D., Davis-Smith, T., Maliszewski, C. R., Clark, E. A., Smith, C. A., Grabstein, D. H., Cosman, D., and Spriggs, M. K., 1992, Molecular and biological characterization of a murine ligand for CD40, *Nature* **357:** 80–82.

Aversa, G., Punnonen, J., Cocks, B. G., de Waal Malefyt, R., Vega, F., Jr., Zurawski, S. M., Zurawski, G., and de Vries, J. E., 1993, An interleukin-4 (IL-4) mutant protein inhibits both IL-4 and IL-13-induced human irnrnunoglobulin G_4 (IgG_4) and IgE synthesis and B cell proliferation: Support for a common component shared by IL-4 and IL-13 receptors, *J. Exp. Med.* **178:**2213–2218.

Azuma, M., Ito, D., Yagita, H., Okamura, K., Phillips, J. H., Lanier, L. L., and Somoza, C., 1993, B70 antigen is a second ligand for CTLA-4 and CD28, *Nature* **366:**76–79.

Bazan, J. F., 1990, Haemopoietic receptors and helical cytokines, *Immunol Today* **11:**350–354.

Briere, F., Bridon, J. M., Servet, C., Rousset, F., Zurawski, G., and Banchereau, J., 1993, Interleukin 10 and interleukin 13 as B cell growth and differentiation factors, *J. Nouv. Fr. Res. Hematol.* **35:**233–235.

Brown, K. D., Zurawski, S. M., Mosmann, T. R., and Zurawski, G., 1989, A family of small inducible proteins secreted by leukocytes are members of a new superfamily that includes leukocytes and fibroblast-derived inflammatory agents, growth factors, and indicators of various activation processes, *J. Immunol.* **142:** 679–687.

Chatelein, R., Varkila, K., and Coffman, R. L., 1992, IL4 induces Th2 responses in *Leishmania major* infected mice, *J. Immunol.* **148:**1182–1187.

Cherwinski, H. M., Schumacher, J. H., Brown, K. D., and Mosmann, T. R., 1987, Two types of mouse helper T cell clone. III. Further differences in lymphokine synthesis between Thl and Th2 clones revealed by RNA hybridization, functionally monospecific bioassays, and monoclonal antibodies, *J. Exp. Med.* **166:**1229–1244.

Cocks, B. G., de Waal Malefyt, R., Galizzi, J.-P., de Vries, J. E., and Aversa, G., 1993, IL-13 induces proliferation and differentiation of human B cells activated by the CD40 ligand, *Int. Immunol.* **5:**657–663.

Coffman, R. L., Varkila, K., Scott, P., and Chateleln, R., 1991, The role of cytokines in the differentiation of CD4$^+$ T cell subsets *in vivo*, *Immunol. Rev.* **123:**189–207.

Colotta, F., Re, F., Muzio, M., Polentarutti, N., Minty, A., Caput, D., Ferrara, P., and Mantovani, A., 1994, Interleukin-13 induces expression and release of interleukin-1 decoy receptor in human polymorphonuclear cells, *J. Biol. Chem.* **269:**12403–12406.

DeFrance, T., Carayon, P., Billian, G., Guillemot, J.-C., Minty, A., Caput, D., and Ferrara, P., 1994, Interleukin-13 is a B cell stimulating factor, *J. Exp. Med.* **179:**135–143.

Denich, K., Borlin, P., O'Hanley, P., Howard, M. C., and Heath, A. W., 1993, The effects of expression of murine interleukin-4 by Aro A-*Salmonella typhimurium*: Persistence, immune response and the inhibition of macrophage killing, *Infect. Immun.* **61:**4818–4827.

Denis, M., and Ghadrian, E., 1994, Interleukin 13 and interleukin 4 protect bronchoalvealoar macrophages against infection with human immunodeficiency virus type-l, *AIDS Res. Human Retroviruses* **10:**795–802.

de Waal Malefyt, R., Figdor, C., Huijbens, R., Mohan-Peterson, S., Bennett, B., Culpepper, J., Dang, W., Zurawski, G., and de Vries, J. E., 1993, Effects of IL-13 on phenotype, cytokine production, and cytotoxic function of human monocytes, *J. Immunol.* **151:**6370–6381.

Doherty, T. M., Kastelein, R., Menon, S., Andrade, S., and Coffman, R. L., 1993, Modulation of murine macrophage function by interleukin-13, *J. Immunol.* **151:**7151–7160.

Fior, R., Vita, N., Raphael, M., Minty, A., Maillot, M. C., Crevon, M. C., Caput, D., Biberfeld, P., Ferrara, P., Galanaud, P., and Emilie, D., 1994, Interleukin-13 gene expression by malignant and EBV-transformed human B-lymphocytes, *Eur. Cytokine Network* **5**:593–600.

Freeman, G. J., Gribben, J. G., Boussiotis, V. A., Ng, V. W., Restivo, V. A., Lombard, L. A., Gray, G. S., and Nadler, L. M., 1993, Cloning of B7-2; a CTLA-4 counterreceptor that co-stimulates human T cell proliferation, *Science* **262**:904–911.

Heinzel, F. P., Scoenhaut, D. S., Rerko, R. M., Rosser, L. E., and Gately, M. K., 1993, Recombinant interleukin 12 cures mice infected with *Leishmania major*, *J. Exp. Med.* **177**:1505–1509.

Hsieh, C-S., Macatonia, S. E., Tripp, C. S., Wolf, S. F., O'Garra, A., and Murphy, K. M., 1993, Development of Thl $CD4^+$ T cells through IL-12 produced by *Listeria*-induced macrophages, *Science* **260**:547–549.

Jacobsen, S. E. W., Okkenhaug, C., Veiby, O. P., Caput, D., Ferrara, P., and Minty, A., 1994, Interleukin-13: Novel role in direct regulation of proliferation and differentiation of primitive hematopoietic progenitor cells, *J. Exp. Med.* **180**:75–82.

Kawabe, T., Naka, T., Yoshida, K., Tanaka, K., Fujiwara, H., Suematsu, S., Yoshida, N., Kishimoto, T., and Kikutani, H., 1994, The immune response in CD40 deficient mice: Impaired immunoglobulin class switching and germinal center formation, *Immunity* **1**:167–178.

Kondo, M., Takeshita, T., Higuchi, M., Nakamura, M., Sudo, T., Nishikawa, S.-I., and Sugamura, K., 1994, Functional participation of the IL-2 receptor γ chain in IL-7 receptor complexes, *Science* **263**:1453–1454.

Kuhn R., Rajewsky, K., and Muller, W., 1991, Generation and analysis of interleukin-4 deficient mice, *Science* **254**:707–710.

Lakkis, F. G., and Cruet, E. N., 1993, Cloning of rat interleukin-13 (IL-13) cDNA and analysis of IL-13 gene expression in experimental glomerulonephritis, *Biochem. Biophys. Res. Commun.* **197**:612–618.

LeGros, G., Ben-Sasson, S. Z., Rader, R., Finkelman, F. D., and Paul, W. E., 1990, Generation of interleukin 4 producing cells *in vitro*, and *in vivo*: IL2 and IL4 are required for *in vitro* generation of IL4 producing cells, *J. Exp. Med.* **172**:921–929.

Lehn, M., Weiser, W. Y., Engelhorn, S., Gillis, S., and Remold, H. G., 1989, IL4 inhibits H_2O_2 production and anti-leishmanial capacity of human cultured monocytes mediated by IFN gamma, *J. Immunol.* **143**:3020–3024.

Lin, J-X., Migone, T-S., Tsang, M., Friedmann, M., Weatherbee, J. A., Zhou, L., Yamauchi, A., Bloom, E. T., Nietz, J., John, S., and Leonard, W. J., 1995, The role of shared receptor motifs and common STAT proteins in the generation of cytokine pleiotropy and redundancy by IL2, IL4, IL7, IL13 and IL15, *Immunity* **2**:331–339.

Magazin, M., Guillemot, J. C., Vita, N., and Ferrara, P., 1994, Interleukin 13 is a monocyte chemoattractant, *Eur. Cytokine Network* **5**:397– 400.

McKenzie, A. N. J., Culpepper, J. A., de Waal Malefyt, R., Briere, F., Punnonen, J., Aversa, G., Sato, A., Dang, W., Cocks, B. G., Menon, S., de Vries, J. E., Banchereau, J., and Zurawski, G., 1993a, Interleukin-13, a novel T-cell-derived cytokine that regulates human monocyte and B cell function, *Proc. Natl. Acad. Sci. U.S.A.* **90**:3735–3739.

McKenzie, A. N. J., Li, X., Largaespada, D. A., Sato, A., Kaneda, A., Zurawski, S. M., Doyle, E. L., Francke, U., Copeland, N. G., Jenkins, N. A., and Zurawski, G., 1993b, Structural comparison and chromosomal localization of the human and mouse IL-13 genes, *J. Immunol.* **150**:5436–5444.

Minty, A., Chalon, P., Derocq, J.-M., Dumont, X., Guillemot, J.-C., Kaghad, K., Labit, C., Leplatois, P., Liauzun, P., Miloux, B., Minty, C., Casellas, P., Loison, G., Lupker, J., Shire, D., Ferrara, P., and Caput, D., 1993, Interleukin-13: A novel human lymphokine regulating inflammatory and immune responses, *Nature* **362**:248–250.

Montanier, L. J., Doyle, A. G., Collin, M., Herbein, G., Illei, P., James, W., Minty, A., and Caput, D., 1993, Interleukin 13 inhibits human immunodeficiency virus type-1 production in primary blood derived human macrophages *in vitro*, *J. Exp. Med.* **178**:743–747.

Morgan, J. G., Dolganov, G. M., Robbins, S. E., Hinton, L. M., and Lovett, M., 1992, The selective isolation of novel cDNAs encoded by the regions surrounding the human interleukin 4 and 5 genes, *Nucleic Acids Res.* **20**:5173–5179.

Mosmann, T. R., and Coffman, R. L., 1989, THl and TH2 cells: Different patterns of lymphokine secretion lead to different functional properties, *Annu. Rev. Immunol.* **7**:145–173.

Noguchi, M., Nakamura, Y., Russell, S. M., Ziegler, S. F., Tsang, M., Cao, X., and Leonard, W. J., 1993, Interleukin-2 receptor γ chain: A functional component of the interleukin-7 receptor, *Science* **262**:1877–1880.

Obiri, N. I., Debinski, W., Leonard, W. J., and Puri, R. K., 1995, Receptor for interleukin 13, *J. Biol. Chem.* **270**:8797–8804.

Punnonen, J., Aversa, G., Cocks, B. G., McKenzie, A. N. J., Menon, S., Zurawski, G., de Waal Malefyt, R., and de Vries, J. E., 1993, Interleukin-13 induces interleukin-4-independent IgG4 and IgE synthesis and CD23 expression by B cells, *Proc. Natl. Acad. Sci. U.S.A.* **90:**3730–3735.

Reiner, S. L., Zheng, S., Wang, Z. E., Stowring, L., and Locksley, R. M., 1994, *Leishmania* promastigotes evade interleukin-12 (IL-12) induction by macrophages and stimulate a broad range of cytokines from $CD4^+$ T cells during initiation of infection, *J. Exp. Med.* **179:**447–456.

Renard, N., Duvert, V., Banchereau, J., and Saeland, S., 1994, Interleukin 13 inhibits the proliferation of normal and leukemic human B cell precursors, *Blood* **84:**2253–2260.

Swain, S. L., Weinberg, A. D., English, M., and Huston, G., 1990, IL4 directs the development of Th2 like helper-effectors, *J. Immunol.* **145:**3796–3806.

te Velde, A. A., Klomp, J. P., Yard, B. A., de Vries, J. E., and Figdor, C. G., 1988, Modulation of phenotypic and functional properties of human peripheral blood monocytes by IL-4, *J. Immunol.* **140:**1548–1554.

Van der Weld, T., Kopf, M., Kohler, G., and Langhorne, J., 1994, The immune response to *Plasmodium chabawdi* malaria in interleukin 4 deficient mice, *Eur. J. Immuol.* **24:**2285–2293.

Vercelli, D., Jabara, H. H., Lee, B.-L., Woodland, N., Geha, R. S., and Leung, D. Y. M., 1988, Human recombinant interleukin-4 induces FcεR/CD23 on normal human monocytes, *J. Exp. Med.* **167:**1406–1416.

Vita, N., Lefort, S., Laurent, P., Caput, P., and Ferrara, P., 1995, Characterisation and comparison of the interleukin 4 receptor on several cell types, *J. Biol. Chem.* **270:**3512–3517.

Walter, M. R., Cook, W. J., Zhoa, B. G., Cameron, R. P., Jr., Ealick, S. E., Walter, R. L., Jr., Reichert, P., Nagabhushan, T. L., Trotta, P. P., and Bugg, C. E., 1992, Crystal structure of recombinant human interleukin-4, *J. Biol. Chem.* **267:**20371–20376.

Wang, Z. E., Reiner, S. L., Zheng, S., Dalton, D. K., and Locksley, R. M., 1994, $CD4^+$ effector cells default to the Th2 pathway in interferon-γ-deficient mice infected with *Leishmania major*, *J. Exp. Med.* **179:**1367–1371.

Wynn, T. A., Eltoum, I., Oswald, I. P., Cheever, A. W., and Sher, A., 1994, Endogenous interleukin-12 (IL-12) regulates granuloma formation induced by eggs of *Schistosoma mansoni*, and exogenous IL-12 both inhibits and prophylactically immunizes against egg pathology, *J. Exp. Med.* **179:**1551–1561.

Zurawski, S. M., Vega, F., Jr., Huyghe, B., and Zurawski, G., 1993, Receptors for interleukin-13 and interleukin-4 are complex and share a novel component that functions in signal transduction, *EMBO J.* **12:**3899– 3905.

Zurawski, S. M., Chomarat, P., Djossou, O., Bidault, C., McKenzie, A. N. J., Miossec, P., Banchereau, J., and Zurawski, G., 1995, The primary binding subunit of the human interleukin-4 receptor is also a component of the interleukin-13 receptor, *J. Biol. Chem.* **270:**13869–13878.

Chapter 3

The Tumor-Necrosis-Factor-Related Superfamily of Ligands and Receptors

David Cosman

1. INTRODUCTION

This review attempts to summarize recent progress in the discovery of an extensive family of ligands and receptors structurally related to tumor necrosis factor (TNF) and its receptors. The structures of these molecules are discussed, together with what is known of their biology. Because TNF has been studied extensively for a number of years and comprehensively reviewed (Vassilli, 1992; Tracey and Cerami, 1994), this review concentrates on the more recently discovered members of the family and refers to the TNF system for comparison. Likewise, although the low-affinity receptor for nerve growth factor (NGF) is a member of the TNFR family, NGF is not structurally related to TNF and is not discussed. The reader is referred to recent reviews by Bradshaw *et al.* (1993) and Eide *et al.* (1993) for more information on NGF and its homolgues, the neurotropins.

2. THE MEMBERS OF THE FAMILY

The currently described members of the TNF/TNFR ligand/receptor family are listed in Table I. The first three, TNF-α, LT-α, and LT-β, and their receptors TNFRp80, TNFRp60, and TNFR-RP (LT-βR), show a complex pattern of cross-binding that will be discussed in detail later. The other family members, however, do not show any such promiscuity of binding, and the one-ligand/one-receptor principle applies. The other pairs of receptors and

David Cosman Department of Molecular Biology, Immunex Research and Development Corporation, Seattle, Washington 98101.

Blood Cell Biochemistry, Volume 7: Hematopoietic Cell Growth Factors and Their Receptors, edited by A. D. Whetton and J. Gordon. Plenum Press, New York, 1996.

Table I
The TNF-Related Ligands and Their Receptors

Receptor (Alternative Names)	Ligand (Alternative Names)
TNFRp75 (TNFRp80, TNFRII)	TNF-α[a], LT-α (TNF-β[a])
TNFRp55 (TNFRp60, TNFRI)	TNF-α, LT-α (TNF-β)
TNFR-RP (LT-βR)	LT-β (p33)
CD40	CD40L (gp39, TRAP, T-BAM)
CD30 (Ki-1 antigen)	CD30L
CD27	CD27L (CD70)
OX40 (ACT35, 106)	OX40L (gp34)
4-1BB (ILA)	4-1BBL
Fas (APO-1, CD95)	FasL (gld protein)
NGFR	(NGF)[b]

[a]In this chapter, the term TNF will be used when describing properties shared between TNF-α and LT-α (TNF-β), which both bind to TNFRp75 and TNFRp55.
[b]NGF is not structurally related to the other ligands.

ligands are CD40/CD40L, CD30/CD30L, CD27/CD27L, OX40/OX40L, 4-1BB/4-1BBL, and Fas/FasL. In many cases, alternative names for these molecules exist (listed in Table I), but this review uses the most common designations. In addition, several poxviruses have been shown to encode secreted proteins with strong homology to the extracellular domains of the TNF receptors that are capable of binding TNF (Smith *et al.*, 1991; Upton *et al.*, 1991). These genes have presumably been acquired by the viruses from host genome during evolution and are used to antagonize the anti-viral action of TNF during infection of the host.

3. DISCOVERY OF THE FAMILY MEMBERS: A HISTORICAL PERSPECTIVE

The first of the family members to be cloned was the low-affinity NGF receptor (Johnson *et al.*, 1986), which was recognized as having a novel structure with repeated cysteine-rich domains within the extracellular portion of the type I membrane glycoprotein (discussed in detail below). The next molecule to be cloned was CD40 (Stamenkovic *et al.*, 1989), and the homology to the extracellular domain of NGFR was immediately recognized. CD40, like many of the members of the receptor family, was initially discovered via the generation of monoclonal antibodies (mAb) possessing interesting biological activities. CD40 was found to be expressed on B cells and certain epithelial carcinomas, and anti-CD40 antibodies were shown to activate B cells (reviewed by Clark and Lane, 1991, and discussed in detail below). Also in 1989, a cDNA for an inducible T-cell gene called 4-1BB was isolated (Kwon and Weissman, 1989) but at that time was not recognized as having homology with the NGFR. The recognition that this was a gene family came in 1990 with the cloning of the two TNF receptors, p75 and p55, and OX40 (Smith *et al.*, 1990; Loetscher *et al.*, 1990; Schall *et al.*, 1990; Mallet *et al.*, 1990). OX40, like CD40, was discovered by the generation of a mAb that recognized rat T helper cells and could costimulate their proliferation (Paterson *et al.*, 1987). The family of receptors was further extended by the cloning of CD27 (Camerini *et al.*, 1991), Fas (Itoh *et al.*, 1991), and CD30 (Dürkop *et al.*, 1992). All three of these proteins were also discovered by the properties of the mAb that recognized them. Antibodies to CD27, like those to OX40, costimulated T-cell proliferation (van Lier *et al.*, 1987); antibodies to Fas had the unusual property of inducing cell death via an apoptotic mechanism (Trauth *et al.*, 1989; Yonehara *et al.*, 1989), and antibodies to CD30 had been used for more than a decade as clinical markers for various lymphomas, particularly Hodgkin's disease (Schwab *et al.*, 1982). Finally, the TNFR-related protein (TNFR-RP) was discovered by accident as an expressed gene mapping to human chromosome 12p13 (Baens *et al.*, 1993).

The discovery of a family of cell-surface molecules, some of which were receptors for known cytokines (TNF, NGF), coupled with the functional activity of antibodies to many of these proteins, led to speculation that natural ligands with cytokine-like activity would exist for all the family members. In order to prove this, it was necessary to develop techniques for identifying and cloning the ligands. A successful strategy was developed for the cloning of the CD40L (Armitage *et al.*, 1992) and subsequently used to identify the ligands for CD30 (Smith *et al.*, 1993), CD27 (Goodwin *et al.*, 1993a), 4-1BB (Goodwin *et al.*, 1993b), Fas (Suda *et al.*, 1993), and OX40 (Baum *et al.*, 1994; Godfrey *et al.*, 1994). In each case, recombinant fusion proteins were constructed that linked the extracellular domain of the receptor to the Fc region of an immunoglobulin (Ig). This fusion protein functioned as a surrogate antibody that could be used in flow cytometry experiments to stain cells that expressed the ligand on their surface and then to clone the ligand via binding to mammalian cells transfected with pools of cDNAs in an expression vector. If necessary, the Fc fusion protein could also be used in fluorescence-activated cell-sorting experiments to enrich for a population of cells expressing higher levels of ligand before constructing a cDNA library (Armitage *et al.*, 1992; Suda *et al.*, 1993).

The success of this strategy was facilitated by the fact that all the ligands for this family of receptors turned out to by type II membrane glycoproteins like TNF-α but unlike LT-α. LT-α was initially characterized as a secreted protein (Pennica *et al.*, 1984), but later work found a cell-surface form of the molecule anchored to the cell surface by association with a membrane glycoprotein, p33 (Browning *et al.*, 1991), which was subsequently purified, cloned, shown to belong to the TNF-related ligand family, and named LT-β (Browning *et al.*, 1993). Subsequently, LT-β was shown to bind to the TNFR-RP (Crowe *et al.*, 1994).

4. RECEPTOR STRUCTURES

The TNF-receptor-related molecules are all type I membrane glycoproteins. The characteristic feature of this family is the presence in the extracellular domain of a variable number of cysteine-rich domains about 40 amino acids in length (Figure 1). Six of the cysteine residues within these domains show a characteristic spacing pattern and are the most strikingly conserved amino acids, but other amino acid sequence similarities can be recognized between domains to give overall homologies in the range of 25% between family members or between domains in a single protein. The cysteine-rich domains of the receptors are responsible for ligand binding. In most family members, there is a "spacer"

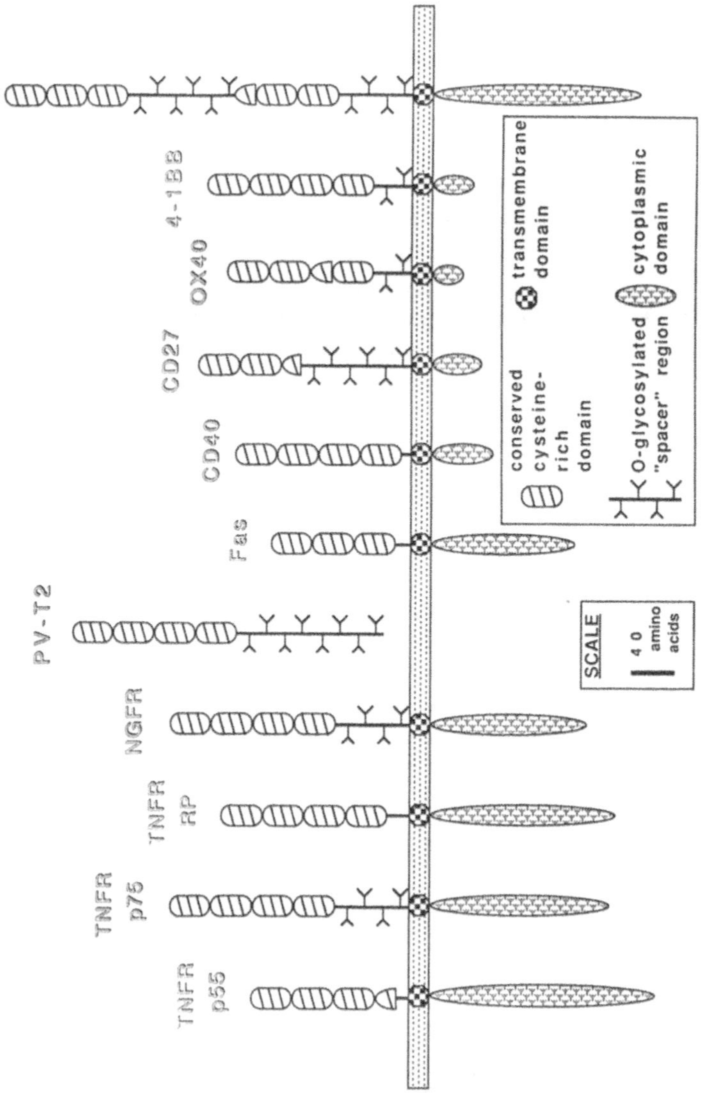

FIGURE 1. Schematic representation of the structures of the TNFR-related family of type I membrane glycoproteins. Note that PV-T2 represents a secreted, poxvirus-encoded, TNF-binding protein. The indicated features of the proteins are discussed in the text.

region between the transmembrane domain and the cysteine-rich domains. These spacer regions are rich in proline, serine, and threonine residues and, where studied, are heavily O-glycosylated. CD30 has an atypical structure in that there is a partial duplication of the cysteine-rich and spacer regions.

The cytoplasmic domains of the receptors are of highly variable length and contain no obvious sequence motifs characteristic of kinases or phosphatases of any kind. Although in general there is little similarity between the cytoplasmic amino acid sequences of the receptors, there are a few limited sequence homologies that have been recognized, such as those between Fas and TNFRp55 (Itoh and Nagata, 1993) and 4-1BB and CD27 (Gravestein *et al.*, 1993). The best-characterized homology is within the "death domain" shared by TNFRp55 and Fas. This domain is responsible for the induction of apoptotic cell death and to some extent is interchangeable between the two receptors (Itoh and Nagata, 1993; Tartaglia *et al.*, 1993).

5. LIGAND STRUCTURES

As mentioned above, the TNF-related ligands (with the exception of LT-α) are type II membrane proteins with a relatively short N-terminal cytoplasmic domain and an extracellular domain containing a region of about 150 amino acids with homology to TNF-α (Figure 2). Percentage amino acid identities between family members in this region vary from 12% to 29%. TNF-α is found in both cell-surface and proteolytically cleaved soluble forms (Kriegler *et al.*, 1988; Mohler *et al.*, 1994); LT-α is found as a secreted protein and also complexed to LT-β on the membrane (discussed below). The x-ray crystallographic structures of the soluble forms of both molecules have been determined and found to be very similar (Eck and Sprang, 1989; Jones *et al.*, 1989; Eck *et al.*, 1992). Both molecules are found as trimers, and each monomer is composed primarily of β-strands. Two sheets formed by eight antiparallel β-strands are arranged in a sandwich structure described as a β-jellyroll. The amino acid homology between ligand family members is highest within the β-strand regions, including the residues that are involved in intersubunit contacts in TNF-α and LT-α. These homologies together with more detailed modeling studies (Peitsch and Jongeneel, 1993) strongly suggest that the ligand family members will share a similar β-sandwich structure and will likely form trimers or other multimers.

6. RECEPTOR–LIGAND STRUCTURE

The x-ray crystallographic structure of LT-α bound to the extracellular domain of the TNFRp55 has been determined (Banner *et al.*, 1993). The LT-α trimer interacts with three molecules of receptor. Each receptor monomer consists of an elongated, slightly bent structure with the conserved cysteine-rich domains stacked on top of each other. Each of the six conserved cysteines is involved in a disulfide bond: C1–C2, C3–C5, and C4–C6. Each receptor monomer binds to a groove formed by the interface between LT-α monomers and therefore makes contact with two ligand monomers. Conversely, each LT-α monomer binds two receptor molecules, but there is no contact between receptor molecules in the regions that bind ligands. This structure suggests a model in which the trimeric ligand

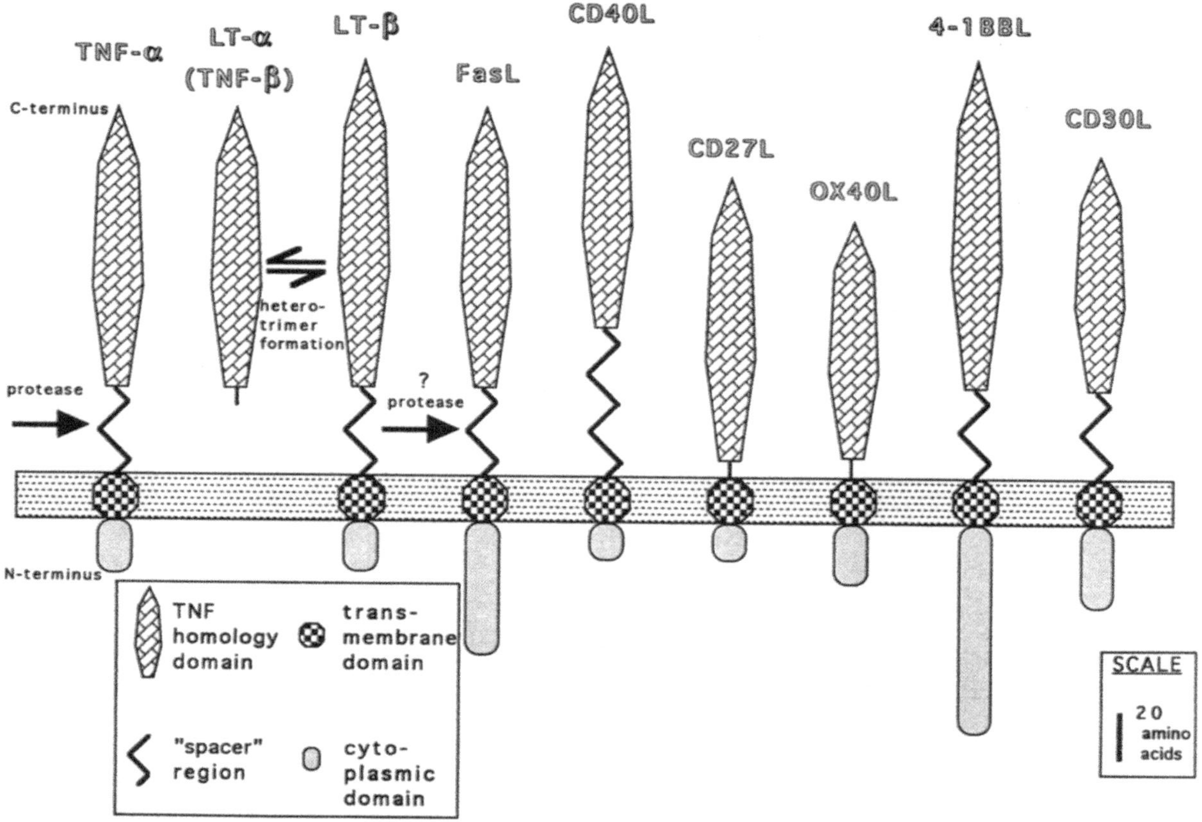

FIGURE 2. Schematic representation of the structures of the TNF-related ligand family. All members of type II membrane glycoproteins with the exception of LT-α, which can be found either as a secreted homotrimer or as a heterotrimeric complex with membrane-bound LT-β.

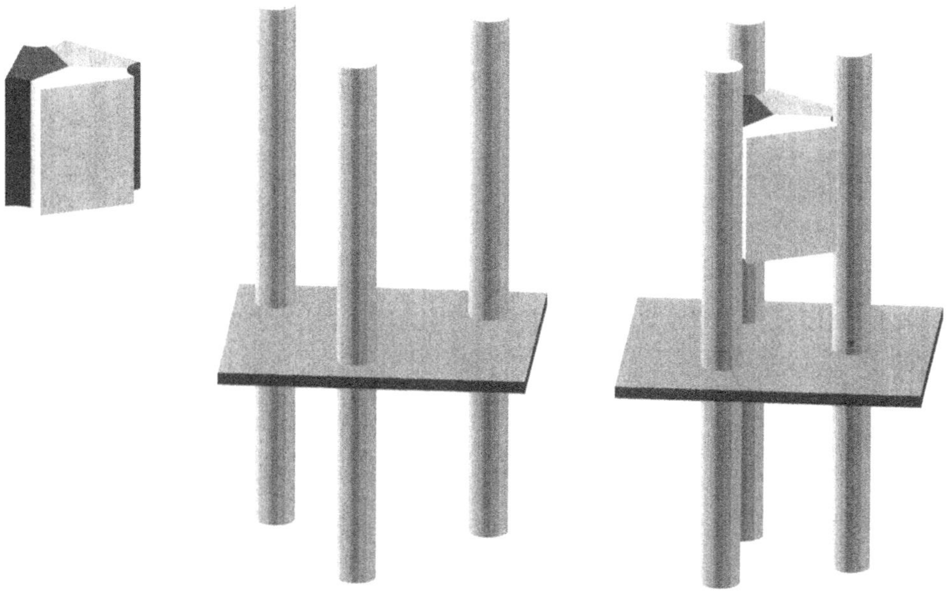

FIGURE 3. Schematic representation of an LT-α homotrimer binding to its receptor. As discussed in the text, each receptor monomer (shown as a rod) binds to the interface between ligand monomers (shown as a groove between differently shaded monomers). Binding of the trimeric ligand to three receptor monomers causes clustering and brings together the cytoplasmic domains of the receptors to indicate signal transduction.

induces clustering of the receptors (shown schematically in Figure 3). Receptor–receptor interactions within the transmembrane or cytoplasmic domains would trigger signal transduction, as has been suggested for other receptor families such as the tyrosine kinases and hematoprotein receptors (Ullrich and Schlessinger, 1990; Cosman, 1993). Consistent with this model is the observed agonistic activity of many mAb directed against the TNFR family members. These antibodies are able to signal only when they are cross-linked and can therefore cause receptor clustering. It is thought that other TNF-related ligands bind to their receptors in a similar manner to LT-α, but this has yet to be demonstrated.

7. GENOMIC ORGANIZATION AND CHROMOSOMAL LOCATION

Thus far, the genomic intron–exon structure has been elucidated for five members of the TNFR family: NGFR (Johnson *et al.*, 1986), TNFRp55 (Fuchs *et al.*, 1992), CD27 (Loenen *et al.*, 1992), CD40 (Grimaldi *et al.*, 1992), and 4-1BB (Kwon *et al.*, 1994). As expected, the genes show clear evidence of evolutionary relatedness. Unlike the Ig superfamily, in which each structurally conserved Ig-like domain is usually found on a single exon, in the TNFR family the structural unit, consisting of about 40 amino acids and containing six conserved cysteine residues, is most commonly interrupted by an intron, usually between cysteines 4 and 5. As with many type I membrane proteins, the signal sequences and transmembrane domains are usually found on their own distinct exons.

Table II
Chromosomal Locations of the Genes for the TNF-Related Ligands and Their Receptors

Ligands	Mouse	Human	Receptors	Mouse	Human
TNF-α	17 (MHC)[a]	6 (MHC)	TNFRp75	4 distal	1p36
LT-α	17 (MHC)	6 (MHC)	CD30		1p36
LT-β	17 (MHC)	6 (MHC)	OX40	4 distal	1p36
CD27L	(17 central)[b]	19p13	4-1BB	4 distal	(1p36)[b]
4-1BBL	17 central	19p13	TNFRp55	6 distal	12p13
FasL	1 distal	(1q25)[b]	TNFR-RP		12p13
OX40L	1 distal	1q25	CD27	(6 distal)[b]	12p13
CD30L	4 central	9q32	CD40	2 distal	20q12
CD40L	X proximal	Xq26	Fas	19 distal	10q24

[a]MHC refers to a location within the major histocompatibility complex for the very closely linked TNF-α, LT-α, and LT-β genes.
[b]Parentheses indicate locations inferred in one species from close linkage in the other but not yet formally confirmed. The information is compiled from the following references: LT-β (Browning *et al.*, 1993), CD27L (Goodwin *et al.*, 1993a), 4-1BBL (Goodwin *et al.*, 1993b), Alderson *et al.*, 1994), FasL (Seldin *et al.*, 1988), OX40L (Baum *et al.*, 1994), CD30L (Smith *et al.*, 1993), CD40L (Allen *et al.*, 1993), TNFRp75 (Baker *et al.*, 1991; Goodwin *et al.*, 1991), CD30 (Fonatsch *et al.*, 1992), OX40 (Latza *et al.*, 1994), 4-1BB (Goodwin *et al.*, 1993b), TNFRp55 (Baker *et al.*, 1991; Goodwin *et al.*, 1991), TNFR-RP (Baens *et al.*, 1993), CD27 (Loenen *et al.*, 1992), CD40 (Grimaldi *et al.*, 1992; Lafage-Pochitaloff *et al.*, 1994), Fas (Watanabe-Fukunaga *et al.*, 1992a; Lichter *et al.*, 1992).

Further evidence of evolutionary relatedness comes from the close genetic linkage of several of the genes (Table II). Among the receptor genes, TNFRp75, CD30, 4-1BB, and OX40 are clustered, as are TNFRp55, TNFR-RP, and CD27. For the ligand genes, three clusters are known: TNF-α, LT-α, and LT-β; CD27L and 4-1BBL; and FasL together with OX40L.

8. CD40L/CD40

CD40, as mentioned above, was discovered as a 50-kDa glycoprotein recognized by mAb and expressed on B cells, B-cell precursors, and epithelial cells (reviewed by Clark and Lane, 1991; Banchereau *et al.*, 1994; see also Chapter 10). Cross-linked antibodies to CD40 were found to stimulate B cells in a variety of assays: increased class II major histocompatibility (MHC) antigen expression, elevated levels of the important costimulatory molecule B7, induction of homotypic adhesion via increased expression of CD23 and intracellular adhesion molecule type 1 (ICAM-1), costimulation of proliferation when used in conjunction with phorbol esters or anti-IgM, and costimulation of Ig secretion together with IL-4 (Clark and Ledbetter, 1986; Gordon *et al.*, 1988; Jabara *et al.*, 1990; Barrett *et al.*, 1991 Björck *et al.*, 1993; Ranheim and Kipps, 1993). Apoptosis of germinal center centrocytes in culture was prevented by stimulation with anti-CD40 (Liu *et al.*, 1989). In addition to these short-term effects, it was found that human B cells could be maintained in long-term culture in the presence of L cells expressing Fc receptors to which anti-CD40 was bound. Either IL-4 or IL-10 was also required for B-cell expansion in these cultures (Banchereau *et al.*,

1991; Banchereau and Rousset, 1991). Different Ig isotypes were produced, depending on the cytokine used: IgM, IgG_1, IgG_2, IgG_3, and IgE were promoted by IL-10, whereas IL-4 stimulated IgG_4 and IgE production (Rousset *et al.*, 1991, 1992).

These dramatic effects of anti-CD40 on B-cell function, together with the discovery that CD40 was homologous to NGFR (Stamenkovic *et al.*, 1989), stimulated the search for a natural ligand for CD40. As described above, a CD40·Fc fusion protein (Fanslow *et al.*, 1992) was used to detect low-level expression of a counterstructure on murine EL-4 thymoma cells. Several rounds of fluorescence-activated cell sorting were used to enrich for cells expressing higher levels of CD40·Fc binding, and a CD40L cDNA was isolated by direct expression cloning (Armitage *et al.*, 1992). A human homologue was found using cross-species hybridization or polymerase chain reaction (PCR) and found to share 78% amino acid identity with the murine CD40L (Spriggs *et al.*, 1992; Hollenbaugh *et al.*, 1992). As described above, the predicted structure of the CD40L is that of a type II membrane glycoprotein with homology to TNF in the C-terminal portion of the extracellular domain. This region contains one N-linked glycosylation site and four cysteine residues (five in the human). The spacer region between the transmembrane and TNF homology domains is the longest of all the TNF-related ligands. Its function is unknown—it is not essential for binding to CD40.

The CD40L was found to be expressed predominantly on activated $CD4^+T$ cells but also on a small proportion of $CD8^+T$ cells (Spriggs *et al.*, 1992). The natural and recombinant proteins migrate as a single band of 33 kDa on SDS polyacrylamide gel electrophoresis (SDS-PAGE) under reducing or nonreducing conditions (Armitage *et al.*, 1992). Despite this evidence against covalently linked multimerization, CD40L is believed to function as a trimer or dimer, based both on modeling studies (Peitsch and Jongeneel, 1993) and the necessity to cross-link CD40 to generate a signal. The function of recombinant CD40L has been tested by using it as a cell-surface protein on transfected cells or as a soluble molecule that has been trimerized by fusion to a modified leucine zipper motif or to the extracellular domain of CD8. In all cases, the recombinant ligand possesses all the biological activities predicted from previous studies using anti-CD40 antibodies (Armitage *et al.*, 1992, 1993a; Hollenbaugh *et al.*, 1992; Spriggs *et al.*, 1992; Grabstein *et al.*, 1993; Holder *et al.*, 1993; Lane *et al.*, 1993, Maliszewski *et al.*, 1993).

Expression of CD40L has also been demonstrated on mast cells and basophils, and these cells have been shown to stimulate IgE synthesis from B cells (Gauchat *et al.*, 1993), supporting an essential role for CD40 and IL-4 in production of IgE. On $CD4^+T$ cells, the most potent stimulus for induction of CD40L expression is the use of phorbol ester together with ionomycin, whereas a more physiological stimulus through the T-cell receptor induces CD40L expression to a lesser extent (Lane *et al.*, 1992; Spriggs *et al.*, 1992). However, the signal through the T-cell receptor can be augmented by antibodies to CD28 and CD2 (Klaus *et al.*, 1994; Kwekkeboom *et al.*, 1994), molecules whose ligands (the B7 family and LFA-3) are expressed by B cells. Because CD40L increases B7 expression on B cells (Kennedy *et al.*, 1994; Yellin *et al.*, 1994), the model that emerges is a reciprocal stimulation between T and B cells that enhances antigen presentation by the B cells (increased MHC class II expression), increases costimulatory function (elevated B7 expression), and increases CD40L expression by the T cells, thus maximizing the strength of T–B cell interactions. Cyclosporin A is able to inhibit the induction of CD40L, which may be an important component of its immunosuppressive capabilities (Fuleihan *et al.*, 1994; Klaus *et al.*, 1994).

At the same time, mechanisms exist to limit or terminate the expression of CD40L on activated T cells during T–B cell interactions. B cells can down-regulate CD40L mRNA levels in T cells and shed soluble CD40 molecules that block CD40L interaction with cell-surface CD40 receptors (van Kooten *et al.*, 1994). Additionally, interaction of CD40L with B-cell-expressed CD40 leads to internalization of CD40L by the T cell (Yellin *et al.*, 1994).

The presence of multiple mechanisms for regulation of CD40L expression and activity suggests a pivotal role for CD40L–CD40 interaction in T–B cell collaboration leading to B-cell proliferation and antibody production. It was already well known that this signal from activated helper T cells to B cells required cell–cell contact as well as cytokine production (reviewed in Parker, 1993; Clark and Ledbetter, 1994). The requirement for T cells could be met by using fixed T cells or T-cell membrane preparations. Under these circumstances, CD40·Fc or an antibody to CD40L could completely block B-cell proliferation and Ig production (Lederman *et al.*, 1992; Noelle *et al.*, 1992). However, several groups have observed incomplete blocking of T-cell-mediated help for B cells by these reagents in cultures where the T cells are activated by antibodies to CD3, suggesting that although CD40L–CD40 interactions are pivotal, other cell-surface molecules may play a role under certain circumstances (Klaus *et al.*, 1994; Kwekkeboom *et al.*, 1994; Nishioka and Lipsky, 1994).

Definitive proof of the importance of the role of CD40L *in vivo* came from the discovery that an inherited human immunodeficiency disease was the result of mutations in the CD40L gene. When the CD40L gene was mapped to Xq26 (Graf *et al.*, 1992; Allen *et al.*, 1993), it was immediately suggested as a candidate gene for hyper-IgM syndrome (HIGM), which had previously been mapped to this locus. In this disease (reviewed by Conley, 1992), affected male children suffer recurrent bacterial infections after birth. These patients have normal number of B cells and normal or elevated levels of IgM and IgD but have no germinal centers and low to absent IgG, A, and E. Studies have shown that these patients are unable to mount an effective high-affinity antibody response to infection (Conley, 1992). CD40L defects were linked to HIGM in several ways. Activated T cells from HIGM patients showed no detectable staining with CD40·Fc, although a polyclonal antibody to CD40L detected CD40L expression in cells from some patients (Allen *et al.*, 1993; Aruffo *et al.*, 1993; DiSanto *et al.*, 1993; Fuleihan *et al.*, 1993; Korthäuer *et al.*, 1993). More definitively, CD40L cDNAs from HIGM patients were sequenced and found to contain a variety of mutations. Introduction of these mutations into CD40L expression vectors abrogated the *in vitro* ability of the ligand to stimulate B-cell proliferation and Ig production (Allen *et al.*, 1993; Aruffo *et al.*, 1993). The mutations, which consisted of single amino acid substitutions as well as deletions, resided within the TNF-homology region of the extracellular domain, the part known to be essential for receptor binding. An exceptional mutation resulted in the introduction of a charged amino acid into the hydrophobic transmembrane domain and prevented cell-surface expression of CD40L (Allen *et al.*, 1993; Aruffo *et al.*, 1993; DiSanto *et al.*, 1993; Korthäuer *et al.*, 1993). HIGM was proven to be a T-cell defect, as B cells from patients responded normally to stimulation with wild-type CD40L *in vitro*. These studies on HIGM patients comfirmed the many *in vitro* studies on CD40L function and demonstrated the critical requirement of CD40L–CD40 interactions for isotype switching and affinity maturation in the germinal center. However, despite potent effects of CD40L on B-cell proliferation *in vitro*, it appears that other pathways can substitute *in vivo*, leading to normal B-cell numbers in HIGM patients.

Recently, murine models for HIGM syndrome have been created in which the genes for CD40 (Kawabe *et al.*, 1994) or CD40L (Renshaw *et al.*, 1994; Xu *et al.*, 1994) have been disrupted. Studies utilizing these mice, as well as experiments in which normal mice were treated with blocking antibody to CD40L (Foy *et al.*, 1993, 1994), have shown that CD40L–CD40 interactions are required for both primary and secondary antibody responses to thymus-dependent antigens as well as for germinal center formation. However, the antibody response to thymus-independent antigens, which is primarily IgM, is unaffected by the absence of CD40L or CD40. This finding provides an explanation for the normal to elevated levels of IgM found in HIGM patients. Immunization of HIGM patients with a thymus-dependent antigen has been shown to generate a depressed primary antibody response (Nonoyama *et al.*, 1993).

However, HIGM patients have some other abnormalities that seem unlikely to be related to defective stimulation of B cells by CD40L. Many patients are neutropenic, and infection by intracellular pathogens such as *Pneumocystis carinii* are common (Conley, 1992). Resistance to such organisms is considered to be mediated by T cells rather than by antibody. These observations suggest that CD40L may have important effects on other cell types, and indeed, several examples of such effects have been demonstrated *in vitro*. In the presence of IL-1 and interferon-γ (IFNγ), CD40 was shown to stimulate production of GM-CSF by thymic epithelial cells (Galy and Spits, 1992). CD40, which is present at low levels on monocytes, is up-regulated by IFNγ, GM-CSF, and IL-3 (Alderson *et al.*, 1993a). These cells can then be triggered by recombinant CD40L to produce IL-6, IL-8, and TNF-α. Independent of treatment with these cytokines, CD40L can stimulate strong tumoricidal activity of monocytes against a melanoma target (Alderson *et al.*, 1993a). Recombinant CD40L can also stimulate a variety of responses in T cells. These include production of IL-2, IFNγ, and TNF-α, up-regulation of IL-2Rα (CD25), CD69, and CD40L itself, and costimulation of proliferation when used in conjunction with agents such as phytohemagglutinin (PHA), concanavalin A (ConA), or anti-CD3 (Armitage *et al.*, 1993b; Fanslow *et al.*, 1994). This T-cell proliferation was largely IL-2 independent. CD40L is also able to stimulate thymic γδ T cells in a similar manner (Ramsdell *et al.*, 1994a). Interestingly, evidence has recently been presented that signaling may also occur through the CD40L molecule on T cells, using recombinant CD40 expressed on transfected cells to costimulate T-cell proliferation (Cayabyab *et al.*, 1994).

Another cell type known to respond to CD40L is B-cell precursors. These cells express CD40 and respond to CD40L in the presence of IL-3 by proliferation and increased CD23 expression (Saeland *et al.*, 1992, 1993). However, no differentiation to mature surface IgM^+ B cells was observed. Because HIGM patients have normal B-cell numbers, other pathways must be able to substitute any role of CD40L in B-cell ontogeny.

Some of the non-B-cell effects of CD40L might account for the neutropenia and intercellular infections found in HIGM patients. Interestingly, preliminary data on mice in which the CD40 gene has been disrupted show that, in addition to the expected defects in germinal center formation and affinity maturation, animals kept under non-pathogen-free conditions have lower neutrophil levels than normal littermates (Kawabe *et al.*, 1994). This experimental system should allow exploration of the mechanisms involved and testing of the CD40-deficient mice for their susceptibility to infection by intracellular pathogens. A possible clinical utility of CD40L–CD40 antagonists is suggested by the blocking of collagen-induced arthritis in mice treated with an antibody to CD40L (Durie *et al.*, 1993).

There may also be another role for CD40L–CD40 agonists in the clinic: stimulation through CD40 is inhibitory to the growth of certain B-cell lymphomas *in vitro* and *in vivo* in SCID mouse models (Heath *et al.*, 1993; Funakoshi *et al.*, 1994). This observation contrasts with the ability of CD40 signaling to prevent apoptosis induced by antigen receptor cross-linking in other B-cell lymphomas (Valentine and Licciardi, 1992; Tsubata *et al.*, 1993).

Taken together, the currently available data on CD40L–CD40 suggest that this ligand–receptor pair plays a crucial and pivotal role in the generation of the humoral response and possibly has even wider significance in B-cell development and cell-mediated immunity.

9. OX40L/OX40

OX40 was first recognized in the rat by generation of a mAb that recognized activated T helper cells (Paterson *et al.*, 1987). This antibody could costimulate T-cell proliferation and was used to isolate a cDNA for rat OX40 (Mallet *et al.*, 1990). Mouse OX40 was cloned by cross-hybridization (Calderhead *et al.*, 1993) and shown to have the same limited distribution pattern as the rat protein. Human OX40 was isolated two ways—by cross-hybridization to the murine cDNA (Baum *et al.*, 1994) and by expression cloning using two mAb (ACT 35 and 106) directed to a human antigen that was also specifically expressed on activated T cells (Godfrey *et al.*, 1993; Latza *et al.*, 1994). Human and rodent proteins shared 63% amino acid identity. OX40.Fc fusion proteins were then used to detect and clone the murine and human OX40L cDNAs (Baum *et al.*, 1994; Godfrey *et al.*, 1994) from murine S49.1 lymphoma and human MSAB B lymphoblastoid cell lines. Sequence analysis of the OX40L cDNAs revealed that the human sequence had been previously characterized as gp34, a cell-surface glycoprotein whose expression was induced in T cells upon HTLV-1 infection (Tozawa *et al.*, 1988; Miura *et al.*, 1991). OX40L is the smallest of the TNF ligand family with respect to primary amino acid sequence (Figure 2). Recombinant OX40L was shown to costimulate T-cell proliferation and induce production of IL-2 and IL-4 (Baum *et al.*, 1994; Godfrey *et al.*, 1994). Expression of OX40 and OX40L is largely restricted to activated T cells, although the proteins are also expressed on HTLV-1-transformed cell lines, and the OX40L is found on PMA- and ionomycin-stimulated B lymphoblastoid cells and THP-1 cells (Baum *et al.*, 1994; Godfrey *et al.*, 1994). At the mRNA level, transcripts of both receptor and ligand are found in multiple tissues, but further work will be necessary to see if cell-surface protein can be expressed by nonlymphoid cells. One report used mouse OX40.Fc constructed to characterize a 70-kDa protein expressed on peritoneal exudate B cells (Calderhead *et al.*, 1993). The molecular weight and expression pattern of this protein are quite different from OX40L/gp34, and its identity as a physiological ligand for OX40 remains to be proven.

10. 4-1BBL/4-1BB

The original identification of 4-1BB was as a transcript induced upon murine T-cell activation (Kwon and Weissman, 1989). The sequence of the cDNA encoded a putative transmembrane protein, but its homology to the TNFR family was not noted until later (Smith *et al.*, 1990). The 4-1BB protein can be detected with a mAb as a 30-kDa glycoprotein on activated T cells and thymocytes that exists in monomeric and dimeric forms (Pollok *et al.*, 1993). The protein was shown to be capable of binding to a tyrosine kinase, lck, via a cytoplasmic consensus sequence that is also found in CD4 and CD8 (Kim *et al.*, 1993). Antibodies to 4-1BB costimulate T-cell proliferation (Pollok *et al.*, 1993), but a direct role of lck in this activity has not been demonstrated.

The murine 4-1BBL was detected using a 4-1BB.Fc fusion protein and expression cloned from an EL-4 thymoma cDNA library (Goodwin *et al.*, 1993b). The protein is the largest of the TNF-related ligands in both the cytoplasmic and extracellular domains, and the natural and recombinant proteins are glycosylated disulfide-linked homodimers of around 100 kDa (Goodwin *et al.*, 1993b).

The human 4-1BB receptor, like that of the mouse, was identified as a cDNA from activated T cells (Schwarz *et al.*, 1993) and also by cross-hybridization with the murine probe (Alderson *et al.*, 1994). The two molecules share 60% amino acid identity. An antibody to the receptor detected expression on activated T cells, EBV-transformed B-cell lines, and resting and activated peripheral blood monocytes (Alderson *et al.*, 1994). Because attempts to isolate human 4-1BBL by cross-hybridization with the mouse probe had failed, a human 4-1BB.Fc fusion protein was used to expression clone the human ligand from a T-cell library (Alderson *et al.*, 1994). The mouse and human ligands share only 36% amino acid identity—much lower than the average 70–80% identity found with other members of the family. No cross-species ligand–receptor interactions can be detected (Alderson *et al.*, 1994). The human and murine ligands are able to costimulate T-cell proliferation, and this is the only biological activity so far demonstrated. However, the broad expression pattern of the ligand, at least at the mRNA level, suggests other functions. Ligand mRNA has been found in bone marrow and thymic stromal cells, activated macrophages, EBV-transformed B cells, a neuroblastoma cell line, a megakaryoblastic leukemia cell line, and a variety of tissues. One report describes the use of a 4-1BB-alkaline phosphatase fusion protein to detect 4-1BBL expression on B-cell lymphomas and activated splenic B cells (Pollok *et al.*, 1994). Recombinant, membrane-bound 4-1BB expressed on insect cells was shown to costimulate B-cell proliferation in conjunction with anti-μ antibodies, suggesting that, like CD40L, 4-1BBL is able to function as a signal-transducing molecule (Pollok *et al.*, 1994). A 4-1BB.Ig fusion protein has also been shown to bind to extracellular matrix proteins, but the functional significance of this binding, if any, is unclear (Chalupny *et al.*, 1992).

11. LYMPHOTOXIN β (LT-β)/LT-βR

Lymphotoxin β was discovered when LT-α, known to be a secreted protein, was found on the cell surface of a T-cell hybridoma. Investigation showed that LT-α was anchored to the membrane by association with a 22-kDa membrane glycoprotein (Browning *et al.*, 1991). Subsequent purification and cloning of this glycoprotein revealed the structure of a type II membrane protein that is a member of the TNF ligand family (Browning *et al.*, 1993). Interestingly, the chromosomal location of the LT-β gene is extremely close to the TNF-α and LT-α genes.

Because LT-α is a homotrimeric protein, it was assumed that LT-α and LT-β would form heterotrimers of both 2 : 1 and 1 : 2 stoichiometry. For LT-α, the receptor binding sites are located at the cleft formed at the interface of two monomers. The LT-α–LT-β heterotrimers would have several types of monomer–monomer interface: α–α, α–β, and β–β. The heterotrimers might therefore be able to bind to more than one receptor. This idea was supported by studies showing that TNFRp55 or p75 could bind to LT-α or a heterotrimer of two molecules of LT-α and one molecule of LT-β but could not bind to LT-β or a heterotrimer of two molecules on LT-β and one molecule of LT-α (Crowe *et al.*, 1994). Conversely, it was discovered that TNFR-RP could bind to LT-β on the 2 : 1 LT-β–LT-α heterotrimer but not to LT-α or the 2 : 1 LT-α–LT-β heterotrimer (Crowe *et al.*, 1994). The TNFR-RP may be the physiological LT-β receptor. Biological evidence does exist for a lymphotoxin receptor, as LT-α knockout mice show a different phenotype from mice with disruptions of the TNFRp75 or p55 genes. The LT-α knockout mice have no peripheral lymph nodes or Peyer's patches and disordered architecture of the spleen (De Togni *et al.*, 1994), whereas disruption of the TNFRp75 or p55 genes produces no such effects. Whether TNFR-RP or LT-β knockout mice would have the same phenotype as LT-α knockout mice remains to be seen.

The ability of LT-α and LT-β to form heterotrimers is the first example of such behavior in the TNF-related ligand family. Clearly, this adds greatly to the potential biological complexity of the ligand–receptor interaction and raises the possibility that heteromeric receptor complexes may exist.

12. CD30L/CD30

Antibodies to CD30 have been used extensively for more than a decade as clinical markers for various lymphomas (Schwab *et al.*, 1982). Despite the large number of papers published in this regard, very little is known about its biological function. The Reed–Sternberg cells of Hodgkin's disease usually express high levels of CD30, as do a subset of non-Hodgkin's lymphoma known as large-cell anaplastic lymphoma (LCAL) (Stein *et al.*, 1985). CD30 is also found on a population of other lymphomas such as Burkitt's and adult T-cell leukemia/lymphoma as well as on a variety of tumors of mesenchymal and epithelial origin (Stein *et al.*, 1985; Mechterscheimer and Moller, 1990). CD30 is also expressed on some normal cell populations such as subsets of activated T and B cells (Stein *et al.*, 1985; Ellis *et al.*, 1993), cultured NK cells (Cambiaggi *et al.*, 1993), and monocytes differentiated *in vitro* (Andreesen *et al.*, 1989).

The CD30 molecule is by far the largest of the TNFR family and is detected as a 120-kDa cell-surface glycoprotein (Froese *et al.*, 1987). There have been reports of an independently synthesized intracellular form of CD30 that is a 57-kDa protein (Hansen *et al.*, 1989), but the molecular cloning of CD30 (Dürkop *et al.*, 1992) gave no clue as to how such a protein could be derived. The structure of CD30 predicted from the cDNA sequence is that of a membrane glycoprotein with a duplication of the characteristic cysteine-rich and serine-threonine-rich domains of the TNFR family (Figure 1) (Dürkop *et al.*, 1992).

A human CD30.Fc fusion protein was used to detect expression of CD30L on an activated murine T helper cell clone, and the CD30L cDNA was cloned by direct expression (Smith *et al.*, 1993). The human homologue was then isolated by cross-species hybridization, again from an activated T-cell library (Smith *et al.*, 1993). The molecules have a typical TNF family structure: five cysteine residues and five (human) or six (mouse) N-linked glycosylation sites in the extracellular domain. The mouse and human amino acid sequences share 72% identity. The protein is detected as a 40-kDa band under reducing conditions that forms dimers and multimers in the absence of reducing agents (Smith *et al.*, 1993). It is expressed by activated monocytes and T cells.

So far a limited spectrum of biological activities have been demonstrated for recombinant CD30L (expressed on transfected cells) or agonistic antibodies to CD30. These include costimulation of T-cell proliferation accompanied by up-regulation of ICAM-1 expression (Smith *et al.*, 1993) and down-regulation of CD28 expression, increased homotypic adhesion, and decreased cytotoxicity on an NK tumor cell line (Bowen, *et al.*, 1993). Stimulation through CD30 had pleiotropic effects on lymphoma cells, including enhanced Ig production from EBV-transformed lymphoblastoid lines, enhanced proliferation of "T-cell-like" Hodgkin's-derived cell lines, and cytotoxic and cytostatic effects on LCAL-derived cell lines (Smith *et al.*, 1993; Gruss *et al.*, 1994). The functional effects on Hodgkin's-derived cell lines together with the association of this disease with CD30 expression suggest a pathological role for CD30/CD30L, but much remains to be understood.

13. CD27L/CD27

CD27 was defined by mAb as a lymphoid-specific 55-kDa glycoprotein that forms a disulfide-bonded homodimer (van Lier *et al.*, 1987). CD27 is a member of the TNFR family, as shown by cDNA cloning, but contains only $2\frac{1}{2}$-cysteine-rich domains (Camerini *et al.*, 1991). Antibodies to CD27 have been described of both agonistic and antagonistic types. On T cells, antibodies to CD27 have been reported to costimulate T-cell proliferation (van Lier *et al.*, 1987) and to inhibit T-cell proliferation (Sugita *et al.*, 1991) when used in combination with PHA, anti-CD3, or other stimuli. Presumably, the antibodies used recognize different epitopes on the CD27 molecule. Recombinant soluble CD27 has biological effects that mimic those of the inhibitory antibody, suggesting that this antibody prevents binding of endogenous CD27L to CD27 (Agematsu *et al.*, 1994). The inhibitory antibody to CD27 is also able to block pokeweed mitogen (PWM)-induced IgG synthesis by B cells (Sugita *et al.*, 1991).

The expression of cell-surface CD27 is induced on T-cell activation (de Jong *et al.*, 1991), and the molecule can be proteolytically cleaved to release the soluble, extracellular domain of the protein (Hintzen *et al.*, 1991a). Hence, increased circulating levels of soluble CD27 are a marker for T-cell activation *in vivo* (Hintzen *et al.*, 1991b). CD27 expression is regulated during T-cell development; thymocytes expressing low levels of CD3 do not express CD27, but as the cells mature to express high levels of CD3, CD27 is expressed (Martorell *et al.*, 1990). $CD45RA^+$ naive peripheral blood T cells express CD27, and activation of these cells through the T-cell receptor greatly up-regulates CD27 expression (Hintzen *et al.*, 1993). However, no CD27 expression is seen on a subpopulation of $CD45RO^+$ T cells that includes fully differentiated, mature effector cells, and CD27 expression cannot be induced in these cells (de Jong *et al.*, 1991). $CD45RO^+$, $CD27^-$ T cells have been shown to be enriched in T_H2 cells (Elson *et al.*, 1994). This distribution pattern of CD27 expression suggests that CD27 function is required more by relatively naive T cells than by completely differentiated effector T cells (Hintzen *et al.*, 1994a; Chapter 8).

Conversely, CD27 is not expressed by naive B cells but only by those cells that have differentiated sufficiently to secrete Ig (Maurer *et al.*, 1992). CD27 is expressed on some NK cells and up-regulated by IL-2 (Sugita *et al.*, 1992). In these cells, the IL-2-induced cytolytic activity can be inhibited with an antagonistic antibody to CD27 (Sugita *et al.*, 1992).

All these functional data using CD27 antibodies implied that there would be a natural ligand for CD27 that would play a role in lymphoid activation. A CD27.Fc fusion protein was used to detect expression of CD27L on an EBV-transformed lymphoblastoid cell line, MP-1, and to expression clone the CD27L cDNA (Goodwin *et al.*, 1993a). The predicted protein structure of the CD27L was that of a typical member of the TNF-related ligand family, with four cysteine residues and two N-linked glycosylation sites in the extracellular domain. The molecular mass of the recombinant protein was 29 kDa under reducing conditions.

Subsequently, CD27 was shown to be identical to the previously identified lymphoid glycoprotein CD70 (Bowman *et al.*, 1994; Hintzen *et al.*, 1994b). Antibodies to CD70 were costimulatory for T cells (Bowman *et al.*, 1994), showing that, as for CD40/CD40L and 4-1BB/4-1BBL, signaling can be mediated through both the ligand and the receptor. CD27L/CD70 is expressed on activated T and B cells, and mRNA can be detected in certain monocytic cell lines (Goodwin *et al.*, 1993a; Hintzen *et al.*, 1994c).

The recombinant CD27L, as expected, costimulated T-cell proliferation. This proliferation was shown to be independent of IL-2 (Goodwin *et al.*, 1993a). CD27 also enhanced the generation of cytotoxic T cells as measured in a lectin-mediated killing assay (Goodwin *et al.*, 1993a), suggesting that CD27 may be involved in the functional differentiation of T cells. Antibody to CD27L was shown to block allogeneic B-cell-induced stimulation of T-cell proliferation (Hintzen *et al.*, 1994c), suggesting that CD27L may be used as an important second signal for T-cell recruitment and expansion by antigen-presenting B cells in addition to the B7/CD28/CTLA-4 costimulatory system.

14. FASL/FAS

The isolation of mAb that could kill cell lines led to the discovery of the Fas or APO-1 antigen (Trauth *et al.*, 1989; Yonehara *et al.*, 1989). The antibody-induced cell death was shown to be mediated by induction of apoptosis (active or programmed cell death; see Chapter 7) in the target cell. The Fas antibodies recognized a 50-kDa membrane glycoprotein that was cloned and shown to belong to the TNFR family (Itoh *et al.*, 1991). This finding was especially interesting because one of the well-known properties of TNF is its ability to kill certain cell lines. Comparison between the cytoplasmic domains of Fas and the TNFRs showed a region of 65 amino acids with 28% amino acid identity between Fas and TNFRp55. Mutational analysis demonstrated that this region was essential for cytotoxicity in both molecules (Itoh and Nagata, 1993; Tartaglia *et al.*, 1993). However, evidence has been presented that the two receptors use at least partially distinct signaling pathways to mediate apoptosis (Wong and Goeddel, 1994). The extreme C-terminal 15 amino acids of the Fas protein appear to negatively regulate cytotoxicity, as their deletion enhanced Fas-mediated killing (Itoh and Nagata, 1993).

The Fas antigen is expressed on a wide variety of cell types, including thymocytes, T cells, B cells, monocytes, and neutrophils. It is expressed on more mature primed or memory ($CD45RO^+$) T cells and surface IgD^- B cells rather than on their naive ($CD45RO^-$ or IgD^+) counterparts (Miyawaki *et al.*, 1992; Iwai *et al.*, 1994). However, Fas expression can be induced in naive T and B cells on mitogenic stimulation (Miyawaki *et al.*, 1992). Fas is also expressed in tissues such as liver, heart, lung, and ovary as well as many cultured cell lines of diverse origin (Watanabe-Fukunaga *et al.*, 1992a). Expression of Fas does not always correlate with sensitivity to Fas-mediated killing. For example, although peripheral blood B cells 1 day after stimulation with PWM were uniformily positive for Fas expression, they were resistant to Fas-mediated killing. However, longer culture with PWM resulted in sensitivity to Fas (Daniel and Krammer, 1994). Similarly, freshly isolated T cells that express Fas are resistant to killing but acquire sensitivity after several days in culture (Owen-Schaub *et al.*, 1992). Nevertheless, at least some primary cells are sensitive to Fas-mediated killing: injection of Fas antibodies into mice cause rapid hepatotoxicity (Ogasawara *et al.*, 1993). Although most research on Fas has involved its ability to induce apoptosis, it was recently found that Fas antibodies were costimulatory to freshly isolated T cells or thymocytes (Alderson *et al.*, 1993b). Thus, Fas shares the dual role played by other members of the TNFR family: they are able to send stimulatory or inhibitory signals depending on the target cell population.

The understanding of Fas biology was facilitated by the discovery that mutations in the Fas gene were the genetic lesions responsible for the phenotype of the *lpr/lpr* mouse, a model for spontaneous autoimmune disease (Watanabe-Fukunaga *et al.*, 1992b). Two different defects have been analyzed. In the *lpr* cg allele, there is a mutation in the cytoplasmic domain of Fas that abolishes Fas-mediated cytotoxicity, and in the other known *lpr* allele there has been an insertion of a transposable element into an intron of the Fas gene, leading to a severe decrease in Fas mRNA levels (Watanabe-Fukunaga *et al.*, 1992b; Adachi *et al.*, 1993). Cells from *lpr/lpr* mice are consequently resistant to Fas antibody-mediated killing. How does a Fas defect lead to autoimmune disease? It was hypothesized initially that Fas was essential for negative selection of potentially autoreactive T cells during their development in the thymus. However, negative selection appeared to be normal in *lpr/lpr* mice (Singer *et al.*, 1989). More recently, evidence has been provided that normal Fas function helps to regulate self-reactive lymphocytes in the periphery and is involved in the phenomenon of activation-induced cell death of lymphocytes (AICD) (Alderson *et al.*, 1993b; Bossu *et al.*, 1993; Kabelitz *et al.*, 1993; Russell *et al.*, 1993; Daniel and Krammer, 1994; Gillette-Ferguson and Sidman, 1994; Green and Scott, 1994; Krammer *et al.*, 1994; Singer and Abbas, 1994). Activation-induced cell death may also function to limit the extent of an immune response on prolonged antigenic stimulation as the cells switch from a state in which Fas mediates a stimulatory signal to a state where Fas engagement causes death. Clearly, lack of a functional Fas response would lead to the overaccumulation of lymphocytes, some of which would be autoreactive. This is the observed phenotype of lymphocytes in *lpr/lpr* mice.

Given the important biological role of Fas, the identification of a natural ligand for Fas was the subject of considerable interest. Mice homozygous for the *gld* mutation demonstrate the same autoimmune phenotype as *lpr/lpr* mice, and it was proposed that the *lpr* and *gld* gene products might bear a receptor–ligand relationship (Allen *et al.*, 1990). Support for this idea came from the discovery that Ca^{2+}-independent toxicity, mediated either by a rat–mouse T cell hybridoma or by PMA- and ionomycin-activated, antigen-specific, normal peritoneal exudate lymphocytes, was impaired by the *lpr* and *gld* mutations. Specifically, targets from *lpr/lpr* mice were not susceptible to this type of killing, and lymphocytes from *gld/gld* mice were unable to act as Ca^{2+}-independent, cytotoxic effector cells (Rouvier *et al.*, 1993). Activated T cells from *gld/gld* mice were also shown to be unable to bind a Fas.Fc fusion protein or to kill target cells by a Fas-mediated mechanism (Ramsdell *et al.*, 1994b). The identification of a cell source for a functional FasL led immediately to its cDNA cloning. The rat–mouse T-cell hybridoma was sorted for increased Fas.Fc binding and used to construct a library from which the rat FasL cDNA was isolated (Suda *et al.*, 1993). As predicted, the molecule belonged to the TNF ligand family and indeed showed closer homology to TNF than to any other family member. Transfection into COS cells generated both membrane-bound and soluble forms of FasL, as is found for TNF-α, and the recombinant molecule was able to trigger apoptosis. As expected, the mouse FasL was proven to be the *gld* gene product (Lynch *et al.*, 1994; Takahashi *et al.*, 1994) by direct sequencing of wild-type and *gld*-derived FasL cDNAs. The *gld*-derived FasL contains a point mutation in the C-terminal part of the TNF-homology domain that completely abrogates function. Apart from T cell, FasL mRNA is expressed in testis, small intestine, lung, and kidney.

The identification of Fas/FasL-mediated T-cell cytotoxicity was intriguing. How did this relate to classical T-cell cytotoxicity directed against tumor cells and virally infected cells? Studies using *lpr/lpr* and *gld/gld* mice together with mice in which the perforin gene had been disrupted showed that perforin played an essential role in classical T-cell cytotoxicity (Kägi *et al.*, 1994a), but not in Fas-mediated cytotoxicity (Kägi *et al.*, 1994b; Kojima *et al.*, 1994). No evidence was found for a third mechanism of cytotoxicity. The ability of the Fas–FasL interaction to kill syngeneic lymphocytes suggested that this pathway of cytotoxicity might play a regulatory role in controlling the immune response of peripheral lymphocytes (Kägi *et al.*, 1994a; Vignaux and Golstein, 1994). Fas–FasL interactions also play a major role in cytotoxicity mediated by $CD4^+$ T cells (Stalder *et al.*, 1994). This has been suggested as a mechanism to eliminate nonprofessional antigen-presenting cells.

15. CONCLUSION

Although TNF has been long recognized as a central mediator of immune activation and inflammation, the discovery of a family of molecules that is structurally related to TNF and its receptors has been comparatively recent. Despite this, good progress has been made in understanding the important roles of several family members in blood cell development, activation, and control. It has been particularly helpful to have preexisting, spontaneously occurring mutations in ligand or receptor genes (*lpr/lpr* mice and HIGM patients), and a start has been made on creating mice in which genes for one or more of the family members has been disrupted (knockouts in TNFRp55, TNFRp75, LT-α, CD40, CD40L: Pfeffer *et al.*, 1993; Rothe *et al.*, 1993; De Togni *et al.*, 1994; Kawabe *et al.*, 1994; Renshaw *et al.*, 1994; Xu *et al.*, 1994). It is to be hoped that knockout mice for all the family members will soon be available and will be as revealing as those currently existing.

Already it is clear that TNF–TNFR family members participate in many interactions between blood cell types, particularly T-cell–B-cell, T-cell–monocyte, and T-cell–T-cell interactions. In several cases we have seen that the interaction is reciprocal, with signaling through both "ligand" and "receptor" molecules. We have also seen that the interactions can be stimulatory or inhibitory depending on the cell type or activation state. A future challenge will be understanding the switch in signaling pathways that must occur when a cell makes a transition from receiving a stimulatory to an inhibitory signal through triggering the same receptor. Other common elements observed following interaction of TNF family ligands and receptors are stimulation of cytokine secretion, up-regulation of adhesion molecules, activation antigens, and costimulatory molecules that function to amplify and spread the stimulatory signal. Clearly, mechanisms also exist for limiting the signal, such as shedding of soluble receptors following activation.

Although there is overlap in the expression pattern of TNF-related ligands and receptors (for example, all family members are expressed on activated T cells, and all ligands can costimulate T-cell proliferation), there are differences in the distribution of the molecules and in their kinetics of induction and their requirements for induction. It will be challenging to dissect their different roles. Currently, little is known about the intracellular signaling pathways used by these receptors, but it is very likely that they each will activate more than one pathway and will be able to synergize with each other and with other cytokines and costimulatory molecules.

The ligand–receptor family is also very interesting from a structural perspective. Although it is clear that there is a common theme of multimeric ligands leading to receptor clustering, it will be interesting to determine if all the family members use the three-ligand–three-receptor paradigm of TNF-α and LT-α.

Finally, some of the receptor molecules have a broader tissue distribution outside the hematopoietic system (TNFR, Fas, CD40), and their function on these other cell types is worthy of attention.

ACKNOWLEDGMENTS. I thank Anne C. Bannister for excellent editorial assistance, Sarah Soutter for artwork, and many colleagues at Immunex for helpful discussions, especially Mark Alderson, Richard Armitage, Peter Baum, Bill Fanslow, Ray Goodwin, Hans-Jürgen Gruss, David Lynch, Charlie Maliszewski, Fred Ramsdell, Craig Smith, Melanie Spriggs, and Mike Widmer.

16. REFERENCES

Adachi, M., Watanabe-Fukunaga, R., and Nagata, S., 1993, Aberrant transcription caused by the insertion of an early transposable element in an intron of the Fas antigen gene of *lpr* mice, *Proc. Natl. Acad. Sci. U.S.A.* **90:**1756–1760.

Agematsu, K., Kobata, T., Sugita, K., Freeman, G. J., Beckmann, M. P., Schlossman, S. F., and Morimoto, C., 1994, Role of CD27 in T cell immune response. Analysis by recombinant soluble CD27, *J. Immunol.* **153:**1421–1429.

Alderson, M. R., Armitage, R. J., Tough, T. W., Strockbine, L., Fanslow, W. C., and Spriggs, M. K., 1993a, CD40 expression by human monocytes: Regulation by cytokines and activation of monocytes by the ligand for CD40, *J. Exp. Med.* **178:**669–674.

Alderson, M. R., Armitage, R. J., Maraskovsky, E., Tough, T. W., Roux, E., Schooley, K., Ramsdell, F., and Lynch, D. H., 1993b, Fas transduces activation signals in normal human T lymphocytes, *J. Exp. Med.* **178:**2231–2235.

Alderson, M. R., Smith, C. A., Tough, T. W., Davis-Smith, T., Armitage, R. J., Falk, B., Roux, E., Baker, E., Sutherland, G. R., Din., W. S., and Goodwin, R. G., 1994, Molecular and biological characterization of human 4-1BB and its ligand, *Eur. J. Immunol.* **24:**2219–2227.

Allen, R. C., Armitage, R. J., Conley, M. E., Rosenblatt, H., Jenkins, N. A., Copeland, N. G., Bedell, M. A., Edelhoff, S., Disteche, C. M., Simoneaux, D. K., Fanslow, W. C., Belmont, J., Spriggs, M. K., 1993, CD40 ligand gene defects responsible for X-linked hyper-IgM syndrome, *Science* **259:**990–993.

Allen, R. D., Marshall, J. D., Roths, J. B., and Sidman, C. D., 1990, Differences defined by bone marrow transplantation suggest that *lpr* and *gld* are mutations of genes encoding an interacting pair of molecules, *J. Exp. Med.* **172:**1367–1375.

Andreesen, R., Burger, W., Löhr, G. W., and Bross, K. J., 1989, Human macrophages can express the Hodgkin's cell-associated antigen Ki-1 (CD30), *Am. J. Pathol.* **134:**187–192.

Armitage, R. J., Fanslow, W. C., Strockbine, L., Sato, T. A., Clifford, K. N., Macduff, B. M., Anderson, D. M., Gimpel, S. D., Davis-Smith, T., Maliszewski, C. R., Clark, E. A., Smith, C. A., Grabstein, K. H., Cosman, D., and Spriggs, M. K., 1992, Molecular and biological characterization of a murine ligand for CD40 *Nature* **357:**80–82.

Armitage, R. J., Macduff, B. M., Spriggs, M. K., and Fanslow, W. C., 1993a, Human B cell proliferation and Ig secretion induced by recombinant CD40 ligand are modulated by soluble cytokines, *J. Immunol.* **150:**3671–3680.

Armitage, R. J., Tough, T. W., Macduff, B. M., Fanslow, W. C., Spriggs, M. K., Ramsdell, F., and Alderson, M. R., 1993b, CD40 ligand is a T cell growth factor, *Eur. J. Immunol.* **23:**2326–2331.

Aruffo, A., Farrington, M., Hollenbaugh, D., Li, X., Milatovich, A., Nonoyama, S., Bajorath, J., Grosmaire, L. S., Stenkamp, R., Neubauer, M., Roberts, R. L., Noelle, R. J., Ledbetter, J. A., Francke, U., and Ochs, H. D., 1993, The CD40 ligand, gp39, is defective in activated T cells from patients with X-linked hyper-IgM syndrome, *Cell* **72:**291–300.

Baens, M., Chaffanet, M., Cassiman, J.-J., van den Berghe, H., and Marynen, P., 1993, Construction and evaluation of a hncDNA library of human 12p transcribed sequences derived from a somatic cell hybrid, *Genomics* **16:**214–218.

Baker, E., Chen, L. Z., Smith, C. A., Callen, D. F., Goodwin, R., and Sutherland, G.R., 1991, Chromosomal location of the human tumor necrosis factor receptor genes, *Cytogenet. Cell Genet.* **57:**117–118.

Banchereau, J., and Rousset, F., 1991, Growing human B lymphocytes in the CD40 system, *Nature* **353:**678–679.

Banchereau, J., de Paoli, P., Vallé, A., Garcia, E., and Rousset, F., 1991, Long-term human B cell lines dependent on interleukin-4 and antibody to CD40, *Science* **251:**70–72.

Banchereau, J., Bazan, F., Blanchard, D., Briere, F., Galizzi, J.-P., van Kooten, C., Liu, Y. J., Rousset, F., and Saeland, S., 1994, The CD40 antigen and its ligand, *Annu. Rev. Immunol.* **12:**881–992.

Banner, D. W., D'Arcy, A., Janes, W., Gentz, R., Schoenfeld, H.-J., Broger, C., Loetscher, H., and Lesslauer, W., 1993, Crystal structure of the soluble human 55 kd TNF receptor-human TNFβ complex: Implications for TNF receptor activation, *Cell* **73:**431–445.

Barrett, T. B., Shu, G., and Clark, E. A., 1991, CD40 signaling activates CD11a/CD18 (LFA-1) mediated adhesion in B cells, *J. Immunol.* **146:**1722–1729.

Baum, P. R., Gayle, R. B., Ramsdell, F., Srinivasan, S., Sorensen, R. A., Watson, M. L., Seldin, M. F., Baker, E., Sutherland, G. R., Clifford, K. N., Alderson, M. R., Goodwin, R. G., and Fanslow, W. C., 1994, Molecular characterization of murine and human OX40/OX40 ligand systems: Identification of a human OX40 ligand as the HTLV-1 regulated protein gp34, *EMBO J.* **13:**3992–4001.

Björck, P., Elenström-Magnusson, C., Rosén, A., Severinson, E., and Paulie, S., 1993, CD23 and CD21 function as adhesion molecules in homotypic aggregation of human B lymphocytes, *Eur. J. Immunol.* **23:**1771–1775.

Bossu, P., Singer, G. G., Andres, P., Ettinger, R., Marshak-Rothstein, A., and Abbas, A. K., 1993, Mature $CD4^+$ T lymphocytes from MRL/lpr mice are resistant to receptor-mediated tolerance and apoptosis, *J. Immunol.* **151:**7233–7239.

Bowen, M. A., Olsen, K. J., Cheng, L., Avila, D., and Podack, E. R., 1993, Functional effects of CD30 on a large granular lymphoma cell line, YT. Inhibition of cytotoxicity, regulation of CD28 and IL-2R, and induction of homotypic aggregation, *J. Immunol.* **151:**5896–5906.

Bowman, M. R., Crimmins, M. A. V., Yetz-Aldape, J., Kriz, R., Kelleher, K., and Herrmann, S., 1994, The cloning of CD70 and its identification as the ligand for CD27, *J. Immunol.* **152:**1756–1761.

Bradshaw, R. A., Blundell, T. L., Lapatto, R., McDonald, N. Q., and Murray-Rust, J., 1993, Nerve growth factor revisited, *Trends Biochem. Sci.* **18:**48–52.

Browning, J. L., Androlewicz, M. J., and Ware, C. F., 1991, Lymphotoxin and an associated 33-kDa glycoprotein are expressed on the surface of an activated human T cell hybridoma, *J. Immunol.* **147:**1230–1237.

Browning, J. L., Ngam-ek, A., Lawton, P., DeMarinis, J., Tizard, R., Chow, E. P., Hession, C., O'Brien-Greco, B., Foley, S. F., and Ware, C. F., 1993, Lymphotoxin β, a novel member of the TNF family that forms a heteromeric complex with lymphotoxin on the cell surface, *Cell* **72:**847–856.

Calderhead, D. M., Buhlmann, J. E., van den Eertwegh, A. J. M., Claassen, E., Noelle, R. J., and Fell, H.P., 1993,

Cloning of mouse Ox40: A T cell activation marker that may mediate T–B cell interactions, *J. Immunol.* **151:**5261–5271.
Cambiaggi, A., Cantoni, C., Marciano, S., De Totero, D., Pileri, S., Tazzari, P. L., Stein, H., and Ferrini, S., 1993, Cultured human NK cells express the Ki-1/CD30 antigen, *Br. J. Haematol.* **85:**270–276.
Camerini, D., Walz, G., Loenen, W. A. M., Borst, J., and Seed, B., 1991, The T cell activation antigen CD27 is a member of the nerve growth factor/tumor necrosis factor receptor gene family, *J. Immunol.* **147:**3165–3169.
Cayabyab, M., Phillips, J. H., and Lanier, L. L., 1994, CD40 preferentially costimulates activation of $CD4^+$ T lymphocytes, *J. Immunol.* **152:**1523–1531.
Chalupny, N. J., Peach, R., Hollenbaugh, D., Ledbetter, J. A., Farr, A. G., and Aruffo, A., 1992, T-cell activation molecule 4-1BB binds to extracellular matrix proteins, *Proc. Natl. Acad. Sci. U.S.A.* **89:**10360–10364.
Clark, E. A., and Lane, P. J. L., 1991, Regulation of human B-cell activation and adhesion, *Annu. Rev. Immunol.* **9:**97–127.
Clark, E. A., and Ledbetter, J. A., 1986, Activation of human B cells mediated through two distinct cell surface differentiation antigens, Bp35 and Bp50, *Proc. Natl. Acad. Sci. U.S.A.* **83:**4494–4498.
Clark, E. A., and Ledbetter, J. A., 1994, How B and T cells talk to each other, *Nature* **367:**425–428.
Conley, M. E., 1992, Molecular approaches to analysis of X-linked immunodeficiencies, *Annu. Rev. Immunol.* **10:**215–238.
Cosman, D., 1993, The hematopoietin receptor superfamily, *Cytokine* **5:**95–106.
Crowe, P. D., VanArsdale, T. L., Walter, B. N., Ware, C. F., Hession, C., Ehrenfels, B., Browning, J. L., Din, W. S., Goodwin, R. G., and Smith, C. A., 1994, A lymphotoxin-β-specific receptor, *Science* **264:**707–710.
Daniel, P. T., and Krammer, P. H., 1994, Activation induces sensitivity toward APO-1 (CD95)-mediated apoptosis in human B cells, *J. Immunol.* **152:**5624–5632.
de Jong, R., Loenen, W. A. M., Brouwer, M., can Emmerik, L., de Vries, E. F. R., Brost, J., and van Lier, R. A. W., 1991, Regulation of expression of CD27, a T cell-specific member of a novel family of membrane receptors, *J. Immunol.* **146:**2488–2494.
De Togni, P., Goellner, J., Ruddle, N. H., Streeter, P. R., Fick, A., Mariathasan, S., Smith, S. C., Carlson, R., Shornick, L. P., Strauss-Schoenberger, J., Russell, J. H., Karr, R., and Chaplin, D. D., 1994, Abnormal development of peripheral lymphoid organs in mice deficient in lymphotoxin, *Science* **264:**703–707.
DiSanto, J. P., Bonnefoy, J. Y., Gauchat, S. F., Fischer, A., and de Saint Basile, G., 1993, CD40 ligand mutations in X-linked immunodeficiency with hyper-IgM, *Nature* **361:**541–543.
Durie, F. H., Fava, R. A., Foy, T. M., Aruffo, A., Ledbetter, J. A., and Noelle, R. J., 1993, Prevention of collagen-induced arthritis with an antibody to gp39, the ligand for CD40, *Science* **261:**1328–1330.
Dürkop, H., Latza, U., Hummel, M., Eitelbach, F., Seed, B., and Stein, H., 1992, Molecular cloning and expression of a new member of the nerve growth factor receptor family that is characteristic for Hodgkin's disease, *Cell* **68:**421–427.
Eck, M. J., and Sprang, S. R., 1989, The structure of tumor necrosis factor-β at 2.6 Å resolution. Implications for receptor binding, *J. Biol. Chem.* **264:**17595–17605.
Eck, M. J., Ultsch, M., Rinderknecht, E., de Vos, A. M., and Sprang, S. R., 1992, The structure of human lymphotoxin (tumor necrosis factor-β) at 1.9 Å resolution, *J. Biol. Chem.* **267:**2119–2122.
Eide, F. F., Lowenstein, D. H., and Reichardt, L. F., 1993, Neurotrophins and their receptors—current concepts and implications for neurologic disease, *Exp. Neurol.* **131:**200–214.
Ellis, T. M., Simms, P. E., Slivnick, D. J., Jack, H.-M., and Fisher, R. I., 1993, CD30 is a signal-transducing molecule that defines a subset of human activated $CD45RO^+$ T cells, *J. Immunol.* **151:**2380–2389.
Elson, L. H., Shaw, S., Van Lier, R. A. W., and Nutman, T. B., 1994, T cell subpopulation phenotypes in filarial infections: CD27 negativity defines a population greatly enriched for T_h2 cells, *Int. Immunol.* **6:**1003–1009.
Fanslow, W. C., Anderson, D., Grabstein, K. H., Clark, E. A., Cosman, D., and Armitage, R. J., 1992, Soluble forms of CD40 inhibit biologic responses of human B cells, *J. Immunol.* **149:**655–660.
Fanslow, W. C., Clifford, K. N., Seaman, M., Alderson, M. R., Spriggs, M. K., Armitage, R. J., and Ramsdell, F., 1994, Recombinant CD40 ligand exerts potent biologic effects on T cells, *J. Immunol.* **152:**4262–4269.
Fonatsch, C., Latza, U., Dürkop, H., Rieder, H., and Stein, H., 1992, Assignment of the human CD30 (Ki-1) gene to 1p36, *Genomics* **14:**825–826.
Foy, T. M., Shepherd, D. M., Durie, F. H., Aruffo, A., Ledbetter, J. A., and Noelle, R. J., 1993 *In vivo* CD40–gp39 interactions are essential for thymus-dependent humoral immunity. II. Prolonged suppression of the humoral immune response by an antibody to the ligand for CD40, gp39, *J. Exp. Med.* **178:**1567–1575.
Foy, T. M., Laman, J. D., Ledbetter, J. A., Aruffo, A., Claassen, E., and Noelle, R. J., 1994, gp39–CD40

interactions are essential for germinal center formation and the development of B cell memory, *J. Exp. Med.* **180:**157–163.

Froese, P., Lemke, H., Gerdes, J., Havsteen, B., Schwarting, R., Hansen, H., and Stein, H., 1987, Biochemical characterization and biosynthesis of the Ki-1 antigen in Hodgkin-derived and virus-transformed human B and T lymphoid cell lines, *J. Immunol.* **139:**2081–2087.

Fuchs, P., Strehl, S., Dworzak, M., Himmler, A., and Ambros, P. F., 1992, Structure of the human TNF receptor 1 (p60) gene (TNFR1) and localization to chromosome 12p13, *Genomics* **13:**219–224.

Fuleihan, R., Ramesh, N., Loh, R., Jabara, H., Rosen, R. S., Chatila, T., Fu, S. M., Stamenkovic, I., and Geha, R. S., 1993, Defective expression of the CD40 ligand in X chromosome-linked immunoglobulin deficiency with normal or elevated IgM, *Proc. Natl. Acad. Sci. U.S.A.* **90:**2170–2173.

Fuleihan, R., Ramesh N., Horner A., Ahern, D., Belshaw, P. J., Alberg, D. G., Stamenkovic, I., Harmon, W., and Geha, R. S., 1994, Cyclosporin A inhibits CD40 ligand expression in T lymphocytes, *J. Clin. Invest.* **93:**1315–1320.

Funakoshi, S., Longo, D. L., Beckwith, M., Conley, D. K., Tsarfaty, G., Tsarfaty, I., Armitage, R. J., Fanslow, W. C., Spriggs, M. K., and Murphy, W. J., 1994, Inhibition of human B-cell lymphoma growth by CD40 stimulation, *Blood* **83:**2787–2794.

Galy, A. H. M., and Spits, H., 1992, CD40 is functionally expressed on human thymic epithelial cells, *J. Immunol.* **149:**775–782.

Gauchat, J.-F., Henchoz, S., Mazzei, G., Aubry, J.-P., Brunner, T., Blasey, H., Life, P., Talabot, D., Flores-Romo, L., Thompson, J., Kishi, K., Butterfield, J., Dahinden, C., and Bonnefoy, J.-Y., 1993, Induction of human IgE synthesis in B cells by mast cells and basophils, *Nature* **365:**340–343.

Gillette-Ferguson, I., and Sidman, C. L., 1994, A specific intercellular pathway of apoptotic cell death is defective in the mature peripheral T cells of autoimmune *lpr* and *gld* mice, *Eur. J. Immunol.* **24:**1181–1185.

Godfrey, W. R., Buck, D., Harara, M. A., and Engleman, E. G., 1993, Molecular cloning of a cDNA encoding the human homolog of the rat OX40 antigen (abstr.), *Tissue Antigen* **42:**253.

Godfrey, W. R., Fagnoni, F. F., Harara, M. A., Buck, D., and Engleman, E. G., 1994, Identification of a human OX-40 ligand, a costimulator of $CD4^+$ T cells with homology to tumor necrosis factor, *J. Exp. Med.* **180:** 757–762.

Goodwin, R. G., Anderson, D., Jerzy, R., Davis, T., Brannan, C. I., Copeland, N. G., Jenkins, N. A., and Smith, C. A., 1991, Molecular cloning and expression of the type 1 and type 2 murine receptors for tumor necrosis factor, *Mol. Cell. Biol.* **11:**3020–3026.

Goodwin, R. G., Alderson, M. R., Smith, C. A., Armitage, R. J., VandenBos, T., Jerzy, R., Tough, T. W., Schoenborn, M. A., Davis-Smith, T., Hennen, K., Falk, B., Cosman, D., Baker, E., Sutherland, G. R., Grabstein, K. H., Farrah, T., Giri, J. G., and Beckmann, M. P., 1993a, Molecular and biological characterization of a ligand for CD27 defines a new family of cytokines with homology to tumor necrosis factor, *Cell* **73:**447–456.

Goodwin, R. G., Din, W. S., Davis-Smith, T., Anderson, D. M., Gimpel, S. D., Sato, T. A., Maliszewski, C. R., Brannan, C. I., Copeland, N. G., Jenkins, N. A., Farrah, T., Armitage, R. J., Fanslow, W. C., and Smith, C. A., 1993b, Molecular cloning of a ligand for the inducible T-cell gene 4-1BB: A member of an emerging family of cytokines with homology to tumor necrosis factor, *Eur. J. Immunol.* **23:**2631–2641.

Gordon, J., Millsum, M. J., Guy, G. R., and Ledbetter, J. A., 1988, Resting B lymphocytes can be triggered directly through the CDw40 (Bp50) antigen, *J. Immunol.* **140:**1425–1430.

Grabstein, K. H., Maliszewski, C. R., Shanebeck, K., Sato, T. A., Spriggs, M. K., Fanslow, W. C., and Armitage, R. J., 1993, The regulation of T cell-dependent antibody formation *in vitro* by CD40 ligan and IL-2, *J. Immunol.* **150:**3141–3147.

Graf, D., Korthäuer, U., Mages, H. W., Senger, G., and Kroczek, R. A., 1992, Cloning of TRAP, a ligand for CD40 on human T cells, *Eur. J. Immunol.* **22:**3191–3194.

Gravestein, L. A., Blom, B., Nolten, L. A., de Vries, E., van der Horst, G., Ossendrop, F., Borst, J., and Loenen, W. A. M., 1993, Cloning and expression of murine CD27: Comparison with 4-1BB, another lymphocyte-specific member of the nerve growth factor receptor family, *Eur. J. Immunol.* **23:**943–950.

Green, D. R., and Scott, D. W., 1994, Activation-induced apoptosis in lymphocytes, *Curr. Opin. Immunol.* **6:** 476–487.

Grimaldi, J. C., Torres, R., Kozak, C. A., Chang, R., Clark, E. A., Howard, M., and Cockayne, D. A., 1992, Genomic structure and chromosomal mapping of the murine CD40 gene, *J. Immunol.* **149:**3921–3926.

Gruss, H.-J., Boiani, N., Williams, D. E., Armitage, R. J., Smith, C. A., and Goodwin, R. G., 1994, Pleiotropic effects of the CD30 ligand on CD30-expressing cells and lymphoma cell lines, *Blood* **83:**2045–2056.

Hansen, H., Lemke, H., Bredfeldt, G., Konnecke, I., and Havsteen, B., 1989, The Hodgkin-associated Ki-1 antigen exists in an intracellular and a membrane-bound form, *Z. Hoppe Seylers Biol. Chem.* **370:**409–416.

Heath, A. W., Chang, R., Harada, N., Santos-Argumedo, L., Gordon, J., Hannum, C., Campbell, D., Shanafelt, A. B., Clark, E. A., Torres, R., and Howard, M., 1993, Antibodies to murine CD40 stimulate normal B lymphocytes but inhibit proliferation of B lymphoma cells, *Cell. Immunol.* **152:**468–480.

Hintzen, R. Q., de Jong, R., Hack, C. E., Chamuleau, M., de Vries, E. F. R., ten Berge, I. J. M., Borst, J., and van Lier, R. A. N., 1991a, A soluble form of the human T cell differentiation antigen CD27 is released after triggering of the TCR/CD3 complex, *J. Immunol.* **147:**29–35.

Hintzen, R. Q., van Lier, R. A., Kuijpers, K. C., Baars, P. A., Schaasberg, W., Lucas, C. J., and Polman, C. H., 1991b, Elevated levels of a soluble form of the T cell activation antigen CD27 in cerebrospinal fluid of multiple sclerosis patients, *J. Neuroimmunol.* **35:**211–217.

Hintzen, R. Q., de Jong, R., Lens, S. M. A., Brouwer, M., Baars, P., and van Lier, R. A. W., 1993, Regulation of CD27 expression on subsets of mature T-lymphocytes, *J. Immunol.* **151:**2426–2435.

Hintzen, R. Q., de Jong, R., Lens, S. M. A., and van Lier, R. A. W., 1994a, CD27: Marker and mediator of T-cell activation? *Immunol. Today* **15:**307–311.

Hintzen, R. Q., Lens, S. M. A., Koopman, G., Pals, S. T., Spits, H., and van Lier, R. A. W., 1994b, CD70 represents the human ligand for CD27, *Int. Immunol.* **6:**477–480.

Hintzen, R. Q., Lens, S. M., Beckmann, M. P., Goodwin, R. G., Lynch, D., and van Lier, R. A. W., 1994c, Characterization of the human CD27 ligand, a novel member of the TNF gene family, *J. Immunol.* **152:**1762–1773.

Holder, M. J., Wang, H., Milner, A. E., Casamayor, M., Armitage, R., Spriggs, M. K., Fanslow, W. C., MacLennan, I. C. M., Gregory, C. D., and Gordon, J., 1993, Suppression of apoptosis in normal and neoplastic human B lymphocytes by CD40 ligand is independent of Bcl-2 induction, *Eur. J. Immunol.* **23:**2368–2371.

Hollenbaugh, D., Grosmaire, L. S., Kullas, C. D., Chalupny, N. J., Braesch-Anderson, S., Noelle, R. J., Stamenkovic, I., Ledbetter, J. A., and Aruffo, A., 1992, The human T cell antigen gp39, a member of the TNF gene family, is a ligand for the CD40 receptor: Expression of a soluble form of gp39 with B cell costimulatory activity, *EMBO J.* **11:**4313–4321.

Itoh, N., and Nagata, S., 1993, A novel protein domain required for apoptosis. Mutational analysis of human Fas antigen, *J. Biol. Chem.* **268:**10932–10937.

Itoh, N., Yonehara, S., Ishii, A., Yonehara, M., Mizushima, S.-I., Sameshima, M., Hase, A., Seto, Y., and Nagata, S., 1991, The polypeptide encoded by the cDNA for human cell surface antigen Fas can mediate apoptosis, *Cell* **66:**233–243.

Iwai, K., Miyawaki, T., Takizawa, T., Konno, A., Ohta, K., Yachie, A., Saki, H., and Taniguchi, N., 1994, Differential expression of bcl-2 and susceptibility to anti-Fas-mediated cell death in peripheral blood lymphocytes, monocytes, and neutrophils, *Blood* **84:**1201–1208.

Jabara, H. H., Fu, S. M., Geha, R. S., and Vercelli, D., 1990, CD40 and IgE: Synergism between anti-CD40 monoclonal antibody and interleukin 4 in the induction of IgE synthesis by highly purified human B cells, *J. Exp. Med.* **172:**1861–1864.

Johnson, D., Lanahan, A., Buck, C. R., Sehgal, A., Morgan, C., Mercer, E., Bothwell, M., and Chao, M., 1986, Expression and structure of the human NGF receptor, *Cell* **47:**545–554.

Jones, E. Y., Stuart, D. I., and Walker, N. P. C., 1989, Structure of tumour necrosis factor, *Nature* **338:**225–228.

Kabelitz, D., Pohl, T., and Pechhold, K., 1993, Activation-induced cell death (apoptosis) of mature peripheral T lymphocytes, *Immunol. Today* **14:**338–339.

Kägi, D., Ledermann, B., Bürki, K., Seiler, P., Odermatt, B., Olsen, K. J., Podack, E. R., Zinkernagel, R. M., and Hengartner, H., 1994a, Cytotoxicity mediated by T cells and natural killer cells is greatly impaired in perforin-deficient mice, *Nature* **369:**31–37.

Kägi, D., Vignaux, F., Ledermann, B., Bürki, K., Depraetere, V., Nagata, S., Hengartner, H., and Golstein, P., 1994b, Fas and perforin pathways as major mechanisms of T cell-mediated cytotoxicity, *Science* **265:** 528–530.

Kawabe, T., Naka, T., Yoshida, K., Tanaka, T., Fujiwara, H., Suematsu, S., Yoshida, N., Kishimoto, T., and Kikutani, H., 1994, The immune responses in CD40-deficient mice: Impaired immunoglobulin class switching and germinal center formation, *Immunity* **1:**167–178.

Kennedy, M. K., Mohler, K. M., Shanebeck, K. D., Baum, P. R., Picha, K. S., Otten-Evans, C. A., Janeway, C. A., Jr., and Grabstein, K. H., 1994, Induction of B cell costimulatory function by recombinant murine CD40 ligand, *Eur. J. Immunol.* **24:**116–123.

Kim, Y.-J., Pollok, K. E., Zhou, A., Shaw, A., Bohlen, J. B., Fraser, M., and Kwon, B. A., 1993, Novel T cell antigen 4-1BB associates with the protein tyrosine kinase $p56^{lck1}$, *J. Immunol.* **151:**1255–1262.

Klaus, S. J., Pinchuk, L. M., Ochs, H. D., Law, C.-L., Fanslow, W. C., Armitage, R. J., and Clark, E. A., 1994, Costimulation through CD28 enhances T cell-dependent B cell activation via a CD40-CD40L interaction, *J. Immunol.* **152:** 5643–5652.

Kojima, H., Shinohara, N., Hanaoka, S., Someya-Shirota, Y., Takagaki, Y., Ohno, H., Saito, T., Katayama, T., Yagita, H., Okumura, K., Shinkai, Y., Alt, F. W., Matsuzawa, A., Yonehara, S., and Takayama, H., 1994, Two distinct pathways of specific killing revealed by perforin mutant cytotoxic T lymphocytes, *Immunity* **1:**357–364.

Korthäuer, U., Graf, D., Mages, H. W., Brière, F., Padayachee, M., Malcolm, S., Ugazio, A. G., Notarangelo, L. D., Levinsky, R. J., and Kroczek, R. A., 1993, Defective expression of T-cell CD40 ligand causes X-linked immunodeficiency with hyper IgM, *Nature* **361:**539–541.

Krammer, P. H., Behrmann, I., Daniel, P., Dhein, J., and Debatin, K. M., 1994, Regulation of apoptosis in the immune system, *Curr. Opin. Immunol.* **6:**279–289.

Kriegler, M., Perez, C., DeFay, K., Albert, I., and Lu, S. D., 1988, A novel form of TNF/cachectin is a cell surface cytotoxic transmembrane protein: Ramifications for the complex physiology of TNF, *Cell* **53:**45–53.

Kwekkeboom, J., de Rijk, D., Kasran, A., Barcy, S., de Groot, C., and de Boer, M., 1994, Helper effector function of human T cells stimulated by anti-CD3 mAb can be enhanced by costimulatory signals and is partially dependent on CD40-CD40 ligand interaction, *Eur. J. Immunol.* **24:**508–517.

Kwon, B. S., and Weissman, S. M., 1989, cDNA sequences of two inducible T-cell genes, *Proc. Natl. Acad. Sci. U.S.A.* **86:**1963–1967.

Kwon, B. S., Kozak, C. A., Kim, K. K., and Pickard, R. T., 1994, Genomic organization and chromosomal localization of the T-cell Antigen 4-1BB, *J. Immunol.* **152:**2256–2262.

Lafage-Pochitaloff, M., Herman, P., Birg, F., Galizzi, J. P., Simonetti, J., Mannoni, P., and Banchereau, J., 1994, Localization of the human CD40 gene to chromosome 20, bands q12–q13.2, *Leukemia* **8:**1172–1175.

Lane, P., Traunecker, A., Hubele, S., Inui, S., Lanzavecchia, A., and Gray, D., 1992, Activated human T cells express a ligand for the human B cell-associated antigen CD40 which participates in T cell-dependent activation of B lymphocytes, *Eur. J. Immunol.* **22:**2573–2578.

Lane, P., Brocker, T., Hubele, S., Padovan, E., Lanzavecchia, A., and McConnell, F., 1993, Soluble CD40 ligand can replace the normal T cell-derived CD40 ligand signal to B cells in T cell-dependent activation, *J. Exp. Med.* **177:**1209–1213.

Latza, U., Dürkop, H., Schnittger, S., Ringeling, J., Eitelbach, F., Hummel, M., Fonatsch, C., and Stein, H., 1994, The human OX40 homolog: cDNA structure, expression and chromosomal assignment of the ACT35 antigen, *Eur. J. Immunol.* **24:**677–683.

Lederman, S., Yellin, M. J., Krichevsky, A., Belko, J., Lee, J. J., and Chess, L., 1992, Identification of a novel surface protein on activated $CD4^+$ T cells that induces contact-dependent B cell differentiation (help), *J. Exp. Med.* **175:**1091–1101.

Lichter, P., Walczak, H., Weitz, S., Behrmann, I., and Krammer, P. H., 1992, The human APO-1 (APT) antigen maps to 10q23, a region that is syntenic with mouse chromosome 19, *Genomics* **14:**179–180.

Liu, Y.-J., Joshua, D. E., Williams, G. T., Smith, C. A., Gordon, J., and MacLennon, I. C. M., 1989, Mechanism of antigen-driven selection in germinal centers, *Nature* **342:**929–931.

Loenen, W. A. M., Gravestein, L. A., Beumer, S., Melief, C. J. M., Hagemeijer, A., and Borst, J., 1992, Genomic organization and chromosomal localization of the human *CD27* gene, *J. Immunol.* **149:**3937–3943.

Loetscher, H., Pan, Y.-C. E., Lahm, H.-W., Gentz, R., Brockhaus, M., Tabuchi, H., and Lesslauer, W., 1990, Molecular cloning and expression of the human 55 kd tumor necrosis factor receptor, *Cell* **61:**351–359.

Lynch, D. H., Watson, M. L., Alderson, M. R., Baum, P. R., Miller, R. E., Tough, T., Gibson, M., Davis-Smith, T., Smith, C. A., Hunter, K., Bhat, D., Din, W., Goodwin, R. G., and Seldin, M. F., 1994, The mouse Fas-ligand gene is mutated in *gld* mice and is part of a TNF family gene cluster, *Immunity* **1:**131–136.

Maliszewski, C. R., Grabstein, K., Fanslow, W. C., Armitage, R., Spriggs, M. K., and Sato, T. A., 1993, Recombinant CD40 ligand stimulation of murine B cell growth and differentiation: Cooperative effects of cytokines, *Eur. J. Immunol.* **23:**1044–1049.

Mallet, S., Fossum, S., and Barclay, A. M., 1990, Characterization of the MRC OX40 antigen of activated CD4 positive T lymphocytes—a molecule related to nerve growth factor receptor, *EMBO J.* **9:**1063–1068.

Martorell, J., Rojo, I., Vilella, R., Martinez-Caceres, E., and Vives, J., 1990, CD27 induction on thymocytes, *J. Immunol.* **145:**1356–1363.

Maurer, D., Fischer, G. F., Fae, I., Majdic, O., Stuhlmeier, K., von Jeney, N., Holter, W., and Knapp, W., 1992, IgM and IgG but not cytokine secretion is restricted to the $CD27^+$ B lymphocyte subset, *J. Immunol.* **148:**3700–3705.

Mechtersheimer, G., and Moller, P., 1990, Expression of Ki-1 antigen (CD30) in mesenchymal tumors, *Cancer* **66:**1732–1737.

Miura, S., Ohtani, K., Numata, N., Niki, M., Ohbo, K., Ina, Y., Gojobori, T., Tanaka, Y., Tozawa, H., Nakamura, M., and Sugamura, K., 1991, Molecular cloning and characterization of a novel glycoprotein, gp34, that is specifically induced by the human T-cell leukemia virus type I transactivator $p40^{tax}$, *Mol. Cell. Biol.* **11:**1313–1325.

Miyawaki, T., Uehara, T., Nibu, R., Tsuji, T., Yachie, A., Yonehara, S., and Taniguchi, N., 1992, Differential expression of apoptosis-related Fas antigen on lymphocyte subpopulations in human peripheral blood, *J. Immunol.* **149:**3753–3758.

Mohler, K. M., Sleath, P. R., Fitzner, J. N., Cerretti, D. P., Alderson, M., Derwar, S. S., Torrance, D. S., Otten-Evans, C., Greenstreet, T., Weerawarna, K., Kronheim, S. R., Peterson, M., Gerhart, M., Kozlosky, C. J., March, C. J., and Black, R. A., 1994, Protection against a lethal dose of endotoxin by an inhibitor of tumour necrosis factor processing, *Nature* **370:**218–220.

Nishioka, Y., and Lipsky, P. E., 1994, The role of CD40–CD40 ligand interaction in human T cell–B cell collaboration, *J. Immunol.* **153:**1027–1036.

Noelle, R. J., Roy, M., Shepherd, D. M., Stamenkovic, I., Ledbetter, J. A., and Aruffo, A., 1992, A 39-kDa protein on activated helper T cells binds CD40 and transduces the signal for cognate activation of B cells, *Proc. Natl. Acad. Sci. U.S.A.* **89:**6550–6554.

Nonoyama, S., Hollenbaugh, D., Aruffo, A., Ledbetter, J. A., and Ochs, H. D., 1993, B cell activation via CD40 is required for specific antibody production by antigen-stimulated human B cells, *J. Exp. Med.* **178:**1097–1102.

Ogasawara, J., Watanabe-Fukunaga, R., Adachi, M., Matsuzawa, A., Kasugai, T., Kitamura, Y., Itoh, N., Suda, T., and Nagata, S., 1993, Lethal effect of the anti-Fas antibody in mice, *Nature* **364:**806–809.

Owen-Schaub, L. B., Yonehara, S., Crump, W. L. III, and Grimm, E. A., 1992, DNA fragmentation and cell death is selectively triggered in activated human lymphocytes by Fas antigen engagement, *Cell. Immunol.* **140:** 197–205.

Parker, D. C., 1993, T cell-dependent B cell activation, *Annu. Rev. Immunol.* **11:**331–360.

Paterson, D. J., Jefferies, W. A., Green, J. R., Brandon, M. R., Corthesy, P., Puklavec, M., and Williams, A. F., 1987, Antigens of activated rat T lymphocytes including a molecule of 50,000 M_r detected only on CD4 positive T blasts, *Mol. Immunol.* **24:**1281–1290.

Peitsch, M. C., and Jongeneel, C. V., 1993, A 3-D model for the CD40 ligand predicts that it is a compact trimer similar to the tumor necrosis factors, *Int. Immunol.* **5:**233–238.

Pennica, D., Nedwin, G. E., Hayflick, J. S., Seeburg, P. H., Derynck, R., Palladino, M. A., Kohr, W. J., Aggarwal, B. B., and Goeddel, D. V., 1994, Human tumour necrosis factor: Precursor structure, expression and homology to lymphotoxin, *Nature* **312:**724–729.

Pfeffer, K., Matsuyama, T., Kündig, T. M., Wakeham, A., Kishihara, K., Shahinian, A., Wiegmann, K., Ohashi, P. S., Kronke, M., and Mak, T. W., 1993, Mice deficient for the 55 kd tumor necrosis factor receptor are resistant to endotoxic shock, yet succumb to *L. monocytogenes* infection, *Cell* **73:**457–467.

Pollok, K. E., Kim, Y.-J., Zhou, A., Hurtado, J., Kim, K. K., Pickard, R. T., and Kwon, B. S., 1993, Inducible T cell antigen 4-1BB. Analysis of expression and function, *J. Immunol.* **150:**771–781.

Pollok, K. E., Kim, Y.-J., Hurtado, J., Zhou, Z., Kim, K. K., and Kwon, B. S., 1994, 4-1BB T-cell antigen binds to mature B cells and macrophages, and costimulates anti-μ-primed splenic B cells, *Eur. J. Immunol.* **24:**367–374.

Ramsdell, F., Seaman, M., Clifford, K. N., and Fanslow, W. C., 1994a, CD40 ligand acts as a costimulatory signal for neonatal thymic γδ T cells, *J. Immunol.* **152:**2190–2197.

Ramsdell, F., Seaman, M. S., Miller, R. E., Tough, T. W., Alderson, M. R., and Lynch, D. H., 1994b, *gld/gld* mice are unable to express a functional ligand for Fas, *Eur. J. Immunol.* **24:**928–933.

Ranheim, E. A., and Kipps, T. J., 1993, Activated T cells induce expression of B7/BB1 on normal or leukemic B cells through a CD40-dependent signal, *J. Exp. Med.* **177:**925–935.

Renshaw, B., Fanslow, W. C. III, Armitage, R. J., Campbell, K. A., Liggitt, D., Wright, B., Davison, B., and Maliszewski, C. R., 1994, Humoral immune responses in CD40 ligand deficient mice, *J. Exp Med.* **180:**1889–1900.

Rothe, J., Lesslauer, W., Lötscher, H., Lang, Y., Koebel, P., Kontgen, F., Althage, A., Zinkernagel, R., Steinmetz, M., and Bluethmann, H., 1993, Mice lacking the tumour necrosis factor receptor 1 are resistant to TNF-mediated toxicity but highly susceptible to infection by *Listeria monocytogenes*, *Nature* **364:**798–802.

Rousset, F., Garcia, E., and Banchereau, J., 1991, Cytokine-induced proliferation and immunoglobulin production of human B lymphocytes triggered through their CD40 antigen, *J. Exp. Med.* **173:**705–710.

Rousset, F., Garcia, E., Defrance, T., Péronne, C., Vezzio, N., Hsu, D.-H., Kastelein, R., Moore, K. W., and Banchereau, J., 1992, Interleukin 10 is a potent growth and differentiation factor for activated human B lymphocytes, *Proc. Natl. Acad. Sci. U.S.A.* **89:**1890–1893.

Rouvier, E., Luciani, M.-F., and Golstein, P., 1993, Fas involvement in Ca^{2+}-independent T cell-mediated cytotoxicity, *J. Exp. Med.* **177:**195–200.

Russell, J. H., Rush, B., Weaver, C., and Wang, R., 1994, Mature T cells of autiommune *lpr/lpr* mice have a defect in antigen-stimulated suicide, *Proc. Natl. Acad. Sci. U.S.A.* **90:**4409–4413.

Saeland, S., Moreau, I., Duvert, V., Pandrau, D., and Banchereau, J., 1992, *In vitro* growth and maturation of human B-cell precursors, *Curr. Top. Microbiol. Immunol.* **182:**85–94.

Saeland, S., Duvert, V., Moreau, I., and Banchereau, J., 1993, Human B cell precursors proliferate and express CD23 after CD40 ligation, *J. Exp. Med.* **178:**113–120.

Schall, T. J., Lewis, M., Koller, K. J., Lee, A., Rice, G. C., Wong, G. H. W., Gatanaga, T., Granger, G. A., Lentz, R., Raab, H., Kohr, W. J., and Goeddel, D. V., 1990, Molecular cloning and expression of a receptor for human tumor necrosis factor, *Cell* **61:**361–370.

Schwab, U., Stein, H., Gerdes, J., Lemke, H., Kirchner, H., Schaadt, M., and Diehl, V., 1982, Production of a monoclonal antibody specific for Hodgkin and Sternberg–Reed cells of Hodgkin's disease and a subset of normal lymphoid cells, *Nature* **299:**65–67.

Schwarz, H., Tuckwell, J., and Lotz, M., 1993, A receptor induced by lymphocyte activation (Ila)—a new member of the human nerve-growth-factor tumor-necrosis-factor receptor family, *Gene* **134:**295–298.

Seldin, M. F., Morse, H. C. III, Reeves, J. P., Scribner, C. L., LeBoeuf, R. C., and Steinberg, A. D., 1988, Genetic analysis of autoimmune *gld* mice. I. Identification of a restriction fragment length polymorphism closely linked to the *gld* mutation within a conserved linkage group, *J. Exp. Med.* **167:**688–693.

Singer, G. G., and Abbas, A. K., 1994, The Fas antigen is involved in peripheral but not thymic deletion of T lymphocytes in T cell transgenic mice, *Immunity* **1:**365–371.

Singer, P. A., Balderas, R. S., and McEvilly, R. J., 1989, Tolerance-related Vβ clonal deletions in normal $CD4^-8^-$, $TCR\text{-}\alpha/\beta^+$ and abnormal *lpr* and *gld* cell populations, *J. Exp. Med.* **170:**1869–1877.

Smith, C. A., Davis, T., Anderson, D., Solam, L., Beckmann, M. P., Jerzy, R., Dower, S. K., Cosman, D., and Goodwin, R. G., 1990, A receptor for tumor necrosis factor defines an unusual family of cellular and viral proteins, *Science* **248:**1019–1023.

Smith, C. A., Davis, T., Wignall, J. M., Din, W. S., Farrah, T., Upton, C., McFadden, G., and Goodwin, R. G., 1991, T2 open reading frame from the shope fibroma virus encodes a soluble form of the TNF receptor, *Biochem. Biophys. Res. Commun.* **176:**335–342.

Smith, C. A., Gruss, H.-J., Davis, T., Anderson, D., Farrah, T., Baker, E., Sutherland, G. R., Brannan, C. I., Copeland, N. G., Jenkins, N. A., Grabstein, K. H., Gliniak, B., McAlister, I. B., Fanslow, W., Alderson, M., Falk, B., Gimpel, S., Gillis, S., Gillis, S., Din, W. S., Goodwin, R. G., and Armitage, R. J., 1993, CD30 antigen, a marker for Hodgkin's lymphoma, is a receptor whose ligand defines an emerging family of cytokines with homology to TNF, *Cell* **73:**1349–1360.

Spriggs, M. K., Armitage, R. J., Strockbine, L., Clifford, K. N., Macduff, B. M., Sato, T. A., Maliszewski, C. R., and Fanslow, W. C., 1992, Recombinant human CD40 ligand stimulates B cell proliferation and Immunoglobulin E secretion, *J. Exp. Med.* **176:**1543–1550.

Stalder, T., Hahn, S., and Erb, P., 1994, Fas antigen is the major target molecule for $CD4^+$ T cell-mediated cytotoxicity, *J. Immunol.* **152:**1127–1133.

Stamenkovic, I., Clark, E. A., and Seed, B., 1989, A B-lymphocyte activation molecule related to the nerve growth factor receptor and induced by cytokines in carcinomas, *EMBO J.* **8:**1403–1410.

Stein, H., Mason, D. Y., Gerdes, J., O'Connor, N., Wainscoat, J., Pallesen, G., Gatter, K., Falini, B., Delsol, G., Lemke, H., Schwarting, R., and Lennert, K., 1985, The expression of the Hodgkin's disease associated antigen Ki-1 in reactive and neoplastic lymphoid tissue: Evidence that Reed–Sternberg cells and histiocytic malignancies are derived from activated lymphoid cells, *Blood* **66:**848–858.

Suda, T., Takahashi, T., Golstein, P., and Nagata, S., 1993, Molecular cloning and expression of the Fas ligand, a novel member of the tumor necrosis factor family, *Cell* **75:**1169–1178.

Sugita, K., Torimoto, Y., Nojima, Y., Daley, J. F., Schlossman, S. F., and Morimoto, C., 1991, The 1A4 molecule (CD27) is involved in T cell activation, *J. Immunol.* **147**:1477–1483.

Sugita, K., Robertson, M. J., Torimoto, Y., Ritz, J., Schlossman, S. F., and Morimoto, C., 1992, Participation of the CD27 antigen in the regulation of IL-2-activated human natural killer cells, *J. Immunol.* **149**:1199–1203.

Takahashi, T., Tanaka, M., Brannan, C. I., Jenkins, N. A., Copeland, N. G., Suda, T., and Nagata, S., 1994, Generalized lymphoproliferative disease in mice, caused by a point mutation in the Fas ligand, *Cell* **76:** 969–976.

Tartaglia, L. A., Ayres, T. M., Wong, G. H. W., and Goeddel, D. V., 1993, A novel domain within the 55 kd TNF receptor signals cell death, *Cell* **74**:845–853.

Tozawa, H., Andoh, S., Takayama, Y., Tanaka, Y., Lee, B., Nakamura, H., Hayami, M., and Hinuma, Y., 1988, Species-dependent antigenicity of the 34-kDa glycoprotein found on the membrane of various primate lymphocytes transformed by human T-cell leukemia virus type-I (HTLV-I) and simian T-cell leukemia virus (STLV-I), *Int. J. Cancer* **41**:231–238.

Tracey, K. J., and Cerami, A., 1994, Tumor necrosis factor—a pleiotropic cytokine and therapeutic target, *Annu. Rev. Med.* **45**:491–503.

Trauth, B. C., Klas, C., Peters, A. M. J., Matzku, S., Möller, P., Falk, W., Debatin, K.-M., and Krammer, P. H., 1989, Monoclonal antibody-mediated tumor regression by induction of apoptosis, *Science* **245**:301–305.

Tsubata, T., Wu, J., and Honjo, T., 1993, B-cell apoptosis induced by antigen receptor crosslinking is blocked by a T-cell signal through CD40, *Nature* **364**:645–648.

Ullrich, A., and Schlessinger, J., 1990, Signal transduction by receptors with tyrosine kinase activity, *Cell* **61:** 203–212.

Upton, C., Macen, J. L., Schreiber, M., and McFadden, G., 1991, Myxoma virus expresses a secreted protein with homology to the tumor necrosis factor receptor gene family that contributes to viral virulence, *Virology* **184**:370–382.

Valentine, M. A., and Licciardi, K. A., 1992, Rescue from anti-IgM-induced programmed cell death by the B cell surface proteins CD20 and CD40, *Eur. J. Immunol.* **22**:3141–3148.

van Kooten, C., Gaillard, C., Galizzi, J.-P., Hermann, P., Fossiez, F., Banchereau, J., and Blanchard, D., 1994, B cells regulate expression of CD40 ligand on activated T cells by lowering the mRNA level and through the release of soluble CD40, *Eur. J. Immunol.* **24**:787–792.

van Lier, R. A. W., Borst, J., Vroom, T. M., Klein, H., Van Mourik, P., Zeijlemaker, W. P., and Melief, C. J. M., 1987, Tissue distribution and biochemical and functional properties of Tp55 (CD27), a novel T cell differentiation antigen, *J. Immunol.* **139**:1589–1596.

Vassalli, P., 1992, The pathophysiology of tumor necrosis factors, *Annu. Rev. Immunol.* **10**:411–452.

Vignaux, F., and Golstein, P., 1994, Fas-based lymphocyte-mediated cytotoxicity against syngeneic activated lymphocytes: A regulatory pathway? *Eur. J. Immunol.* **24**:923–927.

Watanabe-Fukunaga, R., Brannan, C. I., Itoh, N., Yonehara, S., Copeland, N. G., Jenkins, N. A., and Nagata, S., 1992a, The cDNA structure, expression, and chromosomal assignment of the mouse Fas antigen, *J. Immunol.* **148**:1274–1279.

Watanabe-Fukunaga, R., Brannan, C. I., Copeland, N. G., Jenkins, N. A., and Nagata, S., 1992b, Lymphoproliferation disorder in mice explained by defects in Fas antigen that mediates apoptosis, *Nature* **356**:314–317.

Wong, G. H. W., and Goeddel, D. V., 1994, Fas antigen and p55 TNF receptor signal apoptosis through distinct pathways, *J. Immunol.* **152**:1751–1755.

Xu, J., Foy, T. M., Laman, J. D., Elliott, E. A., Dunn, J. J., Waldschmidt, T. J., Elsemore, J., Noelle, R. J., and Flavell, R. A., 1994, Mice deficient for the CD40 ligand, *Immunity* **1**:423–431.

Yellin, M. J., Sinning, J., Covey, L. R., Sherman, W., Lee, J. J., Glickman-Nir, E., Sippel, K. C., Rogers, J., Cleary, A. M., Parker, M., Chess, L., and Lederman, S., 1994, T lymphocyte T cell-B cell-activating molecule/CD40-L molecules induce normal B cells or chronic lymphocytic leukemia B cells to express CD80 (B7/BB-1) and enhance their costimulatory activity, *J. Immunol.* **153**:666–674.

Yonehara, S., Ishii, A., and Yonehara, N., 1989, A cell-killing monoclonal antibody (anti-Fas) to a cell surface antigen co-downregulated with the receptor of tumor necrosis factor, *J. Exp. Med.* **169**:1747–1756.

Chapter 4

Cellular Signaling Events in B Lymphocytes

Margaret M. Harnett

1. INTRODUCTION

The B lymphocyte is the principal cellular mediator of the specific humoral immune response to infection. B cells arise from pluripotent stem cells in the bone marrow via a number of well-defined precursor B-cell phenotypes in a complex process that integrates signals generated by a number of immunoregulatory receptors on several cell types, often in specialized environments (Cushley and Harnett, 1993). Once B cells emerge into the periphery, selection of the appropriate clones by antigen can lead to (1) cellular anergy and/or apoptosis of immature B cells, (2) activation and proliferation of mature B cells, or (3) rescue of germinal center cells from cell suicide followed by differentiation into antibody-secreting plasma cells or memory B cells (reviewed by Cushley and Harnett, 1993). The mechanism by which the antigen receptors (mIg) on B cells can elicit these different biological responses in a maturation-state-dependent manner is one of the central problems in B-cell differentiation yet to be resolved. However, it is clear that many of the signaling events involved in these biological responses are regulated by both cytokines (reviewed by O'Garra *et al.*, 1988) and membrane-associated ligands on helper T cells (reviewed by Clark and Lane, 1991). Characterization of the B-cell receptors that can process these signals has generated a unifying model of (interacting) coreceptor complexes (e.g., CD19/CD21, CD22, CD38, and CD40) and cytokine receptors (e.g., IL-4R) that can modulate mIg-mediated B-cell activation (reviewed by Harnett, 1994).

Margaret M. Harnett Division of Biochemistry and Molecular Biology, University of Glasgow, Glasgow G12 8QQ, United Kingdom.

Blood Cell Biochemistry, Volume 7: Hematopoietic Cell Growth Factors and Their Receptors, edited by A. D. Whetton and J. Gordon. Plenum Press, New York, 1996.

2. THE B-LYMPHOCYTE ANTIGEN RECEPTOR

The B-cell antigen receptor complex (BCR) comprises a clonotypic antigen-binding component, membrane immunoglobulin (mIg), and its accessory transducing molecules, Igα and Igβ (reviewed by Cambier *et al.*, 1994; Pleiman *et al.*, 1994). Although all classes of immunoglobulin can be expressed as antigen receptors on B cells, only IgM and IgD are expressed on mature naive B cells and, indeed, on the vast majority (>90%) of all peripheral B cells. Analysis of μm and δm heavy-chain structure reveals that each heavy chain forms a single transmembrane domain with a charged cytoplasmic tail of only three amino acids, KVK (Kyte and Doolittle, 1982; Cambier *et al.*, 1994). Clearly, this cytoplasmic domain is too small to possess any intrinsic catalytic activity, indicating that the mIg molecule must be associated with accessory signal-transducing molecules in order to transmit antigen-induced transmembrane signals. Two glycoproteins, expressed as a disulfide-bonded heterodimer, are noncovalently associated with the mIg complex (Campbell and Cambier, 1990; Chen *et al.*, 1990; Hombach *et al.*, 1990; Parkhouse, 1990). Both of these mIg-associated proteins have been demonstrated to be encoded by B-cell-specific genes. The Igβ chain (CD79b) encoded by the B29 gene (Hermanson *et al.*, 1988) has a molecular mass of 38 kDa and is a transmembrane glycoprotein that can associate with all classes of mIg (Reth, 1992). The Igα chain (CD79a), also a transmembrane glycoprotein and encoded by the B-cell-specific *mb*-1 gene (Sakaguchi *et al.*, 1988), although differently glycosylated, also associates with all mIg isotypes (Venkitaraman *et al.*, 1991). The stoichiometry of the complex formed between the clonotypic chains of the receptor and the Igα–Igβ heterodimer remains to be defined, but a ratio of 1 : 2 has been suggested by the bilateral symmetry of mIg.

Roles have been inferred for Igα and Igβ in antigen receptor assembly and intracellular trafficking, internalization, and signal transduction. Early evidence for a role for Igα in receptor transport was provided by transfection studies in the J_{558} plasmacytoma. Transfection of μ_m plasmids resulted in synthesis of the μ_m heavy chain but no expression of mIgM at the surface of the cells: surface expression could be achieved only following cotransfection of J_{558} cells with *mb*-1 (i.e., Igα) and μ_m constructs (Hombach *et al.*, 1990), suggesting that Igα is mandatory for surface expression of IgM.

Mutational analysis of mIg has provided a possible molecular mechanism for the requirement of Igα and Igβ in antigen receptor trafficking: a polar patch within the membrane-spanning region, -TTAST-, has been identified that signals ER retention unless Igα and Igβ are also expressed to facilitate transport to the cell surface (Williams *et al.*, 1990; Hombach *et al.*, 1990), suggesting that interactions between the Igαβ heterodimer and this polar patch of the transmembrane region are required for receptor assembly and transport to the cell surface. Interestingly, IgD (Venkitaraman *et al.*, 1991; Reth, 1992) and certain IgG subclasses can be efficiently expressed at the cell surface in the absence of the Igα protein (Venkitaraman *et al.*, 1991; Reth, 1992).

Both Igα and Igβ are single membrane-spanning glycoproteins with sizable intracellular domains (61 and 48 amino acids, respectively). Chimeric receptors of CD8, IgM, or FcγRII expressing the cytoplasmic tails of Igα and Igβ are effective signal transducers (Cambier *et al.*, 1994; Kim *et al.*, 1993). Importantly, Igα and Igβ chimeras can differentially activate specific BCR-coupled signaling events, suggesting some degree of compartmentalization of signaling functions to Igα and Igβ. Although the cytoplasmic domains of Igα and Igβ do not possess any intrinsic catalytic activity, they do, however, possess motifs that are also found in the cytoplasmic domains of a variety of multimeric receptors including the TCR ζ and CD3 γ, δ, and ε complex components (Cambier *et al.*, 1994; Reth, 1992). This motif consists of the conserved sequence D/E-X_7-D/E-X_2-Y-X_2-L/I-X_7-Y-X_2-L/I and has been termed the antigen receptor homology 1 (ARH-1) (or ARAM in T cells, Weiss, 1993) motif (Cambier *et al.*, 1994). There is now abundant evidence that ARH-1 motifs recruit cytoplasmic signal transducers and are thus capable of transducing signals leading to the activation of protein tyrosine phosphorylation and calcium mobilization (Cambier *et al.*, 1994).

Phosphorylation of the tyrosine residues within the ARH-1 motif appears to be the critical event regulating signal transduction via these accessory molecules: even conservative mutations (e.g., Y to P) abrogate receptor-mediated signaling (Pleiman *et al.*, 1994). Interestingly, however, although the Ig$\alpha\beta$ heterodimers appear to play a major role in the transduction of BCR-directed signals, recent analysis of μ_m mutants indicates that point mutations in a polar region of the C-terminal portion of the transmembrane domain of μ_m can abrogate antigen-induced signal transduction without disrupting Igα or Igβ association (Cambier *et al.*, 1994). Taken together with the recent reevaluation of mIg structure (Cambier *et al.*, 1994) by the algorithm of Klein *et al.* (1985), which predicts an 11-amino-acid cytoplasmic tail for μ_m, these data implicate a novel role for the C-terminal region of the μ_m in receptor-mediated signal transduction. Thus, in addition to the role of antigen binding, the clonotypic receptor may also couple to a signal transducer(s) important for antigen-driven B-cell activation.

3. SIGNAL TRANSDUCTION VIA THE B-LYMPHOCYTE ANTIGEN RECEPTOR

The earliest signaling events following ligation of the BCR include the activation of multiple protein tyrosine kinases (PTK) and the tyrosine phosphorylation of Igα and Igβ ARH-1 motifs. The PTKs activated following BCR ligation include the Src-related PTKs Fyn, Lyn, Blk, and Lck and the more distantly related kinase Syk (Gold *et al.*, 1990; Burkhardt *et al.*, 1991; Hutchcroft *et al.*, 1991, 1992; Cambier *et al.*, 1994; Pleiman *et al.*, 1994). Src-related kinases associate with the BCR in unstimulated cells via interaction of their N-terminal domains with the nonphosphorylated Igα (Cambier *et al.*, 1993; Pleiman *et al.*, 1994). Tyrosine phosphorylation of the ARH-1 motifs leads to a dramatic increase in Src-related PTK binding and activity *in vitro*. This is associated with a reorientation and derepression of the kinase such that enhanced PTK recruitment is mediated via interactions of the Src homology domain 2 (SH2) regions of the kinase with the phosphorylated tyrosine residues of the ARH-1 motif (Pleiman *et al.*, 1994).

Although the data are controversial, genetic and biochemical studies (Hutchcroft *et al.*, 1992; Kolanus *et al.*, 1993; Takata *et al.*, 1994; Pleiman *et al.*, 1994) indicate that the primary signal following antigen binding is likely to be activation and autophosphorylation of the PTK Syk, which is already associated with the resting receptor complex, perhaps via the μ_m C-terminal transmembrane polar patch. Autophosphorylation of Syk results in the binding (via SH2 domains) and activation of a Src-related PTK(s), which in turn, tyrosine phosphorylates the ARH-1 motifs in the cytoplasmic domains of Igα and Igβ (Reth, 1992; Pleiman *et al.*, 1994). Tyrosine phosphorylation of the ARH-1 motifs induces not only recruitment of additional Syk but also reorientation, enhanced binding, and activation of ARH-1-associated Src-family PTKs, Blk, Fyn, Lck, and Lyn, leading to BCR association of a number of signal transducers and adaptors, including PLC-γ and Shc (Reth, 1992; Cambier *et al.*, 1994; Pleiman *et al.*, 1994).

The phospholipase C (PLC)-mediated hydrolysis of phosphatidylinositol-4,5-bisphosphate (PIP_2) generates inositol trisphosphate (IP_3) and diacylglycerol (DAG) (reviewed by Cambier *et al.*, 1994). The IP_3 mobilizes intracellular stores of calcium (leading to activation of calcium–calmodulin kinase II, CamKII), and DAG is an activator of the pivotal signaling enzyme, protein kinase C (PKC). The PKC isoforms expressed in B cells include α, β, δ, ζ and η, of which α, β, and δ are DAG regulated (Mischak *et al.*, 1991). Interestingly, although homologous recombination studies in a chicken B lymphoma (Takata *et al.*, 1994) indicate that Syk, in cooperation with Lyn, plays a pivotal role in the regulation of PIP_2 hydrolysis and calcium mobilization (Takata *et al.*, 1994), G-protein antagonist and reconstitution studies indicate that PTK-mediated regulation of PLC-γ appears to be additionally and proximally regulated by a pertussis-toxin-sensitive G protein(s) (reviewed by Harnett and Rigley, 1992; Melamed *et al.*, 1992)

Downstream events, such as the activation of the phosphoinositide 3-kinase (PI-3-K) and Ras-MAPK (mitogen-activating protein kinase) signaling cascades, also appear to be driven by the Src-PTK-mediated recruitment of signal transducers such as Shc, GTPase-activating protein (GAP), MAPK, and PI-3-K via interaction with their SH2, SH3, and N-terminal domains (Kim *et al.*, 1993; Pleiman *et al.*, 1993, 1994). In B cells, Ras activation appears to be regulated not only by the ubiquitous Grb2–Sos system (Pleiman *et al.*, 1994) but also by Vav, a hematopoietic GTP-exchange factor that is rapidly tyrosine-phosphorylated following ligation of the BCR (Bustelo *et al.*, 1992; Cambier *et al.*, 1994). The downstream targets of these signaling cascades remains relatively obscure, but CamKII and PKC have been implicated in the modulation of ets-1 (Fisher *et al.*, 1991) and egr-1 (Seyfert *et al.*, 1990) transcription factor activities, and PIP_3, the product of PI-3-K activity, activates PKCζ, which can regulate the transcription factor NF-κB (Nakanishi *et al.*, 1993; Diaz-Meco *et al.*, 1993). Finally, MAPK has been implicated as a key regulatory element of cellular responses by its targeting of protein kinases (e.g., Rsk-S6 kinase), other signal transducers (e.g., $cPLA_2$), and regulators of gene expression such as *c-jun*, *c-fos*, and *c-myc* (Avruch *et al.*, 1994).

4. CORECEPTORS AND MODULATION OF ANTIGEN RECEPTOR SIGNALING

Most of above studies delineating the signaling pathways coupled to BCR activation have employed anti-Ig reagents to polyclonally mimic antigen ligation of B-cell antigen receptors. It is still a matter of debate how these effects of anti-Ig relate to those of soluble T-cell-dependent antigens, which are likely to be rather poor cross-linking agents and do not cause substantial B-cell activation in the absence of T cells. However, evidence from studies using hapten–protein conjugates suggests that T-dependent antigens may also produce a limited inositol phospholipid turnover leading to abortive activation of B cells. This abortive entry into the cell cycle may be important because the resultant highly efficient internalization of antigen (Ag) followed by proteolysis and reexpression of Ag-derived peptides in association with MHC class II molecules is likely to promote B–T cell cooperation and "cognate" recognition of Ag (Noelle and Snow, 1990). Secondary intercellular interactions between coreceptor molecules may then generate "bidirectional" signaling in which the T helper cell is stimulated to proliferate and release cytokines that, in turn, act to amplify the B-cell response.

4.1. CD22: An Intrinsic Component of BCR Signaling?

Antigen receptor signaling can be modulated by a variety of coreceptor molecules including CD22 (reviewed by Law *et al.*, 1994), CD19/21 (reviewed by Tedder *et al.*, 1994), and CD40 (Durie *et al.*, 1994), which act to lower the threshold of antigen required for activation by recruiting additional Src and Syk PTKs, PLC, and PI-3-K activities to amplify the magnitude of signals generated via mIg. Interestingly, only $CD22^+$ cells can signal via the BCR, suggesting that CD22 and BCR crosstalk is required for mIg-derived signals (Law *et al.*, 1994). Indeed, studies demonstrating coimmunoprecipitation of mIg and CD22 have postulated that CD22, which is a 135-kDa B-cell-specific glycoprotein of the Ig superfamily, may be an intrinsic component of the BCR. However, CD22 is probably better described as a coreceptor, as it does not cocap with mIg. It should also be noted that only 1–2% of the available CD22 coimmunoprecipitates with the BCR, suggesting that CD22 may also possess independent signaling functions. Interestingly, cross-linking of CD22 does not by itself induce calcium mobilization or proliferation. CD22, however, is rapidly tyrosine phosphorylated following mIg cross-linking (Cambier *et al.*, 1994; Schulte *et al.*, 1992; Leprince *et al.*, 1993). Ligation of chimeric receptors expressing the cytoplasmic tail of Igβ is sufficient to induce tyrosine phosphorylation of CD22, which contains a number of Y-X-X-L sequences similar to the structure of the ARH-1 motifs, suggesting that such crosstalk may be achieved via antigen-directed clustering of Igβ, Lyn, and CD22 (Cambier *et al.*, 1994; Stamenkovic and Seed, 1990; Torres *et al.*, 1992; Law *et al.*, 1994).

Phosphorylation of CD22 is now known to induce recruitment of PI-3-K, Lyn, Syk, PLC-γ, protein tyrosine phosphatase (PTP)-1C, and p120. p120 is at present unidentified, but it is not the well-characterized 120-kDa substrate of Src-related kinases, GAP, focal adhesion kinase (FAK), or the p110 catalytic subunit of PI-3-kinase. It is, however, found in association with Syk and PLC-γ and, as such, may provide a "bridge" for BCR–CD22 cross talk (Law *et al.*, 1994; Justement *et al.*, 1994). CD22 is also known to be an adhesion molecule that binds to sialic-acid-containing glycoproteins to mediate B-cell aggregation with other hematopoietic cells. Indeed, CD22-dependent adhesion between B and T cells is mediated via the sialic acid groups of CD45 expressed at the T-cell surface. Interestingly, CD22 has also recently been shown to coimmunoprecipitate with CD45 from B cells, suggesting that CD45 may physically associate with and regulate tyrosine phosphorylation of CD22 (Justement *et al.*, 1994).

4.2. CD19/CD21: Reducing the Threshold of B-Cell Activation

The CD19/CD21 complex is a multimeric receptor (CD19, CD21, TAPA, and Leu-13) on B cells that can reduce the threshold of mIg-mediated B-cell activation by two orders of magnitude (Cambier *et al.*, 1994; Carter and Fearon, 1992). CD21 (which is the complement receptor CR2) is a receptor for iC3b, C3dg, CD23, IFNα, and EBV. This has led to the proposal that B cells that encounter Ag in the form of immune complexes containing complement components or bound to cells expressing CD23 may be more effectively stimulated by such cross-linking of mIg–CD21 than via simple ligation of the BCR (Cambier *et al.*, 1994).

CD21, which is a 140-kDa single membrane-spanning molecule with a cytoplasmic tail of 34 amino acids, does not appear to have the capacity to transduce signals. However, two other chains of this multireceptor complex, CD19 (gp95, a transmembrane glycoprotein of the Ig superfamily) and TAPA (CD81, a 20-kDa serpentine protein of four membrane-spanning domains), which contain extended cytoplasmic structures (the cytoplasmic domain of CD19 contains 243 amino acids), are thought to transduce CD21-mediated signals (Cambier *et al.*, 1994; Bradbury *et al.*, 1991; Zhou *et al.*, 1991; Tedder *et al.*, 1983; Oren *et al.*, 1990). Interestingly, although CD21 and CD19 are effectively B-cell-specific antigens, TAPA-1 and Leu-13 are more ubiquitous and may serve as signal transducers for many multimeric receptors (Cambier *et al.*, 1994). CD19 contains nine Tyr residues within its cytoplasmic domain, several of which are characteristic of binding sites for SH2 domains of cytoplasmic signal transducers. Indeed, CD19 is highly tyrosine phosphorylated following mIg ligation and recruits the Src-PTKs Lyn and Yes, PI-3-K, and GAP via interaction with their SH2 domains. In addition, Fyn and PLC-γ can also associate with CD19. However, because CD19 does not contain consensus binding motifs for either Fyn or PLC-γ, recruitment is likely to be indirect and may involve bridging via one or more of the other signal transducers (Tedder *et al.*, 1994).

Clearly, mIg-mediated tyrosine phosphorylation of CD19 can generate additional recruitment sites for key signal transducers containing SH2 domains and thereby enhance mIg-coupled signals. It should be noted, however, that the signaling capacity of CD19 is not dependent on the coexpression of mIg, and indeed, ligation of CD19 can induce transmembrane signals such as tyrosine phosphorylation and calcium mobilization (Tedder *et al.*, 1994). These findings may have physiological relevance because CD19 is functionally active on B-cell precursors that lack cell surface expression of mIg. Moreover, CD19 is rapidly tyrosine phosphorylated following ligation of CD40 or CD72. Together, these findings suggest that CD19 not only may be important for antigen-specific responses but also is likely to be central to other signal transduction systems (Tedder *et al.*, 1994).

Interestingly, ligation of the CD19/CD21 complex prior to mIg ligation abrogates signaling via the BCR, perhaps via desensitization of mIg-coupled signals (Rijkers *et al.*, 1990). Moreover, this form of negative signaling does not appear to be mediated by the CD19 cytoplasmic tail but rather via the CD19 transmembrane domain, which is important for interactions with TAPA. Indeed, a role for TAPA in mediating negative signals associated with the CD19/21 complex is consistent with the finding that TAPA was first identified as the target of an antiproliferative antibody, TAPA-1, which inhibited cellular proliferation *in vitro* (Tedder *et al.*, 1994).

4.3. CD20: A Regulator of B-Lymphocyte Cell Cycle Progression?

The BCR coreceptor CD20 is a 33-kDa membrane-embedded protein that has four transmembrane domains and a Ser/Thr-rich cytoplasmic domain containing multiple potential phosphorylation sites. CD20 is expressed as a homodimer or homotetramer on the cell surface and generally associates with a number of other other cell-surface proteins, so it may form a multimeric receptor complex (reviewed by Tedder and Engel, 1994). This structure suggested that CD20 might function as either a membrane transporter or an ion channel; indeed, transfection studies established that CD20 itself forms a calcium channel (Tedder and Engel, 1994). Although ligation of CD20 can also induce activation of Ser/Thr and Tyr kinases, it is thought that CD20 may act to regulate BCR-mediated cell cycle progression by altering cell-cycle-dependent calcium homeostasis: because cell cycle progression requires precise levels of $[Ca^{2+}]_i$, the finding that CD20 is itself a calcium channel and can be regulated by PKC and CamKII provides a rationale for CD20-mediated modulation (both positive and negative signals) of B-cell activation (reviewed by Tedder and Engel, 1994).

4.4. CD45: Coordinating PTK and PTPase Signals during B-Cell Activation

The importance of coordinated PTPase and PTK-mediated signals in BCR–coreceptor interactions (reviewed by Justement *et al.*, 1994) is underscored by the finding that expression of the protein tyrosine phosphatase, CD45, is required for full BCR signaling and that Igα/β, CD19/CD21, and CD22 are all substrates for CD45. Moreover, B cells purified from "CD45 knockout" mice are unresponsive to BCR ligation as judged by their lack of proliferation following culture with mitogenic anti-Ig (Cambier *et al.*, 1994; Kishihara *et al.*, 1993). These findings have led to the proposal that CD45 may act by maintaining Igαβ, CD19, and CD22 in a dephosphorylated state in resting cells in order to optimize signaling via the BCR and also to prevent aberrant activation of B cells. However, prior ligation of CD45 abrogates subsequent mIg signaling. This apparent discrepancy may now be reconciled by recent evidence suggesting that CD45 may also positively modulate B-cell antigen receptor signaling by preferentially dephosphorylating (a Tyr residue at a regulatory site in the C-terminal domain) and activating (BCR-associated) Lyn kinase. Thus, CD45 can be viewed both as a positive and as a negative regulator of BCR signaling, depending on the phosphorylation status of the mIg-associated signal transducers (Justement *et al.*, 1994).

4.5. MHC Class II Molecules: Crosstalk among MHC, mIg, and IL-4 Receptors

The activation/maturation status of the B cell can dramatically influence the intracellular signals and biological responses resulting from BCR–coreceptor crosstalk. For example, whereas MHC class II-mediated cyclicAMP generation and nuclear translocation of PKC are associated with inhibitory or apoptotic signals in resting B cells (Cambier *et al.*, 1987; reviewed by Scholl and Geha, 1994), ligation of Class II molecules on B cells preactivated by anti-IgM and IL-4 induces PTK (Fgr and Lyn)-mediated activation of PLC-γ, mobilization of calcium, activation of PKC, and ultimately cell proliferation (Cambier *et al.*, 1991; Scholl and Geha, 1994). The IL-4R is coupled to a transient PLC activation followed by a sustained generation of cyclicAMP in human B cells (Finney *et al.*, 1990). Although these signals are not observed in murine B cells, IL-4 can reduce the threshold of mIg-mediated PKC activation (Harnett *et al.*, 1991). In addition, the IL-4R is also coupled to the PTK-mediated activation of PI-3-K, possibly via a novel docking protein, 4PS (reviewed by Keegan *et al.*, 1994). The IL-4 regulation of gene expression appears to involve Jak1- and Jak3-mediated tyrosine phosphorylation of STATs, such as IL-4NAF and Stf-IL-4 (Johnston *et al.*, 1994; Witthuhn *et al.*, 1994).

4.6. CD40: A Life-or-Death Signal?

CD40 plays a central role in T-cell-dependent proliferation and differentiation of B cells: ligation of CD40, which is a member of the TNF-receptor superfamily, can (1) drive resting cells into cycle, (2) reduce the threshold required for mIg signaling, (3) rescue germinal center (GC) cells from apoptosis, and (4) induce growth arrest or apoptosis of transformed cell lines and lymphomas (reviewed by Durie *et al.*, 1994; see Chapter 7). The molecular mechanisms underlying these biological responses are not fully understood but appear to be quite distinct: although PTK (*Lyn*) activation appears to be important for all CD40-mediated signals and mandatory for rescue of GC cells (Knox and Gordon, 1993), CD40 appears to be additionally coupled to cyclicAMP generation in resting cells and to PLC-γ and PI-3-K in activated B cells (Durie *et al.*, 1994).

4.7. CD38: A Signal for B-Cell Survival?

A novel B-cell coreceptor, CD38, has recently been described and appears to play a pivotal role in the regulation of both proliferation and rescue from apoptosis in B cells (Harada *et al.*, 1993; Santos-Argumedo *et al.*, 1993; Howard *et al.*, 1993; Malavasi *et al.*, 1994). The molecular mechanisms underlying CD38-mediated activation are, as yet, poorly understood, but it is clear that lymphocyte proliferation stimulated by anti-CD38 is preceded by activation of PTK activity (Kirkham *et al.*, 1994) and a CD38-mediated calcium influx that appears to be independent of IP_3 generation and regulation (Santos-Argumedo *et al.*, 1993; Howard *et al.*, 1993). In addition to its role in facilitating proliferation of B cells, CD38 stimulation (1) rescues germinal center or mIgM-activated B cells from apoptosis and (2) demonstrates costimulatory or antagonistic effects with IL-4 in an activation-state-dependent manner (Santos-Argumedo *et al.*, 1993; Howard *et al.*, 1993; Malavasi *et al.*, 1994). The signaling events underlying these differential responses are not yet known, but it has recently been shown that CD38 can associate with mIg, CD19, or CD21 (Malavasi *et al.*, 1994).

Intense interest in the precise role of CD38 has recently been generated following expression cloning and sequencing of this activation antigen. CD38 is a 42-kDa type II transmembrane glycoprotein with a short N-terminal cytoplasmic tail (Harada *et al.*, 1993; Malavasi *et al.*, 1994). Intriguingly, the amino acid sequence of CD38 bears a strong homology to that of the ADP-ribosyl cyclase of the mollusc *Aplysia*. ADP-ribosyl cyclase converts NAD^+ into cyclic ADP-ribose (cADPR), an endogeneous second messenger that can induce the mobilization of intracellular calcium, independently of IP_3, via ryanodine receptors (Santos-Argumedo *et al.*, 1993; Howard *et al.*, 1993; Malavasi *et al.*, 1994; Galione *et al.*, 1991). Indeed, CD38 has been shown to exhibit ADP-ribosyl cyclase activity (Howard *et al.*, 1993; Malavasi *et al.*, 1994). Clearly, elucidation of how this ectoenzyme activity induces the biological responses mediated by CD38 will identify a novel calcium-dependent signaling pathway in B-cell biology.

4.8. Fcγ Receptors: Modulating the Immune Response

In an ongoing immune response, activated B cells are subject to negative regulation by circulating immunoglobulin, which is important in feedback inhibition of the Ab response.

Moreover, resting B cells that encounter antigen in the form of immune complexes may not be activated. In B lymphocytes, such negative feedback regulation of activation and Ab secretion is mediated by the 40-kDa FcγRIIb receptor, CD32 (Hunziker *et al.*, 1990). Fcγ is a member of the Ig superfamily and is composed of two Ig-like domains, a single membrane-spanning region, and a 44- to 76-amino-acid cytoplasmic tail. Cross-linking of mIg and CD32 abrogates mIg-coupled signaling, resulting in abortive activation following antigenic stimulation. However, the negative influence of occupied CD32 can be overcome in B cells by IL-4 (O'Garra *et al.*, 1987).

The biochemical mechanism underlying the FcR-dependent inhibition of mIg signaling has been at least partially elucidated. Cross-linking of mIg and CD32 dramatically inhibits mIg-coupled IP_3 generation and calcium mobilization, resulting in abortive B-cell activation (Bijsterbosch and Klaus, 1985). Mechanistically, this effect is explained in terms of an uncoupling of the Ag receptor complex from the G protein regulating PLC activation (Rigley and Harnett, 1990). Although the mechanism leading to this uncoupling is unknown, one possibility is that engaging CD32 could disrupt receptor–G-protein contact by modifying the receptor contact site with G protein, perhaps by phosphorylation (Rall and Harris, 1987; Sullivan *et al.*, 1987). Alternatively, it has been shown previously that capping of mIg also leads to cocapping of FcR, suggesting that the uncoupling might be brought about by physical dissociation, perhaps by disrupting the mIg–accessory transducing molecules complex. Interleukin 4 reverses the capacity of FcR to inhibit stimulatory signals delivered via the Ag receptor (O'Garra *et al.*, 1987), and although the molecular basis for this effect has not been elucidated, it is clear that IL-4 does not overcome the inhibition of PIP_2 hydrolysis in murine B cells (O'Garra *et al.*, 1987).

5. CYTOKINES AND B-LYMPHOCYTE SIGNALING

B-lymphocyte maturation, activation, growth, and differentiation are regulated by cytokines. In the bone marrow, early B-cell maturation (mIg^- B cells) relies on cytokines derived mainly from stromal cells (e.g., IL-7 and IL-11) and, to a lesser extent, from T cells (reviewed by Callard, 1990; also see Chapter 9). In contrast, most of the cytokines involved in the Ag-dependent phase of B-cell proliferation and differentiation are derived from helper T lymphocytes: for example, whereas IL-2 is involved in regulating growth of cycling B cells and accelerating Ab secretion (as is IL-6), other cytokines such as IL-4, IL-13, and IFNγ direct heavy-chain isotype switching (reviewed by Cushley and Harnett, 1993, 1994). Indeeed, B-cell proliferation and differentiation can be influenced by a wide range of cytokines including IFN-α,β,γ, TNF-α,β, IL-1–4, IL-5 (murine B cells), IL-6, IL-7, IL-10, IL-11, IL-13 (human B cells), IL-15, and TNF-β (Burke *et al.*, 1993). Interestingly, certain of these cytokines can interact synergistically on B cells, but others, such as IL-4 and IFNγ or IL-4 and IL-2, appear to be mutually inhibitory (Cushley and Harnett, 1993, 1994).

Cloning and characterization of many of these cytokine receptors has recently suggested molecular mechanisms for these interactive biological responses: most of the receptors for such cytokines have been found to belong to the hematopoietin receptor superfamily (HRS) (Bazan, 1990; Kishimoto *et al.*, 1994; Chapter 1), which generally comprise multichain complexes consisting of a ligand-specific receptor(s) and an accessory signal transducer(s) common to a cytokine subfamily. For example, although the cytokines IL-6, LIF, and IL-11 interact with cytokine-specific binding chains, they mediate their effects via a common signal transducer, gp130, providing a rationale for the finding that these cytokines mediate similar functions on various tissues (see Chapter 1). Likewise, common signal transducer molecules termed βc and γc have recently been identified for the IL-3, IL-5, GM-CSF, and IL-2, IL-4, IL-7, IL-9, IL-13 cytokine subfamilies, respectively (reviewed by Kishimoto *et al.*, 1994; see Chapter 1). Utilization of such shared signal transducers is likely to underpin the observed functional pleiotropy, redundancy, and antagonism of cytokines in many systems.

5.1. Hematopoietin Receptor Signaling in B Cells

The HRS receptors are transmembrane glycoproteins possessing two extracellular domains (N and C), a short transmembrane region, and a large cytoplasmic tail (Kishimoto *et al.*, 1994). Receptor homology is restricted to the extracellular region containing some 200 amino acids and characterized by four conserved cysteine residues in the N-terminal portion and a very highly conserved pentapeptide motif (Trp-Ser-X-Trp-Ser, the WSxWS motif) located close to the transmembrane region in the primary structure. The intracellular domain is of variable size and shows little conservation within the HRS (Kishimoto *et al.*, 1994). In terms of activation of cellular signaling pathways, the intracellular domains of HRS receptors display no motifs, which indicates that they possess catalytic domains. The ability of the receptor to mediate a mitogenic response is, however, apparently dependent on a limited membrane-proximal region of the cytoplasmic domain that contains box 1 and box 2 motifs, which show limited conservation among members of the HRS family (Murakami *et al.*, 1991). This region is essential for cytokine-mediated coupling to the activation of nonreceptor tyrosine kinases.

Although, some HRS receptors recruit Src-related PTKs (e.g., IL-2R), many of these receptors appear to be coupled to the recruitment and activation of one or more Janus kinases (Jaks) (Stahl and Yancopoulos, 1993). The Jak family (Jak1–3, TYK2) are tyrosine kinases of approx 130 kDa that contain a canonical tyrosine kinase domain at their C terminal. In addition, immediately N-terminal to this kinase domain, they appear to contain a nonactive kinase-like region. Interestingly, these kinases do not contain any SH2 or SH3 domains, and so the molecular mechanisms underpinning receptor association with the membrane-proximal region of the cytoplasmic domain are, at present, unknown. Jak kinases play a central role in cell signaling by activating cytoplasmic proteins termed STATs (signal transducers and activators of transcription), which are translocated to the nucleus, bind response elements, and hence stimulate transcription. Although the best-characterized Jak–STAT systems are those coupled to the IFNs, cytokines that regulate gene expression through tyrosine phosphorylation of the ISGF3α complex (p91 and p113: Darnell *et al.*, 1994), which translocates to the nucleus to activate gene expression (either alone or in conjunction with the 48-kDa ISGF3γ), many other cytokines, including those such as IL-2, IL-4, and IL-6 that modulate B-cell responses, appear to utilize similar systems (Kishimoto *et al.*, 1994; Johnston *et al.*, 1994; Witthuhn *et al.*, 1994; Russell *et al.*, 1994; Miyazaki *et al.*, 1994). Interestingly, there is good evidence that ligation of several of the cytokine receptors of the HRS family can also (alternatively) regulate gene expression via (Src-related?) tyrosine kinase-mediated activation of the Ras–Raf–MAPkinase cascade, resulting in *c-fos* and AP-1 activation (Satoh *et al.*, 1991; Kishimoto *et al.*, 1994).

5.1.1. The IL-2 Receptor

The IL-2 receptor is a multimeric complex comprising α, β, and γ subunits. The IL-2-specific ligand-binding chain, IL-2α, is a 55-kDa glycoprotein structurally unrelated to any known receptor family. The IL-2α, also termed Tac antigen or CD25, is inducible in B cells, notably by IL-4 in human tonsillar B cells and by IL-5 in murine B cells (reviewed by Cushley and Harnett, 1993, 1994). The IL-2β (75 kDa) and -γ (64 kDa) are members of the HRS family, constitutively expressed on B cells, and are responsible for receptor internalization and signal transduction (Kishimoto *et al.*, 1994). The essential role of γ in signal transduction has recently been demonstrated by the genetic localization of the human XSCID immunodeficiency to the IL-2Rγ gene (Noguchi *et al.*, 1993). Interestingly, the IL-2γ chain has also been shown to be involved in signal transduction via IL-4, IL-7, and IL-15 receptors. Together with the recent demonstration that IL-2-like effects can be transduced via the βγ subunits by the novel cytokine IL-15 (Giri *et al.*, 1994), these findings may go some way toward explaining not only the mutually antagonistic effects of IL-2 and IL-4 but also the observed normal immune responses of IL-2-deficient mice (Kundig *et al.*, 1993).

There is now good evidence to show that the IL-2 receptor is coupled, via IL-2Rβ, to the activation of both Src-related PTKs (Lyn in a pre-B-cell line, Torigoe *et al.*, 1992) and Jak PTKs (Jak 1 and 3, Russell *et al.*, 1994; Miyazaki *et al.*, 1994). The IL-2Rβ has two functional domains: a C-terminal domain that recruits PTKs, leading to *c-fos* activation, presumably via PTK-mediated induction of the Ras–Raf–MAPK signaling cascade (Downward *et al.*, 1992; Turner *et al.*, 1991) and a membrane-proximal Ser-rich domain that transduces IL-2-coupled proliferative signals via a novel *c-myc* pathway (Shibuya *et al.*, 1992). These signaling events also appear to be important in the IL-2 up-regulation of IgM secretion that results from an increased biosynthesis of J-chain protein (Tigges *et al.*, 1989). The IL-2 up-regulation of J-chain biosynthesis occurs at the transcriptional level (Tigges *et al.*, 1989) and is controlled by a B-cell-specific IL-2-regulated *trans*-acting factor, NF-J_B, acting on a *cis* promoter element, J_B, which is located at position -75 to -45 in the J-chain gene promoter (Lansford *et al.*, 1992).

5.1.2. The IL-4 Receptor

Interleukin 4 elicits a range of effects on B-cell differentiation, including sustaining the growth of pre-B cells, promoting early activation of quiescent B lymphocytes, and increasing expression of MHC class II antigens and of low-affinity Fc_ε receptors. In addition, IL-4 mediates directed isotype switching in mitogen-activated B-cell blasts (reviewed by Cushley and Harnett, 1993, 1994). Interestingly, in human B cells, IL-13 has similar effects to IL-4 except that IL-4 is more potent in terms of up-regulating IgM, MHC class II, CD23, and CD25 expression (Zurawski and de Vries, 1994). In addition, IL-4 and IL-13 similarly promote the proliferative responses of anti-IgM- or anti-CD40-primed B cells. Interleukin 13 has also been shown to direct class switching to IgE in an IL-4-independent manner. Furthermore, although IL-13 is two- to fivefold less effective than IL-4 at inducing class switching to IgE, there is no additivity or synergy with these two cytokines, suggesting that IL-4 and IL-13 are coupled to common signaling pathways. In contrast, murine B cells do not respond to IL-13, perhaps suggesting that these cells lack a functional IL-13 ligand-binding receptor (Zurawski and de Vries, 1994). These findings may now be reconciled by receptor cloning studies, which show that the IL-4 receptor is also a multimeric HRS receptor comprising, to date, an IL-4-specific 140-kDa glycoprotein ligand-binding receptor of the HRS family and the common HRS signal-transducing subunit, γc, shared by the IL-2, IL-4, IL-7, IL-9, IL-13, and IL-15 receptor complexes (Kishimoto *et al.*, 1994).

The most compelling data currently available for IL-4 signaling in B cells come from studies of human tonsillar B lymphocytes. There, IL-4 stimulates the generation of inositol trisphosphate, DAG, and a rise in intracellular Ca^{2+} immediately after ligand binding. This is followed after several minutes by a sustained rise in intracellular cyclicAMP levels (Finney *et al.*, 1990). Although, in murine B cells, IL-4 does not appear to induce the release of inositol phosphates, mobilize calcium, translocate protein kinase C to the plasma membrane, or, indeed, modify signaling via the phosphoinositide pathway coupled to mIg (Justement *et al.*, 1986; Mizuguchi *et al.*, 1986; Klaus and Harnett, 1990), IL-4 synergizes with anti-Ig to promote translocation of PKC to membranes (Harnett *et al.*, 1991). In addition, the IL-4 receptor is coupled to the activation of PTKs, because culture of murine B cells with IL-4 leads to the appearance of a 75-kDa tyrosine-phosphorylated protein in the membranes of such lymphocytes (McGarvie and Cushley, 1989a). Furthermore, a 42-kDa protein of unknown identity is phosphorylated in isolated membranes in the presence of IL-4 (Justement *et al.*, 1986; McGarvie and Cushley, 1989b). A similar protein kinase substrate has been reported in human tonsillar B cells, although the apparent molecular mass in this model is 38 kDa (Finney *et al.*, 1991).

Interestingly, although IL-4R is also coupled to the PTK-mediated activation of PI-3-K, possibly via a novel docking protein, 4PS, which has recently been cloned and identified as IRS-2 (Sun *et al.*, 1995) (reviewed by Keegan *et al.*, 1994), IL-4, unusually for an HRS cytokine, does not stimulate Ras activation (Satoh *et al.*, 1991). This may reflect the finding that IL-4 regulation of gene expression appears to involve Jak1- and Jak3-mediated tyrosine phosphorylation of STATs such as IL-4NAF and STF-IL-4 (Johnston *et al.*, 1994; Witthuhn *et al.*, 1994). Interleukin 4 up-regulation of MHC class II expression appears to occur at the molecular level by a specific DNA binding protein, NF-BRE. The NF-BRE recognition element is far upsteam of the promoter region of class II α chains (Boothby *et al.*, 1989) and is distinct from the conserved W, X, and Y boxes found in the 5′ promoter region of all MHC class II genes.

5.1.3. The IL-6 Receptor

Interleukin 6 is a highly pleiotropic cytokine that elicits a range of responses in many different cell types, including B and T cells, hepatocytes, neural cells, and fibroblasts (reviewed by Cushley and Harnett, 1993; Kishimoto *et al.*, 1994). The IL-6 receptor complex comprises two distinct transmembrane glycoproteins, gp80 and gp130, both of which are members of the HRS. The gp80 component binds ligand but is, by itself, unable to activate cellular signaling (Taga *et al.*, 1989). However, the binding of IL-6 to gp80 promotes binding of gp80 to the second component of the receptor complex, gp130, and the transmission of cellular signaling information (Hibi *et al.*, 1990). Moreover, a soluble, secreted form of gp80 can bind IL-6, and this complex retains its capacity to interact, via its extracellular domains, with gp130 and provide stimulatory signals to sensitive cells (Taga *et al.*, 1989). Interestingly, the soluble IL-6 receptor is an agonistic receptor, an apparently unique property among the soluble cytokine receptors, because other well-characterized secreted cytokine receptors, for example IL-4 (Mosley *et al.*, 1989) and IL-7 (Goodwin *et al.*, 1990), behave as antagonistic soluble receptors. Interleukin 6 mediates its effects on gene expression, at least in part, via NF-IL-6, a leucine zipper transcription factor that is activated via MAPK-mediated Thr phosphorylation of NF-IL-6 just upstream of its DNA-binding domain. Presumably, IL-6 is coupled to the activation of NF-IL-6 via PTK-mediated activation of the Ras/Raf/MAPK signaling cascade (Satoh *et al.*, 1991; Nakajima *et al.*, 1993). Recent evidence also suggests that IL-6 may additionally regulate gene expression by coupling to a Jak–STAT signaling pathway (Kishimoto *et al.*, 1994).

5.1.4. The IL-7 Receptor

Interleukin 7 is a stromal-cell-derived cytokine that is active on B-cell precursors and certain immature T-cell subsets (Henney, 1989; Cushley and Harnett, 1993). Like many of the other B-cell-directed cytokines, IL-7 mediates its effects via a multimeric HRS receptor complex consisting of an IL-7-specific 75-kDa ligand-binding HRS transmembrane glycoprotein and the common signal-transducing subunit, γc (Kishimoto *et al.*, 1994). Studies on fetal thymocytes and leukemic precursor B-cell lines indicated that the IL-7 receptor was coupled to PIP_2 hydrolysis via PTK-mediated activation of PLC-γ1 (Uckun *et al.*, 1991a,b). However, in striking contrast to these results, a recent report demonstrates that although the IL-7 receptor does indeed mediate tyrosine phosphorylation, it is not coupled to PLC-γ1 phosphorylation or, indeed, to any inositol phosphate generation or calcium mobilization in thymocytes, mature T cells, or prepre-B cells (from a patient with acute lymphoblastic leukemia) (Roifman *et al.*, 1992). This major discrepancy may reflect differential coupling of the IL-7 receptor during differentiation and oncogenic transformation.

5.2. Receptors of the Ig-like Superfamily: The IL-1 Receptor

Interleukin 1 is a highly pleiotropic cytokine, influencing the growth and differentiation of a wide range of cell types, including B and T cells (reviewed by Dinarello *et al.*, 1989). The B-cell IL-1 receptor (type II) (McMahan *et al.*, 1991) has now been cloned. The B-cell-type receptor has a large extracellular region with three Ig-like domains, a single transmembrane region, and a small 29-amino-acid cytoplasmic tail (McMahan *et al.*, 1991). Although the IL-1 receptor is structurally unrelated to the classical seven-transmembrane G-protein receptors, it appears to be linked to a pertussis-toxin-sensitive G protein(s) (Mizel, 1990; O'Neill *et al.*, 1990). Following ligand binding, the receptor undergoes receptor-mediated endocytosis, and the receptor–ligand complex may ultimately translocate to the nucleus via an endosomal compartment. Controversy surrounds the precise downstream signal transduction mechanisms activated in IL-1-sensitive cells, as it appears there may be lineage-specific IL-1-triggered signaling pathways (Mizel, 1990; O'Neill *et al.*, 1990). However, the IL-1-driven transcriptional activation in 70Z/3 pre-B cells is mediated by the NF-κB *trans*-acting factor and can be mimicked by exposure of the cells to forskolin via protein kinase A activation. Together with studies indicating that IL-1 receptor activation leads to elevations in cyclicAMP levels in other lymphocyte models, the data from the 70Z/3 pre-B-cell model suggest that IL-1 mediates its effect on NF-κB activity via elevation of cyclicAMP and subsequent activation of protein kinase A (Mizel, 1990).

5.3. The TNF Receptor Superfamily

Both TNF-α and TNF-β (also known as lymphotoxin or LT-α) are secreted by macrophages and T cells, respectively. These cytokines were first characterized by their ability to kill tumor cell lines, but it has become increasingly evident that they are also primary mediators of immune regulation and the inflammatory response. Indeed, TNF has been implicated as a mediator in the immunopathology of septic shock, autoimmune disorders, and graft-versus-host disease. In contrast to their role in mediating tumor cell death, TNF-α and TNF-β can promote myeloid progenitor cell proliferation and growth inhibition (see Chapters 6 and 11), and B-cell proliferation and have been shown to synergize with IL-2, IL-4, and IFN-γ in supporting B-cell proliferation (reviewed by Cushley and Harnett, 1994). In addition, TNF can enhance spontaneous proliferation of B-CLL cells. Indeed, many lymphoblastoid cell lines secrete TNF, and it has been postulated that TNF may act as an autocrine growth factor for these cells (Cushley and Harnett, 1994).

Cloning of the receptors for TNF has identified two distinct receptors, p75 and p55. More importantly, it has led to the discovery of a novel cytokine receptor superfamily (e.g., TNF-R, CD30, CD40, NGF, OX-40, Fas, and CD27) (see Chapter 3), many of whose members appear to be intimately involved in the processes regulating the balance between cell growth and cell death. B cells express several of these receptors, including CD30, CD27, CD40, and TNF-R. Interestingly, their cytokine ligands of these receptors appear to exist mainly as membrane-bound cell surface proteins, as exemplified by the CD40–CD40L (gp39) interaction formed during T–B cell cooperation (Durie *et al.*, 1994). These cytokine ligands are type II single membrane-spanning proteins of a prototypic pro-TNF-α structure: all cytokine family homology is restricted to the extracellular C-terminal receptor binding region. In contrast, despite considerable variation in size, the receptors are all type 1 transmembrane proteins characterized by varying numbers of a canonical motif of cysteine-rich pseudorepeats, each containing about six cysteines and 40 amino acids. Their cytoplasmic domains are of variable size (46 to 221 amino acids) but generally lack family sequence homology, suggesting that these receptors are coupled to quite different signaling pathways, a hypothesis that is probably consistent with the widely different biological responses induced following receptor ligation.

Although the signal events coupled to these receptors are poorly defined, it is clear that some of the receptors are differentially coupled to distinct signaling pathways in a maturation-state-dependent manner, and this may go some way to explaining the differential biological respones resulting from ligation of these receptors on cells of a single lineage but varying stage of differentiation. For example, CD40 plays a central role in T-cell-dependent proliferation and differentiation of B cells: ligation of CD40 can (1) drive resting cells into cycle, (2) reduce the threshold required for mIg signaling, (3) rescue germinal center (GC) cells from apoptosis, and (4) induce growth arrest or apopotosis of transformed cell lines and lymphomas (reviewed by Durie *et al.*, 1994).

Although the molecular mechanisms underlying these biological responses are not fully understood, they appear to be quite distinct: PTK (*Lyn*) activation appears to be important for all CD40-mediated signals and mandatory for rescue of GC cells (Durie *et al.*, 1994); however, CD40 appears to be additionally coupled to cyclicAMP generation in resting cells and to PLC-γ and PI-3-K in activated B cells (Knox and Gordon, 1993).

Interestingly, at least some of these receptors appear to be coupled to a novel signaling pathway, the sphingomyelin pathway, particularly when transducing differentiation-inducing, antiproliferative, or apoptotic signals. For example, TNF-α can couple to the sphingomyelin pathway in several biological sytems, including the pre-B-cell line 70Z/3, following transfection of the p55 TNF-αR. This pathway involves activation of a sphingomyelinase (SMase), which hydrolyzes sphingomyelin to produce phosphocholine and ceramide. Ceramide is a putative lipid second messenger that can activate a 97-kDa membrane-bound Ser/Thr kinase belonging to the class of proline-directed kinases that also includes the signal transducers MAPK and cdc2. In addition, ceramide can also activate a cytosolic okadaic-acid-sensitive phosphatase of the 2A class. Although ceramide has been implicated in the transduction of antiproliferative signals, it has also been shown to be involved in the transduction of activation signals (Smith *et al.*, 1994).

Interestingly, this pathway also appears to play a role in signaling via other classes of cytokine receptors, for example, IL-1β (fibroblasts, EL4 T cells) and IFNγ receptors (in monocyte differentiatiation of HL60 cells) (Kolesnick and Golde, 1994). There are, however, differences in the regulation of these pathways. Although PLA_2 activation and arachidonic acid generation has recently been proposed as a mediator of TNF-α-directed SMase activation (Jayadev *et al.*, 1994), the IL-1β receptors appear tightly and directly coupled to SMase activation. Together, with the potential of differentially activated levels of ceramide-dependent kinase and phosphatase activation, these lineage- and/or cytokine-specific differences in SMase regulation may go some way to explaining the differential biological responses of cell death versus cell activation observed follwing ceramide generation. Although the precise details of coupling are unknown, TNF-α-mediated ceramide generation leads to the down-regulation of *c-myc* and to the activation of MAPK and NF-κB translocation in a Raf-kinase-dependent manner.

6. FUTURE PERSPECTIVES

Elucidation of the signaling pathways underlying B-lymphocyte activation has identified a wide range of novel signaling pathways and receptor–effector coupling mechanisms. The major areas of B-cell signaling yet to be resolved include the elucidation of the molecular mechanisms underlying B-cell differentiation and the differential coupling of the antigen receptors to the generation of apoptotic tolerogenic signals or activation/survival signals. Resolution of these questions should identify novel signaling pathways, provide strategies for combating infection and immune dysfunctions such as autoimmune disease, leukemia, and lymphoma, and contribute greatly to our understanding of the molecular mechanisms underpinning cell growth, differentiation, and death.

7. REFERENCES

Avruch, J., Zhang, X.-F., and Kyriakis, J. M., 1994, Raf meets ras: Completing the framework of a signal transduction pathway, *Trends Biochem. Sci.* **19**:279–283.

Bazan, J. F., 1990, Haemopoietic receptors and helical cytokines, *Immunol. Today* **11**:350–354.

Bijsterbosch, M. K., and Klaus, G. G. B., 1985, Crosslinking of surface immunoglobulin and Fc receptors on B lymphocytes inhibits B cell stimulation of inositol phospholipid brakdown via the antigen receptors, *J. Exp. Med.* **162:**1825–1836.

Boothby, M., Gravallesse, E., Liou, H-C., and Glimcher, L. H., 1989, A DNA binding protein regulated by IL-4 and by differentiation in B cells, *Science* **242:**1559–1562.

Bradbury, L., Kansa, G., Levy, S., Evans, R. L., and Tedder, T. F., 1991, The CD19/CD21 transducing signal complex of B lymphocytes includes the target of antiproliferative antibody-1 and Leu 13 molecules, *J. Immunol.* **149:**2841–2850.

Burke, F., Naylor, M. S., Davies, B., and Balkwill, F., 1993, The cytokine wallchart, *Immunol. Today* **14:**165–169.

Burkhardt, A. L., Brunswick, M., Bolen, J. B., and Mond, J. J., 1991, Anti-immunoglobulin stimulation of B lymphocytes activates *src*-related protein tyrosine kinases, *Proc. Natl. Acad. Sci. U.S.A.* **88:**7410–7414.

Bustelo, X. R., Ledbetter, J. A., and Barbacid, M., 1992, Product of *vav* protooncogene defines a new class of tyrosine protein kinase substrates, *Nature* **356:**68–71.

Callard, R. E., 1990, *Cytokines and B Lymphocytes*, Academic Press, London.

Cambier, J. C., Newell, M. K., Justement, L. B., McGuire, J. C., Leach, K. L., and Chen, Z. Z., 1987, Ia binding ligands and cAMP stimulate nuclear translocation of PKC in B lymphocytes, *Nature* **327:**629–632.

Cambier, J. C., Morrison, D. C., Chien, M. M., and Lehmann, K. R., 1991, Modelling of T cell contact-dependent B cell activation. Il-4 and antigen receptor ligation primes quiescent B cells to mobilise calcium in response to Ia crosslinking, *J. Immunol.* **146:**2075–2082.

Cambier, J. C., Bedzyk, W., Campbell, K., Chien, N. C., Friedrich, J., Harwood, A., Jensen, W., Pleiman, C. M., and Clark, M. R., 1993, The B cell antigen receptor: Structure and function of primary, secondary, tertiary, and quaternary components, *Immunol. Rev.* **1320:**1–22.

Cambier, J. C., Pleiman, C., and Clark, M. R., 1994, Signal transduction by the B cell antigen receptor and its coreceptors, *Annu. Rev. Immunol.* **12:**457–486.

Campbell, K. S., and Cambier, J. C., 1990, B lymphocyte antigen receptors (mIg) are non-covalently associated with a disulphide-linked, inducibly phosphorylated glycoprotein complex, *EMBO J.* **9:**441–448.

Carter, R. H., and Fearon, D. T., 1992, CD19: Lowering the threshold for antigen receptor stimulation of B lymphocytes, *Science* **256:**105–107.

Chen, J., Stall, A. M., Herzenberg, L. A., and Herzenberg, L. A., 1990, Differences in glycoprotein complexes associated with IgM and IgD on normal murine B cells potentially enable transduction of different signals, *EMBO J.* **9:**2117–2124.

Clark, E. A., and Lane, P. J. L., 1991, Regulation of human B cell activation and adhesion, *Annu. Rev. Immunol.* **9:**97–128.

Cushley, W., and Harnett, M. M., 1993, Cellular signalling mechanisms in B lymphocytes, *Biochem. J.* **292:**313–332.

Cushley, W., and Harnett, M. M., 1994, Regulation of B lymphocyte growth and differentiation by soluble mediators, in *Handbook of B and T Lymphocytes* (C. Snow, ed.), pp.389–420. Academic Press, Orlando, FL.

Darnell, J. E., Jr., Kerr, I. M., and Stark, G. R., 1994, Jak–STAT pathways and transcriptional activation in response to IFNs and other extracellular signalling proteins, *Science* **264:**1415–1421

Diaz-Meco, M. T., Berra, E., Municio, M. M., Sanz, L., Lozano, J., Dominguez, L., Diaz-Golpe, V., Lain de Lera, M. T., Alcami, J., Paya, C. V., Arenzana-Seisdedos, F., Virelizier, J.-L., and Moscat, J., 1993, A dominant negative protein kinase C ζ subspecies blocks NF-κB activation, *Mol. Cell. Biol.* **13:**4770–4775.

Dinarello, C. A., Clark, B. D., Puren, A. J., Savage, N., and Rosoff, P. M., 1989, The interleukin 1 receptor, *Immunol. Today* **10:**49–51.

Downward, J., Graves, J., and Cantrell, D. A., 1992, The regulation and function of p21ras in T lymphocytes, *Immunol. Today* **13:**89–92.

Durie, F. H., Foy, T. M., Masters, S. R., Lama, J. D., and Noelle, R. J., 1994, The role of CD40 in the regulation of humoral and cell mediated immunity, *Immunol. Today* **15:**406–411.

Finney. M., Guy, G., Michell, R. H., Gordon, J., Dugas, B., Rigley, K. P., and Callard, R. E., 1990, Interleukin 4 activates human B lymphocytes via transient inositol lipid hydrolysis and delayed cAMP generation, *Eur. J. Immunol.* **20:**151–156.

Finney, M. J., Michell, R. H., Gillis, S., and Gordon, J., 1991, Regulation of the interleukin 4 signals in human B lymphocytes, *Biochem. Soc. Trans.* **19:**287–291.

Fisher, C. L., Ghysdael, J., and Cambier, J. C., 1991, Ligation of membrane Ig leads to calcium-mediated phosphorylation of the protooncogene product, Ets-1, *J. Immunol.* **146:**1743–1749.

Galione, A., Lee, H. C., and Busa, W. B., 1991, Ca2^{+}-induced Ca^{2+} release in sea urchin egg homogenates: Modulation by cyclic ADP-ribose, *Science* **253:**1143–1146.

Giri, J. G., Ahdieh, M., Eisenman, J., Shanebeck, K., Grabstein, K., Kumacki, S., Namen, A., Park, L. S., Cosman, D., and Anderson, D., 1994, Utilisation of the β and γ chains of the IL-2 receptor by the novel cytokine IL-15, *EMBO J.* **13:**2822–2830.

Gold, M. R., Law, D. A., and De Franco, A. L., 1990, Stimulation of protein tyrosine phosphorylation by the B lymphocyte antigen receptor, *Nature* **345:**810–813.

Goodwin, R. G., Friend, D., Ziegler, S. F., Jerzy, R., Falk, B. A., Gimpel, S., Cosman, D., Dower, S. K., March, C. J., Namen, A. E., and Park, L. S., 1990, Cloning of the human and murine IL-7 receptors: Demonstration of a soluble form and homology to a new receptor superfamily, *Cell* **60:**941–951.

Harada, N., Santos-Argumedo, L., Chang, R., Grimaldi, J. C., Lund, F. E., Heath, A. W., Parkhouse, R. M. E., and Howard, M., 1993, Expression cloning of a cDNA encoding a novel murine B cell activation marker: homology to human CD38, *J. Immunol.* **151:**3111–3118.

Harnett, M. M., 1994, Antigen receptor signalling: From the membrane to the nucleus, *Immunol. Today* **15:**1–3.

Harnett, M. M., and Rigley, K. P., 1992, The role of G-proteins versus protein tyrosine kinases in the regulation of lymphocyte activation, *Immunol. Today* **13:**482–486.

Harnett, M. M., Holman, M. J., and Klaus, G. G. B., 1991, Il-4 promotes anti-Ig-mediated PKC translocation and reverses phorbol ester-mediated PKC downregulation in murine B cells, *J. Immunol.* **147:**3831–3835.

Henney, C. S., 1989, Interleukin 7: Effects on early events in lymphopoiesis, *Immunol Today* **10:**170–173.

Hermanson, G. G., Eisenberg, D., Kincade, P. W., and Wall, R., 1988, B29: A member of the immunoglobulin superfamily exclusively expressed on B-lineage cells, *Proc. Natl. Acad. Sci. U.S.A.* **85:**6890–6894.

Hibi, M., Murakami, M., Saito, M., Hirano, T., Taga, T., and Kishimoto, T., 1990, Molecular cloning and expression of an IL-6 signal transducer, gp130, *Cell* **63:**1149–1157.

Hombach, J., Tsubata, T., Leclercq, L., Stappert, H., and Reth, M., 1990, Molecular components of the B cell antigen receptor complex of the IgM class, *Nature* **343:**760–762.

Howard, M., Grimaldi, J. C., Bazan, J. F., Lund, F. E., Santos-Argumedo, L., Parkhouse, Parkhouse, R. M. E., Walseth, T. F., and Lee, H. C., 1993, Formation and hydrolysis of cyclic ADP-ribose is catalysed by lymphocyte antigen CD38, *Science* **262:**1056–1059.

Hunziker, W., Koch, T., Whitney, S. A., and Mellman, I., 1990, Fc receptor phosphorylation during receptor mediated control of B cell activation, *Nature* **345:**628–632.

Hutchcroft, J. E., Harrison, M. L., and Geahlen, R. L., 1991, B lymphocyte activation is accompanied by phosphorylation of a 72 kDa protein tyrosine kinase, *J. Biol. Chem.* **266:**14846–14849.

Hutchcroft, J. E., Harrison, M. L., and Geahlen, R. L., 1992, Association of the 72-kDa protein tyrosine kinase PTK72 with the B cell antigen receptor, *J. Biol. Chem.* **267:**8613–8619.

Jayadev, S., Linardicx, C. M., and Hannun, Y. A., 1994, Identification of arachidonic acid as a mediator of sphingomyelin hydrolysis in response to TNF-α, *J. Biol. Chem.* **269:**5757–5763.

Johnston, J. A., Kawamura, M., Kirken, R. A., Chen, Y.-Q., Blake, T. B., Shibuya, K., Ortaldo, J. R., McVicar, D. W., and O'Shea, J. J., 1994, Phosphorylation and activation of the Jak-3 Janus kinase in response to interleukin-2, *Nature* **370:**153–157.

Justement, L. B., Chan, Z., Harris, L., Ransom, J., Sandoval, V., Smith, C., Rennick, D., Roehm, N., and Cambier, J. C., 1986, BSFl induces membrane protein phosphorylation but not phosphoinositide metabolism, calcium mobilisation, PKC translocation or membrane depolarisation in resting B lymphocytes, *J. Immunol.* **137:**456–464.

Justement, L. B., Brown, V. K., and Lin, J., 1994, Regulation of B cell activation by CD45: A question of mechanism, *Immunol. Today* **15:**399–406.

Keegan, A. D., Nelms, K., Wang, L.-M., Pierce, J. H., and Paul, W. E., 1994, Interleukin 4 receptor: Signalling mechanisms, *Immunol. Today* **15:**423–432.

Kim, K. M., Alber, G., Weiser, P., and Reth, M., 1993, Differential signalling through the Ig-α and Ig-β components of the B cell antigen receptor, *Eur. J. Immunol.* **23:**911–916.

Kirkham, P. A., Santos-Argumedo, L., Harnett, M. M., and Parkhouse, R. M. E., 1994, Murine B cell activation via CD38 and protein tyrosine phosphorylation, *Immunology* **83:**513–516.

Kishihara, K., Penninger, J., Wallace, V. A., Kundiz, T. M., Kazuhiro, K., Wakeham, A., Timms, E., Pfeffer, K., Ohashi, P. S., Thomas, M. L., Furlonger, C., Paige, C. J., and Mak, T. W., 1993, Normal B lymphocyte development but impaired T cell maturation in CD45-exon6 protein tyrosine phosphatase-deficient mice, *Cell* **74:**143–146.

Kishimoto, T., Taga, T., and Akira, S., 1994, Cytokine signal transduction, *Cell* **76:**253–262.

Klaus, G. G. B., and Harnett, M. M., 1990, Crosstalk betwen B cell sIg and IL-4 receptor: The role of PKC and calcium mediated systems, *Eur J. Immunol.* **20:**2301–2307.

Klein, P., Kanehisa, M., and Delisi, C., 1985, The detection and classification of membrane-spanning proteins, *Biochim. Biophys. Acta* **815:**468–476.

Knox, K. A., and Gordon, J., 1993, Protzin tyrosine phosphorylation is mandatory for CD40-mediated rescue of germinal centre B cells from apoptosis, *Eur. J. Immunol.* **23:**2578–2584.

Kolanus, W., Romeo, C., and Seed, B., 1993, T cell activation by clustered tyrosine kinases, *Cell* **74:**171–183.

Kolesnick, R., and Golde, D. W., 1994, The sphingomyelin pathway in tumor necrosis factor and Il-1 signalling, *Cell* **77:**325–328.

Kundig, T. M., Schorle, H., Bachmann, M. F., Hengartner, H., Zinkernagel, R. M., and Horak, I., 1993, Immune responses in interleukin-2-deficient mice, *Science* **262:**1059–61.

Kyte, J., and Doolittle, R. F., 1982, A simple method for displaying the hydropathic character of a protein, *J. Mol. Biol.* **57:**105–132.

Lansford, R. D., McFadden, H. J., Siu, S. T., Cox, J. S., Cann, G. M., and Koshland, M. E., 1992, A promoter element that exerts positive and negative control of the interleukin-2 responsive J-chain gene, *Proc. Natl. Acad. Sci. U.S.A.* **89:**5966–5970.

Law, C-L., Sidorenko, S. P., and Clark, E. A., 1994, Regulation of lymphocyte activation by the cell surface molecule CD22, *Immunol. Today* **15:**442–450.

Leprince, C., Draves, K. E., Geahlen, R. L., Ledbetter, J. A., and Clark, E. A., 1993, CD22 associates with the human sIgM/B cell antigen receptor complex, *Proc. Natl. Acad. Sci. U.S.A.* **90:**3236–3240.

Malavasi, F., Funaro, A., Roggero, S., Horenstein, A., Calosso, L., and Mehta, K., 1994, Human CD38: A glycoprotein in search of a function, *Immunol. Today* **15:**95–98.

McGarvie, G. M., and Cushley, W., 1989a, Appearance of a novel plasma membrane phosphoprotein following culture of resting B lymphocytes with recombinant interleukin 4, *Immunol. Lett.* **22:**221–226.

McGarvie, G. M., and Cushley, W., 1989b, The effect of recombinant IL-4 upon protein kinase activation associated with murine and human B lymphocyte plasma membranes, *Cellular Signalling* **1:**447–460.

McMahan, C. J., Slack, J. L., Mosley, B., Cosman, D., Lupton, S. D., Brunton, L. L., Grubin, C. E., Wignall, J. M., Jenkins, N. A., Brannan, C. I., Copeland, N. G., Huebner, K., Croce, C. M., Cannizzarro, L. A., Benjamin, D., Dower, S. K., Spriggs, S. K., and Sims, J. E., 1991, A novel IL-1 receptor, cloned from B cells by mammalian expression, is expressed in many cell types, *EMBO J.* **10:**2821–2832.

Melamed, I., Wang, G., and Roifman, C. M., 1992, Antigen receptor-mediated protein tyrosine kinase activity is regulated by a pertusssis toxin-sensitive G-protein, *J. Immunol.* **149:**169–174.

Mischak, H., Kolch, W., Goodnight, J., Davidson, W. F., Rapp, U., Rose-Joh, S., and Mushinski, J. F., 1991, Expression of protein kinase C genes in haemopoietic cells is cell type and B cell differentiation stage specific, *J. Immunol.* **174:**3981–3987.

Miyazaki, T., Kawahara, A., Fujii, H., Nakagawa, Y., Minami, Y., Liu, Z-J., Oishi, I., Silvennoinen, O., Witthuhn, B. A., Ihle, J. N., and Taniguchi, T., 1994, Functional activation of Jak1 and Jak3 by selective association with Il-2 receptor subunits, *Science* **266:**1045–1047.

Mizel, S. B., 1990, Cyclic AMP and interleukin 1 signal transduction, *Immunol. Today* **11:**390–391.

Mizuguchi, J., Beaven, M. A., Ohara, J., and Paul, W. E., 1986, PMA inhibits anti-IgM mediated signalling in resting B cells, *J. Immunol.* **137:**2215–2220.

Mosley, B., Beckmann, M. P., March, C. J., Idzerda, R. L., Gimpel, S. D., VandenBos, T., Friend, D., Alpert, A., Anderson, D., Jackson, J., Wignall, J. M., Smith, C., Gallis, B., Sims, J. E., Urdal, D., Widmer, M. B., Cosman, D., and Park, L. S., 1989, The murine IL-4 receptor: Molecular cloning and characterisation of secreted and membrane-bound forms, *Cell* **59:**335–348.

Murakami, M., Narazaki, M., Hibi, M., Yawata, H., Yasukawa, K., Hamaguchi, M., Taga, T., and Kishimoto, T., 1991, Critical cytoplasmic region of the IL-6 signal transducer, gp130 is conserved in the cytokine receptor family, *Proc. Natl. Acad. Sci. U.S.A.* **88:**11349–11353.

Nakajima, T., Kinoshita, S., Sasagawa, T., Sasaki, K., Naruto, M., Kishimoto, T., and Akira, S., 1993, Phosphorylation at Thr-235 by a Ras-dependent MAPK cascade is essential for transcription factor NF-IL6, *Proc. Natl. Acad. Sci. U.S.A.* **90:**2207–2211.

Nakanishi, H., Brewer, K. A., and Exton, J. H., 1993, Activation of the ζ isozyme of protein kinase C by phosphatidylinositol 3,4,5-trisphosphate, *J. Biol. Chem.* **268:**13–16.

Noelle, R. J., and Snow, E. C., 1990, Cognate interactions between helper T cells and B cells, *Immunol. Today* **11:**361–368.

Noguchi, M., Yi, H., Rosenblaytt, H. M., Filipovitch, A. H., Adelstein, S., Modi, W. S., McBride, O. W., and

Leonard, W. J., 1993, Interleukin-2 receptor g chain mutation results in X-linked severe combined immunodeficiency in humans, *Cell* **73:**147–157.

O'Garra, A., Rigley, K. P., Holman, M., McLaughlin, J. B., and Klaus, G. G. B., 1987, B cell stimulatory factor-1 reverses Fc receptor-mediated inhibition of B lymphocyte activation, *Proc. Natl. Acad. Sci. U.S.A.* **84:**6254–6258.

O'Garra, A., Umland, S., DeFrance, T., and Christiansen, J., 1988, "B cell factors" are pleiotropic, *Immunol. Today* **9:**45–48.

O'Neill, L. A. J., Bird, T. A., and Saklatvala, J., 1990, Interleukin 1 signal transduction, *Immunol. Today* **11:**392–394.

Oren, R., Takahashi, S., Doss, C., Levy, R., and Levy, S., 1990, TAPA-1, the target of an antiproliferative antibody, defines a new family of transmembrane proteins, *Mol. Cell. Biol.* **10:**4007–4015.

Parkhouse, R. M. E., 1990, Three B cell surface molecules associating with membrane immunoglobulin, *Immunology* **69:**289–302.

Pleiman, C. M., Clark, M. R., Winitz, S., Coggeshall, K., Johnson, G. L., and Cambier, J. C., 1993, Mapping of sites on src-family protein tyrosine kinases $p55^{blk}$, $p59^{fyn}$ and $p56^{lyn}$ which interact with the effector molecules PLC-g2, MAP kinase, GAP, PI-3-kinase, *Mol. Cell. Biol.* **13:**5877–5887.

Pleiman, C. M., D'Ambrosio, D., and Cambier, J. C., 1994, The B cell antigen receptor complex: Structure and signal transduction, *Immunol. Today* **15:**393–399.

Rall, T., and Harris B. A., 1987, Identification of the lesion in the stimulatory GTP-binding protein of the uncoupled S49 lymphoma, *FEBS Lett.* **224:**365–371.

Reth, M., 1992, Antigen receptors on B lymphocytes, *Annu. Rev. Immunol.* **10:**97–121.

Rigley, K. P., and Harnett, M. M., 1990, Receptor signalling in B lymphocytes, in *Cytokines and B lymphocytes* (R. E. Callard, ed.), pp. 39–63, Academic Press, Orlando.

Rijkers, G. T., Griffioen, A. W., Zegers, B. J. M., and Cambier, J. C., 1990, Ligation of mIg leads to inactivation of the signal transducing ability of mIg, CD19, CD21 and B cell gp95, *Proc. Natl. Acad. Sci. U.S.A.* **87:** 9766–9770.

Roifman, C. M., Wang, G., Freedman, M., and Pan, Z., 1992, IL-7 receptor mediates tyrosine phosphorylation but does not activate the phosphatidylinositol-phospholipase C-γ1 pathway, *J. Immunol.* **148:**1136–1142.

Russell, S. M., Johnston, J. A., Noguchi, M., Kawamura, M., Bacon, C. M., Friedman, M., Berg, M., McVicar, D., Witthuhn, B. A., Silvennoinen, O., Goldman, A. S., Schmalsteig, F. C., Ihle, J. N., O'Shea, J. J., and Leonard, W. J., 1994, Interaction of IL-2Rb and gc chains with Jak1 and Jak3: Implications for XSCID and XCID, *Science* **266:**1042–1045.

Sakaguchi, N., Kashiwamura, S., Kimoto, M., Thalmann, P., and Melchers, F., 1988, B lymphocyte lineage-restricted expression of *mb*-1, a gene with CD3-like structural properties, *EMBO J* **7:**3457–3464.

Santos-Argumedo, L., Teixeira, C., Preece, G., Kirkham, P., and Parkhouse, R. M. E., 1993, A molecule on the surface of murine B lymphocytes mediating activation and protection from apoptosis via calcium channels, *J. Immunol.* **151:**175–187.

Satoh, T., Nakafuku, M., Miyajima, A., and Kaziro, Y., 1991, Involvement of rasp21 protein in signal transduction pathways from IL-2, IL-3 and GM-CSF, but not from IL-4, *Proc. Natl. Acad. Sci. U.S.A.* **88:**3314–3318.

Scholl, P. R., and Geha, R. S., 1994, MHC class II signalling in B cell activation, *Immunol. Today* **15:**418–423.

Schulte, R. J., Campbell, M. A., Fischer, W. H., and Sefton, B. M., 1992, Tyrosine phosphorylation of CD22 during B cell activation, *Science* **258:**1001–1004.

Seyfert, V. L., McMahon, S. B., Glenn, W. D., Yellen, A. J., Sukhatme, V. P., Cao, X., and Monroe, J. G., 1990, Methylation of an immediate-early inducible gene as a mechanism for B cell tolerance induction, *Science* **250:**197–250.

Shibuya, H., Yoneyama, M., Ninomiya-Tsuji, J., Matsumoto, K., and Taniguchi, T., 1992, IL-2 and EGF receptors stimulate the haematopoietic cycle via signalling pathways: Demonstration of a novel role for *c-myc*, *Cell* **70:**57–67.

Smith, C. A., Farrah, T., and Goodwin, R. G., 1994, The TNF receptor superfamily of cellular and viral proteins: Activation, costimulation and death, *Cell* **76:**959–962.

Stahl, and Yancopoulos, 1993, The alphas, betas and kinases of cytokine receptor complexes, *Cell* **74:**587–589.

Stamenkovic, I., and Seed, B., 1990, The B cell antigen CD22 mediates monocyte and erythrocyte adhesion, *Nature* **345:**74–77.

Sullivan, K. A., Tyler Miller, R., Masters, S. B., Beiderman, B., Heidaman, W., and Bourne, H. R., 1987, Identification of receptor contact site involved in receptor-G protein coupling, *Nature* **330:**758–760.

Sun, X. J., Wang, L.-M., Zhang, Y., Yenush, L., Meyers, M. G., Glasheen, E., Lane, W. S., Pierce, J. H., and Whitz, M. F., 1995, Role of IRS-2 in insulin and cytokine signalling, *Nature* **377**:173–177.

Taga, T., Hibi, M., Hirata, Y., Yamasaki, K., Yasukawa, K., Matsuda, T., Hirano, T., and Kishimoto, T., 1989, Interleukin-6 triggers the association of its receptor with a possible signal transducer, gp130, *Cell* **58**:573–581.

Takata, M., Sabe, H., Hata, A., Inazu, T., Homma, Y., Nukada, T., Yamamura, H., and Kurosaki, T., 1994, Tyrosine kinases lyn and syk regulate B cell receptor-coupled Ca^{2+} mobilisation through distinct pathways, *EMBO J.* **13**:1341–1349.

Tedder, T. F., Clement, L. T., and Coopel, M. D., 1983, Expression of C3d receptors during human B cell differentiation: Immunofluorescence analysis with the HB-5 monoclonal antibody, *J. Immunol.* **133:** 678–683.

Tedder, T. F., Zhou, L.-J., and Engel, P., 1994, The CD19/CD21 signal transduction complex of B lymphocytes, *Immunol. Today* **15**:437–442.

Tedder, T. F., and Engel, P., 1994, CD20: a regulator of cell-cycle progression of B lymphocytes, *Immunol. Today* **15**:442–450.

Torigoe, T., Saragovi, H. U., and Reed, J. C., 1992, Interleukin-2 regulates the activity of the lyn protein tyrosine kinase in a B cell line, *Proc. Natl. Acad. Sci. U.S.A.* **89**:2674–2678.

Tigges, M. A., Casey, L., and Koshland, M. E., 1989, Mechanism of IL-2 signalling: mediation of different outcomes by a single receptor and transduction pathway, *Science* **243**:781–786.

Torres, R. M., Law, D. L., Santos-Argumedo, L., Kirkham, P. A., Grabstein, K., Parkhouse, R. M. E., and Clark, E. A., 1992, Identification and characterisation of the murine homologue of CD22, a lymphocyte restricted adhesion molecule, *J. Immunol.* **149**:2641–2649.

Turner, B., Rapp, U., App, H., Greene, M., Dobashi, K., and Reed, J., 1991, Interleukin-2 induces tyrosine phosphorylation and activation of p72-74 raf-1 kinase in a T cell line, *Proc. Natl. Acad. Sci. U.S.A.* **88**:1227–1231.

Uckun, F. M., Tuel-Ahlgren, L., Obuz, V., Smith, R., Dibirdik, I., Honson, M., Langlie, M.-C., and Ledbetter, J. A., 1991a, Interleukin-7 receptor engagement stimulates tyrosine phosphorylation, inositol phospholpid turnover, proliferation and selective differentiation to the CD4 lineage by human fetal thymocytes, *Proc. Natl. Acad. Sci. U.S.A.* **88**:6323–6327.

Uckun, F. M., Dibirdik, I., Smith, R., Tuel-Ahlgren, L., Langlie, C.-M., Schieven, G. L., Waddick, K. G., Hanson, M., and Ledbetter, J. A., 1991b, IL-7 ligation stimulates tyrosine phosphorylation, inositol phospholipid turnover and clonal proliferation of human B cell precursors, *Proc. Natl. Acad. Sci. U.S.A.* **88**:3589–3593

Venkitaraman, A. R., Williams, G. T., Dariavach, P., and Neuberger, M. S., 1991, The B cell antigen receptor of the five immunoglobulin classes, *Nature* **352**:777–781.

Weiss, A., 1993, T cell antigen receptor signal transduction - A tale of tails and cytoplasmic protein tyrosine kinases, *Cell* **73**:209–212.

Williams, G. T., Venkitaraman, A., Gilmore, D., and Neuberger, M., 1990, The sequence of the μ transmembrane segment determines the tissue specificity of the transport of immunoglobulin M to the cell surface, *J. Exp. Med.* **171**:947–952.

Witthuhn, B. A., Silvennoinen, O., Miura, O., Lai, K. S., Cwik, C., Liu, E. T., and Ihle, J. N., 1994, Involvement of the Jak-3 Janus kinase in signalling by interleukins 2 and 4 in lymphoid and myeloid cells, *Nature* **370:** 153–157.

Zhou, L.-J., Ord, D. C., Hughes, A. L., and Tedder, T. F., 1991, Structure and domain organisation of the CD19 antigen of human, mouse and guinea pig B lymphocytes, *J. Immunol.* **147**:1424–1432.

Zurawski, G., and de Vries, J. E., 1994, Interleukin 13, an interleukin 4-like cytokine that acts on monocytes and B cells but not on T cells, *Immunol. Today* **15**:19–26.

Chapter 5

Cellular and Molecular Aspects of Myeloid Cell Proliferation and Development

Caroline A. Evans and Andrew Pierce

1. INTRODUCTION

Stem cells give rise to all the different types of blood cell. These pluripotent cells undergo self-renewal and differentiation to produce more developmentally restricted cells, which proliferate and develop to form mature cells. Hematopoietic stem cells are defined by their ability to establish long-term reconstitution of hematopoiesis when transplanted to irradiated recipient animals (Till and McCulloch, 1961). Committed progenitor cells have a restricted developmental capacity; e.g., granulocyte–macrophage colony-forming cells (GM-CFC) or erythroid burst-forming units (BFU-e) can form only neutrophil, macrophage, or erythroid cells, respectively. Hematopoiesis occurs in the microenvironment of adult bone marrow where proliferation and differentiation of hematopoietic stem and progenitor cells occurs in close contact with the bone marrow stromal cells and the associated extracellular matrix (ECM) (Adams and Watt, 1993; Yoder and Williams, 1995). These interactions have been termed the local area network (LAN).

The interactions among hematopoietic cells, bone marrow stromal cells, and the surrounding ECM are complex. A scheme of hematopoiesis is shown in Figure 1. Hematopoiesis is regulated in part by hematopoietic growth factors, which are presented to progenitor cells and elicit their biological effects by interaction with specific receptors expressed on the surface of target cells, resulting in activation of intracellular signaling events (Ihle *et al.*, 1994; Pearson *et al.*, 1995) and ultimately regulation of gene transcription promoting survival, proliferation, and development. Stromal cells, including resident bone

Caroline A. Evans and Andrew Pierce Leukaemia Research Fund Cellular Development Unit, Department of Biochemistry and Applied Molecular Biology, UMIST, Manchester M60 1QD, United Kingdom.

Blood Cell Biochemistry, Volume 7: Hematopoietic Cell Growth Factors and Their Receptors, edited by A. D. Whetton and J. Gordon. Plenum Press, New York, 1996.

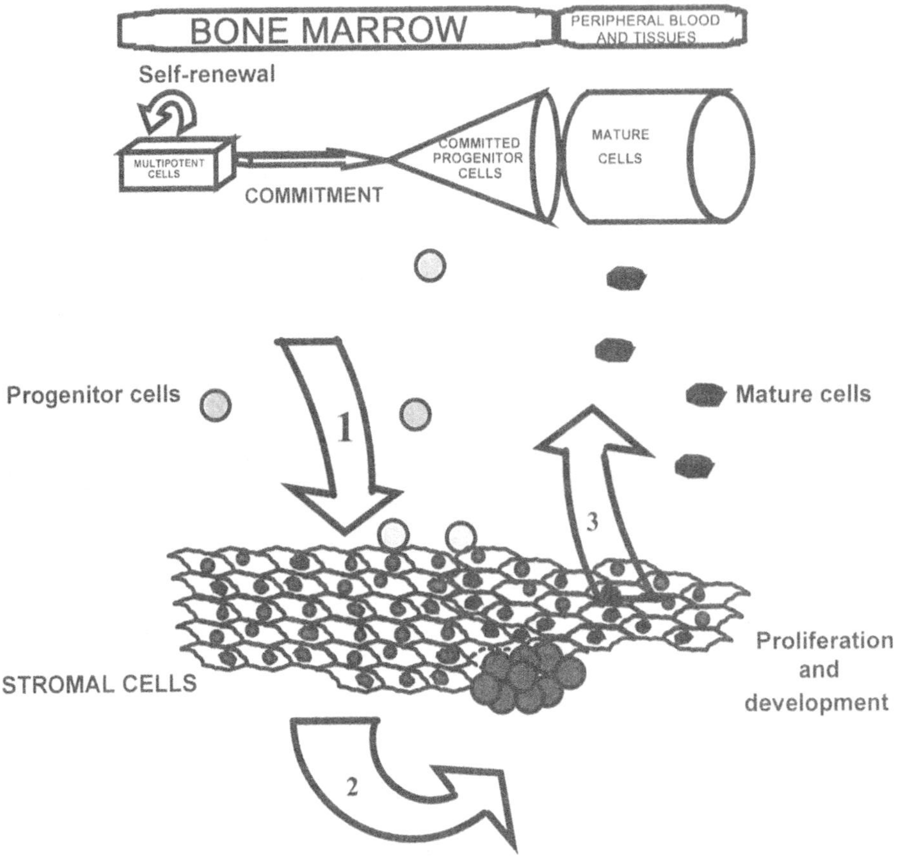

FIGURE 1. Scheme of hematopoiesis. Hematopoietic stem cells are pluripotential progenitor cells and give rise to all of the different types of blood cells. These cells undergo self-renewal and differentiation to produce lineage-restricted (committed) progenitor cells, which proliferate and develop to form mature cells. Pluripotential primitive cells can be purified by a range of techniques. Hematopoiesis occurs in the microenvironment of the adult bone marrow where the progenitor cells interact with stromal cells and ECM (1). Stromal cells synthesize a range of HGFs that regulate hematopoietic cell survival, proliferation, and differentiation (see Figure 2). These HGFs exist as soluble and membrane-bound forms, which are presented in association with stromal cells or components of the ECM (2). Progenitor cells form "cobblestone" areas in long-term bone marrow cultures or long-term culture-initiating cell assays. Developmentally mature hematopoietic cells are released from such sites of hematopoietic activity *in vitro* or into the peripheral blood and tissues *in vivo* (3).

marrow macrophages, produce a number of hematopoietic growth factors (HGF), including macrophage colony-stimulating factor (M-CSF), stem cell factor (SCF), granulocyte–macrophage colony-stimulating factor (GM-CSF), granulocyte colony-stimulating factor (G-CSF), interleukin 1 (IL-1), interleukin 6 (IL-6), interleukin 7 (IL-7), and transforming growth factor β (TGF-β) (Gualtieri *et al.*, 1984). These HGF exist in soluble and membrane-bound forms, which are presented to hematopoietic cells as integral proteins in the plasma membrane of specific stromal cells or associated with components of the ECM (Gordon *et al.*, 1987, 1991; Siczkowski *et al.*, 1992). Binding of cytokines to the ECM may limit proteolysis and promote specific HGF–hematopoietic cell interaction (Roberts *et al.*, 1988). The flt3/flk2 ligand, SCF, and M-CSF are examples of cytokines that can be membrane bound; SCF is an example of an HGF that is also produced in a soluble form. Membrane-bound forms of SCF and M-CSF have both been shown to promote the adhesion of cells expressing their cognate receptors, *c-kit* and *c-fms*, respectively (Flanagan *et al.*, 1991; Rettenmier and Roussel, 1988). Thus, HGFR can act to promote adhesion in addition to growth. Hematopoietic progenitor cells can bind directly to the heparin sulfate molecules on the surface of marrow stromal cells (Siczkowski *et al.*, 1992), which thus act both to anchor and to present HGF to hematopoietic cells.

Integrins also play a role in primitive progenitor cell adhesion to stromal cell layers. The counter-receptors VLA-4 and VCAM, VLA-5 and fibronectin, β_2-integrin and ICAM-1 potentiate progenitor cell interactions with stroma (see review by Yoder and Williams, 1995). Human hematopoietic progenitor cells express a number of cellular adhesion molecules (for example, $\alpha5\beta1$, $\beta2$, LFA3). Their function in stromal cell regulation of hematopoiesis is currently unclear, but interaction of these molecules with their counter-receptors stimulates intracellular signaling events in other cell types (Hynes, 1992). Integrins thus have the potential to act in concert with cytokines to regulate hematopoiesis.

Both positive and negative regulators of primitive hematopoietic cell proliferation are produced by stromal cells; for example, MIP-1α and TGF-β act to inhibit primitive progenitor cell cycling (Eaves *et al.*, 1991; Otsuka *et al.*, 1991). The same molecules can also act to both stimulate and inhibit growth. Tumor necrosis factor α (TNF-α) and TGF-β can inhibit proliferation in some instances and also act on highly purified human progenitor cells to promote growth when present in combination with other cytokines (Caux *et al.*, 1990; Ruscetti *et al.*, 1992). The influence of the stromal cell environment on stem cell proliferation is dependent on the balance of signals from growth inhibitors and growth promoters and the particular progenitor cell type in question. The interaction of hematopoietic progenitor cells with stromal cells of the bone marrow and the associated ECM are summarized in Figure 2.

2. ANALYSIS OF HEMATOPOIETIC GROWTH FACTOR FUNCTION

Investigation of the mechanisms by which HGFs exert their biological effects on primitive hematopoietic cells requires the study of purified populations of primitive cells. Initial purification procedures for hematopoietic progenitor cells were based on the physical properties of progenitor cells (reviewed by Spangrude, 1989). The characterization of specific cell-surface markers has now facilitated a biochemical approach. The most commonly used technique is selection for expression of specific cell-surface antigens or marker proteins using monoclonal antibodies. Cells labeled with antibody cells can be separated by a number of methods, including solid-phase immunologic methods, immunomagnetic beads, or flow cytometry. One of the most commonly used antigens is CD34. This is expressed on early progenitor cells, which constitute 0.5–5% of human bone marrow cells (Civin *et al.*, 1990). CD34-positive cells are a heterogeneous cell population composed of both early and lineage-committed progenitor cells. Additional marker proteins are used to identify the more primitive population of cells found within the CD34 compartment. These

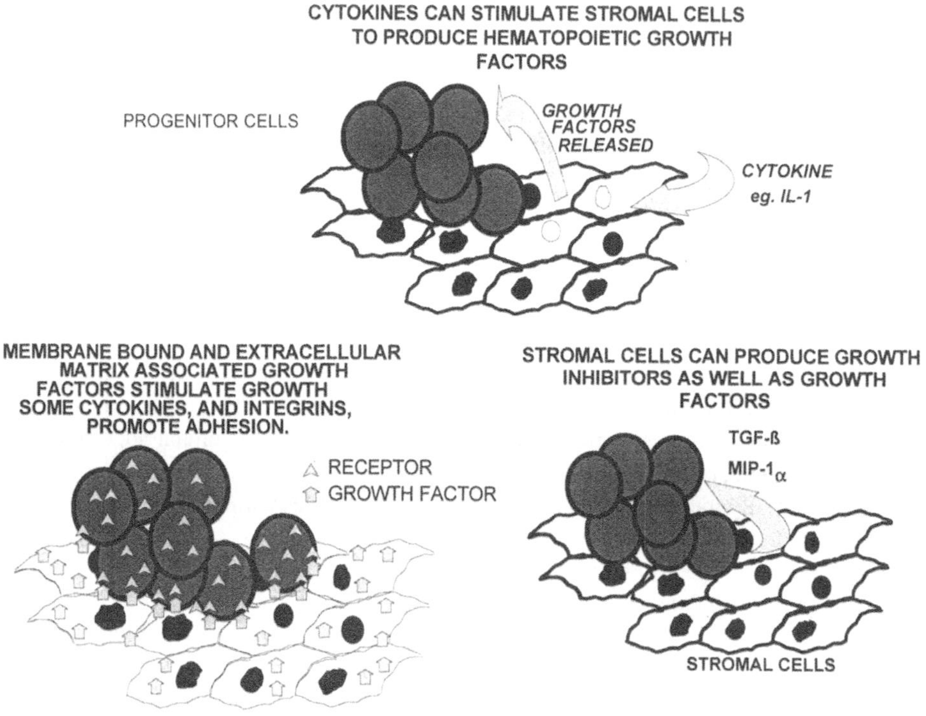

FIGURE 2. Interaction of hematopoietic progenitor cells with stromal cells of the bone marrow and the associated ECM.

include CD38 (Visser and Van Dekkum, 1990; Terstappen *et al.*, 1991; Baum *et al.*, 1992), Thy-1 (Spangrude *et al.*, 1988a), and CD33 (Brandt *et al.*, 1988), which have identified $CD34^+$ early progenitor cells with various phenotypes including $CD34^+/CD38^{low}$, $CD34^+/lin/Thy^+$, and $CD34^+/CD33^-$. There are several markers of murine hematopoietic progenitor cells, which include Thy-1, Sca-1 (Spangrude *et al.*, 1988b), A4.1 (Jordan *et al.*, 1990), and murine CD34 (Krause *et al.*, 1994). Cell populations positive for these markers are heterogeneous and can be subdivided by making use of biological markers such as the mitochondrial and DNA fluorochromes rhodamine 123 (Bertoncello *et al.*, 1991; Visser and de Vries, 1988) and Hoechst 33342 (Baines and Visser, 1983), respectively.

The *in vivo* roles of HGFs in hematopoiesis have also been studied using mice deficient in specific HGFs (reviewed by Metcalf, 1995). In the case of M-CSF and SCF, naturally mutant mice exist in which these HGFs are functionally inactive. An M-CSF-deficient mouse (*op/op*) results from the osteopetrotic mutation, which leads to expression of a truncated, inactive form of M-CSF (Yoshida *et al.*, 1990; Wiktor-Jedrzejczak *et al.*, 1990). The W and Steel strains of mutant mice lack the SCF receptor (encoded by *c-kit*) and fail to express SCF, respectively. The phenotypes of these mice suggest that SCF plays an important role in hematopoietic stem cell, primordial germ cell, and melanocyte survival and proliferation, whereas M-CSF regulates bone maintenance and some macrophage functions. Mice with inactive forms of specific HGF genes, so-called gene knockout mice, have been generated artificially using embryonal stem cells plus an appropriate targeting vector that results in homozygous deletion of part of the HGF gene. The G-CSF knockout mice develop normally but are neutropenic, with a 70–80% reduction in granulocytes, and display a 50% reduction in granulocyte–macrophage colony progenitor cells, maturing granulocytes, and macrophages in the bone marrow. In addition, the mice have reduced ability to mobilize granulocytes (Lieschke *et al.*, 1994a). These data indicate that G-CSF is a major regulator of granulocyte production and function and correlate well with the observed biological effects of G-CSF.

Gene knockout experiments have also demonstrated that HGF can exhibit functional redundancy. For example, the peripheral blood cell counts of the GM-CSF cytokine knockout mice are normal compared to controls and display only slight changes in hematopoietic cell progenitor cell numbers in the bone marrow and spleen (Stanley *et al.*, 1994). The mice, however, had bronchovascular invasion by B lymphocytes and inflamed areas in the lung. These results could also be interpreted as suggesting that GM-CSF plays no role in hematopoiesis but would be at odds with the known biological effects of GM-CSF. Leukemia inhibitory factor (LIF)-deficient mice are also able to survive and maintain normal hematopoiesis, even though LIF is known to act on primitive hematopoietic progenitor cells. Double-knockout mice lacking M-CSF and GM-CSF indicate that, for these HGFs, the effects are combinatorial because the mice both are osteopetrotic and have lung disease (Lieschke *et al.*, 1994b). In the future, data will probably become available for other HGFs.

The knockout approach can also be extended to examine the effects of deleting receptor subunits or signal transduction proteins downstream from the receptors. There are some naturally occurring examples of this, such as the mutation in the IL-3Rα subunit in A/J mice, which prevents colony formation stimulated by IL-3 *in vitro* (Ichihara *et al.*, 1995). The A/J mice, however, display normal hematopoiesis, suggesting a higher level of functional redundancy (see Chapter 1). The cytoplasmic tyrosine phosphatase, SH-PTP1, was identified as playing a major role in HGF-mediated signal transduction (Yi *et al.*, 1993) because it is the product of the "motheaten" locus, and mutations of this gene give rise to a number of hematopoietic abnormalities (Shultz *et al.*, 1993).

2.1. Molecular Mode of Action of Hematopoietic Growth Factors

The HGFs interact with HGFR to mediate their biological effects. There is some evidence of crosstalk between receptors such that one HGF regulates the function of another. This can be at the level of receptor expression; for example, binding of cytokines such as IL-1α, IL-6, or G-CSF to cell surface receptors can influence the expression of receptors for other growth factors such as GM-CSF or IL-3 (S. E. Jacobsen *et al.*, 1992; Testa *et al.*, 1993). In addition, the combinatorial effects of many HGF are often synergistic, both for positive and negative regulators of hematopoiesis (Broxmeyer *et al.*, 1993; F. W. Jacobsen *et al.*, 1994). The mechanism by which synergy is achieved is unclear: two HGFs could potentially stimulate a unique, common, or overlapping sets of signal transduction pathways. With regard to the synergistic promotion of proliferation, both may stimulate proliferation, or, alternatively, one factor could promote survival and the other proliferation. It has been proposed that HGFs such as G-CSF, IL-6, IL-11, IL-12, LIF, and SCF trigger the cycling of primitive progenitor cells, promoting their exit from the resting (G_0) phase of the cell cycle. These cells are then "primed" to respond to a second set of HGFs including IL-3, GM-CSF, and M-CSF (Ogawa, 1993). The HGFs from each of these sets are required to be present continuously to promote proliferation and development of the multipotent cells into mature cells.

The HGFRs can be subdivided into two major subgroups, members of the tyrosine kinase receptor (Ullrich and Schlessinger, 1990) and cytokine receptor (Miyajima, 1992) superfamilies, based on the presence or absence of an intrinsic tyrosine kinase activity. These are reviewed in Chapter 1.

2.2. Tyrosine Kinase Receptors: M-CSF Receptors, SCF Receptors, and Novel flk1 and flk2/flt3 Protein Tyrosine Kinases

Tyrosine kinase receptors are composed of a cytoplasmic tyrosine kinase domain, a single membrane-spanning domain, and a large glycosylated extracellular ligand-binding domain (Hunter and Cooper, 1985). There are four subfamilies, designated class I, II, III, and IV on the basis of structural characteristics. The HGFR for SCF, M-CSF, and the novel flk1 and flk2/flt3 are members of this family and designated class III. They resemble the PDGFR and possess five immunoglobulin repeats, a single transmembrane domain, and a cytoplasmic domain split by a kinase insert domain.

The M-CSF, SCFR, and novel flk1 and flk2/flt3 protein tyrosine kinase receptors are members of the tyrosine kinase receptor superfamily, of which the PDGFR (α and β forms) is the best characterized. These HGFRs resemble the PDGFR in dimerizing on ligand binding and in their activation leading to receptor autophosphorylation by the intrinsic tyrosine kinase and recruitment of signaling proteins via phosphotyrosine-*src* homology region 2 (SH2) domain interactions.

The SCF (also known as Steel or Kit ligand) was identified as the ligand for the protein encoded by the *c-kit* proto-oncogene, which maps to the mouse Steel locus (Witte, 1990); SCF stimulates the survival of progenitor cells when, in combination with other HGFs, it promotes primitive progenitor cell proliferation and development (Migliaccio *et al.*, 1991; Tsuji *et al.*, 1991). The SCF appears to be synthesized in both a soluble and a membrane-bound form (Flanagan *et al.*, 1991), with the membrane-bound form playing a major role in hematopoiesis in the bone marrow (Brannan *et al.*, 1991).

Signal transduction in response to SCF–c-KIT interaction is mediated by c-KIT homodimers. Homodimerization results from the dimeric nature of the ligand intiating a receptor–receptor interaction via their extracellular domain (Lev *et al.*, 1992a,b). SCF stimulates autophosphorylation of c-KIT and binding of signaling proteins, including PI3-kinase and PI-PLCτ (Rottapel *et al.*, 1991; Welham and Schrader, 1992). The interkinase domain of the c-kit receptor binds the p85 subunit of PI3-kinase, and this association is dependent on phosphorylation of tyrosine residues in the kinase insert domain (Lev *et al.*, 1992c). The functional significance of PI-PLCτ is unclear because it is not associated with phosphatidylinositol hydrolysis to generate inositol phosphates (Whetton *et al.*, 1994). The SCF also activates the recently described JAK–STAT pathway, which couples tyrosine kinase activation to transcriptional regulation and is discussed in Chapter 1.

The M-CSF (also known as CSF-1) was characterized as a growth factor for the monocyte/macrophage lineage by its ability to stimulate the formation of monocytic colonies *in vitro* and acts synergistically with other cytokines on primitive hematopoietic cells (Stanley *et al.*, 1986; Mochizuki *et al.*, 1987). The *c-fms* proto-oncogene encodes the receptor for M-CSF. The M-CSFR forms a homodimer, is autophosphosphorylated on ligand binding, and stimulates a range of intracellular signaling events. The p85 subunit of PI3-kinase is activated by c-FMS (Varticovski *et al.*, 1989; Shurtleff *et al.*, 1990) and associates with autophosphorylation sites in the kinase insert domain (Reedijk *et al.*, 1992). The SHC, Grb2, and Sos1 proteins and a 150-kDa tyrosine-phosphorylated protein also bind M-CSFR and form a complex (Lioubin *et al.*, 1994). The M-CSFR also binds and activates members of the *src* family of nonreceptor tyrosine kinases. This interaction is mediated in part by one of the autophosphorylation sites, Tyr^{809} (human), independently of the kinase insert domain (Courtneidge *et al.*, 1993). A series of proliferation-signaling-defective M-CSFR have been generated, and details of what these revealed about the structure–function relationship of the M-CSFR have been reviewed by Roussel (1994).

The novel receptor tyrosine kinases, FLK1 and FLK2 were identified in primitive hematopoietic cells from fetal liver by screening for protein tyrosine kinase domain sequence homology (Matthews *et al.*, 1991a,b); *flk2* was also cloned independently from the placenta and described as *flt3* (Rosnet *et al.*, 1991). The ligand for the orphan *flk2/flt3* receptor has recently been identified, and this, together with the use of chimeric *fms/flk2(flt3)* receptor, has allowed its biological role to be investigated (Hannum *et al.*, 1994; Lyman *et al.*, 1994; Rossner *et al.*, 1994). The *flk2/flt3* receptor is expressed exclusively on a primitive cell subpopulation and acts synergistically with other HGFs to expand the number of primitive progenitor cells and to promote myeloid differentiation (Hannum *et al.*, 1994; Zeigler *et al.*, 1994a; S. E. Jacobsen *et al.*, 1995), the synergistic effects of the FLK2/FLT3 ligand being more pronounced for HGFs binding members of the cytokine receptor superfamily than for tyrosine kinase receptors (S. E. Jacobsen *et al.*, 1995). Transfection of the chimeric *fms/flk2(flt3)* receptor into an IL-3-dependent Ba/F3 cell line abrogated the requirement for IL-3 for survival and proliferation. These effects were associated with ligand-binding-stimulated receptor autophosphorylation and interaction with a number of signaling proteins: PI-PLCτ, $p21^{ras}$GAP, PI3-kinase, SHC, Grb2, vav, fyn, and src. Of these, PI-PLCτ, the p85 subunit of PI3-kinase, SHC, Grb2, src, and fyn were physically associated with the cytoplasmic domain of FLK2/FLT3 (Dosil *et al.*, 1993). The p85 PI3-kinase subunit binds a region outside the kinase insert in the C-terminal end of this receptor (Rottapel *et al.*, 1994). The human homologue of FLK2/FLT3, STK-1, is selectively expressed on $CD34^+$ bone marrow cells and plays a role in proliferation of stem and early progenitor cells (Small *et al.*, 1994).

2.3. Cytokine Superfamily Hematopoietic Growth Factor Receptors

Members of the cytokine receptor superfamily exhibit common structural features including a modified fibronectin type III repeat in the extracellular binding domain, which is comprised of a cysteine motif at the N terminal end and a C-terminal WSxWS box (where x is any amino acid). The characteristics of this superfamily have been described mainly for the erythropoietin receptor (EPOR) and growth hormone receptor. The HGFRs of this family also include the receptors for G-CSF, GM-CSF, IL-2, IL-3, IL-4, IL-5, IL-6, IL-7, and thrombopoietin (TPO).

In general, these receptors bind the cognate ligand with low affinity. A functional high-affinity receptor complex is formed by interaction with an additional receptor subunit(s). Several members of the family such GHR, prolactin, and EPOR are single-chain receptors that function as homodimers. The other cytokine receptors form heterodimers; for these receptors, the subunit with low binding affinity associates with a common subunit. A common subunit, βc, was first described for the IL-3R, GM-CSFR, and IL-5R complexes and other subunits, e.g., gp130 (IL-6R subunit) and γc (IL-2R and IL-7R subunits) have since been identified. These receptors are covered in detail in Chapters 1 and 13. The next section focuses on the receptors for EPO and thrombopoietin, cytokine receptors that promote erythroid and megakaryocyte progenitor cell proliferation, respectively.

2.3.1. The Erythropoietin Receptor

Erythropoietin (EPO) regulates the survival, proliferation, and differentiation of erythroid progenitor cells (Koury, 1992; Liboi *et al.*, 1993) and mediates its effects via interaction with the EPOR. The EPO receptor functions as a homodimer that is formed on ligand binding. Experiments examining the effect of site-directed mutagenesis of amino acid residues predicted to be critical for homodimer formation (Watowich *et al.*, 1994) have confirmed this. The cytoplasmic domain of the EPOR is critical for signal transduction and contains regions associated with positive and negative growth regulation (D'Andrea *et al.*, 1991; Maruyama *et al.*, 1994; Watowich *et al.*, 1994). It is composed of 236 amino acids, of which the membrane-proximal domain ($\leq$103 amino acids) is sufficient for signal transduction promoting EPOR association with Jak2 kinase (via Trp 282 of EPOR), tyrosine phosphorylation of *vav* protein, and expression of *pim-1* serine/threonine kinase (O. Miura *et al.*, 1994). This region is homologous to the β subunit of the IL-2R and gp130 subunit of the IL-6/LIF/oncostatin M receptors (D'Andrea *et al.*, 1989). The C-terminal region of the EPOR has been suggested to be critical for EPO-mediated activation of MAP kinase via the $p21^{ras}$ pathway (Y. Miura *et al.*, 1994). A serine-rich portion of the C-terminal region (approximately 40 amino acids) is subject to negative control and is proposed to interact with the SH-PTP1 tyrosine phosphatase, the product of the "motheaten" locus. The EPOR is also proposed to interact with another tyrosine phosphatase, syp, which is a positive regulator (in contrast to SH-PTP1) and is proposed to propagate the EPOR signal by acting as an adaptor protein between Jak2 kinase and the Grb2 protein, which is a component of the ras/raf-1 signaling pathway (Tauchi *et al.*, 1995).

A truncation mutant of the EPOR that lacks the last 70 amino acids of the C-terminal region is associated with familial erythrocytosis and probably results from the prevention of negative feedback control (de la Chapelle *et al.*, 1993a,b). Similarly, a mutation in the negative regulatory domain of EPOR, which converts Pro 488 to Ser 488, has been described for a patient with sporadic primary polycythemia whose erythroid progenitor cells exhibited increased sensitivity to EPO (Sokol *et al.*, 1994).

2.3.2. The Thrombopoietin Ligand and Its Receptor (c-Mpl)

Thrombopoietin (also known as megakaryocyte growth and development factor, MDGF, and *c-mpl* ligand) plays a major role in the regulation of megakaryotopoiesis, which leads to platelet formation (Kaushansky *et al.*, 1995). Evidence to support this comes from, for example, the observation that mice deficient for the TPO receptor exhibit an 85%

decrease in platelet and megakaryocyte number although the other hematopoietic cell types are present at normal levels (Gurney *et al.*, 1995). Thrombopoietin stimulates the proliferation and development of platelets and megakaryocytes and acts synergistically with EPO, SCF, and IL-11 to enhance megakaryocyte growth (Broudy *et al.*, 1995). The TPO receptor is highly expressed on $CD34^+$ (hematopoietic progenitor cells and cells of the megakaryocytic lineage (Zeigler *et al.*, 1994b; Debili *et al.*, 1995). Two subpopulations of TPO-responsive cells have been proposed, one that is IL-3 responsive but requires TPO for complete development and one that is dependent on TPO for proliferation and development (Kaushansky, 1994).

The TPO receptor is encoded by the *c-mpl* proto-oncogene and is a member of the cytokine receptor superfamily of hematopoietic growth factor receptors. This receptor shows homology to the receptors for EPO and IL-3 and possesses the common structural features of the cytokine receptor superfamily (discussed in Chapter 1), including four conserved cysteine residues and a WSxWS (where x is any amino acid) at the C terminus of the extracellular domain. The viral *mpl* gene, *v-mpl*, encodes a constitutively activated form of c-mpl that transforms murine pro-B and pre-B cells *in vitro* (Fichelson *et al.*, 1995). In common with other cytokine receptors, c-mpl does not possess an intrinsic tyrosine kinase activity, but ligand binding stimulates an increase in cellular phosphotyrosine levels, activates the Jak2 tyrosine kinase (Drachman *et al.*, 1995; Sattler *et al.*, 1995; Tortolani *et al.*, 1995) and TYK2 (Sattler *et al.*, 1995), and stimulates tyrosine phosphorylation of the adaptor protein SHC (Drachman *et al.*, 1995), STAT1, STAT3, and STAT5 (Sattler *et al.*, 1995). Tyrosine phosphorylation of Jak2 is rapid (1–5 min), but its subsequent association with c-mpl (20–60 min) occurs over a relatively longer time course. The kinetics of this Jak2–c-mpl association suggests that Jak2 may not be the initiating kinase in c-mpl signaling (Drachman *et al.*, 1995). Thombopoietin stimulation of Jak2 kinase and SHC phosphorylation leads to activation of Jak2 kinase and association of SHC with Grb2 in human platelets (Miyakawa *et al.*, 1995).

Thrombopoietin signaling has also been studied using chimeric growth hormone–c-mpl receptors composed of the extracellular and transmembrane domains of the growth hormone receptor and deletion mutants of the intracellular domain of c-mpl. This experimental approach has assigned distinct signaling functions to specific regions of the intracellular domain of c-mpl: the proximal region (approximately 60 amino acids) is sufficient for proliferation and mediates activation of the Jak–STAT pathway, but the C-terminal portion is required for tyrosine phosphorylation of the SHC adaptor protein and induction of *c-fos* mRNA (Gurney *et al.*, 1995). Experiments using mutant and wild-type *env–mpl* fusion protein (the transforming agent of myeloproliferative leukaemia retrovirus), which is composed of the cytoplasmic sequence of v-mpl and the extracellular portion of the viral envelope, *env*, protein (Courtois *et al.*, 1995), also indicate that the proximal region of c-mpl is required for the induction of factor-independent proliferation. The proximal region contains the box 1 and box 2 motifs, which are shared with other members of the cytokine receptor superfamily such as EPOR, GM-CSFR, and G-CSFR, suggesting that they share some common signaling mechanisms. By comparison with current models for cytokine receptor activation, c-mpl could potentially function either as a homodimer (like EPOR) or as a heterodimer in association with an additional receptor subunit such as βc (like GM-CSFR). The box1 and 2 motifs of c-MPL also resemble those of the IL-2Rβ and IL-7R (Benit *et al.*, 1994), which form heterodimers with the IL-2 receptor common τ subunit (τ_c)

on ligand binding (Noguchi *et al.*, 1993), raising the possibility that c-MPL may transduce its signal by associating with τ_c. These mechanisms of c-MPL activation clearly require further investigation to determine whether or not specific signals for megakaryocytic development are activated on receptor occupation.

3. SIGNAL TRANSDUCTION PATHWAYS AND LEUKEMIA

Leukemia results from a disregulation in the balance among survival, proliferation, and differentiation of hematopoietic cells that normally functions to maintain a constant supply of blood cells (see Figure 3). These processes are subject to regulation by HGFs, which stimulate a range of cellular signaling events that are subject to negative feedback control, for example, deactivation by SH-PTP1 tyrosine phosphorylation or receptor down-modulation. Perturbation of the signaling pathways resulting from constitutive activation of HGFR (*c-fms* and *c-kit* are proto-oncogenes whose viral counterparts are capable of cell transformation) or downstream signaling molecules potentially results in the delivery of signals leading to suppression of apoptosis, stimulation of proliferation at inappropriate times, and possibly leukaemia (see Chapter 12).

There are several examples of aberrant signaling pathways associated with leukemias and myelodysplastic syndromes (MDS). Mutations of N-*ras* (codons 12, 13, and 61) and *fms* mutation (codons 301 and 969) are proposed to be associated with some cases of AML and MDS. N-*ras* mutation in MDS patient samples is an indicator of poor prognosis and an

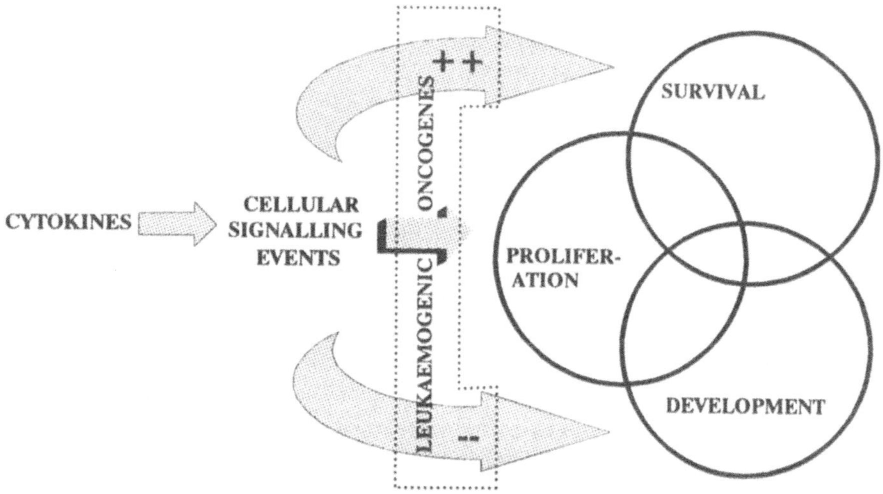

FIGURE 3. Oncogenes and leukemogenesis. The number of hematopoietic cells is determined by the balance between survival, proliferation, and development. Leukemia results from a disruption of these processes, which normally function to maintain a homeostatic supply of blood cells. Proteins encoded by oncogenes can interfere with normal hematopoiesis. Examples include the BCL-2, BCR-ABL, and PML-RARα proteins. BCL-2 and BCR-ABL suppress apoptosis and are associated with follicular lymphoma and chronic myeloid leukemia (CML), respectively, whereas the PML/RARα protein blocks development and is associated with acute promyelocytic leukemia.

increased likelihood of disease progression to leukemia (Paquette *et al.*, 1993). Mutation of N-*ras* is not sufficient alone to cause leukemia (Kubo *et al.*, 1993), but it correlates positively with expression of IL-6 in AML samples, thereby possibly participating in an autocrine loop (Lubbert *et al.*, 1993). Over expression of the *raf-1* proto-oncogene (a serine/threonine kinase that is involved in cellular signaling downstream of $p21^{ras}$) has been detected in erythroleukemic AML samples (Schmidt *et al.*, 1994). It has been proposed that N-*ras* mutation is a "postinitiation" event in AML that contributes to the outgrowth of more malignant clones (Bashey *et al.*, 1992). In addition, the mutational status of N-*ras* and *fms* may be indicative of early secondary leukemic disease in childhood acute lymphoblastic leukemia (Taylor *et al.*, 1995).

Chronic myelomonocytic leukemia is a type of MDS characterized by abnormal clonal myeloid proliferation and progression to AML. A subgroup of cases has a chromosomal translocation, t(5;12)(q333), which leads to expression of a chimeric protein *tel*-PDGFRβ, composed of the tyrosine kinase domain of PDGFRβ and a novel *ets*-like gene, *tel*. This novel protein has oncogenic potential and may mediate the CMML–AML transition (Golub *et al.*, 1994). For a review of the role of hematopoietic growth factors in acute leukemias, see Chapter 12.

Chronic myeloid leukemia (CML) is perhaps the best-defined leukemia in terms of its molecular pathology. It is associated with the Philadelphia chromosome, which results from the reciprocal translocation of *c-abl* on chromosome 9 with the *bcr* locus on chromosome 22 and encodes a constitutively activated form of the *abl* protein tyrosine kinase, a chimeric protein known as BCR-ABL (Lugo *et al.*, 1990). The BCR-ABL tyrosine kinase exists as two major proteins with molecular masses of 210 KDa or 185 KDa, depending on the breakpoint within the *bcr* gene (see Figure 4). These chimeric proteins both contain identical *abl* sequences but differ in the number of *bcr*-encoded N-terminal amino acids (927 or 426 amino acids of BCR, respectively). The tyrosine kinase activity of the *bcr-abl* proteins correlates positively with their transforming potential and is higher for the p185 than the p210 form (McLaughlin *et al.*, 1989). Experiments with transgenic *bcr-abl*-expressing mice and mice whose hematopoietic system has been reconstituted with *bcr-abl*-transfected cells show that this protein has an essential role in the molecular pathology of Ph-positive leukemias (Hariharan *et al.*, 1989; Daley *et al.*, 1990; Elefanty *et al.*, 1990; Heisterkamp *et al.*, 1990; Kelliher *et al.*, 1990; Honda *et al.*, 1995).

Myeloid leukemia is a disease of the hematopoietic progenitor cell population that is characterized by a disruption or blockade in normal myeloid development (Champlin and Golde, 1985); CML is a two-stage disease with chronic and acute phases. The initial chronic phase is characterized by a marked expansion in the late myeloid cell population, probably through an altered self-renewal/differentiation balance in the cells as a consequence of the expression of *bcr-abl*. The CML progenitor cells have been suggested to possess a subtle defect in maturation while retaining the ability to survive and proliferate, resulting in population expansion and consequently disruption of the normally tightly controlled hierarchical amplification system responsible for myelopoiesis (Wang, 1992). The *abl* oncoprotein confers a survival advantage by suppressing apoptosis (see Chapter 7). Studies of primary hematopoietic cells (McLaughlin *et al.*, 1987; Young and Witte, 1988; Gishizky and Witte, 1992), in contrast to many IL-3-dependent cell lines (Daley and Baltimore, 1988; Laneuville *et al.*, 1992; Carlesso *et al.*, 1994; Sirard *et al.*, 1994), indicate that BCR-ABL does not immediately abrogate growth factor requirements for proliferation, but factor-

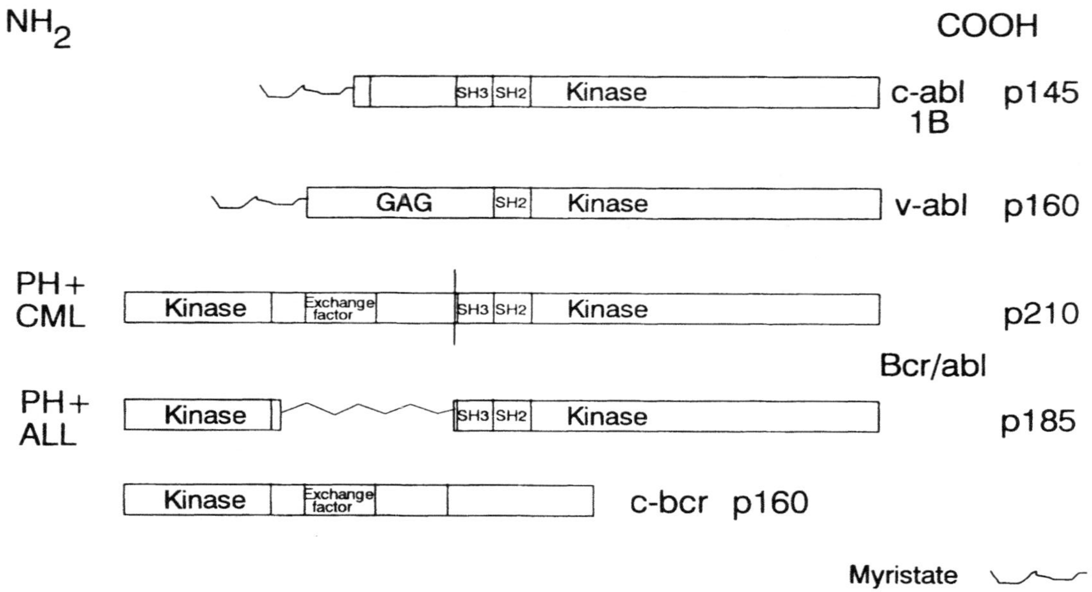

FIGURE 4. A schematic representation of the two forms of *c-abl* and the activated mutants *v-abl* and *bcr-abl*.

independent cells do arise several weeks after activation of the *abl* gene. BCR-ABL abrogates the stromal cell anchorage requirement but not the growth factor requirement for proliferation and does not activate mitogenesis (Renshaw *et al.*, 1995). In addition, CML progenitor cells are unable to respond to MIP1α and thus are refractory to negative regulation by this growth inhibitor (Eaves *et al.*, 1993).

The CML progresses to a fatal acute phase also known as blast crisis, which resembles acute myeloid leukemia (AML) and results in an accumulation of primitive hematopoietic progenitor cells following a blockade in the normal pattern of terminal differentiation. Factor independence arises in a monoclonal fashion (Vogt *et al.*, 1987; Chung *et al.*, 1991), suggesting that *bcr-abl* expression can initiate leukemogenesis but that further genetic changes are required for the disease progression. Evidence exists that the presence of the *bcr-abl* oncogene imposes genetic instability on a cell (Vogt *et al.*, 1987), and conformational changes in p53 (leading to loss of tumor suppressor function) have been observed in progression of the disease (Bi *et al.*, 1994). Because c-ABL undergoes cell-regulated phosphorylation (Kipreos and Wang, 1992), it is plausible that BCR-ABL could potentiate cell cycle progression by mimicking the effects of c-ABL. This remains to be demonstrated. To understand the role of *bcr-abl* in human leukemia, the structure–function relationship of the component BCR and ABL proteins and identification of their intracellular targets have been performed to compare the effects of this signaling protein to those promoted by HGFs.

The c-ABL protein can be divided into four domains, based on homology to other tyrosine kinases. These are an N-terminal variable domain, a kinase regulatory domain (SH2/SH3), a kinase domain, and a large noncatalytic C-terminal domain with no sequence similarity to other tyrosine kinases. The protein is negatively regulated within the cell such that, despite the strong correlation between elevation of kinase activity and oncogenic ability, overexpression of *c-abl* is not transforming (Pendergast *et al.*, 1991). The *c-abl* protein is localized in the cytoplasm and nucleus of both resting and actively dividing cells (Van Etten *et al.*, 1989; Dhut *et al.*, 1991; McWhirter and Wang, 1991) and can interact with both actin filaments (McWhirter and Wang, 1991; Van Etten *et al.*, 1994) and DNA through unique carboxyl-terminal domains (Kipreos and Wang, 1992). Mutational analysis indicates that both an active kinase and nuclear localization domains are important for its biological function. The DNA binding is regulated by serine phosphorylation, which is cell cycle regulated, phosphorylation has no effect on the kinase activity of c-ABL, but *cdc2* kinase-mediated phosphorylation of ABL on serine residues 852 and 853 leads to a loss of DNA-binding activity. The nuclear retinoblastoma gene product (RB) also binds c-ABL and inhibits its kinase activity (Welch and Wang, 1993). Hyperphosphorylation of RB in S phase correlates with the release of c-ABL, which allows activation of its kinase activity. Thus, nuclear c-ABL is an S-phase-active tyrosine kinase, indicating that it may participate directly in regulation of transcription; c-ABL may therefore both modulate the state of the actin cytoskeleton and transduce signals to the nucleus. Unlike the majority of known protein kinases, which are associated with ligand-binding receptors, c-ABL has not been linked to any receptor-mediated signaling pathway. Mutant activated variants of c-abl contribute to lymphoid and myeloid leukemogenesis. These mutations involve activation of ABL PTK by addition of viral *gag* sequences (see Figure 4).

The *bcr* gene product expressed in many cell types is a multifunctional protein. The first exon encodes a unique serine/threonine kinase domain (Maru and Witte, 1991). This

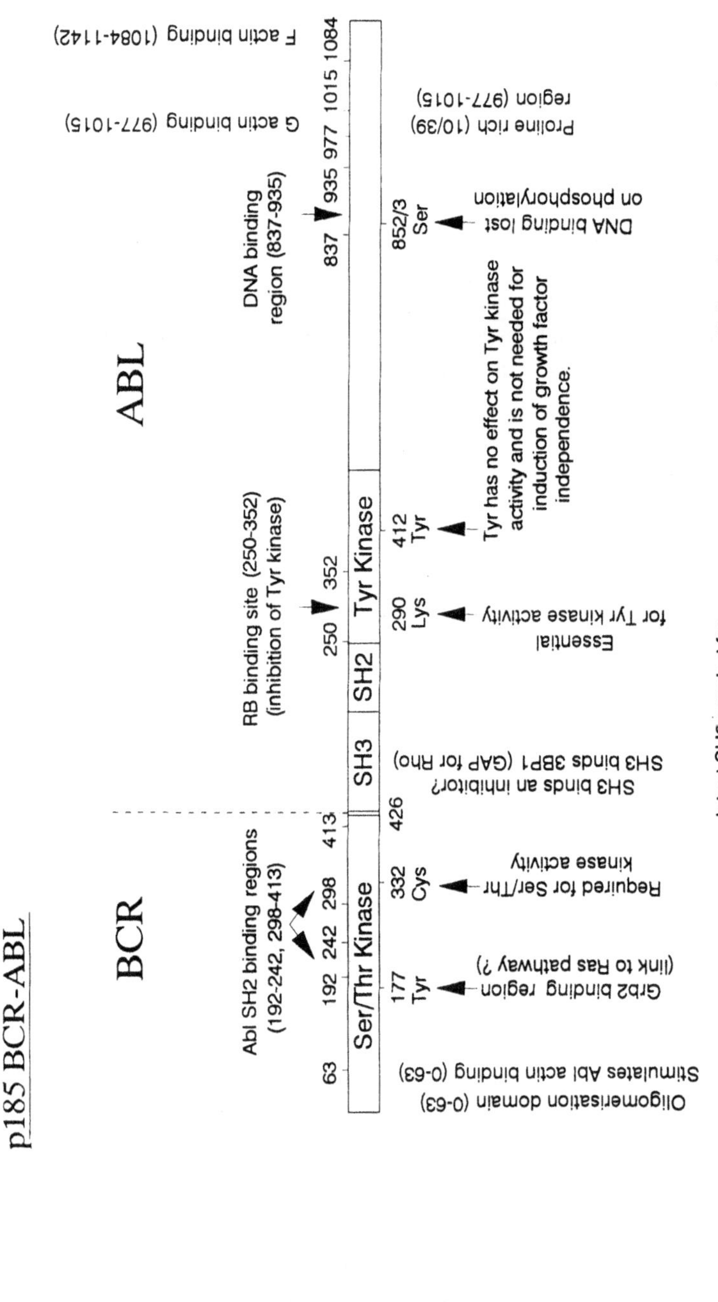

FIGURE 5. A schematic diagram of *bcr-abl* summarizing the data from mutational studies on the structure–functional relationship of various regions in the protein.

auto- and transkinase activity requires a pair of cysteine residues. The first exon also encodes an oligomerization domain (McWhirter *et al.*, 1993). The central region is homologous to several known or putative guanine nucleotide exchange factors including the *vav* and *dbl* proto-oncogene products (Adams *et al.*, 1992). The C-terminal domain has GAP activity *in vitro* toward the *ras*-related product RAC1 (Diekmann *et al.*, 1991). Thus, the *bcr* protein may link the *ras* and *rac* subfamilies of low-molecular-weight G proteins: *bcr* is phosphorylated by the cytoplasmic protein *fes* tyrosine kinase, which leads to the recruitment of the Grb 2 protein coupling it to the *ras* guanine nucleotide exchange factor mSOS, again linking BCR to the *ras* signaling pathway (Maru *et al.*, 1995). BCR also interacts with a member of the 14-3-3 family of proteins, BAP 1 (Reuther *et al.*, 1994). BAP 1 is a substrate for both the *bcr* serine/threonine kinase and is tyrosine-phosphorylated by BCR-ABL. 14-3-3 proteins are essential for cell proliferation and have a role in determining the timing of mitosis in yeast. Through direct binding to bcr the 14-3-3 proteins may link specific signal protein components to mitogenic and cell-cycle control pathways. Hence, c-BCR, with its assorted activities, may function as a point of crosstalk among multiple intracellular signaling pathways. However, little is still known about its biological function.

Although *c-abl* and *c-bcr* both encode proteins with signal transduction potential, their relative contributions to leukemogenesis associated with the chimeric BCR-ABL protein are unclear. Various mutational studies have been undertaken in attempts to address this problem and to attribute specific properties to defined domains of the protein (see Figure 5). Many mutant forms of BCR-ABL transform fibroblasts but not primary hematopoietic cells. As well as demonstrating the need to assess the effects of BCR-ABL in a physiological relevant manner, these findings may reflect a variability in the availability of, or access to, specific substrates.

The subtle biological effects seen in Ph^+ hematopoietic cells (presumably induced by the *bcr-abl* protein tyrosine kinase) therefore resemble, at least in part, signal transduction activated by cytokines (see Chapters 1 and 4). The chronic stimulation of certain signaling pathways by BCR-ABL may lead directly to desensitization of the Ph^+ cell to the effects of growth inhibitors such as MIP-1α or to a decrease in the relative rate of cell death seen when cytokines are removed from Ph^+ cells (when compared to normal progenitor cells; see Chapter 7). The fact that BCR-ABL can mimic actions of cytokines yet leave them HGF dependent for proliferation indicates the complexities of leukemogenesis, which will require a great deal of effort to unravel but may reveal new treatment strategies.

4. REFERENCES

Adams, J., and Watt, F., 1993, Regulation of development and differentiation by ECM, *Development* **117**:1183–1187.

Adams, J. M., Houston, H., Allen, J., Lints, T., and Harvey, R., 1992, The hematopoietically expressed *vav* proto-oncogene shares homology with the *dbl* GDP–GTP exchange factor, the *bcr* gene and a yeast gene (CDC24) involved in cytoskeletal organization, *Oncogene* **7**(4):611–618.

Baines, P., and Visser, J. W., 1983, Analysis and separation of murine bone marrow stem cells by H33342 fluorescence-activated cell sorting, *Exp. Hematol.* **11**(8):701–708.

Bashey, A., Gill, R., Levi, S., Farr, C. J., Clutterbuck, R., Millar, J. L., Pragnell, I. B., and Marshall, C. J., 1992, Mutational activation of the N-*ras* oncogene assessed in primary clonogenic culture of acute myeloid leukemia (AML): Implications for the role of N-*ras* mutation in AML pathogenesis, *Blood* **79**(4):981–989.

Baum, C. M., Weissman, I. L., Tsukamoto, A. S., Buckle, A. M., and Peault, B., 1992, Isolation of a candidate human hematopoietic stem-cell population, *Proc. Natl. Acad. Sci. U.S.A.* **89**(7):2804–2808.

Benit, L., Courtois, G., Charon, M., Varlet, P., Dusanter, F. I., and Gisselbrecht, S., 1994, Characterization of mpl cytoplasmic domain sequences required for myeloproliferative leukemia virus pathogenicity, *J. Virol.* **68**(8): 5270–5274.

Bertoncello, I., Bradley, T. R., Hodgson, G. S., and Dunlop, J. M., 1991, The resolution, enrichment, and organization of normal bone marrow high proliferative potential colony-forming cell subsets on the basis of rhodamine-123 fluorescence, *Exp. Hematol.* **19**(3):174–178.

Bi, S., Barton, C. M., Lemoine, N. R., Cross, N. C., and Goldman, J. M., 1994, Retroviral transduction of Philadelphia-positive chronic myeloid leukemia cells with a human mutant p53 cDNA and its effect on *in vitro* proliferation [see comments], *Exp. Hematol.* **22**(1):95–99.

Brandt, J., Baird, N. Lu, L., Srour, E., and Hoffman, R., 1988, Characterization of a human hematopoietic progenitor cell capable of forming blast cell containing colonies *in vitro*, *J. Clin. Invest.* **82**(3):1017–1027.

Brannan, C. I., Lyman, S. D. Williams, D. E., Eisenman, J., Anderson, D. M., Cosman, D., Bedell, M. A., Jenkins N. A., and Copeland, N.G., 1991, Steel–Dickie mutation encodes a c-kit ligand lacking transmembrane and cytoplasmic domains, *Proc. Natl. Acad. Sci. U.S.A.* **88**(11):4671–4674.

Broudy, V. C., Lin, N. L., and Kaushansky, K., 1995, Thrombopoietin (c-mpl ligand) acts synergistically with erythropoietin, stem cell factor, and interleukin-11 to enhance murine megakaryocyte colony growth and increases megakaryocyte ploidy *in vitro*, *Blood* **85**(7):1719–1726.

Broxmeyer, H. E., Sherry, B., Cooper, S., Lu, L., Maze, R., Beckmann, M. P., Cerami, A., and Ralph, P., 1993, Comparative analysis of the human macrophage inflammatory protein family of cytokines (chemokines) on proliferation of human myeloid progenitor cells. Interacting effects involving suppression, synergistic suppression, and blocking of suppression, *J. Immunol.* **150**:3448–3458.

Carlesso, N., Griffin, J.D., and Druker, J., 1994, Use of a temperature-sensitive mutant to define the biological effects of the p210BCR-ABL tyrosine kinase on proliferation of a factor-dependent murine myeloid cell line, *Oncogene* **9**(1):149–156.

Caux, C., Saeland, S., Favre, C., Duvert, V., Mannoni, P., and Banchereau, J., 1990, Tumor necrosis factor-alpha strongly potentiates interleukin-3 and granulocyte–macrophage colony-stimulating factor-induced proliferation of human $CD34^+$ hematopoietic progenitor cells, *Blood* **75**(12):2292–2298.

Champlin, R. E., and Golde, D. W., 1985, Chronic myelogenous leukaemia: Recent advances, *Blood* **65**:1039–1047.

Chung, S. W., Wong, P. M., Durkin, H., Wu, Y. S., and Petersen, J., 1991, Leukemia initiated by hemopoietic stem cells expressing the *v-abl* oncogene, *Proc. Natl. Acad. Sci. U.S.A.* **88**(4):1585–1589.

Civin, C. I., Strauss, L. C., Fackler, M. J., Trischmann, T. M., Wiley, J. M., and Loken, M.R., 1990, Positive stem cell selection—basic science, *Prog. Clin. Biol. Res.* **333**(387):387–401.

Courtneidge, S. A., Dhand, R., Pilat, D., Twamley, G. M., Waterfield, M. D., and Roussel, M.F., 1993, Activation of Src family kinases by colony stimulating factor-1, and their association with its receptor, *EMBO J.* **12**(3): 943–950.

Courtois, G., Benit, L., Mikaeloff, Y., Pauchard, M., Charon, M., Varlet, P., and Gisselbrecht, S., 1995, Constitutive activation of a variant of the *env-mpl* oncogene product by disulfide-linked homodimerization, *J. Virol.* **69**(5): 2794–2800.

Daley, G. Q., and Baltimore, D., 1988, Transformation of an interleukin 3-dependent hematopoietic cell line by the chronic myelogenous leukemia-specific P210bcr/abl protein, *Proc. Natl. Acad. Sci. U.S.A.* **85**(23):9312–9316.

Daley, G. Q., Van Etten, R.A., and Baltimore, D., 1990, Induction of chronic myelogenous leukemia in mice by the P210*bcr*/*abl* gene of the Philadelphia chromosome, *Science* **247**:824–830.

D'Andrea, A. D., Fasman, G. D., and Lodish, H.F., 1989, Erythropoietin receptor and interleukin-2 receptor beta chain: a new receptor family, *Cell* **58**(6):1023–1024.

D'Andrea, A. D., Yoshimura, A., Youssoufian, H., Zon, L. I., Koo, J. W., and Lodish, H.F., 1991, The cytoplasmic region of the erythropoietin receptor contains nonoverlapping positive and negative growth-regulatory domains, *Mol. Cell Biol.* **11**(4):1980–1987.

Debili, N., Wendling, F., Cosman, D., Titeux, M., Florindo, C., Dusanter, F. I., Schooley, K., Methia, N., Charon, M., Nador, R., Bettaieb, A., and Vainchenker, W., 1995, The Mpl receptor is expressed in the megakaryocytic lineage from late progenitors to platelets, *Blood* **85**(2):391–401.

de la Chapelle, A., Sistonen, P., Lehvaslaiho, H., Ikkala, E., and Juvonen, E., 1993a, Familial erythrocytosis genetically linked to erythropoietin receptor gene, *Lancet* **341**:82–84.

de la Chapelle, A., Traskelin, A. L., and Juvonen, E., 1993b, Truncated erythropoietin receptor causes dominantly inherited benign human erythrocytosis, *Proc. Natl. Acad. Sci. U.S.A.* **90**(10):4495–4499.

Dhut, S., Chaplin, T., and Young, B. D., 1991, Normal *c-abl* gene protein—a nuclear component, *Oncogene* **6:** 1459–1464.

Diekmann, D., Brill, S., Garrett, M. D., Totty, N., Hsuan, J., Monfries, C., Hall, C., Lim, L., and Hall, A., 1991, *Bcr* encodes a GTPase-activating protein for p21 *Rac, Nature* **351:**400–402.

Dosil, M., Wang, S., and Lemischka, I. R., 1993, Mitogenic signalling and substrate specificity of the *Flk2/Flt3* receptor tyrosine kinase in fibroblasts and interleukin 3-dependent hematopoietic cells, *Mol. Cell Biol.* **13**(10):6572–6585.

Drachman, J. G., Griffin, J. D., and Kaushansky, K., 1995, The *c-Mpl* ligand (thrombopoietin) stimulates tyrosine phosphorylation of Jak2, Shc, and *c-Mpl, J. Biol. Chem.* **270**(10):4979–4982.

Eaves, C. J., Cashman, J. D., Kay, R. J., Dougherty, G. J., Otsuka, T., Gaboury, L. A., Hogge, D. E., Lansdorp, P. M., Eaves, A. C., and Humphries, R. K., 1991, Mechanisms that regulate the cell cycle status of very primitive hematopoietic cells in long-term human marrow cultures. II. Analysis of positive and negative regulators produced by stromal cells within the adherent layer, *Blood* **78**(1):110–117.

Eaves, C. J., Cashman, J. D., Wolpe, S. D., and Eaves, A.C., 1993, Unresponsiveness of primitive chronic myeloid leukemia cells to macrophage inflammatory protein 1 alpha, an inhibitor of primitive normal hematopoietic cells, *Proc. Natl. Acad. Sci. U.S.A.* **90**(24):12015–12019.

Elefanty, A. G., Hariharan, I. K., and Cory, S., 1990, *bcr-abl*, the hallmark of chronic myeloid leukemia in man, induces multiple neoplasms in mice, *EMBO J.* **9**(4):1069–1078.

Fichelson, S., Vigon, I., Dusanter, I., Charon, M., Velu, T., Baillou, C., Gisselbrecht, S., and Lemoine, F. M., 1995, *In vitro* transformation of murine pro-B and pre-B cells by *v-mpl*, a truncated form of a cytokine receptor, *J. Immunol.* **154**(4):1577–1586.

Flanagan, J. G., Chan, D. C., and Leder, P., 1991, Transmembrane form of the *kit* ligand growth factor is determined by alternative splicing and is missing in the Sld mutant, *Cell* **64**(5):1025–1035.

Gishizky, M. L., and Witte, O.N., 1992, Initiation of deregulated growth of multipotent progenitor cells by *bcr-abl in vitro*, *Science* **256:**836–839.

Golub, T. R., Barker, G. F., Lovett, M., and Gilliland, D., 1994, Fusion of PDGF receptor beta to a novel *ets*-like gene, *tel*, in chronic myelomonocytic leukemia with t(5;12) chromosomal translocation, *Cell* **77**(2):307–316.

Gordon, M. Y., Riley, G. P., Watt, S. M., and Greaves, M. F., 1987, Compartmentalization of a hematopoietic growth factor (GM-CSF) by glycosaminoglycans in the bone marrow microenvironment, *Nature* **326:** 403–405.

Gordon, M. Y., Atkinson, J., Clarke, D., Dowding, C. R., Goldman, J. M., Grimsley, P. G., Siczkowski, M., and Greaves, M. F., 1991, Deficiency of a phosphatidylinositol-anchored cell adhesion molecule influences hematopoietic progenitor binding to marrow stroma in chronic myeloid leukaemia, *Leukemia* **5**(8):693–698.

Gualtieri, R. J., Shadduck, R. K., Baker, D. G., and Quesenberry, P. J., 1984, Hematopoietic regulatory factors produced in long-term murine bone marrow cultures and the effect of *in vitro* irradiation, *Blood* **64**(2): 516–525.

Gurney, A. L., Kuang, W. J., Xie, M. H., Malloy, B. E., Eaton, D. L., and de Sauvage, S. F., 1995, Genomic structure, chromosomal localization, and conserved alternative splice forms of thrombopoietin, *Blood* **85**(4): 981–988.

Hannum, C., Culpepper, J., Campbell, D., McClanahan, T., Zurawski, S., Bazan, J. F., Kastelein, R., Hudak, S., Wagner, J., and Mattson, J., Luh, J., Duda, G., Martina, N., Peterson, D., Menon, S., Shanafelt, A., Muench, M., Kelner, G,. Namikawa, R., Rennick, D., Roncouolo, M.-G., Zlotnick, A., Rosnet, O, Dubreid, P., Birnbaum, D., and Lee, F., 1994, Ligand for FLT3/FLK2 receptor tyrosine kinase regulates growth of hematopoietic stem cells and is encoded by variant RNAs, *Nature* **368:**643–648.

Hariharan, I. K., Harris, A. W., Crawford, M., Abud, H., Webb, E., Cory, S., and Adams, J. M., 1989, *A bcr-v-abl* oncogene induces lymphomas in transgenic mice, *Mol. Cell Biol.* **9**(7):2798–2805.

Heisterkamp, N., Jenster, G., ten Hoeve, J., Zovich, D., Pattengale, P. K., and Groffen, J., 1990, Acute leukaemia in *bcr/abl* transgenic mice, *Nature* **344:**251–253.

Honda, H., Fujii, T., Takatoku, M., Mano, H., Witte, O. N., Yazaki, Y., and Hirai, H., 1995, Expression of p210*bcr/abl* by metallothionein promoter induced T-cell leukemia in transgenic mice, *Blood* **85**(10):2853–2861.

Hunter, T., and Cooper, J. A., 1985, Protein-tyrosine kinases, *Annu. Rev. Biochem.* **54:**897–930.

Hynes, R., 1992, Integrins: Versatility, modulation and signalling in cell adhesion, *Cell* **69:**11–17.

Ichihara, M., Hara, T., Takagi, M., Cho, L. C., Gorman, D. M., and Miyajima, A., 1995, Impaired interleukin-3

(IL-3) response of the A/J mouse is caused by a branch point deletion in the IL-3 receptor alpha subunit gene, *EMBO J.* **14**(5):939–950.

Ihle, J. N., Witthuhn, B., Tang, B., Yi, T., and Quelle, F. W., 1994, Cytokine receptors and signal transduction, *Baillieres Clin. Haematol.* **7**(1):17–48.

Jacobsen, F. W., Rusten, L. S., and Jacobsen, S. E., 1994, Direct synergistic effects of interleukin-7 on *in vitro* myelopoiesis of human CD34$^+$ bone marrow progenitors, *Blood* **84**(3):775–779.

Jacobsen, S. E., Ruscetti, F. W., Dubois, C. M., Wine, J., and Keller, J. R., 1992, Induction of colony-stimulating factor receptor expression on hematopoietic progenitor cells: Proposed mechanism for growth factor synergism, *Blood* **80**(3):678–687.

Jacobsen, S. E., Okkenhaug, C., Myklebust, J., Veiby, O.P., and Lyman, S.D., 1995, The FLT3 ligand potently and directly stimulates the growth and expansion of primitive murine bone marrow progenitor cells *in vitro*: Synergistic interactions with interleukin (IL)-11, IL-12, and other hematopoietic growth factors, *J. Exp. Med.* **181**(4):1357–1363.

Jordan, C. T., McKearn, J. P., and Lemischka, I. R., 1990, Cellular and developmental properties of fetal hematopoietic stem cells, *Cell* **61**(6):953–963.

Kaushansky, K., 1994, The *mpl* ligand: Molecular and cellular biology of the critical regulator of megakaryocyte development, *Stem Cells Dayt.* **1**(91):91–96.

Kaushansky, K., Broudy, V. C., Lin, N., Jorgensen, M. J., McCarty, J., Fox, N., Zucker, F. D., and Lofton, D. C., 1995, Thrombopoietin, the Mpl ligand, is essential for full megakaryocyte development, *Proc. Natl. Acad. Sci. U.S.A.* **92**(8):3234–3238.

Kelliher, M. A., McLaughlin, J., Witte, O. N., and Rosenberg, N., 1990, Induction of a chronic myelogenous leukemia-like syndrome in mice with *v-abl* and BCR/ABL [published erratum appears in *Proc. Natl. Acad. Sci. U.S.A.* 1990, **87**(22):9072], *Proc. Natl. Acad. Sci. U.S.A.* **87**(17):6649–6653.

Kipreos, E. T., and Wang, J. Y., 1992, Cell cycle-regulated binding of *c-Abl* tyrosine kinase to DNA, *Science* **256:** 382–385.

Koury, M. J., 1992, Programmed cell death (apoptosis) in hematopoiesis, *Exp. Hematol.* **20**(4):391–394.

Krause, D. S., Ito, T., Fackler, M. J., Smith, O. M., Collector, M. I., Sharkis, S. J., and May, W. S., 1994, Characterization of murine CD34, a marker for hematopoietic progenitor and stem cells, *Blood* **84**(3): 691–701.

Kubo, K., Naoe, T., Kiyoi, H., Fukutani, H., Kato, Y., Oguri, T., Yamamori, S., Akatsuka, Y. A., Kodera, Y., and Ohno, R., 1993, Clonal analysis of multiple point mutations in the N-*ras* gene in patients with acute myeloid leukemia, *Jpn. J. Cancer Res.* **84**(4):379–387.

Laneuville, P., Sun, G., Timm, M., and Vekemans, M., 1992, Clonal evolution in a myeloid cell line transformed to interleukin-3 independent growth by retroviral transduction and expression of p210*bcr/abl*, *Blood* **80**(7): 1788–1797.

Lev, S., Yarden, Y., and Givol, D., 1992a, Dimerization and activation of the kit receptor by monovalent and bivalent binding of the stem cell factor, *J. Biol. Chem.* **267**(22):15970–15977.

Lev, S., Yarden, Y., and Givol, D., 1992b, A recombinant ectodomain of the receptor for the stem cell factor (SCF) retains ligand-induced receptor dimerization and antagonizes SCF-stimulated cellular responses, *J. Biol. Chem.* **267**(15):10866–10873.

Lev, S., Givol, D., and Yarden, Y., 1992c, Interkinase domain of kit contains the binding site for phosphatidylinositol 3′ kinase, *Proc. Natl. Acad. Sci. U.S.A.* **89**(2):678–682.

Liboi, E., Carroll, M., D'Andrea, D.A., and Mathey, P. B., 1993, Erythropoietin receptor signals both proliferation and erythroid-specific differentiation, *Proc. Natl. Acad. Sci. U.S.A.* **90**(23):11351–11355.

Lieschke, G. J., Grail, D., Hodgson, G., Metcalf, D., Stanley, E., Cheers, C., Fowler, K. J., Basu, S., Zhan, Y. F., and Dunn, A. R., 1994a, Mice lacking granulocyte colony-stimulating factor have chronic neutropenia, granulocyte and macrophage progenitor cell deficiency, and impaired neutrophil mobilization, *Blood* **84**(6):1737–1746.

Lieschke, G. J., Stanley, E., Grail, E., Hodgson, G., Sinickas, V., Gall, J. A., Sinclair, R. A., and Dunn, A. R., 1994b, Mice lacking both macrophage- and granulocyte–macrophage colony-stimulating factor have macrophages and coexistent osteopetrosis and severe lung disease, *Blood* **84**(1):27–35.

Lioubin, M. N., Myles, G. M., Carlberg, K., Bowtell, D., and Rohrschneider, L. R., 1994, Shc, Grb2, Sosl, and a 150-kilodalton tyrosine-phosphorylated protein form complexes with Fms in hematopoietic cells, *Mol. Cell Biol.* **14**(9):5682–5691.

Lubbert, M., Oster, W., Knopf, H. P., McCormick, F., Mertelsmann, R., and Herrmann, F., 1993, N-*RAS* gene activation in acute myeloid leukemia: Association with expression of interleukin-6, *Leukemia* **7**(12):1948–1954.

Lugo, T. G., Pendergast, A. M., Muller, A. J., and Witte, O. N., 1990, Tyrosine kinase activity and transformation potency of *bcr-abl* oncogene products, *Science* **247**:1079–1082.

Lyman, S. D., Brasel, K., Rousseau, A. M., and Williams, D. E., 1994, The *flt3* ligand: A hematopoietic stem cell factor whose activities are distinct from steel factor, *Stem Cells Dayt.* **1**(99):99–107.

Maru, Y., and Witte, O. N., 1991, The BCR gene encodes a novel serine/threonine kinase activity within a single exon, *Cell* **67**(3):459–468.

Maru, Y., Peters, K. L., Afar, D. E., Shibuya, M., Witte, O. N. and Smithgall, T. E., 1995, Tyrosine phosphorylation of BCR by FPS/FES protein-tyrosine kinases induces association of BCR with GRB-2/SOS, *Mol. Cell Biol.* **15**(2):835–842.

Maruyama, K., Miyata, K., and Yoshimura, A., 1994, Proliferation and erythroid differentiation through the cytoplasmic domain of the erythropoietin receptor, *J. Biol. Chem.* **269**(8):5976–5980.

Matthews, W., Jordan, C. T., Wiegand, G. W., Pardoll, D., and Lemischka, I. R., 1991a, A receptor tyrosine kinase specific to hematopoietic stem and progenitor cell-enriched populations, *Cell* **65**(7):1143–1152.

Matthews, W., Jordan, C. T., Gavin, M., Jenkins, N. A., Copeland, N. G., and Lemischka, I. R., 1991b, A receptor tyrosine kinase cDNA isolated from a population of enriched primitive hematopoietic cells and exhibiting close genetic linkage to *c-kit*, *Proc. Natl. Acad. Sci. U.S.A.* **88**(20):9026–9030.

McLaughlin, J., Chianese, E., and Witte, O. N., 1987, *In vitro* transformation of immature hematopoietic cells by the P210 *BCR/ABL* oncogene product of the Philadelphia chromosome, *Proc. Natl. Acad. Sci. U.S.A.* **84**(18): 6558–6562.

McLaughlin, J., Chianese, E., and Witte, O. N., 1989, Alternative forms of the *BCR-ABL* oncogene have quantitatively different potencies for stimulation of immature lymphoid cells, *Mol. Cell Biol.* **9**(5):1866–1874.

McWhirter, J. R., and Wang, J. Y. J., 1991, Activation of tyrosine kinase and microfilament binding functions of *c-abl* by *bcr* sequences in *bcr/abl* fusion proteins, *Mol. Cell Biol.* **11**:1553–1565.

McWhirter, J. R., Galasso, D. L., and Wang, J. Y., 1993, A coiled-coil oligomerization domain of Bcr is essential for the transforming function of Bcr-Abl oncoproteins, *Mol. Cell Biol.* **13**(12):7587–7595.

Metcalf, D., 1995, The granulocyte–macrophage regulators: Reappraisal by gene inactivation, *Exp. Haematol.* **23**:569–572.

Migliaccio, G., Migliaccio, A. R., Valinsky, J., Langley, K., Zsebo, K., Visser, J. W., and Adamson, J. W., 1991, Stem cell factor induces proliferation and differentiation of highly enriched murine hematopoietic cells, *Proc. Natl. Acad. Sci. U.S.A.* **88**(16):7420–7424.

Miura, O., Miura, Y., Nakamura, N., Quelle, F. W., Witthuhn, B. A., Ihle, J. N., and Aoki, N., 1994, Induction of tyrosine phosphorylation of *Vav* and expression of *Pim-1* correlates with Jak2-mediated growth signaling from the erythropoietin receptor, *Blood* **84**(12):4135–4142.

Miura, Y., Miura, O., Ihle, J. N., and Aoki, N., 1994, Activation of the mitogen-activated protein kinase pathway by the erythropoietin receptor, *J. Biol. Chem.* **269**(47):29962–29969.

Miyajima, A., 1992, Molecular structure of the IL-3, GM-CSF and IL-5 receptors, *Int. J. Cell Clon.* **10**(3):126–134.

Miyakawa, Y., Oda, A., Druker, B. J., Kato, T., Miyazaki, H., Handa, M., and Ikeda, Y., 1995, Recombinant thrombopoietin induces rapid tyrosine phosphorylation of Janus kinase 2 and Shc in human blood platelets, *Blood* **86**:23–27.

Mochizuki, D. Y., Eisenman, J. R., Conlon, P. J., Larsen, A. D., and Tushinski, R. J., 1987, Interleukin 1 regulates hematopoietic activity, a role previously ascribed to hemopoietin 1, *Proc. Natl. Acad. Sci. U.S.A.* **84**(15): 5267–5271.

Noguchi, M., Nakamura, Y., Russell, S. M., Ziegler, S. F., Tsang, M., Cao, X., and Leonard, W. J., 1993, Interleukin-2 receptor gamma chain: A functional component of the interleukin-7 receptor [see comments], *Science* **262**(5141):1877–1880.

Ogawa, M., 1993, Differentiation and proliferation of hematopoietic stem cells, *Blood* **81**(11):2844–2853.

Otsuka, T., Eaves, C. J., Humphries, R. K., Hogge, D. E., and Eaves, A. C., 1991, Lack of evidence for abnormal autocrine or paracrine mechanisms underlying the uncontrolled proliferation of primitive chronic myeloid leukemia progenitor cells, *Leukemia* **5**(10):861–868.

Paquette, R. L., Landaw, E. M., Pierre, R. V., Kahan, J., Lubbert, M., Lazcano, O., Isaac, G., McCormick, F., and Koeffler, H. P., 1993, N-*ras* mutations are associated with poor prognosis and increased risk of leukemia in myelodysplastic syndrome, *Blood* **82**(2):590–599.

Pearson, M. A., Heyworth, C. M., Whetton, A. D., Dexter, T. M., and Owen-Lynch, P. J., 1996, Molecular mechanisms of signal transduction in hematopoietic lineages,in *Hemopoietic Cell Lineages* (N. G. Testa, B. I. Lord, and T. M. Dexter, eds.), M. Dekker, New York (in press).

Pendergast, A. M., Muller, A. J., Havlik, M. H., Clark, R., McCormick, F., and Witte, O. N., 1991, Evidence for regulation of the human ABL tyrosine kinase by a cellular inhibitor, *Proc. Natl. Acad. Sci. U.S.A.* **88**(13): 5927–5931.

Reedijk, M., Liu, X., van der Geer, P., Letwin, K., Waterfield, M. D., Hunter, T., and Pawson, T., 1992, Tyr 721 regulates specific binding of the CSF-1 receptor kinase insert to PI 3′-kinase SH2 domains: A model for SH2-mediated receptor-target interactions, *EMBO J.* **11**(4):1365–1372.

Renshaw, M. W., McWhirter, J. R., and Wang, J. Y., 1995, The human leukemia oncogene *bcr-abl* abrogates the anchorage requirement but not the growth factor requirement for proliferation, *Mol. Cell Biol.* **15**(3):1286–1293.

Rettenmier, C. W., and Roussel, M. F., 1988, Differential processing of colony-stimulating factor 1 precursors encoded by two human cDNAs, *Mol. Cell Biol.* **8**(11):5026–5034.

Reuther, G. W., Fu, H., Cripe, L. D., Collier, R. J., and Pendergast, A. M., 1994, Association of the protein kinases c-Bcr and Bcr-Abl with proteins of the 14-3-3 family [see comments], *Science* **266**:129–133.

Roberts, R., Gallagher, J., Spooncer, E., Allen, T. D., Bloomfield, F., and Dexter, T. M., 1988, Heparin sulphate bound growth factors: A mechanism for stromal cell mediated hematopoiesis, *Nature* **332**:376–378.

Rosnet, O., Marchetto, S., deLapeyriere, O., and Birnbaum, D., 1991, Murine *Flt3*, a gene encoding a novel tyrosine kinase receptor of the PDGFR/CSFIR family, *Oncogene* **6**(9):1641–1650.

Rossner, M. T., McArthur, G. A., Allen, T. D., and Metcalf, D., 1994, Fms-like tyrosine kinase 3 catalytic domain can transduce a proliferative signal in FDC-P1 cells that is qualitatively similar to the signal delivered by c-Fms, *Cell Growth Differ.* **5**(5):549–555.

Rottapel, R., Reedijk, M., Williams, D. E., Lyman, S. D., Anderson, M. D., Pawson, T., and Bernstein, A., 1991, The Steel/W transduction pathway: kit autophosphorylation and its association with a unique subset of cytoplasmic signaling proteins is induced by the Steel factor, *Mol. Cell Biol.* **11**(6):3043–3051.

Rottapel, R., Turck, C. W., Casteran, N., Liu, X., Birnbaum, D., Pawson, T., and Dubreuil, P., 1994, Substrate specificities and identification of a putative binding site for PI3K in the carboxy tail of the murine *Flt3* receptor tyrosine kinase, *Oncogene* **9**(6):1755–1765.

Roussel, M. F., 1994, Signal transduction by the macrophage-colony-stimulating factor receptor (CSF-1R), *J. Cell Sci. Suppl.* **18**(105):105–108.

Ruscetti, F. W., Dubois, C. M., Jacobsen, S. E., and Keller, J. R., 1992, Transforming growth factor beta and interleukin-1: A paradigm for opposing regulation of hematopoiesis, *Baillieres Clin. Haematol.* **5**(3):703–721.

Sattler, M., Durstin, M. A., Frank, D. A., Okuda, K., Kaushansky, K., Ravi, S., and Griffin, J. D., 1995, The thrombopoietin receptor c-MPL activates JAK2 and TYK2 kinases, *Exp. Hematol.* **23**:1040–1048.

Schmidt, C. A., Oettle, H., Ludwig, W. D., Serke, S., Pawlaczyk, P. B., Wilborn, F., Binder, L. T., Huhn, D., and Siegert, W., 1994, Overexpression of the *Raf-1* proto-oncogene in human myeloid leukemia, *Leuk. Res.* **18**(6): 409–413.

Shultz, L. D., Schweitzer, P. A., Rajan, T. V., Yi, T., Ihle, J. N., Matthews, R. J., Thomas, M. L., and Beier, D. R., 1993, Mutations at the murine motheaten locus are within the hematopoietic cell protein-tyrosine phosphatase (Hcph) gene, *Cell* **73**(7):1445–1454.

Shurtleff, S. A., Downing, J. R., Rock, C. O., Hawkins, S. A., Roussel, M. F., and Sherr, C. J., 1990, Structural features of the colony-stimulating factor 1 receptor that affect its association with phosphatidylinositol 3-kinase, *EMBO J.* **9**(8):2415–2421.

Siczkowski, M., Clarke, D., and Gordon, M. Y., 1992, Binding of primitive hematopoietic progenitor cells to marrow stromal cells involves heparin sulfate, *Blood* **80**(4):912–917.

Sirard, C., Laneuville, P., and Dick, J. E., 1994, Expression of *bcr-abl* abrogates factor-dependent growth of human hematopoietic M07E cells by an autocrine mechanism, *Blood* **83**(6):1575–1585.

Small, D., Levenstein, M., Kim, E., Carow, C., Amin, S., Rockwell, P., Witte, L., Burrow, C., Ratajczak, M. Z., Gewirtz, A. M., and Civin, K. I., 1994, *STK-1*, the human homolog of *Flk-2/Flt-3*, is selectively expressed in $CD34^+$ human bone marrow cells and is involved in the proliferation of early progenitor/stem cells, *Proc. Natl. Acad. Sci. U.S.A.* **91**(2):459–463.

Sokol, L., Prchal, J. F., D'Andrea, A., Rado, T. A., and Prchal, J. T., 1994, Mutation in the negative regulatory element of the erythropoietin receptor gene in a case of sporadic primary polycythemia, *Exp. Hematol.* **22**(5): 447–453.

Spangrude, G. J., 1989, Enrichment of murine hematopoietic stem cells: Diverging roads, *Immunol. Today* **10**(10): 344–350.

Spangrude, G. J., Heimfeld, S., and Weissman, I. L., 1988a, Purification and characterization of mouse hematopoietic stem cells [published erratum appears in *Science* 1989, **244:**1030], *Science* **241:**58–62.

Spangrude, G. J., Muller, S. C., Heimfeld, S., and Weissman, I. L., 1988b, Two rare populations of mouse Thy-llo bone marrow cells repopulate the thymus, *J. Exp. Med.* **167**(5)**:**1671–1683.

Stanley, E. R., Bartocci, A., Patinkin, D., Rosendaal, M., and Bradley, T. R., 1986, Regulation of very primitive, multipotent, hemopoietic cells by hemopoietin-1, *Cell* **45**(5)**:**667–672.

Stanley, E., Lieschke, G. J., Grail, D., Metcalf, D., Hodgson, G., Gall, J. A., Maher, D. W., Cebon, J., Sinickas, V., and Dunn, A. R., 1994, Granulocyte/macrophage colony-stimulating factor-deficient mice show no major perturbation of hematopoiesis but develop a characteristic pulmonary pathology, *Proc. Natl. Acad. Sci. U.S.A.* **91**(12)**:**5592–5596.

Tauchi, T., Feng, G. S., Shen, R., Hoatlin, M., Bagby, G. J., Kabat, D., Lu, L., and Broxmeyer, H. E., 1995, Involvement of SH2-containing phosphotyrosine phosphatase Syp in erythropoietin receptor signal transduction pathways, *J. Biol. Chem.* **270**(10)**:**5631–5635.

Taylor, C., McGlynn, H., Carter, G., Baker, A. H., Warren, N., Ridge, S. A., Owen, G., Thompson, E., Thompson, P. W., and Jacobs, A., and Padua, R. A., 1995, RAS and FMS mutations following cytotoxic therapy for childhood acute lymphoblastic leukaemia, *Leukemia* **9**(3)**:**466–470.

Terstappen, L. W., Huang, S., Safford, M., Lansdorp, P. M., and Loken, M. R., 1991, Sequential generations of hematopoietic colonies derived from single nonlineage-committed $CD34^+CD38^-$ progenitor cells, *Blood* **77**(6)**:**1218–1227.

Testa, U., Pelosi, E., Gabbianelli, M., Fossati, C., Campisi, S., Isacchi, G., and Peschle, C., 1993, Cascade transactivation of growth factor receptors in early human hematopoiesis, *Blood* **81**(6)**:**1442–1456.

Till, J. E., and McCulloch, E. A., 1961, A direct measurement of the radiation sensitivity of normal mouse bone marrow cells, *Radiation Res.* **14**(213)**:**383–387.

Tortolani, P. J., Johnston, J. A., Bacon, J. M., McVicar, D. W., Shimosaka, D., Linnekin, D., Longs, D. L., and O'Shea, J. J., 1995, Thrombopoietin induces tyrosine phosphorylation and activation of Janus kinase, JAK2, *Blood* **85:**3444–3452.

Tsuji, K., Zsebo, K. M., and Ogawa, M., 1991, Enhancement of murine blast cell colony formation in culture by recombinant rat stem cell factor, ligand for *c-kit*, *Blood* **78**(5)**:**1223–1229.

Ullrich, A., and Schlessinger, J., 1990, Signal transduction by receptors with tyrosine kinase activity, *Cell* **61**(2)**:** 203–212.

Van Etten, R. A., Jackson, P., and Baltimore, D., 1989, The mouse type IV *c-abl* gene product is a nuclear protein, and activation of transforming ability is associated with cytoplasmic localization, *Cell* **58**(4)**:**669–678.

Van Etten, R. A., Jackson, P. K., Baltimore, D., Sanders, M. C., Matsudaira, P. T., and Janmey, P. A., 1994, The COOH terminus of the c-Abl tyrosine kinase contains distinct F- and G-actin binding domains with bundling activity [published erratum appears in *J. Cell Biol.* 1994, **124**(5)**:**865], *J. Cell Biol.* **124**(3)**:**325–340.

Varticovski, L., Druker, B., Morrison, D., Cantley, L., and Roberts, T., 1989, The colony stimulating factor-1 receptor associates with and activates phosphatidylinositol-3 kinase, *Nature* **342:**699–678.

Visser, J. W., and de Vries, P., 1988, Isolation of spleen-colony forming cells (CFU-s) using wheat germ agglutinin and rhodamine 123 labeling, *Blood Cells* **14**(2–3)**:**369–384.

Visser, J. W., and Van Dekkum, D. W., 1990, Purification of pluripotent hemopoietic stem cells: Past and present, *Exp. Hematol.* **18**(3)**:**248–256.

Vogt, M., Lesley, J., Bogenberger, J. M., Haggblom, C., Swift, S., and Haas, M., 1987, The induction of growth factor-independence in murine myelocytes by oncogenes results in monoclonal cell lines and is correlated with cell crisis and karyotypic instability, *Oncogene Res.* **2**(1)**:** 49–63.

Wang, J. Y. J., 1992, Modelling chronic myeloid leukemia, *Curr. Biol.* **2:**70–72.

Watowich, S. S., Hilton, D. J., and Lodish, H. F., 1994, Activation and inhibition of erythropoietin receptor function: Role of receptor dimerization, *Mol. Cell Biol.* **14**(6)**:**3535–3549.

Welch, P. J., and Wang, J. Y., 1993, A C-terminal protein-binding domain in the retinoblastoma protein regulates nuclear *c-Abl* tyrosine kinase in the cell cycle, *Cell* **75**(4)**:**779–790.

Welham, M. J., and Schrader, J. W., 1992, Steel factor-induced tyrosine phosphorylation in murine mast cells. Common elements withIL-3-induced signal transduction pathways, *J. Immunol.* **149**(8)**:**2772–2782.

Whetton, A. D., Heyworth, C. M., Nicholls, S. E., Evans, C. A., Lord, J. M., Dexter, T. M., and Owen-Lynch, P. J., 1994, Cytokine-mediated protein kinase C activation is a signal for lineage determination in bipotential granulocyte macrophage colony-forming cells, *J. Cell Biol.* **125**(3)**:**651–659.

Wiktor-Jedrzejczak, W., Bartocci, A., Ferrante, A. Jr., Ahmed-Ansari, A., Sell, K. W., Pollard, J. W., and Stanley, E. R., 1990, Total absence of colony-stimulating factor 1 in the macrophage-deficient osteopetrotic (*op/op*) mouse [published erratum appears in *Proc. Natl. Acad. Sci. U.S.A.* 1991, **88**(13):5937], *Proc. Natl. Acad. Sci. U.S.A.* **87**(12):4828–4832.

Witte, O. N., 1990, Steel locus defines new multipotent growth factor, *Cell* **63**(1):5–6.

Yi, T., Mui, A. L., Krystal, G., and Ihle, J. N., 1993, Hematopoietic cell phosphatase associates with the interleukin-3 (IL-3) receptor beta chain and down-regulates IL-3-induced tyrosine phosphorylation and mitogenesis, *Mol. Cell Biol.* **13**(12):7577–7586.

Yoder, M. C., and Williams, D.A., 1995, Matrix molecule interactions with the hematopoietic stem cell, *Exp. Hematol.* **23**:961–967.

Yoshida, H., Hayashi, S., Kunisada, T., Ogawa, M., Nishikawa, S., Okamura, H., Sudo, T., Shultz, L. D., and Nishikawa, S., 1990, The murine mutation osteopetrosis is in the coding region of the macrophage colony stimulating factor gene, *Nature* **345**:442–444.

Young, J. C., and Witte, O. N., 1988, Selective transformation of primitive lymphoid cells by the *bcr/abl* oncogene expressed in long term lymphoid or myeloid cultures, *Mol. Cell Biol.* **8**:4079–4087.

Zeigler, F. C., Bennett, B. D., Jordan, C. T., Spencer, S. D., Baumhueter, S., Carroll, K. J., Hooley, J., Bauer, K., and Matthews, W., 1994a, Cellular and molecular characterization of the role of the *flk-2/flt-3* receptor tyrosine kinase in hematopoietic stem cells, *Blood* **84**(8):2422–2430.

Zeigler, F. C., de Sauvage, F., Widmer, H. R., Keller, A., Donahue, C., Schreiber, R. D., Malloy, B., Hass, P., Eaton, D., and Matthews, W., 1994b, *In vitro* megakaryocytopoietic and thrombopoietic activity of *c-mpl* ligand (TPO) on purified murine hematopoietic stem cells, *Blood* **84**(12):4045–4052.

Chapter 6

Myelosuppressive Cytokines and Peptides

Hal E. Broxmeyer

1. INTRODUCTION

The mature blood cells that fight infection and disease (neutrophilic granulocytes, monocytes/macrophages, lymphocytes, natural killer cells), allow us to breathe (erythrocytes), and prevent us from bleeding to death (platelets), which are thus imperative for maintenance of our health, are produced by hematopoietic stem and progenitor cells (Broxmeyer and Williams, 1988; Broxmeyer, 1992a, 1993). Regulation of blood cell production involves an interacting network of cytokine–cell interactions. Over 50 different cytokines have been identified that can influence blood cell production, at least *in vitro* (Broxmeyer and Williams, 1988; Broxmeyer, 1992a, 1993, 1995a, 1996). This regulation can occur by stimulation, enhancement (augmentation), and/or suppression of the growth and differentiation of hematopoietic stem and progenitor cells. Hematopoietically active cytokines include the colony-stimulating factors (CSF) granulocyte (G)-CSF, macrophage (M)-CSF, GM-CSF, interleukin (IL-3), IL-5, and erythropoietin (Epo), and other interleukins, presently numbering 1 to 17. Non-CSF interleukins include potent costimulating cytokines: steel factor (SLF, or stem cell factor), and flt3/flk2 ligand (L), involved in synergistic stimulation of cell proliferation (Broxmeyer, 1992a,b, 1993, 1995a; Broxmeyer *et al.*, 1991b, 1995a).

Superimposed on this stimulation and enhancement of stimulation, and interactive with these, are a number of cytokines with growth-suppressing activity (Broxmeyer, 1992b). A listing of some of these suppressive cytokines is shown in Table I along with a list of peptides with demonstrated suppressive activity. As the actions *in vitro* and *in vivo* and mechanisms of actions of these negatively acting cytokines are elucidated, they are gaining

Hal E. Broxmeyer Departments of Medicine and Microbiology/Immunology and Walther Oncology Center, Indiana University School of Medicine, Indianapolis, Indiana 46202-5121.

Blood Cell Biochemistry, Volume 7: Hematopoietic Cell Growth Factors and Their Receptors, edited by A. D. Whetton and J. Gordon. Plenum Press, New York, 1996.

Table I
Cytokines and Peptides with Hematopoietic Suppressive Activity

Cytokines	Abbreviations
Macrophage inflammatory protein-1α	MIP-1α
Macrophage inflammatory protein-1β	MIP-1β
Interleukin 8	IL-8
Platelet factor 4	PF4
Monocyte chemotactic peptide 1 (monocyte chemotactic and activating factor)	MCP-1 (MCAF)
Interferon-γ inducible protein 10	IP-10
MIP-related protein 1	MRP-1
MIP-related protein 2	MRP-2
ENA-78	ENA-78
Lactoferrin	LF
H-ferritin	HF
Prostaglandin $E_{1,2}$	$PGE_{1,2}$
Interferons-α, -β, -γ	IFN-α, -β, -γ
Tumor necrosis factors-α, -β	TNF-α, -β
Transforming growth factor β	TGF-β
Inhibin	Inhibin
Vascular endothelial cell growth factor	VEGF
Macrophage-stimulating protein	MSP
Low-molecular-weight inhibitor (~8 kDa)	LMI
Peptides	
(Ala')-Tuftsin-containing 14-mer (NH_2ETVIMKAKPRANFP-COOH)	
pEEDCK	
AcSDKP	

more acceptance as biological regulators with potential physiological significance and are being evaluated, or are being considered for evaluation, in clinical trials as myelosuppressive or myeloprotective agents. There have now been three international conferences devoted entirely to cytokines and peptides with suppressive activity (Najman *et al.*, 1987; Anagnostou *et al.*, 1991; Guigon *et al.*, 1993; Broxmeyer *et al.*, 1991a; Dainiak *et al.*, 1994), although it is clear that these cytokines are similar in respect to most other cytokines in that they have pleiotropic effects. Thus, cytokines that may be suppressive in certain situations, depending on the target cell, the maturation state of the target cell, and other cytokines that may be in the environment of the target cell, may also have stimulating/enhancing activities in other situations. This chapter focuses on the suppressive activities of the cytokines and peptides listed in Table I.

2. ARE SUPPRESSOR CYTOKINES OF RELEVANCE?

Without a dampening mechanism, the unhampered proliferation of cells could result in a hyperproliferative and perhaps leukemic state, whereas in the opposite context, too much suppression could result in a hypoplastic state. Neither type of condition is appropriate for our well-being. Can we envision conditions in which negatively acting cytokines are not involved in the above situations? The answer is yes, but with qualification. For example, one can postulate that positively acting growth factors and cytokines that enhance this stimulation are all that is required. Decreased growth stimulation could occur by decreased production of growth factors, inactivation of growth factors, and/or by decreased sensitivity of target cells to stimulation. There is, however, evidence that some negatively acting cytokines decrease the production or release of growth factors from accessory cells (Broxmeyer *et al.*, 1978; Zucali *et al.*, 1989) or decrease numbers or affinity of growth factor receptors on target cells (S. E. W. Jacobsen *et al.*, 1991b). Overall, it is hard to believe that negatively acting growth factors are not of importance. Many such suppressor cytokines have been identified, isolated, and purified, their production evaluated at the gene level, their target cells of action identified, their intercellular mechanisms of action at least beginning to be elucidated, and their actions *in vivo* demonstrated after administration to animals.

Most importantly, it is crucial to realize that negatively acting cytokines can and do have different cellular sites of action. These actions can be directly on the stem/progenitor cell and/or indirectly on a supporting accessory cell, just as stimulating/enhancing growth factors have been shown to work. What we know of the actions of suppressor molecules comes from studies in which cell separation procedures are used to isolate and enrich stem/progenitor cells or accessory cells. It is known that the CD34 antigen is present on stem and progenitor cells, and antibodies to CD34 antigens have been used to highly enrich these cells. For many years, an action of a cytokine on $CD34^+$ cells was taken as proof that the action was directly on the stem or progenitor cell. However, not all purifications yield pure populations of $CD34^+$ cells, and it was not possible to eliminate the possibility that a few cells, even less than 1% of the population, that were not $CD34^+$ cells were mediating the actions of the cytokines through indirect effects. Most recently, however, it has become clear that not all $CD34^+$ cells, even those expressing the highest density of CD34 antigens, which apparently contain the more immature subsets of stem/progenitor cells (Lu *et al.*, 1993b), are stem and progenitor cells. Although CD34 is a marker for stem and progenitor cells, this antigen is also found on other cells.

One example of this is endothelial cells that are $CD34^+$ (Favaloro, 1993). It has been known for a long time that endothelial cells are capable, both constitutively and after induction, of releasing cytokines, some with stimulating activity (Bagby *et al.*, 1986). Thus, more rigorous proof of actions on stem/progenitor cells requires the study of actions on single isolated $CD34^+$ cells. Many of the negatively acting cytokines shown to inhibit growth of populations of $CD34^+$ cells were subsequently shown to inhibit growth of single isolated $CD34^+$ cells (Lu *et al.*, 1993a), thus confirming their action on stem/progenitor cells without an action of "feeder" cells within the $CD34^+$ population, at least at the initiation of cell culture. This latter qualification is added because, after a cell divides in culture, the daughter cells are not necessarily equivalent to the starting cells, and there is the possibility, if the action of a cytokine is not shown to occur rapidly, that effects can still be considered indirect even when a single isolated cell culture is studied, as the effects may be mediated through the nonstem/progenitor daughter cells.

Recent studies in our laboratory have shown that vascular endothelial cell growth factor (VEGF), whose actions are apparent after a short pulse exposure, has growth-promoting and suppressing activity on very highly purified populations of $CD34^{+++}$ cells, those cells expressing the highest density of CD34 antigens (Broxmeyer *et al.*, 1995b). However, when we analyzed these effects on colony formation by single isolated $CD34^{+++}$

cells, most of the enhancing/suppressing activity was not seen (Broxmeyer *et al.*, 1995b), suggesting that many of the effects we had noted on stem/progenitor cells were mediated indirectly, perhaps through an action on $CD34^{+++}$ endothelial cells.

In a recent review, the hematopoietic colony-stimulating factors (CSF) were reappraised as hematopoietic regulators, based on studies of gene inactivation (Metcalf, 1995). The provisional conclusion was that early *in vitro* studies were correct in identifying CSFs as major regulators of granulocyte–macrophage production. In this context, interleukin 8 (IL-8) has been identified as a negative regulator of hematopoiesis *in vitro* (Broxmeyer *et al.*, 1993a) and *in vivo* (Broxmeyer *et al.*, 1995c), and recent studies have shown that in mice in which the gene for a human homologue of the IL-8 receptor is knocked out by homologous recombination, there is a constitutive hyperproliferative state of the hematopoietic tissue (Cacalano *et al.*, 1994) and enhanced proliferation of progenitor cells (Broxmeyer *et al.*, 1995f). Thus, evidence is becoming available through gene inactivation studies that suppressor cytokines can function as potential major regulators of hematopoiesis.

3. WHAT ARE PRACTICAL USES FOR SUPPRESSOR CYTOKINES?

Two major uses of suppressor cytokines can be currently envisioned. One is as a dampener of hyperproliferation. This assumes that the target cells are not changed in their sensitivity to respond to the effects of the cytokines. In a number of cases, it has been demonstrated through *in vitro* and animal studies that leukemic or virally treated cells do not respond to the regulatory effects of cytokines that suppress the proliferation of normal cells. This has been shown for H-ferritin (Broxmeyer *et al.*, 1981), lactoferrin (LF) (Broxmeyer *et al.*, 1983a), and, to a more heterogeneous degree, for macrophage inflammatory protein (MIP)-1α and other members of the chemokine family (Broxmeyer *et al.*, 1995c). Here, it has been postulated that suppressor molecules may be of use in inhibiting the proliferation of normal progenitor cells but not tumor cells, thus placing the normal progenitors into a more protected phase of the cell cycle (G_0 or G_1) (Broxmeyer, 1995c). This might allow conditioning protocols with chemotherapy to kill the tumor but not the "protected" normal cells (Broxmeyer, 1995b (see Chapter 15).

It is of interest in this regard that treatment of patients with GM-CSF, G-CSF, and PIXY321 (A GM-CSF/IL-3 fusion protein) places progenitor cells into rapid cell cycle (Aglietta *et al.*, 1989; Vadhan Raj *et al.*, 1992; Broxmeyer *et al.*, 1988a, 1989a, 1995d) as long as these CSFs are administered to patients. However, as quickly as 1 day after cessation of the administration of GM-CSF (Aglietta *et al.*, 1989; Vadhan Raj *et al.*, 1992) and of low dosages of PIXY321 (Broxmeyer *et al.*, 1995d), the progenitor cells are placed into a slow or noncycling state below the pre-GM-CSF/PIXY321 baseline levels. This phenomenon of placing cells out of cycle after cessation of administration of a CSF was not seen with G-CSF (Broxmeyer *et al.*, 1994), where the enhanced cycling of progenitors was still noted days after stopping the G-CSF. With higher dosages of PIXY321, it was found that the progenitors were placed into more rapid cycle within the first 8 days of PIXY321 administration, but then the progenitors were put out of cycle while the PIXY321 was still being given (Broxmeyer *et al.*, 1995d). This latter effect was not attributed to the production of neutralizing PIXY321 antibodies (Broxmeyer *et al.*, 1995d). It is believed that the GM-CSF and PIXY321 may be able to be used to place the progenitors in patients into a slow or noncycling state that may be less sensitive to the killing effects of certain chemotherapeutic drugs (Aglietta *et al.*, 1989; Vadhan Raj *et al.*, 1992; Broxmeyer *et al.*, 1995b,d). It is possible that this decreased cycling rate noted with time after administration of GM-CSF and PIXY321 may be related to the release and activity of some of the suppressor cytokines to be described below.

Whether or not this is the case, the possibility exists that some of these suppressor cytokines may be used in the context of myeloprotection: MIP-1α has shown some myeloprotective activity against S-phase cytotoxic drugs in mice (Lord *et al.*, 1992; Dunlop *et al.*, 1992), and two other chemokines, IL-8 and platelet factor 4 (PF4), have shown similar activity (Broxmeyer *et al.*, 1995c). BB10010, an analogue of MIP-1α (Hunter *et al.*, 1995), is currently in clinical trials in Manchester, England and Indianapolis, Indiana as a potential myeloprotective agent. The clinical results from BB10010 use in patients with refractory breast cancer mimic those of the preclinical studies: BB10010 administration is associated in the patients with decreased proliferative rates of myeloid progenitor cells (Broxmeyer *et al.*, 1995g). As we learn more about the actions and mechanisms of actions of suppressor cytokines and their activity *in vivo*, there is optimism that we will be able to utilize these cytokines, or peptides with similar activities, rationally in clinical situations for the benefit of patients with hematologic or other disorders.

4. CHEMOKINE FAMILY OF CYTOKINES

Chemokines are a family of small-molecular-weight inducible heparin-binding proteins that contain either a conserved cysteine-x-cysteine (C-x-C) or a cysteine-cysteine amino acid motif (Schall, 1991; Baggiolini *et al.*, 1994; Michiel, 1993; Clore and Gronenborn, 1995; Van Damme, 1994; Clark-Lewis *et al.*, 1995; M. D. Miller and Krangel, 1992). Members of the chemokine family (Table II) have been implicated in a number of activities associated with attraction and activation of leukocytes with postulated roles as mediators of inflammation. Among the activities of chemokines are their enhancing/suppressing effects on the proliferation of myeloid progenitor cells, with macrophage inflammatory protein (MIP)-1α being the most studied member of this family in this context (Lu *et al.*, 1993a; Gewirtz *et al.*, 1989; Bodine *et al.*, 1991; Clements *et al.*, 1992; Han *et al.*, 1992; Sarris *et al.*, 1993; Broxmeyer *et al.*, 1989b, 1990, 1991c, 1993a,b, 1995c; Mantel *et al.*, 1993; Maltman *et al.*, 1993; Keller *et al.*, 1994; Verfaille *et al.*, 1994; Graham *et al.*, 1990, 1992, 1994; Mayani *et al.*, 1995). The MIP-1α has suppressive activity on early subsets of normal murine and human stem and progenitor cells *in vitro* (Lu *et al.*, 1993a; Broxmeyer *et al.*, 1990, 1991c, 1993a,b, 1995c; Gewirtz *et al.*, 1989; Bodine *et al.*, 1991; Clements *et al.*, 1992; Graham *et al.*, 1990, 1992, 1994; Han *et al.*, 1992; Sarris *et al.*, 1993; Keller *et al.*, 1994; Mayani *et al.*, 1995) and on murine myelopoiesis *in vivo* (Broxmeyer *et al.*, 1995c; Lord *et al.*, 1992; Dunlop *et al.*, 1992; Clemjents *et al.*, 1992; Maze *et al.*, 1992; Cooper *et al.*, 1994).

The first hematopoietic function noted for MIP-1α was as an enhancing activity for progenitors stimulated by GM-CSF and M-CSF (Broxmeyer *et al.*, 1989b, 1990); MIP-1α was subsequently identified as a suppressor molecule for CFU-S and for CFU-A (Graham *et al.*, 1990), the latter cell considered to be an immature progenitor. MIP-1α was subsequently demonstrated to have direct suppressive action on early subsets of myeloid progenitor cells stimulated to proliferate in the presence of two or more growth-stimulating

Table II
Chemokine Family Members

Position-invariant C-C motif: β-subfamily
MIP-1α (macrophage inflammatory protein 1α) = LD78
MIP-1β = Act 2
RANTES (regulated on activation, normal T expressed and secreted)
MCP-1 (monocyte chemotactic peptide 1) = MCAF
MCP-2
MCP-3
MRP-1 (MIP-related protein 1) = C 10
MRP-2
Position-invariant C-x-C motif: α-subfamily
GRO-α = melanoma growth-stimulating factor = KC
GRO-β = MIP-2α
GRO-γ = MIP-2β
NAP-2 (neutrophil-activating peptide 2)
IL-8 (interleukin 2) = NAP-1
PF4 (platelet factor 4)
IP-10 (γ-inducible protein 10)
ENA-78

cytokines such as GM-CSF and steel factor (SLF) (Lu *et al.*, 1993a; Broxmeyer *et al.*, 1990, 1991c), which in the absence of chemokines resulted in synergistic cell proliferation. The limited experience with MIP-1α on the growth *in vitro* of progenitor cells from patients with leukemia suggests heterogeneity in responsiveness to actions of MIP-1α (Broxmeyer *et al.*, 1993b, 1995c; Eaves *et al.*, 1993; Ferrajoli *et al.*, 1994). Based on the suppressive effects of MIP-1α *in vivo* in normal mice and in mice receiving cycle-active chemotherapy (Broxmeyer *et al.*, 1995c; Lord *et al.*, 1992; Dunlop *et al.*, 1992; Clements *et al.*, 1992; Maze *et al.*, 1992; Cooper *et al.*, 1994), it has been suggested that MIP-1α may be of potential use clinically in a setting of myeloprotection during chemotherapy. Phase 1 clinical data using the MIP-1α analogue B10010 (Hunter *et al.*, 1995) demonstrates that this material can put myeloid progenitor cells into a slow or noncycling state (Broxmeyer *et al.*, 1995g). It remains to be determined if this will result in myeloprotection.

Other members of the chemokines family, including MIP-2α (= GRO-β), interleukin 8, platelet factor (PF) 4, monocyte chemotactic and activating factor (MCAF = monocyte chemotactic peptide-1, MCP-1), interferon-γ inducible protein 10 (IP-10), and ENA-78 have also demonstrated suppressive activity on early subsets of stem/progenitor cells *in vitro* (Lu *et al.*, 1993a; Broxmeyer *et al.*, 1993a, 1995c, 1996; Gewirtz *et al.*, 1989; Han *et al.*, 1992; Sarris *et al.*, 1993). Other chemokines, including MIP-1β, MIP-2β (= GRO-γ), GRO-α, neutrophil-activating peptide (NAP)-2, and RANTES (regulated on activation, normal T expressed and secreted) did not manifest this myelosuppressive activity (Broxmeyer *et al.*, 1990, 1991c, 1993a). However, MIP-1β blocked the suppressive activity of MIP-1α (Broxmeyer *et al.*, 1991c, 1993a), and MIP-2β (= GRO-γ) and GRO-α blocked the suppressive effects of PF4 and IL-8 (Broxmeyer *et al.*, 1993a).

An interesting phenomenon associated with suppressive chemokines is that when any two of the active chemokines (MIP-1α, MIP-2α, IL-8, PF4, MCAF, or IP-10) are added together *in vitro* at low concentrations at which neither is active alone, a synergistic suppressive activity is apparent (Broxmeyer *et al.*, 1993a, 1995c, 1996; Sarris *et al.*, 1993). It is not yet clear how this synergism is manifest, but it probably relates to the receptor and intracellular signal transduction activity of these chemokines. Interestingly, it is still only the early progenitors that are responsive to stimulation by multiple (two or more) growth-stimulating/enhancing growth factors and are suppressed by the action of single chemokines and synergistically suppressed by multiple chemokines. Multiple chemokines do not suppress progenitors' response to stimulation by one growth factor (Broxmeyer *et al.*, 1993a, 1995c; Sarris *et al.*, 1993).

Based on previous studies that demonstrated enhancement in the myelosuppressive effects of MIP-1α *in vitro* (Mantel *et al.*, 1993) and *in vivo* (Cooper *et al.*, 1994) after preincubation of the MIP-1α in an acetonitrile (ACN) solution, other members of the chemokine family were tested for suppressive activity after preincubation of these chemokines in ACN (Broxmeyer *et al.*, 1995c). Preincubation of the suppressive chemokines, but not nonsuppressive chemokines such as MIP-1β, MIP-2β, GRO-α, NAP-2, and RANTES, resulted in significantly enhanced suppressive specific activity in which lower concentrations of these chemokines were now active. The ACN-treated chemokines were assessed for activity at concentrations of 5 ng/ml or lower *in vitro* and 8 ng/animal *in vivo* in order to make sure that there was no toxicity associated with carryover of ACN into the assay plates or mice (Broxmeyer *et al.*, 1995c). The very small amount of ACN within the chemokine preparations at these concentrations of chemokines had no significant effect on colony formation, as was apparent from the diluent control. Moreover, this amount of ACN had no effect on nonsuppressive chemokines. The enhanced specific activity of ACN-treated human preparations of MIP-1α, MIP-2α, IL-8, PF4, MCAF, and IP-10 was also apparent when combinations of low concentrations of any two of these six chemokines were assayed. As with these same non-ACN-treated chemokines (Broxmeyer *et al.*, 1993a; Sarris *et al.*, 1993), the ACN-treated chemokines synergized to suppress progenitor cell growth; however, the concentrations of ACN-treated chemokines needed to detect this synergism were lower than for the non-ACN-treated chemokines. Acetonitrile keeps MIP-1α mainly in monomeric form (Mantel *et al.*, 1993; Patel *et al.*, 1993), which is maintained when low concentrations of ACN-treated MIP-1α are added to a PBS solution (Mantel *et al.*, 1993). Many investigators believe that the physiologically relevant and active form of MIP-1α is a monomer (Mantel *et al.*, 1993; Graham *et al.*, 1994; Patel *et al.*, 1993; Avalos *et al.*, 1994). It is possible that the ACN caused dissociation of those chemokines that are known to aggregate, and this may be responsible for the enhanced specific activity of chemokines that have manifested suppressive activity without ACN pretreatment.

There are reports of other chemokines that are monomeric at physiologically relevant concentrations (Paolini *et al.*, 1994), and the use of an analogue of IL-8 that was in monomeric form led some investigators to suggest that the monomer was a functional form of IL-8 for activation of neutrophils (Rajarathnam *et al.*, 1994). Although there is good agreement that the physiological form of MIP 1α is monomeric, others using MIP-1α mutants that can exist as tetramers, dimers, and monomers did not note enhanced specific activity of the monomeric form of MIP-1α compared to the other forms (Graham *et al.*, 1994). A potential problem inherent in the use of muteins to assess specific activity is that changes in the molecule that could confer less aggregation capacity could also elicit changes in the molecule that lessen its intrinsic biological activity for specific functions. Whether the enhanced specific activity noted in our studies (Broxmeyer *et al.*, 1995c; Mantel *et al.*, 1993) with some of the ACN-treated chemokines is a result of decreased aggregation or of other biochemical or biophysical changes remains to be determined. Most importantly, these studies demonstrate the low concentrations at which these chemokines are biologically active both *in vitro* (Mantel *et al.*, 1993) and *in vivo* (Cooper *et al.*, 1994). Those chemokines with suppressive activity *in vitro* (Broxmeyer *et al.*, 1993a; Sarris *et al.*, 1993) have suppressive activity *in vivo* that correlates with their *in vitro* activity (Broxmeyer *et al.*, 1995c; Maze *et al.*, 1992; Cooper *et al.*, 1994).

Previous studies assessing effects of chemokines on purified populations of $CD34^{+++}$ bone marrow cells (Broxmeyer *et al.*, 1990, 1993a; Sarris *et al.*, 1993) and more rigorously at the level of single isolated $CD34^{+++}$ cells (Lu *et al.*, 1993a) strongly suggested that the suppressive effects were direct acting at the level of the hematopoietic stem/progenitor cells themselves. This, of course, does not rule out additional effects mediated by chemokine-activated accessory cells, especially *in vivo*. The mechanisms of action of the chemokines, either alone or in combination, as suppressor molecules of myelopoiesis is only beginning to be evaluated. It is clear that for the suppressive activity of MIP-1α to be manifested, the stem/progenitor cells need to be in a rapidly proliferating state (Lord *et al.*, 1992; Dunlop *et al.*, 1992; Mantel *et al.*, 1993; Maze *et al.*, 1992; Cooper *et al.*, 1994). This was further confirmed by demonstrating that cord blood myeloid progenitors that are in a slow or noncycling state and insensitive to suppression of their proliferation by chemokines (Lu *et al.*, 1993a) can be induced with growth factors in suspension culture *in vitro* into a rapidly cycling phase that places these cells into a state receptive for suppression by the chemokines (Broxmeyer *et al.*, 1995c).

More mechanistic approaches to understanding the action of chemokines requires a cell line model such as the human growth factor-dependent cell line M07e (Hendrie *et al.*, 1991), as it is not yet technically feasible, because of the difficulty of isolating enough such purified cells, to analyze many of the intracellular biochemical and molecular signaling events in primary stem and progenitor cells. M07e cells have receptors for MIP-1α (Mantel *et al.*, 1995; Aronica *et al.*, 1995) and other chemokines (Aronica *et al.*, 1995). M07e colony-forming cells (CFC) respond to suppression by MIP-1α (Broxmeyer *et al.*, 1993b; Mantel *et al.*, 1995; Aronica *et al.*, 1995), and these cells have been used to demonstrate that MIP-1α enhances growth factor-stimulated phosphatidylcholine metabolism and increases cyclic AMP levels, events associated with growth suppression (Mantel *et al.*, 1995; Aronica *et al.*, 1995).

The M07e cells were able to be used as a model for chemokine suppression because these cells respond to the synergistically induced growth proliferation induced by the combination of Steel factor with either GM-CSF or IL-3, and the chemokines suppress the Steel factor enhancement of GM-CSF- or IL-3-stimulated growth (Broxmeyer *et al.*, 1995c; Sarris *et al.*, 1993), essentially duplicating the type of suppressive activity the chemokines manifest on proliferation of normal hematopoietic progenitor cells (Broxmeyer *et al.*, 1990, 1993a, 1995c; Sarris *et al.*, 1993). The MIP-1α enhanced phosphatidylcholine (PC) and phosphocholine turnover rates when cells were synergistically stimulated to proliferate by GM-CSF and Steel factor. Also, MIP-1α, in a dose-dependent fashion, increased levels of cyclic AMP. Exogenous addition of PC or dibutyryl cyclic AMP decreased the percentage of M07e CFC in S phase (Mantel *et al.*, 1995) in a manner similar to that noted with treatment of M07e cells with MIP-1α or the other active chemokines (Broxmeyer *et al.*,

1995c). However, RANTES, a nonsuppressive chemokine (Broxmeyer *et al.*, 1993a, 1995c), did not suppress proliferation of M07e cells and did not increase cyclic AMP levels, but it did enhance the GM-CSF- plus Steel factor-induced turnover of PC. Overall, the results suggest that although PC turnover is probably involved in the suppressive action of MIP-1α, this increased PC turnover alone is not sufficient for suppression. Further studies with the M07e cell line (Aronica *et al.*, 1995) found that MIP-1α, IP-10, PF4, and IL-8, but not the nonsuppressive chemokines MIP-1β or GRO-α, decreased protein synthesis of basal levels by 15% to 20%. The basal levels of protein synthesis in the absence of chemokines were increased by 40% to 65% in the presence of the combination of GM-CSF plus Steel factor, and MIP-1α, IP-10, PF4, and IL-8 were each able to decrease the GM-CSF- plus Steel-factor-stimulated protein synthesis to basal levels. Interestingly MIP-1α and IP-10, but not PF4, IL-8, MIP-1β, or GRO-α, increased cyclic AMP levels in a dose-dependent manner. Moreover, MIP-1α and IP-10, but not PF4, IL-8, MIP-1β, or GRO-α, decreased completely the GM-CSF plus Steel factor synergistically up-regulated phosphorylation of RAF-1 and activation of RAF-1 kinase activity (as measured by phosphorylation of MEK1 MAP-kinase).

Because RAF-1 is inhibited directly by cyclic-AMP-dependent protein kinase A (PKA), the specific PKA inhibitor, PKI, was used to block PKA in M07e cells (Aronica *et al.*, 1995). Pretreatment of cells with PKI had no effect on GM-CSF plus Steel factor activation of RAF-1 kinase activity, but it did block the suppressive activity of MIP-1α and IP-10 on GM-CSF- plus Steel-factor-enhanced RAF-1 kinase activity. Also, PKI blocked the suppressive effect of MIP-1α and IP-10, but not of PF4 or IL-8, on suppression of M07e CFC proliferation (Aronica *et al.*, 1995). The above clearly demonstrated that although the suppressive effects of MIP-1α, IP-10, PF4, and IL-8 were similar at a cellular level, the intracellular signal-transducing effects of MIP-1α and IP-10 associated with suppressive activity could be distinguished from that of PF4 and IL-8. There are a number of different MAP-kinase pathways (Pelech and Charest, 1995; Davis, 1994; Herskowitz, 1995), and it is possible that different chemokines may mediate their suppressive activity on similar or different converging pathways. Moreover, a number of signal transduction pathways such as immediate response genes (Horie and Broxmeyer, 1993) or SH2-domain-interacting intracellular proteins such as SHC (Tauchi *et al.*, 1994c) or the protein phosphatases Syp (Tauchi *et al.*, 1994a,b, 1995) or haematopoietic cell phosphatase (HCP) (Yi and Ihle, 1993; Yi *et al.*, 1993, 1995) may be involved in the negative regulatory effects of the suppressive chemokines (see Chapter 1).

It is important to note that pulsing experiments and continuous addition studies demonstrated that the effects of the chemokines on M07e CFC proliferation are rapidly reversible (Broxmeyer *et al.*, 1995c). Although pulse treatment of M07e cells with some chemokines decreases the percentage of M07e CFC in S phase, it does not decrease actual M07e colony formation, an event only apparent after chemokines have been added for 3 to 4 days to plates containing M07e cells. This is in contrast to studies with primary cells, in which pulse exposure of bone marrow cells to chemokines results in decreased progenitor cell colony formation (Broxmeyer *et al.*, 1991c, 1993a). This rapidly reversible effect may at least partially explain why one laboratory (Avalos *et al.*, 1994) has not reproduced the suppressive effects of MIP-1α on the same factor-dependent murine cell line used in another laboratory (Graham *et al.*, 1993).

Heterogeneity in response of granulocyte–macrophage progenitor cells (CFU-GM) between patients with leukemia to the effect of MIP-1α has been reported (Broxmeyer *et al.*, 1993b; Eaves *et al.*, 1993; Ferrajoli *et al.*, 1994), and this heterogeneity has now been expanded to that of the other chemokines: MIP-2α, IL-8, PF4, MCAF, and IP-10 (Broxmeyer *et al.*, 1995c). Based on the limited number of patients evaluated thus far, this heterogeneity in responsiveness does not appear to be associated with treatment. However, analysis of greater numbers of samples and more mechanistic studies are required before a definitive statement on potential reasons for this heterogeneity can be made. Of extreme interest is the observation that if cells from a patient respond to the suppressive effects of one of the myelosuppressive chemokines, they respond to those of the other suppressive chemokines, and if the cells do not respond to one, they do not respond to other chemokines.

The field of chemokine receptors is still in its infancy. Chemokine receptors have been identified that bind one or more chemokines or that are relatively promiscuous (Graham *et al.*, 1993; Oh *et al.*, 1991; Moser *et al.*, 1991; Holmes *et al.*, 1991; Schumacher *et al.*, 1992; Yamamura *et al.*, 1992; Cerrertti *et al.*, 1993; J. M. Wang *et al.*, 1993; Kelvin *et al.*, 1993; Ahija *et al.*, 1994; Neote *et al.*, 1993a,b, 1994; Chaudhuri *et al.*, 1994), and it is likely that other chemokine receptors will be identified. It remains to be determined if nonresponsive progenitors from some patients with leukemia reflect a lack of one or more chemokine receptors or if this reflects differences in affinity of chemokines for their receptors and/or differences in intracellular signaling events.

In addition to some of the activities shown above for the chemokines, there are reports of IL-8 (Laterveer *et al.*, 1995) and an analogue of MIP-1α (BB10010) (Lord *et al.*, 1995) inducing the mobilization of stem/progenitor cells from the marrow to peripheral blood of mice, effects that could be of value for transplantation purposes should these effects also be apparent in human clinical trials.

The array of members belonging to the chemokine family is by no means complete. A number of laboratories have uncovered chemokines related in sequence to other chemokine members. Examples of these are the MIP-related proteins (MRP)-1 and -2 (Youn *et al.*, 1995). MRP-2 was isolated from a murine macrophage cell line and is composed of 122 amino acids of which the first 21 are a putative signal sequence. The mature protein has an apparent size of 11,600 daltons and is structurally similar to MRP-1 (also identified as C10) and MIP-1α with respective sequence identity at the protein level of 50.8% and 46.3%. Recombinant murine MRP-2 significantly suppressed colony formation by murine and human granulocyte–macrophage (CFU-GM), erythroid (BFU-E), and multipotential (CFU-GEMM) progenitor cells stimulated by combinations of growth factors (Youn *et al.*, 1995).

The redundancy in suppressive activity of chemokines is of interest. It also suggests that it may not be possible, using chemokine knockout mice generated by such procedures as homologous recombination, to prove or disprove the relevancy of one particular chemokine, because other chemokine genes, not knocked out, could easily compensate for the action of the knocked-out gene and its product. A more fruitful approach in this context may be to knock out the gene for the chemokine receptor. As mentioned above, the area of chemokine receptors is not yet clear, and some receptors appear to have the capacity to bind, but not necessarily to activate, these chemokine receptors. There are two IL-8 receptors in humans, but studies with IL-8 receptor knockout mice identified one IL-8 receptor in mice that is a homologue to the two human receptors (Cacalano *et al.*, 1994). Knockout mice lacking this IL-8 receptor demonstrated a hyperproliferation of blood cells (Cacalano *et al.*,

1994), an observation consistent with the negative regulatory role postulated for IL-8 (Broxmeyer *et al.*, 1993a, 1995c). This proliferation enhancement in the IL-8 receptor knockout mice has now been documented for CFU-GM, BFU-E, and CFU-GEMM (Broxmeyer *et al.*, 1995f). Moreover, it has been determined that marrow progenitors from these IL-8 receptor knockout mice respond to the suppressive effects of MIP-1α, PF4, MCAF, and IP-10 but not to IL-8.

A number of orphan chemokine receptors have been identified. These contain sequences consistent with those of other chemokine receptors, but no ligands for these receptors have yet been identified.

In an effort to determine the active portions of certain chemokines with regard to their myelosuppressive and other functional activities, chimeric mutants of IL-8 and PF4 were produced (Daly *et al.*, 1995). Mutation of either of the two DLQ motifs within PF4 resulted in an inactive protein. Changes within the IL-8 dimer interface region inactivated this protein. A class of chimeric mutants consisting of domains of either PF4 and IL-8, GROα and PF4, or GRO-β and PF4 enhanced the suppressive specific activity of the proteins. These chimeric mutants possessed activities that were comparable to or better than the activity observed when wild-type IL-8 and PF4 were added together. In one perturbation of the amino-terminal "ELR" motif, it was found that the highly active IL-8 mutant required both aspartic acid at amino acid residue 4 and either glutamine or asparagine at residue 6. Mutants such as IL-8M1 and IL-8M10, highly active and increased in suppressive specific activity for myeloid progenitors, were significantly reduced in their ability to activate PMN, suggesting that separate mechanisms may exist by which myeloid progenitors and neutrophils are affected by chemokines. In fact, the receptors for IL-8 and PF4 may be different on progenitors and neutrophils.

Some evidence for a link with chemokines and disease is beginning to be uncovered. As an example, cytokine loops involving IP-10 and also interferon-γ has been suggested as an explanation for the epidermotropism of cutaneous T-cell lymphoma (Sarris *et al.*, 1995).

5. LACTOFERRIN

Lactoferrin (LF) is an iron-binding protein (Metz-Boutique *et al.*, 1984) whose three-dimensional structure has been elucidated crystallographically at 3.2 Å (Anderson *et al.*, 1987). It can bind two atoms of iron (Mantel *et al.*, 1994) and has been implicated in normal regulation of myeloid blood cell production *in vitro* (Broxmeyer *et al.*, 1978, 1980, 1986a; Bagby *et al.*, 1989) and *in vivo* (Gentile and Broxmeyer, 1983; Broxmeyer *et al.*, 1987a). Moreover, abnormalities in LF–cell interactions have been associated with progression of leukemia and other hematopoietic disorders (Broxmeyer *et al.*, 1977, 1983a, 1991e). Lactoferrin was first shown to manifest its suppressive activity *in vitro* indirectly, by decreasing the production/release of CSFs (Broxmeyer *et al.*, 1978, 1980; Broxmeyer and Platzer, 1984) or IL-1 (Zucali *et al.*, 1989; Bagby *et al.*, 1986, 1989) from monocytes/macrophages. The LF molecule has a number of interesting characteristics, including the fact that on high-resolution ion-exchange chromatography, the binding of LF to both anionic and cationic media under identical conditions indicated a bipolar charge distribution (Mantel *et al.*, 1994), and multiple forms of LF have been noted in normal and leukemia human granulocytes (Mantel *et al.*, 1994; Furmanski *et al.*, 1989; Furmanski and Li, 1990). Granulocytes are the major hematopoietic cells that store LF. Interestingly, two of the forms appear to express potent RNase activity (Furmanski *et al.*, 1989; Furmanski and Li, 1990). The relationship of these different forms to myelosuppressive activity and the abnormalities of this in leukemia have not been established. One characteristic that has been associated with activity, or more appropriately the loss of activity, is the polymerization of the molecule (Bagby and Bennett, 1982; Bennett *et al.*, 1981). Polymerization was shown to abrogate the capacity of LF to inhibit growth factor release from monocytes/macrophages (Bagby and Bennett, 1982; Bennett *et al.*, 1981). Polymerization was originally shown to be calcium dependent (Bennett *et al.*, 1981) but might in fact be dependent rather on the ionic strength of the solution (Mantel *et al.*, 1994). Inactivation of LF by polymerization has similarities to the reported inactivation of MIP-1α noted on aggregation of MIP-1α samples (Mantel *et al.*, 1993). It was first suggested that LF directly decreased release of CSFs from monocytes/macrophages (Broxmeyer *et al.*, 1978, 1980; Broxmeyer and Platzer, 1984). It was subsequently suggested that LF decreased the release of a monokine or monokines (Bagby *et al.*, 1986, 1989), later identified, at least in part, as IL-1 (Zucali *et al.*, 1989).

Most recently, it has been suggested that LF plays a negative role in GM-CSF expression at the transcriptional level, an effect that may be mediated through IL-1 (Penco *et al.*, 1995). That LF, in some form, can get into the nucleus of monocytes was first suggested by immunoelectron microscopic tracing of LF (Steinmann *et al.*, 1982) and later suggested in K562 cells (Garre *et al.*, 1992). Sequence specificity and transcriptional activation in the binding of LF to DNA have been noted (He and Furmanski, 1995). Abrogation of the action of LF by bacterial lipopolysaccharide (LPS) was shown to be related to the relative concentrations of LF to LPS (Broxmeyer *et al.*, 1978, 1980), effects later shown to result, at least in part, from the formation of LF-LPS complexes (Miyazawa *et al.*, 1991). Lactoferrin can decrease the production or release from monocytes/macrophages of prostaglandin $E_{1,2}$ (Pelus *et al.*, 1979, 1981) and H-ferritin (Broxmeyer *et al.*, 1984, 1985a), two molecules with suppressive activity that will be described below.

Although the original suppressive activities ascribed to LF were indirect ones mediated through monocytes and macrophages, more recent results suggest that LF can also mediate its effects directly on early subsets of stem/progenitor cells (H. E. Broxmeyer, S. Cooper, and L. Lu, unpublished observations). Although LF does not suppress colony formation of human bone marrow cells stimulated by either GM-CSF or erythropoietin (Epo), it does suppress the synergistically induced Steel-factor-enhanced proliferation of CFU-GM stimulated by GM-CSF, and of BFU-E stimulated by Epo. These effects were noted on $CD34^{+++}$ cell populations and at the level of single isolated $CD34^{+++}$ cells.

Lactoferrin has beneficial effects in animal models of virally induced hematopoietic abnormalities (Lu *et al.*, 1987a; Hangoc *et al.*, 1987a; Chen *et al.*, 1987), effects enhanced by pretreatment of mice with interferon-γ (Lu *et al.*, 1991), and it also inhibits growth of solid tumors and development of experimental metastases in mice (Bezault *et al.*, 1994).

6. H-FERRITIN

H-Ferritin is another iron-binding protein that has shown myelosuppressive activity *in vitro* (Broxmeyer *et al.*, 1981, 1982, 1986b, 1991d; Broxmeyer, 1982; Lu *et al.*, 1983) and also in animal models *in vivo* (Broxmeyer *et al.*, 1981, 1988b, 1991d). Abnormalities in H-ferritin–cell interactions have been noted during hematological disorders, including leukemia, where overproduction of H-ferritin and insensitivity of leukemia cells to inhibition by H-ferritin have been shown (Broxmeyer, 1982; Broxmeyer *et al.*, 1981, 1991e). Abnormalities have also been seen in mouse models of hematopoietic disorders, specifically in mice infected with the polycythemia-inducing strain of the Friend virus complex (Lu *et al.*, 1985). The suppressive effects of H-ferritin are directly on the stem/progenitor cells as evaluated using purified populations of cells (Lu *et al.*, 1993a; D. E. Williams *et al.*, 1988). More detailed information of the myelosuppressive activity of H-ferritin is reviewed elsewhere (Broxmeyer, 1989, 1992b,c). Of interest mechanistically, the suppressive activity of H-ferritin appears to reside in the ferroxidase activity of the molecule (Broxmeyer *et al.*, 1991d); the ferroxidase activity of H-ferritin also appears to be involved in the cytoprotective antioxidant strategem of H-ferritin for endothelium (Balla *et al.*, 1992). Specifically, it is the H- and not the L-ferritin-rich molecules that are myelosuppressive (Broxmeyer *et al.*, 1981, 1986b, 1991d). This is in contrast to another biological activity in which both H- and L-rich ferritins suppress antibody production (Morikawa *et al.*, 1994).

7. E-TYPE PROSTAGLANDINS

Prostaglandins are not proteins (Nelson *et al.*, 1982) and, in an exact sense, may not be classified as cytokines, but they do have activities in common with protein cytokines. This includes their capacity to enhance erythroid colony formation (Lu and Broxmeyer, 1985; Lu *et al.*, 1986a, 1987b), an effect related to cell cycling and the Fv-2 mouse locus (Hangoc *et al.*, 1987b), and to suppress granulocyte–macrophage colony formation (Kurland and Moore, 1977; Kurland *et al.*, 1978a,b), an effect most apparent on the macrophage progenitors (Pelus *et al.*, 1979, 1981; N. Williams, 1979; Duma *et al.*, 1988). The suppressive effect was associated with progenitor cell cycling and HLA-DR antigens on the cell surface (Pelus, 1982; Lu *et al.*, 1984), similar to that noted for the suppressive activity of H-ferritin (Broxmeyer, 1982). The $PGE_{1,2}$-mediated inhibition was suggested to be a direct-acting effect on the progenitors (Lu *et al.*, 1987c; D. E. Williams *et al.*, 1988). The PGE synergizes with other negative regulators such as the interferons and tumor necrosis factors (Broxmeyer *et al.*, 1987b; Pelus *et al.*, 1988) to suppress progenitor cell proliferation. Prostaglandin E has a myelosuppressive effect *in vivo* (Gentile and Pelus, 1987; Gentile *et al.*, 1983) that mimics its effects *in vitro* and that at least in part may be related to the induction of suppressor cells in bone marrow and spleen that can mediate inhibition of progenitor cell proliferation (Gentile and Pelus, 1988; Pelus and Gentile, 1988).

8. INTERFERONS

The interferon system is composed of a family of related molecules (Sen and Lengyel, 1992). Interferon-α, -β, and γ have suppressive activity on the proliferation of myeloid progenitor cells (Broxmeyer *et al.*, 1983b, 1985b; Lu *et al.*, 1986b), effects that are in large part apparently direct actions on the progenitors themselves (Lu *et al.*, 1987c). Some of these effects may be indirectly mediated on accessory cells (Ganser 1987), and the indirect and direct suppressive effects of the interferons are not entirely sorted out yet. Of interest and potential physiological relevance are the synergistic suppressive effects that the interferons have in combination with each other (Broxmeyer *et al.*, 1985b) and with other molecules such as the tumor necrosis factors (Broxmeyer *et al.*, 1986c; Pelus *et al.*, 1988), H-ferritin (Broxmeyer *et al.*, 1987b), and PGE (Pelus *et al.*, 1988). The interferons, especially interferon-γ, have a number of effects on the surface of cells, including induction of HLA antigens (Broxmeyer *et al.*, 1986d; Piacibello *et al.*, 1986), and this induction of HLA antigens has been associated with induction of responsiveness of a myeloid cell line to the suppressive influences of LF, H-ferritin, and PGE (Broxmeyer *et al.*, 1986d; Piacibello *et al.*, 1986). Interestingly, PGE counteracts the modulation of HLA antigens on these cells by interferon-γ, and also the enhanced responsiveness to the negative regulatory effects of LF, H-ferritin, and PGE on these cells (Piacibello *et al.*, 1986). Studies are available on the intracellular signaling effects of interferons (Sen and Lengyel, 1992; Darnell *et al.*, 1994; Yip-Schneider *et al.*, 1995), but because the interferons have a plethora of activities, including stimulation of cell proliferation (Hendrie *et al.*, 1991) and the induction of the release of a number of other cytokines including growth factors, it is not clear which intracellular events mediate the suppressive effects of the interferons on the proliferation of myeloid progenitor cells.

9. TUMOR NECROSIS FACTORS

Tumor necrosis factor (TNF)-α, and -β (lymphotoxin) have suppressive activity on myeloid progenitor cells from normal donors and patients with leukemia (Lu *et al.*, 1986b, 1987c; Broxmeyer *et al.*, 1986c; Peetre *et al.*, 1986; Barber *et al.*, 1987a,b; Munker and Koeffler, 1987; Murase *et al.*, 1987; Nagata *et al.*, 1989). Tumor necrosis factor also has a multitude of biological effects, mediated through specific TNF receptors (Beutler and Van Huffle, 1994; Smith *et al.*, 1994; Gruss and Dower, 1995); TNF receptors form a superfamily of cellular and viral proteins involved in activation, costimulation, and death (Smith *et al.*, 1994; see also Chapter 7). Two TNF receptors have been identified: p75 and p55 (Beutler and Van Huffle, 1994; Smith *et al.*, 1994; Gruss and Dower, 1995; see Chapter 3). Differential roles for these two receptors have been reported. For murine hematopoiesis, p55 mediated the TNF suppression of mature progenitors, while p75 was essential for TNF suppression of the more primitive progenitors (F. W. Jacobsen *et al.*, 1994; Chapter 11). With human cells, the p55 and p75 mediated TNF inhibition of primitive progenitors, whereas p55 exclusively mediated enhancement of proliferation by more mature progenitors (Rusten *et al.*, 1994a). The p55 receptor was involved in TNF suppression of stem cell factor (or Steel factor)-enhanced growth, an effect linked to down-regulation of the receptor for stem cell factor, the *c-kit*-encoded protein (Rusten *et al.*, 1994b; Khoury *et al.*, 1994; F. W. Jacobsen *et al.*, 1994, 1995).

10. TRANSFORMING GROWTH FACTOR β

Transforming growth factor (TGF)-β is present in different forms that are latent unless modified (Brown *et al.*, 1990; Miyazono *et al.*, 1993). The crystalline structure of TGF-β_2 has been reported (Schluneger and Grutter, 1992; Daopin *et al.*, 1992), and targeted disruption of the mouse TGF-β_1 gene and a null mutation result in enhanced inflammatory responses and early death (Shull *et al.*, 1992; Kulkarmi *et al.*, 1993), highlighting the significance of this protein. The TGF-β has suppressive activity on normal and leukemic myeloid progenitors *in vitro*, but it also has enhancing activity for myeloid progenitor cell proliferation (S. E. W. Jacobsen *et al.*, 1991b; Ottmann and Pelus, 1988; Sing *et al.*, 1988; Hino *et al.*, 1988; Nara *et al.*, 1989; Kishi *et al.*, 1989; Tessier and Hoang, 1988; Hampson *et al.*, 1989; Ruscetti *et al.*, 1990; Fan *et al.*, 1992; Keller *et al.*, 1988, 1992; Cashman *et al.*, 1990; S. E. W. Jacobsen *et al.*, 1991a, 1994). The inhibitory effect appears to be mainly on the more primitive progenitors (Keller *et al.*, 1988, 1992; S. E. W. Jacobsen *et al.*, 1991a; Cashman *et al.*, 1990), an effect similar to that noted with MIP-1α (Broxmeyer *et al.*, 1990, 1991c, 1993a). Evidence for differential effects of distinct TGF-β isoforms has been reported (S. E. W. Jacobsen *et al.*, 1991a): TGF-β_1 and TGF-β_2 manifested both enhancing and suppressing activity, while TGF-β_3 elicited only suppressive activity, with a potency greater than that of TGF-β_1 and TGF-β_2 (S. E. W. Jacobsen *et al.*, 1991a). In line with the negative regulatory effect of TGF-β on immature progenitors, the primitive progenitors were released from a quiescent state by treatment of cells with antisense oligonucleotides to TGF-β (Hatzfeld *et al.*, 1991).

A substantial amount of information has documented the negative regulatory effects of TGF-β *in vivo* (Roberts and Sporn, 1993; Goey *et al.*, 1989; Jansen *et al.*, 1991; Migdalska *et al.*, 1991; K. L. Miller *et al.*, 1992; Carlino *et al.*, 1992; Lemoli *et al.*, 1992; Grzegorzewski *et al.*, 1994). It has been suggested that TGF-β may be used as a myeloprotective agent in the clinic (Goey *et al.*, 1989; Jansen *et al.*, 1991; Migdalska *et al.*, 1991; Lemoli *et al.*, 1992; Grzegorzewski *et al.*, 1994; Bonewald, 1992).

The suppressive effects of TGF-β are direct acting (Lu *et al.*, 1993a; Keller *et al.*, 1990) and may in part involve the down-modulation of the Steel factor receptor, *c-kit* (de Vos *et al.*, 1993; Dubois *et al.*, 1994; Heinrich *et al.*, 1995). The suppressive effect of TGF-β_1 can be counteracted by basic fibroblast growth factor (Gabrilove *et al.*, 1993).

There is presently more information available on the possible intracellular signal transduction mechanisms mediating the effects of TGF-β than for any other suppressive molecule. Receptors for TGF-β have been characterized, isolated, and studied (ten Dijke *et al.*, 1994; Wrana *et al.*, 1992, 1994; Massague, 1992; Lin *et al.*, 1992; Attisano *et al.*, 1993; Ebner *et al.*, 1993a,b; Franzen *et al.*, 1993; X. F. Wang *et al.*, 1991), and their interactions elucidated. These receptors form a family of transmembrane protein serine/threonine kinases (Massague, 1992). Some of these receptors bind members of the inhibin/activin family, which are described below. Some intracellular sites of TGF-β action include *c-myc* and pRB phosphorylation (Munger *et al.*, 1992), phospholipase C-mediated hydrolysis of phosphatidylcholine (Diaz-Meco *et al.*, 1992), inhibition of cyclin E-dependent kinase (Koff *et al.*, 1993), other G_1 cyclins and cyclin-dependent kinases (Geng and Weinberg, 1993), inhibition of CdK4 (Ewen *et al.*, 1993) and $p15^{INK4B}$ (Hannon and Beach, 1994), inhibition of binding of the transcription factor NF-jun (Sott *et al.*, 1994), and a cyclin-dependent kinase inhibitor p21 through a p53-independent mechanism (Datto *et al.*, 1995).

11. INHIBIN

Inhibin and activin are similar molecules, with inhibin composed of dimers that contain a common α chain and a βA or βB chain, and activins are homodimers or heterodimers of β chains (Broxmeyer *et al.*, 1988c). As mentioned above, these molecules are ligands for some receptors that also bind TGF-β (Massague, 1992; Attisano *et al.*, 1993; Ebner *et al.*, 1993a). Inhibin has suppressive activity (Broxmeyer *et al.*, 1988c; Yu *et al.*,

1987), and activin has enhancing activity *in vitro* (Broxmeyer *et al.*, 1988c), for BFU-E and CFU-GEMM but not for CFU-GM. These effects are indirectly mediated through actions on monocytes and T lymphocytes (Broxmeyer *et al.*, 1988c). The exact mechanism of inhibin suppression of the activin-enhanced erythropoiesis (Broxmeyer *et al.*, 1988c) is not known but may involve the induction of suppressor cytokines from monocytes, T lymphocytes, and/or other accessory cells. The *in vitro* suppressive effects of inhibin (Broxmeyer *et al.*, 1988c; Yu *et al.*, 1987) were substantiated *in vivo* in a mouse model (Hangoc *et al.*, 1992), but because effects were also noted on granulopoiesis, it was clear that at least *in vivo*, inhibition is not restricted to erythropoiesis.

12. VASCULAR ENDOTHELIAL CELL GROWTH FACTOR AND MACROPHAGE-STIMULATING PROTEIN

Vascular endothelial cell growth factor (VEGF) is a ligand for the tyrosine kinase receptor *Flk-1/KDR* and *Flt-1* (De Vries *et al.*, 1992; Quinn *et al.*, 1993; Terman *et al.*, 1992; Millauer *et al.*, 1993) and is considered to be an endothelial-cell-specific mitogen that plays an important role in angiogenesis. Because *Flk-1* mRNA has been detected in primitive and more mature hematopoietic cells (Matthews *et al.*, 1991), recombinant human VEGF was evaluated for its influence on hematopoiesis, which was assayed as *in vitro* colony formation by myeloid progenitor cells from human bone marrow (Broxmeyer *et al.*, 1995b). The VEGF enhanced colony formation by mature subsets of granulocyte–macrophage and erythroid progenitor cells that had been stimulated with a colony-stimulating factor. In contrast, VEGF inhibited colony formation by more immature subsets of granulocyte–macrophage, erythroid, and multipotential progenitor cells synergistically stimulated to proliferate with a colony-stimulating factor and either Steel factor (Broxmeyer *et al.*, 1991b) or the ligand for the Flt-3 receptor tyrosine kinase (Broxmeyer *et al.*, 1995a). The VEGF produced effects similar to those enumerated above on purified $CD34^{++}$ progenitor cells from bone marrow. However, when assessed for effects on single sorted $CD34^{+++}$ cells, VEGF only enhanced or suppressed colony formation by granulocyte–macrophage progenitor cells, and the amplitude of the response was less than that observed when populations of these cells were tested. In the single-cell assays, VEGF had no effect on colony formation by erythroid or multipotential progenitors. This suggested that the effects of VEGF are mediated by both direct and indirect actions on the progenitors but that most of the effects are indirect. Indirect effects are probably mediated through $CD34^{+++}$ endothelial or other accessory cell types. This network and the intracellular signals involved remain to be elucidated for the myeloid effects noted. Additionally, macrophage-stimulating protein (MSP) also has myeloid suppressive activity, which can synergize with that of VEGF and members of the chemokine family. The effects of MSP appear to be indirectly active (Broxmeyer *et al.*, 1996).

13. LOW-MOLECULAR-WEIGHT INHIBITOR AND PEPTIDE MOLECULES

In an effort to determine if LF, H-ferritin, MIP-1α, and PGE might mediate some of their suppressive effects *in vivo* by induction of other suppressor molecules or perhaps one common suppressor mediator, mice and rats were injected with LF, H-ferritin, MIP-1α, or PGE_1. Bone marrow and spleen cells from these injected mice, but not from mice injected with a control diluent, released into conditioned medium *in vitro* an activity that could be purified through a sequence of events into an approximately 8-KDa molecule (low-molecular-weight inhibitor, LMI) capable of suppressing colony formation *in vitro* by granulocyte–macrophage, erythroid, and multipotential progenitors (Gentile *et al.*, 1989; Broxmeyer *et al.*, 1993c; H. E. Broxmeyer, S. Cooper, and C. Mantel, unpublished observations). Although we were unsuccessful in sequencing the small amount of apparently purified LMI, we found that the LMI derived from cells of mice injected with either LF, H-ferritin, MIP-1α, or PGE had the same profile of suppressive activities as a 14-mer peptide (NH_2-ETVIMKAKPRANFP-COOH) (Broxmeyer *et al.*, 1993c; unpublished observations). This (Ala 1)-tuftsin-containing 14-mer peptide had been shown previously to have a more limited range of suppressive activity (Kreisberg *et al.*, 1993) for two factor-responsive macrophage progenitors. Moreover, purified rabbit anti-14-mer peptide, but not anti-MIP-1α, neutralized the suppressive activities *in vitro* of the different LMI preparations and of the 14-mer peptide, suggesting antigenic similarly of the active portion of the different LMI preparations with that of the 14-mer peptide (Broxmeyer *et al.*, 1993c). The different LMI preparations and the 14-mer peptide, but not a scrambled 14-mer, significantly suppressed myelopoiesis in C3H/HeJ mice, suggesting that the LMI was active *in vivo* as well as *in vitro*. The LMI at this time appears to be a common suppressor molecule that mediates, at least in part, the myelosuppressive effects of a number of different cytokines. Although sequencing of the LMI from different sources is needed before identity is definitively established for the different LMI preparations, it is possible that this LMI may be equivalent to the material induced in response to PGE (Gentile and Pelus, 1988; Pelus and Gentile, 1988), as our PGE-induced LMI was neutralized by the 14-mer antibody.

A good deal of information is also available on two other peptides that manifest suppressive activity on hematopoietic stem and progenitor cells. These are the pentapeptide pGlu-Glu-Asp-Cys-Lys(pEEDCK) and the tetrapeptide acetyl-N-Ser-Asp-Lys-Pro(AcSDKP). A recent review has described the source as well as the *in vitro* and *in vivo* activities of these peptides (Guigon and Bonnet, 1995). Interestingly, the pEEDCK is susceptible to oxidation and forms a cysteine-linked dimer that is stimulatory rather than inhibitory. The disulfide bridge of the pEEDCK dimer was removed and replaced with a nonreducible carbon bridge, which prevents the possibility of reduction of the pEEDCK dimer to the inhibitory monomeric form (Pelus *et al.*, 1994), to form the compound SK&F 107647, which was more active than the dimer in enhancing proliferation.

It is the goal of a number of laboratories to identify peptide molecules that will mimic, and thus potentially replace, the use of cytokines for clinical studies on myelosuppression, myeloprotection, and myeloenhancement.

14. CONCLUDING REMARKS

It is clear from the above that the existence of cytokines with myelosuppressive and potential myeloprotective activity is well defined. As we learn more about the production and action of these molecules, and how they interact in the complex network of cell regulation of blood cell production (see Chapters 3, 5, and 11), these molecules, or peptides that may mimic some of their effects, may be the next line of biological molecules to be used clinically to enhance the treatment of blood disorders or lessen cytotoxic effects on the blood system that is inherent in the treatment of tumors.

ACKNOWLEDGMENTS. I wish to thank Rebecca Miller and Shantay Glover for typing the manuscript. Support for the studies reported from my laboratory in the reference section was from grants from the National Institutes of Health and the National Cancer Institute, including the following grants that are presently active: HL 56416, HL 54037, HL 53586, and DK 07519.

15. REFERENCES

Aglietta, M., Piacibello, W., Sanavio, F., Stacchini, A., Apra, F., Schena, M., Mossetti, C., Carnino, F., Caligaris-Cappio, F., and Gavosto, F., 1989, Kinetics of human hemopoietic cells after *in vivo* administration of granulocyte–macrophage colony-stimulating factor, *J. Clin. Invest.* **83:**551–557.

Ahija, S. K., Gao, J. L., and Murphy, P. M., 1994, Chemokine receptors and molecular mimicry, *Immunol. Today* **15:**281–286.

Anagnostou, A., Dainiak, N., and Najman, A. L. (eds.), 1991, *Annals of the New York Academy of Sciences, Vol. 628: Negative Regulators of Hematopoiesis: Studies on Their Nature, Action and Potential Role in Cancer Therapy*, New York Academy of Sciences, New York.

Anderson, B. F., Baker, H. M., Dodson, E. J., Norris, G. E., Rumball, S. V., Waters, J. M., and Baker, E. N., 1987, Structure of human lactoferrin at 3.2-A resolution, *Proc. Natl. Acad. Sci. U.S.A.* **84:**1769–1773.

Aronica, S. M., Mantel, C., Marshall, M. S., Saris, A., Cooper, S., Hague, N., Zhang, X. F., and Broxmeyer, H. E., 1995, Interferon-inducible protein 10 and macrophage inflammatory protein-1α inhibit growth factor stimulation of Raf-1 kinase activity and protein synthesis in a human growth factor-dependent hematopoietic cell line, *J. Biol. Chem.* **270:**21998–22007.

Attisano, L., Carcamo, J., Ventura, F., Weis, F. M. B., Massague, J., and Wrana, J. L., 1993, Identification of human activin and TGFβ type I receptors that form heteromeric kinase complexes with type II receptors, *Cell* **75:** 671–680.

Avalos, B. R., Bartynski, K. J., Elder, P. J., Kotur, M. S., Burton, W. G., and Wilkie, N. M., 1994, The active monomeric form of macrophage inflammatory protein-1α interacts with high- and low-affinity classes of receptors on human hematopoietic cells, *Blood* **84:**1790–1801.

Bagby, G. C., Jr., and Bennett, R. M., 1982, Feedback regulation of granulopoiesis: Polymerization of lactoferrin abrogates its ability to inhibit CSA production, *Blood* **60:**108–112.

Bagby, G. C., Dinarello, C. A., Wallace, P., Wagner, C., Hefeneider, S., and McCall, E., 1986, Interleukin-1 stimulates granulocyte–macrophage colony stimulating activity release by vascular endothelial cells, *J. Clin. Invest.* **78:**1316–1323.

Bagby, G. C., Riga, V. D., Bennett, R. M., Van denbark, A. A., and Garewal, H. S., 1989, Interaction of lactoferrin, monocytes, and T-lymphocyte subsets in the regulation of steady-state granulopoiesis *in vitro*, *J. Clin. Invest.* **68:**56–63.

Baggiolini, M., DeWald, B., and Moser, B., 1994, Interleukin-8 and related chemotactic cytokines-CXC and CC chemokines, *Adv. Immunol.* **55:**97–179.

Balla, G., Jacob, H. S., Balla, J., Rosenberg, M., Nath, K., Apple, F., Eaton, J. W., and Vercellotti, G. M., 1992, Ferritin: A cytoprotective antioxidant stratagem of endothelium, *J. Biol. Chem.* **267:**18148–18153.

Barber, K. E., Crosier, P. S., Gillis, S., and Watson, J. D., 1987a, Human granulocyte–macrophage progenitors and their sensitivity to cytotoxins: Analysis by limiting dilution, *Blood* **70:**1773–1776.

Barber, K. E., Crosier, P. S., and Watson, J. D., 1987b, The differential inhibition of hemopoietic growth factor activity by cytotoxins and interferon-γ, *J. Immunol.* **139:**1108–1112.

Bennett, R. M., Bagby, G. C., and Davis, J., 1981, Calcium-dependent polymerization of lactoferrin, *Biochem. Biophys. Res. Commun.* **101:**88–95.

Beutler, B., and Van Huffle, C., 1994, Unraveling function of the TNF ligand and receptor families, *Science* **264:**667–668.

Bezault, J., Bhimani, R., Wiprovnick, J., and Furmanski, P., 1994, Human lactoferrin inhibits growth of solid tumors and development of experimental metastases in mice, *Cancer Res.* **54:**2310–2312.

Bodine, D. M., Crosier, P. S., and Clark, S. C., 1991, Effects of hematopoietic growth factors on the survival of primitive stem cells in liquid suspension culture, *Blood* **78:**914–920.

Bonewald, L. F., 1992, Can transforming growth factor beta be useful as a protective agent for pluripotent hematopoietic progenitor cells? *Exp. Hematol.* **20:**1249–1251.

Brown, P. D., Wakefield, L. M., Levinson, A. D., and Sporn, M. B., 1990, Physiochemical activation of recombinant latent transforming growth factor-beta's 1, 2, and 3, *Growth Factors* **3:**35–43.

Broxmeyer, H. E., 1982, Relationship of cell-cycle expression of Ia-like antigen determinants on normal and leukemia human granulocyte–macrophage progenitor cells to regulation *in vitro* by acidic isoferritins, *J. Clin. Invest.* **69:**632–642.

Broxmeyer, H. E., 1989, Iron-binding proteins and the regulation of hematopoietic cell proliferation/differentiation, in *Iron, Immunity, Cancer, and Inflammation* (M. deSousa and J. H. Brock, eds.), pp. 199–221, John Wiley & Sons, Chichester.

Broxmeyer, H. E., 1992a, Update: Biomolecule–cell interactions and the regulation of myelopoiesis, in *Concise Reviews in Clinical and Experimental Hematology* (M. J. Murphy, Jr., ed.), pp. 119–149, Alpha Med Press, Dayton, OH.

Broxmeyer, H. E., 1992b, Suppressor cytokines and regulation of myelopoiesis: Biology and possible clinical uses, *Am. J. Pediatr. Hematol./Oncol.* **14:**22–30.

Broxmeyer, H. E., 1992c, H-Ferritin: A regulatory cytokine that down modulates cell proliferation, *J. Lab. Clin. Med.* **120:**367–370.

Broxmeyer, H. E., 1993, Role of cytokines in hematopoiesis, in *Clinical Aspects of Cytokines: Role in Pathogenesis, Diagnosis and Therapy* (J. J. Oppenheim, J. L. Rossio, and A. J. H. Gearing, eds.), pp. 201–206, Oxford University Press, New York.

Broxmeyer, H. E., 1995a, Role of cytokines in hematopoiesis, in *Human Cytokines: Their Role in Disease and Therapy* (B. B. Aggarwal and R. K. Puri, eds.), pp. 27–36, Blackwell Scientific Publications, Cambridge, MA.

Broxmeyer, H. E., 1995b, The cell cycle as therapeutic target, *Cancer Invest.* **13:**617–624.

Broxmeyer, H. E., 1996, Is interleukin 17, an inducible cytokine that stimulates production of other cytokines, merely a redundant player in a sea of other biomolecules? *J. Exp. Med.* **183** (in press).

Broxmeyer, H. E., and Platzer, E., 1984, Lactoferrin acts on I-A and I-E/c antigen-positive subpopulations of mouse peritoneal macrophages in the absence of T-lymphocytes and other cell types to inhibit production of granulocyte–macrophage colony stimulatory factors *in vitro*, *J. Immunol.* **133:**306–314.

Broxmeyer, H. E., and Williams, D. E., 1988, The production of myeloid blood cells and their regulation during health and disease, *CRC Crit. Rev. Oncol./Hematol.* **8:**173–226.

Broxmeyer, H. E., Mendelson, N., and Moore, M. A. S., 1977, Abnormal granulocyte feedback regulation of colony forming and colony stimulating activity-producing cells from patients with chronic myelogenous leukemia, *Leukemia Res.* **1:**3–12.

Broxmeyer, H. E., Smithyman, A., Eger, R. R., Meyers, P. A., and de Sousa, M., 1978, Identification of lactoferrin as the granulocyte-derived inhibitor of colony stimulating activity (CSA)-production, *J. Exp. Med.* **148:**1052–1067.

Broxmeyer, H. E., de Sousa, M., Smithyman, A., Ralph, P., Hamilton, J., Kurland, J. I., and Bognacki, J., 1980, Specificity and modulation of the action of the lactoferrin, a negative feedback regulator of myelopoiesis, *Blood* **55:**324–333.

Broxmeyer, H. E., Bognacki, J., Dorner, M. H., and de Sousa, M., 1981, The identification of leukemia-associated inhibitory activity (LIA) as acidic isoferritins: A regulatory role for acidic isoferritins in the production of granulocytes and macrophages, *J. Exp. Med.* **153:**1426–1444.

Broxmeyer, H. E., Bognacki, J., Ralph, P., Dorner, M. H., Lu, L., and Castro-Malaspina, H., 1982, Monocyte–macrophage derived acidic isoferritins: Normal feedback regulators of granulocyte–macrophage progenitor cells, *Blood* **60:**595–607.

Broxmeyer, H. E., Gentile, P., Bognacki, J., and Ralph, R., 1983a, Lactoferrin, transferrin and acidic isoferritins: Regulatory molecules with potential therapeutic value in leukemia, *Blood Cells* **9:**83–105.

Broxmeyer, H. E., Lu, L., Platzer, E. Feit, C., Juliano, L., and Rubin, B. Y., 1983b, Comparative analysis of the influences of human gamma, alpha and beta interferons on human multipotential (CFU-GEMM), erythroid (BFU-E) and granulocyte–macrophage (CFU-GM) progenitor cells, *J. Immunol.* **131:**1300–1305.

Broxmeyer, H. E., Gentile, P., Cooper, S., Lu, L., Juliano, L., Piacibello, W., Meyers, P. A., and Cavanna, F., 1984, Functional activities of acidic isoferritins and lactoferrin *in vitro* and *in vivo*, *Blood Cells* **10:**397–426.

Broxmeyer, H. E., Juliano, L., Waheed, A., and Shadduck, R. K., 1985a, Release from mouse macrophages of acidic isoferritins that suppress hematopoietic progenitor cells is induced by purified L-cell colony stimulating factors and suppressed by human lactoferrin, *J. Immunol.* **135:**3224–3231.

Broxmeyer, H. E., Cooper, S., Rubin, B. Y., and Taylor, M. W., 1985b, The synergistic influence of human gamma- and alpha-interferons on suppression of hematopoietic progenitor cells is additive with the enhanced sensitivity of these cells to inhibition by interferons at low oxygen tension *in vitro*, *J. Immunol.* **135:**2502–2506.

Broxmeyer, H. E., Bicknell, D. C., Gillis, S., Harris, E. L., Pelus, L. M., and Sledge, G. W., Jr., 1986a, Lactoferrin: Affinity purification from human milk and polymorphonuclear neutrophils using monoclonal antibody (II2C) to human lactoferrin, development of an immunoradiometric assay using II2C, and myelopoietic regulation and receptor-binding characteristics, *Blood Cells* **11:**429–446.

Broxmeyer, H. E., Lu, L., Bicknell, D. C., Williams, D. E., Cooper, S., Levi, S., Salfeld, J., and Arosio, P., 1986b, The influence of purified recombinant human acidic and basic isoferritins on colony formation *in vitro* by granulocyte–macrophage and erythroid progenitor cells, *Blood* **68:**1257–1263.

Broxmeyer, H. E., Williams, D. E., Lu, L., Cooper, S., Anderson, S. L., Beyer, G. S., Hoffman, R., and Rubin, B. Y., 1986c, The suppressive influences of human tumor necrosis factor on bone marrow hematopoietic cells from normal donors and patients with leukemia: Synergism of tumor necrosis factor and gamma interferon, *J. Immunol.* **136:**4487–4495.

Broxmeyer, H. E., Piacibello, W., Juliano, L., Platzer, E., Berman, E., and Rubin, B. Y., 1986d, Gamma interferon induces colony forming cells of the human monoblast cell line U937 to respond to inhibition by lactoferrin, transferrin and acidic isoferritins, *Exp. Hematol.* **14:**35–43.

Broxmeyer, H. E., Williams, D. E., Hangoc, G., Cooper, S., Gentile, P., Shen, R.-N., Ralph, P., Gillis, S., and Bicknell, D. C., 1987a, The opposing action *in vivo* on murine myelopoiesis of purified preparations of lactoferrin and the colony stimulating factors, *Blood Cells* **13:**31–48.

Broxmeyer, H. E., Williams, D. E., Lu, L., Vadhan, S., Cooper, S., Bicknell, D. C., Ralph, P., Gutterman, J., and Tricot, G., 1987b, Biomolecules associated with suppression of myelopoiesis in normal conditions and during myeloid leukemia and other related disorders, in *The Inhibitors of Hematopoiesis*, Vol. 162 (A. Najman, M. Guigon, N. C. Gorin, and J. Y. Mary, eds.), pp. 139–149, John Libbey Eurotext Limited, Paris.

Broxmeyer, H. E., Cooper, S., Williams, D. E., Hangoc G., Gutterman, J. U., and Vadhan, S., 1988a, Effects of purified recombinant human granulocyte–macrophage colony stimulating factor on growth characteristics of bone marrow hematopoietic progenitor cells in patients on a phase I clinical trial, *Exp. Hematol.* **16:**594–602.

Broxmeyer, H. E., Williams, D. E., Geissler, K., Hangoc, G., Cooper, S., Levi, S., and Arosio, P., 1988b, Suppressive effects *in vivo* of purified recombinant human H-subunit (acidic) ferritin on murine myelopoiesis, *Blood* **73:**74–79.

Broxmeyer, H. E., Lu, L., Cooper, S., Schwall, R. H., Mason, A. J., and Nikolics, K., 1988c, Selective and indirect modulation of human hematopoietic progenitor cell proliferation by recombinant human activin and inhibin, *Proc. Natl. Acad. Sci. U.S.A.* **85:**9052–9056.

Broxmeyer, H. E., Cooper, S., and Vadhan-Raj, S., 1989a, Cell cycle status of erythroid (BFU-E) progenitor cells from the bone marrows of patients on a clinical trial with purified recombinant human granulocyte–macrophage colony-stimulating factor, *Exp. Hematol.* **17:**455–459.

Broxmeyer, H. E., Sherry, B., Lu, L., Cooper, S., Carow, C., Wolpe, S. D., and Cerami, A., 1989b, Myelopoietic enhancing effects of murine macrophage inflammatory proteins 1 and 2 *in vitro* on colony formation by murine and human bone marrow granulocyte–macrophage progenitor cells, *J. Exp. Med.* **170:**1583–1594.

Broxmeyer, H. E., Sherry, B., Lu, L., Cooper, S., Oh, K. O., Tekamp-Olson, P., Kwon, B. S., and Cerami, A., 1990, Enhancing and suppressing effects of recombinant murine macrophage inflammatory proteins on colony formation *in vitro* by bone marrow myeloid progenitor cells, *Blood* **76:**1110–1116.

Broxmeyer, H. E., Najman, A., and Dainiak, N., 1991a, Summary of meeting on negative regulators of hematopoiesis, *Exp. Hematol.* **19:**149–152.

Broxmeyer, H. E., Maze, R., Miyazawa, K., Carow, C., Hendrie, P. C., Cooper, S., Hangoc, G., Vadhan-Raj, S., and Lu, L., 1991b, The Kit receptor and its ligand, Steel factor, as regulators of hemopoiesis, *Cancer Cells* **3:**480–487.

Broxmeyer, H. E., Sherry, B., Cooper, S., Ruscetti, F. W., Williams, D. E., Arosio, P., Kwon, B. S., and Cerami, A., 1991c, Macrophage inflammatory protein (MIP)-1β abrogates the capacity of MIP-1α to suppress myeloid progenitor cell growth, *J. Immunol.* **147**:2586–2594.

Broxmeyer, H. E., Cooper, S., Levi, S., and Arosio, P., 1991d, Mutated recombinant human heavy-chain ferritins and myelosuppression *in vitro* and *in vivo*: A link between ferritin ferroxidase activity and biological function, *Proc. Natl. Acad. Sci. U.S.A.* **88**:770–774.

Broxmeyer, H. E., Bicknell, D. C., Cooper, S., Sledge, G., Jr., Williams, D. E., McGuire, W. A., and Coates, T. D., 1991e, Qualitative functional deficiency of affinity purified lactoferrin from neutrophils and patients with chronic myelogenous leukemia, and lactoferrin/H-ferritin cell-interactions in a patient with lactoferrin-deficiency with normal numbers of circulating leukocytes, *Pathobiology* **59**:26–35.

Broxmeyer, H. E., Sherry, B., Cooper, S., Lu, L., Maze, R., Beckmann, M. P., Cerami, A., and Ralph, P., 1993a, Comparative analysis of the suppressive effects of the human macrophage inflammatory protein family of cytokines (chemokines) on proliferation of human myeloid progenitor cells, *J. Immunol.* **150**:3448–3458.

Broxmeyer, H. E., Benninger, L., Hague, N., Hendrie, P., Cooper, S., Mantel, C., Cornetta, K., Vadhan-Raj, S., Sarris, A., and Lu, L., 1993b, Suppressive effects of the chemokine (macrophage inflammatory protein) family of cytokines on proliferation of normal and leukemia myeloid cell proliferation, in *Coloque INSERM, Vol. 229: The Negative Regulation of Hematopoiesis from Fundamental Aspects to Clinical Applications* (M. Guigon, F. M. Lemoine, N. Dainiak, A. Schechter, and A. Najman, eds.), pp. 141–154, John Libbey Eurotext, Paris.

Broxmeyer, H. E., Mantel, C., Cooper, S., Kreisberg, R., and Moore, R. N., 1993c, Administration of macrophage inflammatory protein 1α, lactoferrin or H-ferritin to mice or rats induces a common non-species specific low molecular weight inhibitor active *in vitro* and *in vivo*: Biochemical, biological and antigenic characterization, *Blood* **82:** 181a.

Broxmeyer, H. E., Benninger, L., Patel, S. R., Benjamin, R. S., and Vadhan-Raj, S., 1994, Kinetic responses of human marrow myeloid progenitor cells to *in vivo* treatment of patients with G-CSF is different from that with GM-CSF, *Exp. Hematol.* **22**:100–102.

Broxmeyer, H. E., Lu, L., Cooper, S., Ruggieri, L., Li, Z. H., and Lyman, S. D., 1995a, Flt-3-ligand stimulates/co-stimulates the growth of myeloid stem/progenitor cells, *Exp. Hematol.* **23**:1121–1129.

Broxmeyer, H. E., Cooper, S., Li, Z. H., Lu, L., Song, H-O., Kwon, B. S., Warren, R. E., and Donner, D. B., 1995b, Myeloid progenitor cell regulatory effects of vascular endothelial cell growth factor, a ligand for the tyrosine kinase receptor *Flk-1*, *Int. J. Hematol.* **62**:203–215.

Broxmeyer, H. E., Cooper, S., Hague, N., Benninger, L., Sarris, A., Cornetta, K., Vadhan-Raj, S., Hendrie, P., and Mantel, C., 1995c, Human chemokines: Enhancement of specific activities and effects *in vitro* on normal and leukemic progenitors and a factor dependent cell line and *in vivo* in mice, *Ann. Hematol.* **71**:235–246.

Broxmeyer, H. E., Benninger, L., Cooper, S., Hague, N., Benjamin, R. S., and Vadhan-Raj, S., 1995d, Effects of *in vivo* treatment with PIXY321 (GM-CSF/IL-3 fusion protein) on proliferation kinetics of bone marrow and blood myeloid progenitor cells in patients with sarcoma, *Exp. Hematol.* **23**:335–340.

Broxmeyer, H. E., Cooper, S., Maione, T., Gordon, M., and Daly, T., 1995e, Myeloprotective effects of the chemokines interleukin-8 and platelet factor 4 in a mouse model of cytosine arabinoside (ARA-C) chemotherapy, *Expo. Hematol.* **23**:900.

Broxmeyer, H. E., Cooper, S., Cacalano, G., Hague, N. L., Balish, E., and Moore, M. W., 1995f, Interleukin-8 is a myelosuppressive regulatory cytokine *in vivo*: Evidence from mice lacking the murine IL-8 receptor homolog, *Blood* **86**(Suppl. 1):422.

Broxmeyer, H. E., Hague, N. L., Sledge, G. W., Jr., Rasmussen, H., and Gordon, M. S., 1995g, Suppression of marrow and mobilization of blood myeloid progenitors *in vivo* by BB10010, a genetically engineered variant of human macrophage inflammatory protein (MIP)-1α, in a phase 1 clinical trial in patients with relapsed/refractory breast cancer, *Blood* **86**(Suppl. 1):12a.

Broxmeyer, H. E., Cooper, S., Li, Z.-H., Lu, L., Wang, M-H., Sarris, A., Chang, M.-S., Donner, D. B., and Leonard, E. J., 1996, Macrophage-stimulating protein, a ligand for the RON tyrosine kinase receptor, suppresses myeloid progenitor cell proliferation *in vitro* and synergizes in this effect with vascular endothelial cell growth factor and members of the chemokine family, including ENA-78, *Ann. Hematol.* (in press).

Cacalano, G., Lee, J., Kikly, K., Ryan, A. M., Pitts-Meek, S., Hultgren, B., Wood, W. I., and Moore, M. W., 1994, Neutrophil and B cell expansion in mice that lack the murine IL-8 receptor homolog, *Science* **265**:682–684.

Carlino, J. A., Higley, H. R., Creson, J. R., Avis, P. D., Ogawa, Y., and Ellingsworth, L. R., 1992, Transforming growth factor β1 systematically modulates granuloid, erythroid, lymphoid, and thrombocytic cells in mice, *Exp. Hematol.* **20**:943–950.

Cashman, J. D., Eaves, A. C., Raines, E. W., Ross, R., and Eaves, C. J., 1990, Mechanisms that regulate the cell cycle status of very primitive hematopoietic cells in long-term human marrow cultures. I. Stimulatory role of a variety of mesenchymal cell activators and inhibitory role of TGF-β, *Blood* **75:**96–101.

Cerretti, D. P., Kozlosky, C. J., Bos, T. V., Nelson, N., Gearing, D. P., and Beckmann, M. P., 1993, Molecular characterization of receptors for human interleukin-8, GRO/melanoma growth-stimulatory activity and neutrophil activating peptide-2, *Mol. Immunol.* **30:**359–367.

Chaudhuri, A., Zbrzezna, V., Polyakova, J., Pogo, O., Hesselgesser, J., and Horuk, R., 1994, Expression of the duffy antigen in K562 cells. Evidence that it is the human erythrocyte chemokine receptor, *J. Biol. Chem.* **269:** 7835–7838.

Chen, L. T., Lu, L., and Broxmeyer, H. E., 1987, Effects of purified iron-saturated human lactoferrin on spleen morphology in mice infected with Friend virus complex, *Am. J. Pathol.* **126:**285–292.

Clark-Lewis, I., Kim, K.-S., Rajarathnam, K., Gong, J-H., Dewald, B., Moser, B., Baggiolini, M., and Sykes, B. D., 1995, Structure–activity relationships of chemokines, *J. Leuk. Biol.* **57:**703–711.

Clements, J. M., Craig, S., Gearing, A. J. H., Hunter, M. G., Heyworth, C. M., Dexter, T. M., and Lord, B. I., 1992, Biological and structural properties of MIP-1α expressed in yeast, *Cytokine* **4:**76–82.

Clore, G. M., and Gronenborn, A. M., 1995, Three-dimensional structures of α and β chemokines, *FASEB J.* **9:** 57–62.

Cooper, S., Mantel, C., and Broxmeyer, H. E., 1994, Myelosuppressive effects *in vivo* with very low dosages of monomeric recombinant murine macrophage inflammatory protein-1α, *Exp. Hematol.* **22:**186–193.

Dainiak, N., Guigon, M., Broxmeyer, H. E., Lemoine, F., and Najman, A., 1994, Meeting on negative regulation of hematopoietic cell growth and differentiation, *Exp. Hematol.* **22:**399–405.

Daly, T. J., LaRosa, G. J., Dolich, S., Maione, T. E., Cooper, S., and Broxmeyer, H. E., 1995, High activity suppression of myeloid progenitor proliferation by chimeric mutants of IL-8 and PF4, *J. Biol. Chem.* **270:** 23282–23292.

Daopin, S., Piez, K. A., Ogawa, Y., and Davies, D. R., 1992, Crystal structure of transforming growth factor-β2: An unusual fold for the superfamily, *Science* **257:**369–373.

Darnell, J. E., Jr., Kerr, I. M., and Stark, G. R., 1994, Jak-STAT pathways and transcriptional activation in response to IFNs and other extracellular signaling proteins, *Science* **264:**1415–1421.

Datto, M. B., Li, Y., Panus, J. F., Howe, D. J., Xiong, Y., and Wang, X. F., 1995, Transforming growth factor β induces the cyclin-dependent kinase inhibitor p21 through a p53-independent mechanism, *Proc. Natl. Acad. Sci. U.S.A.* **92:**5545–5549.

Davis, R. J., 1994, MAPKs: New JNK expands the group, *Trends Biochem. Sci.* **19:**470–474.

de Vos, S., Brach, M. A., Asano, Y., Ludwig, W. D., Bettelheim, P., Gruss, H. J., and Herrmann, F., 1993, Transforming growth factor-β1 interferes with the proliferation-inducing activity of stem cell factor in myelogenous leukemia blasts through functional dow-regulation of the *c-kit* proto-oncogene product, *Cancer Res.* **53:**3638–3642.

De Vries, C., Escobedo, J. A., Ueno, H., Houck, K., Ferrara, N., and Williams, L. T., 1992, The fms-like tyrosine kinase, a receptor for vascular endothelial growth factor, *Science* **255:**989–991.

Diaz-Meco, M. T., Dominguez, I., Sanz, L., Municio, M. M., Berra, E., Cornet, M. E., de Herreros, A. G., Johansen, T., and Moscat, J., 1992, Phospholipase C-mediated hydrolysis of phosphatidylcholine is a target of transforming growth factor β1 inhibitory signals, *Mol. Cell. Biol.* **12:**302–308.

Dubois, C. M., Ruscetti, F. W., Stankova, J., and Keller, J. R., 1994, Transforming growth factor-β regulates *c-kit* message stability and cell-surface protein expression in hematopoietic progenitors, *Blood* **83:**3138–3145.

Duma, E. M., Foster, J. S., and Moore, R. N., 1988, Bradykinin sensitization of CSF-1-responsive murine mononuclear phagocyte precursors to prostaglandin E, *J. Immunol.* **141:**3186–3189.

Dunlop, D. J., Wright, E. G., Lorimore, S., Graham, G. J., Holyoake, T., Kerr, D. J., Wolpe, S. D., and Pragnell, I. B., 1992, Demonstration of stem cell inhibition and myeloprotective effects of SCI/rhMIP-1α *in vivo*, *Blood* **79:**2221–2225.

Eaves, C. J., Cashman, J. D., Wolpe, S. D., and Eaves, A. C., 1993, Unresponsiveness of primitive chronic myeloid leukemia cells to macrophage inflammatory protein 1α, an inhibitor of primitive normal; hematopoietic cells, *Proc. Natl. Acad. Sci. U.S.A.* **90:**12015–12019.

Ebner, R., Cheng, R. H., Lawler, S., Zioncheck, T., and Derynck, R., 1993a, Determination of type I receptor specificity by the type II receptors for TGF-β or activin, *Science* **262:**900–902.

Ebner, R., Chen, R. H., Shum, L., Lawler, S., Zioncheck, T. F., Lee, A., Lopez, A. R., and Derynck, R., 1993b,

Cloning of a type I TGF-β receptor and its effect on the TGF-β binding to the type II receptor, *Science* **260:**1344–1348.

Ewen, M. E., Sluss, H. K., Whitehouse, L. L., and Livingston, D. M., 1993, TGFβ inhibition of Cdk4 synthesis is linked to cell cycle arrest, *Cell* **74:**1009–1020.

Fan, K., Ruan, Q., Sensenbrenner, L., and Chen, B., 1992, Transforming growth factor-β1 bifunctionally regulates murine macrophage proliferation, *Blood* **79:**1679–1685.

Favaloro, E. J., 1993, Differential expression of surface antigens on activated endothelism, *Immunol. Cell Biol.* **71:**571–581.

Ferrajoli, A., Talpoz, M., Zipf, T. F., Hirsch-Ginsberg, C., Estey, E., Wolpe, S. D., and Estrov, Z., 1994, Inhibition of acute myelogenous leukemia progenitor proliferation by macrophage inflammatory protein 1-α, *Leukemia* **8:**798–805.

Franzen, P., ten Dijke, P., Ichijo, H., Yamashita, H., Schulz, P., Heldin, C. H., and Miyazono, K., 1993, Cloning of a TGF-β type I receptor that forms a heteromeric complex with the TGF-β type II receptor, *Cell* **75:**681–692.

Furmanski, P., and Li, Z. P., 1990, Multiple forms of lactoferrin in normal and leukemic human granulocytes, *Exp. Hematol.* **18:**932–935.

Furmanski, P., Li, Z. P., Fortuna, M. B., Swamy, C. H. V. B., and Das, M. R., 1989, Multiple molecular forms of human lactoferrin, *J. Exp. Med.* **170:**415–429.

Gabrilove, J. L., Wong, G., Bollenbacher, E., White, K., Kojima, S., and Wilson, E. L., 1993, Basic fibroblast growth factor counteracts the suppressive effect of transforming growth factor-β1 on human myeloid progenitor cells, *Blood* **81:**909–915.

Ganser, A., Carlo-Stella, C., Greher, J., Volkers, B., and Hoelzer, D., 1987, Effect of recombinant interferons alpha and gamma on human bone marrow-derived megakaryocytic progenitor cells, *Blood* **70:**1173–1179.

Garre, C., Bianchi-Scarra, G., Sirito, M., Musso, M., and Ravazzolo, R., 1992, Lactoferrin binding sites and nuclear localization in K562(s) cells, *J. Cell Physiol.* **153:**477–482.

Geng, Y., and Weinberg, R. A., 1993, Transforming growth factor β effects on expression of G_1 cyclins and cyclin-dependent protein kinases, *Proc. Natl. Acad. Sci. U.S.A.* **90:**10315–10319.

Gentile, P., and Broxmeyer, H. E., 1983, Suppression of mouse myelopoiesis by administration of human lactoferrin *in vivo* and the comparative action of human transferrin, *Blood* **61:**982–993.

Gentile, P. S., and Pelus, L. M., 1987, *In vivo* modulation of myelopoiesis by prostaglandin E_2. II. Inhibition of granulocyte–macrophage progenitor cell (CFU-GM) cell-cycle rate, *Exp. Hematol.* **15:**119–126.

Gentile, P. S., and Pelus, L. M., 1988, *In vivo* modulation of myelopoiesis by prostaglandin E_2 IV. Prostaglandin E_2 induction of myelopoietic inhibitor activity, *J. Immunol.* **141:**2714–2720.

Gentile, P., Byer, D., and Pelus, L. M., 1983, *In vivo* modulation of murine myelopoiesis following intravenous administration of prostaglandin E_2, *Blood* **62:**1100–1107.

Gentile, P. S., Mantel, C. R., and Broxmeyer, H. E., 1989, Purification to apparent homogeneity of a new myelopoietic inhibitory factor produced in response to *in vivo* administration of human lactoferrin or recombinant H-subunit ferritin, *Blood* **74:**228a.

Gewirtz, A. M., Calabretta, B., Rucinski, B., Niewiarowski, S., and Xu, W. Y., 1989, Inhibition of human megakaryopoiesis *in vitro* by platelet factor 4 (PF4) and a synthetic COOH-terminal PF4 peptide, *J. Clin. Invest.* **83:**1477–1486.

Goey, H., Keller, J. R., Back, T., Longo, D. L., Ruscetti, F. W., and Wiltrout, R. H., 1989, Inhibition of early murine hemopoietic progenitor cell proliferation after *in vivo* locoregional administration of transforming growth factor-β1, *J. Immunol.* **143:**877–880.

Graham, G. J., Wright, E. G., Hewick, R., Wolpe, S. D., Wilkie, N. M., Donaldson, D., Lorimore, S., and Pragnell, I. B., 1990, Identification and characterization of an inhibitor of haemopoietic stem cell proliferation, *Nature* **344:**442–444.

Graham, G. J., Freshney, M. G., Donaldson, D., and Pragnell, I. B., 1992, Purification and biochemical characterization of human and murine stem cell inhibitors (SCI), *Growth Factors* **7:**151–160.

Graham, G. J., Zhou, L., Weatherbee, J. A., Tsang, M. L. S., Napolitano, M., Leonard, W. J., and Pragnell, I. B., 1993, Characterization of a receptor for macrophage inflammatory protein 1α and related proteins on human and murine cells, *Cell Growth Differ.* **4:**137–146.

Graham, G. J., MacKenzie, J., Lowe, S., Tsang, M. L. S., Weatherbee, J. A., Issacson, A., Medicherla, J., Fang, F., Wilkinson, P. C., and Pragnell, I. B., 1994, Aggregation of the chemokine MIP-1α is a dynamic and reversible phenomenon, *J. Biol. Chem.* **269:**4974–4978.

Gruss, H. J., and Dower, S. K., 1995, Tumor necrosis factor ligand superfamily: Involvement in the pathology of malignant lymphomas, *Blood* **85:**3378–3404.

Grzegorzewski, K., Ruscetti, F. W., Usui, N., Damia, G., Longo, D. L., Carlino, J. A., Keller, J. R., and Wiltrout, R. H., 1994, Recombinant transforming growth factor β1 and β2 protect mice from acutely lethal doses of 5-fluorouracil and doxorubicin, *J. Exp. Med.* **180:**1047–1057.

Guigon, M., and Bonnet, D., 1995, Inhibitory peptides in hematopoiesis, *Exp. Hematol.* **23:**477–481.

Guigon, M., Lemoine, F. M., Dainiak, N., Schechter, A., and Najman, A. (eds.), 1993, *Colloque INSERM, Vol. 229: The Negative Regulation of Hematopoiesis from Fundamental Aspects to Clinical Applications*, pp. 3–492, John Libbey Eurotext, Paris.

Hampson, J., Ponting, I. L. O., Cook, N., Vodinelich, L., Redmond, S., Roberts, A. B., and Dexter, T. M., 1989, The effects of TGFβ on haematopoietic cells, *Growth Factors* **1:**193–202.

Han, Z. C., Maurer, A. M., Bellucci, S., Wan, H. Y., Kroviarski, Y., Bertrand, O., and Caen, J. P., 1992, Inhibitory effects of platelet factor 4 (PF4) on the growth of human erythroleukemia cells: Proposed mechanisms of action of PF4, *J. Lab. Clin. Med.* **120:**645–660.

Hangoc, G., Lu, L., Oliff, A., Hu, W., Bicknell, D. C., Williams, D. E., Gillis, S., and Broxmeyer, H. E., 1987a, Modulation of Friend virus infectivity in mice by purified human lactoferrin and purified recombinant murine interleukin-3, *Leukemia* **1:**762–764.

Hangoc, G., Lu, L., and Broxmeyer, H. E., 1987b, The comparative enhancing effects of prostaglandin E_1 on colony formation by erythroid progenitor (BFU-E) cells from bone marrow of mice differing in the FV-2 locus, *Leukemia Res.* **11:**501–507.

Hangoc, G., Carow, C. E., Schwall, R., Mason, A. J., and Broxmeyer, H. E., 1992, Effects *in vivo* of recombinant human inhibin on myelopoiesis in mice, *Exp. Hematol.* **20:**1243–1246.

Hannon, G. J., and Beach, D., 1994, $p15^{INK4B}$ is a potential effector of TGF-β-induced cell cycle arrest, *Nature* **371:**257–261.

Hatzfeld, J., Li, M. L., Brown, E. L., Sookdeo, H., Levesque, J. P., O'Toole, T., Gurney, C., Clark, S. C., and Hatzfeld, A., 1991, Release of early human hematopoietic progenitors from quiescence by antisense transforming growth factor β1 or Rb oligonucleotides, *J. Exp. Med.* **174:**925–929.

He, J., and Furmanski, P., 1995, Sequence specificity and transcriptional activation in the binding of lactoferrin to DNA, *Nature* **373:**721–724.

Heinrich, M. C., Dooley, D. C., and Keeble, W. W., 1995, Transforming growth factor β1 inhibits expression of the gene products for Steel factor and its receptor (c-kit), *Blood* **85:**1769–1780.

Hendrie, P. C., Miyazawa, K., Yang, Y. C., Langefield, C. D., and Broxmeyer, H. E., 1991, Mast cell growth factor (c-kit ligand) enhances cytokine stimulation of proliferation of human factor dependent cell line, M07e, *Exp. Hematol.* **19:**1031–1037.

Herskowitz, I., 1995, MAP kinase pathways in yeast: For mating and more, *Cell* **80:**187–197.

Hino, M., Tojo, A., Miyazono, K., Urabe, A., and Takaku, F., 1988, Effects of type β transforming growth factors on haematopoietic progenitor cells, *Br. J. Haematol.* **70:**143–147.

Holmes, W. E., Lee, J., Kuang, W. J., Rice, G. C., and Wood, W. I., 1991, Structure and functional expression of a human interleukin-8 receptor, *Science* **253:**1278–1280.

Horie, M., and Broxmeyer, H. E., 1993, Involvement of immediate-early gene expression in the synergistic effects of steel factor in combination with granulocyte–macrophage colony stimulating factor or interleukin-3 on proliferation of a human factor-dependent cell line, *J. Biol. Chem.* **268:**968–973.

Hunter, M. G., Bawden, L., Brotherton, D., Craig, S., Cribbes, S., Czaplewski, L. G., Dexter, T. M., Drummond, A. H., Gearing, A. H., Heyworth, C. M., Lord, B. I., McCourt, M., Varley, P. G., Wood, L. M., Edwards, R. M., and Lewis, P. S., 1995, BB-10010: An active variant of human macrophage inflammatory protein-1α with improved pharmaceutical properties, *Blood* **86:**4400–4408.

Jacobsen, F. W., Rothe, M., Rusten, L., Goeddel, D. V., Smeland, E. B., Veiby, O. P., Slordal, L., and Jacobsen, S. E. W., 1994, Role of the 75-kDa tumor necrosis factor receptor: Inhibition of early hematopoiesis, *Proc. Natl. Acad. Sci. U.S.A.* **91:**10695–10699.

Jacobsen, F. W., Dubois, C. M., Rusten, L. S., Vieby, O. P., and Jacobsen, S. E. W., 1995, Inhibition of stem cell factor-induced proliferation of primitive murine hematopoietic progenitor cells signaled through the 75-kilodalton tumor necrosis factor receptor, *J. Immunol.* **154:**3732–3741.

Jacobsen, S. E. W., Keller, J. R., Ruscetti, F. W., Kondaiah, P., Roberts, A. B., and Falk, L. A., 1991a, Bidirectional effects of transforming growth factor β (TGF-β) on colony-stimulating factor-induced human myelopoiesis *in vitro*: Differential effects of distinct TGF-β isoforms, *Blood* **78:**2239–2247.

Jacobsen, S. E. W., Ruscetti, F. W., Dubois, C. M., Lee, J., Boone, T. C., and Keller, J. C., 1991b, Transforming growth factor-β *trans*-modulates the expression of colony stimulating factor receptors on murine hematopoietic progenitor cells lines, *Blood* **77**:1706–1716.

Jacobsen, S. E. W., Jacobsen, F. W., Fahlman, C., and Rusten, L. S., 1994, TNF-α, the great imitator; role of p55 and p75 TNF receptors in hematopoiesis, *Stem Cells* **12**:111–128.

Jansen, R., Damia, G., Usui, N., Keller, J., Futami, H., Goey, H., Back, T. T., Longo, D. L., Ruscetti, F. W., and Wiltrout, R. H., 1991, Effects of recombinant transforming growth factor-β1 on hematologic recovery after treatment of mice with 5-fluorouracil, *J. Immunol.* **147**:3342–3347.

Keller, J. R., Mantel, C., Sing, G. K., Ellingsworth, L. R., Ruscetti, S. K., and Ruscetti, F. W., 1988, Transforming growth factor β1 selectively regulates early murine hematopoietic progenitors and inhibits the growth of IL-3-dependent myeloid leukemia cell lines, *J. Exp. Med.* **168**:737–750.

Keller, J. R., McNiece, I. K., Sill, K. T., Ellingsworth, L. R., Quesenberry, P. J., Sing, G. K., and Ruscetti, F. W., 1990, Transforming growth factor β directly regulates primitive murine hematopoietic cell proliferation, *Blood* **75**:596–602.

Keller, J. R., Jacobsen, S. E. W., Dubois, C. M., Hestdal, K., and Ruscetti, F. W., 1992, Transforming growth factor-β: A bidirectional regulator of hematopoietic cell growth, *Int. J. Cell Cloning* **10**:2–11.

Keller, J. R., Bartelmez, S. H., Sinicka, E., Ruscetti, F. W., Ortiz, M., Gooya, J. M., and Jacobsen, S. E. W., 1994, Distinct and overlapping direct effects of macrophage inflammatory protein-1α and transforming growth factor β on hematopoietic progenitor/stem cell growth, *Blood* **84**:2175–2181.

Kelvin, D. J., Michiel, D. F., Johnston, J. A., Lloyd, A. R., Springer, H., Oppenheim, J. J., and Wang, J. M., 1993, Chemokines and serpentines: The molecular biology of chemokine receptors, *J. Leukocyte Biol.* **54**: 604–612.

Khoury, E., Andre, C., Pontvert-Delucq, S., Drenou, B., Baillou, C., Guigon, M., Najman, A., and Lemoine, F. M., 1994, Tumor necrosis factor α (TNFα) downregulates *c-kit* protooncogene product expression in normal and acute myeloid leukemia $CD34^+$ cells via p55 TNFα receptors, *Blood* **84**:2506–2514.

Kishi, K., Ellingsworth, L. R., and Ogawa, M., 1989, The suppressive effects of type β transforming growth factor (TGFβ) on primitive murine hemopoietic progenitors are abrogated by interleukin-6 and granulocyte colony-stimulating factor, *Leukemia* **3**:687–691.

Koff, A., Ohtsuki, M., Polyak, K., Roberts, J. M., and Massague, J., 1993, Negative regulation of G1 in mammalian cells: Inhibition of cyclin 3-dependent kinase by TGF-β, *Science* **260**:536–539.

Kreisberg, R., Detrick, M. S., Osmand, A. P., and Moore, R. N., 1993, Cytokine mediation of the suppressive effect of ferritin on colony-stimulating factor-1-dependent monocytopoiesis, *J. Immunol.* **150**:5094–5103.

Kulkarni, A. B., Huh, C. G., Becker, D., Geiser, A., Lyght, M., Flanders, K. C., Roberts, A. B., Sporn, M. B., Ward, J. M., and Karlsson, S., 1993, Transforming growth factor β1 null mutation in mice causes excessive inflammatory response and early death, *Proc. Natl. Acad. Sci. U.S.A.* **90**:770–774.

Kurland, J., and Moore, M. A. S., 1977, Modulation of hemopoiesis by prostaglandins, *Exp. Hematol.* **5**:357–373.

Kurland, J., Bockman, R., Broxmeyer, H. E., and Moore, M. A. S., 1978a, Limitation of excessive myelopoiesis by the intrinsic modulation of macrophage-derived prostaglandin E, *Science* **199**:552–555.

Kurland, J. I., Broxmeyer, H. E., Bockman, R. S., and Moore, M. A. S., 1978b, Role for the monocyte-macrophage colony stimulating factor and prostaglandin E in the positive and negative feedback control of myeloid stem cell proliferation, *Blood* **52**:388–407.

Laterveer, L., Lindley, I. J. D., Hamilton, M. S., Willemze, R., and Fibbe, W. E., 1995, Interleukin-8 induces rapid mobilization of hematopoietic stem cells with radioprotective capacity and long-term myelolymphoid repopulating ability, *Blood* **85**:2269–2275.

Lemoli, R. M., Strife, A., Clarkson, B. D., Haley, J. D., and Gulati, S. C., 1992, TGF-beta 3 protects normal human hematopoietic progenitor cells treated with 4-hydroperoxycyclophosphamide *in vitro*, *Exp. Hematol.* **20**:1252–1256.

Lin, H. Y., Wang, X. F., Ng-Eaton, E., Weinberg, R. A., and Lodish, H. F., 1992, Expression cloning of the TGF-β type II receptor, a functional transmembrane serine/threonine kinase, *Cell* **68**:775–785.

Lord, B. I., Dexter, T. M., Clements, J. M., Hunter, M. A., and Gearing, A. J. H., 1992, Macrophage inflammatory protein protects the cytotoxic effects of hydroxyurea *in vivo*, *Blood* **79**:2605–2609.

Lord, B. I., Woolford, L. B., Wood, L. M., Czaplewski, L. G., McCourt, M., Hunter, M. G., and Edwards, R. M., 1995, Mobilization of early hematopoietic progenitor cells with BB-10010: A genetically engineered variant of human macrophage inflammatory protein-1α, *Blood* **85**:3412–3415.

Lu, L., and Broxmeyer, H. E., 1985, Comparative influences of phytohemagglutin-stimulated leukocyte conditioned medium, hemin, prostaglandin E and low oxygen tension on colony formation by erythroid progenitor cells in normal human bone marrow, *Exp. Hematol.* **13:**989–993.

Lu, L., Broxmeyer, H. E., Meyers, P. A., Moore, M. A. S., and Thaler, H. T., 1983, Association of cell cycle expression of Ia-like antigenic determinants on normal human multipotential (CFU-GEMM) and erythroid (BFU-E) progenitor cells with regulation *in vitro* by acidic isoferritins, *Blood* **61:**250–256.

Lu, L., Pelus, L. M., and Broxmeyer, H. E., 1984, Modulation of expression of Ia-antigens and the proliferation of human erythroid (BFU-E) and multipotential (CFU-GEMM) progenitor cells by prostaglandin E, *Exp. Hematol.* **12:**741–748.

Lu, L., Broxmeyer, H. E., Moore, M. A. S., Sheridan, A. P., and Gentile, P., 1985, Abnormalities in myelopoietic regulatory interactions with acidic isoferritins and lactoferrin in mice infected with Friend virus complex: Associations with altered expression of Ia-antigens on effector and responding cells, *Blood* **65:**91–99.

Lu, L., Pelus, L. M., Broxmeyer, H. E., Moore, M. A. S., Wachter, M., Walker, D., and Platzer, E., 1986a, Enhancement of the proliferation of human marrow erythroid (BFU-E) progenitor cells by prostaglandin E requires the participation of OKT_8 positive T-lymphocytes and is associated with the density expression of major histocompatibility complex class II antigens on BFU-E, *Blood* **68:**126–133.

Lu, L., Welte, K., Gabrilove, J. L., Hangoc, G., Bruno, E., Hoffman, R., and Broxmeyer, H. E., 1986b, Effects of recombinant human tumor necrosis factor alpha, recombinant human interferon-gamma and prostaglandin E on colony formation of human hematopoietic progenitor cells stimulated by natural human pluripotent colony stimulating factor, pluriopoietin-alpha and recombinant erythropoietin in serum-free cultures, *Cancer Res.* **46:**4357–4361.

Lu, L., Hangoc, G., Chen, L. T., Shen, N. R., Oliff, A., and Broxmeyer, H. E., 1987a, The protective influence of lactoferrin on mice infected with the polycythemia inducing strain of the Friend virus complex, *Cancer Res.* **47:**4184–4188.

Lu, L., Pelus, L. M., Piacibello, W., Moore, M. A. S., Hu, W., and Broxmeyer, H. E., 1987b, Prostaglandin E acts at two levels to enhance colony formation *in vitro* by erythroid (BFU-E) progenitor cells, *Exp. Hematol.* **15:** 765–771.

Lu, L., Walker, D., Broxmeyer, H. E., Hoffman, R., Hu, W., and Walker, E., 1987c, Characterization of adult human marrow hematopoietic progenitors highly enriched by two-color sorting with My10 and major histocompatibility (MHC) class II monoclonal antibodies, *J. Immunol.* **139:**1823–1829.

Lu, L., Shen, R-N., Zhou, S. Z., Srivastava, C., Harrington, M. A., Miyazawa, K., Wu, B., Lin, Z-H., Ruscetti, S., and Broxmeyer, H. E., 1991, Synergistic effect of human lactoferrin and recombinant murine interferon-gamma on disease progression in mice infected with the polycythemia-inducing strain of the Friend Virus Complex, *Int. J. Hematol.* **54:**117–124.

Lu, L., Xiao, M., Grigsby, S., Wang, W. X., Wu, B., Shen, R.-N., and Broxmeyer, H. E., 1993a, Comparative effects of suppressive cytokines on isolated single $CD34^{+++}$ stem/progenitor cells from human bone marrow and umbilical cord blood plated with and without serum, *Exp. Hematol.* **21:**1442–1446.

Lu, L., Xiao, M., Shen, R.-N., Grigsby, S., and Broxmeyer, H. E., 1993b, Enrichment, characterization and responsiveness of single primitive $CD34^{+++}$ human umbilical cord blood hematopoietic progenitor with high proliferative and replating potential, *Blood* **81:**41–48.

Maltman, J., Pragnell, I. B., and Graham, G. J., 1993, Transforming growth factor β: Is it a down regulator of stem cell inhibition by macrophage inflammatory protein 1α? *J. Exp. Med.* **178:**925–932.

Mantel, C., Kim, Y. J., Cooper, S., Kwon, B., and Broxmeyer, H. E., 1993, Polymerization of murine macrophage inflammatory protein-1α inactivates its myelosuppressive effects *in vitro*. The active form is monomer, *Proc. Natl. Acad. Sci. U.S.A.* **90:**2232–2236.

Mantel, C., Miyazawa, K., and Broxmeyer, H. E., 1994, Physical characteristics and polymerization during iron saturation of lactoferrin, a myelopoietic regulatory molecule with suppressor activity, in *Lactoferrin: Structure and Function* (T. W. Hutchens, B. Lonnerdal, and S. V. Rumball, eds.), pp. 121–132, Plenum Press, New York.

Mantel, C., Aronica, S., Luo, Z., Marshall, M. S., Kim, Y. J., Cooper, S., Hague, N., and Broxmeyer, H. E., 1995, Macrophage inflammatory protein 1α enhances growth-factor stimulated phospholipid turnover and increases cAMP levels in the human factor-dependent cell line, M07e, *J. Immunol.* **154:**2342–2350.

Massague, J., 1992, Receptors for the TGF-β family, *Cell* **69:**1067–1070.

Matthews, W., Jordan, G. T., Gavin, M., Jenkins, N. A., Copeland, N. G., and Lemischka, I. R., 1991, A receptor tyrosine kinase cDNA isolated from a population of enriched primitive hematopoietic cells and exhibiting close genetic linkage to *c-kit*, *Proc. Natl. Acad. Sci. U.S.A.* **88:**9026–9030.

Mayani, H., Little, M. T., Dragowska, W., Thornberry, G., and Lansdorp, P. M., 1995, Differential effects of the hematopoietic inhibitors MIP-1α, TGF-β and TNF-α on cytokine-induced proliferation of subpopulations of $CD34^+$ cells purified from cord blood and fetal liver, *Exp. Hematol.* **23:**422–427.

Maze, R., Sherry, B., Kwon, B. S., Cerami, A., and Broxmeyer, H. E., 1992, Myelosuppressive effects *in vivo* of purified recombinant murine macrophage inflammatory protein-1 alpha, *J. Immunol.* **149:**1004–1009.

Metcalf, D., 1995, The granulocyte–macrophage regulators: Reappraisal by gene inactivation, *Exp. Hematol.* **23:**569–572.

Metz-Boutique, M. H., Jolles, J., Mazurier, J., Schoentgen, F., Legrand, D., Spik, G., Montreuil, J., and Jolles, P., 1984, Human lactotransferrin: Amino acid sequence and structural comparisons with other transferrins, *Eur. J. Biochem.* **145:**659–676.

Michiel, D., 1993, Chemokines: The missing link, *Biotechnology* **11:**739.

Migdalska, A., Molineux, G., Demuynck, H., Evans, G. S., Ruscetti, F., and Dexter, T. M., 1991, Growth inhibitory effects of transforming growth factor-β1 *in vivo*, *Growth Factors* **4:**239–245.

Millauer, B., Wizigmann-Voos, S., Schnurch, H., Martinez, R., Moller, N., Risau, W., and Ullrich, A., 1993, High affinity VEGF binding and developmental expression suggest Flk-1 as a major regulator of vasculogenesis and angiogenesis, *Cell* **72:**835–846.

Miller, K. L., Carlino, J. A., Ogawa, Y., Avis, P. D., and Carroll, K. G., 1992, Alterations in erythropoiesis in TGF-β1-treated mice, *Exp. Hematol.* **20:**951–956.

Miller, M. D., and Krangel, M. S., 1992, Biology and biochemistry of the chemokines: A family of chemotactic and inflammatory cytokines, *Crit. Rev. Immunol.* **12:**17–46.

Miyazawa, K., Mantel, C., Lu, L., Morrison, D. C., and Broxmeyer, H. E., 1991, Lactoferrin–lipopolysaccharide interactions: Effects on lactoferrin binding to monocyte/macrophage differentiated HL-60 cells, *J. Immunol.* **146:**723–729.

Miyazono, K., Ichijo, H., and Heldin, C. H., 1993, Transforming growth factor-β: Latent forms, binding proteins and receptors, *Growth Factors* **8:**11–22.

Morikawa, K., Oseko, F., and Morikawa, S., 1994, H- and L-rich ferritins suppress antibody production, but not proliferation, of human B lymphocytes *in vitro*, *Blood* **83:**737–743.

Moser, B., Schumacher, C., Tscharner, V. V., Clark-Lewis, I., and Baggiolini, M., 1991, Neutrophil-activating peptide 2 and gro/melanoma growth-stimulatory activity interact with neutrophil-activating peptide/ interleukin-8 receptors on human neutrophils, *J. Biol. Chem.* **266:**10666–10671.

Munger, K., Pietenpol, J. A., Pittelkow, M. R., Holt, J. T., and Moses, H. L., 1992, Transforming growth factor β1 regulation of *c-myc* expression, pRB phosphorylation, and cell cycle progression in keratinocytes, *Cell Growth Differ.* **3:**291–298.

Munker, R., and Koeffler, P., 1987, *In vitro* action of tumor necrosis factor on myeloid leukemia cells, *Blood* **69:**1102–1108.

Murase, T., Hotta, T., Saito, H., and Ohno, R., 1987, Effect of recombinant human tumor necrosis factor on the colony growth of human leukemia progenitor cells and normal hematopoietic progenitor cells, *Blood* **69:**467–472.

Nagata, K., Tohda, S., Suzuki, T., and Nara, N., 1989, Effects of recombinant human tumor necrosis factor on the self-renewal capacity of leukemic blast progenitors in acute myeloblastic leukemia, *Leukemia* **3:**626–630.

Najman, A., Guigon, M., Gorin, N-C., and Mary, J-Y. (eds.), 1987, *Colloque INSERM, Vol. 162: The Inhibitors of Hematopoiesis*, pp. 3–354, John Libbey Eurotext, London.

Nara, N., Tohda, S., Nagata, K., Suzuki, T., and Yamashita, Y., 1989, Inhibition of the *in vitro* growth of blast progenitors from acute myeloblastic leukemia patients by transforming growth factor-β (TGF-β), *Leukemia* **3:**572–577.

Nelson, N. A., Kelly, R. C., and Johnson, R. A., 1982, Prostaglandins and the arachidonic acid cascade, *Chem. Eng. News* **60:**30–44.

Neote, K., Darbonne, W., Oges, J., Horuk, R., and Schall, T. J., 1993a, Identification of a promiscuous inflammatory peptide receptor on the surface of red blood cells, *J. Biol. Chem.* **268:**12247–12249.

Neote, K., DiGregorio, D., Mak, J. Y., Horuk, R., and Schall, T. J., 1993b, Molecular cloning, functional expression, and signalling characteristics of a c-c chemokine receptor, *Cell* **72:**415–425.

Neote, K., Mak, J. Y., Kolakowski, L. F., and Schall, T. J., 1994, Functional and biochemical analysis of the cloned duffy antigen: identity with the red blood cell chemokine receptor, *Blood* **84:**44–52.

Oh, K. O., Zhou, Z., Kim, K. K., Samanta, H., Fraser, M., Kim, Y. J., Broxmeyer, H. E., and Kwon, B. S., 1991, Identification of cell surface receptors for murine macrophage inflammatory protein-1α, *J. Immunol.* **147:**2978–2983.

Ottmann, O. G., and Pelus, L. M., 1988, Differential proliferation effects of transforming growth factor-β on human hematopoietic progenitor cells, *J. Immunol.* **140:**2661–2665.

Paolini, J. F., Willard, D., Consler, T., Luther, M., and Krangel, M. S., 1994, The chemokines IL-8, monocyte chemoattractant protein-1, and I-309 are monomers at physiologically relevant concentrations, *J. Immunol.* **153:**2704–2717.

Patel, S. R., Evans, S., Dunne, K., Knight, G. C., Morgan, P. J., Varley, P. G., and Craig, S., 1993, Characterization of the quaternary structure and conformational properties of the human stem cell inhibitor protein LD78 in solution, *Biochemistry* **32:**5466–5471.

Peetre, C., Gulberg, U., Nilsson, E., and Olsson, I., 1986, Effects of recombinant tumor necrosis factor on proliferation and differentiation of leukemic and normal hemopoietic cells *in vitro*, *J. Clin. Invest.* **78:**1694–1700.

Pelech, S. L., and Charest, D. L., 1995, MAP kinase-dependent pathways in cell cycle control, *Prog. Cell Cycle Res.* Vol. 1 (L. Meijer, S. Guidet, and L. Tung, eds.), pp. 33–52, Plenum Press, New York.

Pelus, L. M., 1982, Association between colony forming units-granulocyte macrophage expression of Ia-like (HLA-DR) antigen and control of granulocyte and macrophage production, *J. Clin. Invest.* **70:**568–578.

Pelus, L. M., and Gentile, P. S., 1988, *In vivo* modulation of myelopoiesis by prostaglandin E_2. III. Induction of suppressor cells in marrow and spleen capable of mediating inhibition of CFU-GM proliferation, *Blood* **71:**1633–1640.

Pelus, L. M., Broxmeyer, H. E., Kurland, J. I., and Moore, M. A. S., 1979, Regulation of macrophage and granulocyte proliferation: Specificities of prostaglandin E and lactoferrin, *J. Exp. Med.* **150:**277–292.

Pelus, L. M., Broxmeyer, H. E., and Moore, M. A. S., 1981, Regulation of human myelopoiesis by prostaglandin E and lactoferrin, *Cell Tissue Kinet.* **14:**515–526.

Pelus, L. M., Ottmann, O. G., and Nocka, K. H., 1988, Synergistic inhibition of human marrow granulocyte-macrophage progenitor cells by prostaglandin E and recombinant interferon-α, -β, and -γ and an effect mediated by tumor necrosis factor, *J. Immunol.* **140:**479–484.

Pelus, L. M., King, A. G., Broxmeyer, H. E., Marsh, P. L., Petteway, S. R., and Bhatnager, P. K., 1994, *In vivo* modulation of hematopoiesis by a novel hematoregulatory peptide, *Exp. Hematol.* **22:**239–247.

Penco, S., Pastroino, S., Bianchi-Scarra, G., and Garre, C., 1995, Lactoferrin down-modulates the activity of the granulocyte macrophage colony-stimulating factor promoter in interleukin-1β stimulated cells, *J. Biol. Chem.* **270:**12263–12268.

Piacibello, W., Rubin, B. Y., and Broxmeyer, H. E., 1986, Prostaglandin E counteracts modulation of the gamma interferon induction of Ia antigens on U937 cells and induction of responsiveness of U937 colony forming cells to suppression by lactoferrin, transferrin, acidic isoferritins and prostaglandin E, *Exp. Hematol.* **14:** 44–50.

Quinn, T. P., Peters, K. G., De Vries, C., Ferrara, N., and Williams, L. T., 1993, Fetal liver kinase 1 is a receptor for vascular endothelial growth factor and is selectively expressed in vascular endothelium, *Proc. Natl. Acad. Sci. U.S.A.* **90:**7533–7537.

Rajaranthnam, K., Sykes, B. D., Kay, C. M., Dewald, B., Geiser, T., Baggiolini, M., and Clark-Lewis, I., 1994, Neutrophil activation by monomeric interleukin-8, *Science* **264:**90–92.

Roberts, A. B., and Sporn, M. B., 1993, Physiological actions and clinical applications of transforming growth factor-β (TGF-β), *Growth Factors* **8:**1–9.

Ruscetti, F. W., Jacobsen, S. E., Birchenall-Roberts, M., Broxmeyer, H. E., Engelman, G. L., Dubois, C., and Keller, J. R., 1990, Role of transforming growth factor β1 in regulation of hematopoiesis, *Ann. N.Y. Acad. Sci.* **628:**31–43.

Rusten, L. S., Jacobsen, F. W., Lessiauer, W., Loetscher, H., Smeland, E. B., and Jacobsen, S. E. W., 1994a, Bifunctional effects of tumor necrosis factor α (TNFα) on the growth of mature and primitive human hematopoietic progenitor cells: Involvement of p55 and p75 TNF receptors, *Blood* **83:**3152–3159.

Rusten, L. S., Smeland, E. B., Jacobsen, F. W., Lien, E., Lesslauer, W., Loetscher, H., Dubois, C. M., and Jacobsen, S. E. W., 1994b, Tumor necrosis factor-α inhibits stem cell factor-induced proliferation of human bone marrow progenitor cells *in vitro*, *J. Clin. Invest.* **94:**165–172.

Sarris, A. J., Broxmeyer, H. E., Wirthmueller, U., Karasavvas, N., Cooper, S., Lu, L., Krueger, J., and Ravetch, J. V., 1993, Human interferon inducible protein 10: Expression and purification of recombinant protein demonstrate inhibition of early human hematopoietic progenitors, *J. Exp. Med.* **178:**1127–1132.

Sarris, A., Esgleyes-Ribot, T., Crow, M., Broxmeyer, H. E., Karasavvas, N., Grossman, D., Pugh, W., Grossman, D., Deisseroth, A., and Duvic, M., 1995, Cytokine loops involving interferon-γ and IP-10, a cytokine chemotactic for CD4-positive lymphocytes: An explanation for the epidermo-tropisin of cutaneous T-cell lymphomas, *Blood* **86:**651–658.

Schall, T. J., 1991, Biology of the RANTES/SIS cytokine family, *Cytokine* **3:**165–183.

Schlunegger, M. P., and Grutter, M. G., 1992, An unusual feature revealed by the crystal structure at 2.2A resolution of human transforming growth factor-β2, *Nature* **358:**430–434.

Schumacher, C., Clark-Lewis, I., Baggliolini, M., and Moser, B., 1992, High- and low-affinity binding of GROα and neutrophil-activating peptide 2 to interleukin 8 receptors on human neutrophils, *Proc. Natl. Acad. Sci. U.S.A.* **89:**10542–10546.

Sen, G. C., and Lengyel, P., 1992, The interferon system: A bird's eye view of its biochemistry, *J. Biol. Chem.* **267:**5017–5020.

Shull, M. M., Ormsby, I., Kier, A. B., Pawlowski, S., Diebold, R. J., Yin, M., Allen, R., Sidman, C., Proetzel, G., Calvin, D., Annunziata, N., and Doetschman, T., 1992, Targeted disruption of the mouse transforming growth factor-β1 gene results in multifocal inflammatory disease, *Nature* **359:**693–699.

Sing, G. K., Keller, J. R., Ellingsworth, L. R., and Ruscetti, F. W., 1988, Transforming growth factor β selectively inhibits normal and leukemic human bone marrow cell growth *in vitro*, *Blood* **72:**1504–1511.

Smith, C. A., Farrah, T., and Goodwin, R. G., 1994, The TNF receptor superfamily of cellular and viral proteins: Activation, costimulation and death, *Cell* **76:**959–962.

Sott, C., Dorner, B., Karawajew, L., Herrmann, F., and Brach, M. A., 1994, Transforming growth factor-β relieves stem cell factor-induced proliferation of myelogenous leukemia cells through inhibition of binding of the transcription factor NF-*jun*, *Blood* **84:**1950–1959.

Steinmann, G., Broxmeyer, H. E., de Harven, E., and Moore, M. A. S., 1982, Immuno-electron microscopic tracing of lactoferrin, a regulator of myelopoiesis, into a subpopulation of human peripheral blood monocytes, *Br. J. Haematol.* **50:**75–84.

Tauchi, T., Feng, G.-S., Shen, R., Song, H. Y., Donner, D., Pawson, T., and Broxmeyer, H. E., 1994a, SH2-containing phosphotyrosine phosphatase Syp is a target of p210bcr-abl tyrosine kinase, *J. Biol. Chem.* **269:** 15381–15387.

Tauchi, T., Marshall, M. S., Shen, R., Mantel, C., Feng, G. S., Pawson, T., and Broxmeyer, H. E., 1994b, The ubiquitously expressed Syp phosphatase interacts with *c-kit* and *Grb2* in hematopoietic cells, *J. Biol. Chem.* **269:**25206–25211.

Tauchi, T., Boswell, H. S., Leibowitz, D., and Broxmeyer, H. E., 1994c, Coupling between p210 *bcr-abl* and Shc and Grb-2 adaptor proteins in hematopoietic cells permits growth factor receptor independent-link to Ras activation pathway, *J. Exp. Med.* **179:**167–175.

Tauchi, T., Feng, G. S., Shen, R., Hoatlin, M., Bagby, G. C., Lu, L., and Broxmeyer, H. E., 1995, Involvement of SH2-containing phosphotyrosine phosphatase Syp in erythropoietin receptor signal transduction pathways, *J. Biol. Chem.* **270:**5631–5635.

ten Dijke, P., Yamashita, H., Ichijo, H., Franzen, P., Laiho, M., Miyazono, K., and Heldin, C. H., 1994, Characterization of type I receptors for transforming growth factor-β and activin, *Science* **264:**101–104.

Terman, B. I., Dougher-Vermazen, M., Carrion, M. E., Dimitrov, D., Armellino, D. C., Gospodarowicz, D., and Bohlen, P., 1992, Identification of the KDR tyrosine kinase as a receptor for vascular endothelial cell growth factor, *Biochem. Biophys. Res. Commun.* **187:**1579–1586.

Tessier, N., and Hoang, T., 1988, Transforming growth factor β inhibits the proliferation of the blast cells of acute myeloblastic leukemia, *Blood* **72:**159–164.

Vadhan Raj, S., Broxmeyer, H. E., Hittelman, W. N., Papadopoulos, N. E., Chawla, S. P., Fenoglio, C., Cooper, S., Buescher, E. S., Frenck, R. W., Jr., Holian, A., Perkins, R. C., Scheule, R. K., Gutterman, J. U., Salem, P., and Benjamin, R. S., 1992, Abrogating chemotherapy-induced myelosuppression by recombinant granulocyte–macrophage colony-stimulating factor: Protection at the progenitor cell level, *J. Clin. Oncol.* **10:**1266–1277.

Van Damme, J., 1994, Interleukin-8 and related chemotactic cytokines, in *The Cytokine Handbook* (A. Thomson, ed.), pp. 185–208, Academic Press, New York.

Verfaille, C. M., Catanzarro, P. M., and Li, W. N., 1994, Macrophage inflammatory protein 1α, interleukin-3 and diffusible marrow stromal factors maintain human hematopoietic stem cells for at least eight weeks *in vitro*, *J. Exp. Med.* **179:**643–649.

Wang, J. M., McVicar, D. W., Oppenheim, J. J., and Kelvin, D. J., 1993, Identification of RANTES receptors on human monocytic cells: Competition for binding and desensitization by homologous chemotactic cytokines, *J. Exp. Med.* **177:**699–705.

Wang, X. F., Lin, H. Y., Ng-Eaton, E., Downward, J., Lodish, H. F., and Weinberg, R. A., 1991, Expression cloning and characterization of the TGF-β type III receptor, *Cell* **67**:797–805.

Williams, D. E., Cooper, S., and Broxmeyer, H. E., 1988, The effects of hematopoietic suppressor molecules on the *in vitro* proliferation of purified murine granulocyte–macrophage progenitor cells (CFU-GM), *Cancer Res.* **48**:1548–1550.

Williams, N., 1979, Preferential inhibition of murine macrophage colony formation by prostaglandin E, *Blood* **53**:1089–1094.

Wrana, J. L., Attisano, L., Carcamo, J., Zentella, A., Doody, J., Laiho, M., Wang, X. F., and Massague, J., 1992, TGFβ signals through a heteromeric protein kinase receptor complex, *Cell* **71**:1003–1014.

Wrana, J. L., Attisano, L., Wieser, R., Ventura, F., and Massague, J., 1994, Mechanism of activation of the TGF-β receptor, *Nature* **370**:341–347.

Yamamura, Y., Hattori, T., Ohmoto, Y., and Takatsuki K., 1992, Identification and characterization of specific receptors for the LD78 cytokine, *Int. J. Hematol.* **55**:131–137.

Yi, T., and Ihle, J. N., 1993, Association of hematopoietic cell phosphatase with c-Kit after stimulation with c-Kit ligand, *Mol. Cell. Biol.* **13**:3350–3358.

Yi, T., Mui, A. L. F., Krystal, G., and Ihle, J. N., 1993, Hematopoietic cell phosphatase associates with the interleukin-3 (IL-3) receptor β chain and down-regulates IL-3-induced tyrosine phosphorylation and mitogenesis, *Mol. Cell. Biol.* **13**:7577–7586.

Yi, T., Zhang, J., Miura, O., and Ihle, J. N., 1995, Hematopoietic cell phosphatase associates with erythropoietin (Epo) receptor after Epo-induced receptor tyrosine phosphorylation: Identification of potential binding sites, *Blood* **85**:87–95.

Yip-Schneider, M. T., Horie, M., and Broxmeyer, H. E., 1995, Transcriptional induction of pim-1 protein kinase expression by interferon gamma and post-transcriptional effects upon co-stimulation with Steel factor, *Blood* **85**:3494–3502.

Youn, B. S., Jang, I-K., Broxmeyer, H. E., Cooper, S., Jenkins, N. A., Gilbert, D. J., Copeland, N. G., Elick, T. A., Fraser, M. J., Jr., and Kwon, B. S., 1995, A novel chemokine MRP-2 inhibits colony formation of bone marrow myeloid progenitors, *J. Immunol.* **155**:2661–2667.

Yu, J., Shao, L., Lemas, V., Yu, A. L., Vaughan, J., Rivier, J., and Vale, W., 1987, Importance of FSH-releasing protein and inhibin in erythrodifferentiation, *Nature* **330**:24–31.

Zucali, J. R., Broxmeyer, H. E., Levy, D., and Morse, C., 1989, Lactoferrin decreases monocyte induced fibroblast production of myeloid colony stimulating activity by suppressing monocyte release of interleukin 1, *Blood* **74**:1531–1536.

Chapter 7

Apoptosis in Hematopoiesis and Leukemogenesis

Rachel S. Chapman, Christopher D. Gregory, and Caroline Dive

1. APOPTOSIS

1.1. Introduction

The number of cells in any tissue is dependent on the balance between cell gain by proliferation and cell loss by differentiation and/or cell death (Figure 1). The importance of this balancing act is demonstrated particularly well within the hematopoietic system, where it is thought that in the order of billions of cells are produced every day to replace cell loss. Thus, it is reasonable to suggest that defects in this cell-gain–cell-loss balance are likely to contribute to a number of pathologies of the blood. Enhanced survival, restricted differentiation, and uncontrolled proliferation together conspire to promote the development of lymphomas and leukemias.

For many years scientific interest has been focused on proliferation as the most important process in regulating the balance, death was considered to be merely a passive event. The discovery that upended this philosophy was the report of a morphologically distinctive cell death process and the realization that it was genetically controlled. This cell death process was named apoptosis (Kerr *et al.*, 1972; Wyllie *et al.*, 1980; reviewed by Dive and Wyllie, 1993). We focus here on the role of apoptosis in the development of leukemias and lymphomas, where the suppression of apoptosis (by growth factors or certain oncogenes) results in an expanded population of cells with a survival advantage. As an active cell process, the apoptotic pathway(s) should be amenable to therapeutic intervention; we

Rachel S. Chapman and Caroline Dive Molecular Pharmacology Group, School of Biological Sciences, University of Manchester, Manchester M13 9PT, United Kingdom. **Christopher D. Gregory** Department of Immunology, School of Medicine, University of Birmingham, Birmingham B15 2TT, United Kingdom.

Blood Cell Biochemistry, Volume 7: Hematopoietic Cell Growth Factors and Their Receptors, edited by A. D. Whetton and J. Gordon. Plenum Press, New York, 1996.

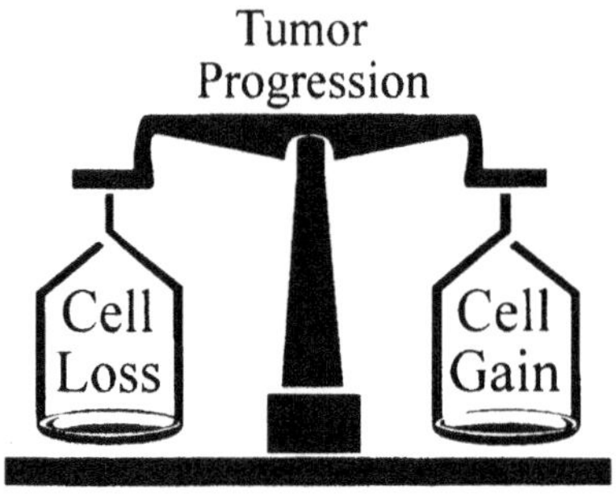

FIGURE 1. The critical balance between cell gain and cell loss determines the size of a tumor cell population.

envisage that new drug targets will be unveiled with increasing knowledge of the molecular regulation of apoptosis.

In this chapter we review the current understanding of the molecular mechanisms that regulate apoptosis (Section 1), we summarize the role of apoptosis in normal hematopoiesis (Section 2), we discuss examples of leukemias and lymphomas in which a failure to control apoptosis is believed to underlie the development of the disease (Section 3), and last, we comment of the suppression of apoptosis as a mechanism of resistance to chemotherapy (Section 4).

1.2. Incidence and Morphology

Apoptosis is a widespread process. In mammalian embryonic development, apoptosis is the mechanism of cell loss in focal deletion of cells to fashion the limbs and occurs during formation of the palate. Apoptosis acts to oppose proliferation during homeostasis of adult tissues. It is essential in the immune system for the deletion of self-reactive B and T lymphocytes (MacDonald and Lees, 1990) and is the mode of cell death in cell-mediated immune killing (Cohen *et al.*, 1985). Apoptosis is responsible for the regression of the prostate following castration (Kerr and Searle, 1973), the involution of breast tissue that occurs after weaning (Walker *et al.*, 1989), and the follicular atresia during the menstrual cycle (Hughes and Gorospe, 1991). Apoptosis is also the cell's physiological response to an adverse environment such as deprivation of nutrients or application of toxic stimuli (Hickman, 1992).

Apoptosis is well defined morphologically (Wyllie, 1992; Figure 2). Apoptotic cells shrink, but intracellular organelles maintain their integrity within the reduced cytoplasm. Cytosolic features include the formation of vacuoles and the slight swelling of mitochondria (Jacobson *et al.*, 1994; Benson *et al.*, 1995). The most obvious changes occur in the nucleus, where chromatin condenses and subsequently fragments. The cell surface develops protuberances (blebs), ultimately forming membrane-bound apoptotic bodies that may or may not contain the condensed chromatin. *In vivo* the apoptotic cells and bodies are recognized and ingested by phagocytes or neighboring cells before the cell spills its contents, thus preventing the development of an inflammatory response (Section 1.6.1). *In vitro*, however, a secondary degradation follows apoptosis in the absence of phagocytes that can be mistaken for a necrotic cell death and is often termed secondary necrosis.

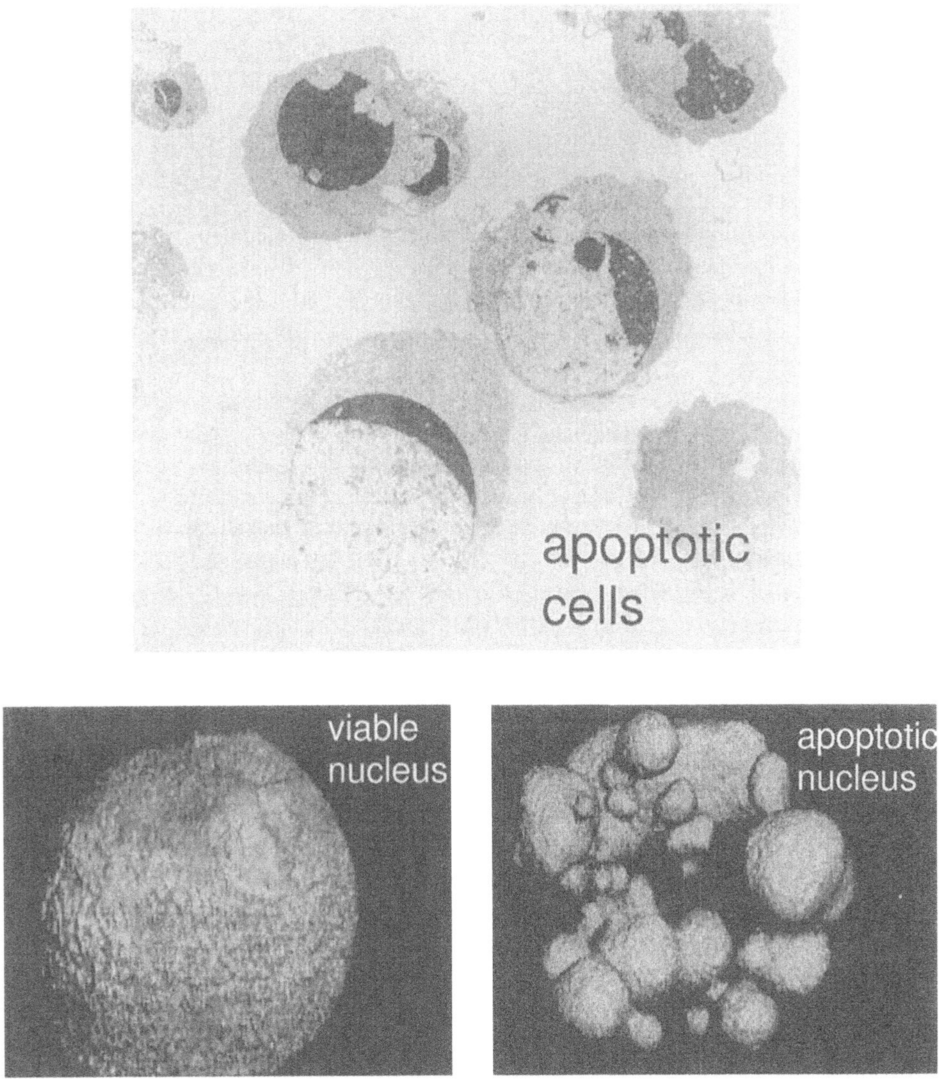

FIGURE 2. The classical morphology of apoptotic cells. The transmission electron micrograph of L1210 leukemia cells deprived of serum shows the classical condensation and fragmentation of the nucleus. Confocal microscopy (with three-dimensional reconstruction) of a viable HL60 cell nucleus (stained with propidium iodide) and a nucleus undergoing apoptosis following serum deprivation also clearly illustrates the dramatic changes within the nucleus (provided with thanks by Dr. Babbetta Marrone, Los Alamos National Flow Cytometry Laboratory, Los Alamos, NM).

1.3. Effectors of Apoptosis

1.3.1. Nonrandom Cleavage of DNA by Endonucleases

The formation of 180-base-pair integer fragments of DNA was reported in initial studies on glucocorticoid-induced apoptosis in immature thymocytes (Wyllie, 1980). This nonrandom cleavage of DNA results from the activation of an endogenous endonuclease(s) that cleaves DNA at internuceosomal linker sites and is typically visualized as a DNA "ladder" on agarose gel electrophoresis of DNA extracted from apoptotic cells (Wyllie, 1980a). The endonuclease(s) was shown to be dependent on calcium and magnesium and inhibited by zinc (Cohen and Duke, 1984) and until recently has generally been accepted as the cardinal feature of apoptosis. Arends *et al.* (1990) suggested that structural changes in the nucleus of apoptotic thymocytes were a direct result of selective nuclease activation in dying cells. Attempts to purify this enzyme have identified NUC18 (Gaido and Cidlowski, 1991), DNase I (Peitsch *et al.*, 1993), and DNase II (Barry and Eastman, 1993) as possible candidates.

However, chromatin condensation occurring during cell death of oligodendrocytes (Ucker *et al.*, 1992) and hepatocytes (Oberhammer *et al.*, 1993a) without endonucleolytic cleavage of DNA into low-molecular-weight fragments led to questions regarding the role of these endonucleases in apoptosis. From field inversion gel electrophoresis, which allows analysis of the integrity of DNA of molecular weight up to 2 megabase pairs, it has been found that cleavage of DNA to 300- and then 50-kilobase fragments is observed before oligonucleosomal fragments are formed in thymocytes (Walker *et al.*, 1991; Sun and Cohen, 1994). These sizes correlate with rosettes (300 kbp) and loops of chromatin (>50 kbp), as had previously been described (Filipski *et al.*, 1990). The appearance of 50-kbp DNA fragments and chromatin condensation is accompanied or immediately preceded by degradation of lamins A and B (Oberhammer *et al.*, 1994). In certain cell types, including MOLT-4 lymphoblastoid cells, classical apoptotic morphology and 50-kilobase fragments were observed with no subsequent oligonucleosomal fragments, indicating that the latter is a dispensable step in apoptosis (Oberhammer *et al.*, 1993b; Beere *et al.*, 1995; Figure 3).

The nature of the endonucleolytic reactions occurring during apoptosis has been further investigated. Zinc completely inhibited the cleavage of DNA into oligonucleosomal fragments in thymocytes induced to undergo apoptosis by dexamethasone but did not

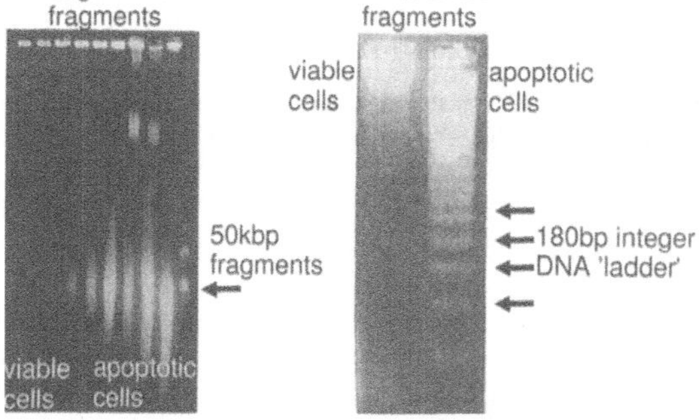

FIGURE 3. Nonrandom cleavage of leukemia cell DNA during apoptosis to generate 50-kbp fragments (detected by pulsed-field gel electrophoresis) and internucleosomal 180-bp integer fragments (detected by conventional gel electrophoresis). Data were kindly provided by Dr. Helen Beere and Dr. Dyfed Evans, School of Biological Sciences, University of Manchester.

prevent cleavage into high-molecular-weight fragments (Brown *et al.*, 1993), suggesting that different enzymes are involved in each stage of DNA degradation. The formation of large DNA fragments (50 kbp) and their subsequent degradation to 50 kbp in thymocytes were found to be magnesium dependent, and the second step was facilitated by, but not dependent on, calcium (Sun and Cohen, 1994). Experiments employing protease inhibitors have shown that serine but not cysteine proteases are in some way involved in activating the endonucleases that mediate internucleosomal cleavage (Weaver *et al.*, 1993).

The nuclear events described above are generally considered to be a hallmark of apoptosis. Together with examination of cell morphology, it is these nuclear events on which many methods to detect apoptosis are based, including many flow cytometric techniques, which are particularly useful for the study of hematopoietic cells (reviewed by Darzynkiewicz *et al.*, 1992). Notably, the TdT assay (and variants thereof) is used to identify and quantify apoptotic cells based on the incorporation by terminal deoxyribonucleotidyl transferase (TdT) of labeled nucleotides into the sites of endonucleolytic cleavage in apoptotic DNA (Gorzyca *et al.*, 1993; Chapman *et al.*, 1995a).

Apoptosis is clearly a strictly controlled coordinated process in which critical events occur in the nucleus, the plasma membrane, and in the cytoplasm. It has been proposed that the control of apoptotic effectors occurs within the cytoplasm and that cytoplasmic molecules set in train the nuclear events observed; this assertion followed experiments using cytoplasts (anucleated cells), which suggested that apoptosis can occur in the absence of a nucleus (Jacobson *et al.*, 1994). Studies with cell-free systems also suggest that cytoplasmic events control the initiation of nuclear fragmentation during apoptosis (Lazebnik *et al.*, 1993). For example, when intact nuclei were added to apoptotic extracts from eggs of *Xenopus laevis* (which were rich in mitochondria), they underwent classical fragmentation and DNA cleavage consistent with apoptosis (Newmeyer *et al.*, 1994). What then are the cytoplasmic effectors of apoptosis?

1.3.2. Apoptotic Proteases

Studies on the nematode *Caenorhabditis elegans* (described in more detail in Section 1.8.1) identified a death-inducing gene *ced-3*, which has sequence homology with the mammalian cysteine protease interleukin-1β-converting enzyme (ICE) (Yuan *et al.*, 1993). This discovery has led to the identification of several similar proteases that play a role in mammalian apoptosis. The ICE cleaves inactive pro-IL-1β to produce the mature form, IL-1β, an inflammatory mediator shown to be released during apoptosis (Hogquist *et al.*, 1991). The active proteolytic site of Ced-3 and ICE is a pentapeptide QACRG in which the central cysteine is critical: this site is conserved among all members of this family of cysteine proteases described to date. Overexpression of *Ice* or *ced-3* in Rat-1 fibroblasts was sufficient to lead to apoptosis. This was dependent on the cysteine protease activity of the enzyme and could be suppressed by overexpression of either the *crmA* gene (a specific inhibitor of ICE) or the *bcl-2* survival gene (Miura *et al.*, 1993; *bcl-2* is discussed in Section 1.8.2).

A novel protease prICE, resembling ICE and Ced-3, was discovered in cytosolic extracts from cells commited to apoptosis (Lazebnik *et al.*, 1994); prICE was subsequently identified as the cysteine protease CPP32/Yama, which is cleaved to an active form that has been renamed apopain (Fernandes-Alnemri *et al.*, 1994; Tewari *et al.*, 1995; Nicholson *et al.*,

1995). Apopain specifically cleaves the nuclear enzyme poly(ADP-ribose) polymerase (PARP) between Asp 216 and Gly 217, a site resembling that cleaved in pro-IL-1β (but it does not cleave pro-IL-1β or other proteins). Inhibition of apopain-mediated PARP cleavage prevented apoptosis. The importance of PARP cleavage for the induction of apoptosis is unclear. PARP activity has been shown to be necessary for the *induction* of apoptosis by hydrogen peroxide (possibly for the detection of damaged DNA; Nosseri *et al.*, 1994). Cleavage of PARP may be necessary at a later stage to prevent the enzyme from repairing DNA strand breaks during apoptosis or deleting energy sources required by the cell to complete the active death process (Lazebnik *et al.*, 1994). PARP cleavage may also be necessary to activate the endonuclease involved in DNA internucleosomal fragmentation (Nicholson *et al.*, 1995). Apopain is insensitive to the protease inhibitors TLCK and TPCK (Lazebnik *et al.*, 1994), but studies employing these inhibitors have demonstrated the importance of proteases acting downstream of apopain (reviewed by Earnshaw, 1995). The formation of 50-kbp DNA fragments and subsequent internucleosomal fragmentation in isolated rat liver nuclei were blocked by the protease inhibitors TPCK and TLCK (Zhivotovsky *et al.*, 1994). Similarly, TLCK inhibited apoptosis in thymocytes, but TPCK only inhibited internucleosomal cleavage, suggesting that protease activity may be cell type specific (Fearnhead *et al.*, 1995).

Ich-1 is another member of the Ced-3/ICE family (L. Wang *et al.*, 1994). Alternative splicing results in the production of two protein products, Ich-1_L and Ich-1_S. Ich-1_L shares homology with both Ced-3 and ICE, and its overexpression is capable of inducing apoptosis, which can be suppressed by the survival gene *bcl-2* but only weakly by *crmA*. Ich-1_S, a truncated version of Ich-1_L, acts to oppose Ich-1_L. The expression of *Ich-1* and *Ice* was detected in the same cells, suggesting that apoptosis may involve a cascade of proteases or the parallel action of these enzymes, but exact relationships among proteases in apoptosis is unclear (L. Wang *et al.*, 1994). In support of parallel or overlapping protease activities is the observation that thymocytes and macrophages from ICE-null mice retained the ability to undergo apoptosis (Li *et al.*, 1995). However, the proteolytic activation of the apopain proenzyme was achieved by addition of ICE (Tewari *et al.*, 1995), which is more suggestive of a protease cascade. The recent identifications of Ich-2 (Kamens *et al.*, 1995) and Mch-2 (Fernandes-Alnemri *et al.*, 1995) suggest that a large family of cysteine proteases exists, possibly operating in different tissues or on different targets during the initiation of apoptosis (Whyte and Evan, 1995).

The studies described above underline the importance of proteolysis steps in the pathway leading to the engagement of apoptosis. It should be noted that cells undergoing apoptosis are not subject to complete proteolytic digestion (Earnshaw, 1995).

1.4. Signal Transduction Events Resulting in Apoptosis

The signals that lead to the induction of apoptosis are beginning to be identified. This is a tough task in view of the asynchronous nature of apoptosis and the nonspecific changes that may occur during the secondary degenerative changes that follow apoptotic cell death *in vitro*. However, several signaling events have been implicated in the regulation of apoptosis, including changes in intracellular calcium concentration ($[Ca^{2+}]_i$) and the generation of ceramide. The studies of induction of apoptosis by Fas/APO-1 and TNF-α have been particularly interesting with respect to cell signaling, and we place some emphasis on this aspect.

1.4.1. Intracellular Calcium

The signaling pathways involved in linking various lethal stimuli to a final common response of apoptosis are poorly understood (reviewed by Dive *et al.*, 1992). Because the endonuclease(s) that cleaves DNA during thymocyte apoptosis was shown to be dependent on calcium (Cohen and Duke, 1984), it had been assumed that a calcium signal was central to the initiation of apoptosis. McConkey *et al.*, (1989) showed that extensive DNA fragmentation was preceded by a sustained increase in average cytosolic calcium concentration, but it has been difficult to understand how a large rise in calcium is consistent with maintenance of plasma membrane integrity in apoptotic cells. HL60 promyelocytic leukemia cells undergoing apoptosis did not exhibit an early increase in calcium levels (Lennon *et al.*, 1992). Moreover, reduction of calcium content in the medium of lymphoid cells treated with dexamethasone appeared to slightly enhance their apoptotic response (Bansal *et al.*, 1990). The role of calcium in apoptosis was further obscured by the observation that calcium ionophores, at concentrations that caused a rapid increase in intracellular calcium, prevented growth factor withdrawal-induced apoptosis of IL-3-dependent cells and stimulated synthesis of the survival factor IL-4 (Rodriguez-Tarduchy *et al.*, 1992). This seems to suggest that calcium signaling in apoptosis is cell type specific and dependent on the lethal stimulus employed.

If a direct influx of extracellular calcium is not involved in the signal transduction pathway of apoptosis, how else might calcium signaling be effected? Withdrawal of IL-3 from an IL-3-dependent hematopoietic cell line was associated with repartitioning of intracellular calcium from nonmitochondrial to mitochondrial pools (Baffy *et al.*, 1993). It was suggested that a decrease in the concentration of calcium in the cytosol, endoplasmic reticulum, or other intracellular compartments may regulate apoptosis. More evidence for the importance of calcium flux between intracellular compartments in apoptosis signaling came from studies on the survival gene *bcl-2* (Section 1.8.2). Bcl-2 reduced calcium efflux through the endoplasmic reticulum membrane in conjunction with the suppression of apoptosis (Lam *et al.*, 1994).

1.4.2. Signaling for Cell Death after Withdrawal of Growth Factors

There are a plethora of data that demonstrate that withdrawal of certain growth factors results in apoptotic cell death in many cell types, and much of these data pertain to hematopoietic cells (see Section 2.1). However, the identification of downstream signaling events following growth factor deprivation is an area of cell signaling that has been little explored with respect to the induction of apoptosis. Recent studies suggest roles for the Ras pathway (Kinoshita *et al.*, 1995) and the level of activtiy of phosphatidyl 3-kinase (Yao and Cooper, 1995).

1.4.3. Fas/Apo-1/CD95 and Tumor Necrosis Factor α

Fas/APO-1 was identified as a cell surface protein following the observations that certain antibodies induced apoptosis in various human cell types (reviewed by Negata and Golstein, 1995). The target of these antibodies, the Fas/APO-1 receptor, was cloned and found to be a member of the tumor necrosis factor (TNF) receptor family. This family includes the low-affinity receptor for nerve growth factor (NGF) and CD40 (see Chapter 3).

Both NGF and CD40 have been reported to provide survival signals to certain types of neurons (Edwards *et al.*, 1991) and to B cells in some situations (see Section 2.2.2), whereas, in contrast, Fas/APO-1 and TNF are associated with death-inducing signaling. The homology throughout this receptor family is found mainly within the extracellular cysteine-rich domains; however, Fas/APO-1 and TNF-R1 also share some similarities within their cytoplasmic regions. A sequence of about 70 amino acids is necessary for Fas or TNF-R1 to transduce an apoptotic signal; this domain has been termed the *death domain*. Interestingly, a similar death domain has been identified within a cytoplasmic protein of *Drosophila* called Reaper (reviewed by Steller, 1995). In this organism, all cell deaths are thought to occur by a common mechanism, and all cell deaths are preceded by *reaper* mRNA expression.

The ligand for Fas/APO-1, termed FasL, has recently been identified (reviewed by Negata and Golstein, 1995) and it is likely that activation of the receptor either by FasL or the antibodies to Fas/APO-1 requires cross-linking of the receptor. The role of Fas/APO-1 in normal B and T cells and the pathologies associated with mutations in Fas/APO-1 and FasL are discussed in Sections 2.2.1 and 2.2.2. TNF-α signaling through TNF-R1 is associated with the ability to generate host defense responses against certain types of pathogen (Pfeffer *et al.*, 1993).

Apoptosis induced by the ligation of Fas/APO-1 is rapid, independent of extracellular calcium, does not require macromolecular synthesis (see Section 1.7), and is only partially suppressed by the overexpression of *bcl-2* (see Section 1.8.2). Apoptosis induced by TNF-α through interaction with TNF-R1 is also rapid and has no requirement for new protein synthesis. Several molecules have now been identified that are thought to act along the death-signaling pathway initiated by FasL (or Fas/APO-1 antibodies) and/or TNF-α (reviewed by Cleveland and Ihle, 1995).

In the yeast two-hybrid assay with the cytoplasmic domain of TNF-R1 or Fas/APO-1 receptor as bait, three related proteins have been identified, namely TRADD, FADD, and RIP (see review by Cleveland and Ihle, 1994). All of these proteins have death domains similar to TNF-R1, Fas/APO-1, and Reaper. TRADD binds to TNF-R1, FADD binds to Fas/APO-1, and RIP binds to both, albeit with differing affinities. Their biochemical function is unclear, although a sequence in RIP suggests that it may act as a protein kinase, a tantalizing thought given that activation of TNF-α or of Fas/APO-1 can result in protein phosphorylations. It is thought that the death domains *per se* do not induce death but act to tether TRADD-like molecules to their respective receptors in order to modulate subsequent cascades of other apoptotic effectors (Cleveland and Ihle, 1995). ICE or ICE-like proteases have been implicated in Fas/APO-1- and TNF-α-driven apoptosis (Tewari and Dixit, 1995; Enari *et al.*, 1995). Overexpression of TRADD, FADD, or RIP can induce apoptosis, and, interestingly, Crm-A, the ICE inhibitor, can suppress this apoptosis.

Another aspect common to FasL-induced apoptosis and that induced by TNF-α is the generation of ceramide, a signaling molecule that has recently been associated with the induction of apoptosis. The precise relationship between the TRADD-like molecules decribed above and the ceramide pathway is not yet clear.

1.4.4. Ceramide Signaling for Apoptosis

Tumor necrosis factor α is principally secreted by macrophages and has a multitide of effects on hematopoietic cells (see Chapter 11) and nonhematopoietic cells in addition to its role described above as a potent inducer of apoptosis. Signaling pathways induced by

TNF-α have been studied intensively and are complex. One known effect of TNF-α is the stimulation of a plasma membrane sphingomyelinase, which hydrolyzes sphingomyelin to generate the second messenger ceramide. Events that result from the production of ceramide include the activation of a protein kinase (ceramide-activated protein kinase, CAPK) and a phosphatase (CAPP) and stimulation of the MAP kinase pathway (probably through Raf kinase). In addition, ceramide promotes the nuclear translocation of the transcription factor NFκB following its release from IκB (reviewed by Kolesnick and Golde, 1994). Whether all or some of these signaling events are involved in TNF-α-induced apoptosis is not yet clear. However, it has been demonstrated that synthetic membrane-permeable analogues of ceramide can induce apoptosis (Obeid *et al.*, 1993; Jarvis *et al.*, 1994a). Morover, ceramide signaling for apoptosis is not restricted to the TNF-α and FasL pathways because ionizing radiation has also been reported to result in generation of ceramide and subsequent apoptosis. Interestingly, activation of PKC (see Section 1.4.3) can suppress ceramide-induced apoptosis (Haimovitz-Friedman *et al.*, 1994).

1.5. Signals for the Suppression of Apoptosis

An increasing body of evidence suggests that all cells must receive survival signals to avoid apoptosis (Raff, 1992). Survival signals are generated, depending on cell type, via receptor binding of soluble growth factors, via cell–cell communication, and via interaction with an extracellular matrix (Figure 4). Hematopoietic growth factors are perhaps the most studied with respect to the suppression of apoptosis, and these are discussed further in Section 2.1. Whether the various survival stimuli share common downstream signaling elements that impinge on and suppress the default apoptotic pathway or whether they signal for survival using independent pathways is not yet clear. A rapidly expanding and widely distributed family of proteins, the Bcl-2 family, function to suppress apoptosis (Reed, 1994; see Section 1.8.2). Despite intensive investigation, their biochemical function(s) is still undefined, although the ratio of hetero- and homodimers within this family appears to dictate cell fate (Oltvai and Korsmeyer, 1994). Signals up- and downstream of Bcl-2-like proteins remain to be identified, although there is some evidence that Bcl-2 is acted on by kinases and phosphatases (Section 1.8.2).

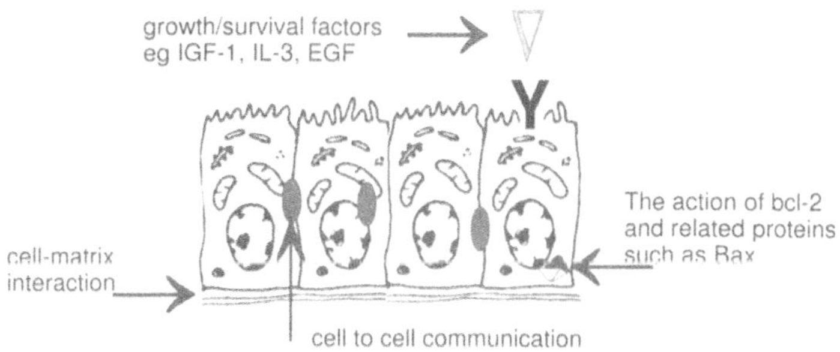

FIGURE 4. Survival signals are required for the suppression of apoptosis. Survival signals can be provided by soluble growth factors, cell–cell contact, interaction with extracellular matrix, and an endogenous survival protein such as Bcl-2. Although the cartoon depicts an epithelial cell interacting with its matrix, it is likely that several types of hematopoietic cells also require some form of stromal interaction.

1.5.1. Extracellular Matrix Interactions

The ability of extracellular matrix to influence apoptosis has been most studied in endothelial and epithelial cells, where absence of any matrix results in cell death by apoptosis (Meredith *et al.*, 1993; Frische and Francis, 1994; Boudreau *et al.*, 1995). The signals generated by the interaction with an extracellular matrix are as yet unknown, but in many instances they are integrin mediated. In addition to dependence on growth factors for survival, B lymphoblasts require direct contact with bone marrow stroma to suppress apoptosis (Manabe *et al.*, 1994). The growth factors within the stromal microenvironment alone were insufficient to support cell survival. Stromal factors crucial for the survival of B cells are thought to be either membrane bound and/or concentrated within the extracellular matrix or integral components of matrix (see also Chapter 5).

1.5.2. Protein Kinase C

Evidence is emerging to suggest that in certain circumstances, PKC activity can suppress apoptosis. Much of the data are based on studies using inhibitors or activators of PKC. Inhibition of PKC by compounds within differing degrees of specificity for PKC resulted in apoptosis in human leukemia cells (Jarvis *et al.*, 1994b); for example, the PKC inhibitor polymixin B induced apoptosis in mature lymphocytes (Lucas *et al.*, 1991). Phorbol esters, which activate the enzyme, have been shown to inhibit apoptosis induced by withdrawal of interleukin 2 from T lymphocytes (Rodriguez-Tarduchy and Lopez-Rivas, 1989). Some growth factors appear to stimulate survival through PKC; IL-1 prevented T-cell receptor-mediated thymocyte apoptosis by a mechanism involving activation of PKC (McConkey *et al.*, 1990), and the suppression of apoptosis by GM-CSF and IL-3 was overcome by inhibition of PKC (Rajotte *et al.*, 1992). In contrast, phorbol ester treatment was shown to induce apoptosis in thymocytes (Kizaki *et al.*, 1989). Evidence for a universal role for PKC in signal transduction pathways that modulate apoptosis is inconclusive. This may be at least partly because of the use of nonspecific inhibitors of the enzyme and the existence of isoforms with differing sensitivities to the inhibitors and phorbol esters.

A role for PKC-β has been proposed in the regulation of apoptosis in two studies in which PKC isoforms, subcellular location, and PKC activity were determined. In the first study, the activation of the v-Abl protein tyrosine kinase (v-Abl PTK, see Section 3.5) resulted in potent suppression of apoptosis in murine mast cells: this was concurrent with an increase in nuclear PKC activity and translocation of PKC-βII to the nucleus; moreover, cell survival was abrogated by the PKC inhibitor calphostin C (Evans *et al.*, 1995). In the second study a phorbol ester that selectively activates PKC-βI accelerated the entry of myelomonocytic cells into apoptosis; here the PKC-βI was cytosolic prior to engagement of apoptosis (Pongracz *et al.*, 1994). These results are intriguing, but clearly more data are required to further define the role of PKC in the regulation of apoptosis.

1.6. Membrane Events in Apoptosis

1.6.1. Phagocytic Clearance of Apoptotic Cells

During apoptosis, a variety of changes occur at the cell surface: contact is lost with neighbors, and key receptors are lost or altered (Wyllie, 1987; Wyllie *et al.*, 1980; Dransfield *et al.*, 1994, 1995). Arguably the most important property of a cell undergoing apoptosis is its ability to be rapidly recognized and phagocytosed by its neighbors or by professional phagocytes. Thus, apoptotic cells are marked at their surface to signal their own swift and efficient disposal. Because, *in vivo*, free apoptotic cells are rarely observed, even at sites of high-rate apoptotic death, it is likely that the efficiency of clearance of apoptotic cells is such that phagocytosis occurs before loss of plasma membrane integrity. The result is the safe packaging of apoptotic cells before leakage of intracytoplasmic or intranuclear constituents and potential histotoxic and inflammatory sequelae. An important contrasting scenario is the catastrophic tissue damage observed as a result of acute necrosis.

How, then, are apoptotic cells marked for clearance? A number of mechanisms have been proposed, some of which have been reviewed recently (Savill *et al.*, 1993). It is perhaps useful to consider the nematode *Caenorhabditis elegans* as a point of reference to the complexity of the clearance question. In this animal no fewer than seven genes are associated with phagocytic clearance of cells undergoing developmental programmed cell death (Ellis *et al.*, 1991). *C. elegans* lacks professional phagocytes, and it may therefore be no surprise that clearance of apoptotic cells by professional phagocytes in mammals operates via multiple molecular mechanisms (Figure 5).

1.6.1a. Vitronectin Receptor. The best-characterized and most widely accepted mechanism through which apoptotic cells are recognized and engulfed by mammalian phagocytes involves the phagocyte surface-associated integrin $\alpha_V\beta_3$, the vitronectin receptor (VNR). The majority of investigations to date have focused on the role of this receptor in the clearance of apoptotic human polymorphonuclear leukocytes *in vitro* by monocyte-derived macrophages (Savill *et al.*, 1993), but the VNR is also functional in clearance of lymphocytes by macrophages and in phagocytosis of neutrophils by fibroblasts (Hall *et al.*, 1994). Although the identity of the component(s) on the apoptotic cell surface with which the VNR interacts is not yet known, it has been established that the adhesive protein, thrombospondin (TSP), binds to $\alpha_V\beta_3$ and forms a molecular "bridge" between apoptotic cell and phagocyte (Savill *et al.*, 1992). Thrombospondin also interacts with an additional phagocyte surface molecule, CD36, which cooperates with $\alpha_V\beta_3$ in mediating phagocytosis of apoptotic cells. It has been shown recently that expression of CD36 in an $\alpha_V\beta_3$-positive

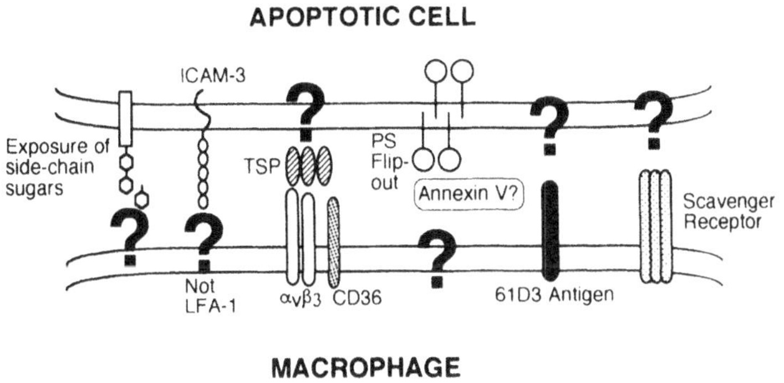

FIGURE 5. Schematic representation of membrane changes in apoptotic cells that are associated with clearance by phagocytes (modified from Savill *et al.*, 1993). TSP, thrombospondin; PS, phosphatidylserine. See text for details.

melanoma cell line after CD36 transfection is sufficient to confer on these cells the ability to phagocytose apoptotic neutrophils, lymphocytes, and fibroblasts (Ren *et al.*, 1995).

1.6.1b. Phagocyte Lectins. Functional activity of the $\alpha_V\beta_3$ VNR on fibroblasts in the clearance of apoptotic neutrophils appears to be complemented by an additional molecular pathway involving a mannose/fucose-specific lectin (Hall *et al.*, 1994). Earlier studies by Wyllie and colleagues suggested that a novel macrophage-associated lectin was involved in the recognition of apoptotic murine thymocytes by peritoneal macrophages (Duvall *et al.*, 1985). These studies indicate that apoptotic cell surface carbohydrate changes provide important signals for recognition. It is perhaps noteworthy in this context that qualitative changes in the adhesive glycoprotein ICAM-3 appear to initiate the involvement of this molecule in recognition of apoptotic human B cells by monocyte-derived macrophages (Flora and Gregory, 1995). Although the physiological ligand for ICAM-3, at least in viable cell–cell interactions, LFA-1, does not participate in apoptotic cell clearance, the possibility that changes in glycosylation pattern provide the basis for the involvement of ICAM-3, a heavily glycosylated molecule, in signaling for clearance has not been discounted. In this scenario, it is conceivable that carbohydrate changes in apoptotic ICAM-3 could produce targets for lectin-like molecules on the macrophage surface.

1.6.1c. Phosphatidylserine. Additional membrane events on apoptotic cells that provide means for recognition and clearance by macrophages involve changes in phospholipid symmetry between the inner and outer leaflets of the plasma membrane. It has been shown that, during apoptosis, the asymmetric distribution of phosphatidylserine (PS), with PS concentrated on the inner leaflet, is lost, and PS appears on the outer surface (Fadok *et al.*, 1992a). The demonstration that PS-containing liposomes inhibit uptake of apoptotic lymphocytes and neutrophils by macrophages pointed to the existence of some kind of PS receptor on the macrophage surface. The nature of this putative receptor has yet to be worked out, although the PS-binding protein annexin V may play a part in this recognition pathway (Homburg *et al.*, 1995). Interestingly, analyses of different populations of macrophages indicate that they utilize the putative PS receptor differentially. Thus, thioglycollate-elicited peritoneal macrophages appear to use the PS recognition mechanism selectively in clearing apoptotic cells, while peripheral blood monocyte- and bone marrow-derived macrophages predominately use the VNR (Fadok *et al.*, 1992b).

1.6.1d. Complexity of Interactions between Phagocytes and Apoptotic Cells. It is clear from Figure 5 that although the interactions at the cell surface between apoptotic cells and phagocytes appear complex, the picture is, as yet, incomplete. No molecular pathway from phagocyte surface to apoptotic cell surface has yet been described in full. Additional phagocyte surface molecules, such as the 61D3 antigen (Flora and Gregory, 1994) and the scavenger receptors of macrophages (Platt and Gordon, 1995), that contribute to recognition and phagocytosis of apoptotic cells at present only add to the complexity of the picture. Future developments in this area are eagerly awaited.

1.6.2. Activation of Tissue Transglutaminase

Tissue transglutaminase activity has been shown to be increased in rat hepatocytes undergoing apoptosis (Fesus *et al.*, 1989). This enzyme cross-links cytoplasmic and membrane proteins and plays a vital role by maintaining the integrity of apoptotic cells prior to phagocytosis, preventing leakage of intracellular enzymes and DNA into surrounding viable tissue.

1.7. Requirement of Macromolecular Synthesis in Apoptosis

Apoptosis is often referred to as programmed cell death, leading to the assumption that new gene transcription and/or translation is necessary for the process to occur. However, apoptosis can be stimulated (Martin *et al.*, 1990), blocked (Williams *et al.*, 1990), or unaffected (Duke *et al.*, 1984) by inhibitors of protein and RNA synthesis. This has led to the development of a model for the genetic program of cell death (Owens and Cohen, 1992). Where apoptosis is dependent on transcription of previously silent genes whose protein products are lethal, the cell is proposed to undergo an "induction" process; that is, gene expression is induced after the apoptotic stimulus. Where inhibition of gene expression or protein synthesis causes apoptosis, a "release" mechanism has been suggested; the death program is constitutively present in the cell but is prevented by an inhibitor with a short half-life, and this is more likely to occur in cells with short life spans to ensure their rapid disposal. If apoptosis is independent of protein or RNA synthesis, it is classed as a "transduction" process and is most likely to be the result of transmembrane signaling.

1.8. Genetic Control of Apoptosis

It now appears that an active program for cell death is highly conserved. Research into active cell death processes in *Caenorhabditis elegans* and *Drosophila* is paving the way in the understanding of the molecular mechanisms that regulate mammalian apoptosis. Apoptosis was originally defined as a morphologically distinct mode of cell death. The genes that regulate the process continue to be identified with exhilarating speed; thus, the interchangeable use of the terms apoptosis and programmed cell death is no longer based on an assumption.

1.8.1. Of Worms and Flies: Providing the Clues

Mention has already been made of the clear genetic program of cell death that has been elucidated in the nematode *Caenorhabditis elegans* (reviewed by Hengartner and Horvitz, 1994), and because this system has led to critical breakthroughs in apoptosis research, it deserves more detailed attention. The genes *ced-3* and *ced-4* act autonomously to cause the programmed cell death of 131 cells within the organism: *ced-4* acts upstream of *ced-3* and requires the presence of *ced-3* to function. The morphology of the dying cells exhibits similarities to mammalian apoptosis. The gene *ced-9* acts upstream of *ced-3* and *ced-4* to suppress the death signal and allow cell survival; *ced-9* is itself regulated, negatively by *egl-1* and positively by the genes *ces-1* and *ces-2*, which encode putative transcription factors and, interestingly, cannot function in a *ced-4*-minus background. Downstream of *ced-3* and *ced-4*, seven *ced* genes act to enable the engulfment of dying celis, and *nuc-1* encodes an endonuclease to degrade the DNA of dying cells. The absence of functional *nuc-1* does not prevent cell death. The *ced-9* is homologous to *bcl-2*, the prototype mammalian gene capable of suppressing apoptosis (Hentgartner and Horvitz, 1994; see Section 1.8.2). Expression of *bcl-2* in *C. elegans* is able to suppress nematode cell deaths (Vaux *et al.*, 1992). As discussed in Section 1.3.2, Ced-3 shares homology with the cysteine protease ICE (Yuan *et al.*, 1993), a finding that spawned the identification of a family of mammalian cysteine proteases involved in apotosis. A mammalian homologue to *ced-4* has yet to be reported.

The Reaper protein, which is of central importance in *Drosphila* apoptosis, has already been mentioned (Section 1.4.3). Reaper is expressed in all cells destined to undergo apoptosis and is thought to be a regulator rather than an effector of apoptosis. The mechanism of this regulation by *reaper* is not defined; it may function to activate death effectors or to inactivate death suppressors (Steller, 1995). Given the powerful technology surrounding *Drosophila* genetics and the plentiful evidence that apoptosis is highly conserved, there is a high probability that new insights into mammalian apoptosis are on the horizon.

1.8.2. The *bcl-2* Family

1.8.2a. *bcl-2*. The gene *bcl-2* was originally identified as a gene at the t(14;18) breakpoint in follicular lymphoma (Section 3.2); its 24-kDa protein product (Bcl-2) is the founding member of a rapidly expanding group of homologous proteins (Oltvai and Korsmeyer, 1994). Bcl-2 suppresses apoptosis induced by many diverse stimuli such as growth factor deprivation (Hockenberry *et al.*, 1990; Nunez *et al.*, 1990), drug treatment (Miyashita and Reed, 1992, 1993; Fisher *et al.*, 1993; Walton *et al.*, 1993), and apoptosis associated with terminal differentiation (Naumovski and Cleary, 1994). Some situations exist, however, where *bcl-2* expression does not increase cell survival. Apoptosis caused by withdrawal of IL-2 from an IL-2-dependent T-cell line or IL-6 from an IL-6-dependent myeloma line could not be prevented by Bcl-2 (Nunez *et al.*, 1990). Clonal deletion of T cells and TNF-α-induced apoptosis were also not inhibited by Bcl-2 (Sentman *et al.*, 1991; Vanhaesebroeck *et al.*, 1993). Therefore, there are Bcl-2-dependent and Bcl-2-independent mechanisms of cell death.

The biochemical function(s) of Bcl-2 is currently unclear and is under intense investigation. Some insight into the role of Bcl-2 has been gained from the study of mice deficient in *bcl-2* expression. Such animals complete embryonic development, but after birth, suffer retarded growth and premature death associated with apoptotic involution of the thymus and spleen and polycystic kidney disease; they also have hypopigmented hair (Veis *et al.*, 1993a). These last two characteristics of the *bcl-2*-null mouse are symptomatic of a defect in the redox pathway, suggesting that the Bcl-2 protein may exert its protective effect via an antioxidant pathway. This theory is further supported by the finding that overexpression of Bcl-2 can suppress lipid peroxidation, and other known antioxidants can suppress apoptosis in an IL-3-dependent murine cell line (Hockenberry *et al.*, 1993). However, Bcl-2 can suppress apoptosis in cells lacking mitochondrial DNA (Jacobson *et al.*, 1993) and in a cellular environment where the oxygen content is less than 1 ppm (Jacobson and Raff, 1995). Thus, the role of Bcl-2 as a protein with an antioxidant function remains controversial.

Further subcellular analysis of Bcl-2 revealed its location at the outer mitochondrial membrane, the nuclear envelope, and endoplasmic reticulum (Monaghan *et al.*, 1992; Krajewski *et al.*, 1993; de Jong *et al.*, 1994). Distribution of the protein is patchy, and it is possibly associated with mitochondrial contact zones and nuclear pores, leading to suggestions that it is involved in protein complexes for transmembrane transport. A report by Ryan *et al.* (1994) adds some weight to this supposition; these authors suggest that Bcl-2 and c-Myc (Section 1.8.4) cooperate to inhibit p53 function by altering the subcellular trafficking of p53 during the cell cycle (p53-mediated apoptosis is discussed in detail in Section 1.8.3). A role for Bcl-2 in subcellular transport was originally supported by structure–

function analysis of Bcl-2, which showed that the carboxyl-terminal domain of the protein was essential for its integration into membranes and that this membrane anchoring was required for the suppression of apoptosis (Nakai *et al.*, 1993; Tanaka *et al.*, 1993; Nguyen *et al.*, 1994). However, further studies have demonstrated that truncated (cytosolic) Bcl-2 is also capable of at least partially suppressing apoptosis and it is motifs at either end of the protein that are required for its survival function (Oltvai *et al.*, 1993; Borner *et al.*, 1994).

Investigations into possible interactions between Bcl-2 and known signaling pathways have produced few results. Bcl-2 has been shown to associate with the *ras*-related protein R-*ras* p23 (Fernandez-Sarabia and Bischoff, 1993) and the serine/threonine-specific Raf-1 kinase (H. G. Wang *et al.*, 1994). WEHI7.2 cells treated with thapsigargin, a highly specific inhibitor of the endoplasmic reticulum (ER)-associated calcium pump, underwent apoptosis that was suppressed by Bcl-2 (Lam *et al.*, 1994). Bcl-2 reduced calcium efflux through the ER membrane, preventing a sustained increase in cytosolic calcium that has been suggested to be important in apoptosis signaling in certain cell types (Section 1.4.1). Further evidence for an effect of Bcl-2 on specific signaling pathways came from studies on murine bone-marrow-derived BAF3 cells. Withdrawal of IL-3 from these cells resulted in apoptosis, which was considerably delayed in cells with forced expression of Bcl-2 (Marvel *et al.*, 1994). These cells arrested in G_1 but were refractory to restimulation into cycle by IL-3; addition of IL-3 failed to induce *cdc2* expression, suggesting that Bcl-2 may suppress apoptosis in part by inhibiting the signaling pathway necessary for cell cycle progression. Regulation of Bcl-2 function has been suggested to occur, at least in part, by phosphorylation. In the scenario in which IL-3 and bryostatin-1 suppressed apoptosis of FDC-P1/ER cells, Bcl-2 was hyperphosphoylated (May *et al.*, 1994). Inhibition of this hyperphosphorylation attenuated their effects on cell survival, and these authors suggest that one regulator of Bcl-2 is PKC. In direct contrast, phosphorylation of Bcl-2 in lymphoid cells attenuated its suppressive effect on apoptosis (Haldar *et al.*, 1995). Thus, many possible functions for Bcl-2 have been suggested, but further evidence is required before any firm conclusions are drawn.

Interaction of *bcl-2* with other genes implicated in the control of apoptosis has been demonstrated. Transgenic mice carrying both *bcl-2* and *c-myc* exhibit enhanced hyperproliferation of pre-B and B cells and have been shown to be more susceptible to the development of tumors than single *bcl-2* or *c-myc* transgenic animals (Strasser *et al.*, 1990; Section 2.2.2). In addition, Bcl-2 protein has been shown to inhibit apoptosis induced by growth factor withdrawal and c-Myc overexpression in Chinese hamster ovary cells (Bissonnette *et al.*, 1992) and in rat-1 fibroblasts (Wagner *et al.*, 1993). Under normal circumstances, c-Myc overexpression provides cells with a growth advantage only in the presence of specific growth factors (Harrington *et al.*, 1994a; Section 1.8.4). It appears that coexpression of Bcl-2 abrogates this requirement for secondary survival signals so that cells that would normally undergo apoptosis in the absence of growth factors not only survive but have the potential to undergo deregulated proliferation.

1.8.2b. Bax. What is clear is that Bcl-2 does not function alone. Extensive studies using the yeast two-hybrid system have suggested that multiple protein–protein interactions can occur within the *bcl-2* family (Oltvai *et al.*, 1993; Sato *et al.*, 1994). Bcl-2 can form homodimers, which are thought to be functionally inert, and also interacts with a number of different proteins. For example, the protein Bax shares extensive homology with Bcl-2, with which it can form heterodimers (Oltvai *et al.*, 1993). In particular, Bcl-2 and Bax share two highly conserved regions named BH1 and BH2 (Bcl-2 homology 1 and 2). These two domains are now known to be involved in the formation of Bcl-2:Bax heterodimers. Mutations within either domain prevented heterodimerization and completely abrogated the suppression of apoptosis by Bcl-2, while Bcl-2 homodimerization still occurred (Yin *et al.*, 1994). However, like Bcl-2, Bax can also form homodimers whose effects on apoptosis oppose that of Bcl-2. Overexpression of Bax has been found to accelerate apoptosis induced by growth factor withdrawal from an IL-3-dependent cell line and abrogate the protective effect of Bcl-2 (Oltvai *et al.*, 1993). Thus, it appears that the decision between survival or death in response to an apoptotic stimulus is determined at least in part by the ratio of Bcl-2 to Bax (Oltvai *et al.*, 1993). It is conceivable that in situations in which Bcl-2 has not suppressed apoptosis, such as those described above, this could be caused not by the existence of Bcl-2-independent mechanisms of cell death but by an insufficient or excessive supply of Bax. This has obvious implications for further studies on the function or mechanism of action of Bcl-2.

The tissue distribution of Bax appears to be more widespread than that of Bcl-2; patterns of Bcl-2 and Bax expression *in vivo* are only partially overlapping, and curiously, Bax protein expression is not confined to tissues characterized by high apoptotic rates (Krajewski *et al.*, 1994a). Immunohistochemical analysis of Bax distribution has shown it to be present in the germinal center of lymph nodes, in contrast to Bcl-2, which was absent in these cells but found in mantle zone lymphocytes (Krajewski *et al.*, 1994a), areas characterized by high and low levels of apoptosis, respectively. Bax was also found in myeloid precursor cells in the bone marrow but not in mature granulocytes (Krajewski *et al.*, 1994a). These cells have very short life spans, so other pathways controlling cell death may be present in these cells.

1.8.2c. *bcl-x*. The gene *bcl-x* is an example of a *bcl-2*-related gene that can regulate apoptosis independently of *bcl-2* (Boise *et al.*, 1993); *bcl-x* exists as two alternatively spliced mRNAs, $bcl\text{-}x_L$ and $bcl\text{-}x_S$. Stable transfection of $bcl\text{-}x_L$ into an IL-3-dependent cell line has been shown to protect the cells from apoptosis induced by growth factor withdrawal to the same extent as *bcl-2*. In contrast, transfection of $bcl\text{-}x_S$ into the same cells inhibited the protective effect of *bcl-2* (Boise *et al.*, 1993). The tissue distribution of *bcl-x* is, in many cases, strikingly different from that of *bcl-2*, and they may regulate apoptosis at different stages of cell differentiation, and their expression may be controlled in a tissue-specific manner (Krajewski *et al.*, 1994b). Of interest is the finding that Bcl-x could prevent cyclosporin-A-induced apoptosis in murine B cells, whereas Bcl-2 could not (Gottschalk *et al.*, 1994).

1.8.2d. Other Bcl-2 Homologues. The list of Bcl-2-like proteins is continually increasing and now includes Mcl-1, whose expression increases during monocyte differentiation of myeloblastic cells (Reynolds *et al.*, 1994); A1, a novel hematopoietic cell-specific early-response gene (Lin *et al.*, 1993); and the virally encoded genes BHRF-1 (Epstein–Barr virus) and LMW5-HL (African swine fever virus).

1.8.2e. Bad. Bad is another heterodimeric partner for Bcl-2 and for Bcl-x_L and exhibits sequence homology limited to the BH1 and BH2 domains (Yang *et al.*, 1995). Bad lacks the hydrophobic C-terminal region of other Bcl-2 homologues and is therefore likely to be cytosolic. Bad has a higher affinity for Bcl-x_L than for Bcl-2 and can reverse the survival function of Bcl-x_L but not that of Bcl-2. A model is emerging to explain how the formation of different heterodimers within the Bcl-2 family might regulate cell fate (Yang *et al.*, 1995; Sato *et al.*, 1994; Figure 6A). In similar vein to Bax:Bcl-2 heterodimerization, the binding of Bax to Bcl-x_L is thought to be required for cell survival. The mechanism whereby Bcl-x_L-mediated survival is abrogated by Bad is thought to depend on the ability of Bad to displace Bax from Bcl-x_L. Bcl-x_S is also thought to operate in this way to displace Bax from Bcl-2 or Bcl-x_L. The result in each case would be the formation of Bax homdimers and the engagement of apoptosis. Whether this model proves to be correct awaits further testing.

1.8.2f. Bak. Three groups identified yet another homologue of bcl-2, named bak (bcl-2-homologous antagonist/killer: Farrow *et al.*, 1995; Chittenden *et al.*, 1995; Kiefer *et al.*, 1995). The Bak protein is widely distributed. It has a membrane anchoring region, and it binds to Bcl-x_L (Farrow *et al.*, 1995) and to Bcl-2 (Chittenden *et al.*, 1995). Bak induces apoptosis when it is overexpressed, countering protection by Bcl-x_L or Bcl-2. Bak has also been shown to bind to the adenoviral protein E1B (19-kD), and at high protein levels Bak can overcome E1B (19-kDa)-mediated cell survival (Farrow *et al.*, 1995).

Clearly, measurement of the protein levels of the members of the bcl-2 family may not be sufficient to understand the survival potential of different cell types given various potentially lethal stimuli, because who binds to whom within this family appears to be of critical importance (Figure 6B).

1.8.2g. Bag-1. Bag-1 is not a Bcl-2 homologue. It does, however, bind to Bcl-2 and modulate its function (Takayama *et al.*, 1995). Fas/APO-1-induced apoptosis is only partially attenuated by Bcl-2; however, cotransfection of *bcl-2* and *bag-1* signficantly increased cell survival, although Bag-1 alone cannot suppress Fas/APO-1-induced apoptosis.

1.9. *p53*: To Check or to Die?

p53 has been named guardian of the genome (Lane, 1992) and is a tumor suppressor gene that, as this title suggests, plays a key role in the control of normal cell growth. This is evidenced by several observations. First, patients with Li Fraumeni syndrome have germline mutations in p53 and exhibit a high frequency of different tumor types (Li *et al.*, 1988); second, mutation or deletion of *p53* occurs in approximately half of all tumors (reviewed by Levine *et al.*, 1994a); and third, *p53* homozygous null mice appear developmentally normal but are predisposed to tumor formation at a very early age and have a very limited survival (Donehouer *et al.*, 1992). Mutations of *p53* are many and varied and can occur by deletion of one or both alleles or a variety of mutations within the gene. What is the molecular basis for the control of cell growth by *p53*?

The p53 protein contains several functional domains including an N-terminal transcriptional transactivation domain, a central DNA-binding domain that confers sequence-specific binding, a nuclear localization signal region, an oligomerization domain, and a C-terminal domain that mediates binding to DNA in a non-sequence-specific fashion (Lin *et al.*, 1994). The p53 protein exists in several conformational forms, which are important with respect to different functions of the protein (Milner, 1995). p53 is a sensor of DNA damage (Kastan *et al.*, 1991). The p53 protein level is elevated by DNA damage (a posttranslational event) and then can induce G_1 growth arrest and/or direct the induction of apoptosis, thus ensuring the prevention of cell proliferation with a damaged genome. Whether the cellular response to damage is cell cycle check or engagement of apoptosis depends on the genetic context in which p53 operates. The mechanistic determinants of this

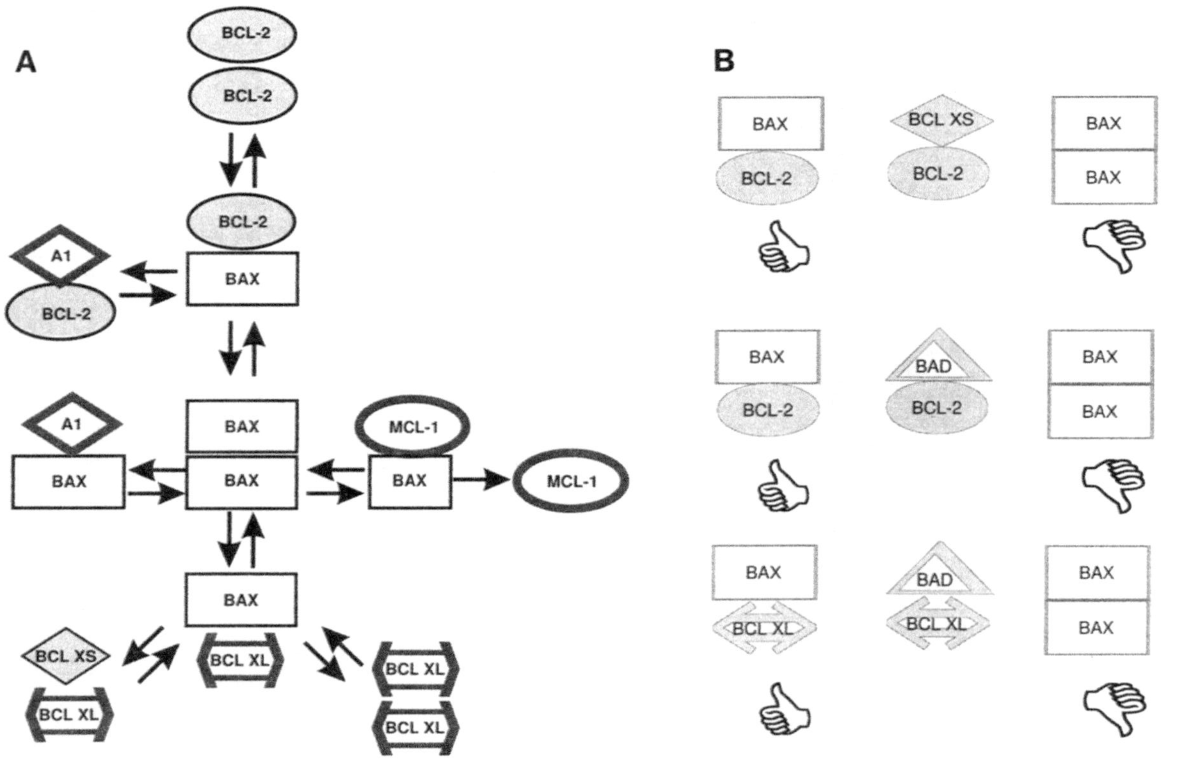

FIGURE 6. Current models for the heterodimerization with the Bcl-2 family (see text for details). (A) In addition to the formation of homodimers, Bcl-2 can heterodimerize with several other proteins within this family. (B) Bax:Bcl-2, Bax:Bcl-x_L heterodimers are thought to promote cell survival (thumbs up). The displacement of Bax from binding to Bcl-2 or Bcl-x_L by either Bcl-x_s or Bad is thought to induce apoptosis (thumbs down).

decision to check or to die are the focus of much current research. As a response to damage, p53 can bind DNA in a sequence-specific manner, resulting in altered transcription from promoters with p53 binding sites (p53 can activate or repress gene transcription) (reviewed by Zambetti and Levine, 1993).

How does p53 sense damaged DNA? DNA strand breaks have been suggested to be the type of DNA lesion required for the stabilization of p53; treatment with the topoisomerase-I inhibitor camptothecin showed that it was not the DNA damage *per se* but the ensuing replication-associated strand breaks that were necessary to induce p53 (Nelson and Kastan, 1994). Studies on the exposure of immature thymocytes to the DNA cross-linking agent cisplatin support this contention: despite no detectable differences in DNA cross links, proliferating but not quiescent thymocytes were induced to undergo apoptosis, implying that a proliferative event was needed to convert DNA cross links to a lesion that could be sensed by p53 (Evans *et al.*, 1994). In addition to strand breaks, the p53 sensor has been shown to recognize insertion/deletion mismatches in DNA (DNA bulges, Lee *et al.*, 1995). Recent studies using cell extracts have shed light on the mechanism by which p53 senses and then responds to DNA damage, and the following model has been proposed (Bayle *et al.*, 1995). Damaged DNA is recognized by p53 via its C terminus (DNA sequence nonspecific), and during this process the conformation of p53 is such that the core DNA sequence-specific binding of p53 in not active. Once bound to the damage site, p53 is phosphorylated and changes conformation. This change in conformation results in detachment from the damage site and activation of DNA sequence specificity. Now p53 can bind to its target sequences (see below) to activate or to repress transcription. The model, although attractive, awaits further testing in intact cells.

The involvement of p53 in halting the cell cycle at the G_1/S boundary has been well characterized. Wild-type p53 is normally present at low levels within cells but is rapidly induced in response to DNA damage and causes G_1 arrest; cells lacking wild-type p53 do not accumulate the protein and do not arrest in G_1 (Kastan *et al.*, 1991; Fritsche *et al.*, 1993). Certain types of *p53* mutation lead to a gain of p53 function, and here the dominant negative mutant *p53* can act as a promoter of cell proliferation. These opposing effects of p53 on cell proliferation are associated with two different conformations of the protein (Milner, 1995). Wild-type *p53* induces the expression of *waf-1/cip-1* (El-Diery *et al.*, 1993; Harper *et al.*, 1993). The product of *waf-1*, p21, is an inhibitor of cyclin-depdendent kinases and hence provides a mechanistic link between *p53* and inhibition of cell proliferation. *p53* has also been suggested to participate in a G_2/M phase block (Stewart *et al.*, 1995), where activation of *waf-1/cip-1* may be involved in the down-regulation of cdc2/cyclin B kinase required for progression through G_2/M. Wild-type (wt) *p53* also transactivates *GADD45* (Kastan *et al.*, 1992), which binds to proliferating cell nuclear antigen (PCNA) (Smith *et al.*, 1994). A PCNA redistribution occurs in the nucleus after DNA damage, possibly to sites of DNA damage, where it stimulates DNA excision repair (Smith *et al.*, 1994). The interaction of Gadd45 and PCNA may facilitate DNA repair and/or decrease the amount of PCNA in replication complcxes, resulting in growth inhibition. Anothcr transcription targct of wild-type *p53* is the *mdm2* gene (Barak *et al.*, 1993). Mdm2 protein binds to *p53* and abrogates its transcriptional activation; this may be an important mechanism for autoregulation of *p53*.

The mechanistic basis of *p53*-mediated apoptosis of DNA damaged cells is less well characterised. *p53* activity alone is capable of inducing apoptosis, as demonstrated when *p53* was expressed as a temperature sensitive mutant in hematopoietic cells (Yonish-Rouach *et al.*, 1991; Ryan *et al.*, 1993). *p53* is also involved in and thought neccessary for the induction of apoptosis following DNA damage in certian circumstances. Mouse thymocytes required p53 for radiation and etoposide-induced apoptosis but not for apoptosis caused by glucocorticoid treatment or T-cell receptor engagement (Clarke *et al.*, 1993; Lowe *et al.*, 1993). Apoptosis was dependent on the amount of functional *p53* present; cells heterozygous for *p53* deletion were partially resistant to radiation and etoposide. Treatment of cells with apoptosis-inducing concentrations of DNA-damaging agents resulted in nuclear accumulation of *p53*, which was a consequence of increased protein stability and not enhanced gene expression (Fritsche *et al.*, 1993). The importance of p53 in the response of tumors to cancer therapy has been corroborated *in vivo* using immunocompromised mice (Lowe *et al.*, 1994a). Tumors expressing p53 contained a high proportion of apoptotic cells and were responsive to treatment with γ-irradiation or adriamycin; in contrast, p53-deficient tumors contained few apoptotic cells and were unresponsive to therapy. Acquired *p53* mutations within p53-expressing tumors were associated with treatment resistance and relapse; the *p53* status of tumors is therefore an important determinant of their chemotherapeutic response (see Section 4). It should be noted that *p53*-independent mechanisms for the induction of DNA damage-induced apoptosis have also been observed (Strasser *et al.*, 1994c).

An important relationship has been demonstrated between expression of Bcl-2 and Bax and the activity of wt p53 (see Section 1.8.2). Transfection of myeloid leukemia (Ml) cells with a temperature-sensitive p53 induced a temperature-dependent decrease in Bcl-2 expression and a simultaneous elevation in Bax expression (Miyashita *et al.*, 1994a). In addition, p53-dependent increased transcription of *bax* was noted after irradiation of Ml cells before the engagment of apoptosis in several cell types with functional wild-type p53 (Zhan *et al.*, 1994). The rapid up-regulation of Bax by p53 is proposed to account for the speed of p53-induced apoptosis. *p53*-independent TGFβ-1-induced Ml leukemia cell death occurs more slowly, and although Bcl-2 expression is reduced following TGFβ-1 administration, no associated increase in Bax expression was demonstrated; note that the protein half-life of Bcl-2 is suggested to be about 24 hr (Selvakumaran *et al.*, 1994). Mice deficient in *p53* have been shown to exhibit elevated Bcl-2 and depleted Bax protein levels in several tissue types (Miyashita *et al.*, 1994a). A negative response element has been identified in the *bcl-2* gene that is thought to mediate the down-regulation of Bcl-2 by p53 (Miyashita *et al.*, 1994b). It is not yet apparent whether the response element mediates this transcriptional down-regulation directly. The hypothesis linking *p53* to the Bax up-regulation is attractive but it is hard to reconcile with recent reports that p53 does not act as a transcriptional activator to engage apoptosis (Caelles *et al.*, 1994). Further controversy in this area is provided by reports that Bcl-2 relieves p53-mediated transcriptional repression (Sabbatini *et al.*, 1995).

1.8.4. *c-myc*

The gene *c-myc* is one of the early response genes rapidly induced in quiescent cells on mitogenic stimulation. Deregulated *c-myc* expression, combined with a block to cell proliferation, potently induces apoptosis in fibroblasts (Evan *et al.*, 1992). Similar results have been obtained using an IL-3-dependent myeloid cell line (Askew *et al.*, 1991). Evan has suggested a model, the dual-signal hypothesis, to explain this observation (Harrington *et al.*,

1994), which argues that c-Myc drives proliferation and apoptosis simultaneously, and the expression of other factors, such as the provision of survival factors and/or the function of proteins such as Bcl-2, determines which outcome is reached. The data to support this model are summarized below. The regions of c-Myc required for induction of apoptosis overlap with regions necessary for cotransformation, autoregulation, and inhibition of differentiation. In a similar fashion to Bcl-2, c-Myc requires a protein partner, Max, with which it forms dimers (Amati *et al.*, 1993). Dimerization is necessary for both the suppression of apoptosis and the mitogenic response, as it facilitates the binding of Myc to DNA, where it acts as a transcription factor. Expression of Bcl-2 has been shown to inhibit *c-myc*-induced apoptosis without affecting the *c-myc* mitogenic function (Bissonnette *et al.*, 1992; Fanidi *et al.*, 1992), providing a mechanistic basis for the oncogenic synergy between these genes. The Bcl-2 homologue Mcl-1 also suppresses apoptosis induced by overexpresion of Myc (Reynolds *et al.*, 1994). A similar role for the suppression of deregulated *c-myc*-induced apoptosis by mutant p53 has been suggested (Lotem and Sachs, 1993a). Where c-Myc expression leads to cell cycle reentry, coexpression of wild-type *p53* induces apoptosis, whereas mutant *p53* allows c-Myc-driven proliferation to occur (Hermeking and Eick, 1994). However, it is not clear whether c-Myc plays a universal role in inducing apoptosis. Dexamethasone-induced apoptosis of human lymphoid leukaemia cells occurred long after a loss of c-Myc protein (Wood *et al.*, 1994).

1.8.5. Genes Expressed during Apoptosis

Several genes described below are regulated during apoptosis, but as yet it is not clear if they drive apoptosis or are expressed as a consequence of apoptosis. In several systems there appears to be induction of early response genes including *c-fos*, *c-myc*, and/or *c-jun* (Smeyne *et al.*, 1993; Ning and Murphy, 1993; Manome *et al.*, 1993; Buttyan *et al.*, 1988). The purpose of this induction is unknown, as the genes are also induced during proliferation and differentiation. Whether it is part of the apoptosis signaling cascade or merely a stress response of cells remains to be determined. The products of *c-fos* and *c-jun* form a stable transcription factor, AP-1, which is activated during apoptosis (Sikora *et al.*, 1993a,b). There is not, however, an indiscriminate activation of transcription factors, as NFκB DNA binding was observed while there was down-regulation of OCT-1 and CREB (Sikora *et al.*, 1993b).

Another gene up-regulated in tissues undergoing apoptosis is the testosterone-repressed prostate message-2 gene (*TRPM-2*), which encodes a highly sulfated cell surface glycoprotein (Buttyan *et al.*, 1989). However, a more detailed study revealed that *TRPM-2* (also known as *clusterin* and *SGP-2*) was not expressed in cells undergoing apoptosis but instead in the neighboring cells that survived (French *et al.*, 1994). Freshly isolated human polymorphonuclear cells undergo rapid apoptosis *in vitro*, which correlated with decreased *TRPM-2* gene expression. The gene could, therefore, have a role in maintaining the viability of specific cells within a regressing tissue.

Human lymphocytes induced to undergo apoptosis by γ-irradiation showed increased levels of ubiquitin mRNA followed by ubiquination of nuclear proteins (Delic *et al.*, 1993). Antisense oligonucleotides to ubiquitin decreased ubiquination and considerably reduced levels of apoptosis, leading to the suggestion that the ubiquitin gene is induced as part of the apoptotic pathway. Ubiquination of nuclear proteins may be involved in the condensation and subsequent fragmentation of chromatin.

2. ROLE OF APOPTOSIS IN NORMAL HEMATOPOIESIS

2.1. Myeloid Lineage Development

2.1.1. Soluble Survival Factors

Many hematopoietic growth factors (survival factors) can promote myeloid cell survival by suppressing apoptosis. This was demonstrated by Williams *et al.* (1990), who showed that withdrawal of IL-3, GM-CSF, or G-CSF from the hematopoeitic precursor cell lines FDCP-1, FDCP-Mix 1, and A4 resulted in apoptosis. Further studies have shown that suppression of apoptosis by survival factors occurs in many cell lineages of the hematopoietic system: viability of erythroid progenitor cells is maintained by erythropoietin (Koury and Bondurant, 1990); GM-CSF (but not TGF-β, G-CSF, IL-3,-4,-6, or -8) provides survival signals to human polymorphonuclear neutrophils (Brach *et al.*, 1992); SCF and IL-3 can both suppress apoptosis in mast cells (Mekori *et al.*, 1993; Iemura *et al.*, 1994); and SCF and leukemia inhibitory factor act on primordial germ cells isolated from mouse embryos to suppress apoptosis (Pesce *et al.*, 1993). Brandt *et al.* (1994) extended this work by examining the relative survival effects of IL-3 and SCF on various classes of human hematopoietic progenitor cells. They found that SCF suppressed apoptosis and promoted the survival of more primitive cells such as pre-CFC, whereas IL-3 played a predominant role in sustaining more differentiated cells.

These *in vitro* results suggest the following model for the regulation of hematopoietic cell number (Koury, 1992). During development from pluripotent stem cells to peripheral blood cells, hematopoietic progenitor cells become exquisitely dependent on growth factors to suppress apoptosis. Normal levels of growth factor allow the survival of some, but not all, progenitor cells. Increased levels of growth factors would increase the number of surviving cells and conseqently that of more mature blood cells. Conversely, a decrease in the levels of growth factors would result in increased apoptosis and lower levels of mature cells.

2.1.2. The Bcl-2 Family

Studies in which apoptosis of multipotent hematopoietic cells was suppressed by the survival gene *bcl-2* have shown that in the absence of growth factors, cells are capable of differentiating into mature cells (Fairbairn *et al.*, 1993). This implies that growth factors are required for the survival of hematopoietic cells, which then follow an intrinsically determined differentiation pathway. Bcl-2 appears to fulfill a physiological role within the hematopoietic system, as it is found in the precursor cells of many hematopoietic lineages but is absent in their mature progeny. It has been suggested that it is involved in conferring longevity to these proliferating cells by suppressing apoptosis (Hockenberry *et al.*, 1991). Up-regulation of Bcl-2 in cell lines has been reported as a consequence of growth factor addition (Deng and Podack, 1993; Otani *et al.*, 1993). Any physiological significance of up-regulation of Bcl-2 by growth factors during hematopoiesis is currently unclear, but it could contribute to the mechanism of growth-factor-mediated expansion of the progenitor cell pool. Part of this regulation by growth factors may be through phosphorylation of Bcl-2 protein itself. As mentioned in Section 1.8.2, treatment of FDC-P1/ER with IL-3 and bryostatin-1 (an activator of PKC) suppressed growth factor deprivation-induced apoptosis and resulted in hyperphosphorylated Bcl-2 (May *et al.*, 1994). The effect of phosphorylation of Bcl-2 on its ability to bind to Bax and other members of the Bcl-2 family, and hence the effect on survival function, remains to be established.

2.2. Lymphoid Lineage Development

2.2.1. T Lymphocytes

The majority of recirculating T cells bear on their surface heterodimeric receptors for antigen (TCR) that are composed of α and β polypeptide chains associated with a TCR signal transduction complex, CD3. Mature T cells recognize protein antigens processed to peptides and presented on cell surface molecules encoded by the polymorphic major histocompatibility complex (MHC). Two functionally distinct subsets of $\alpha\beta$ TCR-expressing T cells have been identified according to their expression of the "coreceptors" CD4 and CD8: $CD4^+$ cells are largely of the helper/inducer type and recognize peptides presented by MHC class II molecules; $CD8^+$ cells function mainly as cytotoxic T cells and recognize peptides presented by MHC class I molecules. Development of the $\alpha\beta$ T-cell lineage takes place in the thymus, a primary lymphoid organ that provides the appropriate microenvironment required for the "education" of immature T cells (see Chapter 8). In the thymus, vast numbers of immature T cells are produced from relatively few precursors, and their developmental pathway follows selection processes that favor maturation of cells bearing TCR-recognizing self MHC molecules. At the same time, the selection mechanisms operate to prevent maturation of T cells that recognize self peptides.

The sequence of changes in TCR gene rearrangement, expression of TCR polypeptide chains, and surface expression of TCR and ancillary molecules during T-cell development in the thymus has been reviewed recently (Kisielow, 1995) and is summarized in Figure 7. Of particular relevance to the present chapter are the mechanisms underlying activation or repression of the apoptotic program in maturing T cells. Establishment of a population of mature T cells with a diverse repertoire of specificities requires a high rate of cell turnover: large numbers of immature T cells are produced (about 5×10^6 per day in the mouse thymus), and the stringent selection processes result in almost all of these cells (more than 95%) undergoing apoptosis (Egerton *et al.*, 1990). Von Boehmer and colleagues described the thymus as an organ that "... selects the useful, neglects the useless, and destroys the harmful" (Von Boehmer *et al.*, 1989). In the context of apoptosis as a default mechanism in developing T cells, as in any other cell type (and immature thymocytes show high-rate spontaneous apoptosis *in vitro*), the preceding statement may be reconsidered: the thymus promotes repression of apoptosis in the useful but prevents or overrides such repression in the harmful; survival signals are not accessible to the useless.

How is thymocyte survival regulated? Cells that do not make a functional TCR ("the useless") may not receive the survival signals necessary to permit them to move to the next stage in their development, and apoptosis ensues by default. In this scenario, synthesis of a functional TCR is a critical determinant of thymocyte fate. Supportive evidence on the role of the TCR in thymocyte survival is provided by studies of mice with *scid* or *rag* mutations. These animals are unable to rearrange their TCR genes, and T-lineage cells die by apoptosis at the progenitor stage. Expression of TCR transgenes, however, promotes T-cell survival in these mice (Cory, 1995).

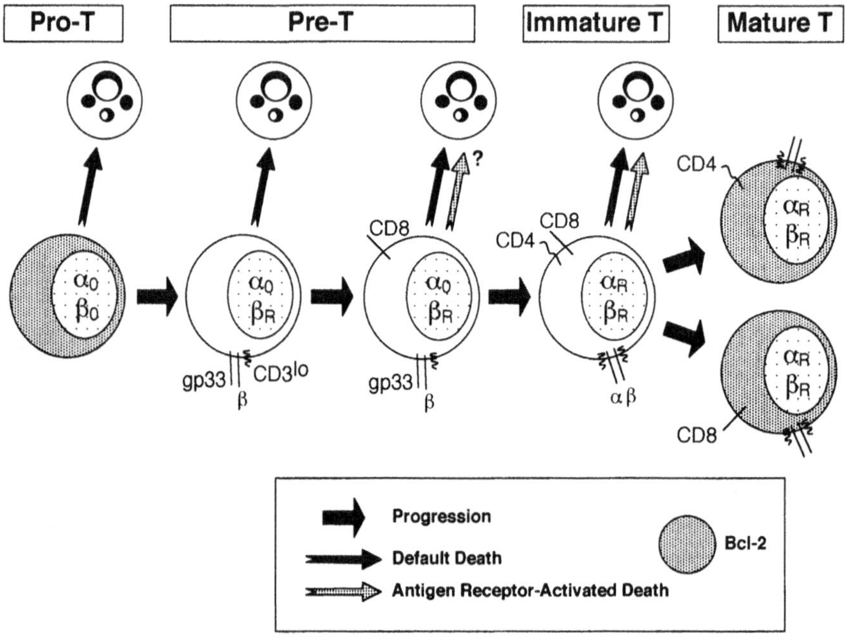

FIGURE 7. Simplified scheme of progression or death during T-cell development in the thymus. Stages of TCR-α and TCR-gene rearrangements (0, germline; R, rearranged) and Bcl-2 protein expression (cytoplasmic shading) are indicated. Surface expression of TCR-α, -β, CD3, surrogate heavy chain (gp33), and coreceptor CD4 and CD8 polypeptides are also shown. Note that the "default" death pathway may be activated by glucocorticoids.

Potentially harmful (i.e., self-reactive) thymocytes may reactivate their apoptotic program as a result of "inappropriate" signal transduction initiated at the TCR: in an immature thymocyte, engagement of the TCR generates an apoptotic signal, whereas in a mature T cell, recognition of antigen by the TCR generates a mitogenic signal (see Kisielow, 1995). The paradox of T-cell development in the thymus is that the TCR promotes both positive and negative selection, i.e., survival and apoptosis, respectively. Thus, self MHC-reactive thymocytes are selected for, and self peptide-reactive cells are selected against. The qualitative differences among survival, apoptotic, and mitogenic signals generated at TCR are beginning to be elucidated. Differences in affinity/avidity of the interactions of TCR, the CD4 and CD8 coreceptors, and availability of thymic ligands for ancillary molecules including the IL-2 receptor, CD2, CD44, LFA-1, and Fas are all likely to contribute to the fate-determining signals.

There is evidence to suggest that intracellular signaling pathways initiated at the TCR diverge during T-cell maturation. For example, positive selection appears to alter the coupling of second-messenger pathways to TCR in developing thymocytes (Malissen and Schmitt-Verhulst, 1993). In addition, differential expression of the transcription factor Nur77 may mark an important point of divergence in TCR-generated intracellular signalling for apoptosis versus proliferation (Liu *et al.*, 1994; Woronicz *et al.*, 1994).

2.2.1a. Bcl-2. The role played by *bcl-2* in T-cell development has been the subject of intense investigation. In thymocytes, the expression of Bcl-2 protein is high at the progenitor (pro-T)-cell and mature T-cell stages (Figure 8) but low in the stages during which selection occurs, in between (Veis *et al.*, 1993; Gratiot-Deans *et al.*, 1994; Moore *et al.*, 1994). In *bcl-2*-knockout mice, T cells appear to develop normally (Nakayama *et al.*, 1993). In *bcl-2*-transgenic mice that express *bcl-2* constitutively in the T-cell lineage, negative selection (deletion of self-reactive cells) continued to occur (Sentman *et al.*, 1991; Strasser *et al.*, 1994a). With regard to positive selection, expression of Bcl-2 was found to be insufficient to enhance mature T-cell numbers, even though levels of immature cells were increased. Thus, although Bcl-2 subserved the survival element of positive selection, the oncoprotein did not support the differentiation element (Strasser *et al.*, 1994a). In *scid* mice, expression of Bcl-2 failed to promote T-cell maturation (Strasser *et al.*, 1994b). Taken together, these results indicate that T-cell development occurs by a route that either circumvents or negates Bcl-2 function. The role played by Bcl-2-homologues in T-cell development will be an important avenue of future investigation.

2.2.1b. Fas. In mice, Fas is expressed on both immature and mature thymocytes, although the apoptosis-inducing activity of Fas appears to be restricted to the immature cells (Nishimura *et al.*, 1995; Ogasawara *et al.*, 1995). Although the function of Fas suggests that interaction with Fas ligand in the thymus could play a role in thymocyte deletion, direct

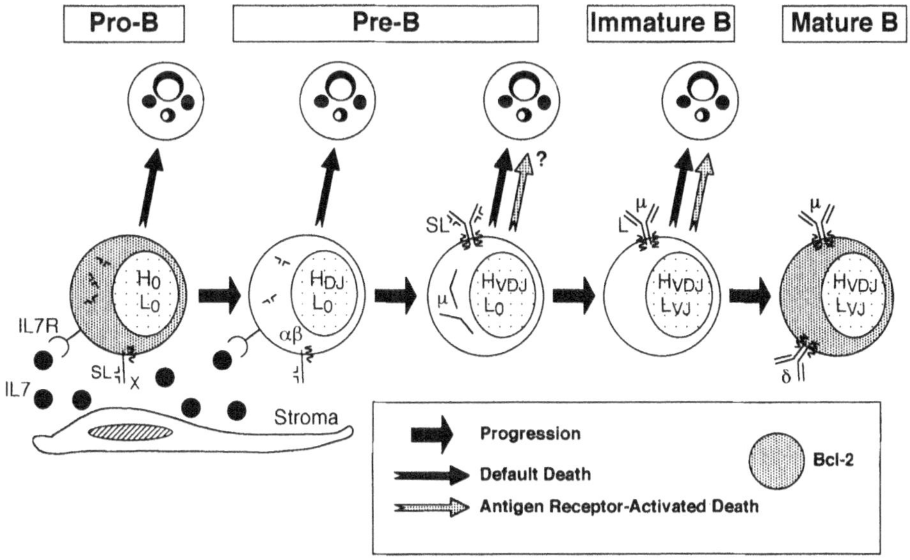

FIGURE 8. Simplified scheme of progression or death during initial maturation of B cells in the bone marrow. Heavy (H)- and light (L)-chain antigen receptor gene rearrangements (0, germline; DJ, VDJ, and VJ, rearrangements of variable (V), diversity (D), and joining (J) gene segments) are indicated. Cytoplasmic and surface expression of surrogate light chains (SL) and μ heavy chains are shown. Bcl-2 protein expression is indicated by cytoplasmic shading, and surface expression of IL7 receptors (IL7R), surrogate heavy chains (X), immunoglobulin-α and -β polypeptides, light chains (L), and γ heavy chains are also shown. Growth/survival factors provided by bone marrow stroma are exemplified by IL7 but are likely to include stromal cell-bound elements in addition.

evidence has not been reported. In mice carrying homozygous *lpr* and *gld* mutations, affecting the *fas* and *fas*-ligand genes, respectively, large numbers of abnormal T cells accumulate in peripheral lymphoid tissues (see Cohen and Eisenberg, 1995).

2.2.2. B Lymphocytes

Just as for the T lineage, maturation of B lymphocytes is critically dependent on the successful synthesis and surface expression of antigen receptors (see Figure 8) as well as hematopoietic growth factors (see Chapters 9 and 10). The antigen receptors of B cells interact with native antigens, and during the production of cells in the bone marrow, self-reactive cells are physically deleted or rendered functionally inactive (anergic). Apoptosis plays a central role in the successful generation of bone marrow B cells with a diverse repertoire of receptors. Thus, B lymphopoiesis involves high cell turnover with about 70–75% of the precursor B (pre-B)-cell pool ($>5 \times 10^7$ cells in the average mouse) being deleted. Cell loss occurs through typical apoptosis, and clearance is mediated by resident macrophages with which developing B cells intimately associate (Osmond, 1994).

Successful maturation of B cells, like that of T cells, follows a "progression-or-death" pathway (Figure 8). Unsuccessful rearrangement of immunoglobulin genes leads to death by default: mice with *rag* and *scid* mutations, which preclude productive Ig gene rearrangement, produce no B cells beyond the progenitor stage (Rolink and Melchers, 1993). Expression of receptors for key growth/survival factors (e.g., IL-7 receptor) and competition for soluble and bone marrow stroma-associated factors are additional elements in successful survival and progression along the B-cell differentiation pathway (reviewed by Gregory, 1995). Elimination of self-reactive cells in the bone marrow appears to occur at two levels: (1) arrest of development such that the transition from immature to mature B cell is blocked and the expression of an array of cell surface molecules involved in B-cell activation and migration is restricted; (2) cell death, which ensues rapidly after developmental arrest (Hartley *et al.*, 1993). A recent study indicates that a transitional phase of differentiation between the immature and mature B-cell stages marks a key point for self-reactive cell deletion after genesis in the bone marrow (Carsetti *et al.*, 1995).

A second important phase of B-cell selection occurs in the germinal centers (GC) of secondary lymphoid follicles (see MacLennan, 1994). These are the locations at which B cells responding to T-dependent (protein) antigens proliferate at high rate and undergo point mutations at Ig V-region genes in order to improve the affinity of antibodies produced against the immunizing antigen. B cells expressing mutant antigen receptors are thought to undergo selection through their ability to (1) compete for antigen held in the form of immune complexes by follicular dendritic cells and (2) interact with CD40-ligand, a tumor necrosis factor family member expressed transiently on activated T cells. In the absence of these and probably additional signals, GC B cells default into apoptosis (Liu *et al.*, 1989, 1992). Furthermore, a self-censoring mechanism, which appears to occur directly as a result of interaction of B cells with soluble antigen, operates to activate apoptosis in self-reactive cells generated unfortuitously as a result of Ig V-region mutation (Shokat and Goodnow, 1995; Pulendran *et al.*, 1995).

2.2.2a. Bcl-2. The physiological mechanisms through which B-cell apoptosis is regulated are only partially understood. Clearly, the antigen receptor is a central moderator of B-cell survival and death responses. The distribution of Bcl-2 during B-cell differentiation suggests a role for the protein in promoting survival (Figure 8). Early progenitor B cells in the bone marrow, as well as mature B cells, express high levels of Bcl-2 protein (Merino *et al.*, 1994). Pre-B and immature B cells in the bone marrow and GC B cells in the follicles are Bcl-2-low or negative. However, the presence of Bcl-2 is neither required for initial B cell development in the bone marrow, nor does it override self-censorship of newly produced B cells (Nakayama *et al.*, 1993; Hartley *et al.*, 1993). In *scid* mice, constitutive expression of Bcl-2 appears to permit normal B-cell survival and differentiation in the absence of antigen receptors. This contrasts markedly with T-cell development in these animals, which, as described above, is not promoted by Bcl-2, and it has been suggested that pro-T cells lack a function necessary for the activity of Bcl-2 that is expressed by pro-B cells (Strasser *et al.*, 1994b).

Bcl-2 appears to play a role in the regulation of survival of mature B cells in the periphery. In transgenic mice, constitutive expression of Bcl-2 in the B-cell lineage results in the accumulation of mature B cells in secondary lymphoid tissues (McDonnell *et al.*, 1989). The prolonged survival of these cells results in the extension of secondary immune responses to T-dependent antigens, development of autoimmune phenomena in some animals, and progression to polyclonal lymphoproliferative lesions (McDonnell *et al.*, 1989, 1990; Nunez *et al.*, 1991; McDonnell and Korsmeyer, 1991; Strasser *et al.*, 1991). Rescue of normal GC B cells (which no longer express Bcl-2) by antigen receptor- and CD40-cross-linking *in vitro* results in reexpression of Bcl-2 (Liu *et al.*, 1991). However, during CD40-ligand-induced rescue, Bcl-2 is not the primary effector of survival but, rather, is switched on in rescued cells several hours after repression of apoptosis (Holder *et al.*, 1993). In this respect, expression of Bcl-2 protein appears to occur as part of a differentiation step, possibly leading to long-term survival.

Can Bcl-2 override self-censorship in the periphery? It has been found that autoreactive Ly-1 ($CD5^+$) B cells (see Tsubata, 1995, for review) can persist in certain cloistered locations, such as the peritoneal cavity. However, when deliberately exposed to antigen, these cells rapidly undergo apoptosis. This response is abrogated by the presence of Bcl-2 (Nisitani *et al.*, 1993). Similarly, self-censorship of GC B cells also appears to be inhibited by Bcl-2 (Shokat and Goodnow, 1995). Therefore, it may be concluded that, whereas Bcl-2 is ineffective in repressing the deletion of self-reactive cells in the bone marrow, the oncoprotein plays an important role in controlling the survival of mature B cells in the periphery. The function of Bcl-2 homologues in regulating B-cell apoptosis is an active area of current investigation, and reports to date indicate that Bcl-x provides critical survival signals not only for B-cell genesis but for development of the hematopoietic system as a whole (Choi *et al.*, 1995; Motoyama *et al.*, 1995).

2.2.2b. Fas. The distribution of Fas on B cells is consistent with the idea that Fas signaling contributes to B-cell selection. In this respect, it has been shown that Fas is high on GC B cells (Möller *et al.*, 1993) and on the transitional (between immature and mature) B cells of the bone marrow (Carsetti *et al.*, 1995). CD40 ligation, an established survival mechanism for GC B cells, has been found to sensitize B cells to Fas-mediated apoptosis in the absence, but not in the presence, of antigen receptor cross-linking. This response may represent a fail-safe mechanism for the elimination of antigen-unreactive ("bystander") cells receiving CD40-ligand-mediated survival signals (Rothstein *et al.*, 1995).

That the Fas/Fas-ligand system plays a role in normal control of B-cell death and survival is also inferred from studies of *lpr* mice. B cells from these animals are defective in their ability to undergo apoptosis *in vitro* and show intrinsic abnormalities *in vivo* such that the autoimmune antibodies characteristically observed in *lpr* mice are only produced by *lpr* (i.e., Fas-negative) B cells. Thus, in radiation chimaeras constructed from mixtures of normal and *lpr* lymphoid cells, only the *lpr* B cells produced autoantibodies (Cohen and Eisenberg, 1995). Recent work suggests a mechanism behind the abnormality inherent in *lpr* B cells. Anergic, self-reactive B cells appear to be deleted by a Fas-dependent mechanism in which death is triggered through interaction of B cells with Fas-ligand-expressing T cells. This mechanism of deletion is defective in *lpr* B cells (Rathmell *et al.*, 1995).

3. ROLE OF APOPTOSIS IN THE DEVELOPMENT OF LEUKEMIAS AND LYMPHOMAS

3.1. Introduction

As described in Section 1, the normal control of the cell number of any hematopoietic cell population is critically dependent on the balance between cell gain and cell loss. Our contention is that the suppression of apoptosis disrupts this control and that this is a likely prelude to the development of several leukemias. The suppression can occur by several mechanisms, including up-regulation of survival genes, increased levels of survival factors, or down-modulation/loss of death-inducing genes. The resulting cells would possess an increased survival advantage and have greater risk of acquiring further genetic mutations leading to disease progression (reviewed in Harrington and Evan, 1994). In the following sections examples are cited in which proteins with known abilites to affect the regulation of apoptosis are implicated in leukemogenesis or lymphomagenesis, and subsequent sections present a more detailed account of specific leukemias and lymphomas in which normal control of the apoptotic pathway is lost.

3.1.1. Role of Bcl-2

Bcl-2 introduction into early pre-B cells *in vitro* suppressed apoptosis induced by growth factor withdrawal (Borzillo *et al.*, 1992). A subset of cells survived indefinitely and spontaneously gave rise to factor-independent variants. Bcl-2 expression is increased in B-lineage acute lymphoblastic leukemia (ALL) cells compared with normal immature B cells; its expression was associated with survival *in vitro* of leukemic B cells in culture even in the absence of bone marrow stromal layers (Campana *et al.*, 1993). This increased survival ability of the cells may contribute to leukemogenesis and explains the ability of these cells to survive outside the bone marrow microenvironment. Bcl-2 expression was also correlated with disease progression in ALL by Maung *et al.* (1994): the number of cells staining positive for Bcl-2 and the intensity of the staining increased in relapsed patients compared with patients at presentation. Bcl-2 is expressed in acute myeloid leukemia (AML), where levels of the protein have been correlated with extended survival of cells *in vitro*, a low complete remission rate after intensive chemotherapy, and a significantly shorter patient survival (Campos *et al.*, 1993). Treatment of AML cells *in vitro* with antisense to *bcl-2* resulted in decreased cell survival and clonogenicity and more effective killing of cells by cytotoxic chemotherapy (Campos *et al.*, 1994).

3.1.2. Mutations in p53

As the most commonly mutated gene in human cancer, *p53* would be expected to play an important role in leukemogenesis (for a more complete review, see Imamura *et al.*, 1994; Prokocimer and Rotter, 1994). *p53* mutations have been detected in L_3-type B-cell acute lymphoblastic leukemia and B-cell chronic lymphocytic leukemia, especially a particularly aggressive stage of its progression known as Richter's transformation (Gaidano *et al.*, 1991; see Section 3.4). In contrast, mutated *p53* is rarely seen in acute lymphoblastic leukemia (Gaidano *et al.*, 1991). Blast cells from patients with acute myeloblastic leukemia expressed different conformations of *p53* (Zhu *et al.*, 1993). Those cells with mutant or no *p53* exhibited autonomous growth *in vitro*, whereas expression of wild-type *p53* in the promoter conformation resulted in non-autonomous growth. *p53* mutation in leukemia therefore appears to be related to progression of the disease and to a worsened prognosis. This has been confirmed in *p53*-deficient mice, where bone marrow progenitor cells were more resistant to apoptosis in low growth factors, irradiation, or heat shock (Lotem and Sachs, 1993b).

As described in Section 1.8.3, an increasing body of evidence suggests that mutations of *p53* prevent the G_1 growth arrest and/or apoptotic response to DNA damage. *p53* mutation has also been reported to suppress apoptosis induced by growth factor withdrawal (Gottlieb *et al.*, 1994; Zhu *et al.*, 1994), thereby allowing growth of cells in limited amounts of growth factors.

3.1.3. Disregulated c-Myc

According to the dual-conflict model for c-Myc action proposed by Fanidi *et al.* (1992), activation of c-Myc alone would lead to apoptosis in the absence of survival signals. However, c-Myc overexpression is sometimes a secondary event in leukemogenesis that takes advantage of the suppression of apoptosis afforded by another oncogene such as *bcl-2*, mutated *p53*, or *bcr-abl*, which allows the mitogenic function of *c-myc* to predominate. This is thought to be the case in blast crisis CML (McCarthy *et al.*, 1984), Burkitt lymphoma (see Section 3.3), and L_3-type B-cell acute lymphoblastic leukemia (Gaidano *et al.*, 1991).

3.1.4. Hematopoietic Survival Factors

The survival factors GM-CSF, G-CSF, and IL-6 suppressed apoptosis induced by several anticancer agents in myeloid leukemic cells (Lotem and Sachs, 1992), and IL-3 induced resistance to x-irradiation, cisplatin, and etoposide (Collins *et al.*, 1992). Increased levels or altered responsiveness to growth factors could therefore result in increased numbers of circulating cells with an increased threshold for a cytotoxic response to drugs. Patients with CML were found to have increased serum levels of GM-CSF and G-CSF (Balleari *et al.*, 1994), and a subtype of human acute pre-B-cell leukemia has a t(5;14) translocation resulting in overexpression of IL-3. In addition to suppressing apoptosis itself, IL-3 has been shown to cooperate with Bcl-2 in inhibiting etoposide-induced apoptosis in a murine pre-B-cell line (Ascaso *et al.*, 1994). Juvenile chronic myeloid leukemia progenitor cells were reported to be hypersensitive to GM-CSF (Emanuel *et al.*, 1991). The clinical use of growth factors to correct therapy-associated myelosuppression must therefore be carefully considered and timed to avoid protection of malignant cells from cytotoxic chemotherapy (Lotem and Sachs, 1992; see also Chapter 15). It has been suggested that use of antibodies to growth factors may be an effective means of increasing responsiveness to chemotherapy by reducing resistance to apoptosis (Sachs and Lotem, 1993). Leukemic cells from some patients with ALL were less dependent on contact with bone marrow stroma for survival than their normal B progenitor cell counterparts (Manabe *et al.*, 1994). This finding supports the hypothesis that leukemic cells can survive free of signals that regulate normal B-cell growth and explains the premature release of leukemic cells and their ability to expand in extramedullary sites. A similar situation appears to exist in CML: CML progenitors were deficient in their adhesion to stromal cells *in vitro* compared with normal cells (Gordon *et al.*, 1987). This could account for their premature release into the blood, which occurs along with suppression of apoptosis by *bcr-abl* (McGahon *et al.*, 1994).

3.2. Follicular Lymphoma

Follicular non-Hodgkin lymphomas (NHL) are tumors arising from germinal center B cells. More than 85% of cases contain t(14;18) translocations (Yunis *et al.*, 1987) causing deregulation of the *bcl-2* oncogene. Whereas few normal germinal center B cells are Bcl-2-positive, the majority of follicular lymphoma cells express high levels of the protein (Pezzella *et al.*, 1990). Follicular lymphomas are indolent, low-grade tumors with few cells in active cycle, lending support to the view that *bcl-2* belongs to a novel class of oncogenes that deregulate cell survival rather than proliferation. Thus, in low-grade follicular lymphoma, there is a gradual accumulation of cells as a result of suppression of apoptosis rather than enhancement of proliferation.

That Bcl-2 contributes in this way to the acquisition of a neoplastic phenotype in B cells is most strongly supported by investigations of transgenic mice that constitutively express Bcl-2 in the B-cell lineage. Accumulation of mature, resting B cells was observed in the lymphoid follicles, and in some animals polyclonal lymphoproliferative lesions progressed to monoclonal high-grade tumors (McDonnell and Korsmeyer, 1991; Strasser *et al.*, 1991). However, the latter were observed only at low incidence, and many of the tumors also contained *c-myc* mutations. Interestingly, the frequent change of human follicular lymphomas from a low-grade indolent form to a high-grade aggressive form also appears to involve mutations of additional cellular oncogenes such as *c-myc* (Gauwerky *et al.*, 1988; De Jong *et al.*, 1988). When *bcl-2* transgenic animals were crossed with mice constitutively expressing *c-myc*, the double transgenic animals developed tumors more rapidly than those expressing *myc* alone (which was more strongly tumorigenic than *bcl-2* alone, see below). These rapidly developing tumors were found to have the phenotype of a lymphoid stem cell or early progenitor cell (Strasser *et al.*, 1990).

In addition to follicular lymphoma, Bcl-2 is also present in the majority of other low-grade NHL that do not contain a t(14;18) translocation, and its presence often correlates with a low frequency of apoptosis. However, in both high- grade and low-grade lymphomas, the presence of Bcl-2 appears to have no bearing on prognosis (Pezzella *et al.*, 1992; Piris *et al.*, 1994; Leoncini *et al.*, 1995). Some low-grade NHL display both high levels of Bcl-2 and a high rate of apoptosis (Leoncini *et al.*, 1995), suggesting that the presence of the oncoprotein is insufficient to promote tumor cell survival. In this respect, it is plausible that the tumor microenvironment, such as the germinal center, provides factors necessary for survival. Thus, the survival and growth of isolated follicular lymphoma cells is potentiated by CD40 ligation (Johnson *et al.*, 1993). Autocrine factors may also be important: soluble autocrine factors appear to be required for the growth of a Bcl-2-expressing follicular lymphoma cell line (Blagosklonny and Neckers, 1995).

Therefore, in follicular lymphoma, it seems reasonable to conclude that in cases in which *bcl-2* is constitutively expressed, the oncogene contributes to the pathogenesis of the disease through its ability to repress apoptosis. Additional mutations and/or environmental signals are also required, however, to maintain and/or advance the malignant status. It is interesting to note that Bcl-2-expressing follicular lymphoma cells may remain susceptible to peptide ligands and antiidiotypic antibodies that induce apoptosis *in vitro* and tumor regression *in vivo* via signals generated at the antigen receptor (Renschler *et al.*, 1994; Vuist *et al.*, 1994).

3.3. Burkitt Lymphoma

The protooncogene *c-myc* is far more potent in promoting tumor progression in the B-cell lineage than *bcl-2*. All transgenic mice expressing Myc consitutively in B cells (using a transgene linked to the immunoglobulin enhancer, E_μ) develop tumors by 12 months (Adams *et al.*, 1985; Harris *et al.*, 1988), whereas in the same period tumors arise in only a small minority of E_μ-*bcl*-2-transgenic animals (Strasser *et al.*, 1993). In man, all cases of the B-cell tumor Burkitt lymphoma (BL) contain a translocation that juxtaposes *c-myc* to an immunoglobulin locus, and the consequent loss of normal regulation of *c-myc* plays a critical role in the outgrowth of malignant cells (Magrath, 1990). B-cell tumors in rodents also contain *myc*–Ig translocations (Cory, 1986).

In view of its pleiotropic functions, on the one hand as a promoter of cell cycle progression and on the other as a driver of apoptosis, the role of *c-myc* in tumor pathogenesis is an intriguing one. In E_μ–*myc* transgenic mice, monoclonal tumors evolve from polyclonal lymphoproliferative lesions of pro- and pre-B cells, indicating that additional changes must be acquired for progression to the malignant state (Adams *et al.*, 1985). Of note is the prelymphomatous phase in these animals, in which the enlarged polyclonal cycling population of B cells does not continue to expand progressively after birth but levels off at 3–5 weeks of age, suggesting that the high proliferation rate of these cells is balanced by a high rate of cell death (Langdon *et al.*, 1986). More recently it has been demonstrated that the rate of apoptosis among B cells in the bone marrow of E_μ–*myc* mice is higher than that of control animals (Osmond *et al.*, 1994). These results may be interpreted according to Myc's dual action in promoting both proliferation and apoptosis: B cells of E_μ–*myc* mice continue to expand at immature phases of development under the influence of *c-myc* so long as survival pathways are fully in place; when conditions do not favor survival, for example, in cells that fail to compete successfully for limited quantities of growth/survival factors available in the microenvironment, *myc* then necessarily drives apoptosis.

High-rate apoptosis, as evidenced by the classical histological picture of the tumor, is a characteristic feature of Burkitt lymphoma (BL): the malignant B cells are interspersed with macrophages laden with apoptotic tumor cells and their debris, producing the typical "starry sky" appearance of BL (O'Conor, 1961; Cooper *et al.*, 1966; Berard *et al.*, 1969). In addition, BL lines that retain the phenotypic features of the parental tumor cells are primed to undergo apoptosis *in vitro* (Gregory *et al.*, 1991). In this respect they behave as germinal center B cells, which they resemble in cell surface phenotype and in their lack of expression of Bcl-2 (Gregory *et al.*, 1987; Henderson *et al.*, 1991; Liu *et al.*, 1991). Burkitt lymphoma cell apoptosis *in vitro* is driven by *c-myc* (Milner *et al.*, 1993) and can be induced at a high rate by a variety of agents and treatments, including limitation of growth/survival factors, exposure to Ca^{2+} ionophore, antiimmunoglobulin treatment, and temperature shock (reviewed by Gregory, 1995). Apoptosis of BL cells can be repressed by serum- and fibroblast-derived growth/survival factors (Gregory *et al.*, 1991; Falk *et al.*, 1993), CD40 ligand (Holder *et al.*, 1993), Epstein–Barr virus (EBV) latent protein expression (Gregory *et al.*, 1991), Bcl-2 (Henderson *et al.*, 1991; Milner *et al.*, 1992), and interferon-α (Milner *et al.*, 1992, 1993). Although EBV promotes survival at least in part through the up-regulation of Bcl-2 by the latent proteins LMP-1 and EBNA-2 (Henderson *et al.*, 1991; Finke *et al.*, 1992), repression of apoptosis by exogenous factors occurs in the absence of Bcl-2 (Milner *et al.*, 1992; Holder *et al.*, 1993; Falk *et al.*, 1993). Contrary to popular belief, c-Myc can be regulated at the protein level in BL cells, and rapid transient down-regulation of Myc protein occurs during both Bcl-2-independent and Bcl-2-dependent rescue from apoptosis (Milner *et al.*, 1993), providing a common route to survival.

In conclusion, although BL cells are successful in tipping the cell-gain/cell-loss population balance in favor of gain, the fully fledged malignant B cells retain the ability to engage their apoptotic program rapidly. The involvement of *myc* in BL cell life and death is reminiscent of its role in tumorigenesis in E_μ–*myc* mice: if survival pathways are in place, the BL cell population expands under the influence of *myc*; if not, *myc*-driven apoptosis ensues. For successful malignant outgrowth, BL cells therefore require environmental factors or additional mutations favoring enhanced survival. The high rate of apoptosis present in BL tumors *in situ* suggests that competition for survival factors or autocrine production of survival factors is an important determinant of BL cell fate.

3.4. Chronic Lymphocytic Leukemia

In B-cell chronic lymphocytic leukemia (B-CLL), the tumor cells are $CD5^+$ (B-1) B cells that accumulate in the peripheral blood (Caligaris-Cappio *et al.*, 1993). Although the origin of these cells is not yet clear, the majority are not actively in cycle, and much evidence suggests that inhibition of apoptosis is likely to contribute to the pathogenesis of the disease. Most B-CLL cells appear to be resting, long-lived cells that express Bcl-2 in their cytoplasm and Fas on their surface (Schena *et al.*, 1993; Mapara *et al.*, 1993). Bcl-2 expression may not be sufficient to suppress apoptosis because the majority of CLL cells die rapidly in culture (Collins *et al.*, 1989). Apoptosis can be inhibited by a variety of cytokines, notably IL-4 and interferons γ and -α (Danescu *et al.*, 1992; Buschle *et al.*, 1993; Chaouchi *et al.*, 1994). Interleukin 4 inhibits apoptosis in the absence of proliferative responses and increases Bcl-2 expression (Danescu *et al.*, 1992). By contrast, interferon-α elicits protection, as in BL cells, by a Bcl-2-independent pathway (Chaouchi *et al.*, 1994). Interferon-γ may sustain cell survival via an autocrine route, because B-CLL cells both produce and respond to the cytokine (Buschle *et al.*, 1993). Contact with endothelial cells has also been reported to promote survival of B-CLL cells (Long *et al.*, 1995).

Therefore, although much is yet to be learned of the oncogenic events that lead to B-CLL, such as the identity of the clonogenic progenitor cell and the nature of the transforming lesion(s), it is likely that inhibition of apoptosis through Bcl-2 expression and additional survival pathways activated by endocrine, paracrine, autocrine, or juxtacrine factors promote accumulation of leukemic cells. Translocations involving *bcl-2* are not a feature of B-CLL, and knowledge of expression patterns of other Bcl-2-family proteins in B-CLL cells will assist in our understanding of their survival mechanisms. Whatever the nature of the survival pathways, they are easily blocked or circumvented because apoptosis can be induced readily in CLL cells by various means. Thus, in addition to factor deprivation, apoptosis of B-CLL cells can be accelerated by inhibition of protein synthesis (Collins *et al.*, 1991), exposure to IL-5 (Craig *et al.*, 1993) or IL-10 (Fluckiger *et al.*, 1994), and cross-linking of surface immunoglobulin (McConkey *et al.*, 1991; Kobayashi *et al.*, 1993) or Fas antigen (Mapara *et al.*, 1993).

3.5. Chronic Myeloid Leukemia

Chronic myeloid leukemia (CML) accounts for 20–30% of cases of leukemia and has an incidence of one case per 100,000 persons per year. The disease progresses through three phases: chronic phase, accelerated phase, and blast crisis. During chronic phase there is an expansion of the myeloid compartment of the blood, but the leukemic cells still retain the capacity to differentiate. After (on average) 3–4 years, patients pass through an accelerated phase into blast crisis, where additional genetic changes often become apparent (described below). Blast crisis is characterizsed by increasing numbers of immature cells in the peripheral blood and bone marrow and resembles an acute leukemia; at this stage, the blast cells can have a myeloid, lymphoid, erythroid, megakaryocytic, or undifferentiated phenotype (Koeffler and Golde, 1981). The mechanism by which the leukemic cells have a growth advantage is not yet clear, as they do not have an accelerated rate of proliferation (Chervenick and Boggs, 1968). There is no increased production of growth factors by CML cells or decreased levels of inhibitory cytokines (Otsuka *et al.*, 1991). However, CML cells exhibit a decreased response to one of these inhibitors, MIP-1α (Eaves *et al.*, 1993; see also Chapter 6), which is not a result of down-regulation of receptors for the cytokine but possibly of modification of a downstream signaling event (Chasty *et al.*, 1995).

The cytogenetic hallmark of CML is the formation of the Philadelphia chromosome, a shortened chromosome 22 resulting from a reciprocal translocation with chromosome 9 (reviewed by Daley and Ben-Neriah, 1991). The result of this translocation is the formation of a chimeric *bcr-abl* gene. The *bcr-abl* encodes a 210-kDa protein in which the N-terminal domain of protooncogene *c-abl* is replaced by the first two domains of *bcr* (breakpoint cluster region). The ability of c-Abl to bind DNA is lost along with the tightly controlled regulation of the *c-abl*-encoded PTK, and the resulting *bcr-abl* PTK is cytoplasmic and constitutively activated (see Chapter 5).

Both *bcr-abl* and *v-abl* (the viral homologue of *c-abl*, which also encodes a cytoplasmic and constitutively active PTK) are capable of transforming various hematopoietic cells *in vitro* (Kipreos *et al.*, 1987; Chung *et al.*, 1988; McLaughlin *et al.*, 1987; Daley and Baltimore, 1988). Transformation is accompanied by growth factor independence (the loss of an apoptotic response to factor deprivation) in some cell lines with no evidence for autocrine production of growth factors (Chung *et al.*, 1988; Daley and Baltimore, 1988). However, factor-independent growth does not necessarily accompany transformation. Transfection of multipotent progenitor cells with *bcr-abl* resulted in factor-dependent growth of colonies that were capable of multilineage differentiation, but factor-independent colonies evolved over a period of time and eventually dominated the culture (Gishizky and Witte, 1992). These results are consistent with the apparently "preleukemic" chronic phase of CML and the observation that expression of Bcr-Abl is not sufficient for full oncogenic behavior; clonal outgrowth toward a fully malignant phenotype occurs in some clonal lines, probably reflecting the progression of the disease *in vivo* (McLaughlin *et al.*, 1987).

The ability of either *v-abl* or *bcr-abl* to transform cells is dependent on the tyrosine kinase activity of the protein (Kipreos *et al.*, 1987; Kabarowski *et al.*, 1994). The ability of Bcr-Abl to bind to actin is also important for transformation of cells (McWhirter and Wang, 1993). Another essential component for transformation of cells by either *v-abl* or *bcr-abl* is *c-myc* (Sawyers *et al.*, 1992). Myc could be acting downstream of *bcr-abl* or *v-abl*, consistent with the observation that *c-myc* expression is increased by both *v-abl* (Cleveland *et al.*, 1989) and *bcr-abl* (Sawyers *et al.*, 1992).

Like Bcl-2, Abl PTK can suppress apoptosis induced by growth factor withdrawal and by drug treatment (Evans *et al.*, 1993; Bedi *et al.*, 1994; Chapman *et al.*, 1994; McGahon *et al.*, 1994). The potential importance of this in leukemogenesis and the development of drug resistance is discussed below. Unlike Bcl-2, however, Abl PTK has a known function as a signaling molecule, and several candidate signaling molecules have been identified that may be involved in the pathway for the suppression of apoptosis; in this respect, the involvement of PKC (see Section 1.5.2), Shc, Ras, and possibly MAP kinase deserve further investigation (Owen *et al.*, 1993; Evans *et al.*, 1995; Owen-Lynch *et al.*, 1995; Sawyers *et al.*, 1995; Pendergast *et al.*, 1993; Puil *et al.*, 1994; Tauchi *et al.*, 1994; Kabarowski *et al.*, 1994).

An interesting study by Laneuville *et al.* (1992) showed that an interleukin-3-dependent cell line transfected with $p210^{bcr\text{-}abl}$ was phenotypically and genetically unstable when compared with control vector-transfected cells over a period of 6 months. This genomic instability and clonal evolution are reflected clinically in the progression of CML into blast crisis, where several genetic changes in addition to *bcr-abl* can be detected. Some of these genetic changes described below involve several genes whose ability to influence apoptosis has been discussed earlier in this chapter. A study of 242 patients with CML in blast crisis revealed that clonal evolution was most frequently associated with development of a double Philadelphia chromosome, trisomy 8, or isochrome 17q, and these factors were significantly associated with a poor survival of the patient (Kantarjian *et al.*, 1987). The *c-myc* oncogene is located on the long arm of chromosome 8; trisomy 8 has, in some instances, been associated with enhanced expression of *c-myc* (Blick *et al.*, 1987) which may promote proliferation if apoptosis is blocked (Section 1.8.4). The formation of isochrome 17q results in the loss of the short arm of chromosome 17p, the location of the tumor suppressor gene *p53*.

In contrast to chronic-phase CML, where *p53* expression is usually normal, blast crisis can be associated with loss and/or mutation of *p53* (Ahuja *et al.*, 1989; Mashal *et al.*, 1990). Mutation in p53 may prevent the engagement of apoptosis after exposure to DNA-damaging drugs (Section 1.8.3). A study on a particular patient with CML revealed a deletion of 17p and a point mutation of the remaining single allele of *p53* appearing concomitantly with the onset of blast crisis and then disappearing with the establishment of a second chronic phase, suggesting a temporal relationship between this abnormality and the progression of the disease (Foti *et al.*, 1991). The expression of *bcl-2* (the prototype survival gene, which can suppress drug-induced apoptosis or growth factor deprivation-induced apoptosis; Section

1.8.2) has also been noted in peripheral blood cells from patients in blast crisis of chronic myeloid leukemia (CML) (Delia *et al.*, 1992). The ability of Abl PTK to suppress drug-induced apoptosis is discussed in detail in Section 4 below.

4. SUPPRESSION OF APOPTOSIS: A MECHANISM OF ANTICANCER DRUG RESISTANCE

Many mechanisms of drug resistance have been described at the level of the drug–target interaction (Kessel, 1989). The observation that disparate anticancer agents induce apoptosis led to the suggestion that a novel mechanism of drug resistance that occurs downstream of the primary drug–target interaction and its immediate biochemical sequelae is the suppression of the apoptotic pathway itself (Dive and Hickman, 1991; Figure 9). This was first shown to be the case in studies of cell lines that overexpressed Bcl-2 (Fisher *et al.*, 1993; Kameski *et al.*, 1993; Oliver *et al.*, 1993) and subsequently in cell lines with mutant *p53* (Lowe *et al.*, 1993) and activated v-Abl PTK (Chapman *et al.*, 1995b). In each case, the exposure to drugs resulted in no differences in detectable drug-induced damage compared to cells without these oncogenes, but drug-induced apoptosis was suppressed. These studies have enormous implications with respect to the future design of cancer chemotherapy strategies. The vital role of apoptosis in hematopoiesis and thus the potential for dramatic effect on the cell-loss/cell-gain balance when apoptosis is suppressed should be considered when one is thinking about chemotherapy for leukemias.

One approach is the use of antisense oligodeoxynucleotides to directly inhibit the

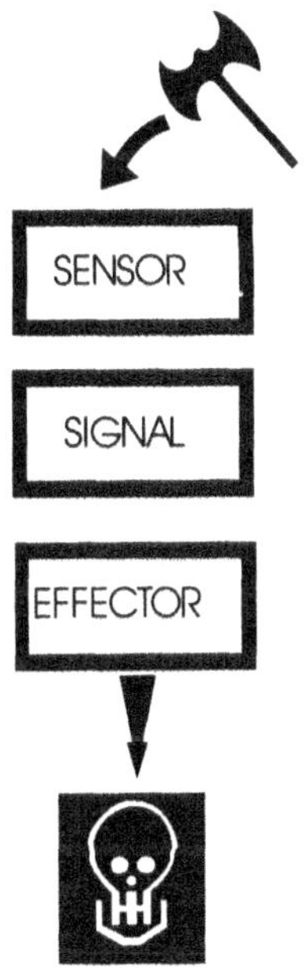

FIGURE 9. The induction of apoptosis after cellular damage (the axe represents all types of anticancer drug) depends on the sensing of the damage and the response of the cell to activate irreversibly the apoptosis effector machinery (see text for details).

expression of an apoptosis-suppressing gene such as *bcl-2* (Campos *et al.*, 1994) or *bcr-abl* (Szczylik *et al.*, 1991). It would presumably be more difficult to achieve selective toxicity of leukemic cells with antisense to *bcl-2*, as their normal counterparts in the bone marrow appear to need expression of this gene to survive. Redundancy within the Bcl-2 family must exist, however, as evidenced by the birth of the Bcl-2-null mouse. Another strategy that may tip the balance in favor of leukemic cell death involves reintroduction of an apoptosis-inducing gene such as wild-type *p53* (or the conversion of mutant to a wild-type conformation). This has produced some promising results *in vitro* (Cheng *et al.*, 1992), where introduction of wild-type p53 into a human T-leukemia cell line that contained no p53 resulted in a reduced growth rate and suppressed colony formation *in vitro* and abrogation of tumorigenicity in nude mice *in vivo*. However, the caveat to this approach is the possibility that wild-type and fully functioning p53 may still not lead to apoptosis if the pathway is still blocked by, for example, Bcl-2 or Bcr-Abl.

A controversial issue surrounds the treatment with DNA-damaging agents of leukemic cells that express an apoptosis-suppressing gene or genes. If DNA damage is not repaired and apoptosis is suppressed, surviving cells may possess the potential to proliferate subsequently with a damaged genome. This unfortunate chain of events may only need to occur in a small number of cells, which could acquire further mutations and develop into a more aggressive disease. A case in point is the recent demonstration that the alkylating agent N-ethyl-N-nitrosourea causes mutations within codon 248 of *p53* (Hussain *et al.*, 1994), one of the "hotspot" mutation regions (Levine *et al.*, 1994b), leading to the controversial contention that the such drugs may even contribute to disease progression in certain contexts.

If the mechanism whereby Bcr-Abl PTK and other leukemogenic oncogenes, perhaps in tandem with growth factors and matrix interactions within the bone marrow, together elicit a survival advantage and a drug-resistant phenotype, continued efforts to understand the signal transduction pathways involved in the associated survival signaling pathways is likely to lead to improved chemotherapy for these leukemias.

5. REFERENCES

Adams, J. M., Harris, A. W., Pinkert, C. A., Corcoran, L. M., Alexander, W. S., Cory, S., Palmiter, R. D., and Brinster, R. L., 1985, The *c-myc* oncogene driven by immunoglobulin enhancers induces lymphoid malignancy in transgenic mice, *Nature* **318:**533–538.

Ahuja, H., Bar-Eli, M., Advani, S. H., Benchimol, S., and Cline, M. J., 1989, Alterations in the *p53* gene and the clonal evolution of the blast crisis of chronic myelocytic leukemia, *Proc. Natl. Acad. Sci. U.S.A.* **86:**6783–6787.

Amati, B., Littlewood, T. D., Evan, G. I., and Land, H., 1993, The c-MYC protein induces cell cycle progression and apoptosis through dimerization with MAX, *EMBO J.* **12:**5083–5087.

Arends, M. J., Morris, R. G., and Wyllie, A. H., 1990, Apoptosis: The role of the endonuclease, *Am. J. Pathol.* **136:**593–608.

Ascaso, R., Marvel, J., Collins, M. K. L., and Lopez-Rivas, A., 1994, Interleukin-3 and Bcl-2 cooperatively inhibit etoposide-induced apoptosis in a murine pre-B cell line, *Eur. J. Immunol.* **24:**537–541.

Askew, D. S., Ashmun, R. A., Simmons, B. C., and Cleveland, J. L., 1991, Constitutive *c-myc* expression in an IL-3-dependant myeloid cell line suppresses cell cycle arrest and accelerates apoptosis, *Oncogene* **6:**1915–1922.

Baffy, G., Miyashita, T., Williamson, J. R., and Reed, J. C., 1993, Apoptosis induced by withdrawal of interleukin-3

(IL-3) from an IL-3-dependent hematopoietic cell line is associated with repartitioning of intracellular calcium and is blocked by enforced bcl-2 oncoprotein production, *J. Biol. Chem.* **268:**6511–6519.
Balleari, E., Bason, C., Visani, G., Gobbi, M., Ottaviani, E., and Ghio, R., 1994, Serum levels of granulocyte–macrophage colony-stimulating factor and granulocyte colony-stimulating factor in treated patients with chronic myelogenous leukemia in chronic phase, *Haematologica* **79:**7–12.
Bansal, N., Houle, A. G., and Melnykovych, G., 1990, Dexamethasone-induced killing of neoplastic cells of lymphoid derivation: Lack of early calcium involvement, *J. Cell. Physiol.* **143:**105–109.
Barak, Y., Juven, T., Haffner, R., and Oren, M., 1993, *mdm2* expression is induced by wild type p53 activity, *EMBO J.* **12:**461–468.
Barry, M. A., and Eastman, A., 1993, Identification of deoxyribonuclease II as an endonuclease involved in apoptosis, *Arch. Biochem. Biophys.* **300:**440–450.
Bayle, J. H., Elenbaas, B., and Levine, A. J., 1995, The carboxyl-terminal domain of the p53 protein regulates sequence-specific DNA binding through its nonspecific nucleic acid-binding activity, *Proc. Natl. Acad. Sci. U.S.A.* **92:**5729–5733.
Bedi, A., Zehnbauer, B. A., Barber, J. P., Sharkis, S. J., and Jones, R. J., 1994, Inhibition of apoptosis by BCR-ABL in chronic myeloid leukemia, *Blood* **83:**2038–2044.
Beere, H. M., Chresta, C. M., Alejo-Herberg, A., Skladanowski, A., Dive, C., Larsen, A. K., and Hickman, J. A., 1995, Investigation of the mechanism of higher order chromatin fragmentation observed in drug-induced apoptosis, *Mol. Pharmacol.* **47:**986–996.
Benson, R. S. P., Heer, S., Dive, C., and Watson, A. J. M., 1995, Characterization of cell volume loss in CEM-C7A cells during dexamethasone-induced apoptosis, *Am. J. Physiol.* (in press).
Berard, C., O'Conor, G. T., Thomas, L. B., and Torloni, H., 1969, Histopathological definition of Burkitt's tumour. *Bull. WHO* **40:**601–607.
Bissonnette, R. P., Echeverri, F., Mahboubi, A., and Green, D. R., 1992, Apoptotic cell death induced by *c-myc* is inhibited by *bcl-2*, *Nature* **359:**552–554.
Blagosklonny, M. V., and Neckers, L. M., 1995, The role of bcl-2 protein and autocrine growth factors in a human follicular lymphoma-derived B-cell line, *Eur. Cytokine Network* **6:**21–27.
Blick, M., Romero, P., Talpaz, M., Kurzrock, R., Shtalrid, M., Andersson, B., Trujillo, J., Beran, M., and Gutterman, J., 1987, Molecular characteristics of chronic myelogenous leukemia in blast crisis, *Cancer Genet. Cytogenet.* **27:**349–356.
Boise, L. H., Gonzalez-Garcia, M., Postema, C. E., Ding, L., Lindsten, T., Turka, L. A., Mao, X., Nunez, G., and Thompson, C. B., 1993, *bcl-x* a *bcl-2*-related gene that functions as a dominant regulator of apoptotic cell death, *Cell* **74:**1–20.
Borner, C., Martinou, I., Mattamnn, C., Irmler, M., Schaerer, E., Martinou, J. C., and Tschopp, J., 1994, The protein bcl-2alpha does not require membrane attachment, but two conserved domains to suppress apoptosis, *J. Cell. Biol.* **126:**1059–1068.
Borzillo, G. V., Endo, K., and Tsujimoto, Y., 1992, *Bcl-2* confers growth and suvival advantage to interleukin 7-dependant early pre-B cells which become factor independant by a multistep process in culture, *Oncogene* **7:**869–876
Boudreau, N., Sympson, C. J., Werb, Z., and Bissell, M. J., 1995, Suppression of ICE and apoptosis in mammary epithelial cells by extracellular matrix, *Science* **267:**891–893.
Brach, M. A., deVos, S., Gruss, H. J., and Herrmann, F., 1992, Prolongation of survival of human polymorphonuclear neutrophils by granulocyte–macrophage colony-stimulating factor is caused by inhibition of programmed cell death, *Blood* **80:**2920–2924.
Brandt, J. E., Bhalla, K., and Hoffman, R., 1994, Effects of interleukin-3 and *c-kit* ligand on the survival of various classes of human hematopoietic progenitor cells, *Blood* **83:**1507–1514.
Brown, D. G., Sun, X. M., and Cohen, G. M., 1993, Dexamethasone-induced apoptosis involves cleavage of DNA to large fragments prior to internucleosomal fragmentation, *J. Biol. Chem.* **268:**3037–3039.
Buschle, M., Campana, D., Carding, S. R., Richard, C., Hoffbrand, A. V., Brenner, M. K., 1993, Interferon gamma inhibits apoptotic cell death in B cell chronic lymphocytic leukaemia, *J. Exp. Med.* **177:**213–218.
Buttyan, R., Zakeri, Z., Lockshin, R., and Wolgemuth, D., 1988, Cascade induction of *c-fos*, *c-myc*, and heat shock 70K transcripts during regression of the rat ventral prostate gland, *Mol. Endocrinol.* **2:**650–657.
Buttyan, R., Olsson, C. A., Pintar, J., Chang, C., Bandyk, M., Ng, P., and Sawczuk, I. S., 1989, Induction of the *TRPM-2* gene in cells undergoing programmed death, *Mol. Cell. Biol.* **9:**3473–3481.

Caelles, C., Helmberg, A., and Karin, M., 1994, *p53*-dependent apoptosis in the absence of transcriptional activation of *p53*-target genes, *Nature* **370**:220–223.

Caligaris-Cappio, F., Gottardi, D., Alfarano, A., Stacchini, A., Gregoretti, M. G., Ghia, P., Bertero, M. T., Novarino, A., and Bergui, L., 1993, The nature of the B lymphocyte in B-chronic lymphocytic leukemia, *Blood Cells* **19**:601–613.

Campana, D., Coustan-Smith, E., Manabe, A., Buschle, M., Raimondi, S. C., Behm, F. G., Ashmun, R., Arico, M., Biondi, A., and Pui, C. H., 1993, Prolonged survival of B-lineage acute lymphoblastic leukemia cells is accompanied by overexpression of *bcl-2* protein, *Blood* **81**:1025–1031.

Campos, L., Rouault, J. P., Sabido, O., Oriol, P., Roubi, N., Vasselon, C., Archimbaud, E., Magaud, J. P., and Guyotat, D., 1993, High expression of BCL-2 protein in acute myeloid leukemia cells is associated with poor response to chemotherapy, *Blood* **81**:3091–3096.

Campos, L., Sabido, O., Rouault, J. P., and Guyotat, D., 1994, Effects of *bcl-2* antisense oligodeoxynucleotides on *in vitro* proliferation and survival of normal marrow progenitors, *Blood* **84**:595–600.

Carsetti, R., Kohler, G., and Lamers, M. C., 1995, Transitional B-cells are the target of negative selection in the B-cell compartment, *J. Exp. Med.* **181**:2129–2140.

Chaouchi, N., Wallon, C., Taieb, J., Auffredou, M. T., Tertian, G., Lemoine, F. M., Delfraissy, J. F., and Vazquez, A., 1994, Interferon-alpha-mediated prevention of *in vitro* apoptosis of chronic lymphocytic-leukemia B cells—role of *bcl-2* and *c-myc*, *Clin. Immunol. Immunopathol.* **73**:197–204.

Chapman, R. S., Whetton, A. D., and Dive, C., 1994, The suppression of drug-induced apoptosis by activation of v-ABL protein tyrosine kinase, *Cancer Res.* **54**:5131–5137.

Chapman, R. S., Chresta, C. M., Herberg, A. A., Beere, H. M., Heer, S., Whetton, A. D., Hickman, J. A., and Dive, C., 1995a, Flow cytometric analysis of apoptosis in drug resistant and drug sensitive leukaemic cells; further characterisation of the *in situ* terminal deoxynucleotidyl transferase (TdT) assay, *Cytometry* **20**:245–256.

Chapman, R. S., Chresta, C. M., Whetton, A. D., and Dive, C., 1995b, Characterisation of v-Abl protein tyrosine kinase mediated suppression of drug-induced apoptosis, *Mol. Pharmacol.* **48**:334–343.

Chasty, R. C., Lucas, G. S., Owen-Lynch, P. J., Pierce, A., and Whetton, A. D., 1995, CD34 enriched cells from patients with chronic myeloid leukemia express receptors for but are nonresponsive to macrophage inflammatory protein-1 alpha, *Blood* **86**:4270–4277.

Cheng, J., Yee, J. K., Yeargin, J., Friedmann, T., and Haas, M., 1992, Suppression of acute lymphoblastic leukemia by the human wild-type *p53* gene, *Cancer Res.* **52**:222–226.

Chervenick, P. A., and Boggs, D. R., 1968, Granulocyte kinetics in chronic myelocytic leukaemia, *Semin. Haematol.* **1**:24–35.

Chittenden, T., Harrington, E. A., O'Connor, R., Flemington, C., Lutz, R. J., Evan, G. I., and Guild, B., 1995, Induction of apoptosis by the Bcl-2 homologue Bak, *Nature* **374**:733–736.

Choi, M. S. K., Boise, L. H., Gottschalk, A. R., Quintans, J., Thompson, C. B., and Klaus, G. G. B., 1995, The role of bcl-x(L) in CD40-mediated rescue from anti-mu-induced apoptosis in WEHI-231 B-lymphoma-cells, *Eur. J. Immunol.* **25**:1352–1357.

Chung, S. W., Ruscetti, S., and Wong, P. M. C., 1988, Formation of factor-independent hematopoietic multilineage colonies after Abelson virus infection, *Blood* **71**:973–977.

Clarke, A. R., Purdie, C. A., Harrison, D. J., Morris, R. G., Bird, C. C., Hooper, M. L., and Wyllie, A. H., 1993, Thymocyte apoptosis induced by p53-dependent and independent pathways, *Nature* **362**:849–852.

Cleveland, J. L., and Ihle, J. N., 1995, Contenders in Fas/TNF death signaling, *Cell* **81**:479–482.

Cleveland, J. L., Dean, M., Rosenberg, N., Wang, J. Y. J., and Rapp, U. R., 1989, Tyrosine kinase oncogenes abrogate IL-3 dependence of muring myeloid cells through signaling pathways involving *c-myc*: Conditional regulation of *c-myc* transcription by temperature sensitive v-Abl, *Mol. Cell. Biol.* **9**:5685–5695.

Cohen, J. J., and Duke, R. C., 1984, Glucocorticoid activation of a calcium-dependent endonuclease in thymocyte nuclei leads to cell death, *J. Immunol.* **132**:38–42.

Cohen, J. J., Duke, R. C., Chervenak, R., Sellins, K. S., and Olson, L. R., 1985, DNA fragmentation in targets of CTL: An example of programmed cell death in the immune system, *Adv. Exp. Med. Biol.* **184**:493–508.

Cohen, P. L., and Eisenberg, R. A., 1995, Fas/Apo-1: A cell surface receptor that signals apoptosis, in: *Apoptosis and the Immune Response* (C. D. Gregory, ed.), pp. 169–186, Wiley-Liss Inc., New York.

Collins, M. K. L., Marvel, J., Malde, P., and Lopez-Rivas, A., 1992, Interleukin 3 protects murine bone marrow cell from apoptosis induced by DNA damaging agents, *J. Exp. Med.* **176**:1043–1051.

Collins, R. J., Verschuer, L. A., Harmon, B. V., Prentice, R. L., Pope, J. H., and Kerr, J. F. R., 1989, Spontaneous

programmed death (apoptosis) of B-chronic lymphocytic leukemia cells following their culture *in vitro*, *Br. J. Haematol.* **71**:343–350.
Collins, R. J., Harmon, B. V., Souvlis, T., Pope, J. H., and Kerr, J. F. R., 1991, Effects of cycloheximide on B-chronic lymphocytic leukemic and normal lymphocytes *in vitro*—induction of apoptosis, *Br. J. Cancer* **64**:518–522.
Cooper, E. H., Frank, G. L., and Wright, D. H., 1966, Cell proliferation in Burkitt tumours, *Eur. J. Cancer* **2:** 377–384.
Cory, S., 1986, Activation of cellular oncogenes in hematopoietic cells by chromosome translocation, *Adv. Cancer Res.* **47**:189–234.
Cory, S., 1995, Regulation of lymphocyte survival by the *bcl-2* gene family, *Annu. Rev. Immunol.* **13**:513–543.
Craig, V. A., Mainoufowler, T., and Prentice, A. G., 1993, Effect of interleukin-4 (IL4) and dexamethasone on apoptosis and *bcl2* expression in B-chronic lymphocytic-leukemia (B-CLL) in culture, *Exp. Hematol.* **21:** 1090–1096.
Daley, G. Q., and Baltimore, D., 1988, Transformation of an interleukin 3-dependant hematopoietic cell line by the chronic myelogenous leukemia-specific p210bcr/abl protein, *Proc. Natl. Acad. Sci. U.S.A.* **85**:9312–9316.
Daley, G. Q., and Ben-Neriah, Y., 1991, Implicating the *bcr/abl* gene in the pathogenesis of Philadelphia chromosome-positive human leukemia, *Adv. Cancer Res.* **57**:151–184.
Dancescu, M., Rubio-Trujillo, M., Biron, G., Bron, D., Delespesse, G., and Sarfati, M., 1992, Interleukin 4 protects chronic lymphocytic leukemic B cells from death by apoptosis and upregulates Bcl-2 expression, *J. Exp. Med.* **176**:1319–1326.
Darzynkiewicz, Z., Bruno, S., Del Bino, G., Gorczyca, W., Hotz, M. A., Lassota, P., and Traganos, F., 1992, Features of apoptotic cells measured by flow cytometry, *Cytometry* **13**:795–808.
De Jong, D., Voetdijk, B. M. H., Beverstock, G. C., Vanommen, G. J. B., Willemze, R., and Kluin, P. M., 1988, Activation of the *c-myc* oncogene in a precursor B-cell blast crisis of follicular lymphoma, presenting as composite lymphoma, *N. Engl. J. Med.* **318**:1373–1378.
De Jong, D., Prins, F. A., Mason, D. Y., Reed, J. C., van Ommen, G. B., and Kluin, P. M., 1994, Subcellular localisation of the BCL-2 protein in malignant and normal lymphoid cells, *Cancer Res.* **54**:256–260.
Delia, D., Aiello, A., Soligo, D., Fontanella, E., Melani, C., Pezzella, F., Pierotti, M. A., and Porta, G. D., 1992, Bcl-2 proto-oncogene expression in normal and neoplastic human myeloid cells, *Blood* **79**:1291–1298.
Delic, J., Morange, M., and Magdelenat, H., 1993, Ubiquitin pathway involvement in human lymphocyte gamma-irradiation-induced apoptosis, *Mol. Cell. Biol.* **13**:4875–4883.
Deng, G., and Podack, E. R., 1993, Suppression of apoptosis in a cytotoxic T-cell line by interleukin 2-mediated gene transcription and deregulated expression of the proto-oncogene *bcl-2*, *Proc. Natl. Acad. Sci. U.S.A.* **90**:2189–2193.
Dive, C., and Hickman, J. A., 1991, Drug–target interactions: Only the first step in the commitment to a programmed cell death, *Br. J. Cancer* **64**:192–196.
Dive, C., and Wyllie, A. H., 1993. Apoptosis and cancer chemotherapy, in *Frontiers in Pharmacology* (J. A. Hickman and T. R. Tritton, eds.), pp. 21–56, Blackwell Scientific, Oxford.
Dive, C., Evans, C. A., and Whetton, A. D., 1992, Induction of apoptosis—new targets for cancer chemotherapy, *Semin. Cancer Biol.* **3**:417–427.
Donehouer, L. A., Harvey, M., Slagle, B. L., McArthur, M. J., Montgomery, C. A., Jr., Butel, J. S., and Bradley, A., 1992, Mice deficient for *p53* are developmentally normal but susceptible to spontaneous tumours, *Nature* **356**:215–221.
Dransfield, I., Buckle, A. M., Savill, J. S., McDowall, A., Haslett, C., and Hogg, N., 1994, Neutrophil apoptosis is associated with a reduction in CD16 (Fc Riii) expression, *J. Immunol.* **153**:1254–1263.
Dransfield, I., Stocks, S. C., and Haslett, C., 1995, Regulation of cell-adhesion molecule expression and function associated with neutrophil apoptosis, *Blood* **85**:3264–3273.
Duke, R. C., Chervenak, R., and Cphen J. J., 1984, Endogenous endonuclease-induced DNA fragmentation: An early event in cell mediated cytolysis, *Proc. Natl. Acad. Sci. U.S.A.* **80**:6361–6365.
Duvall, E., Wyllie, A. H., and Morris, R. G., 1985, Macrophage recognition of cells undergoing programmed cell death (apoptosis), *Immunology* **56**:351–358.
Earnshaw, W. C., 1995, Nuclear changes in apoptosis, *Curr. Opin. Cell. Biol.* **7**:337–343.
Eaves, C. J., Cashman, J. D., Wolpe, S. D., and Eaves, A. C., 1993, Unresponsiveness of primitive chronic myeloid leukemia cells to macrophage inflammatory protein 1alpha, an inhibitor of primitive normal hematopoietic cells, *Proc. Natl. Acad. Sci. U.S.A.* **90**:12015–12019.

Edwards, S. N., Buckmaster, A. E., and Tolkovsky, A. M., 1991, The death programme in cultured sympathetic neurones can be suppressed at the post-translational level by nerve growth factor, cyclic AMP, and depolarisation, *J. Neurochem.* **57**:2140–2143.

Egerton, M., Scollay, R., and Shortman, K., 1990, Kinetics of mature T-cell development in the thymus, *Proc. Natl. Acad. Sci. U.S.A.* **87**:2579–2582.

El-Deiry, W. S., Tokino, T., Velculesco, V. E., Levy, D. B., Parsons, R., Trent, J. M., Lin, D., Mercer, W. E., Kinzler, K. W., and Vogelstein, B., 1993, WAF1, a potential mediator of *p53* tumour suppression, *Cell* **75**:817–825.

Ellis, R. E., Jacobson, D. M., and Horvitz, H. R., 1991, Genes required for the engulfment of cell corpses during programmed cell death in *Caenorhabditis elegans*, *Genetics* **129**:79–94.

Emanuel, P. D., Bates, L. J., Castleberry, R. P., Gualtieri, R. J., and Zuckerman, K. S., 1991, Selective hypersensitivity to granulocyte–macrophage colony stimulating factor by juvenile chronic myeloid leukaemia haematopoietic progenitors, *Blood* **77**:925–929.

Enari, M., Hug, H., and Nagata, S., 1995, Involvement of an ICE-like protease in Fas mediated apoptosis, *Nature* **375**:78–81.

Evan, G. I., Wyllie, A. H., Gilbert, C. S., Littlewood, T. D., Land, H., Brooks, M., Waters, C. M., Penn, L. Z., and Hancock, D. C., 1992, Induction of apoptosis in fibroblasts by *c-myc* protein, *Cell* **69**:119–128.

Evans, C. A., Owen-Lynch, P. J., Whetton, A. D., and Dive, C., 1993, Activation of Abelson tyrosine kinase activity is associated with suppression of apoptosis in hemopoietic cells, *Cancer Res.* **53**:1735–1738.

Evans, C. A., Lord, J. M., Owen-Lynch, P. J., Johnson, G., Dive, C., and Whetton, A. D., 1995, Suppression of apoptosis by v-ABL protein tyrosine kinase is associated with nuclear translocation and activation of protein kinase C in an interleukin-3 dependent haemopoietic cell line, *J. Cell. Sci.* **108**:2591–2598.

Evans, D. L., Tilby, M., and Dive, C., 1994, Differential sensitivity to the induction of apoptosis by cisplatin in proliferating and quiescent immature rat thymocytes is independent of the levels of drug accumulation and DNA adduct formation, *Cancer Res.* **54**:1596–1603.

Fadok, V. A., Voelker, D. R., Campbell, P. A., Cohen, J. J., Bratton, D. L., and Henson, P. M., 1992a, Exposure of phosphatidylserine on the surface of apoptotic lymphocytes triggers specific recognition and removal by macrophages, *J. Immunol.* **148**:2207–2216.

Fadok, V. A., Savill, J. S., Haslett, C., Bratton, D. L., Doherty, D. E., Campbell, P. A., and Henson, P. M., 1992b, Different populations of macrophages use either the vitronectin receptor or the phosphatidylserine receptor to recognize and remove apoptotic cells, *J. Immunol.* **149**:4029–4035.

Fairbairn, L. J., Cowling, G., Reipert, B. M., and Dexter, T. M., 1993, Suppression of apoptosis allows differentiation and development of a multipotent hemopoietic cell line in the absence of added growth factors, *Cell* **74**:1–10.

Falk, M. H., Hultner, L., Milner, A., Gregory, C. D., and Bornkamm, G. W., 1993, Irradiated fibroblasts protect Burkitt lymphoma cells from apoptosis by a mechanism independent of *bcl-2*, *Int. J. Cancer* **55**:485–491.

Fanidi, A., Harrington, E. A., and Evan, G. I., 1992, Cooperative interaction between *c-myc* and *bcl-2* proto-oncogenes, *Nature* **359**:554–556.

Farrow, S. N., White, J. H. M., Martinou, I., Raven, T., Pun, K. T., Grinham, C. J., Martinou, J. C., and Brown, R., 1995, Cloning of a *bcl-2* homologue by interaction with adenovirus E1B 19K, *Nature* **374**:731–733.

Fearnhead, H. O., Rivett, A. J., Dinsdale, D., and Cohen, G. M., 1995, A pre-existing protease is a common effector of thymocyte apoptosis mediated by diverse stimuli, *FEBS Lett.* **357**:242–246.

Fernandez-Alnemri, T., Litwack, G., and Alnemri, E. S., 1994, CPP32, a novel human apoptotic protein with homology to *Caenorhabditis elegans* cell death protein Ced-3 and mammalian interleukin-1B-converting enzyme, *J. Biol. Chem.* **49**:30761–30764.

Fernandez-Alnemri, T., Litwack, G., and Alnemri, E. S., 1995, *Mch2*, a new member of the apoptotic *ced-3/Ice* cystein protease gene family, *Cancer Res.* **55**:2737–2742.

Fernandez-Sarabia, M. J., and Bischoff, J. R., 1993, Bcl-2 associates with the *ras*-related protein R-*ras* p23, *Nature* **366**:274–275.

Fesus, L., Thomazy, V., Autuori, F., Ceru, M. P., Taresa, E., and Piacentini, M., 1989, Apoptotic hepatocytes become insoluble in detergents and chaotropic agents as a result of transglutaminase action, *FEBS Lett.* **245**:150–154.

Filipski, J., Leblanc, J., Youdale, T., Sikorska, M., and Walker, P. R., 1990, Periodicity of DNA folding in higher order chromatin structures, *EMBO J.* **9**:1319–1327.

Finke, J., Fritzen, R., Ternes, P., Trivedi, P., Bross, K. J., Lange, W., Mertelsmann, R., and Dolken, G., 1992, Expression of *bcl-2* in Burkitts lymphoma cell lines—induction by latent Epstein–Barr virus genes, *Blood* **80**:459–469.

Fisher, T. C., Milner, A. E., Gregory, C. D., Jackman, A. L., Aherne, G. W., Hartley, J. A., Dive, C., and Hickman, J. A., 1993, Bcl-2 modulation of apoptosis induced by anticancer drugs: Resistance to thymidylate stress is independant of classical resistance pathways, *Cancer Res.* **53:**3321–3326.

Flora, P. K., and Gregory, C. D., 1994, Recognition of apoptotic cells by human macrophages—inhibition by a monocyte/macrophage-specific monoclonal-antibody, *Eur. J. Immunol.* **24:**2625–2632.

Flora, P. K., and Gregory, C. D., 1995, Recognition pathways in the interaction of macrophages with apoptotic B cells, in *Leucocyte Typing V* (Sclossman, et al., eds.), pp. 1675–1677, Oxford University Press, Oxford.

Fluckiger, A. C., Durand, I., and Banchereau, J., 1994, Interleukin-10 induces apoptotic cell death of B-chronic lymphocytic-leukemia cells, *J. Exp. Med.* **179:**91–99.

Foti, A., Ahuja, H. G., Allen, S. J., Koduru, P., Schuster, M. W., Schulman, P., Bar-Eli, M., and Cline, M. J., 1991, Correlation between molecular and clinical events in the evolution of chronic myelocytic leukemia to blast crisis, *Blood* **77:**2441–2444.

French, L. E., Wohlwend, A., Sappino, A. P., Tschopp, J., and Schifferli, J. A., 1994, Human clusterin gene expression is confined to surviving cells during *in vitro* programmed cell death, *J. Clin. Invest.* **93:**877–884.

Frische, S. M., and Francis, H., 1994, Disruption of epithelial cell matrix interactions induces apoptosis, *J. Cell. Biol.* **124:**619–626.

Fritsche, M., Haessler, C., and Brandner, G., 1993, Induction of nuclear accumulation of the tumor-suppressor protein p53 by DNA-damaging agents, *Oncogene* **8:**307–318.

Gaidano, G., Ballerini, P., Gong, J. Z., Inghirami, G., Neri, A., Newcomb, E. W., Magrath, I. T., Knowles, D. M., and Dalla-Favera, R., 1991, *p53* mutations in human lymphoid malignancies: Association with Burkitt lymphoma and chronic lymphocytic leukemia, *Proc. Natl. Acad. Sci. U.S.A.* **88:**5413–5417.

Gaido, M. L., and Cidlowski, J. A., 1991, Identification, purification, and characterisation of a calcium-dependent endonuclease (NUC18) from apoptotic rat thymocytes, *J. Biol. Chem.* **266:**18580–18585.

Gauwerky, C. E., Hoxie, J., Nowell, P. C., and Croce, C. M., 1988, Pre-B-cell leukemia with a t(8–14) and a t(14–18) translocation is preceded by follicular lymphoma, *Oncogene* **2:**431–435.

Gishizky, M. L., and Witte, O. N., 1992, Initiation of deregulated growth of multipotent progenitor cells by *bcr-abl in vitro*, *Science* **256:**836–839.

Gorczyca, W., Gong, J., Ardelt, B., Traganos, F., and Darzynkiewicz, Z., 1993, The cell cycle related differences in susceptibility of HL-60 cells to apoptosis induced by various antitumour agents, *Cancer Res.* **53:**3186–3192.

Gordon, M. Y., Dowding, C. R., Riley, G. P., Goldman, J. M., and Greaves, M. J., 1987, Altered adhesive interactions with marrow stroma of hematopoietic progenitor cells in chronic myeloid leukemia, *Nature* **328:**342–344.

Gottlieb, E., Haffner, R., von Ruden, T., Wagner, E. F., and Oren, M., 1994, Down regulation of wild-type *p53* activity interferes with apoptosis of IL-3 dependent hematopoietic cells following IL-3 withdrawal, *EMBO J.* **13:**1368–1374.

Gottschalk, A. R., Boise, L. H., Thompson, C. B., and Quintans, J., 1994, Identification of immunosuppressant-induced apoptosis in a murine B-cell line and its prevention by *bcl-x* but not *bcl-2*, *Proc. Natl. Acad. Sci. U.S.A.* **91:**7350–7354.

Gratiot-Deans, J., Merino, R., Nunez, G., and Turka, L. A., 1994, Bcl-2 expression during T-cell development—early loss and late return occur at specific stages of commitment to differentiation and survival, *Proc. Natl. Acad. Sci. U.S.A.* **91:**10685–10689.

Gregory, C. D., 1995, Control of apoptosis in human B cells: Implications for neoplasia, in *Apoptosis and the Immune Response* (C. D. Gregory, ed.), pp. 257–311, Wiley-Liss Inc., New York.

Gregory, C. D., Dive, C., Henderson, S., Smith, C. A., Williams, G. T., Gordon, J., and Rickinson, A. B., 1991, Activation of Epstein–Barr virus latent genes protects human B cells from death by apoptosis, *Nature* **349:**612–614.

Gregory, C. D., Tursz, T., Edwards, C. F., Tetaud, C., Talbot, M., Caillou, B., Rickinson, A. B., Lipinski, M., 1987, Identification of a subset of normal B cells with a Burkitt's lymphoma (BL)-like phenotype, *J. Immunol.* **139:**313–318.

Haimovitz-Friedman, A., Kan, C. C., Ehleiter, D., Persaud, R. S., McLoughlin, M., Fuks, Z., and Kolesnick, R. N., 1994, Ionizing radiation acts on cellular membranes to generate ceramide and initiate apoptosis, *J. Exp. Med.* **180:**525–535.

Haldar, S., Jena, N., and Croce, C. M., 1995, Inactivation of Bcl-2 by phosphorylation, *Proc. Natl. Acad. Sci. U.S.A.* **92:**4507–4511.

Hall, S. E., Savill, J. S., Henson, P. M., and Haslett, C., 1994, Apoptotic neutrophils are phagocytosed by fibroblasts with participation of the fibroblast vitronectin receptor and involvement of a mannose/fucose-specific lectin, *J. Immunol.* **153:**3218–3227.

Harper, J. W., Adami, G. R., Wei, N., Keyomarsi, K., and Elledge, S. J., 1993, The p21 Cdk-interacting protein CIP1 is a potent inhibitor of G_1 cyclin-dependent kinases, *Cell* **75:**805–816.

Harrington, E. A., and Evan, G. I., 1994, Oncogenes and cell death, *Curr. Opin. Gen. Dev.* **4:**120–129.

Harrington, E. A., Bennett, M. R., Fanidi, A., and Evan, G. I., 1994, c-Myc induced apoptosis in fibroblasts is inhibitied by specific cytokines, *EMBO J.* **13:**3286–3295.

Harris, A. W., Pinkert, C. A., Crawford, M., Langdon, W. Y., Brinster, R. L., and Adams, J. M., 1988, The E_μ-*myc* transgenic mouse—a model for high-incidence spontaneous lymphoma and leukemia of early B-cells, *J. Exp. Med.* **167:**353–371.

Hartley, S. B., Cooke, M. P., Fulcher, D. A., Harris, A. W., Cory, S., Basten, A., and Goodnow, C. C., 1993, Elimination of self-reactive B lymphocytes proceeds in two stages—arrested development and cell death, *Cell* **72:**325–335.

Henderson, S., Rowe, M., Gregory, C., Croomcarter, D., Wang, F., Longnecker, R., Kieff, E., and Rickinson, A., 1991, Induction of *bcl-2* expression by Epstein–Barr virus latent membrane protein-1 protects infected B-cells from programmed cell death, *Cell* **65:**1107–1115.

Hengartner, M. O., and Horvitz, H. R., 1994, *C. elegans* cell survival gene *ced-9* encodes a functional homolog of the mammalian proto-oncogene *bcl2*, *Cell* **76:**665–676.

Hermeking, H., and Eick, D., 1994, Mediation of c-Myc-induced apoptosis by *p53*, *Science* **265:**2091–2093.

Hickman, J. A., 1992, Apoptosis induced by anticancer drugs, *Cancer Metab. Rev.* **11:**121–139.

Hockenberry, D., Nunez, G., Milliman, C., Schreiber, R. D., and Korsmeyer, S. J., 1990, Bcl-2 is an inner mitochondrial membrane protein that blocks programmed cell death, *Nature* **348:**334–336.

Hockenberry, D., Zutter, M., Hickey, W., Nahm, M., and Korsmeyer, S. J., 1991, BCL2 protein is topographically restricted in tissues characterized by apoptotic cell death, *Proc. Natl. Acad. Sci. U.S.A.* **88:**6961–6965.

Hockenberry, D. M., Oltvai, Z. N., Yin, X., Milliman, C. L., and Korsmeyer, S. J., 1993, Bcl-2 functions in an antioxidant pathway to prevent apoptosis, *Cell* **75:**241–251.

Hogquist, K. A., Nett, M. A., Unanue, E. R., and Chaplin, D. D., 1991, Interleukin is processed and released during apoptosis, *Proc. Natl. Acad. Sci. U.S.A.* **88:**8485–8489.

Holder, M. J., Wang, H., Milner, A. E., Casamayor, M., Armitage, R., Spriggs, M. K., Fanslow, W. C., Maclennan, I. C. M., Gregory, C. D., and Gordon, J., 1993, Suppression of apoptosis in normal and neoplastic human B-lymphocytes by CD40 ligand is independent of Bcl-2 induction, *Eur. J. Immunol.* **23:**2368–2371.

Homburg, C. H. E., Dehaas, M., Vondemborne, A., Verhoeven, A. J., Reutelingsperger, C. P. M., and Roos, D., 1995, Human neutrophils lose their surface Fc Riii and acquire annexin-V binding sites during apoptosis *in vitro*, *Blood* **85:**532–540.

Hughes, F. M., and Gorospe, W. C., 1991, Biochemical identification of apotposis (programmed cell death) in granulosa cells: Evidence for a potential mechanism underlying follicular atresia, *Endocrinology* **129:**2415–2422.

Hussain, S. P., Aguilar, F., and Cerutti, P., 1994, Mutagenesis of codon 248 of the human *p53* tumor suppressor gene by N-ethyl-N-nitrosourea, *Oncogene* **9:**13–18.

Iemura, A., Tsai, M., Ando, A., Wershil, B., and Galli, S. J., 1994, The *c-kit* ligand, stem cell factor, promotes mast cell survival by suppressing apoptosis, *Am. J. Pathol.* **144:**321–328.

Imamura, J., Miyoshi, I., and Koeffler, H. P., 1994, *p53* in hematologic malignancies, *Blood* **84:**2414–2421.

Jacobson, M. D., and Raff, M. C., 1995, Programmed cell death and bcl-2 protection in very low oxygen, *Nature* **374:**814–816.

Jacobson, M. D., Burne, J. F., King, M. P., Miyashita, T., Reed, J. C., and Raff, M. C., 1993, Bcl-2 blocks apoptosis in cells lacking mitochondrial DNA, *Nature* **361:**365–369.

Jacobson, M. D., Burne, J. F., and Raff, M. C., 1994, Programmed cell death and Bcl-2 protection in the absence of a nucleus, *EMBO J.* **13:**1899–1910.

Jarvis, W. D., Kolesnick, R. N., Fornari, F. A., Traylor, R. S., Gewirtz, D. A., and Grant, S., 1994a, Induction of apoptotic DNA damage and cell death by activation of the sphingomyelin pathway, *Proc. Natl. Acad. Sci. U.S.A.* **91:**73–77.

Jarvis, W. D., Turner, A. J., Povriack, L. F., Traylor, R. S., and Grant, S., 1994b, Induction of apoptotic DNA fragmentation and cell death in HL-60 human promyelocytic leukemia cells by pharmacological inhibitors of protein kinase C, *Cancer Res.* **54:**1707–1714.

Johnson, P. W. M., Watt, S. M., Betts, D. R., Davies, D., Jordan, S., Norton, A. J., and Lister, T. A., 1993, Isolated follicular lymphoma cells are resistant to apoptosis and can be grown *in vitro* in the CD40 stromal cell system, *Blood* **82:**1848–1857.

Kabarowski, J. H. S., Allen, P. B., and Weidemann, L. M., 1994, A temperature sensitive p210 BCR-ABL mutant defines the primary consequences of BCR-ABL tyrosine kinase expression in growth factor dependent cells, *EMBO J.* **13:**5887–5895.

Kamens, J., Paskind, M., Hugunin, M., Talanian, R. V., Allen, H., Banach, D., Bump, N., Hackett, M., Johnston, C. G., Li, P., Mankovich, J. A., Terranove, M., and Ghayur, T., 1995, Identification and characterization of ICH-2, a novel member of the interleukin-1B-converting enzyme family of cystein proteases, *J. Biol. Chem.* **270:**15250–15256.

Kamesaki, S., Kamesaki, H., Jorgensen, T. J., Tanizawa, A., Pommier, T., and Crossman, J., 1993, Bcl-2 protein inhibits etoposide induced apoptosis throught its effects subsequent to topoisomerase II-induced DNA strand breaks and repair, *Cancer Res.* **53:**4251–4256.

Kantarjian, H. M., Keating, M. J., Talpaz, M., Walters, R. S., Smith, T. L., Cork, A., McCredie, K. B., and Freireich, E. J., 1987, Chronic myelogenous leukaemia in blast crisis, *Am. J. Med.* **83:**445–454.

Kastan, M. B., Onyekwere, O., Sidransky, D., Vogelstein, B., and Craig, R. W., 1991, Participation of *p53* in the cellular response to DNA damage, *Cancer Res.* **51:**6304–6311.

Kerr, J. F. K., and Searle, J., 1973, Deletion of cells by apoptosis during castration-induced involution of the rat prostate, *Virchows Arch. Biol.* **13:**87–102.

Kerr, J. F. K., Wyllie, A. H., and Currie, A. R., 1972, Apoptosis: A basic biological phenomenon with wide ranging implications in tissue kinetics, *Br. J. Cancer* **26:**239–257.

Kessel, D. (ed.), 1989, *Resistance to Antineoplastic Drugs*, CRC press, Boca Raton, FL.

Kiefer, M. C., Brauer, M. J., Powers, V. C., Wu, J. J., Umansky, S. R., Tomei, L. D., and Barr, P. J., 1995, Modulation of apoptosis by the widely distributed Bcl-2 homologue Bak, *Nature* **374:**736–739.

Kinoshita, T., Yokota, T., Arai, K., and Miyajima, A., 1995, Regulation of Bcl-2 expression by oncogenic Ras protein in hematopoietic cells, *Oncogene* **10:**2207–2212.

Kipreos, E. T., Lee, G. J., and Wang, J. Y. J., 1987, Isolation of temperature-sensitive tyrosine kinase mutants of *vabl* oncogene by screening with antibodies for phosphotyrosine, *Proc. Natl. Acad. Sci. U.S.A.* **84:**1345–1349.

Kisielow, P., 1995, Apoptosis in intrathymic T-cell development, in *Apoptosis and the Immune Response* (C. D. Gregory, ed.), pp. 13–53, Wiley-Liss, New York.

Kizaki, H., Tadakuma, T., Odaka, C., Muramatsu, J., and Ishimura, Y., 1989, Activation of a suicide process of thymocytes through DNA fragmentation by calcium ionophores and phorbol esters, *J. Immunol.* **143:**1790–1794.

Kobayashi, R., Rassenti, L. Z., Meisenholder, G., Carson, D. A., and Kipps, T. J., 1993, Autoantigen inhibits apoptosis of a human B cell leukemia that produces pathogenic rheumatoid factor, *J. Immunol.* **151:**7273–7283.

Koeffler, H. P., and Golde, D. W., 1981, Chronic myelogenous leukemia—new concepts, *New Engl. J. Med.* **304:**1201–1207.

Kolesnick, R., and Golde, D. W., 1994, The sphingomyelin pathway in tumour necrsosi factor and interleukin-1 signalling, *Cell* **77:**325–328.

Koury, M. J., 1992, Programmed cell death (apoptosis) in hematopoiesis, *Exp. Hematol.* **20:**391–394.

Koury, M. J., and Bondurant, M. C., 1990, Erythropoietin retards DNA breakdown and prevents programmed cell death in erythroid progenitor cells, *Science* **248:**378–381.

Krajewski, S., Tanaka, S., Takayama, S., Schibler, M. J., Fenton, W., and Reed, J. C., 1993, Investigation into the subcellular distribution of the *bcl-2* oncoprotein: Residence in the nuclear envelope, endoplasmic reticulum, and outer mitochondrial membranes, *Cancer Res.* **53:**4701–4714.

Krajewski, S., Krajewski, M., Shabaik, A., Miyashita, T., Wang, H. G., and Reed, J. C., 1994a, Immunohistochemical determination of *in vivo* distribution of Bax, a dominant inhibitor of Bcl-2, *Am. J. Pathol.* **145:**1323–1336.

Krajewski, S., Krajewski, M., Shabaik, A., Wang, H. G., Irie, S., Fong, L., and Reed, J. C., 1994b, Immunohistochemical analysis of *in vivo* patterns of Bcl-X expression, *Cancer Res.* **54:**5501–5507.

Lam, M., Dubyak, G., Chen, L., Nunez, G., Miesfeld, R. L., and Distelhorst, C. W., 1994, Evidence that BCL-2 represses apoptosis by regulating endoplasmic reticulum-associated Ca^{2+} fluxes, *Proc. Natl. Acad. Sci. U.S.A.* **91:**6569–6573.

Lane, D. P., 1992, *p53*, guardian of the genome, *Nature* **358:**15–16.

Laneuville, P., Sun, G., Timm, M., and Vekemans, M., 1992, Clonal evolution in a myeloid cell line transformed to interleukin-3 independant growth by retroviral transduction and expression of $p210^{bcr/abl}$, *Blood* **80:**1788–1797.

Langdon, W. Y., Harris, A. W., Cory, S., and Adams, J. M., 1986, The *c-myc* oncogene perturbs B-lymphocyte development in E_{μ}*-myc* transgenic mice, *Cell* **47:**11–18.

Lazebnik, Y. A., Cole, S., Cooke, C. A., Nelson, W. G., and Earnshaw, W. C., 1993, Nuclear events of apoptosis *in vitro* in cell-free mitotic extracts: A model system for analysis of the active phase of apoptosis, *J. Cell. Biol.* **123:**7–22.

Lazebnik, Y. A., Kaufmann, S. H., Desnoyers, S., Poirier, G. G., and Earnshaw, W. C., 1994, Cleavage of poly(ADP-ribose) polymerase by a proteinase with properties like ICE, *Nature* **371:**346–347.

Lee, S., Elenbaas, B., Levine, A., and Griffith, J., 1995, p53 and its 14kDa C-terminal domain recognize primary DNA damage in the form of insertion/deletion mismatches, *Cell* **81:**1013–1020.

Lennon, S. V., Kilfeather, S. A., Hallett, M. B., Campbell, A. K., and Cotter, T. G., 1992, Elevations in cytosolic free Ca^{2+} are not required to trigger apoptosis in human leukaemia cells, *Clin. Exp. Immunol.* **87:**465–471.

Leoncini, L., Delvecchio, M. T., Spina, D., Megha, T., Barbini, P., Sabattini, E., Pileri, S., Tosi, P., Kraft, R., Laissue, J. A., and Cottier, H., 1995, Presence of the bcl-2 protein and apoptosis in non-Hodgkin lymphomas with diffuse growth-pattern, *Int. J. Cancer* **61:**826–831.

Levine, A. J., Chang, A., Dittmer, D., Notterman, D. A., Silver, A., Thorn, K., Welsh, D., and Wu, M., 1994a, The *p53* tumor suppressor gene, *J. Lab. Clin. Med.* **123:**817–823.

Levine, A. J., Perry, M. E., Chang, A., Silver, A., Dittmer, D., Wu, M., and Welsh, D., 1994b, The 1993 Walter Hubert lecture: The role of the *p53* tumour-suppressor gene in tumorigenesis, *Br. J. Cancer* **69:**409–416.

Li, F. P., Fraumeni, J. F., Mulvihill, J. J., Blattner, W. A., Dreyfus, M. G., Tucker, M. A., and Miller, R. M., 1988, A cancer family syndrome in twenty four kindreds, *Cancer Res.* **48:**4358–5362.

Li, P., Allen, H., Banerjee, S., Franklin, S., Herzog, L., Johnston, C., McDowell, J., Paskind, M., Rodman, L., Salfeld, J., Towne, E., Tracey, D., Wardwell, S., Wei, F. Y., Wong, W., Kamen, R., and Seshadri, T., 1995, Mice deficient in IL-1B-converting enzyme are defective in production of mature IL-1B and resistant to endotoxic shock, *Cell* **80:**401–411.

Lin, E. Y., Orlofsky, A., Berger, M. S., and Prystowsky, M. B., 1993, Characterisation of A1, a novel hemopoietic-specific early resonse gene with sequence similarity to *bcl-2*, *J. Immunol.* **151:**1979–1988.

Lin, J., Chen, J., Elenbaas, B., and Levine, A. J., 1994, Several hydrophobic amino acids in the p53 amino-terminal domain are required for transcriptional activation, binding to mdm-2 and the adenovirus 5 E1B 55kd protein, *Genes. Dev.* **8:**1235–1246.

Liu, Y. J., Joshua, D. E., Williams, G. T., Smith, C. A., Gordon, J., and Maclennan, I. C. M., 1989, Mechanism of antigen-driven selection in germinal centers, *Nature* **342:**929–931.

Liu, Y. J., Mason, D. Y., Johnson, G. D., Abbot, S., Gregory, C. D., Hardie, D. L., Gordon, J., and Maclennan, I. C. M., 1991, Germinal center cells express Bcl-2 protein after activation by signals which prevent their entry into apoptosis, *Eur. J. Immunol.* **21:**1905–1910.

Liu, Y. J., Johnson, G. D., Gordon, J., and Maclennan, I. C. M., 1992, Germinal centers in T cell-dependent antibody responses, *Immunol. Today* **13:**17–21.

Liu, Z. G., Smith, S. W., McLaughlin, K. A., Schwartz, L. M., and Osborne, B. A., 1994, Apoptotic signals delivered through the T-cell receptor of a T-cell hybrid require the immediate-early gene *nur77*, *Nature* **367:**281–284.

Long, B. W., Witte, P. L., Abraham, G. N., Gregory, S. A., and Plate, J. M. D., 1995, Apoptosis and interleukin-7 gene-expression in chronic B-lymphocytic leukemia-cells, *Proc. Natl. Acad. Sci. U.S.A.* **92:**1416–1420.

Lotem, J., and Sachs, L., 1992, Hematopoietic cytokines inhibit apoptosis induced by transforming growth factor beta1 and cancer chemotherapy compounds in myeloid leukemia cells, *Blood* **80:**1750–1757.

Lotem, J., and Sachs, L., 1993a, Hematopoietic cells from mice deficient in wild-type *p53* and more resistant to induction of apoptosis by some agents, *Blood* **82:**1092–1096.

Lotem, J., and Sachs, L., 1993b, Regulation by *bcl-2, c-myc* and *p53* of susceptibility to induction of apoptosis by heat shock and cancer chemotherapy compounds in differentiation-competent and -defective myeloid leukemia cells, *Cell Growth Differ.* **4:**41–47.

Lowe, S. W., Schmitt, E. M., Smith, S. W., Osborne, B. A., and Jacks, T., 1993, *p53* is required for radiation-induced apoptosis in mouse thymocytes, *Nature* **362:**847–849.

Lowe, S. W., Bodis, S., McClatchey, A., Remington, L., Ruley, H. E., Fisher, D. E., Housman, D. E., and Jacks, T., 1994a, *p53* status and the efficacy of cancer therapy *in vivo*, *Science* **266:**807–810.

Lucas, M., Solano, F., and Sanz, A., 1991, Induction of programmed cell death (apoptosis) in mature lymphocytes, *FEBS Lett.* **279:**19–20.

MacDonald, H. R., and Lees, R. K., 1990, Programmed cell death of autoreactive thymocytes, *Nature* **343:** 642–644.

MacLennan, I. C. M., 1994, Germinal centers, *Annu. Rev. Immunol.* **12:**117–139.

Malissen, B., and Schmitt-Verhulst, A. M., 1993, Transmembrane signalling through the T-cell-receptor–CD3 complex, *Curr. Opin. Immunol.* **5:**324–333.

Manabe, A., Murti, K. G., Coustan-Smith, E., Kumagai, M., Behm, F. G., Raimondi, S. C., and Campana, D., 1994, Adhesion-dependent survival of normal and leukemic human B lymphoblasts on bone marrow stromal cells, *Blood* **83:**758–766.

Manome, Y., Datta, R., Taneja, N., Shafman, T., Bump, E., Hass, R., Weichselbaum, R., and Kufe, D., 1993, Coinduction of *c-jun* gene expression and internucleosomal DNA fragmentation by ionizing radiation, *Biochemistry* **32:**10607–10613.

Mapara, M. Y., Bargou, R., Zugck, C., Dohner, H., Ustaoglu, F., Jonker, R. R., Krammer, P. H., and Dorken, B., 1993, Apo-1 mediated apoptosis or proliferation in human chronic B-lymphocytic leukemia—correlation with *bcl-2* oncogene expression, *Eur. J. Immunol.* **23:**702–708.

Martin, S. J., Lennon, S. V., Bonham, A. M., and Cotter, T. G., 1990, Induction of apoptosis (programmed cell death) in human leukaemic HL-60 cells by inhibition of RNA or protein synthesis, *J. Immunol.* **145:**1859–1867.

Marvel, J., Perkins, G. R., Lopez-Rivas, A., and Collins, M. K. L., 1994, Growth factor starvation of *bcl-2* overexpressing murine bone marrow cells induced refractoriness to IL-3 stimulation of proliferation, *Oncogene* **9:**1117–1122.

Mashal, R., Shtalrid, M., Talpaz, M., Kantarjian, H., Smith, L., Beran, M., Trujillo, J., Gutterman, J., and Deisseroth, A., 1990, Rearrangement and expression of *p53* in the chronic phase and blast crisis of chronic myelogenous leukemia, *Blood* **75:**180–189.

Maung, Z. T., MacLean, F. R., Reid, M. M., Pearson, A. D. J., Procter, S. J., Hamilton, P. J., and Hall, A. G., 1994, The relationship between *bcl-2* expression and response to chemotherapy in acute leukemia, *Br. J. Haematol.* **88:**105–109.

May, W. S., Tyler, G. P., Ito, T., Armstrong, D. K., Qatsha, K. A., and Davidson, N. E., 1994, Interleukin-3 and bryostatin-1 mediate hyperphosphorylation of *bcl-2* in association with suppression of apoptosis, *J. Biol. Chem.* **269:**26865–26870.

McCarthy, D. M., Goldman, J. M., Rassool, F. V., Graham, S. V., and Birnie, G. D., 1984, Genomic alterations involving the *c-myc* proto-oncogene locus during the evolution of a case of chronic granulocytic leukaemia, *Lancet* **2:**1362–1365.

McConkey, D. J., Nicotera, P., Hartzell, P., Bellomo, G., Wyllie, A. H., and Orrenius, S., 1989, Glucocorticoids activate a suicide process in thymocytes through an elevation of cytosolic calcium concentration, *Arch. Biochem. Biophys.* **259:**365–370.

McConkey, D. J., Hartzell, P., Chow, S. C., Orrenius, S., and Jondal, M., 1990, Interleukin 1 inhibits T cell receptor-mediated apoptosis in immature thymocytes, *J. Biol. Chem.* **265:**3009–3011.

McConkey, D. J., Aguilar-Santelises, M., Hartzell, P., Eriksson, I., Mellstedt, H., Orrenius, S., and Jondal, M., 1991, Induction of DNA fragmentation in chronic B-lymphocytic leukemia cells, *J. Immunol.* **146:**1072–1076.

McDonnell, T. J., and Korsmeyer, S. J., 1991, Progression from lymphoid hyperplasia to high-grade malignant lymphoma in mice transgenic for the t(14, 18), *Nature* **349:**254–256.

McDonnell, T. J., Deane, N., Platt, F. M., Nunez, G., Jaeger, U., McKearn, J. P., and Korsmeyer, S. J., 1989, Bcl-2-immunoglobulin transgenic mice demonstrate extended B-cell survival and follicular lymphoproliferation, *Cell* **57:**79–88.

McDonnell, T. J., Nunez, G., Platt, F. M., Hockenberry, D., London, L., McKearn, J. P., and Korsmeyer, S. J., 1990, Deregulated *bcl-2*-immunoglobulin transgene expands a resting but responsive immunoglobulin-M and immunoglobulin-D-expressing B-cell population, *Mol. Cell. Biol.* **10:**1901–1907.

McGahon, A., Bissonnette, R. P., Schmitt, M., Cotter, K. M., Green, D. R., and Cotter, T. G., 1994, *BCR-ABL* maintains resistance of chronic myelogenous leukemia cells to apoptotic cell death, *Blood* **83:**1179–1187.

McLaughlin, J., Chianese, E., and Witte, O. N., 1987, *In vitro* transformation of immature hematopoietic cells by the P210 *BCR/ABL* oncogene product of the Philadelphia chromosome, *Proc. Natl. Acad. Sci. U.S.A.* **84:**6558–6562.

McWhirter, J. R., and Wang, J. Y. J., 1993, An actin-binding function contributes to transformation by the Bcr-Abl oncoprotein of Philadelphia chromosome-positive human leukaemias, *EMBO J.* **12:**1533–1546.

Mekori, Y. A., Oh, C. K., and Metcalfe, D. D., 1993, IL-3-dependent murine mast cells undergo apoptosis on removal of IL-3, *J. Immunol.* **151:**3775–3784.

Meredith, J. E., Fazeli, B., and Schwartz, M., 1993, The extracellular matrix as a cell survival factor, *Mol. Biol. Cell* **4:**953–961.

Merino, R., Ding, L. Y., Veis, D. J., Korsmeyer, S. J., and Nunez, G., 1994, Developmental regulation of the Bcl-2 protein and susceptibility to cell death in B lymphocytes, *EMBO J.* **13:**683–691.

Milner, A. E., Johnson, G. D., and Gregory, C. D., 1992, Prevention of programmed cell death in Burkitt lymphoma cell lines by *bcl-2*-dependent and *bcl-2*-independent mechanisms, *Int. J. Cancer* **52:**636–644.

Milner, A. E., Grand, R. J. A., Waters, C. M., and Gregory, C. D., 1993, Apoptosis in Burkitt-lymphoma cells is driven by *c-myc*, *Oncogene* **8:**3385–3391.

Milner, J., 1995, Flexibility: the key to *p53* function, *Trends Biochem. Sci.* **20:**49–51.

Miura, M., Zhu, H., Rotello, R., Hartweig, E. A., and Yuan, J., 1993, Induction of apoptosis in fibroblasts by IL-1B-converting enzyme, a mammalian homolog of the *C. elegans* cell death gene *ced-3*, *Cell* **75:**653–660.

Miyashita, T., and Reed, J. C., 1992, *bcl-2* gene transfer increases relative resistance of S49.1 and WEHI7.2 lymphoid cells to cell death and DNA fragmentation induced by glucocorticoids and multiple chemotherapeutic drugs, *Cancer Res.* **52:**5407–5411.

Miyashita, T., and Reed, J. C., 1993, Bcl-2 oncoprotein blocks chemotherapy-induced apoptosis in a human leukemia cell line, *Blood* **81:**151–157.

Miyashita, T., Harigai, M., Hanada, M., and Reed, J. C., 1994a, Identification of a *p53*-dependent negative response element in the *bcl-2* gene, *Cancer Res.* **54:**3131–3135.

Miyashita, T., Krajewski, S., Krajewski, M., Wang, H. G., Lin, H. K., Liebermann, D., Hoffman, B., and Reed, J. C., 1994b, Tumor suppressor *p53* is a regulator of *bcl-2* and *bax* gene expression *in vitro* and *in vivo*, *Oncogene* **9:**1799–1805.

Möller, P., Henne, C., Leithauser, F., Eichelmann, A., Schmidt, A., Bruderlein, S., Dhein, J., Krammer, P. H., 1993, Co-regulation of the APO-1 antigen with intercellular adhesion molecule 1 (CD54) in tonsillar B cells and coordinate PTO, *Blood* **81:**2067–2075.

Monaghan, P., Robertson, D., Amos, T. A. S., Dyer, M. J. S., Mason, D. Y., and Greaves, M. F., 1992, Ultrastructural localization of BCL-2 protein, *J. Histochem. Cytochem.* **40:**1819–1825.

Moore, N. C., Anderson, G., Williams, G. T., Owen, J. J. T., and Jenkinson, E. J., 1994, Developmental regulation of *bcl-2* expression in the thymus, *Immunology* **81:**115–119.

Motoyama, N., Wang, F. P., Roth, K. A., Sawa, H., Nakayama, K., Nakayama, K., Negishi, I., Senju, S., Zhang, Q., Fujii, S., and Loh, D. Y., 1995, Massive cell death of immature hematopoietic cells and neurons in *bcl-x*-deficient mice, *Science* **267:**1506–1510.

Nakai, M., Takeda, A., Cleary, M. L., and Endo, T., 1993, The bcl-2 protein is inserted into the outer membrane but not the inner membrane of rat liver mitochondria *in vitro*, *Biochem. Biophys. Res. Commun.* **196:**233–239.

Nakayama, K., Nakayama, K., Negishi, I., Kuida, K., Shinkai, Y., Louie, M. C., Fields, L. E., Lucas, P. J., Stewart, V., Alt, F. W., and Loh, D. Y., 1993, Disappearance of the lymphoid system in Bcl-2 homozygous mutant chimeric mice, *Science* **261:**1584–1588.

Negata, S., and Golstein, P., 1995, The *fas* death factor, *Science,* **267:**1449–1455.

Nelson, W. G., and Kastan, M. B., 1994, DNA strand breaks: The DNA template alterations that trigger *p53*-dependent DNA damage response pathways, *Mol. Cell Biol.* **14:**1815–1823.

Newmeyer, D. D., Farschon, D. M., and Reed, J. C., 1994, Cell-free apoptosis in *Xenopus* egg extracts: Inhibition by Bcl-2 and requirement for an organelle fraction enriched in mitochondria, *Cell* **79:**353–364.

Nguyen, M., Branton, P. E., Walton, P. A., Oltvai, Z. N., Korsmeyer, S. J., and Shore, G. C., 1994, Role of membrane anchor domain of Bcl-2 in suppression of apoptosis caused by E1B-defective adenovirus, *J. Biol. Chem.* **269:**16521–16524.

Nicholson, D. W., Ali, A., Thornberry, N. A., Vaillancourt, J.P., Ding, C. K., Gallant, M., Gareau, Y., Griffin, P. R., Labelle, M., Lazebnik, Y. A., Munday, N. A., Raju, S. M., Smulson, M. E., Yamin, T. T., Yu, V. L., and Miller, D. K., 1995, Identification and inhibition of the ICE/CED-3 protease necessary for mammalian apoptosis, *Nature* **376:**37–43.

Ning, Z. Q., and Murphy, J. J., 1993, Calcium ionophore-induced apoptosis of human B cells is preceeded by the induced expression of early response genes, *Eur. J. Immunol.* **23:**3369–3372.

Nishimura, Y., Ishii, A., Kobayashi, Y., Yamasaki, Y., and Yonehara, S., 1995, Expression and function of mouse Fas antigen on immature and mature T-cells, *J. Immunol.* **154:**4395–4403.

Nisitani, S., Tsubata, T., Murakami, M., Okamoto, M., and Honjo, T., 1993, The *bcl-2* gene product inhibits clonal deletion of self-reactive B lymphocytes in the periphery but not in the bone marrow, *J. Exp. Med.* **178:**1247–1254.

Nosseri, C., Coppola, S., and Ghibelli, L., 1994, Possible involvement of poly(ADP-ribosyl) polymerase in triggering stress-induced apoptosis, *Exp. Cell. Res.* **212:**367–373.

Nunez, G., London, L., Hockenberry, D., Alexander, M., McKearn, J. P., and Korsmeyer, S. J., 1990, Deregulated Bcl-2 gene expression selectively prolongs survival of growth factor deprived hemopoietic cell lines, *J. Immunol.* **144:**3602–3610.

Nunez, G., Hockenbery, D., McDonnell, T. J., Sorensen, C. M., and Korsmeyer, S. J., 1991, Bcl-2 maintains B-cell memory, *Nature* **353:**71–73.

Obeid, L. M., Linardic, C. M., Karolak, L. A., and Hannun, Y. A., 1993, Programmed cell death induced by ceramide, *Science* **259:**1769–1771.

Oberhammer, F., Fritsch, G., Schmied, M., Pavelka, M., Printz, D., Purchio, T., Lassmann, H., and Schulte-Hermann, R., 1993a, Condensation of the chromatin at the membrane of an apoptotic nucleus is not associated with activation of an endonuclease, *J. Cell. Sci.* **104:**317–326.

Oberhammer, F., Wilson, J. W., Dive, C., Morris, I. D., and Hickman, J. A., 1993b, Apoptotic death in epithelial cells: cleavage of DNA to 300 and/or 50 kilobase fragments prior to or in the absence of internucleosomal fragmentation, *EMBO J.* **12:**3679–3684.

Oberhammer, F., Hochegger, K., Froschl, G., Tiefenbacher, R., and Pavelka, M., 1994, Chromatin condensation during apoptosis is accompaied by degradation of lamin A+C, without enhanced activation of *cdc2* kinase, *J. Cell. Biol.* **126:**827–837.

O'Conor, G.T., 1961, Malignant lymphoma in African Children. II. Pathological entity, *Cancer* **14:**270–283.

Ogasawara, J., Suda, T., and Nagata, S., 1995, Selective apoptosis of $CD4^+CD8^+$ thymocytes by the anti-*fas* antibody, *J. Exp. Med.* **181:**485–491.

Oliver, F. J., Marvel, J., Collins, M. K. L., and Lopez-Rivas, A., 1993, *Bcl-2* oncogene protects a bone marrow-derived pre-B cell line from 5′-fluor,2′-deoxyuridine-induced apoptosis, *Biochem. Biophys. Res. Commun.* **194:**126–132.

Oltvai, Z. N., and Korsmeyer, S. J., 1994, Checkpoints of duelling dimers foil death wishes, *Cell* **79:**189–192.

Oltvai, Z. N., Milliman, C. L., and Korsmeyer, S. J., 1993, *Bcl-2* heterodimerises *in vivo* with a conserved homolog, *bax*, that accelerates programmed cell death, *Cell* **74:**609–619.

Osmond, D. G., Ricovargas, S., Valenzona, H., Fauteux, L., Liu, L., Janani, R., Lu, L., and Jacobsen, K., 1994, Apoptosis and macrophage-mediated cell deletion in the regulation of B-lymphopoiesis in mouse bone-marrow, *Immunol. Rev.* **142:**209–230.

Otani, H., Erdos, M., and Leonard, W. J., 1993, Tyrosine kinase(s) regulate apoptosis and *bcl-2* expression in a growth factor-dependent cell line, *J. Biol. Chem.* **268:**22733–22736.

Otsuka, T., Eaves, C. J., Humphries, R. K., Hogge, D. E., and Eaves, A. C., 1991, Lack of evidence for abnormal autocrine or paracrine mechanisms underlying the uncontrolled proliferation of primitive chronic myeloid leukemia progenitor cells, *Leukemia* **5:**861–868.

Owen, P. J., Musk, P., Evans, C. A., and Whetton, A. D., 1993, Cellular signaling events elicited by *v-abl* associated with growth factor independence in an IL-3 dependent cell line, *J. Biol. Chem.* **268:**15696–15703.

Owen-Lynch, P. J., Wong, A. K. Y., and Whetton, A. D., 1995, v-Abl mediated apoptotic suppression is associated with SHC phosphorylation without concomitant MAP kinase activation, *J. Biol. Chem.* **270:**5956–5962.

Owens, G. P., and Cohen, J. J., 1992, Identification of genes involved in programmed cell death, *Cancer Metab. Rev.* **11:**149–156.

Peitsch, M. C., Polzar, B., Stephan, H., Crompton, T., MacDonald, H. R., Mannherz, H. G., and Tschopp, J., 1993, Characterization of the endogenous deoxyribonuclease involved in nuclear DNA degradation during apoptosis (programmed cell death), *EMBO J.* **12:**371–377.

Pendergast, A. M., Quilliam, L. A., Cripe, L. D., Bassing, C. H., Dai, Z., Li, N., Batzer, A., Rabun, K. M., Der, C. J., Schlessinger, J., and Gishizky, M. L., 1993, BCR-ABL-induced oncogenesis is mediated by direct interaction with the SH2 domain of the GRB-2 adaptor protein, *Cell* **75:**175–185.

Pesce, M., Farrace, M. G., Piacentini, M., Dolci, S., and De Felici, M., 1993, Stem cell factor and leukemia inhibitory factor promote primordial germ cell survival by suppressing programmed cell death (apoptosis), *Development* **118:**1089–1094.

Pezzella, F., Tse, A. G. D., Cordell, J. L., Pulford, K. A. F., Gatter, K. C., and Mason, D. Y., 1990, Expression of the *bcl-2* oncogene protein is not specific for the 14–18 chromosomal translocation, *Am. J. Pathol.* **137:** 225–232.

Pezzella, F., Jones, M., Ralfkiaer, E., Ersboll, J., Gatter, K. C. and Mason, D. Y., 1992, Evaluation of bcl-2 protein expression and 14–18 translocation as prognostic markers in follicular lymphoma, *Br. J. Cancer* **65:**87–89.

Pfeffer, K., Matsuyama, T., Kundig, T. M., Wakeham, A., Kishihara, T., Shahinian, A., Wiegmann, K., Ohashi, P. S., Kronke, M., and Mak, T. M., 1993, Mice defiecient for the 55 kD tumour necrosis factor receptor are resistant to endotoxic shock, yet succumb to *L. monocytogenes* infection, *Cell* **73:**457–467.

Piris, M. A., Pezzella, F., Martinezmontero, J. C., Orradre, J. L., Villuendas, R., Sanchezbeato, M., Cuena, R., Cruz, M. A., Martinez, B., Garrido, M. C., Gatter, K., Aiello, A., Delia, D., Giardini, R., and Rilke, F., 1994, P53 and bcl-2 expression in high-grade B-cell lymphomas—correlation with survival-time, *Br. J. Cancer* **69:** 337–341.

Platt, N., and Gordon, S., 1995, Role of murine scavenger receptor in the recognition of apoptotic thymocytes by macrophages, *J. Cell. Biochem. [Suppl.]* **19B:**300.

Pongracz, J., Johnson, G. D., Crocker, J., Burnett, D., and Lord, J. M., 1994, The role of protein kinase C in myeloid cell apoptosis, *Biochem. Soc. Trans.* **22:**593–597.

Prokocimer, M., and Rotter, V., 1994, Structure and function of *p53* in normal cells and their aberrations in cancer cells: Projection on the hematologic lineages, *Blood* **84:**2391–2411.

Puil, L., Liu, J., Gish, G., Mbamalu, G., Bowtell, D., Pelicci, P. G., Arlinghaus, R., and Pawson, T., 1994, Bc-Abl oncoproteins bind directly to activators of the Ras signalling pathway, *EMBO J.* **13:**764–773.

Pulendran, B., Kannourakis, G., Nouri, S., Smith, K. G. C., and Nossal, G. J. V., 1995, Soluble antigen can cause enhanced apoptosis of germinal-center B cells, *Nature* **375:**331–334.

Raff, M. C., 1992, Social controls on cell survival and cell death, *Nature* **356:**397–400.

Rajotte, D., Haddad, P., Haman, A., Cragoe E. J., Jr., and Hoang, T., 1992, Role of protein kinase C and the Na^+/H^+ antiporter in suppression of apoptosis by granulocyte–macrophage colony-stimulating factor and interleukin-3, *J. Biol. Chem.* **267:**9980–9987.

Rathmell, J. C., Cooke, M. P., Ho, W. Y., Grein, J., Townsend, S. E., Davis, M. M., and Goodnow, C. C., 1995, CD95 (*fas*) dependent elimination of self-reactive B-cells upon interaction with CD(+) T-cells, *Nature* **376:**181–184.

Reed, J. C., 1994, BCL-2 and the regulation of programmed cell death, *J. Cell. Biol.* **124:**1–6.

Ren, Y., Silverstein, R. L., Allen, J., and Savill, J., 1995, CD36 gene-transfer confers capacity for phagocytosis of cells undergoing apoptosis, *J. Exp. Med.* **181:**1857–1862.

Renschler, M. F., Bhatt, R. R., Dower, W. J., and Levy, R., 1994, Synthetic peptide ligands of the antigen binding receptor induce programmed cell death in a human B cell lymphoma, *Proc. Natl. Acad. Sci. U.S.A.* **91:**3623–3627.

Reynolds, J. E., Yang, T., Qian, L., Jenkinson, J. D., Zhou, P., Eastman, A., and Craig, R. W., 1994, Mcl-1, a member of the Bcl-2 family, delays apoptosis induced by c-Myc overexpression in chinese hamster ovary cells, *Cancer Res.* **54:**6348–6352.

Rodriguez-Tarduchy, G., and Lopez-Rivas, A., 1989, Phorbol esters inhibit apoptosis in IL-2-dependent T lymphocytes, *Biochem. Biophys. Res. Commun.* **164:**1069–1075.

Rodriguez-Tarduchy, G., Malde, P., Lopez-Rivas, A., and Collins, M. K. L., 1992, Inhibition of apoptosis by calcium ionophores in IL-3 dependant bone marrow cells is dependant upon production of IL-4, *J. Immunol.* **146:**1416–1422.

Rolink, A., and Melchers, F., 1993, Generation and regeneration of cells of the B-lymphocyte lineage, *Curr. Opin. Immunol.* **5:**207–217.

Rothstein, T. L., Wang, J. K. M., Panka, D. J., Foote, L. C., Wang, Z. H., Stanger, B., Cui, H., Ju, S. T., and Marshakrothstein, A., 1995, Protection against fas-dependent Th1-mediated apoptosis by antigen receptor engagement in B-cells, *Nature* **374:**163–165.

Ryan, J. J., Danish, R., Gottlieb, C. A., and Clarke, M. F., 1993, Cell cycle analysis of *p53*-induced cell death in murine erythroleukemia cells, *Mol. Cell. Biol.* **13:**711–719.

Ryan, J. J., Prochownik, E., Gottlieb, C. A., Apel, I. J., Merino, R., Nunez, G., and Clarke, M. F., 1994, *c-myc* and *bcl-2* modulate p53 function by altering p53 subcellular trafficking during the cell cycle, *Proc. Natl. Acad. Sci. U.S.A.* **91:**5878–5882.

Sachs, L., and Lotem, J., 1993, Control of programmed cell death in normal and leukemic cells: New implications for therapy, *Blood* **82:**15–21.

Sato, T., Hanada, M., Bodrug, S., Irie, S., Iwama, N., Boise, L. H., Thompson, C. B., Golemis, E., Fong, L., Wang, H. G., and Reed, J. C., 1994, Interactions among members of the bcl-2 protein family analyzed with a yeast two-hybrid system, *Proc. Natl. Acad. Sci. U.S.A.* **91:**9238–9242.

Savill, J., Hogg, N., Ren, Y., and Haslett, C., 1992, Thrombospondin cooperates with CD36 and the vitronectin receptor in macrophage recognition of neutrophils undergoing apoptosis, *J. Clin. Invest.* **90:**1513–1522.

Savill, J., Fadok, V., Henson, P., and Haslett, C., 1993, Phagocyte recognition of cells undergoing apoptosis, *Immunol. Today* **14:**131–137.

Sawyers, C. L., Callahan, W., and Witte, O. N., 1992, Dominant negative MYC blocks transformation by *ABL* oncogenes, *Cell* **70:**901–910.

Sawyers, C. L., McLaughlin, J., and Witte, O. N., 1995, Genetic requirement for *ras* in the transformation of fibroblasts and hematopoietic cells by the *bcr-abl* oncogene, *J. Exp. Med.* **181:**307–313.

Schena, M., Gottardi, D., Ghia, P., Larsson, L. G., Carlsson, M., Nilsson, K., and Caligariscappio, F., 1993, The role of *bcl-2* in the pathogenesis of B-chronic lymphocytic leukemia, *Leukemia Lymphoma* **11:**173–179.

Selvakumaran, M., Lin, H. K., Miyashita, T., Wang, H. G., Krajewski, S., Reed, J. C., Hoffman, B., and Liebermann, D., 1994, Immediate early up-regulation of *bax* expression by *p53* but not *TGFB1*: A paradigm for distinct apoptotic pathways, *Oncogene* **9:**1791–1798.

Sentman, C. L., Shutter, J. R., Hockenberry, D., Kanagawa, O., and Korsmeyer, S. J., 1991, *bcl-2* inhibits multiple forms of apoptosis but not negative selection in thymocytes, *Cell* **67:**879–888.

Shokat, K. M., and Goodnow, C. C., 1995, Antigen-induced B-cell death and elimination during germinal-center immune-responses, *Nature* **375:**334–338.

Sikora, E., Grassilli, E., Bellesia, E., Troiano, L., and Franceschi, C., 1993a, Studies of the relationship between cell proliferation and cell death. III. AP-1 DNA-binding activity during concanavalin A-induced proliferation or dexamethasone-induced apoptosis of rat thymocytes, *Biochem. Biophys. Res. Commun.* **192:**386–391.

Sikora, E., Grassilli, E., Radziszewska, E., Bellesia, E., Barbieri, D., and Franceschi, C., 1993b, Transcription factors DNA-binding activity in rat thymocytes undergoing apoptosis after heatshock or dexamethasone treatment, *Biochem. Biophys. Res. Commun.* **197:**709–715.

Smeyne, R. J., Vendrell, M., Hayward, M., Baker, S. J., Miao, G. G., Schilling, K., Robertson, L. M., Curran, T., and Morgan, J. I., 1993, Continuous *c-fos* expression precedes programmed cell death *in vivo*, *Nature* **363:** 166–169.

Smith, M. L., Chen, I. T., Zhan, Q., Bae, I., Chen, C. Y., Gilmer, T. M., Kastan, M. B., O'Connor, P. M., and Fornace A. J., Jr., 1994, Interaction of the *p53*-regulated protein Gadd45 with proliferating cell nuclear antigen, *Science* **266:**1376–1380.

Steller, H., 1995, Mechanisms and genes of cellular suicide, *Science* **267:**1445–1449.

Stewart, N., Hicks, G. G., Paraskevas, F., and Mowat, M., 1995, Evidence for a second cell cycle block at G_2/M by *p53*, *Oncogene* **10:**109–115.

Strasser, A., Harris, A. W., Bath, M. L., and Cory, S., 1990, Novel primitive tumours induced in transgenic mice by cooperation between *myc* and *bcl-2*, *Nature* **348:**331–333.

Strasser, A., Whittingham, S., Vaux, D. L., Bath, M. L., Adams, J. M., Cory, S., and Harris, A. W., 1991, Enforced *bcl-2* expression in B-lymphoid cells prolongs antibody responses and elicits autoimmune disease, *Proc. Natl. Acad. Sci. U.S.A.* **88:**8661–8665.

Strasser, A., Harris, A. W., and Cory, S., 1993, E_μ–*bcl-2* transgene facilitates spontaneous transformation of early pre-B and immunoglobulin-secreting cells but not T cells, *Oncogene* **8:**1–9.

Strasser, A., Harris, A. W., von Boehmer, H., and Cory, S., 1994a, Positive and negative selection of T-cells in T-cell receptor transgenic mice expressing a *bcl-2* transgene, *Proc. Natl. Acad. Sci. U.S.A.* **91:**1376–1380.

Strasser, A., Harris, A. W., Corcoran, L. M., and Cory, S., 1994b, Bcl-2 expression promotes B-lymphoid but not T-lymphoid development in *scid* mice, *Nature* **368:**457–460.

Strasser, A., Harris, A. W., Jacks, T., and Cory, S., 1994c, DNA damage can induce apoptosis in proliferating lymphoid cells via *p53*-independent mechanisms inhibitable by *bcl-2*, *Cell* **79:**329–339.

Sun, X. M., and Cohen, G. M., 1994, Mg^{2+}-dependent cleavage of DNA into kilobase pair fragments is responsible for the initial degradation of DNA in apoptosis, *J. Biol. Chem.* **269:**14857–14860.

Szczylik, C., Skorski, T., Nicolaides, N. C., Manzella, L., Malaguarnera, L., Venturelli, D., Gewirtz, A. M., and Calabretta, B., 1991, Selective inhibition of leukemia cell proliferation by BCR-ABL antisense oligodeoxynucleotides, *Science* **253:**562–565.

Takayama, S., Sato, T., Krajewski, S., Kochel, K., Irie, S., Millan, J. A., and Reed, J. C., 1995, Cloning and functional analysis of BAG-1: A novel Bcl-2-binding protein with anti-cell death activity, *Cell* **80:**279–284.

Tanaka, S., Saito, K., and Reed, J. C., 1993, Structure-function analysis of the BCL-2 oncoprotein, *J. Biol. Chem.* **268:**10920–10926.

Tauchi, T., Boswell, H. S., Leibowitz, D., and Broxmeyer, H. E., 1994, Coupling between p210*^bcr-abl^* and Shc and

Grb-2 adaptor proteins in hematopoietic cells permits growth factor receptor-independent link to *ras* activation pathway, *J. Exp. Med.* **179:**167–175.

Tewari, M., Dixit, V. M., 1995, Fas- and tumor necrosis factor induced apoptosis is inhibited by the pox virus crmA gene product, *J. Biol. Chem.* **270:**3255–3260.

Tewari, M., Quan, L. T., O'Rourke, K., DesNoyers, S., Zheng, Z., Beidler, D. R., Poirier, G. G., Salvesen, G. S., and Dixit, V. M., 1995, Yama/CPP32B, a mammalian homolog of CED-3, is a crmA-inhibitable protease that cleaves the death substrate poly(ADP-ribose) polymerase, *Cell* **81:**801–809.

Tsubata, T., 1995, B-1 B cells (Ly-1 or $CD5^+$ B cells), apoptosis and autoimmunity, in *Apoptosis and the Immune Response* (C. D. Gregory, ed.), pp. 217–256, Wiley-Liss, New York.

Ucker, D. S., Obermiller, P. S., Eckhart, W., Apgar, J. R., Berger, N. A., and Meyers, J., 1992, Genome digestion is a dispensible consequence of physiological cell death mediated by cytotoxic T-lymphocytes, *Mol. Cell. Biol.* **12:**3060–3069.

Vanhaesebroeck, B., Reed, J. C., Valck, D. D., Grooten, J., Miyashita, T., Tanaka, S., Beyaert, R., Van Roy, F., and Fiers, W., 1993, Effect of *bcl-2* proto-oncogene expression on cellular sensitivity to tumor necrosis factor-mediated cytotoxicity, *Oncogene* **8:**1075–1081.

Vaux, D. L., Weissman, I. L., and Kim, S. K., 1992, Prevention of programmed cell death in *Caenorhabditis elegans* by human *bcl-2*, *Science* **258:**1955–1957.

Veis, D. J., Sorensen, C. M., Shutter, J. R., and Korsmeyer, S. J., 1993a, Bcl-2-deficient mice demonstrate fulminant lymphoid apoptosis, polycystic kidneys, and hypopigmented hair, *Cell* **75:**229–240.

Veis, D. J., Sentman, C. L., Bach, E. A., and Korsmeyer, S. J., 1993b, Expression of the Bcl-2 protein in murine and human thymocytes and in peripheral T-lymphocytes, *J. Immunol.* **151:**2546–2554.

Von Boehmer, H., Teh, H.-S., and Kisielow, P., 1989, The thymus selects the useful, neglects the useless and destroys the harmful, *Immunol. Today* **10:**57–61.

Vuist, W. M., Levy, R., and Maloney, D. G., 1994, Lymphoma regression induced by monoclonal anti-idiotypic antibodies correlates with their ability to induce Ig signal transduction and is not prevented by tumor expression of high levels of bcl 2 protein, *Blood* **83:**899–906.

Wagner, A. J., Small, M. B., and Hay, N., 1993, Myc-mediated apoptosis is blocked by ectopic expression of Bcl-2, *Mol. Cell. Biol.* **13:**2432–2440.

Walker, N. I., Bennett, R. E., and Kerr, J. F. K., 1989, Cell death by apoptosis during involution of the lactating breast in mice and rats, *Am. J. Pathol.* **185:**19–32.

Walker, P. R. C., Smith, T., Youdale, J., Leblanc, J. F., Whitfield, J. F., and Sikorska, M., 1991, Topoisomerase II-reactive drugs induce apoptosis in thymocytes, *Cancer Res.* **51:**1078–1085.

Walton, M. I., Whysong, D., O'Connor, P. M., Hockenberry, D., Korsmeyer, S. J., and Kohn, K. W., 1993, Constitutive expression of human *bcl-2* modulates nitrogen mustard and camtothecin induced apoptosis, *Cancer Res.* **53:**1853–1861.

Wang, H. G., Miyashita, T., Takayama, S., Sato, T., Torigoe, T., Krajewski, S., Tanaka, S., Hovey, L., Troppmair, J., Rapp, U. R., and Reed, J. C., 1994, Apoptosis regulation by interaction of Bcl-2 protein and raf-1 kinase, *Oncogene* **9:**2751–2756.

Wang, L., Miura, M., Bergeron, L., Zhu, H., and Yuan, J., 1994, *Ich-1*, an *Ice/ced-3*-related gene, encodes both positive and negative regulators of programmed cell death, *Cell* **78:**739–750.

Weaver, V. M., Lach, B., Walker, P. R., and Sikorska, M., 1993, Role of proteolysis in apoptosis: Involvement of serine proteases in internucleosomal DNA fragmentation in thymocytes, *Biochem. Cell. Biol.* **71:**488–500.

Whyte, M., and Evan, G., 1995, The last cut is the deepest, *Nature* **376:**17–18.

Williams, G. T., Smith, C. A., Spooncer, E., Dexter, T. M., and Taylor, D. R., 1990, Haemopoietic colony stimulating factors promote cell survival by suppressing apoptosis, *Nature* **343:**76–79.

Wood, A. C., Waters, C. M., Garner, A., and Hickman, J. A., 1994, Changes in *c-myc* expression and the kinetics of dexamethasone-induced programmed cell death (apoptosis) in human lymphoid leukaemia cells, *Br. J. Cancer* **69:**663–669.

Woronicz, J. D., Calnan, B., Ngo, V., and Winoto, A., 1994, Requirement for the orphan steroid receptor nur77 in apoptosis of T-cell hybridomas, *Nature* **367:**277–281.

Wyllie, A. H., 1980, Glucocorticoid-induced thymocyte apoptosis is associated with endogenous endonuclease activation, *Nature* **284:**555–556.

Wyllie, A. H., 1987, Apoptosis: Cell death under homeostatic control, *Arch. Toxicol. [Suppl.]* **11:**3–10.

Wyllie, A. H., 1992, Apoptosis and the regulation of cell numbers in normal and neoplastic tissues: An overview, *Cancer and Metastasis Reviews* **11:**95–103.

Wyllie, A. H., Kerr, J. F. R., and Currie, A. R., 1980, Cell death: The significance of apoptosis, *Int. Rev. Cytology* **68:**251–306.

Yang, E., Zha, J., Jockei, J., Boise, L. H., Thompson, C. B., and Korsmeyer, S. J., 1995, Bad, a heterodimeric partner for Bcl-x_L and bcl-2, displaces bax and promotes cell death, *Cell* **80:**285–291.

Yao, R., and Cooper, G. M., 1995, Requirement for phosphatidylinositol-3 kinase in the prevention of apoptosis by nerve growth factor, *Science* **267:**2003–2006.

Yin, X. M., Oltval, Z. N., and Korsmeyer, S. J., 1994, BH1 and BH2 domains of Bcl-2 are required for inhibition of apoptosis and heterodimerization with Bax, *Nature* **369:**321–323.

Yonish-Rouach, D., Resnitzky, D., Lotem, J., Sachs, L., Kimchi, A., and Oren, M., 1991, Wild-type *p53* induces apoptosis of myeloid leukaemia cells that is inhibited by interleukin-6, *Nature* **352:**345–347.

Yuan, J., Shaham, S., Ledoux, S., Ellis, H. M., and Horvitz, H. R., 1993, The *C. elegans* cell death gene *ced-3* encodes a protein similar to mammalian interleukin-1B-converting enzyme, *Cell* **75:**641–652.

Yunis, J. J., Frizzera, G., Oken, M. M., McKenna, J., Theologides, A., and Arnesen, M., 1987, Multiple recurrent genomic defects in follicular lymphoma—a possible model for cancer, *N. Engl. J. Med.* **316:**79–84.

Zambetti, G. P., and Levine, A. J., 1993, A comparison of the biological activitites of wild-type and mutant *p53*, *FASEB J.* **7:**855–865.

Zhan, Q., Fan, S., Bae, I., Guillouf, C., Liebermann, D. A., O'Connor, P. M., and Fornace A. J., Jr., 1994, Induction of *bax* by genotoxic stress in human cells correlates with normal p53 status and apoptosis, *Oncogene* **9:**3743–3751.

Zhivotovsky, B., Wade, D., Gahm, A., Orrenius, S., and Nicotera, P., 1994, Formation of 50kbp chromatin fragments in isolated liver nuclei is mediated by protease and endonuclease activation, *FEBS Lett.* **351:** 150–154.

Zhu, Y. M., Bradbury, D., and Russell, N., 1993, Expression of different conformations of p53 in the blast cells of acute myeloblastic leukaemia is related to *in vitro* growth characteristics, *Br. J. Cancer* **68:**851–855.

Zhu, Y. M., Bradbury, D. A., and Russell, N. H., 1994, Wild-type *p53* is required for apoptosis induced by growth factor deprivation in factor-dependent leukaemic cells, *Br. J. Cancer* **69:**468–472.

Chapter 8

Human T-Cell Differentiation and Cytokine Regulation

Mike Salmon, Darrell Pilling, Clair Mappin, and Arne N. Akbar

1. INTRODUCTION

T lymphocytes play an important role in regulating the behavior of other cells and the nature of immune responses. This is mostly by the action of secreted cytokines, but T cells are also constantly affected by signals derived from themselves and the microenvironment around them. It is difficult to separate these processes in a meaningful way, so in this review we discuss the coordination of the signals delivered and received by T cells in regulating immunity.

2. REGULATION OF T-CELL CYTOKINE PRODUCTION

Fine control of immune reactions by T lymphocytes appears to be achieved largely by precise combinations of cytokines delivered to the right place at the right time. For this to be possible, T cells must have the capacity to separate the production of different lymphokines, either in time or in space. The major cytokines produced by T cells are summarized in Table I (Salmon and Akbar, 1993), though not all of these molecules are produced exclusively or even principally by T lymphocytes. Perhaps the clearest separation to be achieved both on theoretical grounds and in observed reality is between interferon-γ (IFNγ) and interleukin 4 (IL-4). These molecules have essentially counteractive effects on other cells and on their own production: IFNγ is mainly associated with cellular immune reactions and delayed

Mike Salmon, Darrell Pilling, and Clair Mappin Department of Rheumatology, The University of Birmingham, Birmingham B15 2TT, United Kingdom. **Arne N. Akbar** Department of Clinical Immunology, The Royal Free Hospital, London NW3 2PF, United Kingdom.

Blood Cell Biochemistry, Volume 7: Hematopoietic Cell Growth Factors and Their Receptors, edited by A. D. Whetton and J. Gordon. Plenum Press, New York, 1996.

Table I
Cytokines[a]

Cytokine	Length (amino acids)	Principal actions
Interleukin 2	133	T-cell growth factor. Also activates NK cells; regulates B-cell function
Interleukin 3	134–140	Growth factor for bone marrow stem cells, also activation of mast cells and basophils
Interleukin 4	120	B-cell growth and differentiation factor
Interleukin 5	115	Eosinophil-differentiating factor. In mice, IL-5 is also an important B-cell-differentiating factor
Interleukin 6	189	B cell growth and differentiation, T cell costimulation; induction of acute-phase response
Interleukin 8	72	Neutrophil chemotaxis and activation
Interleukin 9	126	Helper T-cell growth factor
Interleukin 10	160	Inhibits IL-2 and IFN production
Interferon-γ	146	Macrophage activation, class II MHC regulation, B-cell differentiation; promotes NK-cell function
Tumor necrosis factor α and β	α 157 β 171	T-cell growth factor, macrophage activation, endothelial cell activation, and fever induction
Transforming growth factor β	112	Macrophage activation; lymphocyte inhibition; regulates B-cell function

[a]Cytokines produced by human T cells and some of their characteristic effects. The latter should be taken with a pinch of salt, because most cytokines have effects that are both redundant (the same as other molecules) and pleotropic (many and various). Cytokines appear to act together in a network fashion, and the effects of different cytokine combinations are very difficult to predict.

hypersensitivity, and IL-4 seems to be involved mostly in regulating antibody production by B cells and is particularly important for synthesis of IgE. A number of possible models have been proposed to account for T-cell cytokine segregation, and these fall broadly into three categories:

1. Fixed segregation in which distinct subsets of cells produce various combinations of cytokines.
2. Transient segregation, which is associated with differentiation or environmental control.
3. Direct transcriptional regulation of cytokine genes by environmental signals, producing the appropriate molecules from pluripotent cells

The constraints applied to and by the production of cytokines are essential for limiting the potential diversity of cytokine interactions within the network. We discuss each of these three main strategies in more detail and then suggest that each of these influences on the regulation of T-cell cytokine production play a part and can be reconciled into a unified model.

2.1. Fixed Segregation

In 1986, Mosmann *et al.* described a panel of mouse T-cell clones that fell into two distinct categories (Mosmann *et al.*, 1986; Kurt-Jones *et al.*, 1986). T helper 1 (Th1) clones synthesized IL-2, IFNγ, and other cytokines that act predominantly on the cellular arm of the immune response, while Th2 clones produce IL-4 and IL-5, which mainly regulate B-cell immunity. More recently, IL-10 has been shown to segregate into Th2 clones in mice and is believed to play a major role in their development; IL-10 in humans seems rather more equivocal. Other cytokines such as IL-3, granulocyte–macrophage colony-stimulating factor (GM-CSF), and tumor necrosis factor (TNF) seem to show no clear segregation and are produced equally by both sets of clones.

Subsequently, it became apparent that many murine T-cell clones fall into a third category, which has been termed Th0 (Firestein *et al.*, 1989). These cells make all or at least most cytokines tested with very little selectivity. Nevertheless, the Th1/Th2 segregation has become the most widely accepted model for lymphokine segregation.

In fact, it has proved considerably more difficult to produce either Th1 or Th2 clones of human T cells. The majority of reports have consistently failed to show such a pattern of lymphokine segregation (Salmon *et al.*, 1989; Palliard *et al.*, 1988), and in most of those that have done so, the segregation is only partial (Plant, 1990), reflecting quantitative rather than qualitative differences in the molecules produced.

Probably the best evidence for a physiological role for the Th1/Th2 model comes from allergy (Seder and Paul, 1994) and from parasitic disease models such as *Leishmania* (Wang *et al.*, 1994). In both systems, Th2-like IL-4-producing cells have been isolated. In *Leishmania* and leprosy, where polarized forms of the disease exist, reflecting different types of immune response, both Th1-like and Th2-like cells have been observed (Wang *et al.*, 1994; Cillari *et al.*, 1991).

2.2. The Effect of Environmental Signals

The presence of specific cytokines in the microenvironment during stimulation or initial priming seems to be a major factor in regulating T-cell lymphokine production. It is not entirely clear to what extent this reflects an influence on the permanent differentiation of T cells or a direct regulatory effect on gene expression. For example, IL-4 present during the initial stimulation of naive $CD4^+$ T cells from T-cell receptor transgenic mice promotes the production of high levels of IL-4 on restimulation (Hsieh *et al.*, 1993; Seder *et al.*, 1993), whereas IL-12 promotes the production of IFNγ (Seder *et al.*, 1992; Hsieh *et al.*, 1992). In humans, similar effects of IL-4 and IL-12 have been shown when these cytokines were included in cultures of peripheral blood T cells before cloning, but the discrimination of Th1 and Th2 cells appears much less clear-cut than in mice. For example, the presence of IL-4 before cloning may shift the differentiation of PPD-specific T cells from IL-2 and IFNγ production toward T cell clones able to produce IL-4 and IL-5 in addition to IL-2 and IFNγ (Maggi *et al.*, 1992). The presence of IL-12 during cloning of human T cells induced the differentiation of clones able to produce high levels of IFNγ, but with the same pattern of IL-4 production observed in clones generated in cultures without IL-12 (Manetti *et al.*, 1994). Interleukin-10 has also been shown to have a role in influencing T-cell lymphokine production, inhibiting IFNγ production by T cells through an action on accessory cells (Vieira *et al.*, 1991) and directly inhibiting T-cell IL-2 production (De Waal Malefyt *et al.*, 1993). Recent studies have suggested a role for the costimulatory molecules B-7.1 (CD80) and B-7.2 (CD86) in directing development of T helper cell responses. Continuous stimulation with B-7.1 favors IFNγ production, and B-7.2 enhances IL-4 production (reviewed by Reiner and Seder, 1995).

In vitro cloning produces T cells with a highly differentiated phenotype and altered biological function. Such cells *in vivo* have a diminished capacity to produce sufficient IL-2 to support their own survival and proliferation, markedly reduced expression of *bcl-2*, but increased expression of Fas (Salmon *et al.*, 1994; see also Chapter 3). Consequently, they tend to die by apoptosis. Cloning is likely to select cells capable of survival under these conditions or to require provision of exogenous survival factors. The regulation of cytokine production by early primed cells may well differ from that of clones in extended culture, as we discuss later.

If Thl and Th2 clones with fixed heterogeneity of lymphokine production are produced *in vivo*, how do they arise? Several possibilities have been suggested. Virgin T cells on first challenge with antigen make only IL-2 (Salmon *et al.*, 1989). These cells may then develop into pluripotential Th0 cells. The final differentiation event into a Thl or Th2 state would take many cell cycles and depend on a particular set of environmental signals. Alternatively, the virgin T cells may themselves be precommitted, requiring a number of cycles before they can effectively suppress the "forbidden" genes (Figure 1) (Male *et al.*, 1992). Neither of these models is entirely satisfactory.

2.3. Transient Segregation: The Progressive Differentiation of Primed T Cells

Human virgin T cells make high levels of IL-2 if stimulated, but little else. In contrast, stimulated primed T cells produce rather less IL-2 on a unit basis, but they also make all

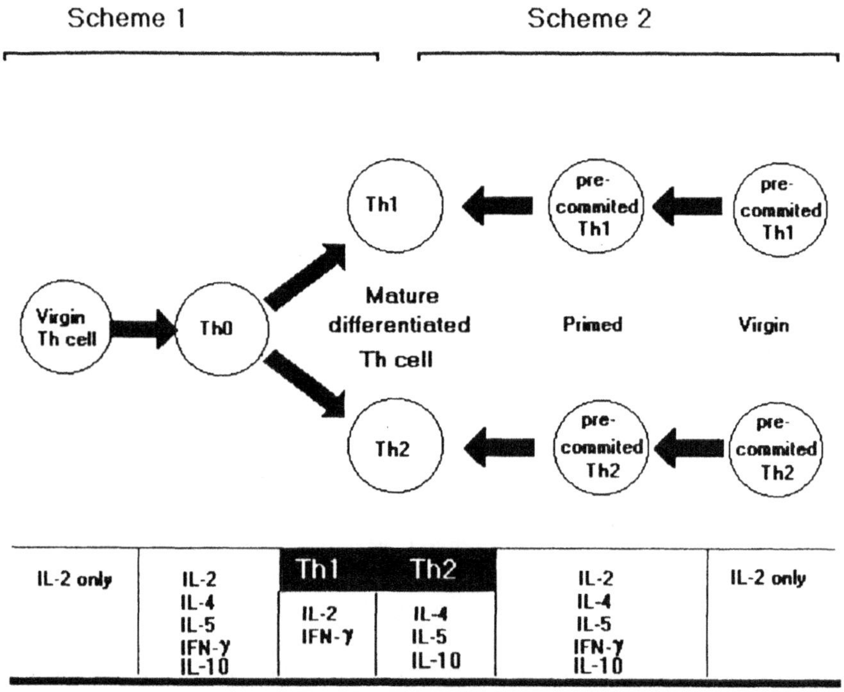

FIGURE 1. Two proposed routes for the *in vivo* development of Thl and Th2 clones. Many variants have been proposed to these basic themes, but the essential difference between these two is the existence or absence in virgin T-cell populations of precommitment. The window of activity of specific cytokines is shown.

of the other T-cell-derived cytokines (Salmon *et al.*, 1989). Because virgin T cells become primed after stimulation, this is a simple model of transient segregation of lymphokine synthesis: nonpermanent states in which T cells produce a restricted range of cytokines. Our recent work suggests that primed T cells also show such a transient pattern of segregation (Salmon *et al.*, 1994).

The differentiation state of human T cells can be identified by their expression of different isoforms of the leucocyte common antigen CD45. Virgin T cells express CD45RA and high levels of CD45RB (B bright); primed T cells express CD45RO and also highly variable levels of CD45RB (Salmon *et al.*, 1994; Mason and Powrie, 1990). Recent studies have suggested that T-cell memory resides in clones of primed CD45RO$^+$ T cells with a short intermitotic half-life rather than long-loved resting cells (Beverley, 1990; Akbar *et al.*, 1991). Until recently, no markers were available to distinguish newly primed cells from long-established clones, which severely hampered attempts to determine the progressive development of immunoregulatory responses. Recently, we described a spectrum of primed CD4$^+$ T cells characterized by an inverse relationship between the expression of CD45RB and CD45RO (Salmon *et al.*, 1994). Primed CD4$^+$ T cells progress through many cycles of division from a CD45RB^{bright}O^{dull} to a CD45RB^{dull}O^{bright} state.

This progressive differentiation defined by the shift from CD45RB^{bright} to CD45RB^{dull} is paralleled by the gradual loss of bcl-2 and gain of Fas expression, two features associated with an increased propensity for apoptosis. At the same time, these highly differentiated CD45RB^{dull} cells selectively lose the capacity to synthesize IL-2, a cytokine that is particularly effective in preventing T-cell apoptosis, and also, to a lesser extent, IFNγ. This leads to the apoptosis of primed CD45RB^{dull} cells when they are stimulated in the absence of exogenous IL-2 (Salmon *et al.*, 1994). This process is likely to provide a major constraint for the persistence of T-cell memory maintained by continuous priming and an important mechanism contributing to the maintenance of T-cell homeostasis and immunoregulatory potential *in vivo*. Although memory itself appears to require the persistence of antigen (Gray and Matzinger, 1991), progression through the spectrum may not. Naive cells require specific antigen to be recruited to the primed population, but their continued cycling is probably maintained by cross-reactive antigens or by third-party responses to which they are recruited nonspecifically.

As cells progress through an increasing number of cycles, they become more and more fragile. Persistent antigen is probably required to replace those cells that are lost to apoptosis with newly recruited T cells from the virgin or resting pool. T cells moving through the differentiation spectrum do not simply become "clonally exhausted" because, as they progressively differentiate, losing the capacity to produce IL-2 and IFNγ, they simultaneously produce higher levels of IL-4 (Salmon *et al.*, 1994). This provides a model for lymphokine segregation in which different combinations of cytokine may be produced as a continuous spectrum by cells in various states of differentiation. More recently, we have studied the influence of environmental cytokines on this basic pattern of differentiation. As we discussed earlier, specific cytokines in the microenvironment during stimulation or initial priming seem to be a major factor in regulating T-cell lymphokine production. However, it has been unclear to what extent this reflects an influence on the permanent differentiation of T cells, producing subsets like the Thl and Th2 divisions or perhaps a direct and transient regulatory influence on gene expression.

We investigated the effects of IL-4, IL-10, and IL-12 on the differentiation of naive human CD4$^+$ CD45RA$^+$ T cells *in vitro*, analyzing the changes in T-cell lymphokine production as the cells differentiated (Mappin *et al.*, 1995). Each cytokine was present during primary stimulation of the naive T cells and was either maintained or removed during subsequent stimulations to see whether its effects on T-cell lymphokine production were persistent (influencing differentiation) or transient (requiring the continued presence of the cytokine). Both IL-12 and IL-4 had persistent effects, leading to the development of high or low IL-2 and IFNγ producers, respectively. Interleukin 4 did not up-regulate its own production, but IL-4 still became the dominant cytokine by suppression of IL-2 and IFNγ. An unexpected but highly consistent finding was that IL-12 also transiently increased IL-10 production, which is possibly a mechanism for limiting its own persistent effects. The most prominent feature of these experiments was that without the addition of exogenous cytokines, T cells differentiated as expected from a Th1-like to a Th2-like state. However, IL-12 appeared to retard the differentiation of cells, maintaining them in an "early" phase, while IL-4 accelerated differentiation into a late phase (Figure 2). Although the early phenotype persists after removal of IL-12, cells will ultimately resume the differentiation process. Interestingly, the increasing susceptibility of cells to apoptosis as they progress through the spectrum was mirrored here: IL-12-treated cells that maintained a Th1-like phenotype were strongly resistant to apoptosis, but the Th2-like cells produced by treating cells with IL-4 were highly susceptible to this process. The model of T-cell differentiation suggested by these new data (Figure 3) allows a reconciliation of the subset/differentiation arguments and points to an understanding of these processes, which may have therapeutic significance. The effects of B-7.1 and B-7.2 acting as costimulatory ligands, as discussed earlier, are likely to superimpose on this spectrum.

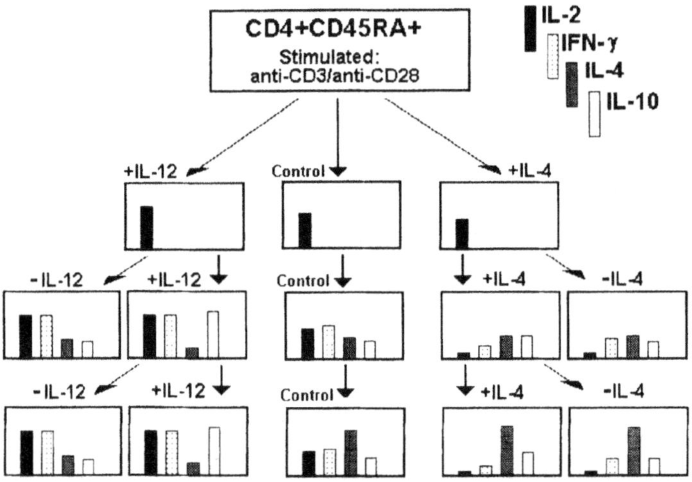

FIGURE 2. The effect of environmental cytokines on the differentiation of human T cells. Each stimulation was performed 14 days after the previous level. Persistence of influence was assessed by removing cytokines during the subsequent round of stimulation. The central vertical thread represents points on the basic pattern of differentiation; IL-4 accelerates differentiation to a much later phase, whereas IL-12 retards it at the level seen after about a week of normal stimulation. An unexpected effect of IL-12 was the transient but dramatic increase in IL-10 production, which appears to represent a negative feedback loop.

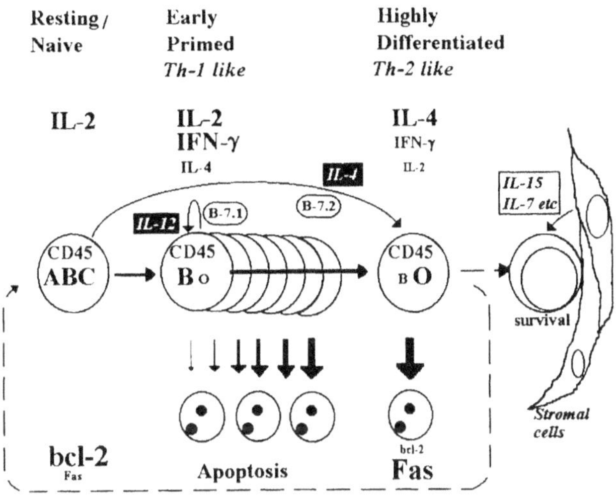

FIGURE 3. The progressive differentiation of human primed T cells. Virgin T cells in the resting pool produce only IL-2 on first stimulation but subsequently differentiate through a Thl-like to a Th2-like state. This does not proceed at a fixed rate but can be either retarded or accelerated by costimulatory ligands and environmental cytokines. As cells progress through the spectrum, they become increasingly susceptible to apoptosis, though they can be sustained at all points by interaction with certain stromal cells including fibroblasts and endothelium. Current evidence favors some degree of reversion to the completely resting ($CD45RA^+$) state, but not reversion to an earlier state within the primed spectrum. Reversion is likely to be a minor phenomenon, allowing replacement of cells taken from the resting pool but not considerable expansion in a mature immune system. The fate of the overwhelming majority of cells within the primed spectrum appears to be death.

3. TRANSCRIPTION REGULATION

The cellular regulation of lymphokine gene expression operates principally at the level of transcription (Taniguchi, 1988). Regulatory elements 5′ (upstream) of the genes for IL-2, IL-4, and IFNγ show regions of close homology, suggesting a degree of coordinated regulation by DNA binding proteins. There are three regions (each roughly 30 base pairs) of about 70% homology between the regulatory sequences of IL-2 and IL-4 genes, and two regions of complete homology exist between the IL-4 and IFNγ 5′ sequences (Arai *et al.*, 1989). Similar but quite distinct homologies are found between IL-2 and IFNγ (Arai *et al.*, 1989). Transcriptional regulation of these genes has been mapped using deletion constructs, gel retardation, and by DNase-1 footprinting (Taniguchi, 1988; Jain *et al.*, 1995). Sites known to bind transcriptionally active factors such as NF-κB, AP-1, and a CD28 response element have been identified. These observations suggest that a limited number of nuclear DNA-binding proteins may interact with both unique and shared binding sites in a fairly complex manner to regulate the expression of each of these lymphokines in the same cells in a coordinated fashion. It has become clear that quantitative regulation of transcription involves complex interactions of a number of DNA-binding proteins acting together (Jain *et al.*, 1995; Maniatis *et al.*, 1987) (Figure 4).

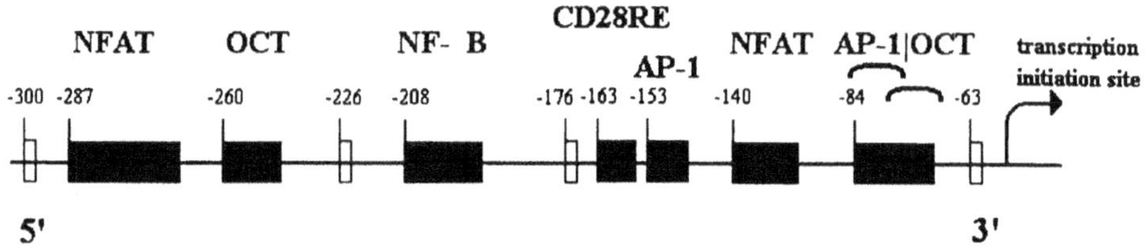

FIGURE 4. Regulatory sequences on the 5′ region of the IL-2 gene (after Jain *et al.*, 1995).

4. THE REGULATION OF GENES TRIGGERED BY CYTOKINES IN T LYMPHOCYTES

Possible outcomes of T-cell activation include cellular proliferation, production of cytokines and their receptors, proliferation, and death. Immediately after the binding of MHC–antigen complex by the TCR, a series of proteins are activated by the phosphorylation of tyrosine residues. Although the TCR/CD3 complex has no intrinsic tyrosine kinase activity, several cytoplasmic protein tyrosine kinases (PTK) are associated with CD3 and the CD4 and CD8 coreceptors. These include $p59^{fyn}$, ZAP-70, and $p56^{lcl}$ (Ledbetter *et al.*, 1993). These enzymes phosphorylate a number of proteins including phospholipase Cγl (PLCγl) and activate $p21^{ras}$.

The critical episode following the initial phosphorylation events associated with TCR-mediated activation is the hydrolysis of phosphatidylinositol-4,5-biphosphate (PIP_2) by PLCγl, releasing the second messengers diacylglycerol (DAG) and inositol-4,5-trisphosphate (IP_3) (Berridge, 1993; see Chapter 4). Diacylglycerol production leads to the activation of protein kinase C (PKC), which is involved in the regulation of many secondary kinase enzymes, such as those on the Ras/Raf/MAP kinase pathway, that lead to the activation of transcription factors such as NFAT and AP-1. The second derivative from PIP_2, IP_3, is involved in the elevation of intracellular calcium levels by the release of Ca^{2+} ions from the endoplasmic reticulum and the control of membrane Ca^{2+} channels. The rise in calcium levels leads to the activation of Ca^{2+}-dependent enzymes including calcineurin, which is vital for the induction of IL-2 gene transcription, through the translocation of NFAT from the cytoplasm to the nucleus.

Signals generated by the interaction of the TCR with an antigenic peptide bound to MHC are generally insufficient for maximal IL-2 release and cellular proliferation. However, a number of cell surface proteins that are incapable of inducing an immune response on their own synergize with signals generated by the TCR for efficient T-cell activation.

The most extensively studied of these molecules is CD28, which, along with CTLA-4, binds to CD80 (B7-1) and CD86 (B7-2) on antigen-presenting cells. CD28 ligation enhances T-cell activation by up-regulating cytokine production, particularly IL-2. This is achieved by the generation of a nuclear binding factor that binds to a 5′ regulatory domain of the IL-2 gene (CD28RE) (Civil *et al.*, 1992), resulting in the stabilization of IL-2 mRNA and prolonging the production of IL-2 protein (Ledbetter *et al.*, 1985). The biochemical signal produced by CD28 ligation is independent of those involving the TCR (Pages *et al.*, 1994). CD28 binds to PI3 kinase, an intracellular enzyme involved in the phosphorylation of phosphatidylinositol lipids. If this association is absent, then T cells do not produce IL-2, fail to proliferate, and probably die by apoptosis.

The signals provided by the TCR and CD28 are required for the effective activation of T cells, but the secretion of IL-2 has a positive feedback on T-cell proliferation. Events in the IL-2 signal pathway include the induction of several transcription factors such as AP-1 and NFAT (see Chapter 4), similar to TCR-mediated signals. However, in common with many cytokine receptors, the IL-2 receptor complex associates with a novel family of tyrosine kinase enzymes called Janus-associated kinases (Jak) (otherwise known as just another kinase). The association of members of the Jak family with cytokine receptors (Table II) has shed light on the observations that certain cytokines have overlapping if not identical effects, while others are specific. There are at least four members of the Jak family: Jak1,2,3 and Tyk2. Jak kinases are associated with the intracellular portions of cytokine receptors and are activated when dimerization of cytokine receptors occurs after ligation by the specific cytokine. This leads to the phosphorylation of a family of cytoplasmic transcription factors called STAT (signal transducer and activator of transcription). The role of Jak family

Table II
Signaling Pathways Operated by Cytokine Receptors[a]

Cytokine receptor family	Family members	Jak family member	STAT family member	References
Growth hormone	GH	Jak2		Witthuhn *et al.* (1993)
	EPO		STAT5	Gronowski *et al.* (1995)
	Prolactin			
	G-CSF			
IL-3	IL-3	Jak2	STAT5	Shual *et al.* (1993)
(Common β chain)	GM-CSF		STAT6	
	IL-5			
IL-6	IL-6	Jak1	STAT1/STAT3	Guschin *et al.* (1995)
(Common gp130)	LIF			Bacon *et al.* (1995)
	IL-11			Jacobson *et al.* (1995)
	IL-12 (?)	Jak2 and Tyk2	STAT4	
IL-2	IL-2 β chain	Jak1	STAT5/STAT6	Miyazaki *et al.* (1994)
(Common γ chain)	IL-2 γ chain	Jak3		Witthuhn *et al.* (1994)
	IL-4	Jak1 and Jak3	STAT6	Beadling *et al.* (1994)
	IL-7	Jak3		Hou *et al.* (1995)
	IL-9			Zeng *et al.* (1994)
	IL-13			
	IL-15	Jak1 and Jak3		
IFNα and IFNβ		Jak1 and Tyk2	STAT1α	Watling *et al.* (1993)
			STAT2	Muller *et al.* (1993)
IFNγ	α Receptor	Jak1	STAT1α	Watling *et al.* (1993)
	β Receptor	Jak2	STAT4	Muller *et al.* (1993)
IL-10		?	STAT1α	Webemordt *et al.* (1994)
			STAT3	

[a]The interaction of different Janus associated kinases (Jak) and STAT proteins (signal transducer and activator of transcription) associated with signaling through cytokine receptors. This exciting new area represents a major breakthrough in understanding the transduction of signals in the immune system, from the binding of surface receptors through the induction of subsequent biological effects.

members in cytokine receptor signal transduction was first elucidated in studies with mutant IFN-nonresponsive cell lines. Interferon-α and IFNβ receptors bind to Jak1 and Tyk2, whereas IFNγ receptors bind Jak1 and Jak2 (Muller *et al.*, 1993; Watling *et al.*, 1993). The interaction of these different Jak family members leads to the activation of different members of the STAT family (Table II). These findings have been extended to many cytokine receptors, including IL-2. The IL-2β and -γ chains associate with Jak1 and Jak3, respectively (Miyazaki *et al.*, 1994). The shared use of the IL-2 receptor γ chain by receptors specific for IL-2, IL-4, IL-7, IL-9, IL-13, and IL-15 may explain some of the apparent redundancy observed among these cytokines (see below).

The phosphorylation of tyrosine residues of STAT proteins leads to the formation of complexes of one or more STATs, which migrate from the cytoplasm to the nucleus. These complexes bind to specific DNA sequences and promote transcription of cytokine-responsive genes (Table II) (Shual *et al.*, 1993). This novel signaling pathway may be essential for the effective function of activated T cells. Indeed, if IL-2 is prevented from binding to its receptor, then activated T cells enter apoptosis. Therefore, the expression of different cytokine receptors and the association of alternate Jak/STAT family members in T cells may greatly affect their function. The STAT family is a good illustration of the complexity of transcriptionally active DNA-binding proteins. The same receptor on different cells may lead to the activation different STAT proteins and regulate expression of different genes. On the other hand, different STAT proteins may regulate the same genes in different cells, operating in concert with other factors.

5. REGULATION OF T-CELL APOPTOSIS BY CYTOKINES THAT BIND TO COMPONENTS OF THE IL-2 RECEPTOR

It has been apparent for some years that activated T cells die by apoptosis if deprived of IL-2 (Salmon *et al.*, 1994; Duke and Cohen, 1986; Akbar *et al.*, 1993). This is associated with reduced expression of the *bcl-2* gene product (Akbar *et al.*, 1993). Recently, we found that the group of cytokines that signal through a common receptor component, which represents the γ chain of the IL-2 receptor, all prevent apoptosis of activated T cells (Akbar *et al.*, 1995). These cytokines include IL-2 itself, IL-4, IL-7, and IL-15. This cytokine-mediated rescue is proliferation independent because cells placed in S-phase block with etoposide are still rescued. Treatment with these cytokines leads to induction of the antiapoptosis genes *bcl-2* and *bcl-x*, but has no effect on expression of *bax* and *bcl-*x_s, which promote apoptosis (see Chapter 7). The most effective rescue cytokine was IL-15, which also shares a β chain with the IL-2 receptor, suggesting an important role for this molecule also. The regulation of apoptosis through Bcl-2 expression by IL-2R γ-chain-reactive cytokines may be a direct phenomenon; the 5′ regulatory sequences of *bcl-2* and *bcl-x* genes contain sequences that suggest a consensus binding site for STAT proteins. A recent report suggests that the same group of IL- 2R γ-chain-reactive cytokines lead to the rescue of resting T cells that have been induced to die by apoptosis following γ irradiation (Boise *et al.*, 1995). This is a little harder to explain because the expression of such receptors by these cells is low or nonexistent, and the rescue appeared to be Bcl-2 independent. These results suggest that alternative pathways exist for the rescue of T cells by cytokines, possibly involving direct influence on the expression of *p53* (see Chapter 7).

6. CONCLUDING REMARKS

The networked interactions of cytokines and T cells in regulating immunity are complex. That segregation of T-cell-derived cytokines is both necessary and apparent has led to the development of several models to account for this phenomenon. A simple binary model like the Th1–Th2 division appears too simplistic to account for complex regulatory events. Although in a number of human systems polarization of responses can be observed, the evidence of signaling and differentiation suggests no good reason why T cells should specialize into subsets producing different combinations of cytokines. A pattern of closely coordinated regulation will mean that T cells produce lymphokines directly related to signals received from their environment. Recently, Kelso *et al.* (1995) showed that when a "Th2" response is induced in mice to the antigen keyhole limpet hemocyanin, cytokines at a population level show clear Th2 segregation. However, studied at the level of single cells, cytokine gene segregation was entirely random, representing all possible combinations. The results favor stochastic regulation and a differentiation mechanism rather than the selection of helper T cell subsets.

7. REFERENCES

Akbar, A. N., Salmon, M., and Janossy, G., 1991, The synergy between naive and memory T cells during activation, *Immunol. Today* **12:**184–188.

Akbar, A. N., Borthwick, N., Salmon, M., Gombert, W., Bofill, M., Shamsadeen, N., Pilling, D., Pett, S., Grundy, J. E., and Janossy, G., 1993, The significance of low *bcl-2* expression by CD45RO T cells in normal individuals and patients with acute viral infections. The role of apoptosis in T cell memory, *J. Exp. Med.* **178:** 427–438.

Akbar, A. N., Borthwick, N., Wickremasinghe, R. G., Panayiotidis, P., Pilling, D., Bofill, M., Krajewski, S., Reed, J. C., Salmon, M., 1996, Interleukin-2 receptor common γ-chain cytokines regulate activated T sell apoptosis in response to growth factor withdrawal: selective induction of anti-apoptotic (bcl-2, bcl-x_L) but not pro-apoptotic (bax, bcl-x_s) gene expression, *Eur. J. Immunology* **26:**294–299.

Airai, N., Nomura, D., Villaret, D., Malefijt, R. D., Seiki, M., Yoshida, M., Minoshima, S., Fukuyama, R., Maekawa, M., Kudoh, J., Shimizu, N., Yokota, K., Abe, E., Yokota, T., Takebe, Y., and Arai, K., 1989, Complete nucleotide sequence of the chromosomal gene for human IL-4 and its expression, *J. Immunol.* **142:** 274–282.

Bacon, C. M., Mcvicar, D. W., Ortaldo, J. R., Rees, R. C., Oshea, J. J., and Johnston, J. A., 1995, Interleukin-12 (IL-12) incudes tyrosine phosphorylation of Jak2 and Tyk2—differential use of Janus family tyrosine kinases by IL-2 and IL-12, *J. Exp. Med.* **181:**399–404.

Beadling, C., Guschin, D., Witthuhn, B. A., Ziemiecki, A., Ihle, J. N., Kerr, I. M., and Cantrell, D. A., 1994, Activation of Jak kinases and STAT proteins by interleukin-2 and interferon-alpha, but not the T-cell antigen receptor, in human T-lymphocytes, *EMBO J.* **13:**5605–5615.

Berridge, M. J., 1993, Inositol trisphosphate and calcium signalling, *Nature* **361:**315–325.

Beverley, P. C. L., 1990, Is T cell memory maintained by cross-reactive stimulation? *Immunol. Today* **11:**203–205.

Boise, L. I. I., Minn, A. J., June, C. H., Lindsten, T., and Thompson, C. B., 1995, Growth factors can enhance lymphocyte survival without committing the cell to undergo cell division, *Proc. Natl. Acad. Sci. U.S.A.* **92:** 5491–5495.

Cillari, E., Milano, S., Dieli, M., Maltese, E., Dirosa, S., Mansueto, S., Salerno, A., and Liew, F. Y., 1991, Reduction in the number of UCHL-1^+ cells and IL-2 production in the peripheral blood of patients with visceral leishmaniasis, *J. Immunol.* **146:**1026–1030.

Civil, A., Geerts, M., Aarden, L. A., and Verweij, C. L., 1992, Evidence for a role of CD28RE as a response element for distinct mitogenic T cell activation signals, *Eur. J. Immunol.* **22:**3041–3043.

De Waal Malefyt, R., Yssel, H., and De Vries, J. E., 1993, Direct effects of IL-10 on subsets of human $CD4^+$ T-cell clones and resting T-cells—specific inhibition of IL-2 production and proliferation, *J. Immunol.* **150:**4754–4765.

Duke R. C., and Cohen, J. J., 1986, IL-2 addiction—withdrawal of growth-factor activates a suicide program in dependent T-cells, *Lymphokine Res.* **5:**289–299.

Firestein, G. S., Roeder, W. D., Laxer, J. A., Townsend, K. S., Weaver, C. T., Horn, J. T., Linton, J., Torbett, B. E., and Glasebrooke, A. L., 1989, A new murine CD4$^+$ T cell subset with an unrestricted cytokine profile, *J. Immunol.* **143:**518–524.

Gray, D., and Matzinger, P., 1991, T-cell memory is short-lived in the absence of antigen, *J. Exp. Med.* **174:** 969–974.

Gronowski, A. M., Zhong, Z., Wen, Z. L., Thomas, M. J., Darnell, J. E., and Rotwein, P., 1995, *In-vivo* growth-hormone treatment rapidly stimulates the tyrosine phosphorylation and activation of STAT3, *Mol. Endocrinol.* **9:**171–177.

Guschin, D., Rogers, N., Briscoe, J., Witthuhn, B., Watling, D., Horn, F., Pellegrini, S., Yasukawa, K., Heinrich, P., Stark, G. R., Ihle, J. N., and Kerr, I. M., 1995, A major role for the protein-tyrosine kinase Jak1 in the Jak/STAT signal-transduction pathway in response to interleukin-6, *EMBO J.* **14:**1421–1429.

Hou, J. Z., Schindler, U., Henzel, W. J., Wong, S. C., and Mcknight, S. L., 1995, Identification and purification of human STAT proteins activated in response to interleukin-2, *Immunity* **2:**321–329.

Hsieh, C. S., Heimberger, A. B., Gold, J. S., O'Garra, A., and Murphy, K. M., 1992, Differential regulation of T-helper phenotype development by interleukin-4 and interleukin-10 in an alpha-beta-T-cell-receptor transgenic system, *Proc. Natl. Acad. Sci. U.S.A.* **89:**6065–6069.

Hsieh, C. S., Macatonia, S. E., Tripp, C. S., Wolf, S. F., Ogarra, A., and Murphy, K. M. N., 1993, Development of Th1 CD4$^+$ T-cells through IL-12 produced by *Listeria*-induced macrophages, *Science* **260:**547–549.

Jacobson, N. G., Szabo, S. J., Webernordt, R. M., Zhong, Z., Schreiber, R. D., Darnell, J. E., and Murphy, K. M. A., 1995, Interleukin-12 signaling in T-helper type-1 (th1) cells involves tyrosine phosphorylation of signal transducer and activator of transcription (STAT)3 and STAT4, *J. Exp. Med.* **181:**1755–1762.

Jain, J., Loh, C., and Rao, A., 1995, Transcriptional regulation of the IL-2 gene, *Curr. Opin. Immunol.* **7:**333–342.

Kelso, A., Groves, P., Troutt, A. B., and Francis, K., 1995, Evidence for the stochastic acquisition of cytokine profile by CD4$^+$ T cells in a T helper type 2-like response *in vivo*, *Eur. J. Immunol.* **25:**1168–1175.

Kurt-Jones, E. A., Hamberg, S., O'Hara, J., Paul, W. E., and Abbas, A. K., 1986, Heterogeneity of helper/inducer T lymphocytes 1. Lymphokine production and lymphokine responsiveness, *J. Exp. Med.* **166:**1774–1787.

Ledbetter, J. A., Martin, P. J., Spooner, C. E., Wofsy, D., Tsu, T. T., Beatty, P. G., and Gladstone, P., 1985, Antibodies to tp67 and tp44 augment and sustain proliferative responses of activated T cells, *J. Immunol.* **135:**2331–2336.

Ledbetter, J. A., Deans, J. P., Aruffo, A., Grosmaire, L. S., Kanner, S. B., Bolen, J. B., and Schieven, G. L., 1993, CD4, CD8, and the role of CD45 in T cell activation, *Curr. Opin. Immunol.* **5:**334–340.

Maggi, E., Parronchi, P., Manetti, R., Simonelli, C., Piccinni, M. P., Rugiu, F. S., Decarli, M., Ricci, M., Romagnani, S., 1992, Reciprocal regulatory effects of IFN-gamma and IL-4 on the *in vitro* development of human Th1 and Th2 clones, *J. Immunol.* **148:**2142–2147.

Male, D., Champion, B., Cooke, A., and Owen, M., 1992, Cytokines, in *Advanced Immunology*, pp. 11.1–11.16, Gower, London.

Manetti, R., Gerosa, F., Giudizi, M. G., Biagiotti, R., Parronchi, P., Piccinni, M. P., Sampognaro, S., Maggi, E., Romagnani, S., and Trinchieri, G., 1994, Interleukin-12 induces stable priming for interferon-gamma (IFN-gamma) production during differentiation of human T-helper (Th) cells and transient IFN-gamma production in established Th2 cell clones, *J. Exp. Med.* **179:**1273–1283.

Maniatis, T., Goodbourne, S., and Fischer, J. A., 1987, Regulation of inducible and tissue specific gene expression, *Science* **236:**1237–1245.

Mappin, C., Pilling, D., Scheel-Toellner, D., Akbar, A. N., and Salmon, M., 1996, IL-12 accelerates and IL-4 retards the process of progressive differentiation in human primed T cells (in press).

Mason, D., and Powrie, F., 1990, Memory CD4$^+$ T cells in man form 2 distinct subpopulations, defined by their expression of isoforms of the leukocyte common antigen CD45, *Immunology* **70:**427–433.

Miyazaki, T., Kawahara, A., Fujii, H., Nakagawa, Y., Minami, Y., Liu, Z. J., Oishi, I., Silvennoinen, O., Witthuhn, B. A., Ihle, J. N., and Taniguchi, T., 1994, Functional activation of *jak1* and *jak3* by selective association with IL-2 receptor subunits, *Science* **266:**1045–1047.

Mosmann, T., Cherwinski, H., Bond, M. W., Giedlin, M. A., Coffman, R. L., 1986, Two types of murine helper T cell clone. I. Definition according to profiles of lymphokine activities and secreted proteins, *J. Immunol.* **136:** 2348–2355.

Muller, M., Briscoe, J., Laxton, C., Guschin, D., Ziemiecki, A., Silvennoinen, O., Harpur, A. G., Barbieri, G., Witthuhn, B. A., Schindler, C., Pellegrini, S., Wilks, A. F., Ihle, J. N., Stark, G. R., and Kerr, I. M., 1993, The protein-tyrosine kinase jak1 complements defects in interferon-alpha/beta and interferon-gamma signal-transduction, *Nature* **366:**129–135.

Pages, F., Ragueneau, M., Rottapel, R., Truneh, A., Nunes, J., Imbert, J., and Olive, D., 1994, Binding of phosphatidylinositol-3-oh kinase to CD28 is required for T cell signaling, *Nature* **369:**327–329.

Palliard, X., De Waal Malefijt, R., Yssel, H., Blanchard, D., Chretien, I., Abrams, J., De Vries, J., and Spits, H., 1988, Simultaneous production of IL-2, IL-4 and IFN-γ by activated human $CD4^+$ and $CD8^+$ T cell clones, *J. Immunol.* **141:**849–855.

Plant, M., 1990, Antigen-specific lymphokine secretory patterns in atopic disease, *J. Immunol.* **144:**4497–500.

Reiner, S. L., and Seder, R. A., 1995, T helper cell differentiation in immune response, *Curr. Opin. Immunol.* **7:**360–366.

Salmon, M., and Akbar, A. N., 1993, T cell cytokines, in *The Immunology of Liver Transplantation* (Neuberger and Adams, eds.), pp. 84–94, Edward Arnold, London.

Salmon, M., Kitas, G. D., and Bacon, P. A., 1989, Production of lymphokine mRNA by $CD45R^+$ and $CD45R^-$ helper T cells from human peripheral blood and by human $CD4^+$ T cell clones, *J. Immunol.* **143:**907–912.

Salmon, M., Pilling, D., Borthwick, N., Viner, N., Bacon, P., Janossy, G., and Akbar, A., 1994, The progressive differentiation of primed T cells is associated with an increasing susceptibility to apoptosis, *Eur. J. Immunol.* **24:**892–899.

Seder, R. A., and Paul, W. E., 1994, Acquisition of lymphokine-producing phenotype by $CD4^+$ T-cells, *Annu. Rev. Immunol.* **12:**635–673.

Seder, R. A., Paul, W. E., Davis, M. M., and Destgroth, B. F., 1992, The presence of interleukin-4 during in vitro priming determines the lymphokine-producing potential of $CD4^+$ T-cells from T-cell receptor transgenic mice, *J. Exp. Med.* **176:**1091–1098.

Seder, R. A., Gazzinelli, R., Sher, A., and Paul, W. E., 1993, Interleukin-12 acts directly on $CD4^+$ T-cells to enhance priming for interferon-gamma production and diminishes interleukin-4 inhibition of such priming, *Proc. Natl. Acad. Sci. U.S.A.* **90:**10188–10192.

Shual, K., Ziemiecki, A., Wilks, A. F., Harpur, A. G., Sadowski, H. B., Gilman, M. Z., and Darnell, J. E., 1993, Polypeptide signaling to the nucleus through tyrosine phosphorylation of Jak and STAT proteins, *Nature* **366:**580–583.

Taniguchi, T., 1988, Regulation of cytokine gene expression, *Annu. Rev. Immunol.* **6:**439–464.

Vieira, P., Dewaalmalefyt, R., Dang, M. N., Johnson, K. E., Kastelein, R., Fiorentino, D. F., Devries, J. E., Roncarolo, M. G., Mosmann, T. R., and Moore, K. W., 1991, Isolation and expression of human cytokine synthesis inhibitory factor cDNA clones—homology to Epstein–Barr virus open reading frame BCRFI, *Proc. Natl. Acad. Sci. U.S.A.* **88:**1172–1176.

Wang, Z. E., Zheng, S. C., Corry, D. B., Dalton, D. K., Seder, R. A., Reiner, S. L., and Locksley, R. M., 1994, Interferon-gamma-independent effects of interleukin-12 administered during acute or established infection due to *Leishmania major*, *Proc. Natl. Acad. Sci. U.S.A.* **91:**12932–12936.

Watling, D., Guschin, D., Muller, M., Silvennoinen, O., Witthuhn, B. A., Quelle, F. W., Rogers, N. C., Schindler, C., Stark, G. R., Ihle, J. N., and Kerr, I. M., 1993, Complementation by the protein-tyrosine kinase Jak2 of a mutant-cell line defective in the interferon-gamma signal-transduction pathway, *Nature* **366:**166–170.

Webernordt, R. M., Greenlund, A. C., Darnell, J. E., and Schreiber, R. D., 1994, IL-10 induces STAT1 and STAT3 activation in cells, *Blood* **84:**A223.

Witthuhn, B. A., Quelle, F. W., Silvennoinen, O., Yi, T. L., Tang, B., Miura, O., and Ihle, J. N., 1993, Jak2 associates with the erythropoietin receptor and is tyrosine-phosphorylated and activated following stimulation with erythropoietin, *Cell* **74:**227–236.

Witthuhn, B. A., Silvennoinen, O., Miura, O., Lai, K. S., Cwik, C., Liu, E. T., and Ihle, J. N., 1994, Involvement of the Jak-3 Janus kinase in signaling by interleukin-2 and interleukin-4 in lymphoid and myeloid cells, *Nature* **370:**153–157.

Zeng, Y. X., Takahashi, H., Shibata, M., and Hirokawa, K., 1994, Jak3 Janus kinase is involved in interleukin-7 signal pathway, *FEBS Lett.* **353:**289–293.

Chapter 9

Hematopoietic Growth Factors Involved in B-Cell Development

Ana Cumano, Barbara L. Kee, Isabelle Godin,
Françoise Dieterlen-Lièvre, and C. J. Paige

1. INTRODUCTION

B lymphocytes are continuously generated throughout life. Like all other members of the hematopoietic system, B cells are derived from multipotent hematopoietic stem cells (Wu *et al.*, 1967). The developmental pathway that leads from multipotent stem cells to committed B lymphocytes is characterized by a series of differentiation steps. Many of these steps can be recognized based on the appearance of proteins such as growth factor receptors or the B-lineage-specific molecules that allow or promote the further development of progenitor cells. These stages of differentiation can also be recognized based on functional assays that have been developed over the last 30 years. Initially, such assay systems relied on the ability of progenitor cells to repopulate the hematopoietic system of a mouse that had previously been subjected to high doses of ionizing radiation. This approach has been crucial for the identification of multipotent stem cells and defining cells, which have long-term reconstituting potential (Dick *et al.*, 1985; Keller *et al.*, 1985; Lemischka *et al.*, 1986). Irradiation/reconstitution experiments have been less useful for identifying intermediate stages in the developmental process. This method is also inadequate for studying either the essential cellular interactions or the growth and differentiation factors that promote hematopoiesis.

Ana Cumano Unité de Biologie Moléculaire du Gène, Institut Pasteur, Paris 75724, France. **Barbara L. Kee and C. J. Paige** The Wellesley Hospital Research Institute, Toronto, Ontario M4Y 1J3, Canada. **Isabelle Godin and Françoise Dieterlen-Lièvre** Institut d'Embryologie du CNRS et du Collège de France, Nogent-sur-Marne, France.

Blood Cell Biochemistry, Volume 7: Hematopoietic Cell Growth Factors and Their Receptors, edited by A. D. Whetton and J. Gordon. Plenum Press, New York, 1996.

Fortunately, parallel progress in the development of *in vitro* culture systems has allowed these questions to be addressed.

In the last few years, we and others (Cumano *et al.*, 1992a; Hirayama *et al.*, 1992; Kee *et al.*, 1994; Godin *et al.*, 1995) have defined *in vitro* culture conditions that support lymphocyte differentiation from uncommitted multipotent hematopoietic precursors. These assays are modifications of single-cell cloning assays developed in the 1970s, for the detection of various progenitor cells. To ensure clonality, such assays rely on either a semisolid medium, such as agar (Paige, 1983) or methylcellulose, or on limiting dilution in 96-well liquid culture plates (Rolink *et al.*, 1991; Cumano and Paige, 1992). Our current method is based on a liquid culture assay that, in limiting dilution conditions or after single-cell manipulation, allows us to quantify precursor cells both at the initiation of culture and as the cultures progress. This strategy also permits us to alter the culture conditions to meet the changing requirements of progenitor cells as they progress through different developmental stages (Cumano *et al.*, 1992a,b; Kee *et al.*, 1994).

In this review we describe recent advances in the development of *in vitro* culture systems that allow the differentiation of immunoglobulin-secreting plasma cells from uncommitted hematopoietic stem cells isolated from different sites during ontogeny. We discuss the growth factor requirements and differentiation potentials of these precursor cells.

2. OVERVIEW OF B-CELL DEVELOPMENT

Hematopoietic precursors are first detectable in the yolk sac at day 7.5 of gestation (Moore and Metcalf, 1970). Later, at the 10- to 12-somite stage (day 8.5 of gestation), they are found in the paraaortic splanchnopleura and tissues derived from this site (Godin *et al.*, 1993; Medvinsky *et al.*, 1993). By day 10 of gestation, the fetal liver becomes a principal site of hematopoiesis. In this organ, around day 12–13 of gestation, the first cells committed to the B lineage appear (Raff *et al.*, 1976; Kamps *et al.*, 1982). The liver remains a site of B-cell production throughout gestation and for a few weeks after birth. Subsequently, the adult bone marrow becomes the only major site of B lymphopoiesis. B cells play a central role in the immune response, principally by producing antibodies. The process of antibody formation is itself a fascinating subject about which numerous reviews have been written (see below). Moreover, monitoring the emergence of antibody molecules has also proven useful for tracking the progression of cells through the B-cell developmental pathway.

2.1. Immunoglobulin Gene Rearrangement

Immunoglobulin (Ig) molecules on the surface of B lymphocytes are formed by the association of two heavy and two light polypeptide chains. The diversity in the heavy chain is initially achieved by the somatic recombination of three gene segments that constitute the variable region of the Ig molecule: the variable (V), the diversity (D), and the joining (J) segments. There are approximately 100–500 variable, 12 diversity, and four joining genes in the heavy chain locus in the mouse (Tonegawa, 1983). The light chain is encoded in two separated loci: the κ and the λ locus. In both light-chain loci rearrangement occurs between a V and a J segment. There are around 100 variable genes in the κ locus and three functional variable genes in the λ locus. Potential diversity is therefore more restricted in the λ than in the κ light-chain. In an adult mouse, 95% of circulating immunoglobulins express the κ light-chain.

Each individual B lymphocyte expresses only one heavy and one light chain gene product, although in most precursor cells, rearrangements of both chromosomes do occur; this phenomenon is called allelic exclusion.

The process of Ig rearrangement generally follows a predictable pattern during B-cell development. Usually, rearrangement starts at the heavy-chain locus by recombination of a D segment to a J segment (a D–J join) (Alt *et al.*, 1984). Typically, both chromosomes undergo D–J joining. Subsequently, a V segment combines with an existing D–J joint to form the immunoglobulin variable region. In the process of recombination, exonuclease activity can delete nucleotides in both ends of the recombination product. Nucleotides can also be inserted (N sequences and P nucleotides) by the terminal deoxynucleotidyl-transferase (TdT) (Alt and Baltimore, 1982) or by complementarity to both ends of the coding strand (Lafaille *et al.*, 1989). These mechanisms confer additional variability to the recombined product. Light-chain rearrangement usually starts after heavy-chain rearrangement is completed. There is, therefore, a well-established order of events in the rearrangement process that, although not absolute, is followed in most B-cell precursors.

Despite its potential to create more than one heavy chain and one light chain, a single B cell generally expresses only one of each, a process called allelic and isotypic exclusion. The molecular basis for allelic exclusion is not yet clear. Probabilistic models of allelic exclusion suggest that the likelihood of two independent productively rearranged heavy-chain genes is relatively low (Cohn and Langman, 1990). However, given the high output of B cells produced each day (estimated to be 5×10^7 per day in the mouse), this model might well allow the emergence of rare B lymphocytes expressing more that one immunoglobulin specificity on the cell surface. Regulatory models have therefore been evoked to explain the apparent absence of such cells. These models also postulate a role for Ig-associated proteins in the developmental process. One model postulates that, as soon as a precursor cell has achieved one productive heavy-chain rearrangement, it expresses the protein on the cell surface together with two invariant chains (λ5 and V-pre-B). The immediate consequence of this expression may be to stop further rearrangement at the locus (Reth *et al.*, 1987). After light-chain rearrangement has produced an appropriate Ig light chain, surface expression of a complete Ig might prevent further rearrangement in this cell.

Although the role of productive Ig rearrangement in the assembly process itself remains controversial, there are certainly interesting experiments suggesting a role for Ig products in the developmental process of lymphopoiesis. Support for this hypothesis comes from the analysis of mice in which the transmembrane part of the immunoglobulin M has been mutated by homologous recombination in embryonic stem cells (μMT) (Kitamura *et al.*, 1991). These mice have a complete block in B-cell development, showing that surface expression of Ig is essential, after V–D–J rearrangement, for progression to the mature stage. Interestingly, allelic exclusion is incomplete in F_1 mice carrying one mutated and one wild-type chromosome, showing that surface expression of Ig is involved in this process (Kitamura and Rajewsky, 1992). Likewise, mice carrying a disrupted λ5 gene show a deficiency in B lymphopoiesis (Kitamura *et al.*, 1992). The precursor population is blocked at the transition from D–J to V–D–J rearrangement (Ehlich *et al.*, 1993). However, unlike μMT −/− mice, λ5 −/− mice have some mature B cells in the spleen. These results shows that although the absence of immunoglobulin heavy-chain leads to an absolute lack of B cells, the absence of λ5 can probably be overcome by some other chain, possibly a dimer of V-pre-B.

After productive rearrangement of heavy and light chains, the developing B cell will express this gene product (IgM) on the cell surface together with the associated receptor molecules, namely, mb1 and B29 (Hombach *et al.*, 1990). The latter are critical components that mediate signal transduction after Ig cross-linking. At this stage of differentiation, the cell has acquired the ability to respond to exogenous stimuli. Mitogens or polyclonal activators for B cells can be used to induce *in vitro* stimulation of these cells.

Lipopolyssacharide (LPS) is one of the most effective mitogens that can promote, with high efficiency, proliferation and differentiation of mature B cells into immunoglobulin-secreting plasma cells.

2.2. Surface Antigens Expressed by B-Cell Precursors at Different Stages of Development

Additional cell surface proteins have proven useful for following B-cell development. In some cases, such as growth factor receptors, their expression is known to be developmentally regulated and to coincide with the stages of B-cell development in which extensive proliferation occurs. In other cases, however, the ligand is unknown or does not correlate with any particular function. Their expression can only be correlated with a developmental stage. Nonetheless, the expression pattern is useful for *in vitro* enrichment.

2.2.1. B220

B220 (CD45R) is an isoform of CD45, a membrane-associated phosphatase, expressed on all B-lineage cells (except for plasma cells) (Kincade *et al.*, 1981) and on a small subpopulation of T cells. It first becomes detectable at the stage of the initial D–J rearrangements. In the adult bone marrow, approximately 20% of the cells are B220$^+$, half of which coexpress Ig on the surface, marking them as B lymphocytes (Figure 1).

2.2.2. CD43

CD43 is a 90- to 120-kDa molecule expressed on the surface of multiple myeloid and lymphoid cells. Pre-B cells in the bone marrow can be partitioned into two populations based on this marker (Hardy *et al.*, 1991). The majority of pre-B cells are CD43 negative, have undergone V–D–J rearrangement in the heavy-chain locus, express cytoplasmic Ig, and are progressing to mature B cells. A small percentage (3–5%), however, express CD43 and are at an earlier stage of differentiation during which cells progress from D–J to V–D–J rearrangements (Figure 1).

2.2.3. BP-1/6C3

BP-1/6C3 is a surface antigen belonging to family of aminopeptidase A (Wu *et al.*, 1990, 1991). The surface expression of this molecule is detectable in a fraction of CD43$^+$

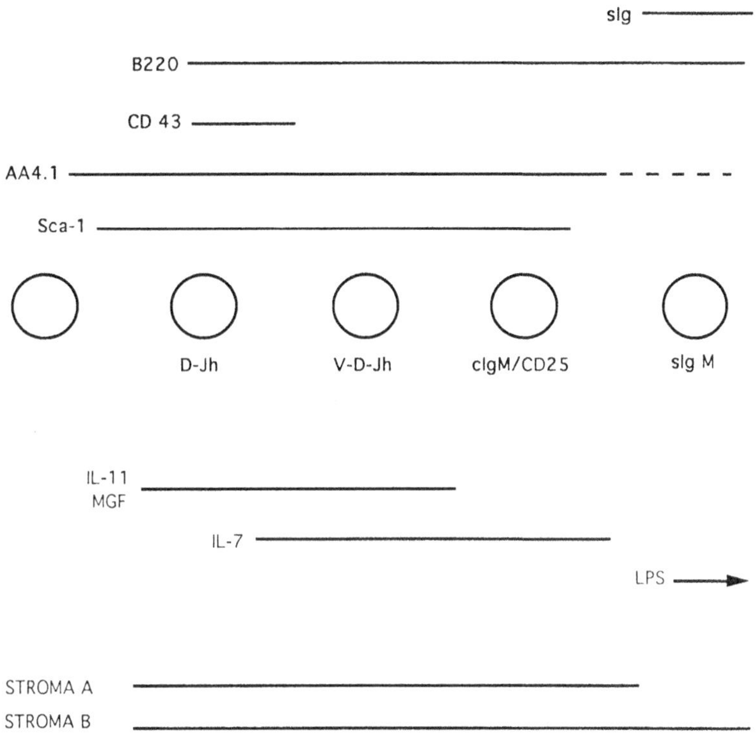

FIGURE 1. B-cell development according to surface markers and growth factor requirements. Scheme of B-cell development from the stage of unrearranged multipotent hematopoietic cell until mature B lymphocyte. Surface markers, growth factor requirements, and stromal cell effects are indicated.

$B220^+$ cells and remains until surface Ig starts to be expressed. The first $BP\text{-}1/6C3^+$ cells are thus detected after B220 expression (Hardy *et al.*, 1991).

2.2.4. CD25

CD25 (IL-2 receptor α; see Chapter 4) is expressed on large cycling $B220^+CD43^+$ and small $B220^+CD43^-$ resting progenitors that do not express *c-kit* (Rolink *et al.*, 1994; Chen *et al.*, 1994; see Chapter 5). Its expression coincides with that of cytoplasmic heavy-chain (cIg).

2.2.5. Sca-1

Sca-1 (Ly6 A) is present on all multipotent hematopoietic stem cells in the bone marrow and in fetal liver as well as in multiple myeloid and lymphoid precursors (Spangrude *et al.*, 1988; Uchida and Weissman, 1992). This marker delineates a population of uncommitted B-cell precursors in fetal liver at day 12 of gestation that can generate B cells and macrophages but not other myeloid cells (Cumano *et al.*, 1992a). Sca-1 is still expressed on pre-B cells that expand in response to IL-7 and kit-ligand *in vitro*.

2.2.6. AA4.1

AA4.1 is a cell surface protein, yet undefined, that is recognized by a monoclonal antibody (AA4.1mAb) (Jordan *et al.*, 1990). It is present in multipotent hematopoietic stem cells in fetal liver but not in bone marrow (Trevisan and Iscove, 1995) and is the first hematopoietic marker detected during ontogeny (Cumano *et al.*, 1993). It is present on multipotent precursors in the yolk sac and in the paraaortic splanchnopleura beginning at the 15 to 25 somite stage (Godin *et al.*, 1995). Expression of AA4 precedes the expression of Sca-1 marker in ontogeny.

3. AN *IN VITRO* ASSAY SYSTEM THAT PROVIDES CONDITIONS FOR MYELOID AND LYMPHOID CELL DEVELOPMENT FROM SINGLE PRECURSORS

In vitro assays that provide the conditions for B-cell precursors to differentiate to immunoglobulin-secreting plasma cells have been established. These assay systems have revealed many aspects of B-cell development, including (1) the definition of growth factors involved in this process; (2) the quantification of B-cell precursors during embryonic development; (3) the isolation of intermediate stages of differentiation; and (4) establishing the pattern of differential gene expression during the developmental process. There are critical issues, however, that remain unresolved: the nature and site of emergence of the first committed B-cell precursors in embryonic development; the developmental potential of the immediate precursor of a committed B-lineage progenitor; the precise growth factor requirements that promote this transition; the molecular events necessary for B-cell commitment; and the selective mechanisms that operate on pre-B cells in the bone marrow and fetal liver. We discuss some of these problems and the strategies we used to approach them.

3.1. Enrichment and Characterization of B-Cell Progenitors

Because of the low frequency of progenitor cells, it was necessary to find ways to enrich for B-cell precursors. We showed that the AA4.1$^+$ population contains all clonable pre-B cells in mouse fetal liver (Cumano and Paige, 1992). Fetal liver contains a large fraction of committed erythroid cells as well as hepatocyte precursors. B-cell progenitors constitute only 0.5–5% of fetal liver cells, depending on the stage of gestation. Therefore, we used two surface markers to enrich for B-cell precursors in fetal liver: the B220 marker, which is expressed on committed B lineage cells, and the AA4.1 marker, which recognizes multilineage as well as committed B-cell precursors. AA4.1$^+$ fetal liver cells were separated into B220$^+$ and B220$^-$ cells and plated under limiting dilution conditions with S17 stromal cells and Interleukin 7 (IL-7). Clones are allowed to expand and are tested for B-cell progenitors by LPS stimulation followed by detection of Ig secretion (Cumano and Paige, 1992).

At day 16 of gestation the majority of B-cell progenitors express B220. In limiting dilution experiments, one in three of these cells generate clones of B-lineage cells (Table I).

Table I
Numbers of B-Cell Precursors in Enriched Populations of Bone Marrow and Fetal Liver[a]

Cell phenotype	Percentage of cells recovered after enrichment	Number of B-cell precursors/ organ (frequency)	Heavy-chain uncommitted (%)
A BM	10	173,611 (1:27)	9
$B220^+$			
$AA4.1^+$ $B220^-$	0.71	303 (1:110)	89
D 16 FL	5.12	303,000 (1:3)	47
$B220^+$			
$AA4.1^+$ $B220^-$	2.1	47,000 (1:8)	81
D 12 FL	0.6	197 (1:125)	87
$B220^+$			
$AA4.1^+$ $B220^-$	1.5	1,320 (1:42)	89

[a]Fetal liver and bone marrow cells were enriched by panning for cells expressing the B220 or the AA4.1 surface markers. Cells were then cultured under limiting dilution conditions at 60, 20 and 6 cells per well of 96 well plates. For day 16 FL $B220^+$ cells the concentrations were 10, 3 and 1 cells per well and for bone marrow $AA4.1^+$, $B220^-$ cells 180, 60 and 20 cells per well were plated. Cultures were checked by microscope inspection after 5–7 days and single clones of lymphoid appearance identified. Clones were transferred at different time points into S17 cells and LPS. Immunoglobulin secretion was tested 12 days later, by ELISA as previously described (Cumano and Paige, 1992) 20 to 25 individual clones that secreted Ig were tested for allotype expression. For bone marrow, total numbers of precursors correspond to cells isolated from two femurs.

By contrast, the $B220^-AA4.1^+$ cells were less efficient in the clonable pre-B-cell assay. This result is not surprising because AA4.1 marks both myeloid and lymphoid progenitors. When the total numbers of clonable precursors are calculated, $B220^+$ cells contribute to 86% of the 3.5×10^5 fetal liver B-lineage precursors found at this stage of gestation.

In contrast to day 16, fetal liver at day 12 of gestation contains more $B220^-AA4.1^+$ than $B220^+$ precursors. Only 13% of B-cell precursors detected at this stage express B220. There is a more than 100-fold increase in the total precursor numbers detected per fetal liver at day 16 as compared to day 12 of gestation. This increase is predominantly observed in the $B220^+$ compartment (1500-fold increase).

These progenitor cells could be distinguished based on other criteria as well. For example, the B-cell progenitors derived from 12-day fetal liver of Balb/c × C57BL/6 F_1 embryos are uncommitted for Ig heavy-chain allotype expression, because Igh^a and Igh^b are detected by specific mAbs in the supernatant of LPS-activated clones (Sauter and Paige, 1988). Thus, single progenitors gave rise to mature progeny, some of which had undergone productive rearrangement at one allele while others utilized the second allele. By day 16 of gestation, nearly half of the progenitor cells were already allotype restricted. Another developmental difference was found when the interval between the initial cloning step and the emergence of mitogen reactivity was measured. $B220^-$ cells required 8 days in culture before they became LPS reactive. $B220^+$ cells required only 5 days in culture to achieve the same stage (Figure 2).

Most B-cell precursors found in adult bone marrow also express the marker B220. These cells can be isolated and cultured in limiting dilution assays. Clones are thus generated that contain mononuclear cells that subsequently develop the capacity to secrete

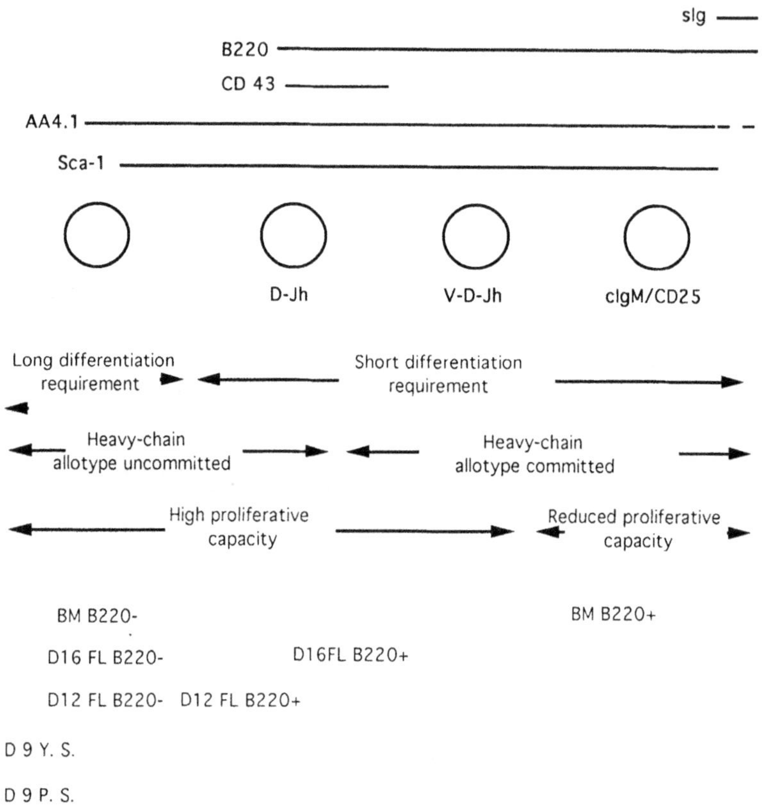

FIGURE 2. B-cell development according to immunoglobulin gene rearrangement and *in vitro* properties. Scheme of B cell development indicating the phenotype and major properties *in vitro* of precursors found in fetal liver, bone marrow, and in day 9 embryos.

immunoglobulin when cultured in the presence of LPS and stromal cells. B-cell progenitors are also found among the $B220^-AA4.1^+$ cells (Table I). They represent a minority (0.2%) of B-cell precursors found in this organ and are probably more immature B-cell precursors. For example, in bone marrow, more than 90% of clonable B-cell precursors are heavy-chain allotype committed (Figure 2), and they are found in the $B220^+$ population. Most $B220^-$ cells are allotype uncommitted. Another major difference between $B220^+$ and $B220^-$ bone marrow cells is found when we measured the capacity to respond to LPS at different time points after culture (Figure 2). Clones from $B220^-$ precursors require 3 days longer than $B220^+$ precursors before they are capable of responding to LPS. While $B220^+$ and $B220^-$ fetal liver-derived clones continue to proliferate and maintain LPS responsivess after 11–12 days in culture, most cells in bone marrow $B220^+$ derived clones die after 9 days and can no longer respond to LPS at day 11 of culture (Figure 2).

From the results described above, we conclude that B lymphopoiesis starts in fetal liver around day 12 of gestation, with the appearance of the first $B220^+$ cells and the first detectable immunoglobulin rearrangements. At this stage, 90% of B-cell precursors are

$B220^-$, heavy-chain Ig uncommitted, and they require more than 8 days in culture to become LPS responsive. Around 10% of clonable precursors are already more advanced in the B-lineage pathway: they express B220 and differentiate within 5 days in culture to LPS-responsive cells; however, they are mostly heavy-chain allotype uncommitted (Figure 3).

As gestation progresses, precursors expand extensively and differentiate from more primitive hematopoietic precursors. At day 16 of gestation, B-lineage-committed $B220^+$ cells (now 85% of total precursors detected) have increased more than three orders of magnitude, while $B220^-$ cells (15%) have only increased 35 times. Half of them are still heavy-chain uncommitted and maintain high proliferative capacity *in vitro* as pre-B cells. A more advanced $B220^+$, heavy-chain-committed precursor now represents 40% of total B-cell precursors (Figure 3).

3.2. Detection of Bipotent Macrophage/B-Cell Precursors in Fetal Liver

$AA4.1^+B220^-$ cells comprise 90% of total B-cell precursors detected in fetal liver at day 12 of gestation. Depletion of $B220^+$ cells and $Mac1^+$ cells (a marker for myeloid precursors and macrophages) revealed the presence of a cell type that developed into macrophages and pre-B cells in the culture conditions previously described. We further enriched for cells with this property using a selection for the Ly6A marker (Sca-1) (Cumano *et al.*, 1992a).

One in three to one in seven $AA4.1^+B220^-/Mac1^-Ly6A^+$ cells could differentiate into macrophages and B lymphocytes but not other myeloid cells. These precursors were incapable of generating other myeloid cells under conditions in which multilineage myeloid precursors could do so. For example, the addition of interleukin 3 (IL-3), SCF, and erythropoietin to S17 cells and IL-7 increased the frequency of myeloid clones detected but failed to change the frequency of macrophage/B-cell clones.

Examination of the macrophages that were derived from single macrophage/B-cell precursors revealed that their Ig loci were in germ line configuration. This demonstrated that the clonogenic cell itself must also have its Ig locus in germ-line configuration. As these progenitors differentiate along the B-cell pathway, their progeny undergo rearrangement of heavy and light chains during the culture period. They differentiate into mature B cells that secrete immunoglobulin when stimulated with LPS. Macrophage cells resemble macrophages obtained from myeloid colonies: they express high levels of lysozyme and c-fms (the receptor for the macrophage growth factor M-CSF), and they stain positively for naphthol AS-D chloroacetate esterase.

The existence of macrophage/B-cell precursors in normal bone marrow has been suggested, based on the growth of both lineages from limited numbers of cells. However, we have been unable to enrich bone marrow cells sufficiently to undertake the single-cell cloning experiments required for proof of their existence. Therefore, although it is possible that they exist in adult bone marrow, their absolute and relative representation could be quite small.

The bipotent nature of the cell type we described is intriguing because it is not compatible with most schemes of hematopoietic development. Experiments are in progress to determine whether these cells also are able to give rise to T cells. Whether or not the bipotential cell is a lymphoid/macrophage progenitor or a B lymphoid/macrophage progenitor, the data suggest a lymphoid progenitor that retains a highly restricted myeloid differen-

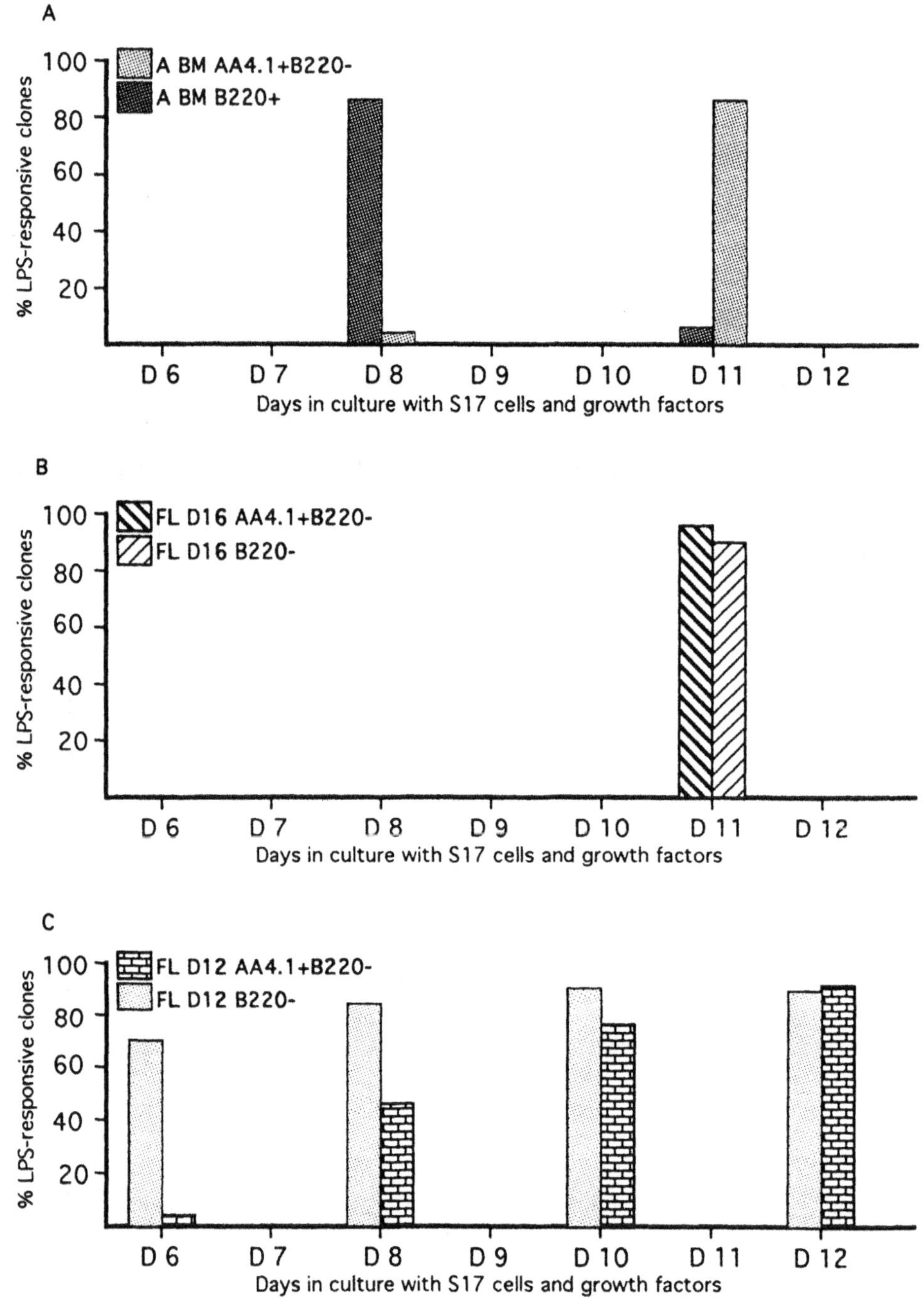

FIGURE 3. Development of LPS-reactive clones from bone marrow and fetal liver cells. Precursors were isolated as described in Table I. 20–40 individual clones were transferred into LPS conditions after 6 to 12 days from the beginning of the precursor cultures. Shown is the percentage of clones responsive to LPS as detected by ELISA. $B220^+$ BM cells generated clones that survived for less than 11 days in culture. By day 11 clones contained a large proportion of dead cells and could no longer respond to LPS, cell death was observed between days 9 and 10.

tiative capacity. It is conceivable that transition to the lymphoid differentiation pathway is not determined by a single event but by a succession of steps that sequentially restrict the differentiation options of a multipotent precursor. If this hypothesis is correct, macrophage differentiation would be the last potential to be lost, maybe because its own commitment program is most compatible with lymphoid or any other pathway of differentiation. Compatible with this idea, a few transformed pre-B-cell lines exist that can generate macrophages *in vitro*.

4. ROLE OF SOLUBLE MEDIATORS AND DIFFERENT STROMAL CELLS IN THE DIFFERENTIATION OF UNCOMMITTED B-CELL PRECURSORS

The establishment of a culture system that supports uncommitted B-cell precursors to differentiate into multilineage hematopoietic cells allows us to examine the role of stromal cells and exogenous growth factors in lymphopoiesis (Kee *et al.* 1994).

4.1. Stromal Cells

The term "stromal cell" is used to designate a heterogeneous group of adherent cells, which include fibroblasts, adventitial reticular cells, and epithelial cells (Collins and Dorshkind, 1987; Hunt *et al.*, 1987; Whitlock *et al.*, 1987; Nishikawa *et al.*, 1988; Pietrangelli *et al.*, 1988). As mentioned before, all *in vitro* culture systems that support B-cell precursor proliferation and differentiation into mature cells that have been developed in the past relied on the presence of stromal cell components. Generally, these were the plastic adherent fraction obtained from fresh cell suspensions of lymphoid organs such as bone marrow, fetal liver, spleen, or thymus. This heterogeneous cell population was later substituted by cell lines.

The mechanism by which stromal cells interact with the hematopoietic precursors is not understood. One possible mechanism of interaction is certainly the production of growth factors. Stromal cells are able to make various cytokines involved in hematopoietic cell differentiation such as IL-7, IL-11, SCF, M-CSF, GM-CSF, and IL-1. On the other hand, *in vitro* lymphopoiesis can be blocked by monoclonal antibodies that inhibit direct interaction between stromal cells and lymphoid cells (Miyake *et al.*, 1990). Direct contact with adhesion molecules on the surface of stromal cells might therefore deliver a direct signal to B-cell precursors (Miyake *et al.*, 1991). We emphasize, however, that from the stage of bipotent macrophage/B-cell precursor until the stage of plasma cell, B-cell differentiation *in vitro* has been obtained with comparable efficiency under condition where no direct contact between stromal cells and lymphoid progenitors occurs. It is known that growth factors are produced locally and that they can be isolated from the extracellular matrix or from the cell surface of stromal cells. It is therefore possible that direct cell interaction is required *in vivo* to guarantee access to growth factors but not *in vitro*, where excess soluble mediators are supplied.

Stromal cells also express high levels of peptidases such as BP-1/6C3, CD10/CALLA, and CD13, which might be responsible for the fast degradation of interleukins produced by stromal cells; this may constitute an important mechanism to regulate hematopoiesis (Witlock *et al.*, 1987; Kee *et al.*, 1992).

Heterogeneity of stromal cells themselves has been found and is the subject of much current interest. Cloned stromal cell lines have been characterized and found to produce different growth factor combinations. They also differ in their ability to support different stages of B-cell differentiation. It is not clear whether these differences mark distinct subsets of stromal cells or are merely the result of variation introduced during isolation procedures *in vitro*. Whether or not different kinds of stromal cells *in vivo* have functionally distinct roles, they serve *in vitro* to help distinguish distinct stages of B-cell development.

We defined two main types of stromal cells based on the capacity to support different stages of B-cell differentiation. Stroma A (Figure 1) is capable of supporting differentiation of uncommitted B-cell precursors into pre-B cells that respond to IL-7. Several stromal cells tested were capable of supporting this phase, namely, NIH3T3 embryonic fibroblasts, BMS 2.2, 95/1.7, S10, and S17 (Cumano *et al.*, 1994). Moreover, it was shown that the role of stroma A in the transition from bipotent precursor to B-lineage-committed cell can be replaced by IL-11 and SCF. It is interesting to note that the attempt to support development of precursor cells at earlier stages of differentiation than macrophage/B-cell precursors in stromal free conditions was unsuccessful, suggesting that other growth factors or direct cell interaction are required for differentiation of more immature cell types.

Later in B-cell development, stroma B (Figure 1) is required to support the transition from the IL-7-responsive cells to LPS-responding cells (Cumano *et al.*, 1990). S17 and S10 were the only tested cell lines capable of supporting this transition in the absence of direct contact. This was achieved by physically separating precursors and stromal cells in an agar matrix. However, in liquid cultures, where cell contact is unavoidable, stroma-A-type cells were also effective in supporting this transition of *in vitro* B-cell differentiation. It is evident from these results that no direct contact between stromal cells and precursors is strictly required for this transition. On the other hand, attempts to replace stromal cells by defined growth factors alone or in combination were unsuccessful. The differences between stromal cells could relate to different rates in production and degradation of yet unidentified growth factors in individual lines. Close contact with stromal cells could also induce the production of some soluble mediator that would be constitutively produced in other lines. Evidence for this possibility has been documented.

4.2. Growth Factors

Recently, we established culture conditions in which the S17 stromal cell layer can be replaced by IL-11 and SCF for the bipotent macrophage/B-cell precursor differentiation (Kee *et al.*, 1994). These culture conditions allowed the characterization of growth requirements for this precursor cell and will provide further insight in the role of interleukins in lineage commitment and gene expression.

4.2.1. Interleukin 7

In 1988 a novel growth factor, interleukin 7 (IL-7), was isolated (Namen *et al.*, 1988). This 25-kDa molecule induces proliferation of bone marrow and fetal-liver-derived pre-B cells. Addition of IL-7 to *in vitro* cultures containing stromal cells and B-cell progenitors significantly increases the proliferation of B-lineage cells, suggesting that it is either produced in limiting amounts by stromal cells or that it is readily degraded (Lee *et al.*, 1989).

The first IL-7-responsive cells can be detected around day 13 of gestation in fetal liver. They express the B220 surface marker, although the presence of this molecule is not required for induction of proliferation.

During expansion in response to IL-7, precursors rearrange the immunoglobulin heavy-chain genes and become allotype committed (Cumano and Paige, 1992). Whether IL-7 plays a direct role in this process or merely provides an opportunity for it to occur by keeping the progenitor cells alive is still unknown. It has been suggested that IL-7 may induce rearrangement of TCR-expressing T cells. Those experiments are, however, difficult to interpret because IL-7 cannot be replaced by other interleukins to support the survival of lymphocyte precursors *in vitro*. An alternative explanation is that once a hematopoietic precursor enters the pathway of B-lineage differentiation, an internal program is set in motion leading to a mature B cell. Survival and possibly a certain number of rounds of cell division are the only requirements for progression of differentiation. The role of IL-7 would be to provide these requirements. As noted previously, B-cell precursors have a limited capacity to proliferate to IL-7. This may be related to maturational events that occur during IL-7-dependent expansion leading to the loss of IL-7 responsiveness itself.

Interleukin-7 supports the proliferation of pre-B lymphocytes but does not promote the transition to LPS responsiveness (Cumano *et al.*, 1990). To differentiate further to the mature LPS-responding stage, IL-7-responsive pre-B cells require the presence of stroma B as defined in Figure 1. After 18–21 hr, maximal frequencies of LPS-responding cells can be obtained. This step of differentiation seems, in contrast to the previous one, independent of cell division.

Interleukin-7 not only is important in B-cell development but also plays a fundamental role in T-cell development; in fact, the thymus is the organ where IL-7 mRNA is most abundant. The role of IL-7 in T- and B-cell development has been studied extensively *in vitro* but also *in vivo*. Elevation of IL-7 levels *in vivo* in IL-7 transgenic mice resulted in increased numbers of T and B lymphocytes and a high incidence of lymphomas (Samaridis *et al.*, 1991; Fisher *et al.*, 1992). Depletion of IL-7 by injections with anti-IL-7 specific mAbs (Grabstein *et al.*, 1993; Sudo *et al.*, 1993) or by inactivation of the high-affinity IL-7 receptor (Peschon *et al.*, 1994) resulted in a decline in peripheral T and B cells. These experiments confirm the fundamental role of IL-7 in lymphocyte precursor proliferation.

4.2.2. Stem Cell Factor

The characterization of soluble mediators involved in hematopoiesis has been often related to the identification of an underlying defect in mutant mice. This was the case for the W and Sl mutations. Both viable forms of these mutations (W/Wv and Sl/Sld) express identical phenotype, characterised by severe anemia, reduced numbers of mast cells, and poor erythropoiesis. The W mutation was later identified as a group of alterations in the *c-kit* protooncogene, expressed on the membrane of most hematopoietic precursors (Chabot *et al.*, 1988; Geissler *et al.*, 1988). It was soon clear that c-kit, a tyrosine kinase, functions as a growth factor receptor and that Sl mice carried a mutation in the ligand for c-kit. The c-kit ligand, also called stem cell factor (SCF), mast cell growth factor, or Steel factor, is a protein of 25 kDa present in soluble form, as a dimer, or membrane bound, and it is expressed in most stromal cell lines from hematopoietic organs (Anderson *et al.*, 1990; Zsebo *et al.*, 1990).

The c-kit ligand (SCF), in conjunction with IL-11, can substitute for the requirements for stromal cells and allow the *in vitro* development of uncommitted B-cell precursors (Figure 4) (Kee *et al.*, 1994). It synergizes with IL-7 to increase the frequency of clonable pre-B cells (Billips *et al.*, 1992). Although in a day-14 fetal liver the increase is around fourfold, a dramatic effect is observed in day-12 fetal liver, where no IL-7-responsive cells can be detected. An exogenous supply of SCF results in 50–100 clonable B-cell precursors per fetal liver. The effect of SCF on B-cell development is still unclear. We favor the hypothesis that the major role of SCF is to support cells at the earliest stages of differentiation, including committed and uncommitted progenitors. This hypothesis is based on the observation that the cells that clone in SCF and IL-7 develop, during ontogeny, 24 hr before

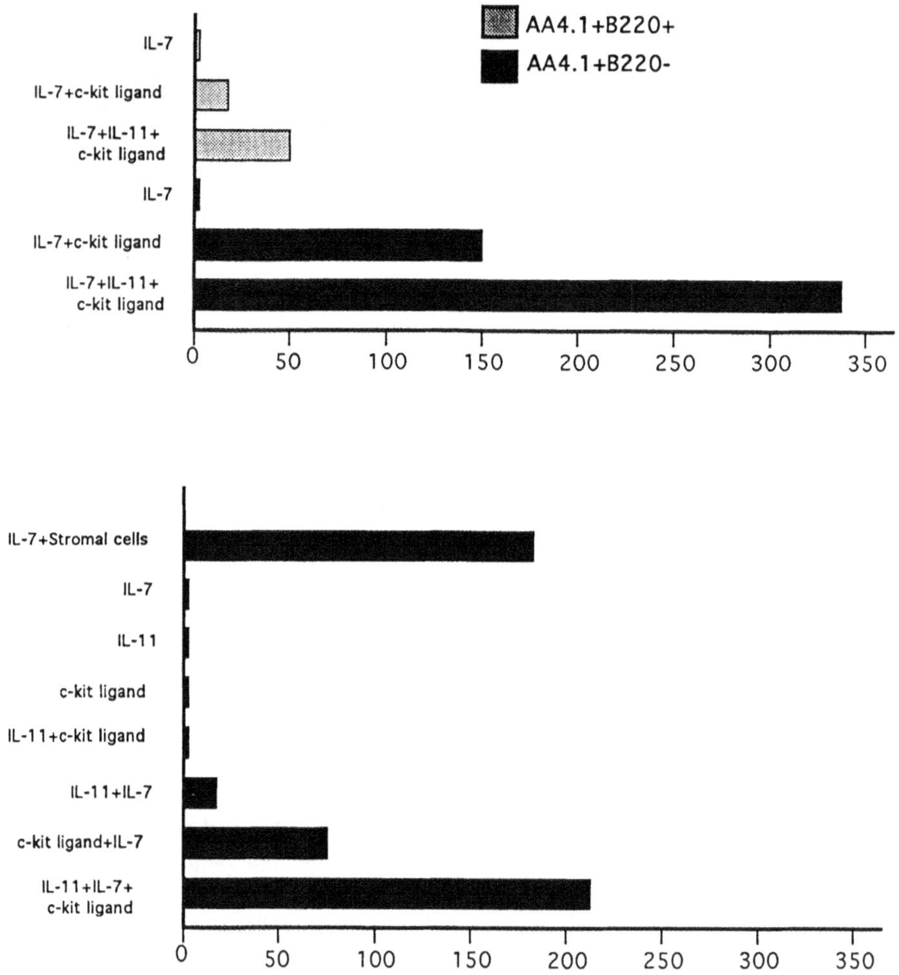

FIGURE 4. Growth requirements of precursors isolated from day 12 fetal liver. Cells were enriched for the expression of AA4.1 and B220 and cloned onto different conditions. Shown is the number of B-cell containing clones per fetal liver, as ascertained by the capacity to secrete Ig in response to LPS.

cells that clone in IL-7 alone. When SCF is added to cultures of IL-7-responsive pre-B cells, a twofold increase in thymidine incorporation is observed, suggesting that it can synergize with IL-7 to induce proliferation. It is also possible, however, that it promotes survival of precursors during commitment to the B-lineage pathway, allowing them to acquire the capacity to respond to IL-7. It is clear that SCF acts on lineage-unrestricted progenitors because the development of B lymphocytes from bipotent B-cell/macrophage precursors is dependent on both IL-11 and SCF at the initiation of the culture (Kee *et al.*, 1994).

W/Wv mice have close to normal numbers of B-cell precursors in the bone marrow. This indicates that SCF is not essential for B-cell development. Other stromal growth factors might replace this factor in the interaction with B-cell precursors.

4.2.3. Interleukin 11

Interleukin 11 is a growth factor produced by stromal cells. It has many functional similarities with interleukin 6 (IL-6), and it was identified as a growth factor capable of supporting proliferation of a plasmacytoma line in the presence of antibodies directed against IL-6 (Paul *et al.*, 1990). Similar to IL-6, it also has pleiotropic effects in hematopoietic precursors, and it appears to influence the cycling stage of resting hematopoietic stem cells. Alone or in combination with IL-7, it has no known effect on committed B-cell precursors that proliferate in response to IL-7. However, in combination with c-kit ligand (SCF) and IL-7, it supports development of macrophage/B-cell precursors isolated from day-12 fetal liver (Figure 4) (Kee *et al.*, 1994). The cloning efficiency of this factor combination is comparable to that obtained with stromal cells. Moreover, the properties of macrophages and B cells derived from these cultures are, in all tested parameters, similar to those of cells derived from stroma cultures.

4.2.4. Other Factors Involved in B-Cell Differentiation

Recently, a 10-kDa molecule produced by stromal cells has been identified; this molecule, insulin-like growth factor 1 (IGF-1), can be produced in a soluble form or together with IGF binding proteins (Landreth *et al.*, 1992). It potentiates the differentiation of B-cell precursors in cIg^+ pre-B cells and synergizes with IL-7 in promoting pre-B-cell proliferation.

Another interleukin that has been involved in B-cell development is interleukin 3 (IL-3). This effect has been controversial. Although IL-3-dependent cell lines of B-cell precursors have been reported, other investigators failed to observe a direct effect of IL-3 in B-cell development. We analyzed IL-3 in colony assays and concluded that the effect on B-cell development was indirect. Although murine IL-3 progenitor assays failed to show a direct role of this growth factor in B-cell development, IL-3 in human pre-B-cell assays directly affects B lymphopoiesis.

4.3. Growth Factor Receptors

The characterization of different interleukins and their functions revealed that their effects were often overlapping. The biological explanation for this observation emerged from the characterization of growth factor receptors. Many are complexes of two or three polypeptide chains that are membrane associated. Some of these receptor elements can be shared among different receptors. The receptors for IL-2, IL-7, and IL-4 share a common chain (γ chain) (Kondo *et al.*, 1994; Nogushi *et al.*, 1993; Chapter 4) directly associated with signal transduction (Kondo *et al.*, 1994). Growth factors with similar target cells share the same chain in their receptor: IL-2R and IL-15R have a common β and α chain (Giri *et al.*, 1994). Both interleukins are involved in T-cell proliferation. However, IL-2 and IL-15 bind differently to different cells, implying that an additional IL-15-specific element might exist.

Similarly, IL-11R, IL-6R, and leukemia inhibitory factor (LIF) have a common component (Chater 1). It appears, therefore, possible, and in some cases it has been confirmed, that different interleukins that share receptor chains can have overlapping effector functions (Ip *et al.*, 1992; Yin *et al.*, 1993). It is still not clear that *in vivo* conditions will allow low-affinity interactions to occur.

5. FIRST SITES WHERE HEMATOPOIETIC PRECURSORS CAN BE DETECTED IN EMBRYONIC DEVELOPMENT

Before fetal liver is seeded by hematopoietic cells at day 10 of gestation, precursors are found in the yolk sac starting at day 8 of gestation (Paige *et al.*, 1979). Hematopoietic cells capable of reconstituting irradiated recipient mice have been detected in the yolk sac between days 8 and 10 of gestation but also in the caudal region of the embryo proper at day 9 of gestation (Tyan and Herzenberg, 1968). More recently, we showed that B-cell precursors that undergo *in vitro* differentiation can be found in the yolk sac and in the embryo proper from the stage of eight to ten somites (8.5 to 9 days of gestation) (Figure 5) (Godin *et al.*, 1995).

The results indicate that before fetal liver, other locations in the mouse embryo proper harbor hematopoietic precursors. A parallel situation has been reported in birds. In a quail–chicken chimeric embryonic environment, it was shown that adult hematopoiesis originates from precursors present in the embryo proper and that the yolk sac had an important role only in the establishment of primitive erythropoiesis (Dieterlen-Lièvre *et al.*, 1981). The precise location of these cells in the embryo proper is in the paraaortic region. The paraaortic splanchnopleura in the mouse is the combination of the mesodermic and endodermic layers. These layers will contribute to the formation of spleen, liver, omentum, and mesentery. In the mouse, the paraaortic splanchnopleura was isolated and grafted under the kidney capsule of severe combine immunodeficient (SCID) mice; B lymphocytes of donor origin were detected, showing that B-cell precursors were present in the grafted tissue (Godin *et al.*, 1993). More recent results show that in the AGM (aorta, gonad, mesonephros) region derived from the splanchnopleura, precursors of day 12 CFU-S (colony-forming units in spleen) and long-term reconstituting cells could be detected (Müller *et al.*, 1994). This result suggests that multipotent hematopoietic stem cells are present in the paraaortic splanchnopleura.

Cells isolated from the yolk sac and from the paraaortic splanchnopleura (P-Sp) were capable of generating B cells *in vitro* under limiting dilution conditions containing S17 stromal cells and IL-7. A detailed analysis of numbers of B-cell precursors obtained at different stages of embryonic development, as shown in Figure 5, indicated that the first

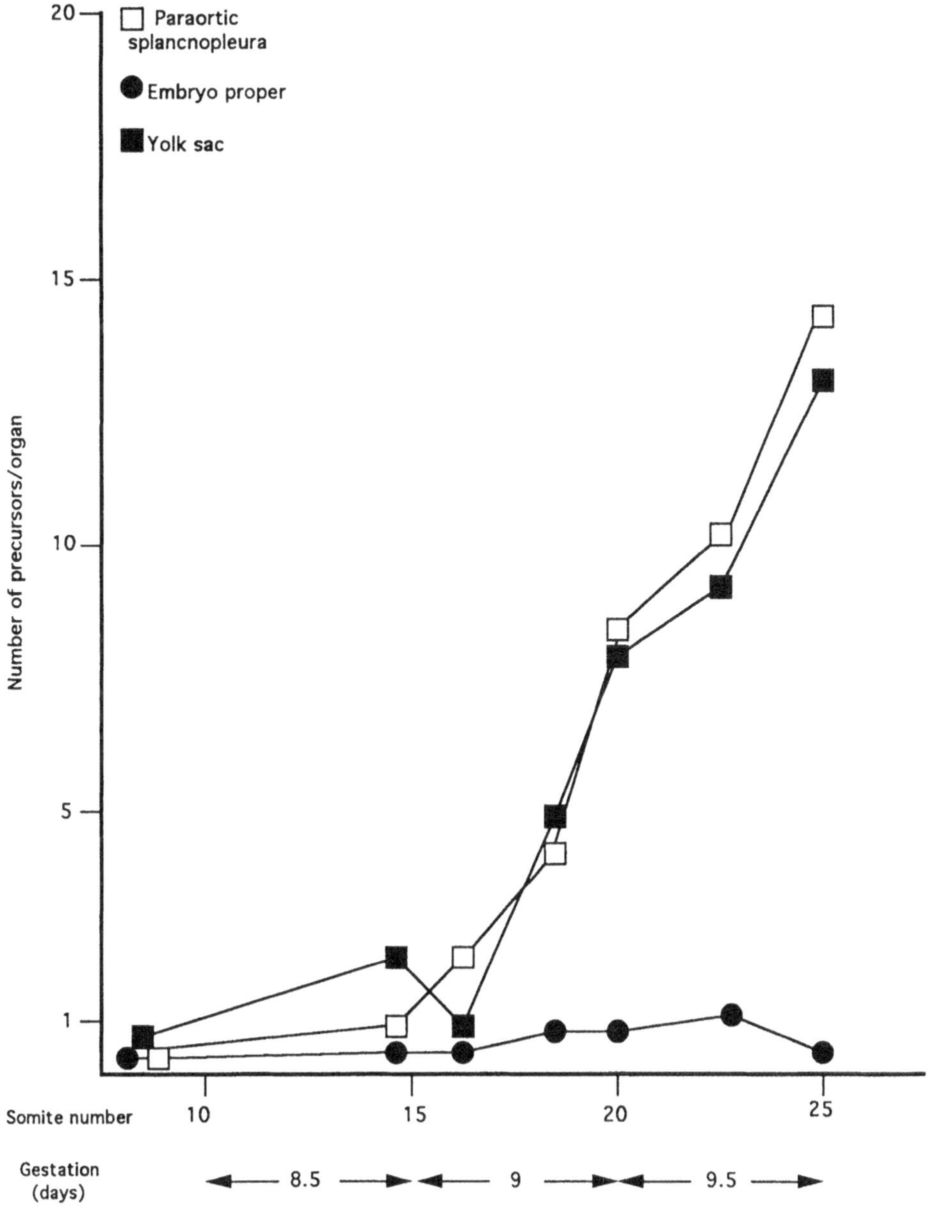

FIGURE 5. Quantification of B-cell precursors in pre-liver embryos. Absolute numbers of B-cell precursors detected in different sites of embryos at various stages of gestation starting at day 8.5. Embryos were isolated and dissected (Godin *et al.*, 1995). Cells were cultured under limiting dilution conditions and lymphoid clones counted.

B-cell precursors could be detected at the stage of eight somites in the P-Sp and in the yolk sac (Godin *et al.*, 1995). At this stage, however, the number of precursors is less than one per embryo. Subsequently, precursors increase rapidly in the yolk sac and in the P-Sp. The remaining cells in the embryo proper were also analyzed, but they failed to generate B lymphocytes.

Analysis of cell surface phenotype of progenitors found in yolk sac and in the para-aortic splanchnopleura indicated that they express AA4.1, similar to the bipotential cells found in fetal liver. However, unlike the fetal liver-derived progenitors, they failed to express Sca1 (Ly6). They also lacked B220 and Mac1 (expressed on B-lineage and myeloid-committed precursors, respectively).

This group of experiments shows that progenitor cells capable of differentiating into B cells arise simultaneously in the yolk sac and in the paraaortic splanchnopleura, starting at the stage of eight to ten somites. It should be emphasized, however, that circulation between the yolk sac and the embryo starts exactly at this time. It is therefore difficult to argue for an independent origin of both hematopoietic precursors. It is clear that the first signs of hematopoiesis occur in the yolk sac with intense primitive erythropoietic activity around day 7.5 of gestation. No erythropoietic activity is observed in the paraaortic splanchnopleura.

A major distinction between the progenitors found at this stage of differentiation and those found in fetal liver was revealed when their progeny was examined. The B-cell precursors detected in the paraaortic splanchnopleura between the stages of 12 and 25 somites are also capable of differentiating into macrophages, megakaryocytes, and granulocyte precursors under the appropriate culture conditions. This property was experimentally demonstrated by cloning cells under limiting dilution conditions or by single-cell manipulation experiments in cultures supplemented with SCF, GM-CSF, and IL-3 in addition to S17 cells and IL-7 (Godin *et al.*, 1995).

It is interesting to note that stromal cells were essential at this stage of differentiation to generate B-lineage cells and could not be replaced by IL-11 and SCF. Direct cell contact or additional growth factor requirements are two possible explanations for the different responses in culture between fetal liver uncommitted bipotent precursors and multipotent cells from embryos at day 9 of gestation.

Although we have not directly tested for self-renewability, it is possible that the precursors detected are the first multipotent hematopoietic stem cells in the embryonic development of the mouse. This possibility finds some support in the observation that the stage of development when the first multipotent precursors can be detected coincides with the stage at which *in vivo* long-term reconstituting cells are first detected. Interestingly, the numbers of multipotent cells detected increased with time until the stage of 30–35 somites, indicating that multipotent precursors undergo expansion without detectable lineage commitment.

6. T-CELL GENERATION FROM MULTIPOTENT PRECURSORS

The characterization of multipotent hematopoietic cells involves the detection of multilineage myeloid and lymphoid cells. The culture conditions described do not allow the differentiation of mature T lymphocytes. Several attempts to detect surface expression of

T-cell receptor or CD3-bearing cells were unsuccessful. To test this capacity, we used an assay system known to provide *in vitro* conditions for T-cell differentiation by using irradiated fetal thymic lobes repopulated *ex vivo* by potential T-cell precursors (Jenkinson *et al.*, 1982). Fetal liver and bone marrow cells can readily generate mature T cells when tested in fetal thymic organ culture assays (FTOC). Single cells from day-9 P-Sp were expanded for 12–14 days with S17 stromal cells, IL-7, and c-kit ligand. After this initial culture period, multipotent clones were identified by microscopic inspection, and 5×10^3 to 10^4 cells were isolated from a single clone and used to repopulate three to six irradiated fetal thymic lobes. The presence of mature T lymphocytes is detected by surface staining with antibodies detecting T-cell-specific markers. All multipotent clones analyzed so far were capable of generating TCRαβ as well as CD4-expressing cells, and clones that are kept in culture for more than 3 weeks lose this capacity (Godin *et al.*, 1995).

This experiment shows not only that the precursors studied are multipotent cells but also that the potential to generate T cells is preserved under these culture conditions. Two main possibilities can be considered. (1) Multipotent cells can undergo self-renewal in culture. It is clear that to obtain multiple lineage cells from a single founder, multipotency must be kept, at least for the first one or two cell divisions. It is possible, however, that multiple rounds of divisions without lineage commitment can occur and even continue in some cells after the first steps of differentiation occur. If this is true, one could consider that multipotent cells repopulate the thymic lobes and generate T cells. It has been demonstrated that cells enriched for long-term reconstituting elements can, on direct thymic injection, generate T lymphocytes. This interpretation implies that the conditions described support the proliferation of multipotent hematopoietic cells. (2) Alternatively, however, we can assume that the cell responsible for thymic repopulation is a committed T-cell precursor. The existence of prethymic T-cell committed precursors has been suggested in bone marrow and more recently in fetal development, but direct evidence for their existence has never been obtained. If this alternative is correct, the culture conditions described here could provide the technical tools to characterize this cell type and its growth and differentiation requirements.

7. PERSPECTIVES

We summarized here the recent advances in culture systems that allow the characterization of growth requirements, the discrete differentiation stages in lymphocyte differentiation from multipotent cells, and their relative representation in hematopoietic organs during ontogeny.

New growth factors involved in hematopoietic development are being found, and their characterization *in vitro* will provide insights in their role in the lineage commitment process. On the other hand, the biochemical study of growth factor receptors (see Chapters 1 and 3) and the regulation of their expression will certainly lead to better understanding of how the differentiation process is regulated. The apparent redundancy in the effect of interleukins in cell proliferation can be clarified once local concentrations of interleukins in hematopoietic organs can be assessed. It is possible that many of the observed effects *in vitro* have no physiological relevance because the required concentrations of interleukins are never attainable in *in vivo* conditions. Finally, the capacity to clone progenitor cells and to follow intermediate steps of differentiation will allow better definition of the mechanism that regulates B-cell development.

ACKNOWLEDGMENTS. We thank P. Kincade, F. Melchers, K. Dorshkind, S. Gillis (Immunex), and D. Donaldson (Genetics Institute) for reagents. C. Furlonger and I. McDermott provided excellent technical assistance. A.C. and B.K. were fellows from the Medical Research Council of Canada. Experiments reported were funded by the National Research Council of Canada, a Terry Fox Program Project Grant, the Medical Research Council of Canada through grants to C.P., and the program Human Capital and Mobility from the E.C. through a grant to A.C.

8. REFERENCES

Alt, F. W., and Baltimore, D., 1982, Joining of immunoglobulin heavy chain gene segments: Implications from a chromosome with evidence of three D–JH fusions, *Proc. Natl. Acad. Sci. U.S.A.* **79:**4118–4122.

Alt, F. W., Yancopoulos, G. D., Blackwell, T. K., Wood, C., Thomas, E., Boss, M., Coffman, R., Rosenberg, N., Tonegawa, S., and Baltimore, D., 1984, Ordered rearrangement of immunoglobulin heavy chain variable region segments, *EMBO J.* **3:**1209–1219.

Anderson, D. M., Lyman, S. D., Baird, A., Wignall, J., Eisenman, J., Rauch, C., March, C. J., Boswell, H. S., Gimpel, S. D., Cosman, D., and Williams, D. E., 1990, Molecular cloning of mast cell growth factor, a hematopoietin that is active in both membrane bound and soluble form, *Cell* **63:**233–243.

Billips, L. G., Petitte, D., Dorshkind, K., Narayanan, R., Chiu, C.-P., and Landreth, K. S., 1992, Differential roles of stromal cells, interleukin-7, and *kit*-ligand in the regulation of B lymphopoiesis, *Blood* **79:**1185–1192.

Chabot, B., Stephenson, D. A., Chapman, V. M., Besmer, P., and Bernstein, A., 1988, The protooncogene *c-kit* encoding a transmembrane tyrosine kinase receptor maps to the mouse *W* locus, *Nature* **335:**88–89.

Chen, J., Ma, F., Young, F., and Alt, F. W., 1994, IL-2 receptor α chain expression during early B lymphocyte differentiation, *Int. Immunol.* **6:**1265–1268.

Cohn, M., and Langman, R. E., 1990, The protecton: The evolutionarily selected unit of humoral immunity, *Immunol. Rev.* **115:**1–131.

Collins, L. S., and Dorshkind, K., 1987, A stromal cell line from myeloid long-term bone marrow cultures can support myelopoiesis and B lymphopoiesis, *J. Immunol.* **138:**1082–1087.

Colon, P. J., Morrissey, R. P., Nordan, K. H., Grabstein, K. H., Prickett, K. S., Reed, S. G., Goodwin, R., Cosman, D., and Namen, A. E., 1989, Murine thymocytes proliferate in direct response to interleukin-7, *Blood* **74:** 1368–1371.

Cumano, A., and Paige, C. J., 1992, Enrichment and characterization of uncommitted B-cell precursors from fetal liver at day 12 of gestation, *EMBO J.* **11:**593–601.

Cumano, A., Dorshkind, K., Gillis, S., and Paige, C. J., 1990, The influence of S17 stromal cells and interleukin-7 on B cell development, *Eur. J. Immunol.* **20:**2183–2189.

Cumano, A., Paige, C. J., Iscove, N. N., and Brady, G., 1992a, Bipotential precursors of B cells and macrophages in murine fetal liver, *Nature* **356:**612–615.

Cumano, A., Furlonger, C., and Paige, C. J., 1992b, Differentiation and characterization of B-cell precursors detected in the yolk sac and embryo body of embryos beginning at the 10–12 somite stage, *Proc. Natl. Acad. Sci. U.S.A.* **90:**6429–6433.

Cumano, A., Kee, B. L., Ramsden, D. A., Marshall, A., Paige, C. J., and Wu, G. E., 1994, Development of B lymphocytes from lymphoid committed and uncommitted progenitors, *Immunol. Rev.* **137:**5–33.

Dick, J. E., Magli, M. C., Huszar, D., Phillips, R. A., and Bernstein, A., 1985, Introduction of a selectable gene into primitive stem cells capable of long term reconstitution of the hemopoietic system of W/W^v mice, *Cell* **42:**71–79.

Dieterlen-Lièvre, F., and Martin, C., 1981, Diffuse intraembryonic hemopoiesis in normal and chimeric avian development, *Dev. Biol.* **88:**180–188.

Ehlich, A., Schaal, S., Gu, H., Kitamura, D., Müller, W., and Rajewsky, K., 1993, Immunoglobulin heavy and light chain genes rearrange independently at early stages of B cell developent, *Cell* **72:**695–704.

Fisher, A. G., Burdet, C., LeMeur, M., Haasner, D., Gerber, P., and Ceredig, R., 1992, Lymphoproliferative desorders in an IL-7 transgenic mouse line, *Leukemia* **2:**566–568.

Geissler, E. N., Ryan, M. A., and Houseman, D. E., 1988, The dominant white spotting locus of the mouse encodes the *c-kit* protooncogene, *Cell* **55:**185–192.

Giri, J., Ahdieh, M., Eisenman, J., Shanabeck, K., Grabstein, K., Kumaki, S., Namen, A., Park, L., Cosman, D., and Anderson, D., 1994, Utilization of the β and the γ chains of the IL-2 receptor by the novel cytokine IL-15, *EMBO J.* **13:**2822–2830.

Godin, I., Garcia-Porrero, J. A., Coutinho, A., Dieterlen-Lièvre, F., and Marcos, M. A. R., 1993, Para-aortic splanchnopleura from early mouse embryos contains B1a cell progenitors, *Nature* **364:**67–70.

Godin, I., Dieterlen-Lièvre, F., and Cumano, A., 1995, Emergence of multipotent hematopoietic cells in the yolk sac and para-aortic splanchnopleura of 8.5 dpc mouse embryos, *Proc. Natl. Acad. Sci. U.S.A.* **92:** 773–777.

Grabstein, K. H., Waldschmidt, T. J., Finkelman, F. D., Hess, B. W., Alpert, A. R., Boiani, N. E., Namen, A. E., and Morrissey, P. J., 1993, Inhibition of murine B and T lymphopoiesis *in vivo* by anti-interlekin 7 monoclonal antibody, *J. Exp. Med.* **178:**257–264.

Hardy, R. R., Carmack, C. E., Shinton, S. A., Kemp, J. D., and Hayakawa, K., 1991, Resolution and characterization of pro-B and pre-pro-B cell stages in normal mouse bone marrow, *J. Exp. Med.* **173:**1213–1225.

Hirayama, F., Shih, J.-P., Awgulewitsch, A., Warr, G. W., Clark, S. S., and Ogawa, M., 1992, Clonal proliferation of murine lymphohemopoietic progenitors in culture, *Proc. Natl. Acad. Sci. U.S.A.* **89:**5907–5911.

Hombach, J., Tsubata, T., Leclercq, L., Stappert, H., and Reth, M., 1990, Molecular components of the B-cell antigen receptor complex of the IgM class, *Nature* **343:**760–762.

Hunt, P., Robertson, D., Weiss, D., Rennick, D., Lee, F., and Witte, O. N., 1987, A single bone marrow-derived stromal cell type supports the *in vitro* growth of early lymphoid and myeloid cells, *Cell* **48:**997–1007.

Ip, N. Y., Nye, S. H., Boulton, T. G., Davis, S., Taga, T., Li, Y., Birren, S. J., Yasukawa, K., Kishimoto, T., Anderson, D. J., Stahl, N., and Yancopoulos, G. D., 1992, CNTF and LIF act on neuronal cells via shared signaling pathways that involve the IL-6 signal transducing receptor component gp130, *Cell* **69:**1121–1132.

Jenkinson, E. J., Fanchi, L. L., Kingston, R., and Owen, J. J. T., 1982, Effect of deoxyguanosine on lymphopoiesis in the developing thymus rudiment *in vitro*: Application in the production of chimeric thymus rudiments, *Eur. J. Immunol.* **12:**583–592.

Jordan, C. T., McKearn, J. T., and Lemischka, I. R., 1990, Cellular and developmental properties of fetal hematopoietic stem cells, *Cell* **61:**953–963.

Kamps, W. A., and Cooper, M. D., 1982, Microenvironmental studies of pre-B and B cell development in human and mouse fetuses, *J. Immunol.* **129:**526–531.

Kee, B. L., Paige, C. J., and Letarte, M., 1992, Characterization of murine CD10, an endopeptidase expressed on bone marrow adherent cells, *Int. Immunol.* **4:**1041–1047.

Kee, B. L., Cumano, A., Iscove, N. N., and Paige, C. J., 1994, Stromal cell independent growth of bipotent B cell-macrophage precursors from murine fetal liver, *Int. Immunol.* **6:**401–407.

Keller, G., Paige, C. J., Gilboa, E., and Wagner, E. F., 1985, Expression of a foreign gene in myeloid and lymphoid cells derived from multipotent heamatopoietic precursors, *Nature* **318:**149–154.

Kincade, P. W., Lee, G., Watanabe, S., and Scheid, M. P., 1981, Antigens displayed on murine B lymphocyte precursors, *J. Immunol.* **127:**2262–2268.

Kitamura, D., and Rajewsky, K., 1992, Targeted disruption of μ chain membrane exon causes loss of heavy-chain allelic exclusion, *Nature* **356:**154–156.

Kitamura, D., Roes, J., Kühn, R., and Rajewsky, K., 1991, A B cell deficient mouse generated through targeted disruption of the membrane exon of the immunoglobulin μ chain, *Nature* **350:**423–426.

Kitamura, D., Kudo, A., Schaal, S., Müller, W., Melchers, F., and Rajewsky, K., 1992, A critical role of λ 5 protein in B cell development, *Cell* **69:**823–831.

Kondo, M., Takeshita, T., Higuchi, M., Nakamura, M., Sudo, T., Nishikawa, S.-I., and Sagamura, K., 1994, Functional participation of the IL-2 receptor γ chain in IL-7 receptor complexes, *Science* **263:**1453–1454.

Lafaille, J. J., DeCloux, A., Bonneville, M., Takagaki, Y., and Tonegawa, S., 1989, Junctional sequences of T cell receptor gd genes: Implications for gd T cell lineages and for a novel intermediate of V–(D)–J joining, *Cell* **59:**859–868.

Landreth, K.S., Narayanan, R., and Dorshkind, K., 1992, Insulin-like growth factor-1 regulates pro-B cell defferentiation, *Blood* **80:**1207–1212.

Lee, G., Namen, A. E., Gillis, S., Ellingsworth, L. R., and Kincade, P. W., 1989, Normal B cell precursors responsive to recombinant murine IL-7 and inhibition of IL-7 activity by transforming growth factor-β, *J. Immunol.* **142:**3875–3883.

Lemischka, I. R., Raulet, D., and Mulligan, R. C., 1986, Developmental potential and dynamic behaviour of hemopoietic stem cells, *Cell* **45:**917–927.

Medvinsky, A. L., Samoylina, N. L., Muller, A. M., and Dzierzak, E. A., 1993, An early pre-liver intraembryonic source of CFU-S in the developing mouse, *Nature* **364:**64–67.

Miyake, K., Medina, K. L., Hayashi, S.-I., Ono, S., Hamaoka, T., and Kincade, P. W., 1990, Monoclonal antibodies to Pgp-1/CD44 block lympho-hemopoiesis in long-term bone marrow cultures, *J. Exp. Med.* **171:**477–488.

Miyake, K., Weissman, I. L., Greenberger, J. S., and Kincade, P. W., 1991, Evidence for a role of the integrin VLA-4 in lympho-hemopoiesis, *J. Exp. Med.* **173:**599–607.

Moore, M. A. S., and Metcalf, D., 1970, Ontogeny of the haemopoietic system: Yolk sac origin of *in vivo* and *in vitro* colony forming cells in the developing mouse embryo, *Br. J. Haematol.* **18:**279–295.

Müller, A. M., Medvinsky, A., Strouboulis, J., Grosveld, F., and Dzierzak, E., 1994, Development of hematopoietic stem cell activity in the mouse embryo, *Immunity* **4:**291–302.

Namen, A. E., Lupton, S., Hjerrild, K., Wignall, J., Mochizuki, D. Y., Schmierer, A., Mosley, B., March, C. J., Urdal, K., Gillis, S., Cosman, D., and Goodwin, R. G., 1988, Stimulation of B cell progenitors by cloned murine interleukin 7, *Nature* **333:**571–573.

Nishikawa, S.-I., Ogawa, M., Nishikawa, S., Kunisada, T., and Kodama, H., 1988, B lymphopoiesis on stromal cell clone: Stromal cell clones acting on different stages of B cell differentiation, *Eur. J. Immunol.* **18:**1767–1771.

Noguchi, M., Nakamura, Y., Russel, S., Ziegler, S. F., Tsang, M., Cao, X., and Leonard, W. J., 1993, Interleukin-2 receptor γ chain: A functional component of the interleukin-7 receptor, *Science* **262:**1877–1880.

Paige, C. J., 1983, Surface immunoglobulin-negative B-cell precursors detected by formation of antibody-secreting colonies in agar, *Nature* **302:**711–713.

Paige, C. J., Kincade, P. W., Moore, M. A. S., and Lee, G., 1979, The fate of fetal and adult B-cell progenitors grafted into immunodeficient CBA/N mice, *J. Exp. Med.* **150:**548–563.

Paul, S. R., Bennett, F., Calvetti, J. A., Kelleher, K., Wood, C. R., O'Hara, R. M., Leary, A. C., Sibley, B., Clark, S. C., Williams, D. A., and Yang, Y.-C., 1990, Molecular cloning of a cDNA encoding interleukin 11, a stromal cell-derived lymphopoietic and hematopoietic cytoquine, *Proc. Natl. Acad. Sci. U.S.A.* **87:**7512–7516.

Peschon, J. J., Morrissey, P. J., Grabstein, K. H., Ramsdell, F. J., Maraskovsky, E., Gliniak, B. C., Park, L. S., Ziegler, S. F., Williams, D. E., Ware, C. B., Meyer, J. D., and Davison, B. L., 1994, Early lymphocyte expansion is severely impaired in interleukin 7 receptor-deficient mice, *J. Exp. Med.* **180:**1955–1960.

Pietrangeli, C. E., Hayashi, S.-I., and Kincade, P., 1988, Stromal cell lines which support lymphocyte growth: Characterization, sensitivity to radiation and responsivness to growth factors, *Eur. J. Immunol.* **18:**863–872.

Raff, M. C., Megson, M., Owen, J. J. T., and Cooper, M. D., 1976, Early production of intracellular IgM by B-lymphocyte precursors in mouse, *Nature* **259:**224–226.

Reth, M. G., Petrac, E., Wiese, P., Lobel, L., and Alt, F., 1987, Activation of Vk gene rearrangement in pre-B cells follows the expression of membrane-bound immunoglobulin heavy chains, *EMBO J.* **6:**3299–3305.

Rolink, A., Kudo, A., Karasuyama, H., Kikuchi, Y., and Melchers, F., 1991, Long-term proliferating early pre-B cell lines and clones with the potential to develop to surface Ig-positive, mitogen reactive B cells *in vitro* and *in vivo*, *EMBO J.* **10:**327–336.

Rolink, A., Grawunder, U., Winkler, T. H., Karasuyama, H., and Melchers, F., 1994, IL-2 receptor a chain (CD25, TAC) expression defines a crucial stage in pre-B cell development, *Int. Immunol.* **6:**1257–1264.

Russel, S. M., Keegan, A. D., Harada, N., Nakamura, Y., Noguchi, M., Leland, P., Friedmann, M. C., Miyajima, A., Puri, R. K., Paul, W. E., and Leoanrd, W. J., 1993, Interleukin-2 receptor γ chain: A functional component of the interleukin-4 receptor, *Science* **262:**1880–1883.

Samaridis, J., Casorati, G., Traunecker, A., Igesias, A., Gutierrez, J., Muller, U., and Palacios, R., 1991, Development of lymphocytes in interleukin 7 transgenic mice, *Eur. J. Immunol.* **21:**453–460.

Sauter, H., and Paige, C. J., 1988, B cell progenitors have different growth requirements before and after immunoglobulin heavy chain commitment, *J. Exp. Med.* **168:**1511–1514.

Spangrude, G. J., Heimfeld, S., and Weissman I. L., 1988, Purification and characterization of mouse hematopoietic stem cells, *Science* **241:**58–62.

Sudo, R., Nishikawa, S., Ohno, M., Akiyama, N., Tamakoshi, M., Yoshida, H., and Nishikawa, S.-I., 1993, Expression and function of interleukin 7 receptor in murine lymphocytes, *Proc. Natl. Acad. Sci. U.S.A.* **90:** 9125–9129.

Tonegawa, S., 1983, Somatic generation of antibody diversity, *Nature* **302:**575–577.

Trevisan, M., and Iscove N. N., 1995, Phenotypic analysis of murine long-term hemopoietic reconstituting cells quantitated competitively in vivo and comparison with more advanced colony-forming progeny, *J. Exp. Med.* **181:**93–103.

Tyan, M., and Herzenberg, L. A., 1968, Studies on the ontogeny of the mouse immune system. II. Immunoglobulin-producing cells, *J. Immunol.* **101:**446–452.

Uchida, N., and Weissman, I., 1992, Searching for hematopoietic stem cells: Evidence that Thy-1.1^{lo}Lin^{-}Sca-1^{+} cells are the only stem cells in C57BL/Ka-Thy1.1 bone marrow, *J. Exp. Med.* **175:**175–184.

Whitlock, C. A., Tidmarch, G. F., Muller-Sieberg, C., and Weissman, I. L., 1987, Bone marrow stromal cell lines with lymphopoietic activity express high levels of a pre-B neoplasia associated molecule, *Cell* **48:**1009–1021.

Wu, A. M., Till, J. E., Siminovitch, L., and McCulloch, E. A., 1967, Cytological evidence for a relationship between normal hemopoietic colony-forming cells and cells of the lymphoid system, *J. Exp. Med.* **127:**455–463.

Wu, Q., Lahti, J. M., Air, G. M., Burrows, P. D., and Cooper, M. D., 1990, Molecular cloning of the murine BP-1/6C3 antigen: A member of the zinc-dependent metallopeptidase family, *Proc. Natl. Acad. Sci. U.S.A.* **87:**993–997.

Wu, Q., Li, L., Cooper, M. D., Pierres, M., and Gorvel, J. P., 1991, Aminopeptidase A activity of the murine B-lymphocyte differentiation antigen BP-1/6C3, *Proc. Natl. Acad. Sci. U.S.A.* **88:**676–680.

Yin, T., Yaga, T., Tsang, M. L.-S., Yasukawa, D., Kishimoto, T., and Yang, Y.-C., 1993, Involvment of IL-6 signal transducer gp130 in IL-11 mediated signal transduction, *J. Immunol.* **151:**2555–2561.

Zsebo, K. M., Williams, D. A., Geissler, E. N., Broudy, V. C., Martin, F. H., Atkins, H. L., Hsu, R.-Y., Birkett, N. C., Okino, K. H., Murdock, D. C., Jacobsen, F. W., Langley, K. E., Smith, K. A., Takeishi, T., Cattanach, B. M., Galli, S. J., and Sugges, S. V., 1990, Stem cell factor is encoded at the *Sl* locus of the mouse and is the ligand for *c-kit* tyrosine kinase receptor, *Cell* **63:**213–224.

Chapter 10

Regulation of Peripheral B-Cell Growth and Differentiation

Jacques Banchereau, Pierre Garrone, and Yong-Jun Liu

1. INTRODUCTION

The life histories of B lymphocytes and T lymphocytes share many similar features. They derive from immature precursor cells and use similar genes and strategy to encode, assemble, and select their surface antigen receptors. T-cell development in the thymus involves a series of stringent positive and negative selection processes that ensure (1) that the generated T-cell receptors distinguish self and nonself antigen and (2) the dependence on the recognition of self-MHC to respond to nonself antigens. B-cell development within the bone marrow also involves a series of selection processes that ensure that the generated B-cell antigen receptors directly recognize nonself antigen without the involvement of self-MHC. The current views of primary T-cell and B-cell lymphopoiesis in the thymus and bone marrow, respectively, are generated by two considerations: (1) each stage of gene rearrangement and the selection of the host cells correspond to a particular stage of cellular differentiation characterized by surface molecules that collectively define a cell subset; (2) a particular differentiation stage/cell subset is localized within a particular anatomic compartment of the thymus or bone marrow. Thus, the whole differentiation pathway seems to follow a specific migration pattern through different functional compartments of thymus or bone marrow, where they interact with different stromal cells/accessory cells. These two experimental approaches, together with gene cloning, should ultimately permit the isolation of the regulatory molecules that drive the cells from one differentiation stage to the next during the primary B-cell and T-cell developments.

Jacques Banchereau, Pierre Garrone, and Yong-Jun Liu Schering-Plough, Laboratory for Immunological Research, 69571 Dardilly, France.

Blood Cell Biochemistry, Volume 7: Hematopoietic Cell Growth Factors and Their Receptors, edited by A. D. Whetton and J. Gordon. Plenum Press, New York, 1996.

It is only recently that similar approaches have been used to dissect peripheral B-cell (Kelsoe, 1994; Liu and Banchereau, 1995; Pascual *et al.*, 1994) and T-cell (McHeyzer-Williams and Davis, 1995) development. Peripheral B-cell development takes place in peripheral lymphoid tissues such as lymph nodes, spleen, and mucosal-associated lymphoid tissues. It is characterized by antigen-driven clonal expansion, selection, somatic hypermutation in IgV region genes, and isotype switch in Ig heavy-chain constant-region genes. This chapter discusses the regulation of peripheral B-cell development including the *in vivo* sites of B-cell activation, the characterization of mature B-cell subsets, and the response of B cells to extracellular signals.

2. A SCHEMATIC VIEW OF ANTIGEN-INDUCED T-CELL-DEPENDENT B-CELL ACTIVATION: IMMUNOPOIESIS

The final stage of primary B-cell development in the bone marrow (lymphopoiesis) is the generation of large numbers of mature B cells that carry surface immunoglobulin M (sIgM) and sIgD antigen receptors (Melchers *et al.*, 1994). These naive B cells continuously migrate into the peripheral lymphoid tissues, passing through the T cell and interdigitating cell-rich extrafollicular areas, where they may die or pursue their journey into the primary follicles where they become incorporated into the peripheral B-cell pool (MacLennan and Gray, 1986).

Peripheral lymphoid tissues provide basic structures to facilitate antigen trapping and interactions between lymphocytes and antigen-presenting cells. All peripheral lymphoid tissues are organized into at least two basic structures, a T-cell-rich interfollicular area and a B-cell-rich follicle through which naive B cells percolate. The antigen-presenting cells within these two compartments are quite distinct: interdigitating dendritic cells (IDC) of hematopoietic origin (Steinman, 1991) are found mainly in T-cell-rich extrafollicular areas (Inaba *et al.*, 1984; Steinman, 1991), and follicular dendritic cells (FDC) of mesenchymal origin (Schriever and Nadler, 1992) are localized mainly in the B-cell follicles (Kosco, 1991; Schriever and Nadler, 1992; Tew *et al.*, 1990). In thymus-dependent humoral immune responses, antigen-specific naive B cells are activated first in extrafollicular areas in association with T cells and IDC, leading to the formation of so-called foci (Jacob *et al.*, 1991a; Liu *et al.*, 1991a). They undergo clonal expansion, and most of them differentiate into short-lived plasma cells. However, some of the activated cells are recruited into the B-cell follicles without undergoing terminal differentiation. The activated B cell blasts colonizing the follicles undergo high rate proliferation, yielding centroblasts (Kroese *et al.*, 1987; Liu *et al.*, 1991a; MacLennan, 1994; Zhang *et al.*, 1988) that form the dark zones of germinal centers (Fliedner *et al.*, 1964; Liu *et al.*, 1991b). Centroblasts undergo somatic mutation in their IgV region genes (Berek *et al.*, 1991; Jacob *et al.*, 1991b; McHeyzer-Williams *et al.*, 1993; Pascual *et al.*, 1994). They give rise to nonproliferating centrocytes, which undergo positive affinity selection in the light zone of germinal centers (Liu *et al.*, 1989). The FDC play a key role in B-cell positive selection through the immune complexes they bind via Fc receptors as well as complement receptors. Then, high-affinity centrocytes undergo differentiation into either memory B cells (Coico *et al.*, 1983; Klaus *et al.*, 1980; Weiss and Rajewsky, 1990) or plasma cells (Kosco et al., 1989; Tew *et al.*, 1992). There is evidence suggesting that Ig isotype switching occurs within germinal centers (Kraal *et al.*, 1982). For more than a decade, we and many others have been concentrating our efforts on identifying the cells and molecules that control these various steps of B-lymphocyte immunopoiesis.

3. CHARACTERIZATION OF HUMAN TONSILLAR B-CELL SUBSETS

3.1. Surface IgD and CD38 Define Four Subpopulations of Tonsillar B Cells

Human tonsils represent an ideal secondary lymphoid tissue for understanding antigen-driven T-cell-dependent B-cell maturation inasmuch as they are the site of ongoing immune responses and therefore display all stages of B-cell differentiation including germinal center (GC) reactions.

As shown by immunohistology, IgD and CD38 are mutually exclusive markers that define naive B cells in the follicular mantle as IgD^+CD38^- cells and GC B cells as IgD^-CD38^+ cells. However, a flow cytometry analysis of tonsillar B cells with anti-IgD-FITC and anti-CD38-PE reveals two additional B-cell subsets (Figure 1B). One subset is IgD^-CD38^-, double negative, and the other is IgD^+CD38^+, double positive. As summarized and discussed here, we have carried out detailed phenotypic, cell cycle, survival, anatomic, functional, and immunoglobulin gene analysis on these four subsets of B cells. Note that the Bm abbreviations used in this review stand for "mature B cell subset."

3.2. IgD^+CD38^- B Cells Are Naive Follicular Mantle B Cells Composed of $CD23^-$ (Bm1) and $CD23^+$ (Bm2) Subsets

IgD^+CD38^- naive B cells can be isolated from high-density tonsillar B cells by depletion of $CD38^+$, IgG^+, and IgA^+ B cells with immunomagnetic beads or by direct FACS sorting of IgD^+CD38^- B cells. These cells express all the antigens that can be detected on follicular mantles of tonsil sections, including IgM, CD5, CD23, CD44, and Bcl-2 protein (Table I). In contrast, all antigens that are restricted to GC, including CD10, CD71, CD77, Fas/APO-1/CD95, and Ki67 nuclear antigen, are negative on IgD^+CD38^- naive B cells. They are small resting B cells based on morphological and cell cycle analysis. They do not undergo rapid spontaneous apoptosis during *in vitro* culture, in sharp contrast with GC B cells (Liu *et al.*, 1989). They do not induce proliferation of allogeneic CD4 T cells or allospecific T-cell lines. They fail to rapidly up-regulate the key costimulating molecules B7.1/CD80 (Gimmi *et al.*, 1991; Linsley *et al.*,1991) and B7.2/CD86 (Azuma *et al.*, 1993; Freeman *et al.*, 1993; Hathcock *et al.*, 1993) during cognate T–B interactions. Ligation of their antigen receptor and CD40 antigen results in an up-regulation of B7.1 and B7.2 that is slower and lower than that observed with the memory B cells. The IgD^+CD38^- B cells can be further separated into $CD23^-$ (Bm1) and $CD23^+$ (Bm2) subsets (Figure 1). Analysis of IgV_H genes shows that these two subsets of B cells have not undergone somatic mutation, confirming that they are naive B cells (Pascual *et al.*, 1994).

Studies in the mouse system have suggested that some bone marrow IgD^+IgM^+ B cells are recruited into the peripheral B-cell pool after selection, possibly by an idiotypic network or environmental antigen (Chies *et al.*, 1994; Gu *et al.*, 1991; Schittek and Rajewsky, 1992), thus making it very difficult to distinguish truly naive B cells from ligand-selected B cells. The IgD^+CD38^- follicular mantle B cells isolated from human tonsils correspond to the

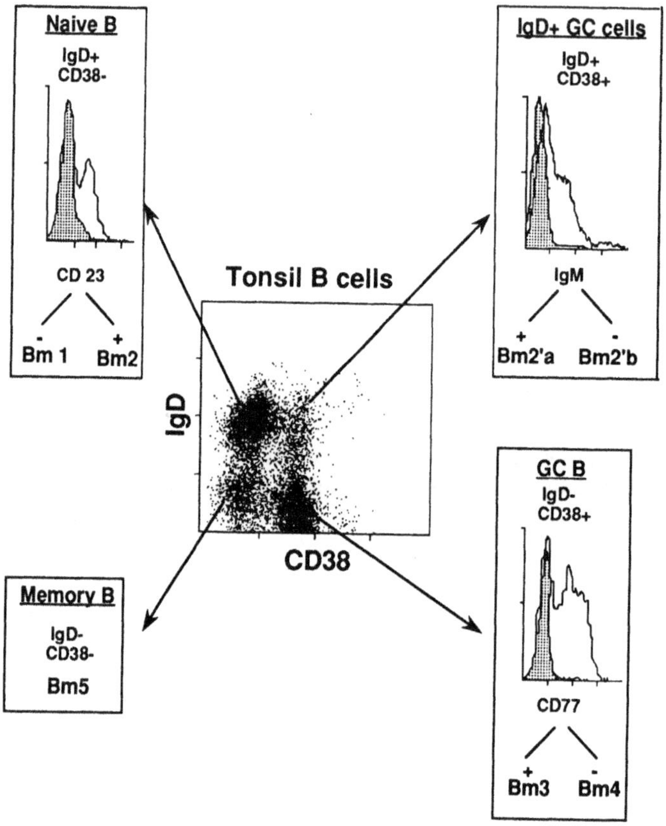

FIGURE 1. Isolation of seven tonsil B-cell subsets by FACS sorting. Tonsillar B cells can be sorted into four subsets based on anti-IgD and anti-CD38 double-immunofluorescence staining. Then $sIgD^+CD38^-$ naive B cells are separated into a $CD23^-$ subset (Bml) and a $CD23^+$ subset (Bm2). The $sIgD^+CD38^+$ cells are separated into a $sIgM^+$ subset (Bm2′a) and a $sIgM^-$ subset (Bm2′b). The $sIgD^+CD38^+$ germinal center B cells are separated into $CD77^+$ centroblasts (Bm3) and $CD77^-$ centrocytes (Bm4). The $sIgD^-CD38^-$ B cells (Bm5) represent memory B cells.

naive B cells, as they are unmutated IgM^+IgD^+ B cells. Because CD23 is an antigen that can be transiently up-regulated on IgD^+ B cells early after activation *in vitro* (Banchereau and Rousset, 1992), it is possible that $IgD^+CD38^-CD23^-$ cells (Bml) represent early immigrated naive B cells, and $IgD^+CD38^-CD23^+$ cells (Bm2) represent ligand-selected B cells.

3.3. IgD^+CD38^+ B Cells (Bm2′) Contain a Subset of IgM^+ (Bm2′a) Germinal Center Founder Cells and a Subset of IgM^- (Bm2′b) B Cells with Extensively Mutated IgV Genes

IgD^+CD38^+ B cells represent 5–15% of tonsillar B cells and can be isolated by FACS sorting (Liu *et al.*, 1996a). Detailed phenotypic analysis using three-color flow cytometry and *in situ* immunocytology shows that these IgD^+CD38^+ B cells have acquired all the

Table I
Properties and Relationship of B-Cell Subpopulations from Human Tonsils

	IgD^+CD38^- Bm1 + 2 naive B cells	IgD^+CD38^+ Bm2′a germinal center precursors	IgD^-CD38^+ Bm3 + 4 germinal center B cells	IgD^-CD38^- Bm5 memory B cells
CD23	30%+	–	–	–
CD39	+	?	–	+
CD44	+	Low	Low	+
CD10	–	+	+	–
CD71	–	+	+	–
CD77	–	50%+	40%+	–
IgM	+	+	10%+	5%+
IgG	–	–	+	+
Bcl-2	+	–	–	+
Fas	–	+	+	Low
Ki67	–	+	+	–
Somatic mutation	–	–	+	+
Switching activity	–	–	+	–
Location	Follicular mantle	Germinal center	Germinal center	Epithelium

features of GC B cells (Table I): (1) they are positive for all surface markers identified so far on GC B cells, including CD10, CD71, and CD77; (2) they lack most markers expressed on follicular mantle B cells, such as CD5, CD23, and CD44; (3) they express Fas but not Bcl-2 and are sensitive to rapid spontaneous apoptosis during *in vitro* culture; (4) most of them are proliferating B cells as indicated by expression of the proliferation-related nuclear antigen Ki67 or the high DNA content as analyzed by Hoechst 33342 staining. Double immunoenzymatic staining permits to show that these $IgD^+CD38^+Ki67^+$ B cells are located within the dark zone of GC.

Interestingly, IgD^+CD38^+ B cells can be further separated into an IgM^+ subset (Bm2′a) and an IgM^- subset (Bm2′b) (Figure 1). IgV_H gene analysis shows that the IgM^+ subsets contain germ line IgV genes, while the IgM^- subsets have accumulated a distinct high load of mutations in the IgV-δ transcripts (36–77 mutations per transcript). This suggests that the $sIgM^+IgD^+CD38^+$ B cells (Bm2′a) may represent the founder cells of germinal centers. These cells, which are prone to apoptosis, may therefore be selected before entering into high-rate proliferation, somatic mutations, and affinity selection. Such a selection mechanism of early GC B cells has indeed been demonstrated in immunized mice (Jacob *et al.*, 1993; McHeyzer-Williams *et al.*, 1993).

The hypermutated $sIgM^-IgD^+CD38^+$ B cells (Bm2′b) may correspond to cells generated by two mechanisms: (1) activated $sIgD^+$ naive B cells that have been sequestered within the germinal center have undergone many rounds of proliferation and somatic mutation; (2) $sIgD^+$ memory B cells have been recycling through the GCs (Kepler and Perelson, 1993). The absence of comparable rate of mutation in γ transcripts of germinal center and memory B cells from the same tonsil indicate that the $sIgM^-IgD^+CD38^+$ B cells may not switch to the downstream isotype. Such a switching inability could be explained by deletion of the C_μ locus, through homologous recombination between $\Sigma\mu$ and $\Sigma\delta$ sequences (Gilliam *et al.*, 1984; Yasui *et al.*, 1989), which could remove the switch region (S_μ) required for further isotype switch, thus leading to a $sIgM^-IgD^+$ phenotype. This would also result in the deletion of a putative somatic mutation silencer, thus contributing to enhanced somatic mutation in IgD transcripts. The functional role of this unique B-cell population remains to be understood.

3.4. The Germinal Center IgD^-CD38^+ B Cells Can Be Further Separated into $CD77^+$ Centroblasts (Bm3) and $CD77^-$ Centrocytes (Bm4)

IgD^-CD38^+ germinal center B cells can be isolated directly by FACS sorting or by negative depletion of IgD^+ and $CD39^+$ B cells (incuding naive and memory B cells as well as plasma cells) with immunomagnetic beads. IgD^-CD38^+ GC B cells express CD10, CD71, Fas, and the nuclear Ki67 antigen. They do not express CD5, CD23, CD39, CD44, or Bcl-2 (Table I). During *in vitro* culture, they undergo a rapid apoptosis, which can be slowed down by triggering their antigen receptors or CD40 (Liu *et al.*, 1989). The IgD^-CD38^+ GC B cells can be further separated into $CD77^+$ centroblasts (Bm3) and $CD77^-$ centrocytes (Bm4) using FACS sorting (Figure 1). Sequence analysis of their IgV_H genes shows that both subsets of GC B cells have undergone somatic mutation in their IgM and IgG transcripts (Pascual *et al.*, 1994). When compared with centroblasts (Bm3), centrocytes (Bm4) show a significant increase in the ratios of replacement mutations versus silent mutations in CDRs of both IgM and IgG transcripts (Pascual *et al.*, 1994). This indicates a progressive antigen-driven selection during the differentiation from centroblasts to centrocytes. Immunoglobulin-sterile transcripts $I\gamma$, $I\alpha$, and $I\varepsilon$ as well as switch circles can be selectively detected within the GC centrocyte Bm4 subset. This provides direct molecular evidence that isotype switching is occurring within the GC after somatic mutation has occured (Liu *et al.*, 1996b).

3.5. IgD^-CD38^- B Cells Are Memory B Cells that Colonize Mucosal Epithelium and Act as Strong Antigen-Presenting Cells

IgD^-CD38^- B cells can be isolated through negative selection by depleting IgD^+ B cells and $CD38^+$ B cells, using either FACS sorting or magnetic beads. These cells express neither the GC markers CD10, CD71, and CD77 nor the follicular mantle B-cell markers CD5 and CD23. All of them express CD20, CD39, and CD44 (Lagresle *et al.*, 1993); a high proportion express high levels of sIgG, and only a few display sIgA. They do not express the proliferation-related nuclear Ki67 antigen, thus indicating their resting status. All express Bcl-2; only a few have low levels of Fas, and, accordingly, they do not undergo rapid spontaneous apoptosis during *in vitro* culture. On polyclonal activation *in vitro* with IL-2, IL-10, and CD40 ligand (CD40-L), these cells, isolated from tetanus-toxoid-vaccinated donors, produce large amounts of anti-tetanus-toxoid IgG antibodies, whereas IgD^+CD38^- naive B cells do not. Sequence analysis of their IgV_H genes shows that they have undergone somatic mutation (Pascual *et al.*, 1994). These properties qualify IgD^-CD38^- tonsillar B cells to be memory B cells.

Human tonsillar memory B cells have two unique features (Liu *et al.*, 1995). First, immunohistological analysis shows that IgD$^-$CD38$^-$ memory B cells are located mainly within the mucosal epithelium but not in the follicles. Second, memory B cells but not naive B cells can induce proliferation of allo-CD4$^+$ T cells, indicating that memory B cells can present antigen directly to T cells. The difference in the capacity of antigen presentation between memory B cells and naive B cells is associated with the rapid up-regulation of B7.1 and B7.2 on memory B cells during the cognate interaction with the allospecific T-cell line. Thus, the epithelial localization and the potent antigen-presenting capacity of memory B cells may explain the rapid and robust secondary humoral immune responses.

3.6. CD38^{++}CD20$^-$ Tonsillar Plasma Cells Undergo Rapid Apoptosis *in Vitro* that Is Prevented by Contact with Bone Marrow Stroma Cells

Plasma cells represent the final stage of B-cell differentiation. Because of their high level of CD38 expression (Ling *et al.*, 1987), plasma cells can be isolated by sorting CD38^{++}CD20$^-$ cells from tonsillar populations that have previously been enriched in plasma cells through BSA gradients. CD38^{++}CD20$^-$ plasma cells express CD19, CD40, CD24, CD37, CD39, CD74, and HLA-DR. They do not express CD3, CD14, CD28, CD49d, CD56, or HLA-DQ. The purified CD38^{++}CD20$^-$ cells have a typical morphology of plasma cells, contain high levels of intracellular Ig, and release IgG and IgA in roughly equivalent levels. During *in vitro* culture, these plasma cells undergo rapid apoptosis unless cultured on bone marrow stromal cells (Merville *et al.*, 1996). The bone marrow stroma permits the survival of plasma cells through cell–cell interactions that are presently uncharacterized.

4. *IN VITRO* RESPONSES OF B CELLS TO SIGNALS THROUGH ANTIGEN RECEPTORS, CD40, FAS, AND CYTOKINE RECEPTORS

Over the past decade, numerous studies have been performed with the aim of understanding the role of cytokines in the growth and differentiation of human B cells activated *in vitro* through different means. The following section summarizes the *in vitro* effects of recombinant cytokines on purified human B lymphocytes triggered through: (1) their antigen receptors, (2) CD40 antigen, and (3) Fas antigen.

4.1. Functional Consequences of Antigen Receptor Engagement

Purified mature B lymphocytes undergo limited DNA synthesis when activated through their antigen receptors (BCR) by immobilized anti-Ig antibodies (Banchereau and Rousset, 1992; Jelinek and Lipsky, 1987). Particles of *Staphylococcus aureus* strain Cowan (SAC) also act as a strong polyclonal activator of B cells, which involves BCR cross-linking and possibly the engagement of other B-cell surface molecules.

Interleukin 2 (Mingari *et al.*, 1984), IL-3 (Xia *et al.*, 1992), IL-4 (Defrance *et al.*, 1987), IL-10 (Rousset *et al.*, 1992), IL-13 (McKenzie *et al.*, 1993), IFNγ (Defrance *et al.*, 1986; Romagnani *et al.*, 1986), TNF-α and -β (Kehrl *et al.*, 1987a,b), and high-molecular-weight BCGF (IL-14?) (Ambrus *et al.*, 1993) have been shown to enhance to some extent the

DNA synthesis of BCR-activated B cells. Interleukin 2 represents one of the most efficient cytokines, particularly after several days of culture. IL-2 and IL-10 can induce SAC-activated B cells to secrete IgG, IgA, and IgM (Defrance *et al.*, 1988a; Rousset *et al.*, 1992), whereas IL-4 only induces the production of low levels of IgM and IgG (Defrance et al., 1988b). Under these conditions, B cells triggered through their BCR do not undergo isotype switch in response to cytokines.

4.2. Functional Consequences of CD40 Engagement

4.2.1. Historical Considerations

The limited B-cell growth and differentiation observed following BCR cross-linking led to the proposal that the engagement of BCR may essentially represent an activation and/or a selection signal for B cells. Such a hypothesis implies that other B-cell surface molecules are involved in the massive expansion and differentiation of the rare antigen-specific B cells *in vivo*. In particular, models were established in which activated T cells were shown to induce considerable B-cell proliferation and Ig secretion (Parker, 1993). Although cytokines produced by activated T cells play a key role in the growth and differentiation of B cells, the contact between T cells and B cells cannot be replaced by T-cell supernatants. Thus, fixed activated T cells or their membranes can induce B cells to proliferate, and addition of cytokines can further enhance growth and can induce Ig secretion. These studies led to the identification of CD40 as a key antigen involved in B-cell growth and differentiation. CD40, a member of the TNF-α receptor family, is a receptor for CD40-L, an activation antigen of T cells and a member of the TNF-α family (Banchereau *et al.*, 1994; Spriggs *et al.*, 1993).

4.2.2. Proliferation of B Lymphocytes

Anti-CD40 antibodies were identified by their ability to costimulate (with either anti-IgM antibodies or phorbol esters) the proliferation of purified B cells (Clark and Ledbetter, 1986; Gordon *et al.*, 1987; Ledbetter *et al.*, 1987; Vallé *et al.*, 1989). Culture of resting B cells in the CD40 system (anti-CD40 antibody presented by a fibroblastic cell line transfected with FcγRII/CD32) or on CD40-L-transfected cells results in strong and long-lasting B-cell proliferation (Armitage *et al.*, 1993; Banchereau *et al.*, 1991; Cocks *et al.*, 1993; Gauchat *et al.*, 1993; Grabstein et al., 1993; Hollenbaugh *et al.*, 1992; Spriggs *et al.*, 1992). These culture conditions allow the proliferation of various B-cell subpopulations including naive $sIgD^+$ $sIgM^+$ B cells, germinal center and memory B cells, and $CD5^+$ and $CD5^-$ B cells (Defrance *et al.*, 1992a). Certain soluble anti-CD40 antibodies and soluble trimeric forms of CD40-L can induce DNA replication of B cells.

Addition of IL-4 to B cells cultured in the CD40 system or over CD40-L-transfected cells results in their sustained proliferation (Armitage *et al.*, 1993; Banchereau *et al.*, 1991; Defrance *et al.*, 1992a; Rousset *et al.*, 1991; Spriggs *et al.*, 1992) and allows the generation of factor-dependent long-term normal B-cell lines. B-cell clones of several hundred cells can be obtained under these conditions. Interleukin 13, which shares homology with IL-4 (McKenzie *et al.*, 1993; Minty *et al.*, 1993), can also induce an intense proliferation of B cells stimulated in the CD40 system or with CD40-L-transfected cells (Cocks *et al.*, 1993;

see Chapter 2). Cells cultured in the CD40 system with IL-4 express CD19, CD20, CD40, sIg, high levels of CD23, and HLA class II antigens (Galibert et al., 1994).

Both viral and human IL-10 enhance the proliferation of CD40-activated B cells and appear to be almost as efficient as IL-4 during early cultures (Rousset *et al.*, 1992). The combination of IL-4 and IL-10 results in additive proliferation. Interleukin 2 poorly enhances the proliferation of CD40-activated B cells. B cells cultured in IL-10 differ microscopically from those cultured in IL-4 in that loose aggregates are observed early on, which then yield cultures mostly composed of single large cells. Interleukin 10 up-regulates the expression of CD25/Tac on anti-CD40-activated B cells, and accordingly, addition of IL-2 strongly enhances B-cell proliferation (Fluckiger *et al.*, 1993)

4.2.3. CD40 Triggering Turns on Expression of Functional Fas

The ligation of the Fas molecule by functional antibodies or by its ligand (Fas-L) induces death of sensitive cells through an apoptotic process (Krammer *et al.*, 1994; Nagata and Golstein, 1995). Although all germinal center B cells express Fas, only a few memory B cells express it at low levels and naive B cells do not express Fas at all (Liu *et al.*, 1995; Lagresle *et al.*, 1995). Activation of resting B-cell populations that are composed of both naive and memory B cells through their CD40 results in prompt expression of Fas. In contrast, engagement of the BCR does not turn on Fas expression on B cells (Garrone et al., 1995). Addition of a functional anti-Fas antibody to cultures of CD40-activated B cells results in an inhibition of cell proliferation that can be seen after 3 to 4 days of culture. This late inhibition results from delayed acquisition of sensitivity to Fas-mediated programmed cell death (apoptosis). Neither recombinant cytokines (IL-2, IL-4, IL-10, IFNγ) nor activated T-cell supernatants could counteract Fas-induced death of CD40-activated B cells and down-regulate CD40-induced Fas expression. The delayed functional response to Fas ligation after CD40 activation may represent a way to limit the size of a specific B-cell clone that is generated during T/B cell interactions.

4.2.4. Differentiation of B Lymphocytes

B cells cultured in the CD40 system or with CD40-L-transfected cells produce marginal amounts of immunoglobulins (Armitage *et al.*, 1993; Grabstein *et al.*, 1993; Maliszewski *et al.*, 1993; Rousset *et al.*, 1991). However, B cells cultured in the CD40 system together with SAC particles produce large amounts of IgM, IgG, and IgA without IgE (Defrance *et al.*, 1992b). Naive sIgD$^+$ sIgM$^+$ B cells secrete only IgM, whereas sIgD$^-$sIgM$^\pm$ memory B cells secrete IgG and IgA and lower amounts of IgM. The surface antigen that is specifically engaged by SAC particles to turn on the differentiation of CD40 activated B cells remains unidentified.

Interestingly, stimulation of blood B lymphocytes with soluble monoclonal antibody to CD40 induces germ-line transcription of all heavy-chain isotypes except $\gamma 4$ and ε, thus suggesting that CD40 triggering participates in early steps of switch recombination (Jumper *et al.*, 1994).

Addition of IL-4 or IL-13 to CD40-activated B cells results in a slight increase in the production of IgM and IgG and in the secretion of IgE following isotype switching (Armitage *et al.*, 1993; Jabara *et al.*, 1990; Punnonen *et al.*, 1993; Rousset *et al.*, 1991).

Addition of IL-10 to CD40-activated B lymphocytes results in the production of considerable amounts of IgM, IgG, and IgA without IgE as a result of a differentiation into plasma cells (Armitage *et al.*, 1993; Rousset *et al.*, 1992). Interleukin 10 induces CD40-activated total tonsil B cells to secrete IgG_1, IgG_2, and IgG_3. CD40-activated naive $sIgD^+sIgM^+$ B cells were found to secrete principally IgM but also IgG_1 and IgG_3 in response to IL-10, indicating that this cytokine acts as a switch factor for certain IgG subclasses (Brèire *et al.*, 1994). CD40-activated naive $sIgD^+sIgM^+$ B cells cultured with IL-10 also produce low levels of IgA, and addition of TGF-β induces large amounts of IgA while inhibiting IgM and IgG production (Defrance *et al.*, 1992b). The TGF-β is likely to be an IgA switch factor both in human and in mouse (Coffman *et al.*, 1989; Islam *et al.*, 1991). Thus, the engagement of CD40 on B cells turns on their isotype switching machinery, the specificity of which is subsequently provided by cytokines.

4.2.5. Role of CD40 in the Maturation of B-Cell Subsets

Our accumulated knowledge of the phenotype of human B-cell subsets and plasma cells and about the molecular regulation of human B-cell activation *in vitro* allowed us to determine the signals controlling the differentiation of naive B cells into germinal center B cells and that of germinal center B cells into either memory B cells or plasma cells.

4.2.5a. Prolonged BCR Triggering Primes CD40-Activated B Cells to Undergo Apoptosis. A combination of CD40 triggering and BCR triggering synergizes for strong B-cell proliferation/expansion over short-term cultures (up to 5 days). The observed cellular expansion is comparable to that observed with CD40 triggering and IL-4 (Wheeler *et al.*, 1993; C. van Kooten, unpublished data). However, after 5 days of culture, all activated B cells quickly enter into apoptosis. Phenotypic analysis at this time shows that many IgD^+ B cells have acquired some germinal center B-cell markers including CD38, Fas, and carboxypeptidase M (de Saint-Vis *et al.*, 1995; Galibert *et al.*, 1996). Thus, naive B cells acquired one of function features of GC B cells, the propensity of undergoing apoptosis, after dual triggering of BCR and CD40.

4.2.5b. CD40 Triggering Directs the Differentiation of Germinal Center B Cells Toward Memory B Cells Rather Than Plasma Cells. Inasmuch as GC B cells are $CD38^+CD20^+$, memory B cells are $CD38^-CD20^+$, and plasma cells are $CD38^+CD20^-$, the levels of CD38 and CD20 can be used to follow the *in vitro* differentiation of GC B cells into either plasma cells or memory B cells. Germinal center B cells can be successfully cultured on CD40-L-transfected L cells with IL-2 and IL-10, a system that mimics the *in vivo* situation when high-affinity GC B cells have picked up antigen from follicular dendritic cells and are presenting it to CD40-L-expressing GC T cells (Arpin *et al.*, 1995; Lagresle *et al.*, 1993). Germinal center B cells, rescued from apoptotic cell death, grow in exponential phase after 3 days of culture as $CD38^+CD20^+$ B-cell blasts. These $CD38^+CD20^+$ cells acquire a $CD38^-CD20^+$ memory B-cell phenotype after a further 3 days of culture on CD40-L-transfected L cells with IL-2 and IL-10. These cells secrete little Ig and have the ability to proliferate in response to triggering through antigen receptor and CD40. In contrast, the $CD38^+CD20^+$ B blasts become terminally differentiated plasma cells after a further 3 days of culturing with IL-2 and IL-10 in the absence of CD40-L signal. Thus, CD40-L plays a key role in directing germinal center B cells toward the memory B-cell pathway rather than the plasma cell pathway.

5. MOLECULAR CONTROL OF B-CELL IMMUNOPOIESIS: A SYNTHESIS

It is always difficult and potentially misleading to try to give *in vivo* biological significance to phenomena that have been observed *in vitro*. This is particularly difficult in the case of human immunology, as studies in humans are limited by ethical considerations. Nevertheless, we try to associate the data obtained *in vitro* from purified human B lymphocytes with those obtained by immunohistological analysis of secondary lymphoid organs from humans or antigen-challenged animals such as mice or rats.

B-cell immunopoiesis can be divided into four different stages, which occur in distinct anatomic areas (Kroese *et al.*, 1990; Liu *et al.*, 1992; MacLennan and Gray, 1986): (1) primary and secondary B-cell proliferation and differentiation into plasma blasts in extrafollicular areas; (2) blast/centroblast proliferation in the germinal center dark zone; (3) antigen-driven selection of mutated B cells, proliferation of selected B cells, and their differentiation into either memory cells or plasma blasts in the light zone; and (4) differentiation of plasma blasts into plasma cells in the lymphoid organ medullary cords, bone marrow, or mucosal lamina propria (Figure 2).

5.1. Extrafollicular Reaction

At the site of tissue injury, pathogens/antigens are captured by dendritic cells (DC) (called Langerhans cells in the epidermis) (Steinman, 1991). The antigen-loaded dendritic cells migrate through lymphatics (they are now called veiled cells) into the secondary lymphoid organs (e.g., draining lymph nodes). The arrival of antigen-loaded dendritic cells (called interdigitating dendritic cells at that level) into the paracortical T-cell-rich areas of the lymph nodes results in the shutting of the valvules of the efferent lymphatics. This permits the accumulation of lymphocytes from afferent lymphatics and the selection of antigen-specific T and B cells (Inaba *et al.*, 1984).

There is no evidence that dendritic cells present antigen to B lymphocytes directly. An attractive hypothesis is that dendritic cells may present unprocessed antigen to B cells and may present processed antigen to T cells. Dendritic cells may use surface Fc receptors to carry the native antigens in the form of immune complexes and use surface lectins (mannose receptor or DEC 205) and CD1a molecule to take up antigen for processing. The interaction between T lymphocytes and IDC will result in specific T-cell activation. Such activated T cells express activation antigens and secrete cytokines (Bogen *et al.*, 1993). The CD40-L and other activation antigens on T cells may directly induce B cells to proliferate and differentiate in an IL-2-dependent fashion (Bogen *et al.*, 1993). The generated plasma blasts will migrate into the medullary cords following loss of certain adhesion molecules and acquisition of novel ones. Naive $IgM^+IgD^+CD23^\pm$ B cells (Bml and Bm2) will generate plasma cells, producing IgM primary antibody responses. At that stage, naive B cells may undergo isotype switch without somatic mutations, as shown in the murine system (Jacob *et al.*, 1991a,b; Jacob and Kelsoe, 1992; McHeyzer-Williams *et al.*, 1993). Memory Ig-D^-CD38^- B cells (Bm5) expressing either sIgG or sIgA or sIgE will generate plasma cells producing IgG, IgA, and IgE secondary antibody responses. CD40/CD40-L interactions may not be mandatory to initiate the primary B-cell reaction in extrafollicular T-cell areas, as human and mice displaying altered CD40/CD40-L interactions are able to mount primary IgM responses (Callard *et al.*, 1993; Durie *et al.*, 1994; Foy *et al.*, 1994; Kawabe *et al.*, 1994;

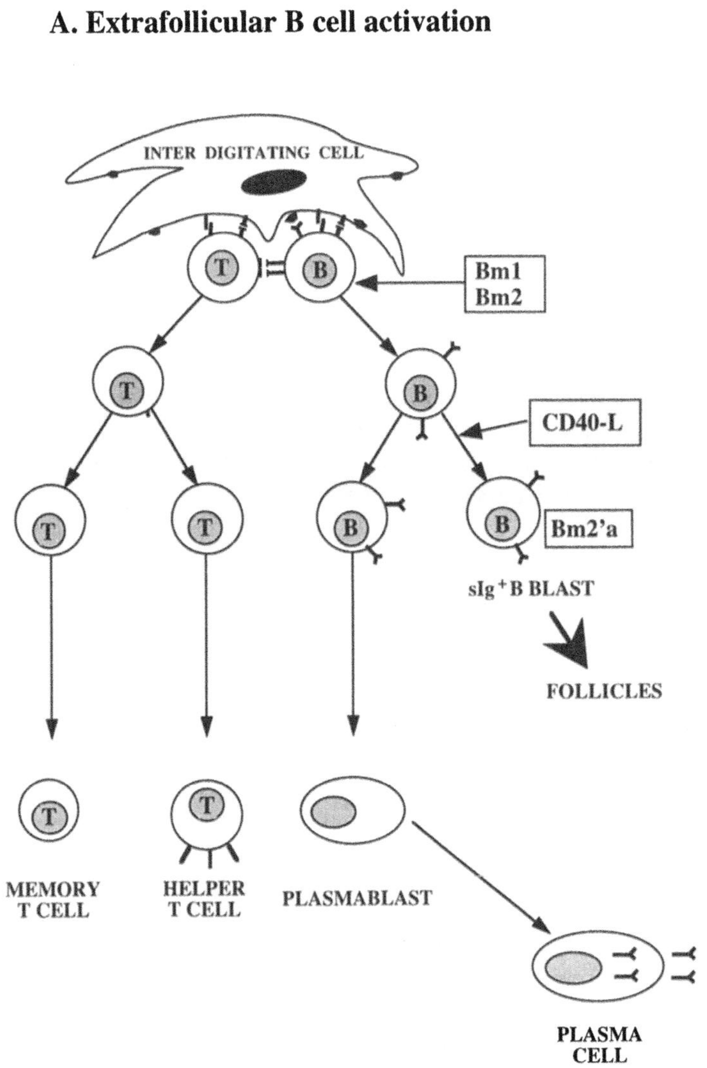

FIGURE 2. Schematic representation of antigen-driven T-cell-dependent B-cell activation (A) and germinal center reaction (B) in secondary lymphoid organs (for details see text). CD40-L indicates where CD40/CD40-L interactions are occurring. B low is B lymphocyte that has undergone somatic mutation and whose BCR has an affinity equal to or lower than that of the initial antigen-specific B cell. B high has a BCR of higher affinity. B auto indicates that after somatic mutation, the BCR has acquired the ability to recognize autoantigens. TBM, tingible body macrophage; T, T helper cells; FDC, follicular dendritic cells.

Xu *et al.*, 1994). The CD40-independent activation of B cells may involve other members of the TNF/TNF-receptor superfamilies such as TNF-α itself (Aversa *et al.*, 1993; Macchia *et al.*, 1993) or lectins such as CD72 (Gordon, 1994). This extrafollicular reaction also results in the generation of $IgD^+IgM^+CD38^+$ B-cell blasts (Bm2′a) and T-cell blasts that will play a key role in the generation of germinal centers (Jacob and Kelsoe, 1992). *In vitro*

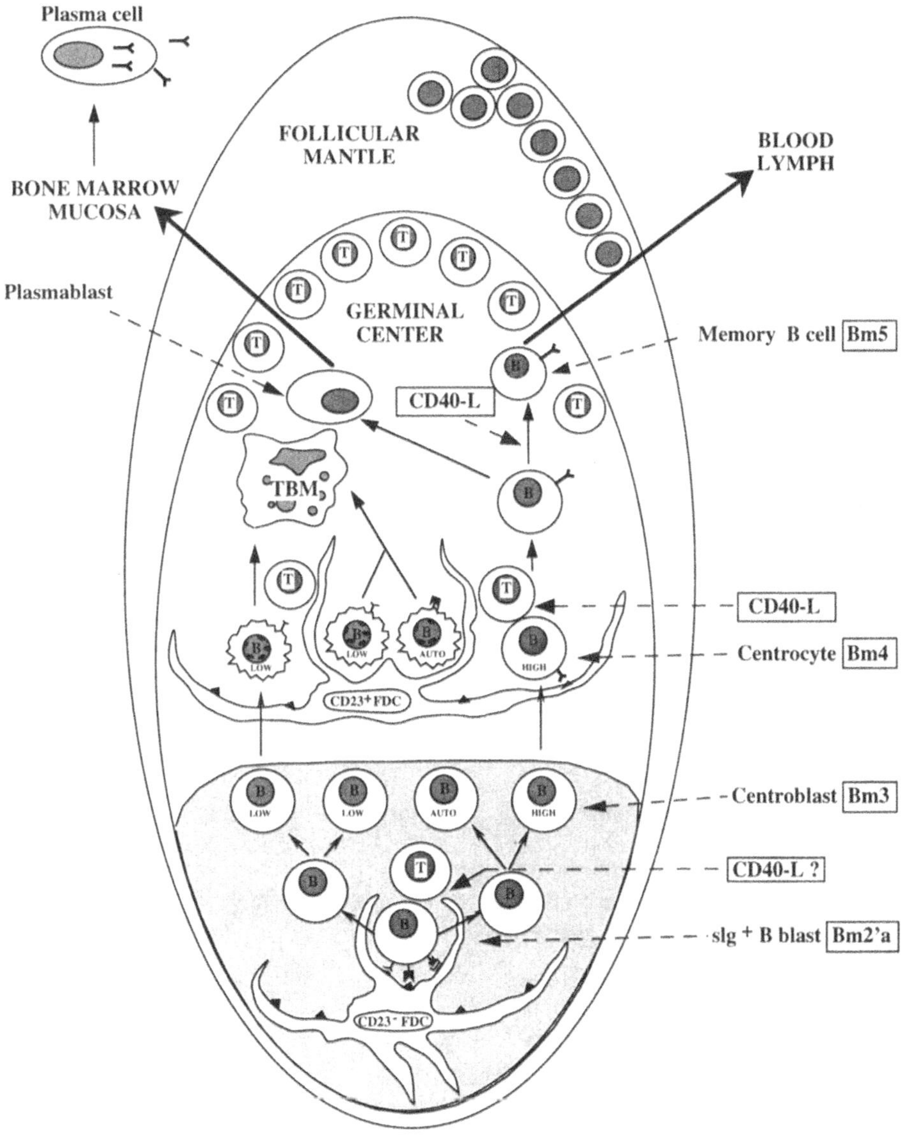

FIGURE 2. (*Continued*)

and *in vivo* studies lead to the conclusion that BCR and CD40 antigen triggering are required to induce B cells to enter into a germinal center reaction. This dual triggering *in vitro* permits the generation of cells with a germinal center phenotype. An altered CD40/CD40-L interaction *in vivo*, through either anti-CD40-L antibodies or through a genetic alteration of either CD40 or CD40-L, results in an inability to generate germinal centers or to mount a B-cell memory response. Finally, T-cell-independent antigens induce B-cell activation *in vivo*, mostly without inducing the formation of germinal centers.

It is also possible that memory B cells localized in epithelia capture antigen in place of dendritic cells and migrate into the T-cell-rich areas to present processed antigen to antigen-specific T cells (Liu *et al.*, 1995). T cells may then be activated to proliferate and differentiate into helper effectors that will induce B-cell growth and differentiation as described above.

5.2. Proliferation of Blasts and Centroblasts in the Dark Zone

The $IgD^+IgM^+CD38^+$ B-cell blasts (Bm2′a) generated in the extrafollicular areas migrate, after acquisition of specific adhesion molecules, into primary follicles. At that level, $CD38^+CD77^+$ centroblasts (Bm3) undergo extensive proliferation, and a few blasts give rise to $\approx 10^4$ cells within 3 days, imposing a cell cycle time of 6–10 hr (Kroese *et al.*, 1987; Liu *et al.*, 1991a). This stage is characterized by the occurrence of somatic mutations randomly over the Ig V-region genes following presently undetermined signaling pathways (Berek *et al.*, 1991; Jacob *et al.*, 1991b; Küppers *et al.*, 1993). The molecular mechanisms involved in this proliferation remain largely unknown. They may involve the production of autocrine high-molecular-weight BCGF as deduced from the possible autocrine role of this molecule in lymphoma cell proliferation (Ambrus *et al.*, 1993). Furthermore, FDC processes may provide these centroblasts with proliferation signals yet to be determined. The few T cells that are found within the germinal center dark zone may also play a critical role in B-cell proliferation and the triggering of somatic mutations. The generated centroblasts further differentiate into nonproliferating $IgD^-CD38^+CD77^-$ centrocytes (Bm4) that form the light zone of the germinal centers.

5.3. Selection and Differentiation of B Lymphocytes in the Light Zone

In the basal light zone, centrocytes will undergo affinity selection based on their ability to compete for the antigens present in the form of immune complexes on FDC processes. Three scenarios may happen according to the specificity and affinity provided to the BCR by the somatic mutations.

5.3.1. The Generated Centrocytes Display BCRs That Have Acquired a Higher Affinity for the Elicitating Antigen

The centrocytes with high-affinity BCR will thus retrieve the antigen from the complexes, process it, and then present it to the antigen-specific T cells (Gray *et al.*, 1991; Kosco *et al.*, 1989). These T cells are likely to represent the progeny of antigen-specific T cells that have proliferated during the extrafollicular reaction that preceded the germinal center reaction (Fuller *et al.*, 1993). This T-cell/B-cell dialogue permits the expansion of the uniquely produced centrocytes that express high-affinity BCR. This dialogue involves CD40/CD40-L interactions as shown by the presence of CD40-L-positive T cells within germinal centers (Casamayor-Palleja *et al.*, 1995; Lederman *et al.*, 1992). The importance of this CD40/CD40-L interaction has also been demonstrated *in vivo* by the administration of soluble CD40 (a weak antagonist to CD40-L), which results in the blocking of the generation of B-cell memory while the formation of germinal centers is not affected (Gray *et al.*, 1994).

The CD40/CD40-L interaction is most likely to be crucial at that level in turning on the isotype switch machinery the specificity of which is given by cytokines such as IL-4/IL13, IL-10, or TGF-β. In fact, it is economic for the immune system to set up an isotype switch after selection of high-affinity mutants rather than at the same time (e.g., in the dark zone) or even earlier (e.g., during the extrafollicular reaction). Indeed, the sequencing of many human IgG transcripts from sorted B-cell populations has shown their V genes to be heavily mutated, whereas IgM transcripts displayed both germ-line and mutated V genes (Pascual *et al.*, 1994). Furthermore, germinal center $CD4^+$ T cells isolated according to CD57 expression display IL-4 and IL-10, which represent switch factors toward IgG_4, IgE, IgG_1, IgG_3, and IgA (Butch *et al.*, 1993). The proliferating centrocytes will mature into memory B cells in response to prolonged CD40 triggering and into plasma blasts when CD40 signal is limiting and when IL-10 is available (Arpin *et al.*, 1995; Dechanet *et al.*, 1995; Merville *et al.*, 1995; Rousset *et al.*, 1995).

In all cases the high affinity B cell clone expansion will be limited because the prolonged contact between the antigen-specific T cell and the antigen-specific B cell may result in cell death as a consequence of protracted Fas/Fas-L interactions (Garrone *et al.*, 1995). In fact, such a mechanism prevents the possibility that immune responses may be dominated by only a few B-cell clones that would prevent the building of a repertoire vast enough to match the extreme diversity of potentially harmful antigens. Yet, it is interesting to stress that BCR-triggered B cells are less sensitive to Fas/Fas-L-dependent B-cell death than B cells that have not engaged their BCR (Rothstein *et al.*, 1995; Lagresle *et al.*, 1995). This would give antigen-triggered B cells a survival advantage when compared to non-antigen-triggered B cells that may interact with the activated T cell in a noncognate fashion.

5.3.2. The Generated Centrocytes Display BCRs That Have Acquired an Affinity for Autoantigens

A very large repertoire of autoantigens is likely to be present within the germinal center microenvironment. The centrocytes that have acquired BCR binding to such antigens will undergo protracted BCR triggering in the absence of specific T-cell help (autoreactive T cells are normally deleted in the thymus). As demonstrated *in vitro*, these cells will enter apoptosis. In keeping with this, recent *in vivo* studies have indeed demonstrated that injection of high doses of soluble antigen induces apoptosis (and phagocytosis by tingible body macrophages) of antigen-specific B cells within germinal centers (Han *et al.*. 1995; Pulendran *et al.*, 1995; Shokat and Goodnow, 1995). The B cells that will escape this selection stage, e.g., by interacting in a noncognate fashion with activated T cells that express CD40-L and secrete IL-4, will be rescued and will thus represent the source of high-affinity autoantibodies. Such an event is, however, unlikely to result in autoimmunity as long as autoreactive T cells, necessary for launching a secondary humoral response, do not escape thymic selection.

5.3.3. The Generated Centrocytes Display BCRs That Have Lost Any Antigen-Binding Capacity

Such cells will receive no signals from their BCR or from their CD40 antigen, as no antigen-specific T cell will be encountered. We propose that these B cells will be eliminated by entering into apoptosis either spontaneously or following noncognate interaction with activated T cells expressing Fas-L. In keeping with this, the upper light zone is filled with activated T cells. These cells may express Fas-L and thus may represent an ultimate selection gate to kill nonselected B cells in a Fas/Fas-L-dependent interaction.

5.4. Plasma Cell Differentiation

The last important stage to consider in the B-cell natural history is their differentiation into plasma cells. This occurs in three anatomic sites, including the medullary cords of lymph nodes, the mucosal lamina propria, and the bone marrow. It has been estimated that this latter organ produces two-thirds of daily production of IgG and IgA and 20–30% of daily IgM secretion. In fact, fairly little is known regarding the molecular control of plasma cell differentiation. Interleukin 6 has been claimed to play a crucial role in the development of myelomas and plasmacytomas (Kishimoto *et al.*, 1992). Plasma cells of bone marrow can differentiate over bone marrow stroma in an IL-6-dependent fashion and following VLA-4–fibronectin interactions (Roldan *et al.*, 1992). Yet, highly purified plasma cells from tonsils interact with fibroblasts using other presently uncharacterized molecules. Plasma blasts generated by anti-CD3-activated T-cell stimulation of B lymphocytes differentiate over bone marrow stroma or synoviocytes, and addition of IL-3 and/or IL-10 further enhances Ig secretion (Merville *et al.*, 1995). Indeed, stimulatory effects of IL-3 on the growth of myeloma progenitors and the growth and differentiation of normal B cells have been reported earlier (Bergui *et al.*, 1989; Tadmori *et al.*, 1989; Xia *et al.*, 1992).

6. PERSPECTIVES

The past few years have witnessed immense progress in the understanding of antigen-induced B-cell immunopoiesis, a multistage event occurring in many independent anatomic locations throughout the organism and involving different cell types and many different molecules. Among the major questions that remain to be addressed are (1) the nature of CD40/CD40-L-independent T/B cell interactions that permit the activation, proliferation, and differentiation of naive B cells as observed in the hyper-IgM syndrome, (2) the nature of the molecules involved in interactions of B-cells with dendritic cells, (3) the mechanisms initiating and controlling the massive proliferation of B-cell blasts and centroblasts in the dark zone, (4) the cell surface trigger whose engagement leads to the activation of the somatic mutation machinery, (5) a characterization of the molecular complexes involved in somatic mutations and isotype switch, and (6) the mechanisms controlling the migration of the B cells at their various stages of differentiation. This understanding of normal B-cell development should ultimately allow an elucidation of the mechanisms leading to lymphomagenesis/myelomagenesis and autoimmunity, which may then lead to the design of specific therapies.

ACKNOWLEDGMENTS. We would like to thank our many collaborators over the past 10 years who permitted us, through their work and discussions, to write this chapter. C. Arpin, F. Brière, D. Capra, O. de Bouteiller, L. Galibert, S. Lebecque, H. Martinez-Valdez, V. Pascual, and F. Rousset have contributed greatly to the most recent experiments. We thank N. Courbière for superb editorial help.

7. REFERENCES

Ambrus, J. L., Pippin, J., Joseph, A., Xu, C., Blumenthal, D., Tamayo, A., Claypool, K., McCourt, D., Srikiatchatochorn, A., and Ford, R. J., 1993, Identification of a cDNA for a human high-molecular-weight B-cell growth factor, *Proc. Natl. Acad. Sci. U.S.A.* **90:**6330–6334.

Armitage, R. J., Macduff, B. M., Spriggs, M. K., and Fanslow, W. C., 1993, Human B cell proliferation and Ig secretion induced by recombinant CD40 ligand are modulated by soluble cytokines, *J. Immunol.* **150:**3671–3680.

Arpin, C., Dechanet, J., van Kooten, K., Merville, P., Grouard, G., Brière, F., Banchereau, J., and Liu, Y.-J., 1995, *In vitro* generation of memory B cells and plasma cells, *Science* **268:**720–722.

Aversa, G., Punnonen, J., and de Vries, J.E., 1993, The 26-kD transmembrane form of tumor necrosis factor α on activated $CD4^+$ T cell clones provides a costimulatory signal for human B cell activation, *J. Exp. Med.* **177:**1575–1585.

Azuma, M., Ito, D., Yagita, H., Okumura, K., Phillips, J. H., Lanier, L. L., and Somoza, C., 1993, B70 antigen is a second ligand for CTLA-4 and CD28, *Nature* **366:**76–79.

Banchereau, J., and Rousset, F., 1992, Human B lymphocytes: Phenotype, proliferation and differentiation, *Adv. Immunol.* **52:**125–251.

Banchereau, J., de Paoli, P., Vallé, A., Garcia, E., and Rousset, F., 1991, Long term human B cell lines dependent on interleukin 4 and antibody to CD40, *Science* **251:**70–72.

Banchereau, J., Bazan, F., Blanchard, D., Brire, F., Galizzi, J.-P., van Kooten, C., Liu, Y.-J., Rousset, F., and Saeland, S., 1994, The CD40 antigen and its Ligand, *Annu. Rev. Immunol.* **12:**881–922.

Berek, C., Berger, A., and Apel, M., 1991, Maturation of immune responses in germinal centers, *Cell* **67:**1121–1129.

Bergui, L., Schena, M., Gaidano, G., Riva, M., and Caligaris-Cappio, F., 1989, Interleukin 3 and interleukin 6 synergistically promote the proliferation and differentiation of malignant plasma cell precursors in multiple myeloma, *J. Exp. Med.* **170:**613–618.

Bogen, S. A., Fogelman, I., and Abbas, A. K., 1993, Analysis of IL-2, IL-4, and IFN-γ-producing cells in situ during immune responses to protein antigens, *J. Immunol.* **150:**4197–4205.

Brière, F., Servet-Delprat, C., Bridon, J.-M., Saint-Rémy, J.-M., and Banchereau, J., 1994, Human interleukin 10 induces naive $sIgD^+$ B cells to secrete IgG_1 and IgG_3, *J. Exp. Med.* **179:**757–762.

Butch, A. W., Chung, G.-H., Hoffmann, J. W., and Nahm, M. H., 1993, Cytokine expression by germinal center cells, *J. Immunol.* **150:**39–47.

Callard, R. E., Armitage, R. J., Fanslow, W. C., and Spriggs, M. K., 1993, CD40 ligand and its role in X-linked hyper-IgM syndrome, *Immunol. Today* **14:**559–564.

Casamayor-Palleja, M., Khan, M., and MacLennan, I. C. M., 1995, A subset of $CD4^+$ memory T cells contains preformed CD40 Ligand that is rapidly but transiently expressed on their surface after activation through the T cell receptor complex, *J. Exp. Med.* **181:**1293–1301.

Chies, J. A. B., Lembezat, M. P., and Freitas, A. A., 1994, Entry of B lymphocytes into the persistent cell pool in non-immunized mice is not accompanied by somatic mutation of V_H genes, *Eur. J. Immunol.* **24:**1657–1664.

Clark, E. A., and Ledbetter, J. A., 1986, Activation of human B cells mediated through two distinct cell surface differentiation antigens, Bp35 and Bp50, *Proc. Natl. Acad. Sci. U.S.A.* **83:**4494–4498.

Cocks, B. G., de Waal-Malefyt, R., Galizzi, J.-P., de Vries, J. E., and Aversa, G., 1993, IL-13 induces proliferation and differentiation of human B cells activated by the CD40 ligand, *Int. Immunol.* **5:**657–663.

Coffman, R. L., Lebman, D. A., and Shrader, B., 1989, Transforming growth factor β specifically enhances IgA production by lipopolysaccharide-stimulated murine B lymphocytes, *J. Exp. Med.* **170:**1039–1045.

Coico, R. F., Bhogal, B. S., and Thorbecke, G. J., 1983, Relationship of germinal centers in lymphoid tissue to immunologic memory. IV. Transfer of B cell memory with lymph node cells fractionated according to their receptors for peanut agglutinin, *J. Immunol.* **131:**2254–2257.

Dechanet, J., Merville, P., Durand, I., Banchereau, J., and Miossec, P., 1995, The ability of synoviocytes to support terminal differentiation of activated B cells may explain plasma cell accumulation in rheumatoid synovium, *J. Clin. Invest.* **95:**456–463.

Defrance, T., Aubry, J. P., Vanbervliet, B., and Banchereau, J., 1986, Human interferon-γ acts as a B cell growth factor in the anti-IgM antibody costimulatory assay but has no direct B cell differentiation activity, *J. Immunol.* **137:**3861–3867.

Defrance, T., Vanbervliet, B., Aubry, J.P., Takebe, Y., Arai, N., Miyajima, A., Yokota, T., Lee, T., Arai, K., de Vries, J. E., and Banchereau, J., 1987, B cell growth-promoting activity of recombinant human interleukin-4, *J. Immunol.* **139:**1135–1141.

Defrance, T., Vanbervliet, B., Aubry, J. P., and Banchereau, J., 1988a, Interleukin 4 inhibits the proliferation but not the differentiation of activated human B cells in response to interleukin 2, *J. Exp. Med.* **168:**1321–1337.

Defrance, T., Vanbervliet, B., Pène, J., and Banchereau, J., 1988b, Human recombinant IL-4 induces activated B lymphocytes to produce IgG and IgM, *J. Immunol.* **141:**2000–2005.

Defrance, T., Vanbervliet, B., Brière, F., Durand, I., Rousset, F., and Banchereau, J., 1992a, Interleukin 10 and transforming growth factor β cooperate to induce anti-CD40-activated naive human B cells to secrete immunoglobulin A, *J. Exp. Med.* **175:**671–682.

Defrance, T., Vanbervliet, B., Durand, I., Briolay, J., and Banchereau, J., 1992b, Proliferation and differentiation of human $CD5^+$ and $CD5^-$ B cell subsets activated through their antigen receptors or CD40 antigens, *Eur. J. Immunol.* **22:**2831–2839.

de Saint-Vis, B., Cupillard, L., Pandrau-Garcia, D., Ho, S., Renard, N., Grouard, G., Duvert, V., Thomas, X., Galizzi, J.P., Banchereau, J., and Saeland, S., 1995, Distribution of carboxypeptidase-M on lymphoid and myeloid cells parallels the other zinc-dependent proteases CD10 and CD13, *Blood* **86:**1098–1105.

Durie, F. H., Foy, T. M., Masters, S. R., Laman, J. D., and Noelle, R. J., 1994, The role of CD40 in the regulation of humoral and cell-mediated immunity, *Immunol. Today* **15:**406–410.

Fliedner, T. M., Kress, M., Cronkite, E. P., and Robertson, J. S., 1964, Cell proliferation in germinal centre of the rat spleen, *Ann. N.Y. Acad. Sci.* **113:**578–588.

Fluckiger, A.-C., Garrone, P., Durand, I., Galizzi, J. P., and Banchereau, J., 1993, IL-10 upregulates functional high affinity IL-2 receptors on normal and leukemic B lymphocytes, *J. Exp. Med.* **178:**1473–1481.

Foy, T. M., Laman, J. D., Ledbetter, J. A., Aruffo, A., Claassen, E., and Noelle, R. J., 1994, gp39–CD40 interactions are essential for germinal center formation and the development of B cell memory, *J. Exp. Med.* **180:**157–163.

Freeman, G. F., Gribben, J. G., Boussiotis, V. A., Ng, J. W., Restivo, V. A. Jr., Lombard, L. A., Gray, G. S., and Nadler, L. M., 1993, Cloning of B7-2: A CTLA-4 counterreceptor that costimulates human T cell proliferation, *Science* **262:**909–911.

Fuller, K. A., Kanagawa, O., and Nahm, M. H., 1993, T cells within germinal centers are specific for the immunizing antigen, *J. Immunol.* **151:**4505–4512.

Galibert, L., Durand, I., Rousset, F., and Banchereau, J., 1994, CD40 activated surface IgD positive lymphocytes constitute the long term IL-4 dependent proliferating B cell pool, *J. Immunol.* **152:**22–29.

Galibert, L., Burdin, N., de Saint-Vis, B., Garrone, P., van Kooten, C., Banchereau, J., and Rousset, F., 1996, CD40 and BCR dual triggering of resting B lymphocytes turns on a germinal center phenotype, *J. Exp. Med.* **183:**77–85.

Garrone, P., Neidhardt, E. M., Garcia, E., Galibert, L., van Kooten, C., and Banchereau, J., 1995, Fas ligation induces apoptosis of CD40-activated human B lymphocytes, *J. Exp. Med.* **182:**1265–1273.

Gauchat, J.-F., Aubry, J.-P., Mazzei, G., Life, P., Jomotte, T., Elson, G., and Bonnefoy, J.-Y., 1993, Human CD40-ligand: Molecular cloning, cellular distribution and regulation of expression by factors controlling IgE production, *FEBS* **3:**259–266.

Gilliam, A. C., Shen, A., Richards, J. E., Blattner, F. R., Mushinski, J. F., and Tucker, P. W., 1984, Illegitimate recombination generates a class switch from C mu to C delta in an IgD-secreting plasmacytoma, *Proc. Natl. Acad. Sci. U.S.A.* **81:**4164–4168.

Gimmi, C. D., Freeman, G. J., Gribben, J. G., Sugita, K., Freedman, A. S., Morimoto, C., and Nadler, L. M., 1991,

B-cell surface antigen B7 provides a costimulatory signal that induces T cells to proliferate and secrete interleukin 2, *Proc. Natl. Acad. Sci. U.S.A.* **88:**6575–6579.

Gordon, J., 1994, B-cell signalling via the C-type lectins CD23 and CD72, *Immunol. Today* **15:**411–417.

Gordon, J., Millsum, M. J., Guy, G. R., and Ledbetter, J. A., 1987, Synergistic interaction between interleukin 4 and anti-Bp50 (CDw40) revealed in a novel B cell restimulation assay, *Eur. J. Immunol.* **17:**1535–1538.

Grabstein, K. H., Maliszewski, C. R., Shanebeck, K., Sato, T. A., Sprigg, M. K., Fanslow, W. C., and Armitage, R. J., 1993, The regulation of T cell-dependent antibody formation in vitro by CD40 ligand and IL-2, *J. Immunol.* **150:**3141–3147.

Gray, D., Kosco, M., and Stockinger, B., 1991, Novel pathways of antigen presentation for the maintenance of memory, *Int. Immunol.* **3:**141–148.

Gray, D., Dullforce, P., and Jainandunsing, S., 1994, Memory B cell development but not germinal center formation is impaired by in vivo blockade of CD40–CD40 ligand interaction, *J. Exp. Med.* **180:**141–155.

Gu, H., Tarlington, D., Müller, W., Rajewsky, K., and Förster, I., 1991, Most peripheral B cells in mice are ligand selected, *J. Exp. Med.* **173:**1357–1371.

Han, S.-H., Zheng, B., Dal Porto, J., and Kelsoe, G., 1995, In situ studies of the primary immune response to (4-hydroxy-3-nitrophenyl) acetyl IV. Affinity-dependent, antigen-driven B cell apoptosis in germinal centers as a mechanism for maintaining self-tolerance, *J. Exp. Med.* **182:**1635–1644.

Hathcock, K. S., Laszlo, G., Dickler, H. B., Bradshaw, J., Linsley, P., and Hodes, R. J., 1993, Identification of an alternative CTLA-4 ligand co-stimulatory for T cell activation, *Science* **262:**905–907.

Hollenbaugh, D., Grosmaire, L. S., Kullas, C. D., Chalupny, N. J., Braesch-Andersen, S., Noelle, R. J., Stamenkovic, I., Ledbetter, J. A., and Aruffo, A., 1992, The human T antigen gp39, a member of the TNF gene family, is a ligand for the CD40 receptor: expression of a soluble form of gp39 with B cell co-stimulatory activity, *EMBO J.* **11:**4313–4321.

Inaba, K., Witmer, M. D., and Steinman, R. M., 1984, Clustering of dendritic cells, helper T lymphocytes, and histocompatible B cells, during primary antibody responses *in vitro*, *J. Exp. Med.* **160:**858–876.

Islam, K.B., Nilsson, L., Sideras, P., Hammarstrm, L., and Smith, C.I.E., 1991, TGF-β1 induces germ-line transcripts of both IgA subclasses in human B lymphocytes, *Int. Immunol.* **3:**1099–1106.

Jabara, H. H., Fu, S. M., Geha, R. S., and Vercelli, D., 1990, CD40 and IgE: Synergism between anti-CD40 monoclonal antibody and interleukin 4 in the induction of IgE synthesis by highly purified human B cells, *J. Exp. Med.* **172:**1861–1864.

Jacob, J., and Kelsoe, G., 1992, *In situ* studies of the primary immune response to (4-hydroxy-3-nitrophenyl) acetyl II. A common clonal origin for periarteriolar lymphoid sheath-associated foci and germinal centers, *J. Exp. Med.* **176:**679–687.

Jacob, J., Kassir, R., and Kelsoe, G., 1991a, *In situ* studies of the primary immune response to (4-hydroxy-3-nitrophenyl) acetyl. I. The architecture and dynamics of responding cell populations, *J. Exp. Med.* **173:**1165–1175.

Jacob, J., Kelsoe, G., Rajewsky, K., and Weiss, U., 1991b, Intraclonal generation of antibody mutants in germinal centres, *Nature* **354:**389–392.

Jacob, J., Przylepa, J., Miller, C., and Kelsoe, G., 1993, *In situ* studies of the primary immune response to (4-hydroxyl-3-nitrophenyl) acetyl. III. The kinetics of V region mutation and selection in germinal center B cells, *J. Exp. Med.* **178:**1293–1307.

Jelinek, D., and Lipsky, P. E., 1987, Regulation of human B lymphocyte activation, proliferation and differentiation, *Adv. Immunol.* **40:**1–59.

Jumper, M. D., Splawski, J. B., Lipsky, P. E., and Meek, K., 1994, Ligation of CD40 induces sterile transcripts of multiple Ig heavy chain isotypes in human B cells, *J. Immunol.* **152:**438–445.

Kawabe, T., Naka, T., Yoshida, K., Tanaka, T., Fujiwara, H., Suematsu, S., Yoshida, N., Kishimoto, T., and Kikutani, H., 1994, The immune response in CD40-deficient mice: Impaired immunoglobulin class switching and germinal center formation, *Immunity* **1:**167–178.

Kehrl, J. H., Alvarez-Mon, M., Delsing, G. A., and Fauci, A. S., 1987a, Lymphotoxin is an important T cell-derived growth factor for human B cells, *Science* **238:**1144–1146.

Kehrl, J. H., Miller, A., and Fauci, A. S., 1987b, Effect of tumor necrosis factor α on mitogen-activated human B cells, *J. Exp. Med.* **166:**786–791.

Kelsoe, G., 1994, B cell diversification and differentiation in the periphery, *J. Exp. Med.* **180:**5–6.

Kepler, T.B., and Perelson, A.S., 1993, Cyclic re-entry of germinal center B cells and the efficiency of affinity maturation, *Immunol. Today* **14:**412–415.

Kishimoto, T., Akira, S., and Taga, T., 1992, Interleukin-6 and its receptor: A paradigm for cytokines, *Sciences* **258:**593–597.

Klaus, G. G. B., Humphrey, J. H., Kunkl, A., and Dongworth, D. W., 1980, The follicular dendritic cell: Its role in antigen presentation in the generation of immunological memory, *Immunol. Rev.* **53:**3–28.

Kosco, M. H., 1991, Cellular interactions during the germinal centre response, *Res. Immunol.* **142:**245–248.

Kosco, M. H., Burton, G. F., Kapasi, Z. F., Szakal, A. K., and Tew, J. G., 1989, Antibody-forming cell induction during an early phase of germinal centre development and its delay with ageing, *Immunology* **68:**312.

Kraal, G., Weissman, I. L., and Butcher, E. C., 1982, Germinal centre B cells: Antigen specificity and charges in heavy chain class expression, *Nature* **298:**377.

Krammer, P. H., Dhein, J., Walczak, H., Behrmann, I., Mariani, S., Matiba, B., Fath, M., Daniel, P. T., Knipping, E., Westendorp, M. O., Stricker, K., Bäumler, C., Hellbardt, S., Germer, M., Peter, M. E., and Debatin, K.-M., 1994, The role of APO-1-mediated apoptosis in the immune system, *Immunol. Rev.* **142:**175–191.

Kroese, F. G. M., Wubbena, A. S., Seijen, H. G., and Nieuwenhuis, P., 1987, Germinal centers develop oligoclonally, *Eur. J. Immunol.* **17:**1069–1072.

Kroese, F. G. M., Timens, W., and Nieuwenhuis, P., 1990, Germinal center reaction and B lymphocytes: Morphology and function, in *Current Topics in Pathology: Reaction Patterns of the Lymph Node* (E. Grundman and E. Vollmer, eds.), pp. 103–148, Springer-Verlag, Berlin.

Küppers, R., Zhao, M., Hansmann, M.-L., and Rajewsky, K., 1993, Tracing B cell development in human germinal centres by molecular analysis of single cells picked from histological sections, *EMBO J.* **12:**4955–4967.

Lagresle, C., Bella, C., and Defrance, T., 1993, Phenotypic and functional heterogeneity of the IgD$^-$ B cell compartment: Identification of two major tonsillar B cell subsets, *Int. Immunol.* **5:**1259–1268.

Lagresle, C., Bella, C., Daniel, T., Krammer, P. H., and Defrance, T., 1995, Regulation of germinal center B cell differentiation. Role of the human APO-1/Fas (CD95) molecule, *J. Immunol.* **154:**5746–5756.

Ledbetter, J. A., Shu, G., Gallagher, M., and Clark, E. A., 1987, Augmentation of normal and malignant B cell proliferation by monoclonal antibody to the B cell-specific antigen BP50 (CDw40), *J. Immunol.* **138:** 788–794.

Lederman, S., Yellin, M. J., Inghirami, G., Lee, J. J., Knowles, D. M., and Chess, L., 1992, Molecular interactions mediating T-B lymphocyte collaboration in human lymphoid follicles. Roles of T cell-B cell-activating molecule (5c8 antigen) and CD40 in contact-dependent help, *J. Immunol.* **149:**3817–3826.

Ling, N. R., MacLennan, I. C. M., and Mason, D., 1987, B cell and plasma cell antigens: New and previously defined clusters, in *Leucocyte Typing III* (A. J. McMichael, ed.), pp. 302–335, Oxford University, Press, Oxford.

Linsley, P. S., Brady, W., Grosmaire, L., Aruffo, A., Damle, N. K., and Ledbetter, J. A., 1991, Binding of the B cell activation antigen B7 to CD28 costimulates T cell proliferation and interleukin 2 mRNA accumulation, *J. Exp. Med.* **173:**721–730.

Liu, Y. J., and Banchereau, J., 1996, Human peripheral B cell subsets, in *Handbook of Experimental Immunology* (D. Weir, C. Blackwell and L. Hersenberg, eds.), Blackwell Scientific Publications, Oxford (in press).

Liu, Y. J., Joshua, D. E., Williams, G. T., Smith, C. A., Gordon, J., and MacLennan, I. C. M., 1989, Mechanisms of antigen-driven selection in germinal centres, *Nature* **342:**929–931.

Liu, Y.-J., Mason, D. Y., Johnson, G. D., Abbot, S., Gregory, C. D., Hardie, D. L., Gordon, J., and MacLennan, I. C. M., 1991a, Germinal center cells express bcl-2 protein after activation by signals which prevent their entry into apoptosis, *Eur. J. Immunol.* **21:**1905–1910.

Liu, Y.-J., Zhang, J., Lane, P. J. L., Chan, E. Y.-T., and MacLennan, I. C. M., 1991b, Sites of specific B cell activation in primary and secondary responses to T cell-dependent and T cell-independent antigens, *Eur. J. Immunol.* **21:**2951-2962.

Liu, Y.-J., Johnson, G. D., Gordon, J., and MacLennan, I. C. M., 1992, Germinal centres in T-cell-dependent antibody responses, *Immunol. Today* **13:**17–21.

Liu, Y.-J., Barthelemy, C., de Bouteiller, O., Arpin, C., Durand, I., and Banchereau, J., 1995, Memory B cells from human tonsils colonize mucosal epithelium and directly present antigen to T cells by rapid upregulation of B7.1 and B7.2, *Immunity* **2:**239–248.

Liu, Y.-J., Arpin C., de Bouteiller, O., Guret, C., Banchereau, J., Martinez-Valdez, H., and Lebecque, S., 1996a, Sequential triggering of apoptosis somatic mutation and isotype switch during germinal center development, *Seminar Immunol.* **8:**169–177.

Liu, Y.-J., Malisan, F., de Bouteiller, O., Guret, C., Lebecque, S., Banchereau, J., Mills, F. C., Max, E. E., and Martinez-Valdez, H., 1966b, Within germinal centers isotype switching of immunoglobulin genes occurs after onset of somatic mutation, *Immunity* **4:**241–250.

Macchia, D., Almerigogna, F., Parronchi, P., Ravina, A., Maggi, E., and Romagnani, S., 1993, Membrane tumour necrosis factor-α is involved in the polyclonal B-cell activation induced by HIV-infected human T cells, *Nature* **363:**464–466.

MacLennan, I. C. M., 1994, Germinal Centers, *Annu. Rev. Immunol.* **12:**117–139.

MacLennan, I. C., and Gray, D., 1986, Antigen-driven selection of virgin and memory B cells, *Immunol. Rev.* **91:**61–85.

Maliszewski, C. R., Grabstein, K., Fanslow, W. C., Armitage, R., Spriggs, M. K., and Sato, T. A., 1993, Recombinant CD40 ligand stimulation of murine B cell growth and differentiation: Cooperative effects of cytokines, *Eur. J. Immunol.* **23:**1044–1049.

McHeyzer-Williams, M. G., and Davis, M. M., 1995, Antigen-specific development of primary and memory T cells *in vivo*, *Science* **268:**106–111.

McHeyzer-Williams, M. G., McLean, M. J., Lalor, P. A., and Nossal, G. J. V., 1993, Antigen-driven B cell differentiation *in vivo*, *J. Exp. Med.* **178:**295–307.

McKenzie, A. N. J., Culpepper, J. A., de Waal Malefyt, R., Brière, F., Punnonen, J., Aversa, G., Sato, A., Dang, W., Cocks, B.G., Menon, S., de Vries, J. E., Banchereau, J., and Zurawski, G., 1993, Interleukin-13, a novel T cell-derived cytokine that regulates human monocyte and B cell function, *Proc. Natl. Acad. Sci. U.S.A.* **90:**3735–3739.

Melchers, F., Haasner, D., Grawunder, U., Kalberer, C., Karasuyama, H., Winkler, T., and Rolink, A. G., 1994, Roles of IgH and L chains and of surrogate H and L chains in the development of cells of the B lymphocyte lineage, *Annu. Rev. Immunol.* **12:**209–225.

Merville, P., Dechanet, J., Grouard, G., Durand, I., and Banchereau, J., 1995, T cell-induced B cell blasts differentiate into plasma cells when cultured on bone marrow stroma with IL3 and IL10, *Int. Immunol.* **7:** 636–643.

Merville, P., Dechanet, J., Desmoulière, A., Durand, I., de Bouteiller, O., Garrone, P., Banchereau, J., and Liu, Y.-J., 1996, Bcl-2 positive tonsillar plasma cells are rescued from prompt apoptosis by bone marrow fibroblasts, *J. Exp. Med.* **183:**227–236.

Mingari, M. C., Gerosa, F., Carra, G., Accolla, R. S., Moretta, A., Zubler, R. H., Waldmann, T. A., and Moretta, L., 1984, Human interleukin-2 promotes proliferation of activated B cells via surface receptors similar to those of activated T cells, *Nature* **312:**641–643.

Minty, A., Chalon, P., Derocq, J.-M., Dumont, X., Guillemot, J.-C., Kaghad, M., Labit, C., Leplatois, P., Liauzun, P., Miloux, B., Minty, C., Casellas, P., Loison, G., Lupker, J., Shire, D., Ferrara, P., and Caput, D., 1993, Interleukin-13 is a new human lymphokine regulating inflammatory and immune responses, *Nature* **362:** 248–250.

Nagata, S., and Golstein, P., 1995, The Fas death factor, *Science* **267:**1449–1456.

Parker, D. C., 1993, T cell-dependent B cell activation, *Annu. Rev. Immunol.* **11:**331–360.

Pascual, V., Liu, Y.-J., Magalski, A., de Bouteiller, O., Banchereau, J., and Capra, J. D., 1994, Analysis of somatic mutation in five B cell subsets of human tonsil, *J. Exp. Med.* **180:**329–339.

Pulendran, B., Kannourakis, G., Nouri, S., Smith, K. G. C., and Nossal, G. J. V., 1995, Soluble antigen can cause enhanced apoptosis of germinal-centre B cells, *Nature* **375:**331–334.

Punnonen, J., Aversa, G. G., Cocks, B. G., McKenzie, A. N. J., Menon, S., Zurawski, G., de Waal Malefyt, R., and de Vries, J. E., 1993, Interleukin-13 induces interleukin-4-independent IgG4 and IgE synthesis and CD23 expression by human B cells, *Proc. Natl. Acad. Sci. U.S.A.* **90:**3730–3734.

Roldan, E., Garcia-Pardo, A., and Brieva, J. A., 1992, VLA-4-fibronectin interaction is required for the terminal differentiation of human bone marrow cells capable of spontaneous and high rate immunoglobulin secretion, *J. Exp. Med.* **175:**1739–1747.

Romagnani, S., Giudizi, M. G., Biagiotti, R., Almerigogna, F., Mingari, M. G., Maggi, E., Liang, M. C., and Moretta, L., 1986, B cell growth factor activity of interferon γ. Recombinant human interferon γ promotes proliferation of anti-Ig activated human B lymphocytes, *J. Immunol.* **136:**3513–3516.

Rothstein, T. L., Wang, J. K. M., Panka, D. J., Foote, L. C., Wang, Z., Stanger, B., Cui, H., Ju, S.-T., and Marshak-Rothstein, A., 1995, Protection against Fas-dependent Th1-mediated apoptosis by antigen receptor engagement in B cells, *Nature* **374:**163–165.

Rousset, F., Garcia, E., and Banchereau, J., 1991, Cytokine-induced proliferation and immunoglobulin production of human B lymphocytes triggered through their CD40 antigen, *J. Exp. Med.* **173:**705–710.

Rousset, F., Garcia, E., Defrance, T., Péronne, C., Vezzio, N., Hsu, D. H., Kastelein, R., Moore, K. W., and Banchereau, J., 1992, Interleukin 10 is a potent growth and differentiation factor for activated human B lymphocytes, *Proc. Natl. Acad. Sci. U.S.A.* **89:**1890–1893.

Rousset, F., Peyrol, S., Garcia, E., Vezzio, N., Andujar, M., Grimaud, J.-A., and Banchereau, J., 1995, Long-term cultured CD40-activated B lymphocytes differentiate into plasma cells in response to IL-10 but not IL-4, *Int. Immunol.* **7**:1243–1253.

Schittek, B., and Rajewsky, K., 1992, Natural occurrence and origin of somatically mutated memory B cells in mice, *J. Exp. Med.* **176**:427–438.

Schriever, F., and Nadler, L.M., 1992, The central role of follicular dendritic cells in lymphoid tissues, *Adv. Immunol.* **51**:243–284.

Skokat, K. M., and Goodnow, C. C., 1995, Antigen-induced B-cell death and elimination during germinal-centre immune responses, *Nature* **375**:334–338.

Spriggs, M. K., Armitage, R. J., Strockbine, L., Clifford, K. N., Macduff, B. M., Sato, T. A., Maliszewski, C. R., and Fanslow, W. C., 1992, Recombinant human CD40 ligand stimulates B cell proliferation and immunoglobulin E secretion, *J. Exp. Med.* **176**:1543–1550.

Spriggs, M. K., Fanslow, W. C., Armitage, R. J., and Belmont, J., 1993, The biology of the human ligand for CD40, *J. Clin. Immunol.* **13**:373–380.

Steinman, R. M., 1991, The dendritic cell system and its role in immunogenicity, *Annu. Rev. Immunol.* **9**:271–296.

Tadmori, W., Feingersh, D., Clark, S. C., and Choi, Y. S., 1989, Human recombinant IL-3 stimulates B cell differentiation, *J. Immunol.* **142**:1950–1955.

Tew, J. G., Kosco, M. H., Burton, G. F., and Szakal, A. K., 1990, Follicular dendritic cells as accessory cells, *Immunol. Rev.* **117**:185–211.

Tew, J. G., DiLosa, R.-M., Burton, G. F., Kosco, M. H., Kupp, L. I., Masuda, A., and Szakal, A. K., 1992, Germinal centers and antibody production in bone marrow, *Immunol. Rev.* **126**:99–112.

Vallé, A., Zuber, C. E., Defrance, T., Djossou, O., De Rie, M., and Banchereau, J., 1989, Activation of human B lymphocytes through CD40 and interleukin 4, *Eur. J. Immunol.* **19**:1463–1467.

Weiss, U., and Rajewsky, K., 1990, The repertoire of somatic antibody mutants accumulating in the memory compartment after primary immunization is restricted through affinity maturation and mirrors that expressed in the secondary response, *J. Exp. Med.* **172**:1681–1689.

Wheeler, K., Pound, J. D., Gordon, J., and Jefferis, R., 1993, Engagement of CD40 lowers the threshold for activation of resting B cells via antigen receptor, *Eur. J. Immunol.* **23**:1165–1168.

Xia, X., Li, L., and Choi, Y. S., 1992, Human recombinant IL-3 is a growth factor for normal B cells, *J. Immunol.* **148**:491–497.

Xu, J., Foy, T.M., Laman, J. D., Elliott, E. A., Dunn, J. J., Waldschmidt, T. J., Elsemore, J., Noelle, R. J., and Flavell, R. A., 1994, Mice deficient for the CD40 ligand, *Immunity* **1**:423–431.

Yasui, H., Akahori, Y., Hirano, M., Yamada, K., and Kurosawa, Y., 1989, Class switch from μ to δ is mediated by homologous recombination between σμ and Σμ sequences in human immunoglobulin gene loci, *Eur. J. Immunol.* **19**:1399–1403.

Zhang, J., MacLennan, I. C. M., Liu, Y.-J., and Lane, P. J. L., 1988, Is rapid proliferation in B centroblasts linked to somatic mutation in memory B cell clones? *Immunol. Lett.* **18**:297–299.

Chapter 11

In Vitro Regulation of Dendritic Cell Development and Function

Christophe Caux and Jacques Banchereau

1. SUMMARY

Human and murine dendritic cells (DC) can now be generated *in vitro* in large numbers by culturing progenitor cells. The establishment of such culture systems will allow clarification of the relationship between different DC populations and monocytes. In all studies, granulocyte–macrophage colony-stimulating factor (GM-CSF) has been required for the generation of DC, suggesting a crucial role for this hematopoietic growth factor during the ontogeny of the DC lineage. Those DC generated *in vitro* represent interesting models for studying the function of DC. Such studies established that immature DC precursors are able to phagocytose particles much more efficiently than the mature DC. These DC generated *in vitro* induce, after antigen loading and injection into mice, antigen-specific immune responses. Human DC generated by culturing monocytes with GM-CSF and IL-4 capture macromolecules through high-rate macropinocytosis. Human DC can also be generated by culturing $CD34^+$ hematopoietic progenitors in the presence of GM-CSF, and further addition of TNF-α results in a strong increase of DC yield. These cultured human DC are able to prime naive T cells, a phenomenon that involves interactions between CD80/CD86 on DC and CD28 on T cells. Interestingly, DC express a functional CD40 antigen whose triggering up-regulates expression of CD80 and CD86 and induces cytokine secretion.

Actually, CD40/CD40L interaction between DC and T cells (see Chapter 3) might represent a very early event, interruption of which might lead to prevention of T-cell priming. Finally, DC generated *in vitro* interact directly with naive B cells that are activated

Christophe Caux and Jacques Banchereau Schering-Plough, Laboratory for Immunological Research, 69571 Dardilly, France.

Blood Cell Biochemistry, Volume 7: Hematopoietic Cell Growth Factors and Their Receptors, edited by A. D. Whetton and J. Gordon. Plenum Press, New York, 1996.

through their CD40 antigen, leading to enhanced growth, differentiation (IgM production), and preferential isotype switch toward IgA. Thus, in the extrafollicular area of lymphoid organs, in addition to prime naive T cells, DC might also provide costimulatory factors involved in the initiation of the primary B-cell response and in the development of humoral immunity.

In vivo administration of DC generated *in vitro* and loaded with antigen may ultimatly represent a powerful immunotherapy of presently uncontrolled infectious diseases and tumors.

2. INTRODUCTION

Dendritic cells represent a trace population of cells expressing high levels of MHC class II products and with an unusual dendritic shape. They are professional antigen-presenting cells (APC), which are required for the priming of naive T lymphocytes. They comprise a system that occupies (a) discrete areas of nonlymphoid tissues with the Langerhans cells of epithelium (skin, mucosa, lung) and the interstitial DC of heart and kidney, (b) the circulation with the afferent lymph veiled cells and the peripheral blood DC, the T-cell-rich areas within secondary lymphoid organs, where they are the interdigitating cells (IDC), and (c) the thymic medulla (Metlay *et al.*, 1989; Streilein and Grammer, 1989; Hart and McKenzie, 1990; Steinman, 1991; Knight and Stagg, 1993; Williams *et al.*, 1994, for reviews). The various DC, which are thought to represent different steps of maturation of the same lineage, are interconnected by defined pathways of circulation (Steinman, 1991). In the periphery, DC such as Langerhans cells (LC) capture antigens (Inaba *et al.*, 1986; Romani *et al.*, 1989b; Streilein and Grammer, 1989; Puré *et al.*, 1990) and then migrate via the lymphatics and home to the T-cell-rich areas of secondary lymphoid organ (IDC) (Macatonia *et al.*, 1987; Austyn *et al.*, 1988; Fossum, 1988; Kupiec-Weglinski *et al.*, 1988; Kripke *et al.*, 1990; Larsen *et al.*, 1990). There, they present processed antigen to naive T cells and generate an antigen-specific primary T-cell response (Inaba *et al.*, 1990b; Sornasse *et al.*, 1992; Liu and MacPherson, 1993). During their migration from periphery to draining lymph nodes, DC are thought to undergo, under yet undefined signals, modulations of phenotype and functions including loss of antigen processing and acquisition of accessory function (Schuler and Steinman, 1985; Romani *et al.*, 1989b; Streilein and Grammer, 1989; Larsen *et al.*, 1990).

Defining the mechanisms of DC action is important, as it may ultimately permit us to understand how T cells are primed. This knowledge should allow the manipulation of immune responses at the early sensitization phase of immunity: down-regulation in case of autoimmunity or transplantation and enhancement in case of infectious diseases or cancer. In addition, understanding DC physiology has implications in the context of human immunodeficiency virus infections inasmuch as DC are known to be infected *in vivo* (Knight *et al.*, 1991; Langhoff *et al.*, 1991; Cameron *et al.*, 1992, 1994; Dezutter-Bambuyant and Schmitt, 1993; Weissman *et al.*, 1995).

Although DC are known to originate from bone marrow, the conditions that direct their growth and differentiation have been deciphered only relatively recently. GM-CSF appears to act as a central factor for DC development in both mice and humans. Because DC are difficult to isolate from tissues, their *in vitro* generation represents an important step toward the understanding of DC physiology and their potential use in immunotherapy.

After a review of the litterature on the dendritic cell system *in vivo*, we concentrate our attention on the *in vitro* culture conditions that allow the generation of DC from hematopoietic progenitors or from circulating mononuclear cells. The functional properties of these cells generated *in vitro* are detailed.

3. THE DENDRITIC CELL SYSTEM

3.1. Bone Marrow Origin of Dendritic Cells

The fact that they express the leukocyte common antigen (CD45) argues for their hematopoietic origin, but direct evidence comes from irradiation reconstitution experiments following bone marrow transplantation. Dendritic cells from mouse spleen (Steinman *et al.*, 1974), mouse epidermis (Katz *et al.*, 1979), rat lymph (Pugh *et al.*, 1983), and human epidermis (Pelletier *et al.*, 1984; Volc-Platzer *et al.*, 1984) are derived from bone marrow progenitors, as shown by their expression of donor MHC molecules in recipients of marrow transplants.

3.2. Nomenclature and Distribution

Many tissues contain a trace cell type, the DC, which in each site share features of morphology and function, especially the ability to capture antigens and initiate T-cell-mediated immunity.

3.2.1. Epithelial Langerhans Cells

Epidermal Langerhans cells are the best characterized nonlymphoid DC. The LC are localized in basal and suprabasal layers of the epidermis (Birbeck *et al.*, 1961). They have also been described in other stratified epithelia such as mouth, esophagus, and lung (De Fraissinette *et al.*, 1988; Holt *et al.* 1990; Schon-Hegrad *et al.*, 1991; Holt, 1993). In the human, 460 to 1000 LC are found per square centimeter of skin (Rowden, 1981). Langerhans cells have a multilobulated nucleus and are characterized by cytoplasmic structures formed by double membrane joinings named Birbeck granules (Birbeck *et al.*, 1961; Wolff, 1967; Sagebiel and Reed, 1968). The role of Birbeck granules is still a matter of debate, with three major hypotheses presently considered: (1) Birbeck granules could be structures derived from the Golgi apparatus with a secretory role (Wolff and Schreiner, 1970); (2) they could be formed by plasma membrane invagination, suggesting a role in endocytosis (Takahashi and Hashimoto, 1985; Takigawa *et al.*, 1985; Bartosik, 1992); (3) they might be formed from endosomes with a role in intracellular trafficking (Hanau *et al.*, 1987; Stössel *et al.*, 1990).

Human LC are characterized by the expression of the CD1a antigen (Fithian *et al.*, 1981; van de Rijn *et al.*, 1984), which is otherwise found only on thymocytes. CD1a is a nonpolymorphic class-I-like molecule associated with β2-migroglobulin (Terhorst *et al.*, 1981), of unknown function. However, a recent study described $CD4^{-}CD8^{-}$ αβ T-cell clones recognizing antigens from *Mycobacterium tuberculosis* in CD1b or CD1c (Beckman *et al.*, 1994). The antigen presented in CD1b appeared to be a mycolic acid, a lipid from the mycobacteria cell wall, suggesting that the function of this family of molecules (CD1a to

CD1e) might be to present to $CD4^-CD8^-$ T-cell clones nonpeptidic antigenic determinants specifically borne by some pathogens.

Langerhans cells are the only cell type in the healthy epidermis that express MHC class II.

3.2.2. Interstitial Dendritic Cells

Initially characterized in heart (Hart and Fabre, 1981; Prickett *et al.*, 1988; Larsen *et al.*, 1990; Spencer and Fabre, 1990) by mAbs against HLA class II or CD45, DC have subsequently been identified in the interstitium of most organs and tissues (liver, kidney, pancreas, ureter, bladder, thyroid, gut, dermis) (Hart and Fabre, 1981; Klinkert *et al.*, 1982; Steiniger *et al.*, 1984; Stern and Nawroth, 1986; Holt *et al.*, 1988; Prickett *et al.*, 1988; Nicod *et al.*, 1989; Hart and McKenzie, 1990; Pavli *et al.*, 1990; Lenz *et al.*, 1993; Nestle *et al.*, 1993). The interstitial DC sit in the extravascular areas, and their dendritic processes extend into the interstitium for a considerable distance away from the cell body. In these tissues, the DC are loosely bound to blood vessels and possibly lymphatics.

3.2.3. Circulating Dendritic Cells

Dendritic cells are found in different biological fluids including the peripheral blood, the afferent lymph, and synovial fluid (Freudenthal and Bhardwaj, 1990).

3.2.3a. Afferent Lymph. Afferent but not efferent lymph contains leukocytes termed "veiled" cells in rabbit (Knight *et al.*, 1982), pig (Drexhage *et al.*, 1979), rat (Mason *et al.*, 1981; Pugh *et al.*, 1983), mouse (Rhodes and Agger, 1987), sheep (Bujdoso *et al.*, 1989; Harkiss *et al.*, 1990), and human (Spry *et al.*, 1980; Brand *et al.*, 1993). Like DC, veiled cells have a low buoyant density. Veiled cells may come either from the interstitium and epithelium of many organs or from a pool of cells that leaves the blood, moves through nonlymphoid tissues, and enters the lymph. In rat, veiled cells can be obtained in relatively large numbers in the thoracic duct after ablation of mesenteric lymph nodes (Pugh *et al.*, 1983). In humans, by means of microsurgical lymph cannulation, skin lymph veiled cells can be recovered from the late phase of an elicitation reaction to a hapten. They express CD1a but lack Birbeck granules (Brand *et al.*, 1993).

3.2.3b. Blood. In human blood, although DC are the main APC (Freudenthal and Steinman, 1990; Thomas *et al.*, 1993), they represent less than 0.1% of leukocytes. The DC display a high motility, continually extending and retracting large lamellipodia or "veils," a property that is not shared by any other cell in blood. Dendritic cells that display a mature phenotype and morphology are, *in vivo*, in rosettes with T cells and can be isolated on the basis of CD83 expression (Weissman *et al.*, 1995; Zhou and Tedder, 1995). However, in more conventional protocols of purification involving T-cell depletion, freshly isolated DC appear as medium-sized lymphocytes that lack typical DC morphology and phenotype (Young and Steinman, 1988; O'Doherty *et al.*, 1993; Thomas *et al.*, 1993; Zhou and Tedder, 1995). After 2 days of culture, blood DC mature into cells displaying a typical DC phenotype, morphology, and function (O'Doherty *et al.*, 1993). Two subsets of blood DC have been identified on the basis of CD11c expression: $CD11c^-$ cells (also $CD45RO^-$) are dependent on exogenous factors for their maturation, while $CD11c^+$ cells (also $CD45RO^+$) are capable of spontaneous maturation (O'Doherty *et al.*, 1994). It is hypothesized that the

$CD11c^-$ cells are bone-marrow-derived precursors to tissue DC, such as epidermal LC, whereas the $CD11c^+$ cells arise from tissue where they have been activated by antigen and are en route to the spleen or lymph nodes. These $CD11c^-$ and $CD11c^+$ subsets might respectively correspond to the $CD14^{low}$ $CD33^+$ and $CD14^{low}$ $CD33^{bright}$ subsets described independently (Thomas *et al.*, 1993).

3.2.4. Thymic Dendritic Cells

Large, bone-marrow-derived, MHC class-II-rich DC are found in the thymus medulla (Barclay and Mayrhofer, 1981; Guillemot *et al.*, 1984). These cells correspond to the interdigitating cells that had been identified by electron microscopy (Kaiserling *et al.*, 1974) as well as the DC that are harvested after enzymatic digestion of thymus (Duijvestijn *et al.*, 1983; Kyewski *et al.*, 1986). The thymic DC are about twice as large as spleen DC, with an overall comparable phenotype (Crowley *et al.*, 1989; Inaba *et al.*, 1990a). Within the thymus, mice DC express the α and β chains of the CD8 antigen (Vremec *et al.*, 1992) and are believed to contribute to the deletion of developing T cells with self-reacting potential (Jenkinson *et al.*, 1985; Fairchild and Austyn, 1990).

3.2.5. Dendritic Cells from the T-Cell-Rich Areas of Secondary Lymphoid Organs

"Interdigitating cells" (IDC) (Veldman and Kaiserling, 1980), which display the morphology and phenotype of DC (Dijkstra, 1982; Witmer and Steinman, 1984), can be identified in sections of T-cell-rich areas of secondary lymphoid organs such as lymph nodes (Van Nieuwkerk *et al.*, 1992), tonsils (Hart and McKenzie, 1988), spleens (Steinman, 1991), and Peyer's patches (Spalding *et al.*, 1983; Wilders *et al.*, 1983). The spleen IDC presently represent the best-studied DC in mice (Steinman, 1991) and can be purified according to CD11c expression (Crowley *et al.*, 1990). In sections of spleen, nests of IDC lie in the periphery of the T-cell-rich areas interrupting the marginal zone of macrophages (Agger *et al.*, 1990; Metlay *et al.*, 1990). As both T cells and antigens leave the arterial tree via the marginal zone, IDC are positioned as "doorways" through which T cells pass on entry into the periarterial sheaths. The IDC from tonsils have been purified and characterized in humans (Hart and McKenzie, 1988).

3.3. Phenotype of Human Dendritic Cells

In the human, because they are relatively easily obtained, peripheral blood DC represent the most studied DC population with regard to phenotype (Table I) (Freudenthal and Steinman, 1990; Thomas *et al.*, 1993; O'Doherty *et al.*, 1993, 1994; Thomas and Lipsky, 1994; Zhou and Tedder, 1995). The phenotype of DC isolated from human tonsils (Hart and McKenzie, 1988), epidermis (Romani *et al.*, 1989b; Teunissen *et al.*, 1990), and dermis (Lenz *et al.*, 1989; Nestle *et al.*, 1993) has also been reported. Consistent with their strong antigen-presenting cell function, DC express all MHC class II proteins at high levels. They also display surface invariant chain (CD74) (Freudenthal and Steinman, 1990; Puré *et al.*, 1990). Polymorphic MHC class I products are also abundant and can be used to monitor either the repopulation of DC in bone marrow chimera (Lenz *et al.*, 1989) or their movement

Table I
Phenotype of Human Dendritic Cells[a]

	Fresh PBL DC	Cultured PBL DC	Fresh LC	Cultured LC	Dermal DC organ culture	Monocyte derived	CD34 derived
MHC							
HLA class I	+++	+++	+	++	+++	++	++
HLA DR	++	++++	++	+++	+++	++	+++
HLA DP	++	+++	++	+++		++	++
HLA DQ	++	+++	++	+++	+++	+	++
CD74	+	+ or ++	++	+	++	+	++
T cell							
CD1a	−	−	++	−/+	−/++	+	++
CD1b	−	−	−	−	−	+	+
CD1c	tr	++	+	−	−/++	+	++
CD2	− and sp +	++ or −	−		−		+
CD3	−	−	−	−	−	−	−
CD4	++	− or tr	+	tr/++	tr/++	+	++
CD5	− and sp ++	++ and sp −	−	tr	++	+	
CD7	− and sp +	−	−		−		−
CD8	−	−	−	−	−		−
CD99		+					
B cell							
CD19	−	−	−	−	−	−	−
CD20	−	−	−	−	−	−	−
CD24	tr	−	−	+	+	−/+	tr
CD40	+	+	+	++	++	++	++
CD72	tr	− or +					tr
CD78		+++					+++
CD82		+					
CD83		++	++				++
Myeloid, plat.							
CD9		+					
CD13	− or ++	++	−		++		++
CD14	− or tr	−	tr	−	− and sp ++	−/tr	tr
CD15	−	−	−		−	−	−
CD15s		+++	+				++
CD17		+/−					
CD33	+	++	++	++	++	+	+
CD34	−	−		−	−		−
CD36	tr	+	−		+/+++		+
CD41/61		++					
CD42a		+/−					
CD63		+/−					
CD65	−	−					
Leukocytes							
CD25	−	+	−	+	tr/+	−	−/tr
CD38		+					+
CD39		+					+
CD43	++	+					++
CD45	++	+	++	++	++		++

Table I (*Continued*)

	Fresh PBL DC	Cultured PBL DC	Fresh LC	Cultured LC	Dermal DC organ culture	Monocyte derived	CD34 derived
CD45RO	− and sp +	+ and sp −				++	++
CD45RA	+ and sp −	−				−	−
CD48		++					++
CD53		+					
CD55		++					
CD59		++					++
CD70	−	−					−
CD98		+/−					
Adhesion							
CD11a	++	++	−/+	++	++	++	++
CD11c	− and sp +	++	++	+	++	++	++
CD18	++	++	++	+	++		++
CD29	++	++	+				++
CD31		+/−					
CD44		+++				++	++
CD49		++(e)					++(d,e)
CD50		++					++
CD54	++	++	−/+	++	++	++	++
CD58	++	++	−/+	++	+	++	++
CD80	− or tr	+/++		+	++	+	++
CD86		+++					++
CTLA4-Ig			−/+	++	++		++
CD102		+					
CD105		+++					
Comp. Receptors							
CD16	− or +		−		−	tr	tr
CD32	+	+/−	++	+	+	+	++
CD64	tr	−	−		−	−	−
CD11b	− and sp +	−	++		++	++	− and sp +
CD21		−			−		−
CD35		−	−		− and sp +		−
Others							
CD68		−	−		−	tr	+ and sp −
S100			++				++
Factor XIIIa			−		++		
E-cadherin			++				− and sp +
c-fms	−						−
Lag			++	−	−	−	− and sp +

[a]Phenotype (from studies referred to in the text) of human fresh or cultured dendritic cells isolated either from peripheral blood, epidermis, or dermis (see Section 3.2), and phenotype of dendritic cells generated *in vitro* from peripheral blood monocytes cultured in presence of GM-CSF + IL-4 (see Section 4.2) or $CD34^+$ hematopoietic progenitor cells cultured in presence of GM-CSF + TNF-α (see Section 4.3). −, lack of expression; + to ++++, intensity of expression of positive cells; tr, expression at trace level; sp, expression on subpopulation; or, variation from studies to studies.

during transplantation (Codner *et al.*, 1990). CD1 molecules are not encoded in the MHC locus but are MHC class-I-like and may act as a restriction element for nonpeptidic determinants (Beckman *et al.*, 1994). CD1a and CD1c are abundant on LC in skin (Fithian *et al.*, 1981; Ray *et al.*, 1989), and CD1b on afferent lymph DC (Bujdoso *et al.*, 1989). CD1c expression has been observed on peripheral blood DC by some (Zhou and Tedder, 1995) but not all investigators (O'Doherty *et al.*, 1993).

Although DC fail to express most T, B, NK, or myelomonocytic lineage-restricted antigens, they do express some lineage-associated surface structures. CD2 and CD4 are present on peripheral blood DC (O'Doherty *et al.*, 1994; Zhou and Tedder, 1995), but CD4 is lost on culture (O'Doherty *et al.*, 1993). CD5 has been reported as positive on circulating DC (O'Doherty *et al.*, 1994; Thomas and Lipsky, 1994; Zhou and Tedder, 1995) and dermal DC (Lenz *et al.*, 1993) but not on LC (Lenz *et al.*, 1993). CD40 is expressed on all DC that have been studied to date (Hart and McKenzie, 1988; Lenz *et al.*, 1993; Nestle *et al.*, 1993; O'Doherty *et al.*, 1993, 1994). The B-cell antigens CD78 and CD82 have also been identified on blood DC (Zhou and Tedder, 1995). The CD83 antigen, borne by germinal center B cells, is expressed on DC *in situ* (Zhou *et al.*, 1992) and on mature blood DC that are in the form of rosettes with T cells (Weissman *et al.*, 1995). After overnight culture of peripheral blood cells, CD83 is up-regulated on immature DC, thereby identifying a uniform population of DC. Thus, CD83 represents an interesting marker for the purification of cultured DC (Zhou and Tedder, 1995). The expression of CD72 has been described on fresh DC (O'Doherty *et al.*, 1994) but not on cultured DC (Zhou and Tedder, 1995). Dendritic cells lack the T-cell markers CD3, CD6, CD7, and CD8 as well as the B-cell markers CD19 and CD20. They display certain surface antigens commonly found on myeloid cells such as CD13 and CD33 but lack most myeloid markers, including CD14, CD15, CD68, and the M-CSF-R. CD36, a molecule involved in phagocytosis (Ren *et al.*, 1995), has been described on circulating DC but not on LC. CD15S (sialyl Lewis x) has been reported to be expressed on LC (Tabata *et al.*, 1993; Ross *et al.*, 1994) as well as on cultured blood DC (Zhou and Tedder, 1995).

Concerning adhesins and accessory molecules, resting DC harbor CD11a, CD11c, CD18, CD29, CD44, CD49e, CD54, CD58, and low levels of the CD28 ligands CD80/B7-1 and CD86/B7-2 (Freudenthal and Steinman, 1990; Thomas *et al.*, 1993; O'Doherty *et al.*, 1993, 1994; Thomas and Lipsky, 1994; Zhou and Tedder, 1995). Concerning Fc and complement receptors, DC express CD32/FcγRII (Schmitt *et al.*, 1990; O'Doherty *et al.*, 1994; Zhou and Tedder, 1995), and CD16/FcγRIII has been documented in a single study (Thomas and Lipsky, 1994). The expression of complement receptor CR3 (CD11b) is documented on LC (De Panfilis *et al.*, 1989) but remains a matter of debate for blood DC (O'Doherty *et al.*, 1994; Thomas and Lipsky, 1994; Zhou and Tedder, 1995). The DC express the phosphotyrosine phosphatase CD45, fresh cells expressing mainly CD45RA and cultured cells displaying mainly CD45RO (O'Doherty *et al.*, 1993, 1994). Finally, LC have some surface or intracytoplasmic molecules not found on other dendritic cell types such as CD1a (Fithian *et al.*, 1981), E-cadherin (Tang *et al.*, 1993), and the Langerhans cell granule-associated antigen (Lag) (Kashihara *et al.*, 1986). Also, dermal DC (dendrocytes) have been reported to specifically express the coagulation factor XIIIa (Nestle *et al.*, 1993). Culturing DC or LC results in up-regulation of CD25, CD40, and CD28 ligands (CD80 and CD86) and loss of Birbeck granules on LC (Romani *et al.*, 1989b; O'Doherty *et al.*, 1993; Symington *et al.*, 1993).

In summary, DC are characterized by the lack of most lineage-associated antigens

(CD3, CD8, CD19, CD20, CD14, CD15, CD56, CD57) and the expression of high levels of MHC products and accessory molecules such as CD11a, CD54, CD58, and, after culture, high levels of CD80, CD86, and CD83.

3.4. Functions of Dendritic Cells

3.4.1. Motility

Motility represents one of the important characteristics of DC inasmuch as they are constantly changing shape by developing lamellipodia and veils (Freudenthal and Steinman, 1990).

3.4.2. Antigen Capture

Although described as professional APC, DC were, until recently, considered to display poor endocytic and phagocytic capacities (Steinman *et al.*, 1974). Yet, LC were shown able to display all steps of receptor-mediated endocytosis (Hanau *et al.*, 1987) and to phagocytose particles of relatively large size such as latex beads, viruses (Barfoot *et al.*, 1989), bacteria (Schuler *et al.*, 1983; Reis e Sousa *et al.*, 1993), and intracellular parasites such as *Leishmania major* (Moll, 1993). The pinocytic activity of DC has been reported to be at least as high as that of other APC (Levine and Chain, 1992). Antigen uptake by afferent lymph DC can also occur in the form of immune complexes (Harkiss *et al.*, 1990). Finally, molecules such as the mannose receptor, with multiple lectin-like domains have been reported to be expressed on DC (Sallusto *et al.*, 1995; Nussenzweig *et al.*, 1995). Such molecules might be involved in uptake of bacterial antigen with specific glycosylations and, thus, might represent a first line of discrimination between self and nonself.

3.4.3. Antigen Processing

Whereas freshly isolated LC have numerous acidic vacuoles, presumably endosomes, cultured mouse and human epidermal LC display some acidic lysosomes but few acidic endocytic vesicles (Stössel *et al.*, 1990). Also, freshly isolated LC actively synthesize MHC class II products and display a lysosomal compartment enriched in MHC class II (Kleijmeer *et al.*, 1994), whereas cultured LC appear to specifically lose MHC class II synthesis (Shimada *et al.*, 1987; Witmer-Pack *et al.*, 1988). Furthermore, fresh but not cultured LC express high levels of cytoplasmic invariant chain (Puré *et al.*, 1990). Finally, antigen-pulsed DC retain processed antigen for long periods in an immunogenic form (Inaba *et al.*, 1990b; Puré *et al.*, 1990).

3.4.4. Primary T-Cell Responses

The most commonly used functional assessment of DC is their remarkable ability to drive allogeneic and syngeneic mixed lymphocyte reactions (MLR). In doing this, DC are about 100-fold more efficient than any other APC population including B cells and monocytes (Steinman and Witmer, 1978; Crow and Kunkel, 1982; Van Voorhis *et al.*, 1983). Although $CD4^+$ T-cell reactivity is most often analyzed, DC can also stimulate $CD8^+$ T cells to enter into an allogeneic reaction, though higher numbers of APC are required

(Inaba *et al.*, 1987; Young and Steinman, 1990). This *in vitro* activation of naive CD8$^+$ T cells by DC is independent of CD4$^+$ T cells (Elbe *et al.*, 1994), an observation that is in accordance with *in vivo* examples of CTL development in apparent absence of CD4 help (Buller *et al.*, 1987; Rahemtulla *et al.*, 1991).

The use of superantigens has further demonstrated the efficiency of DC in primary T-cell activation. Ten- to 50-fold fewer DC than either monocytes or B cells are required to induce T cells to proliferate in response to various superantigens (Bhardwaj *et al.*, 1992, 1994b). In addition, DC can induce T cells to respond to femtomolar concentrations of superantigens, but B cells and monocytes necessitate the use of ten- to 100-fold more superantigen (Bhardwaj *et al.*, 1992, 1993).

Recently, the availability of mice expressing transgenic T-cell receptors has allowed the analysis of the capacity of DC to induce a primary antigen-specific T-cell response to soluble antigens *in vitro*. In such systems, DC appear to be 100- to 300-fold more efficient than any other APC (Croft *et al.*, 1992; Macatonia *et al.*, 1993b).

In vivo models have shown that B cells, which can activate memory T cells, are unable to prime naive T cells (Ronchese and Hausmann, 1993). In keeping with this, B cells were further shown to induce tolerization of naive T cells (Fuchs and Matzinger, 1992). In addition, SCID mice reconstituted with T cells but not B cells exhibit functional T-cell priming. The capacity of DC to prime T cells *in vivo* has been directly demonstrated in cell transfer experiments. Thus, mouse DC pulsed *in vitro* with protein antigen induced, upon reinjection into foot-pad or blood or upon intratracheal instillation, MHC-restricted antigen-specific T-cell responses (Inaba *et al.*, 1990b; Havenith *et al.*, 1993; Levin *et al.*, 1993; Liu and MacPherson, 1993). Also, mouse DC pulsed with HIV envelope-derived peptides have been shown to induce CD8$^+$ cytotoxic responses *in vivo* (H. Takahashi *et al.*, 1993).

All those studies outline the crucial and probably unique capacity of DC to induce antigen-specific priming of naive T cells.

3.4.5. Primary Antibody Responses

The need for spleen-adherent cells in primary antibody response (Mosier, 1967) led to the discovery of the essential function of DC in such responses (Inaba *et al.*, 1983). In case of hapten carrier-conjugates, DC sensitize carrier-specific T cells, which can subsequently interact in a cognate fashion with hapten-specific B cells (Inaba and Steinman, 1985). Antigen-activated B cells finally proliferate and differentiate in response to membrane interactions and released cytokines. These cellular interactions occur in discrete aggregates that include DC, B cells, and T cells (Inaba *et al.*, 1984; Inaba and Steinman, 1987). Studies performed *in vivo*, further demonstrated the T-cell priming (Inaba *et al.*, 1990b,c) to be followed by the appearance of antigen specific immunoglobulin in serum (Francotte and Urbain, 1985; Sornasse *et al.*, 1992; Berg *et al.*, 1994; Flamand *et al.*, 1994). However, Ig productions could be detected only after a second challenge with soluble antigen a few days after DC injection (Sornasse *et al.*, 1992; Liu and MacPherson, 1993). Antigen-pulsed B cells are unable to trigger such responses.

3.4.6. T-Cell Maturation

In TcR transgenic mice, DC were initially described as being neutral with regard to the maturation of newly primed naive T cells into either Thl or Th2 *in vitro* (Hsieh *et al.*, 1993;

Macatonia *et al.*, 1993b). The commitment toward Th1 was dependent on the presence of activated macrophages as a source of IL-12, and the commitment toward Th2 was dependent on the presence of IL-4 produced by memory T cells. However, more recent experiments that used a similar system but devoid of memory T cells (i.e., devoid of IL-4-producing cells) demonstrated that DC rather favored the maturation of naive T cells into the Th1 pathway. This commitment toward Th1 was dependent on the endogenous production of IL-12 by DC (Macatonia *et al.*, 1995). In the same line, *in vivo* priming in mice with antigen-pulsed DC further substantiated a commitment toward the Th1 pathway inasmuch as specific IgG2a immunoglobulins (a signature of of IFNγ production, a Th1 product) were preferentially produced (Sornasse *et al.*, 1992; De Becker *et al.*, 1994).

3.4.7. Secondary T-Cell Responses

Dendritic cells are able to present soluble antigen to established T-cell clones or to memory T cells primed *in vivo* (Romani *et al.*, 1989a; Harkiss *et al.*, 1990; Aiba and Katz, 1991; De Bruijn *et al.*, 1992). However, their efficiency remains to be compared to that of other APC.

Following viral infection, DC were shown, in mice (Nonacs *et al.*, 1992) and humans (Bhardwaj *et al.*, 1994a), to be able to induce, *in vitro*, secondary influenza-virus-specific $CD8^+$ CTL responses in absence of $CD4^+$ T cells.

3.5. Migration and Turnover of Dendritic Cells

3.5.1. From the Bone Marrow to the Periphery

Newly generated DC migrate from the bone marrow to the nonlymphoid tissues, presumably through the bloodstream (Figure 1). In particular, in the lung, following inhalation of *Moraxella catarrhalis* organisms, DC accumulate rapidly ($\leq$1 hr) in the airway epithelium (McWilliam *et al.*, 1994). Also, in the skin, intradermal injection of GM-CSF induces local accumulation of LC within 48–72 hr (Kaplan *et al.*, 1992). Local production of GM-CSF has also been demonstrated to be associated with the distribution of DC *in vivo* in the lung (Tazi *et al.*, 1993). These accumulations of DC are thought to represent recruitment of circulating DC previously generated in the bone marrow, though local production from circulating precursors could not be excluded.

3.5.2. From the Periphery to the Secondary Lymphoid Organs

Entry of DC into the spleen occurs from circulating blood (Austyn *et al.*, 1988; Kupiec-Weglinski *et al.*, 1988): DC administered intravenously migrate through the liver and then proceed via the afferent lymphatics to the celiac lymph nodes (Fossum, 1988), but few, if any, enter the lymph nodes directly from blood. Several experiments suggest that DC leave the nonlymphoid organs through the afferent lymph. First, during intestinal *Salmonella* infection, specific antigens are present on DC in lymph draining the intestines (Mayrhofer *et al.*, 1986). Second, after subcutaneous injection of antigen, DC in lymph draining the site of injection could present the antigen to sensitized T cells (Bujdoso *et al.*, 1989). Finally, allogeneic skin transplantation models (Macatonia *et al.*, 1987; Kripke *et al.*, 1990; Larsen *et al.*, 1990) as well as injection of radiolabeled DC (Fossum, 1988) or *Leishmania*-infected

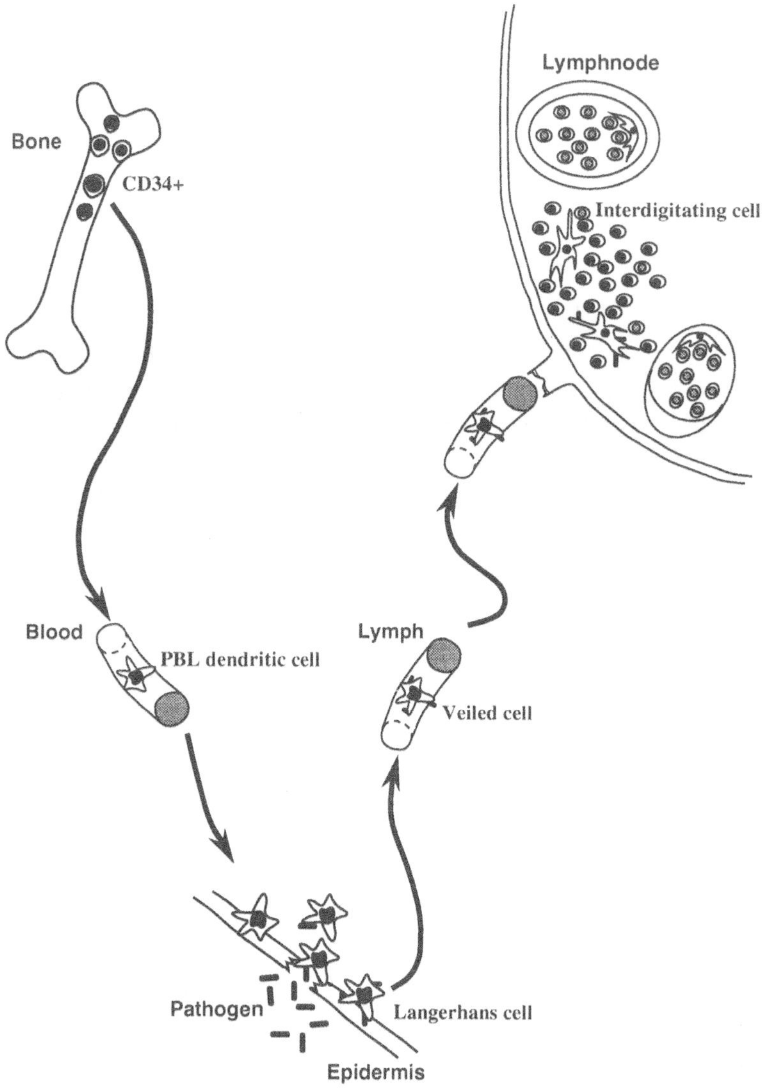

FIGURE 1. Migration of dendritic cells *in vivo*. Newly generated dendritic cells migrate from the bone marrow to the nonlymphoid tissues presumably through the bloodstream (peripheral blood dendritic cells). During an injury, interstitial dendritic cells (such as Langerhans cells in the epidermis) capture the antigen and then, under microenvironmental signals (such as TNF-α production in the dermis) leave the nonlymphoid tissues through the afferent lymph (veiled cells). Then, dendritic cells enter into lymph nodes, where they home to the T-cell-rich area (interdigitating cells) and induce an antigen-specific primary T-cell response. Few dendritic cells exit lymph nodes, possibly because they may be either programmed to die after antigen presentation or destroyed by the afferent immune response.

LC (Moll *et al.*, 1993) into footpads or dermis, respectively, demonstrated that DC enter into the lymph nodes through the afferent lymph. The local production of TNF-α in the dermis has been suggested to represent one potential mechanism involved in the migration of epidermal LC to lymph nodes (Cumberbatch and Kimber, 1992, 1995). Few DC exit lymph nodes, possibly because they may be either programmed to die after antigen presentation or destroyed by the efferent immune response.

3.5.3. Turnover

Bone marrow reconstitution of irradiated mice or rats has permitted an estimation of DC turnover. The LC in skin (Katz *et al.*, 1979; Krueger *et al.*, 1983; Chen *et al.*, 1986) and DC in heart (Hart and Fabre, 1981; Leszczynski *et al.*, 1985) have a relatively slow turnover with a half-life ≥15 days. In contrast, DC of the epithelial airway (Holt *et al.*, 1994) or of the gut wall (Pugh *et al.*, 1983; Fossum, 1989a) undergo a rapid turnover with a half-life ≤2 days.

3.6. Maturation

During their migration, after antigen loading, DC are thought to undergo changes in phenotype and function (Larsen *et al.*, 1990; Streilein *et al.*, 1990). Fresh LC can process native proteins with great efficiency and subsequently present the processed peptides to memory T cells. Conversely, cultured LC as well as IDC of lymphoid organs are relatively inefficient in antigen processing but strongly efficient in naive T-cell activation (Schuler and Steinman, 1985; Inaba *et al.*, 1986; Shimada *et al.*, 1987; Heufler *et al.*, 1988; Romani *et al.*, 1989a,b; Streilein and Grammer, 1989; Puré *et al.*, 1990). In this respect, epidermal LC, when compared to cultured LC or IDC, express relatively low levels of MHC class II molecules and accessory molecules such as B7 (Romani *et al.*, 1989b; Teunissen *et al.*, 1990; Dai *et al.*, 1993; Hart *et al.*, 1993; Kraal *et al.*, 1993; O'Doherty *et al.*, 1993; Symington *et al.*, 1993; Inaba *et al.*, 1994; Larsen *et al.*, 1992, 1994). Such differences between fresh and cultured LC are considered as reflecting a maturation of these cells in response to factors such as GM-CSF, which is secreted by keratinocytes (Witmer-Pack *et al.*, 1987; Heufler *et al.*, 1988; Koch *et al.*, 1990). The maturation of LC that occurs *in vitro* in culture is viewed as a counterpart to the physiological events that occur *in vivo* during migration of LC from skin to draining lymph nodes (Larsen *et al.*, 1990).

3.7. Dendritic Cells and Tolerance

As sentinels of the immune system, DC are known mainly for their unique capacity to initiate immune responses. However, their localization in peripheral tissues such as the epithelial airway or the gut wall appears in contrast with the well-known oral tolerance (Rubin *et al.*, 1981; MacDonald, 1983; Gautam and Battisto, 1985; Mowat, 1987). In such tissues, the function of DC might be repressed by other cell types such as the alveolar macrophages in the lung (Holt *et al.*, 1988, 1993). Alternatively, DC, under the influence of microenvironmental signals (lytic cell death, cell stress) (Matzinger, 1994; Ibrahim *et al.*, 1995), might control the outcome of the immune response: immunity versus tolerance. In this respect, the liver, with its relatively high DC content, has long been recognized as the most tolerogenic whole organ (Calne *et al.*, 1969; Starzl *et al.*, 1965, 1993) that, across mouse strain combinations, can be transplanted successfully without host immunosuppression (Qian *et al.*, 1994). The DC precursors isolated from mouse liver, even when cultured with GM-CSF ($\pm$ TNF or IFNγ), express low levels of MHC class II antigen and are inefficient in stimulating allogeneic naive T cells (Lu *et al.*, 1994). These results suggest a possible linkage between hepatic tolerogenicity and the presence, within the liver, of MHC class II low-DC progenitors that resist conventional attempts to induce maturation *in vitro*. Yet, those DC precursors could be induced to differentiate after prolonged exposure to GM-CSF in the presence of type I collagen (Lu *et al.*, 1994).

Thus, microenvironmental signals controlling DC maturation are congruent with the possibility that the migration of immature DC from normal liver and perhaps other organs may explain their inherent tolerogenicity.

On the other hand, DC might also be involved in tolerance through the activation of T cells with a regulatory function. In this context, it has been shown that DC from pancreatic lymph nodes can protect nonobese diabetic (NOD) mice from diabetes. This protection occurs through the generation of a transferable regulatory T-cell population (Clare-Salzler *et al.*, 1992).

4. GENERATION OF DENDRITIC CELLS *IN VITRO*

Progress in the generation of DC from mouse and human precursors has recently been accomplished. This progress has led to the establishment of culture systems that allow the *in vitro* generation of large numbers of DC.

4.1. From Mouse Precursor Cells

Although it was known early that DC arise from proliferating precursors *in vivo* (Fossum, 1989b), it has been difficult to identify progenitors in culture. Inaba *et al.* (1992b) found that aggregates of growing DC develop in cultures of mouse blood that are supplemented with GM-CSF but not with other CSFs. The DC precursors derive from the Ia negative and nonadherent cell fraction. Comparable culture systems have also been applied to bone marrow cells (Inaba *et al.*, 1992a; Scheicher *et al.*, 1992). The starting population is composed of MHC class II negative precursors, and the majority of nonadherent, newly formed granulocytes are removed by gentle washes during the first 2–4 days of culture. The proliferating clusters that are loosely attached remain in culture. At day 4–6, clusters can be resuspended, and on reculturing, by about 2 weeks, large numbers of single, MHC class-II-rich DC are being released into the medium. These cells, which proliferate within the aggregates, lack certain antigenic markers that are found on mature DC. However, in pulse-chase protocols, the [^{3}H]TdR-labeled progeny exhibit many typical DC features, including abundant MHC class II. Culturing blood or bone marrow progenitors for 2 weeks in semisolid culture medium permitted identification of DC in 1–2% of the colonies in association with granulocytes and monocytes. This demonstrates that granulocytes, monocytes, and DC can develop from a common MHC class II$^-$ progenitor under the aegis of GM-CSF (Inaba *et al.*, 1993a).

Marrow progenitors represent a major source of DC because more than 5×10^6 DC can develop within 1 week from precursors within the large hindlimb bones of a single animal.

4.2. From Human Peripheral Blood Monocytes

Attempts to induce monocytes to differentiate into DC have been going on for about a decade. In pioneering studies, Knight *et al.* (1986) described monocytes that had acquired a veiled and dendritic appearance after separation. Subsequently, a high proportion of elutriated monocytes (40–60%) were found to differentiate into cells with a dendritic morphology following a pulse with metrizamide (Kabel *et al.*, 1989). Furthermore, monocytes purified by adherence were found to express CDla and loose CD14 after culture in low-serum medium (Rossi *et al.*, 1992). In keeping with this, monocytes were induced to express CDla after treatment with GM-CSF plus IL-4 (Porcelli *et al.*, 1992) or GM-CSF alone (Kasinerk *et al.*, 1993). This phenomenon is observed only with monocytes and not with any other cell or selected cell lines, and is produced only with GM-CSF out of a panel of cytokines (Kasinrerk *et al.*, 1993). These findings have recently been confirmed and extended further (Romani *et al.*, 1994; Sallusto and Lanzavecchia, 1994). With GM-CSF and IL-4, adherent blood mononuclear cells yield DC lines that can be maintained in culture for several weeks and retain the antigen-capturing and -processing capacity characteristic of immature DC. These cells demonstrate a typical dendritic morphology, express high levels of MHC class I and class II molecules, CD1, CD32/FcγRII, CD40, CD80, CD44, and CD54, and lack CD14 (Table I). Those DC most likely derive from the monocyte population.

4.3. From Human CD34$^+$ Hematopoietic Progenitor Cells

4.3.1. In Association with GM-CSF or IL-3, TNF-α Induces Development of Dendritic Cells from CD34$^+$ Cells

CD34$^+$ hematopoietic progenitor cells (HPC), isolated from cord blood or bone marrow mononuclear cells, can be induced to proliferate in response to several cytokines and their combinations. Tumor necrosis factors (α and β) strongly potentiate the proliferation of CD34$^+$ HPC induced by either IL-3 or GM-CSF (Caux *et al.*, 1990; Backx *et al.*, 1991; Moore, 1991; Jacobsen *et al.*, 1992; Reid *et al.*, 1992). Indeed, such stimulatory effects of TNF-α have also been observed with acute myeloid leukemias (Hoang *et al.*, 1989; Salem *et al.*, 1990; see Chapter 12). Within 8 days, the number of cells generated in TNF-α-supplemented liquid cultures of CD34$^+$ HPC is increased three to eight times when compared to cultures carried out with IL-3 or GM-CSF alone. Among cells generated after 12 days with IL-3/GM-CSF and TNF-α, both adherent and nonadherent cells with a dendritic morphology and delicate membrane projections are observed (Figure 2A–D) (Caux *et al.*, 1992). Of the cells generated with GM-CSF alone, 5–15% express CDla, a molecule borne by LC and thymocytes but not by freshly isolated HPC. Adding TNF-α to GM-CSF cultures increases the proportion of CDla$^+$ cells up to 70%, yielding 10 to 30 × 10^6 CDla$^+$ cells from 10^6 CD34$^+$ HPC. All the cells expressing CDla display a dendritic morphology, and the DC are characterized by a high level of MHC class II expression. Under electron microscopy, HPC-derived CDla$^+$ cells exhibit a lobulated nucleus and a villous surface with dendritic projections. Furthermore, one of five CDla$^+$ cells shows Birbeck granules, which are characteristic of LC (Figure 2E–I). Although IL-3 alone does not permit generation of significant numbers of CDla$^+$ cells, the combination IL-3 plus TNF-α yields 11–17% CDla$^+$ cells. The strong growth induced by IL-3 plus TNF-α permits the generation of as many CDla$^+$ cells as does GM-CSF plus TNF-α (Caux *et al.*, 1996).

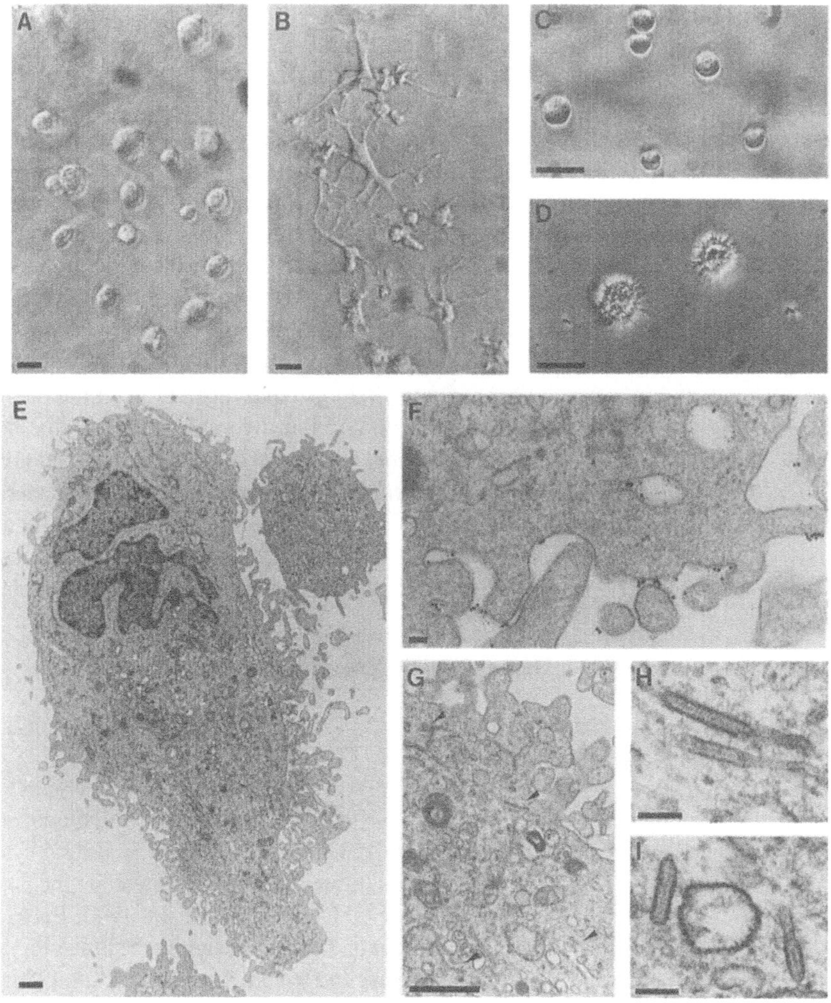

FIGURE 2. Morphology of cells generated from CD34$^+$ HPC in response to GM-CSF and GM-CSF plus TNF-α. CD34$^+$ HPC were cultured for 12 days with GM-CSF alone (panels A and C) or GM-CSF plus TNF-α (panels B and D to I). Panels A and B: Adherent cells (bar = 10 μm). Panels C and D: nonadherent cells (bar = 10 μm) examined under phase-contrast microscopy. Panels E to I: Transmission electron microscopy. Panel E: CDla (10 nm) immunogold labeling (bar = 1 μm). Panel F: CDla (5 nm) and HLA-DR (17 nm) immunogold double labeling (bar = 100 nm). Panel G: Portion of panel E at higher magnification; arrows indicate Birbeck granules (bar = 1 μm). Panels H and I: Birbeck granules from panel E (bar = 100 nm).

CDla$^+$ cells (generated with GM-CSF plus TNF-a or IL-3 plus TNF-α) differ from CD14$^+$ cells (generated with GM-CSF alone or IL-3 plus TNF-α) by (1) dendritic morphology, (2) higher expression of CDla, CDlc, CD4, CD40, adhesion molecules (CDllc, CD54, CD58), MHC class II molecules, and CD28 ligands (CD80 and CD86), (3) lack of Fc receptor FcγRI (CD64) and complement receptor CR1 (CD35) expression (Caux *et al.*,

1992, Caux *et al.*, 1996). CD1a$^+$ are CD45RO$^+$ and express the myeloid markers CD13, CD15s (sialyl Lewis X), CD33, CD36, but not CD15 (Lewis X) or the c-fms (M-CSFR). Importantly, they express the DC-specific marker CD83. They also express the B-cell marker CD78 and a large panel of accessory molecules (CD11a, CD18, CD48, CD59). CD1a$^+$ cells express S100 and are heterogeneous for the expression of CD68, E-cadherin, and the Lag antigen (see Table I). Thus, CD1a$^+$ cells express antigens previously identified on DC isolated from either tissues or blood. Among all CSFs tested, only GM-CSF and IL-3, in association with TNF-α or TNF-β, were able to support DC development from CD34$^+$ progenitors. Moreover, adding SCF to GM-CSF plus TNF-α increases by three- to sixfold the yield of CD1a$^+$ cells at early time points (days 5–7) (Szabolcs *et al.*, 1995; C. Caux, unpublished results).

The requirement of TNF-α for the generation of DC from human CD34$^+$ HPC has been established by several groups (Reid *et al.*, 1992; Santiago-Schwarz *et al.*, 1992; Romani *et al.*, 1994). Furthermore, in semisolid cultures, DC were shown to arise within single colonies together with monocytes/macrophages, suggesting the existence of a common precursor cell (Reid *et al.*, 1990, 1992; Santiago-Schwarz *et al.*, 1992). For optimal generation of DC, TNF-α appears to be required only during the first days of culture (Santiago-Schwarz *et al.*, 1992; C. Caux, unpublished results), and it probably acts through up-regulation of the GM-CSF receptor (Santiago-Schwarz *et al.*, 1993; Sato *et al.*, 1993).

4.3.2. Identification of Two Dendritic Cell Subsets

Although most cells are CD1a$^+$CD14$^-$ after 12 days of culture, two cell subsets, one CD1a$^+$CD14$^-$ and one CD1a$^-$CD14$^+$, can be clearly distinguished after 5–6 days. These two subsets, sorted at day 5, have no proliferative capacity but mature at day 14 into CD1a$^+$CD14$^-$ cells that express high levels of MHC class II and DC characteristic molecules (CD80, CD86, CD83, CD58). Of interest, the CD1a$^+$CD14$^-$ precursors differentiate into cells that contain Birbeck granules (BG), while the CD1a$^-$CD14$^+$ precursors lead to CD1a$^+$ DC-lacking BG characterized by the expression of IL-10 and the coagulation factor XIIIa. TNF-α is required for the emergence of the precursors at day 5 but not for the subsequent maturation of the two subsets (day 5 to 14) (Caux *et al.*, 1995b).

4.3.3. Tumor Necrosis Factor α Modulates Hematopoiesis at Multiple Levels

The enhancing effect of TNF-α on myelopoiesis is a direct effect and is caused by the recruitment of early hematopoietic progenitor cells hyporesponsive to IL-3 (Figure 3) (Caux *et al.*, 1993). Probably, TNF-α up-regulates the β chain of the IL-3/GM-CSF receptor complex (Sato *et al.*, 1993), thus inducing CSF responsiveness of pluripotent progenitors (phase 1). Then, TNF-α blocks both erythropoiesis (phase 2) and granulopoiesis (Backx *et al.*, 1991; Jacobsen *et al.*, 1992), the latter by inhibiting granulocytic differentiation of uncommitted progenitors (phase 3) and proliferation of immature granulocytic cells (phase 4) (Caux *et al.*, 1991). Concomitantly, TNF-α specifically induces the generation of functionally mature DC (phase 5 and 6).

All those effects of TNF-α are mediated through the p55 TNF receptor (Jacobsen *et al.*, 1994; C. Caux, unpublished results; Rusten *et al.*, 1994). Other studies have demonstrated that TNF-α may inhibit the recruitment of primitive progenitors in response to combinations

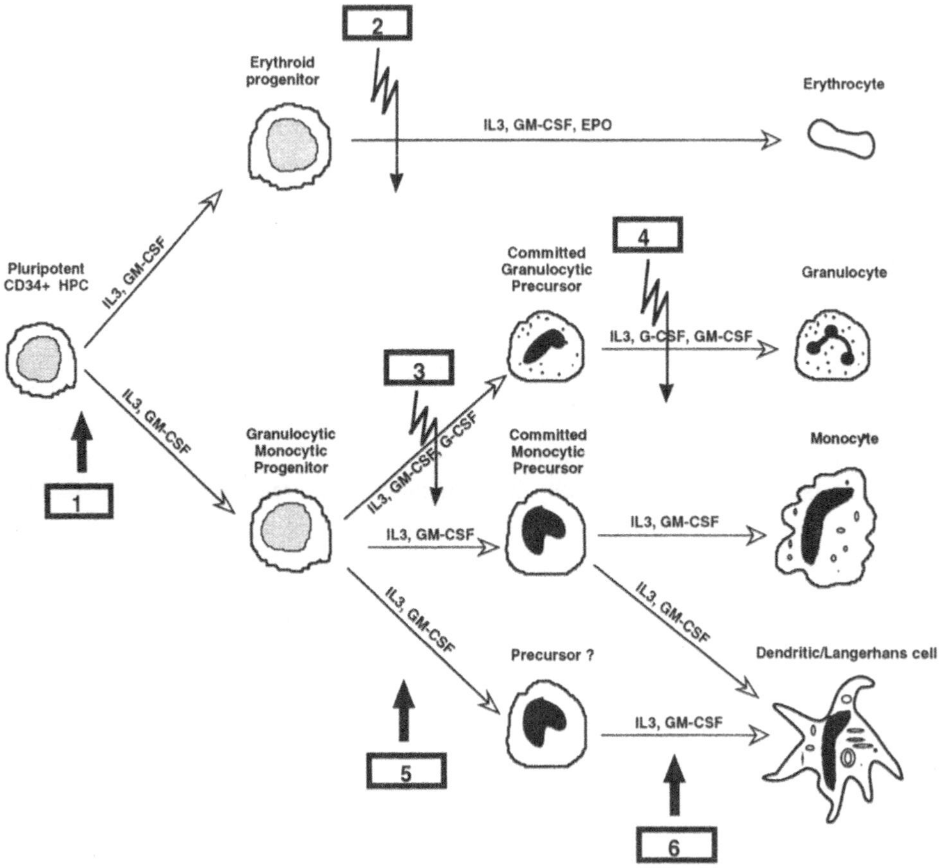

FIGURE 3. Tumor necrosis factor α modulates human hematopoiesis at multiple levels. Probably, TNF-α up-regulates the β chain of the IL-3/GM-CSF receptor complex, thus inducing CSF responsiveness of pluripotent progenitors (phase 1). Then, TNF-α blocks both erythropoiesis (phase 2) and granulopoiesis, the latter by inhibiting granulocytic differentiation of uncommitted progenitors (phase 3) and proliferation of immature granulocytic cells (phase 4). Concomitantly, TNF-α specifically induces the generation of functionally mature DC (phases 5 and 6).

of factors including SCF, an effect transduced through the p75 TNF receptor (Jacobsen *et al.*, 1994; Rusten *et al.*, 1994). Thus, TNF-α may maintain very early progenitors in a quiescent states.

4.4. Dendritic Cell Lines

Though the generation of DC from hematopoietic precursors yields significant numbers of cells, thus avoiding the tedious purification procedure that would be needed for a trace population, it is still time consuming and costly in terms of reagents that are required at all stages of the culture. Therefore, the generation of continuous dendritic cell lines represents an interesting alternative. Accordingly, several groups have recently reported the successful establishment of murine and avian cell lines with DC characteristics. Such DC cell lines, with a DC phenotype and the capacity to prime naive T cells *in vivo*, have been generated by infecting fetal mouse spleen cells using a recombinant retrovirus transducing the *v-myc*^{MH2} oncogene (Paglia *et al.*, 1993). In chicken, a conditional v-Rel estrogen receptor fusion protein, v-RelER, causes estrogen-dependent *v-rel*-specific transformation of bone marrow cells. Following inactivation of v-RelER oncoprotein activity, cells differentiate into DC as judged by morphological and functional criteria (Boehmelt *et al.*, 1995).

Others have established, from fetal mouse skin, growth factor (GM-CSF, IL-2)-dependent, MHC class-II-negative cell lines (Elbe *et al.*, 1994; Xu *et al.*, 1995). Those MHC class-II-negative cell lines can prime *in vitro* allogeneic CD8$^+$ naive T cells in absence of CD4$^+$ T cells.

4.5. Role of GM-CSF in the Development of Dendritic Cells

In vitro studies in mice and humans concur in the conclusion that GM-CSF plays a key role in the development of DC. In humans, the importance *in vivo* of GM-CSF in dendritic cell ontogeny is further demonstrated by its intradermal administration, which results in a progressive local accumulation of CD1a$^+$, Birbeck-granule-positive cells (Kaplan *et al.*, 1992). The Langerhans cell infiltration peaks after 3 days, suggesting that these cells originate from attracted progenitors maturing locally rather than from the recruitment of surrounding LC. The role of GM-CSF in dendritic cell ontogeny therefore complements its known stimulatory activity on freshly isolated LC (Witmer-Pack *et al.*, 1987; Heufler *et al.*, 1988). Yet, mice whose GM-CSF gene has been inactivated by homologous recombination display normal levels of LC and DC (Dranoff *et al.*, 1994). This suggests that other factors may act in lieu of GM-CSF, and IL-3 appears to be a real candidate as long as it is associated to TNF-α, in humans at least. It will be interesting to examine the functional status of DC in mice whose IL-3 and GM-CSF genes are jointly inactivated. The lack of CD1a on interstitial dendritic cell from numerous tissues indicates a possibly GM-CSF-independent ontogeny inasmuch as DC generated with GM-CSF from either hematopoietic progenitors or monocytes always express CD1a.

4.6. Different Dendritic Cell Populations May Originate from Different Progenitor Cells

In vitro and *in vivo* studies in mice and humans presently lead us to the conclusion that several dendritic cell subsets may originate from at least three different progenitors, including G–M–DC progenitors, T–B–DC progenitors, and circulating monocytes (Figure 4).

4.6.1. G–M–DC Progenitors

Separation of bone-marrow- or cord-blood-derived CD34$^+$ hematopoietic progenitors into subsets at various stages of their growth *in vitro* has led to the identification of a pluripotent G–M–DC progenitor that branches out into G–M progenitors or DC progenitors that can subsequently yield mature DCs. Indeed, these growing DC precursors can

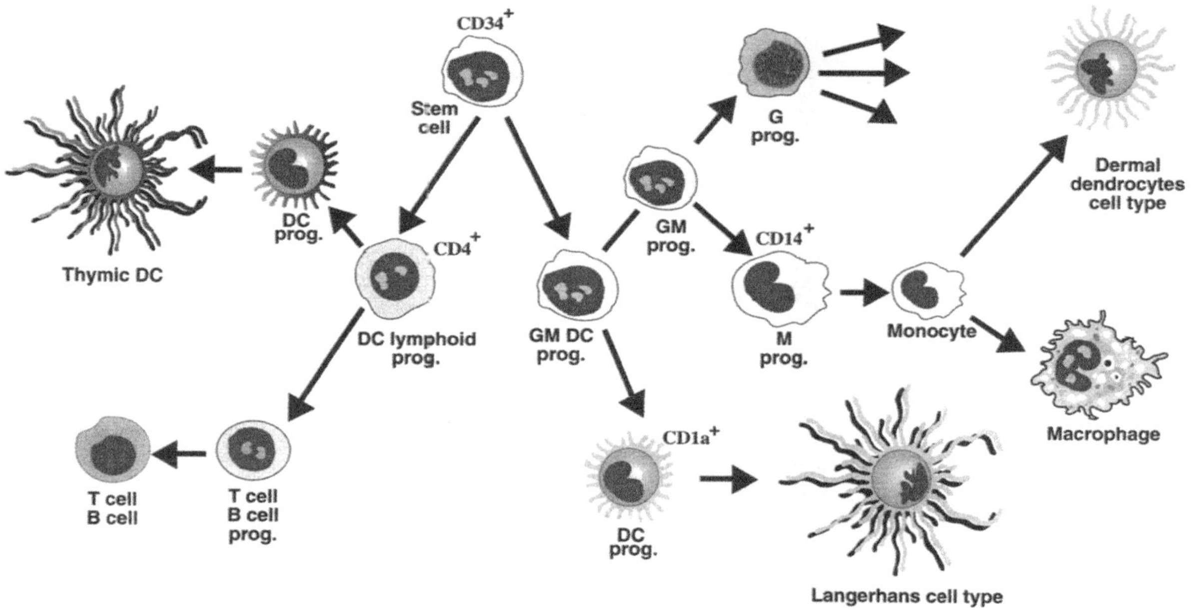

FIGURE 4. Different dendritic cell populations may originate from different progenitor cells. *In vitro* and *in vivo* studies in mice and humans permit the suggestion that several dendritic cell subsets may originate from at least three different progenitors. A progenitor cell common for granulocytes, monocytes, and dendritic cells (G–M–DC), identified in semisolid medium, further differentiates into several lineage-specific precursors. DC-specific precursors ($CD1a^+$) lead to Langerhans-cell-type DC, and M precursors ($CD14^+$) lead to either dermal-type DC or macrophages as a function of microenvironmental signals. Another precursor for monocytes/macrophages, dependent on M-CSF might also exist (not illustrated here). Finally, the existence of a third progenitor for thymic DC has also been demonstrated in mice. This progenitor displays T and B lymphoid, but no myeloid, differentiation potential.

even be separated early during the culture into two populations according to exclusive expression of either CD1a or CD14, both of them leading to CD1a$^+$ cells lacking CD14. Yet these two mature populations differ in their exclusive expression of Birbeck granule on one side (CD1a precursors) and the expression of IL-10 and the coagulation factor XIIIa on the other side (CD14 precursors). These cells appear to correspond, respectively, to epidermal LC and dermal DC (dendrocytes) whose functional specialization remains unclear.

Recent experiments with mice whose *rel-B* gene has been disrupted further illustrate the different origin/relationship of dendritic cell subsets inasmuch as these animals display epidermal LC but no DC in their thymus and spleen (Burkly *et al.*, 1995; Weih *et al.*, 1995).

4.6.2. T–B–DC Progenitors

A thorough analysis of cellular populations within the mouse thymus has permitted the identification of progenitors that can differentiate into either T cells or DC when injected in thymus lobes or into B cells following homing into the spleen after intravenous injection (Ardavin *et al.*, 1993). In no case have these progenitors been found to differentiate into the myeloid pathway. Such T–B–DC progenitors have not yet been identified in humans.

4.6.3. Monocytes

In response to GM-CSF plus IL-4, monocytes can be induced to differentiate into cells that express high levels of CD1a and are able to induce an alloreactive response. Yet, in our hands, these cells display only a partial dendritic cell phenotype and may be closest to dermal DC or CD14-precursor-derived DC in view of the lack of Birbeck granules. Indeed, it is tempting to speculate that dendrocytes may originate from monocytes that have entered tissues and encountered IL-4/IL-13 released by tissue mast cells.

Although monocytes entering into tissues can also differentiate into macrophages, it remains unknown whether it is the same monocyte subset that yields either a macrophage or a dendritic cell or whether lineage-committed monocytes can be distinguished within the circulation. In this context, mice with an abnormal M-CSF gene (*op/op* mice) lack some subsets of monocytes/macrophages in some organs (peritoneal cavity, blood) but not others (secondary lymphoid organs, liver, lung) and display normal DC (Wiktor-Jedrzejczak *et al.*, 1992; K. Takahashi *et al.*, 1993; Witmer-Pack *et al.*, 1993), thus supporting the existence of different lineage-committed monocytes.

5. FUNCTIONS OF DENDRITIC CELLS GENERATED *IN VITRO*

5.1. From Mouse Precursor Cells

The DC generated in response to GM-CSF are potent stimulators of MLR and, after injection into footpads, can home to the T-dependent areas of lymph nodes (Inaba *et al.*, 1992b). Furthermore, DC were shown to internalize particles at an early stage in their development from proliferating progenitors. Progenitors that had internalized *Bacillus Calmette-Guérin* organisms differentiated into DC that were able to present efficiently mycobacterial antigens to T cells both *in vitro* and *in vivo* (Inaba *et al.*, 1993b). Furthermore, the ability of bone-marrow-derived DC to present a peptide derived from E7 human papilloma virus-16 protein has been evaluated *in vivo* in mice. Dendritic cells pulsed with peptide permitted tumor regression induced by mouse embryonic cells transformed with E7 (Mayordomo *et al.*, 1995).

5.2. From Human Peripheral Blood Cells

The DC generated from mononuclear blood cells in response to GM-CSF plus IL-4 are powerful stimulators in MLR and are capable of activating cord-blood naive T cells (Romani *et al.*, 1994; Sallusto and Lanzavecchia, 1994). These DC are as efficient as antigen-specific B cells in presenting tetanus toxoid (TT) to specific T-cell clones (Sallusto and Lanzavecchia, 1994). Their efficiency in antigen presentation is further enhanced by immune complexes, which are captured through FcR-mediated endocytosis. Those DC possess a specialized mechanism, macropinocytosis, that allows them, by fluid-phase pinocytosis, to internalize soluble antigens with high efficiency (Sallusto and Lanzavecchia, 1995). In contrast to what has been found with other cell types, macropinocytosis in those DC is constitutive and allows an accumulation of macromolecules that is linear in time. A single DC can accumulate, within a single hour, an amount of marker that is contained in four times its own volume of extracellular fluid. The marker taken up by the macropinosomes is delivered rapidly into large compartments rich in MHC class II molecules. Incubation of these cultured DC with TNF-α, soluble CD40L, IL-1, or lipopolysaccharide (LPS) leads to an irreversible down-regulation of macropinocytosis and a disappearance of the MHC class II compartment. This is associated with an increased expression of surface MHC antigens, adhesion receptors, and costimulatory molecules. Concomitantly, levels of FcγRII decrease significantly. The functional consequences are an increase in T-cell stimulatory capacity in MLR but a tenfold decrease in presentation of soluble antigen and a 100-fold decrease in presentation of immune complexes (Sallusto and Lanzavecchia, 1994, 1995).

5.3. From Human CD34$^+$ Hematopoietic Progenitor Cells

5.3.1. Interactions between Dendritic Cells and T Cells

As few as ten CD1a$^+$ cells, generated by culture of CD34$^+$ HPC in presence of GM-CSF plus TNF-α, induce a strong proliferation of allogeneic CD4$^+$ T cells, whereas CD1a$^-$ cells are poor APC (Caux *et al.*, 1992, 1996). DC were also found to induce significant proliferation of syngeneic CD4$^+$ cord blood T cells, but at a magnitude 20 times lower than that observed for allogeneic CD4$^+$ T cells. Allogeneic CD4$^+$ T cells form rosettes with CD1a$^+$ DC within 24 hr and take on a blast morphology after 96 hr of coculture (Figure 5). In addition, naive CD4$^+$ T cells are induced to proliferate in the presence of very low numbers of syngeneic DC and low concentration of superantigens. Allogeneic CD8$^+$ T lymphocytes are also able to proliferate when cultured in the presence of DC generated *in vitro*. The proliferation of CD8$^+$ T cells is weaker than that of CD4$^+$ T cells when cultured with DC alone but reaches comparable levels in the presence of cytokines such as IL-2, IL-4, or IL-7. Accordingly, allospecific CD4$^+$ or CD8$^+$ T-cell lines can be generated by repeated culturing

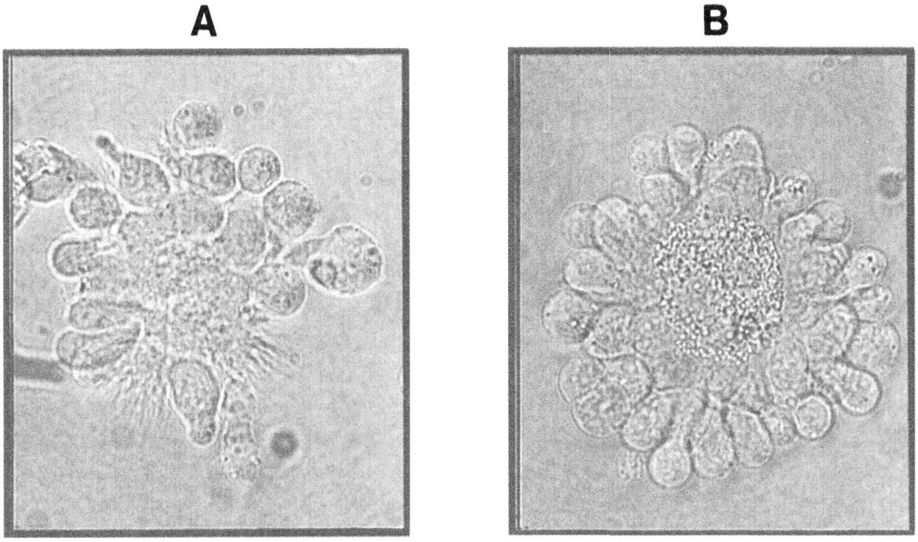

FIGURE 5. Phase-contrast morphology of $CD34^+$ HPC-derived DC interacting with allogeneic $CD4^+$ T cells. GM-CSF plus TNF-α generated cells after 48 (panel A) and 96 hr (panel B), respectively, of coculture with allogeneic $CD4^+$ T cells (magnification ×400).

of T-cells on DC, and the $CD8^+$ T-cell lines display a high cytotoxic potential in an MHC-restricted manner (Caux *et al.*, 1995a; Wettendorf *et al.*, 1995).

The DC generated from $CD34^+$ HPC are also able to present soluble antigen to MHC-matched tetanus-toxoid-specific T-cell clones (Caux *et al.*, 1995a).

By using immunogold labeling for CDla/HLA-DR, the LC generated *in vitro* were found to internalize membrane complexes through receptor-mediated endocytosis via coated pits, coated vesicles, endosomes, and lysosomes (Caux *et al.*, 1995a). Induced endocytosis involved Birbeck granule structures generated from labeled plasma membrane.

Thus, $CDla^+$ dendritic/LC generated *in vitro* from $CD34^+$ HPC can prime naive T cells as well as capture, internalize, process and present soluble protein to primed T cells. The ability of these DC generated *in vitro* to elicit naive syngeneic T cells to mount primary responses specific for soluble antigens has yet to be demonstrated.

5.3.2. Control of Molecular Interactions Between Dendritic Cells and T Cells

5.3.2a. Role of CD28-CTLA-4/CD80-CD86 in DC/T Cell Interactions. APC express two molecules, B7-1/BB1/CD80 and B7-2/B70/CD86, that share 21% homology and that bind to two different receptors on T cells, CD28 and CTLA-4, whose extracellular domains share homology (Azuma *et al.*, 1993; Boussiotis *et al.*, 1993; Freeman *et al.*, 1993a,b; Hathcock *et al.*, 1993) Coligation of the T-cell receptors together with the CD28 antigens results in high cytokine (IL-2) production and strong T-cell proliferation (Azuma *et al.*, 1993; Freeman *et al.*, 1993b; Linsley and Ledbetter, 1993). The specific functions of

CD28 and CTLA-4, which display unrelated intracytoplasmic domains, remain unclear. Yet recent studies indicate the existence of a third ligand for CTLA-4 that does not bind to CD28 and that functions to induce apoptosis of activated T cells whose CD28 antigen has not been engaged (Gribben *et al.*, 1995).

Although all the DC generated *in vitro* by culturing CD34$^+$ HPC with GM-CSF plus TNF-α express CD80 (B7/BB1), only a subset display CD86 (B70/B7-2) (Caux *et al.*, 1994c). It is unlikely that DC generated *in vitro* display a third ligand for CD28/CTLA-4 because the binding of CTLA4-Ig is completely inhibited by a combination of antibodies (mAbs) against CD80 and CD86. This is in apparent contrast to B cells that have been shown to display such a third ligand (Boussiotis *et al.*, 1993). CD28 triggering appears essential during DC-induced T-cell alloreaction, with CD86 being the major functional DC ligand, inasmuch as most T-cell proliferation is suppressed by mAbs against CD86 but not by mAbs against CD80 (Caux *et al.*, 1994c; Zhou and Tedder, 1995). The critical role of CD86 has also been demonstrated in mice whose DC from various organs display levels of CD86 that are higher than those of CD80. Studies with antibodies also demonstrated a major involvement of CD86 during T-cell activation (Inaba *et al.*, 1994). The functional importance of the CD28 pathway has also been demonstrated *in vivo*, where its blocking results in T cell anergy, prolongation of graft survival, and inhibition of B-cell responses (Lenschow *et al.*, 1992; Linsley *et al.*, 1992; Turka *et al.*, 1992).

The rationale for two CD28/CTLA-4 ligands mediating apparently similar effects on T-cell activation is largely unknown. However, CD80 and CD86 have recently been shown, in a mouse model of experimental encephalomyelitis, to activate differentially the Thl/Th2 developmental pathways, respectively (Kuchroo *et al.*, 1995).

5.3.2b. Role of CD40/CD40L in DC/T Cell Interactions. CD40 is expressed on LC, blood DC as well as interdigitating cells from lymphoid organs (Figure 6) (Hart and McKenzie, 1988; Lenz *et al.*, 1993; Nestle *et al.*, 1993; O'Doherty *et al.*, 1993, 1994). The CD40 antigen, which is of critical importance in T-dependent B-cell growth, differentiation, and isotype switch (Banchereau *et al.*, 1994), is also functional in cells other than mature B cells, including human B lymphocyte progenitors (Saeland *et al.*, 1993), thymic epithelial cells (Galy and Spits, 1992), and monocytes (Alderson *et al.*, 1993). Human dendritic–Langerhans cells generated *in vitro* with the use of GM-CSF plus TNF-α also express CD40 at high density (Caux *et al.*, 1994b). Transferring these cultured DC into a medium devoid of GM-CSF results in their prompt death unless their CD40 antigen is engaged by CD40 ligand (CD40-L)-transfected cells. In addition, CD40 triggering induces changes in morphology and phenotype. Furthermore, CD40-activated DC up-regulate accessory molecules such as CD58, CD80, and CD86 and secrete a limited set of cytokines (TNF-α, IL-8, and MIP-1α). CD40 engagement also turns on DC maturation, as illustrated by up-regulation of CD25 and down-regulation of CDla, which respectively appear and disappear on LC/veiled cells when they enter secondary lymphoid organs to become interdigitating dendritic cells (Romani *et al.*, 1989b; Larsen *et al.*, 1990). Because T cells that are activated by DC up-regulate CD40-L (Caux *et al.*, 1994b), it is likely that CD40 activation of DC represents an important physiological interaction between DC and T cells (see Figure 6).

In keeping with this, T-cell alloreaction induced by blood DC (Zhou and Tedder, 1995) and epidermal LC (Péguet-Navarro *et al.*, 1995) is inhibited by antibodies interrupting CD40/CD40L interactions. Furthermore, patients who have a genetically altered nonfunctional CD40-L (see Chapter 10), present a major T-cell defect as shown by frequent

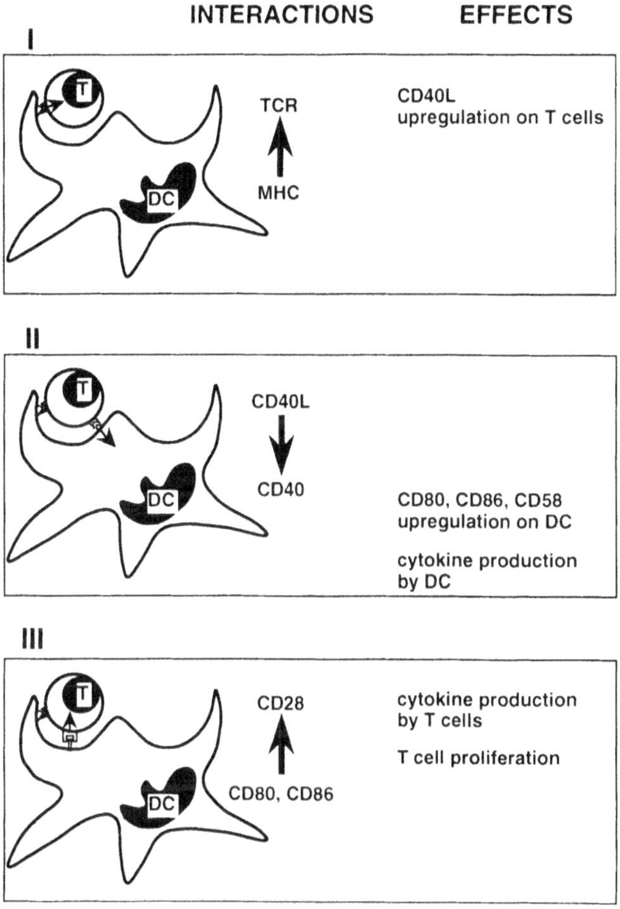

FIGURE 6. A schematic view of early interactions between T cells and dendritic cells. The priming of naive T cells occurs in the T-cell-rich area of secondary lymphoid organs. There, the dendritic cells loaded with antigen, emigrating from nonlymphoid organs through the afferent lymph stream, present the antigen to naive T cells. Processed peptide presented in the MHC groove triggers the T-cell receptor, inducing up-regulation of CD40L on the naive T cell. CD40L, by engaging CD40 on the dendritic cell, leads to CD80 (B7-1) and CD86 (B7-2) up-regulation and cytokine production. Then, the triggering of CD28 on the T cells in the presence of dendritic-cell-derived cofactors (TNF-α), induces full T-cell activation.

Pneumocystis carinii infections (Notarangelo *et al.*, 1992). Finally, *in vivo* administration of CD40/CD40-L antagonist has been shown to block the development of T-cell-dependent collagen-induced arthritis (Durie *et al.*, 1993) as well as graft-versus-host diseases (Durie *et al.*, 1994). Taken together, these data can be interpreted as depending on the interruption of a CD40/CD40-L interaction between T cells and DC rather than between T cells and B cells. A cell-specific disruption of the CD40 gene would make it possible to clarify the role of CD40 on the various cell types.

5.3.2c. Interleukin 10 Is a Strong Inhibitor of DC/T-cell Interactions. Interleukin 10 was discovered for its cytokine synthesis inhibitory activity (Fiorentino *et al.*, 1989; Moore *et al.*, 1990). It inhibits primary and secondary proliferative responses of both $CD4^+$ and $CD8^+$ T cells induced by allogeneic $CD1a^+$ DC (Caux *et al.*, 1994a). Interleukin 10 specifically acts on DC rather than on T cells, as it does not affect the T-cell proliferation induced by EBV lymphoblastoid cell lines (EBV-LCL) or that induced by immobilized anti-CD3. The inhibitory effect of IL-10 does not result from the production of "suppressive" factors by the DC. Rather, the early inhibition of cytokine production (IL-2, GM-CSF, TNF, IFN-γ) (after 24 hr of coculture) may explain the inhibition of the late T-cell DNA synthesis. The inhibitory effect of IL-10 is not restricted to DC generated *in vitro* because IL-10 was also found to inhibit the alloreaction induced by LC purified from skin (Péguet-Navarro *et al.*, 1994).

Thus, regarding their sensitivity to IL-10, human DC are not different from blood mononuclear cells, monocytes, and B cells (Bejarano *et al.*, 1992). In contrast, murine DC, isolated from either spleen or skin, appear able to mount a full alloreactive T-cell proliferation in the presence of IL-10 (Ding and Shevach, 1992; Enk *et al.*, 1993; Macatonia *et al.*, 1993a). Yet, mIL-10 alters interactions between mouse DC and T cells because it reduces the synthesis of IFN-γ induced by spleen DC (Macatonia *et al.*, 1993a) and that of IFN-γ and IL-2 induced by LC (Enk *et al.*, 1993). Furthermore, studies in mice have clearly established that *in vivo* administration of IL-10 can induce transferable tolerance to the hapten-mediated delayed-type hypersensitivity reaction (Enk *et al.*, 1994). Further studies are clearly required to understand the effects of IL-10 on DC.

5.3.3. Interactions between Dendritic Cells and B Cells

Although the T-dependent primary B-cell activation is known to be dependent on DC (Inaba *et al.*, 1983, 1984; Francotte and Urbain, 1985; Inaba and Steinman, 1985; Sornasse *et al.*, 1992; Flamand *et al.*, 1994) and to occur in the extrafollicular area of secondary lymphoid organs, little information is presently available regarding the potential interactions between DC and B cells. We recently engaged in such an analysis using DC generated by culturing $CD34^+$ HPC with GM-CSF plus TNF-α and tonsillar B lymphocytes. In a system in which the T-cell signal is provided by CD40-L-transfected L cells, DC were found to markedly and directly modulate all stages of B-cell growth and differentiation (Dubois *et al.*, 1995). In particular, DC induce a three- to sixfold enhancement of CD40-L-dependent B-cell proliferation in the absence of exogenous cytokines. Furthermore, DC considerably enhance (ten- to 100-fold increase) the secretion of IgG, IgA, and IgM by CD40-activated memory B cells in the absence of exogenous cytokines. However, DC are not able by themselves to induce Ig production by CD40-activated naive $sIgD^+$ B cells. Interestingly, such cocultures of naive B cells and DC become permissive for high levels of IgM secretion in presence of IL-2. Indeed, IL-2 may act on the $CD25^+$ CD40-activated DC inasmuch as CD40-activated B cells do not respond to IL-2. Furthermore, in the presence of IL-10, DC strikingly stimulate CD40-activated naive $sIgD^+$ B cells to express surface IgA and secrete large amounts of this isotype (Fayette *et al.*, 1995). In keeping with this, murine DC have earlier been shown to specifically enhance the secretion of IgA by activated B cells (Spalding and Griffin, 1986; Schrader *et al.*, 1990). Figure 7 summarizes our current views about the possible interactions that may occur between DC and B cells during antigen-induced T-cell-dependent primary humoral immune responses.

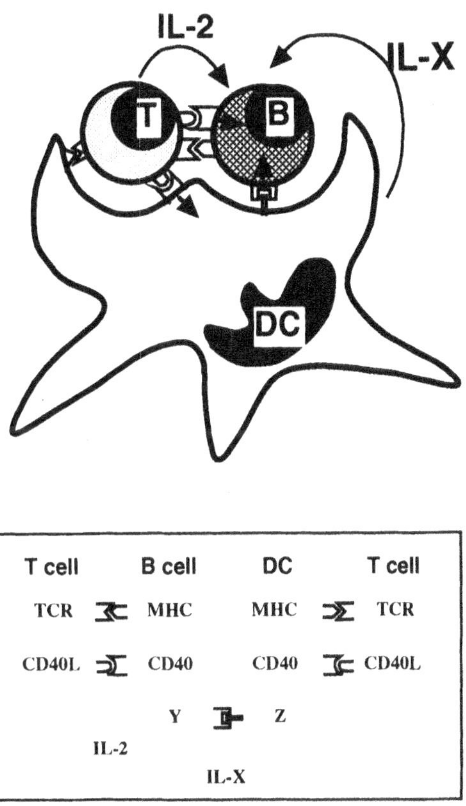

FIGURE 7. A schematic view of possible interactions between dendritic cells and B cells. Humoral responses are initiated in the T-cell-rich area of secondary lymphoid organs consequently to naive-T-cell priming. Thus, dendritic cells might represent a bridge between antigen-specific naive T cells and antigen-specific naive B cells. After priming by a dendritic cell (see Figure 6), naive T cells can then interact in a cognate manner with naive B cells. Following this cognate interaction (CD40L triggering, IL-2 production), naive B cells are then sensitive to signals (both soluble and membrane bound) provided by the dendritic cell activated through CD40 engagement. Under concomitant signals provided by both the primed T cell and the activated dendritic cell, B cells undergo strong proliferation and differentiation, leading to the extrafollicular IgM plasma cell formation observed during primary humoral responses.

6. CONCLUDING REMARKS

The large numbers of DC generated from precursors or monocytes represent a material that should allow easier progress in the field of immunity aiming at characterizing important molecular pathways involved in the regulation of initiation of immune responses and in the modulation of the class of immune responses. The discovery of these pathways is likely to offer new ways to modulate immune responses *in vivo*. In addition, the discovery of the factors controlling DC generation might lead to strategies to control their development *in vivo*. Also, the identification of factors involved in the control of DC maturation might have some implications in the understanding of the role of DC in the induction of tolerance versus immunity. Such possibilities might be of interest either to stimulate immune responses in case of immunodeficiencies, cancers, and severe infectious diseases or to suppress immune responses in case of autoimmunity and transplantation. Finally, the generation of DC *in vitro* should make it feasible to explore the immunogenic potential of those cells in clinical situations, such as the presentation of antigens in resistance to infections and tumors.

ACKNOWLEDGMENTS. We would like to thank our many collaborators over the past 3 years who permitted us, through their work and discussions, to write this chapter. F. Brière, B. de Saint-Vis, C. Dezutter-Dambuyant, B. Dubois, J. Fayette, S. Lebecque, C. Massacrier,

B. Vanbervliet, and M. Wettendorf have contributed greatly to the most recent experiments. We thank N. Courbière for outsanding help and D. Lepot and M. Vatan for invaluable assistance.

7. REFERENCES

Agger, R.. Crowley, M. T., and Witmer-Pack, M. D., 1990, The surface of dendritic cells in the mouse as studied with monoclonal antibodies, *Int. Rev. Immunol.* **6:**89–101.

Aiba, S., and Katz, S. I., 1991, The ability of cultured Langerhans cells to process and present protein antigen is MHC dependent, *J. Immunol.* **146:**2479–2487.

Alderson, M. R., Armitage, R. J., Tough, T. W., Strockbine, L., Fanslow, W. C., and Spriggs, M. K., 1993, CD40 expression by human monocytes: Regulation by cytokines and activation of monocytes by the ligand for CD40, *J. Exp. Med.* **178:**669–674.

Ardavin, C., Wu, L., Li, C. L., and Shortman, K., 1993, Thymic dendritic cells and T cells develop simultaneously in the thymus from a common precursor population, *Nature* **362:**761–763.

Austyn, J. M., Kupiec-Weglinski, J. W., Hankins, D. F., and Morris, P. J., 1988, Migration patterns of dendritic cells in the mouse. Homing to T cell-dependent areas of spleen, and binding within marginal zone, *J. Exp. Med.* **167:**646–651.

Azuma, M., Ito, D., Yagita, H., Okumura, K., Phillips, J. H., Lanier, L. L., and Somoza, C., 1993, B70 antigen is a second ligand for CTLA-4 and CD28, *Nature* **366:**76–79.

Backx, B., Broeders, L., Bot, F. J., and Lowenberg, B., 1991, Positive and negative effects of tumor necrosis factor on colony growth from highly purified normal marrow progenitors, *Leukemia* **5:**66–70.

Banchereau, J., Bazan, F., Blanchard, D., Brière, F., Galizzi, J.-P., van Kooten, C., Liu, Y.-J., Rousset, F., and Saeland, S., 1994, The CD40 antigen and its ligand, *Annu. Rev. Immunol.* **12:**881–922.

Barclay, A. N., and Mayrhofer, G., 1981, Bone marrow origin of Ia-positive cells in the medulla of rat thymus, *J. Exp. Med.* **153:**1666–1671.

Barfoot, R., Denham, S., Gyure, L. A., Hall, J. G., Hobbs, S. M., Jackson, L. E., and Robertson, D., 1989, Some properties of dendritic macrophages from peripheral lymph, *Immunology* **68:**233–239.

Bartosik, J., 1992, Cytomembrane-derived Birbeck granules transport horseradish peroxidase to the endosomal compartment in the human Langerhans cells, *J. Invest. Dermatol.* **99:**53–58.

Beckman, E. M., Porcelli, S. A., Morita, C. T., Behar, S. M., Furlong, S. T., and Brenner, M. B., 1994, Recognition of a lipid antigen by CDl-restricted $\alpha\beta^+$ T cells, *Nature* **372:**691–694.

Bejarano, M.-T., de Waal Malefyt, R., Abrams, J. S., Bigler, M., Bacchetta, R., de Vries, J. E., and Roncarolo, M.-G., 1992, Interleukin 10 inhibits allogeneic proliferative and cytotoxic T cell responses generated in primary mixed lymphocyte cultures, *Int. Immunol.* **4:**1389–1397.

Berg, S. F., Mjaaland, S., and Fossum, S., 1994, Comparing macrophages and dendritic leukocytes as antigen-presenting cells for humoral responses *in vivo* by antigen targeting, *Eur. J. Immunol.* **24:**1262–1268.

Bhardwaj, N., Friedman, S. M., Cole, B. C., and Nisanian, A. J., 1992, Dendritic cells are potent antigen-presenting cells for microbial superantigens, *J. Exp. Med.* **175:**267–273.

Bhardwaj, N., Young, J. W., Nisanian, A. J., Baggers, J., and Steinman, R. M., 1993, Small amounts of superantigen, when presented on dendritic cells, are sufficient to initiate T cell responses, *J. Exp. Med.* **178:** 633–642.

Bhardwaj, N., Bender, A., Gonzalez, N., Bui, L. K., Garrett, M. C., and Steinman, R. M., 1994a, Influenza virus-infected dendritic cells stimulate strong proliferative and cytolytic responses from human $CD8^+$ T cells, *J. Clin. Invest.* **94:**797–807.

Bhardwaj, N., Hodtsev, A.S., Nisanian, A., Kabak, S., Friedman, S. M., Cole, B. C., and Posnett, D. N., 1994b, Human T-cell responses to *Mycoplasma arthritidis*-derived superantigen, *Infect. Immun.* **62:**135–144.

Birbeck, M. S., Breathnach, A. S., and Everall, J. D., 1961, An electron microscopic study of basal melanocytes and high level clear cells (Langerhans' cells) in vitiligo, *J. Invest. Dermatol.* **37:**51–64.

Boehmelt, G., Madruga, J., Dörfler, P., Briegel, K., Schwarz, H., Enrietto, P. J., and Zenke, M., 1995, Dendritic cell progenitor is transformed by a conditional v-Rel estrogen receptor fusion protein v-RelER, *Cell* **80:**341–352.

Boussiotis, V. A., Freeman, G. J., Gribben, J. G., Daley, J., Gray, G., and Nadler, L. M., 1993, Activated human B lymphocytes express three CTLA-4 counterreceptors that costimulate T-cell activation, *Proc. Natl. Acad. Sci. U.S.A.* **90:**11059–11063.

Brand, C. U., Gerber, H. A., Hunziker, T., Schaffner, T., Limat, A., and Brathen, L.R., 1993, Phenotype of Langerhans cells in human afferent skin lymph derived from allergic contact dermatitis, *Exp. Dermatol.* **2:** 274–279.

Bujdoso, R., Hopkins, J., Dutia, B. M., Young, P., and McConnell, I., 1989, Characterization of sheep afferent lymph dendritic cells and their role in antigen carriage, *J. Exp. Med.* **170:**1285–1301.

Buller, M. L., Holmes, K. L., Hugin, A., Frederikson, T. N., and Morse, H. C., 1987, Induction of cytotoxic T-cell response *in vivo* in the absence of CD4 helper cells, *Nature* **328:**77–79.

Burkly, L., Hession, C., Ogata, L., Reilly, C., Marconi, L. A., Olson, D., Tizard, R., Cate, R., and Lo, D., 1995, Expression of *rel*B is required for the development of thymic medulla and dendritic cells, *Nature* **373:** 531–536.

Calne, R. Y., Sells, R. A., Pena, J. R., Davis, D. R., Millard, P. R., Herbertson, B. M., Binns, R. M., and Davis, D. A. L., 1969, Induction of immunological tolerance by porcine liver allografts, *Nature* **233:**472–476.

Cameron, P. U., Freudenthal, P. S., Barker, J. M., Gezelter, S., Inaba, K., and Steinman, R. M., 1992, Dendritic cells exposed to human immunodeficiency virus type-1 transmit a vigorous cytopathic infection to $CD4^+$ T cells, *Science* **257:**383–387.

Cameron, P. U., Lowe, M. G., Crowe, S. M., O'Doherty, U., Pope, M., Gezelter, S., and Steinman, R. M., 1994, Susceptibility of dendritic cells to HIV-1 infection *in vitro*, *J. Leukocyte Biol.* **56:**257–265.

Caux, C., Saeland, S., Favre, C., Duvert, V., Mannoni, P., and Banchereau, J., 1990, Tumor necrosis factor-alpha strongly potentiates interleukin-3 and granulocyte–macrophage colony-stimulating factor-induced proliferation of human $CD34^+$ hematopoietic progenitor cells, *Blood* **75:**2292–2298.

Caux, C., Favre, C., Saeland, S., Duvert, V., Durand, I., Mannoni, P., and Banchereau, J., 1991, Potentiation of early hematopoiesis by tumor necrosis factor-α is followed by inhibition of granulopoietic differentiation and proliferation, *Blood* **78:**635–644.

Caux, C., Dezutter-Dambuyant, C., Schmitt, D., and Banchereau, J., 1992, GM-CSF and TNF-α cooperate in the generation of dendritic Langerhans cells, *Nature* **360:**258–261.

Caux, C., Durand, I., Moreau, I., Duvert, V., Saeland, S., and Banchereau, J., 1993, TNF-α cooperates with IL-3 in the recruitment of a primitive subset of human $CD34^+$ progenitors., *J. Exp. Med.* **177:**1815–1820.

Caux, C., Massacrier, C., Vanbervliet, B., Barthelemy, C., Liu, Y.-J., and Banchereau, J., 1994a, Interleukin-10 inhibits T cell alloreaction induced by human dendritic cells, *Int. Immunol.* **6:**1177–1185.

Caux, C., Massacrier, C., Vanbervliet, B., Dubois, B., van Kooten, C., Durand, I., and Banchereau, J., 1994b, Activation of human dendritic cells through CD40 cross-linking, *J. Exp. Med.* **180:**1263–1272.

Caux, C., Vanbervliet, B., Massacrier, C., Azuma, M., Okumura, K., Lanier, L. L., and Banchereau, J., 1994c, B70/B7-2 is identical to CD86 and is the major functional ligand for CD28 expressed on human dendritic cells, *J. Exp. Med.* **180:**1841–1847.

Caux, C., Massacrier, C., Dezutter-Dambuyant, C., Vanbervliet, B., Jacquet, C., Schmitt, D., and Banchereau, J., 1995a, Human dendritic Langerhans cells generated in vitro from $CD34^+$ progenitors can prime naive $CD4^+$ T cells and process soluble antigen, *J. Immunol.* **155:**5427–5435.

Caux, C., Vanbervliet, B., Massacrier, C., Dubois, B., Dezutter-Dambuyant, C., Schmitt, D., and Banchereau, J., 1995b, Characterization of human $CD34^+$ derived Dendritic/Langerhans cells (D-Lc), in *Dendritic Cells in Fundamental and Clinical Immunology* (J. Banchereau and D. Schmitt, eds.), pp. 1–5, Plenum Press, London.

Caux, C., Vanbervliet, B., Massacrier, C., Durand, I., and Banchereau, J., 1996, Interleukin-3 cooperates with Tumor Necrosis Factor α for the development of human dendritic/Langerhans cells from cord blood $CD34^+$ hematopoietic progenitor cells, *Blood* **87:**2376–2385.

Chen, H., Ma, C., Yuan, J., Wang, Y., and Silvers, W. K., 1986, Occurrence of donor Langerhans cells in mouse and rat chimeras and their replacement in skin grafts, *J. Invest. Dermatol.* **86:**630–633.

Clare-Salzler, M. J., Brooks, J., Chai, A., Van Herle, K., and Anderson, C., 1992, Prevention of diabetes in nonobese diabetic mice by dendritic cell transfer, *J. Clin. Invest.* **90:**741–748.

Codner, M. A., Shuster, B. A., Steinman, R. M., Harper, A. D., LaTrenta, G. S., and Hoffman, L. A., 1990, Migration of donor leukocytes from limb allografts into host lymphoid tissues, *Ann. Plastic Surg.* **25:** 353–359.

Croft, M., Duncan, D. D., and Swain, S. L., 1992, Response of naive antigen-specific $CD4^+$ T cells *in vitro*: Characteristics and antigen-presenting cell requirements, *J. Exp. Med.* **176:**1431–1437.

Crow, M. K., and Kunkel, H. G., 1982, Human dendritic cells: Major stimulators of the autologous and allogeneic mixed leucocyte reactions, *Clin. Exp. Immunol.* **49:**338–346.

Crowley, M., Inaba, K., Witmer, P. M., and Steinman, R. M., 1989, The cell surface of mouse dendritic cells: FACS analyses of dendritic cells from different tissues including thymus, *Cell. Immunol* **118:**108–125.

Crowley, M. T., Inaba, K., Witmer-Pack, M. D., Gezelter, S., and Steinman, R. M., 1990, Use of the fluorescence activated cell sorter to enrich dendritic cells from mouse spleen, *J. Immunol. Methods* **133:**55–66.

Cumberbatch, M., and Kimber, I., 1992, Dermal tumour necrosis factor-alpha induces dendritic cell migration to draining lymph nodes, and possibly provides one stimulus for Langerhans' cell migration, *Immunology* **75:**257–263.

Cumberbatch, M., and Kimber, I., 1995, Tumour necrosis factor-α is required for accumulation of dendritic cells in draining lymph nodes and for optimal contact sensitization, *Immunology* **84:**31–35.

Dai, R., Grammer, S. F., and Streilein, J. W., 1993, Fresh and cultured Langerhans cells display differential capacities to activate hapten-specific T cells, *J. Immunol.* **150:**59–66.

De Becker, G., Sornasse, T., Nabavi, N., Bazin, H., Tielemans, F., Urbain, J., Leo, O., and Moser, M., 1994, Immunoglobulin isotype regulation by antigen-presenting cells *in vivo*, *Eur. J. Immunol.* **24:**1523–1528.

De Bruijn, M. L. H., Nieland, J. D., Harding, C. V., and Melief, C. J. M., 1992, Processing and presentation of intact hen egg-white lysozyme by dendritic cells, *Eur. J. Immunol.* **22:**2347–2352.

De Fraissinette, A., Schmitt, D., Dezutter-Dambuyant, C., Guyotat, D., Zabot, M., and Thivolet, J., 1988, Culture of putative Langerhans cell bone marrow precursors: Characterization of their phenotype, *Exp. Hematol.* **16:**764–768.

De Panfilis, G., Soligo, D., Manara, G. C., Ferrari, C., and Torresani, C., 1989, Adhesion molecules on the plasma membrane of epidermal cells. I. Human resting Langerhans cells express two members of the adherence-promoting CD11/CD18 family, namely, H-Mac-1 (CD11b/CD18) and pg 150,95 (CD11c/CD18), *J. Invest. Dermatol.* **93:**60–69.

Dezutter-Bambuyant, C., and Schmitt, D., 1993, Epidermal Langerhans cells and HIV-1 infection, *Immunol. Lett.* **39:**33–37.

Dijkstra, C. D., 1982, Characterization of nonlymphoid cells in rat spleen, with special reference to strongly Ia-positive branched cells in T-cell areas, *J. Reticuloendothel. Soc.* **32:**167–178.

Ding, L., and Shevach, E., 1992, IL-10 inhibits mitogen-induced T cell proliferation by selectively inhibiting macrophage costimulatory function, *J. Immunol.* **148:**3133–3139.

Dranoff, G., Crawford, A. D., Sadelain, M., Ream, B., Rashid, A., Bronson, R. T., Dickersin, G. R., Bachurski, C. J., Mark, E. L., Whitsett, J. A., and Mulligan, R. C., 1994, Involvment of granulocyte-macrophage colony-stimulating factor in pulmonary homeostasis, *Science* **264:**713–716.

Drexhage, H. A., Mullink, H., de Groot, J., Clark, J., and Balfour, B. M., 1979, A study of cells present in peripheral lymph of pigs with special reference to a type of cell resembling the Langerhans cells, *Cell Tiss. Res.* **202:**407–430.

Dubois, B., Fayette, J., Vanbervlient, B., Banchereau, J., Brière, F., and Caux, C., 1995, Human dendritic cells enhance growth and differentiation of CD40 activated B cells, in *Dendritic Cells in Fundamental and Clinical Immunology* (J. Banchereau and D. Schmitt, eds.), pp. 397–399, Plenum Press, New York.

Duijvestijn, A. M., Schutte, R., Kohler, Y. G., Korn, C., and Hoefsmit, E. C. M., 1983, Characterization of the population of phagocytic cells in thymic cell suspensions. A morphological and cytochemical study, *Cell Tiss. Res.* **231:**313–323.

Durie, F. H., Fava, R. A., Foy, T. M., Aruffo, A., Ledbetter, J. A., and Noelle, R. J., 1993, Prevention of collagen-induced arthritis with an antibody to gp39, the ligand for CD40, *Science* **261:**1328–1330.

Durie, F. H., Aruffo, A., Ledbetter, J., Crassi, K. M., Green, W. R., and Fast, L. D., 1994, Antibody to the ligand of CD40, gp39, blocks the occurence of the acute and chronic forms of graft-vs-host disease, *J. Clin. Invest.* **94:**1333–1338.

Elbe, A., Schleischitz, S., Strunk, D., and Stingl, G., 1994, Fetal skin-derived MHC class I^+, MHC class II^- dendritic cells stimulate MHC class I-restricted responses of unprimed $CD8^+$ T cells, *J. Immunol.* **153:**2878–2889.

Enk, A. H., Angeloni, V. L., Udey, M. C., and Katz, S. I., 1993, Inhibition of Langerhans cell antigen-presenting function by IL-10, *J. Immunol.* **151:**2390–2398.

Enk, A. H., Saloga, J., Becker, D., Mohamadzadeh, M., and Knop, J., 1994, Induction of hapten-specific tolerance by interleukin 10 *in vivo*, *J. Exp. Med.* **179:**1397–1402.

Fairchild, P. J., and Austyn, J. M., 1990, Thymic dendritic cell. Phenotype and function, *Int. Rev. Immunol.* **6:**187–196.

Fayette, J., Dubois, B., Caux, C., Banchereau, J., and Brière, F., 1995, Human dendritic cells can drive CD40-activated sIgD$^+$ B cells to mount mucosal-type humoral response, in *Dendritic Cells in Fundamental and Clinical Immunology* (J. Banchereau and D. Schmitt, eds.), pp. 401–403, Plenum Press, New York.

Fiorentino, D. F., Bond, M. W., and Mosmann, T. R., 1989, Two types of mouse T helper cell. IV. Th2 clones secrete a factor that inhibits cytokine production by Thl clones, *J. Exp. Med.* **170:**2081–2095.

Fithian, E., Kung, P., Goldstein, G., Rubenfeld, M., Fenoglio, C., and Edelson, R., 1981, Reactivity of Langerhans cells with hybridoma antibody, *Proc. Natl. Acad. Sci. U.S.A.* **78:**2541–2544.

Flamand, V., Sornasse, T., Thielemans, K., Demanet, C., Bakkus, M., Bazin, H., Tielemans, F., Leo, O., Urbain, J., and Moser, M., 1994, Murine dendritic cells pulsed *in vitro* with tumor antigen induce tumor resistance *in vivo*, *Eur. J. Immunol.* **24:**605–610.

Fossum, S., 1988, Lymph-borne dendritic leucocytes do not recirculate, but enter the lymph node paracortex to become interdigitating cells, *Scand. J. Immunol.* **27:**97–105.

Fossum, S., 1989a, Dendritic leukocytes: features of their *in vivo* physiology, *Res. Immunol.* **140:**883–891.

Fossum, S., 1989b, The life history of dendritic leukocytes (DL), *Current Topics in Pathology* (O. H. Ivessen ed.), pp. 101–124, Springer-Verlag, Berlin.

Francotte, M., and Urbain, J., 1985, Enhancement of antibody responses by mouse dendritic cells pulsed with tobacco mosaic virus or with rabbit antiidiotypic antibodies raised against a private rabbit idiotype, *Proc. Natl. Acad. Sci. U.S.A.* **82:**8149–8152.

Freeman, G. F., Gribben, J. G., Boussiotis, V. A., Ng, J. W., Restivo, V. A., Jr., Lombard, L. A., Gray, G. S., and Nadler, L. M., 1993a, Cloning of B7-2: A CTLA-4 counterreceptor that costimulates human T cell proliferation, *Science* **262:**909–911.

Freeman, G. J., Borriello, F., Hodes, R. J., Reiser, H., Hathcock, K. S., Laszlo, G., McKnight, A. J., Kim, J., Du, L., Lombard, D. B., Gray, G. S., Nadler, L. M., and Sharpe, A. H., 1993b, Uncovering of functional alternative CTLA-4 counter-receptor in B7-deficient mice, *Science* **262:**907–909.

Freudenthal, P., and Bhardwaj, N., 1990, Dendritic cells in human blood and synovial exudates, *Int. Rev. Immunol.* **6:**103–116.

Freudenthal, P. S., and Steinman, R. M., 1990, The distinct surface of human blood dendritic cells, as observed after an improved isolation method, *Proc. Natl. Acad. Sci. U.S.A.* **87:**7698–7702.

Fuchs, E. J., and Matzinger, P., 1992, B cells turn off virgin but not memory T cells, *Science* **258:**1156–1159.

Galy, A. H. M., and Spits, H., 1992, CD40 is functionnally expressed on human thymic epithelial cells, *J. Immunol.* **149:**775–782.

Gautam, S. C., and Battisto, J. R., 1985, Orally induced tolerance generates an efferently acting suppressor T cell and an acceptor T cell that together downregulate contact sensitivity, *J. Immunol.* **135:**2975–2983.

Gribben, J. G., Freman, G. J., Boussiotis, V. A., Rennert, P., Jellis, C. L., Greenfield, E., Barber, M., Restivo, V. A., Jr., Ke, X., Gray, G.S., and Nadler, L.M., 1995, CRLA4 mediates antigen-specific apoptosis of human T cells, *Proc. Natl. Acad. Sci. U.S.A.* **92:**811–815.

Guillemot, F. P., Oliver, P. D., Peault, B. M., and LeDourain, N. M., 1984, Cells expressing Ia antigen in the avian thymus, *J. Exp. Med.* **160:**1803–1819.

Hanau, D., Fabre, M., Schmitt, D. A., Stampf, J.-L., Garaud, J.-C., Bieber, T., Grosshans, E., Benezra, C., and Cazenave, J.-P., 1987, Human epidermal Langerhans cells internalized by receptor-mediated endocytosis T6 (CDl "NAl/34") surface antigen. Birbeck granules are involved in the intracellular traffic of the antigen, *J. Invest. Dermatol.* **89:**172–177.

Harkiss, G. D., Hopkins, J., and McConnell, I., 1990, Uptake of antigen by afferent lymph dendritic cells mediated by antibody, *Eur. J. Immunol.* **20:**2367–2373.

Hart, D. N. J., and Fabre, J. W., 1981, Demonstration and characterization of Ia-positive dendritic cells in the interstitial connective tissues of rat heart and other tissues, but not brain, *J. Exp. Med.* **154:**347–361.

Hart, D. N. J., and McKenzie, J. L., 1988, Isolation and characterization of human tonsil dendritic cells, *J. Exp. Med.* **168:**157–160.

Hart, D. N. J., and McKenzie, J. L., 1990, Interstitial dendritic cells, *Int. Rev. Immunol.* **6:**128–149.

Hart, D. N., Starling, G. C., Calder, V. L., and Fernando, N. S., 1993, B7/BB-1 is a leucocyte differentiation antigen on human dendritic cells induced by activation, *Immunology* **79:**616–620.

Hathcock, K. S., Laszlo, G., Dickler, H. B., Bradshaw, J., Linsley, P., and Hodes, R. J., 1993, Identification of an alternative CTLA-4 ligand co-stimulatory for T cell activation, *Science* **262:**905–907.

Havenith, C. E., Breedijk, A. J., Betjes, M. G., Calame, W., Beelen, R. H., and Hoefsmit, E. C., 1993, T cell priming *in situ* by intratracheally instilled antigen-pulsed dendritic cells, *Am. J. Respir. Cell. Mol. Biol.* **8:**319–324.

Heufler, C., Koch, F., and Schuler, G., 1988, Granulocyte/macrophage colony-stimulating factor and interleukin 1 mediate the maturation of murine epidermal Langerhans cells into potent immunostimulatory dendritic cells, *J. Exp. Med.* **167:**700–705.

Hoang, T., Levy, B., Onetto, N., Haman, A., and Rodriguez-Cimadevilla, J. C., 1989, Tumor necrosis factor α stimulates growth of the clonogenic cells of acute myeloblastic leukemia in synergy with granulocyte–macrophage colony-stimulating factor, *J. Exp. Med.* **170:**15–26.

Holt, P. G., 1993, Regulation of antigen-presenting cell function(s) in lung and airway tissues, *Eur. Respir. J.* **6:** 120–129.

Holt, P. G., Schon-Hegrad, M. A., and Oliver, J., 1988, MHC class II antigen-bearing dendritic cells in pulmonary tissues of the rat: Regulation of antigen presentation activity by endogenous macrophage populations, *J. Exp. Med.* **167:**262–274.

Holt, P. G., Schon-Hegrad, M. A., and McMenamin, P. G., 1990, Dendritic cells in the respiratory tract, *Int. Rev. Immunol.* **6:**139–149.

Holt, P. G., Oliver, J., McMenamin, C., Bilyk, N., Kraal, G., and Thepen, T., 1993a, The antigen presentation functions of lung dendritic cells are downmodulated *in situ* by soluble mediators from pulmonary alveolar macrophages, *J. Exp. Med.* **177:**397–407.

Holt, P. G., Haining, S., Nelson, D. J., and Sedgwick, J. D., 1994, Origin and steady-state turnover of class II MHC-bearing dendritic cells in the epithelium of the conducting airways, *J. Immunol.* **153:**256–261.

Hsieh, C.-S., Macatonia, S. E., Tripp, C. S., Wolf, S. F., O'Garra, A., and Murphy, K. M., 1993, Development of TH1 $CD4^+$ T cells through IL-12 produced by *Listeria*-induced macrophages, *Science* **260:**547–549.

Ibrahim, M. A. A., Chain, B. M., and Katz, D. R., 1995, The injured cell: The role of the dendritic cell system as a sentinel receptor pathway, *Immunol. Today* **14:**181–186.

Inaba, K., and Steinman, R. M., 1985, Protein-specific helper T lymphocyte formation initiated by dendritic cells, *Science* **229:**475–479.

Inaba, K., and Steinman, R. M., 1987, Dendritic and B cell function during antibody responses in normal and immunodeficient (*xid*) mouse spleen cultures, *Cell. Immunol.* **105:**432–442.

Inaba, K., Granelli-Piperno, A., and Steinman, R. M., 1983, Dendritic cells are critical accessory cells for thymus-dependent antibody responses in mouse and man, *Proc. Natl. Acad. Sci. U.S.A.* **80:**6041–6045.

Inaba, K., Witmer, M. D., and Steinman, R. M., 1984, Clustering of dendritic cells, helper T lymphocytes, and histocompatible B cells, during primary antibody responses *in vitro*, *J. Exp. Med.* **160:**858–876.

Inaba, K., Schuler, G., Witmer, M. D., Valinksy, J., Atassi, B., and Steinman, R. M., 1986, Immunologic properties of purified epidermal Langerhans cells, *J. Exp. Med.* **164:**605–613.

Inaba, K., Young, J. W., and Steinman, R. M., 1987, Direct activation of $CD8^+$ cytotoxic T lymphocytes by dendritic cells, *J. Exp. Med* **166:**182–194.

Inaba, K., Hosono, M., and Inaba, M., 1990a, Thymic dendritic cells and B cells: Isolation and function, *Int. Rev. Immunol.* **6:**117–122.

Inaba, K., Metlay, J. P., Crowley, M. T., and Steinman, R. M., 1990b, Dendritic cells pulsed with protein antigens *in vitro* can prime antigen-specific, MHC-restricted T cells in situ, *J. Exp. Med.* **172:**631–640.

Inaba, K., Metlay, J. P., Crowley, M. T., Witmer-Pack, M., and Steinman, R. M., 1990c, Dendritic cells as antigen presenting cells *in vivo*, *Int. Rev. Immunol.* **6:**197–206.

Inaba, K., Inaba, M., Romani, N., Aya, H., Deguchi, M., Ikehara, S., Muramatsu, S., and Steinman, R. M., 1992a, Generation of large numbers of dendritic cells from mouse bone marrow cultures supplemented with granulocyte/macrophage colony-stimulating factor, *J. Exp. Med.* **176:**1693–1702.

Inaba, K., Steinman, R. M., Pack, M. W., Aya, H., Inaba, M., Sudo, T., Wolpe, S., and Schuler, G., 1992b, Identification of proliferating dendritic cell precursors in mouse blood, *J. Exp. Med.* **175:**1157–1167.

Inaba, K., Inaba, M., Deguchi, M., Hagi, K., Yasumizu, R., Ikehara, S., Muramatsu, S., and Steinman, R. M., 1993a, Granulocytes, macrophages, and dendritic cells arise from a common major histocompatibility complex class II-negative progenitor in mouse bone marrow, *Proc. Natl. Acad. Sci. U.S.A.* **90:**3038–3042.

Inaba, K., Inaba, M., Naito, M., and Steinman, R. M., 1993b, Dendritic cells progenitors phagocytose particulates, including *Bacillus Calmette-Guérin* organisms, and sensitize mice to mycobacterial antigens *in vivo*, *J. Exp. Med.* **178:** 479–488.

Inaba, K., Witmer-Pack, M., Inaba, M., Hathcock, K. S., Sakuta, H., Azuma, M., Yagita, H., Okumura, K., Linsley, P. S., Ikehara, S., Muramatsu, S., Hodes, R. J., and Steinman, R. M., 1994, The tissue distribution of the B7-2 costimulator in mice: Abundant expression on dendritic cells *in situ* and during maturation *in vitro*, *J. Exp. Med.* **180:**1849–1860.

Jacobsen, F. W., Rothe, M., Rusten, L., Goeddel, D. V., Smeland, E. B., Veiby, O. P., Slordal, L., and Jacobsen, E. W., 1994, Role of the 75-kDa tumor necrosis factor receptor: Inhibition of early hematopoiesis, *Proc. Natl. Acad. Sci. U.S.A.* **91:**10695–10699.

Jacobsen, S. E. W., Ruscetti, F. W., Dubois, C. M., and Keller, J. R., 1992, Tumor necrosis factor α directly and indirectly regulates hematopoietic progenitor cell proliferation: Role of colony-stimulating factor receptor modulation, *J. Exp. Med.* **175:**1759–1772.

Jenkinson, E. J., Jhittay, P., Kingston, R., and Owen, J. J. T., 1985, Studies of the role of the thymic environment in the induction of tolerance to MHC antigens, *Transplantation* **39:**331–333.

Kabel, P. J., de Haan-Meulman, M., Voorbij, H. A., Kleingeld, M., Knol, E. F., and Drexhage, H. A., 1989, Accessory cells with a morphology and marker pattern of dendritic cells can be obtained from elutriator-purified blood monocyte fractions. An enhancing effect of metrizamide in this differentiation, *Immunobiology* **179:**395–411.

Kaiserling, E., Stein, H., and Mueller-Hermelink, H. K., 1974, Interdigitating reticulum cells in the human thymus, *Cell Tiss. Res.* **155:**47–55.

Kaplan, G., Walsh, G., Guido, L. S., Meyn, P., Burkhardt, R. A., Abalos, R. M., Barker, J., Frindt, P. A., Fajardo, T. T., Celona, R., and Cohn, Z. A., 1992, Novel responses of human skin to intradermal recombinant granulocyte/macrophage-colony-stimulating factor: Langerhans cell recruitment, keratinocyte growth, and enhanced wound healing, *J. Exp. Med.* **175:**1717–1728.

Kashihara, M., Ueda, M., Horiguchi, Y., Furukawa, F., Hanaoka, M., and Imamura, S., 1986, A monoclonal antibody specifically reactive to human Langerhans cells, *J. Invest. Dermatol.* **87:**602–607.

Kasinrerk, W., Baumruker, T., Majdic, O., Knapp, W., and Stockinger, H., 1993, CD1 molecule expression on human monocytes induced by granulocyte-macrophage colony-stimulating factor, *J. Immunol.* **150:**579–584.

Katz, S. I., Tamaki, K., and Sachs, D. H., 1979, Epidermal Langerhans cells are derived from cells originating in bone marrow, *Nature* **282:**324–326.

Kleijmeer, M. J., Oorschot, V. M., and Geuze, H. J., 1994, Human resident Langerhans cells display a lysosomal compartment enriched in MHC class II, *J. Invest. Dermatol.* **103:**516–523.

Klinkert, W. E. F., Labadie, J. H., and Bowers, W. E., 1982, Accessory and stimulating properties of dendritic cells and macrophages isolated from various rat tissues, *J. Exp. Med.* **156:**1–19.

Knight, S. C., and Stagg, A. J., 1993, Antigen-presenting cell types, *Curr. Opin. Immunol.* **5:**374–382.

Knight, S. C., Balfour, B. M., O'Brien, J., Buttifant, L., Sumerska, T., and Clark, J., 1982, Role of veiled cells in lymphocyte activation, *Eur. J. Immunol.* **12:**1057–1060.

Knight, S. C., Farrant, J., and Bryan, A., 1986, Non-adherent, low density cells from human peripheral blood contain dendritic cells and monocytes, both with veiled morphology, *Immunology* **57:**595–603.

Knight, S. C., Macatonia, S. E., Bedford, P. A., and Patterson, S., 1991, Dendritic cells and HIV infection, in *Accessory Cells in HIV and Other Retroviral Infections* (P. Racz, C. D. Dijkstra, and J. C. Gluckman, eds.), pp. 145–154, Karger, Basel.

Koch, F., Heufler, C., Kmpgen, E., Schneeweiss, D., Böck, G., and Schuler, G., 1990, Tumor necrosis factor alpha maintains the viability of murine epidermal Langerhans cells in culture but, in contrast to granulocyte–macrophage colony-stimulating factor, does not induce their functional maturation, *J. Exp. Med.* **171:**159–172.

Kraal, G., Van Wilsem, E., and Breve, J., 1993, The phenotype of murine Langerhans cells from skin to lymph node, *In Vivo* **7:**203–206.

Kripke, M. L., Munn, C. G., Jeevan, A., Tang, J.-M., and Bucana, C., 1990, Evidence that cutaneous antigen-presenting cells migrate to regional lymph nodes during contact sensitization, *J. Immunol.* **145:**2833–2838.

Krueger, G. G., Daynes, R. A., and Emam, M., 1983, Biology of Langerhans cells: Selective migration of Langerhans cells into allogeneic and xenogeneic grafts on nude mice, *Proc. Natl. Acad. Sci. U.S.A.* **80:**1650–1654.

Kuchroo, V. K., Das, M. P., Brown, J. A., Ranger, A. M., Zamvil, S. S., Sobel, R. A., Weiner, H. L., Nabavi, N., and Glimcher, L. H., 1995, B7-1 and B7-2 costimulatory molecules activate differentially the Thl/Th2 developmental pathways: application to autoimmune disease therapy, *Cell* **80:**707–718.

Kupiec-Weglinski, J. W., Austyn, J. M., and Morris, P. J., 1988, Migration patterns of dendritic cells in the mouse. Traffic from blood, and T cell-dependent and independent entry to lymploid tissues, *J. Exp. Med.* **167:**632–645.

Kyewski, B. A., Fathman, C. G., and Rouse, R. V., 1986, Intrathymic presentation of circulating non-MHC antigens by medullary dendritic cells. An antigen-dependent microenvironment for T cell differentiation, *J. Exp. Med.* **163:**231–246.

Langhoff, E., Terwilliger, E. F., Bos, H. J., Kalland, K. H., Poznansky, M. C., Bacon, O. M. L., and Haseltine,

W. A., 1991, Replication of human immunodeficiency virus type 1 in primary dendritic cell cultures, *Proc. Natl. Acad. Sci. U.S.A.* **88:**7998–8002.

Larsen, C. P., Steinman, R. M., Witmer-Pack, M. D., Hankins, D. F., Morris, P. J., and Austyn, J. M., 1990, Migration and maturation of Langerhans cells in skin transplants and explants, *J. Exp. Med.* **172:**1483–1494.

Larsen, C. P., Ritchie, S. C., Pearson, T. C., Linsley, P. S., and Lowry, R. P., 1992, Functional expression of the costimulatory molecule, B7/BB1, on murine dendritic cell populations, *J. Exp. Med.* **176:**1215–1220.

Larsen, C. P., Ritchie, S. C., Hendrix, R., Linsley, P. S., Hathcock, K. S., Hodes, R. J., Lowry, R. P., and Pearson, T. C., 1994, Regulation of immunostimulatory function and costimulatory molecule (B7-1 and B7-2) expression on murine dendritic cells, *J. Immunol.* **152:**5208–5219.

Lenschow, D. J., Zeng, Y., Thistlethwaite, J. R., Montag, A., Brady, W., Gibson, M. G., Linsley, P. S., and Bluestone, J. A., 1992, Long-term survival of xenogeneic pancreatic islet grafts induced by CTLA4Ig, *Science* **257:**789–792.

Lenz, A., Heufler, C., Rammensee, H. G., Glassl, H., Koch, F., Romani, N., and Schuler, G., 1989, Murine epidermal Langerhans cells express significant amounts of class I major histocompatibility complex antigens, *Proc. Natl. Acad. Sci. U.S.A.* **86:**7527–7531.

Lenz, A., Heine, M., Schuler, G., and Romani, N., 1993, Human and murine dermis contain dendritic cells, *J. Clin. Invest.* **92:**2587–2596.

Leszcynski, R., Renkonen, R., and Hayry, P., 1985, Turnover of dendritic cells in rat heart, *Scand. J. Immunol.* **22:**351–360.

Levin, D., Constant, S., Pasqualini, T., Flavell, R., and Bottomly, K., 1993, Role of dendritic cells in the priming of $CD4^+$ T lymphocytes to peptide antigen *in vivo*, *J. Immunol.* **151:**6742–6750.

Levine, T. P., and Chain, B. M., 1992, Endocytosis by antigen presenting cells: Dendritic cells are as endocytically active as other antigen presenting cells, *Proc. Natl. Acad. Sci. U.S.A.* **89:**8342–8346.

Linsley, P. S., and Ledbetter, J. A., 1993, The role of the CD28 receptor during T cell responses to antigen, *Annu. Rev. Immunol.* **11:**191–212.

Linsley, P. S., Wallace, P. M., Johnson, J., Gibson, M. G., Greene, J. L., Ledbetter, J. A., Singh, C., and Tepper, M. A., 1992, Immunosuppression *in vivo* by a soluble form of the CTLA-4 T cell activation molecule, *Science* **257:**792–795.

Liu, L.M., and MacPherson, G.G., 1993, Antigen acquisition by dendritic cells: intestinal dendritic cells acquire antigen administered orally and can prime naive T cells *in vivo*, *J. Exp. Med.* **177:**1299–1307.

Lu, L., Woo, J., Rao, A. S., Li, Y., Watkins, S. C., Qian, S., Starzl, T. E., Demetris, A. J., and Thomson, A. W., 1994, Propagation of dendritic cell progenitors from normal mouse liver using granulocyte/macrophage colony-stimulating factor and their maturational development in the presence of type-1 collagen, *J. Exp. Med.* **179:**1823–1834.

Macatonia, S. E., Knight, S. C., Edwards, A. J., Griffiths, S., and Fryer, P., 1987, Localization of antigen on lymph node dendritic cells after exposure to the contact sensitizer fluorescein isothiocyanate, *J. Exp. Med.* **166:**1654–1667.

Macatonia, S. E., Doherty, T. M., Knight, S. C., and O'Garra, A., 1993a, Differential effect of interleukin 10 on dendritic cell-induced T cell proliferation and interferon-γ production, *J. Immunol.* **150:**3755–3765.

Macatonia, S. E., Hsieh, C.-S., Murphy, K. M., and O'Garra, A., 1993b, Dendritic cells and macrophages are required for Thl development of $CD4^+$ T cells from αβ TCR transgenic mice: IL-12 substitution for macrophages to stimulate IFN-γ production is IFN-γ-dependent, *Int. Immunol.* **5:**1119–1128.

Macatonia, S. E., Hosken, N. A., Litton, M., Vieira, P., Hsieh, C.-S., Culpepper, J. A., Wysocka, M., Trinchieri, G., Murphy, K. M., and O'Garra, A., 1995, Dendritic cells produce interleukin-12 and direct the development of Thl cells from naive $CD4^+$ T cells, *J. Immunol.* **154:**5071–5079.

MacDonald, T. T., 1983, Immunosuppression caused by antigen feeding II. Suppressor T cells mask Peyer's patch B cell priming to orally administered antigen, *Eur. J. Immunol.* **13:**138–142.

Mason, D. W., Pugh, C. W., and Webb, M., 1981, The rat mixed lymphocyte reaction: Roles of a dendritic cell in intestinal lymph and T cell subsets defined by monoclonal antibodies, *Immunology* **44:**75–87.

Matzinger, P., 1994, Tolerance, danger, and the extended family, *Annu. Rev. Immunol.* **12:**991–1045.

Mayordomo, J. I., Storkus, W. J., Kast, W. M., Zorina, T., DeLeo, A. B., and Lotze, M. T., 1995, Bone marrow-derived dendritic cells serve as potent adjuvants for peptide-based antitumor vaccines, *J. Cell. Biochem. [Supp.]* **21A:**21.

Mayrhofer, G., Holt, P. G., and Papadimitriou, J. M., 1986, Functional characteristics of the veiled cells in afferent lymph from the rat intestine, *Immunology* **58:**379–387.

McWilliam, A. S., Nelson, D., Thomas, J. A., and Holt, P. G., 1994, Rapid dendritic cell recruitment is a hallmark of the acute inflammatory response at mucosal surfaces, *J. Exp. Med.* **179:**1331–1336.

Metlay, J. P., Puré, E., and Steinman, R. M., 1989, Control of the immune response at the level of antigen-presenting cells: A comparison of the function of dendritic cells and B lymphocytes, *Adv. Immunol.* **47:**45–116.

Metlay, J. P., Witmer-Pack, M. D., Agger, R., Crowley, M. T., Lawless, D., and Steinman, R. M., 1990, The distinct leukocyte integrins of mouse spleen dendritic cells as identified with new hamster monoclonal antibodies, *J. Exp. Med.* **171:**1753–1772.

Moll, H., 1993, Epidermal Langerhans cells are critical for immunoregulation of cutaneous leishmaniasis, *Immunol. Today* **14:**383–387.

Moll, H., Fuchs, H., Blank, C., and Rollinghoff, M., 1993, Langerhans cells transport *Leishmania major* from the infected skin to the draining lymph node for presentation to antigen-specific T cells, *Eur. J. Immunol.* **23:**1595–1601.

Moore, K. W., Vieira, P., Fiorentino, D. F., Trounstine, M. L., Khan, T. A., and Mosmann, T. R., 1990, Homology of the cytokine synthesis inhibitory factor (IL-10) to the Epstein–Barr virus gene BCRFI, *Science* **248:** 1230–1234.

Moore, M. A. S., 1991, Clinical implications of positive and negative hematopoietic stem cell regulators, *Blood* **78:**1–19.

Mosier, D. E., 1967, A requirement for two cell types for antibody formation *in vitro*, *Science* **158:**1573–1575.

Mowat, A. M., 1987, The regulation of immune responses to dietary protein antigens, *Immunol. Today* **8:**93–98.

Nestle, F. O., Zheng, X.-G., Thompson, C. B., Turka, L. A., and Nickoloff, B. J., 1993, Characterization of dermal dendritic cells obtained from normal human skin reveals phenotypic and functionally distinctive subsets, *J. Immunol.* **151:**6535–6545.

Nicod, L. P., Lipscomb, M. F., Weissler, J. C., Lyons, C. R., Alberton, J., and Toews, G. B., 1989, Mononuclear cells from human lung parenchyma support antigen-induced T lymphocyte proliferation, *J. Leuk. Biol.* **45:** 336–344.

Nonacs, R., Humborg, C., Tam, J. P., and Steinman, R. M., 1992, Mechanisms of mouse spleen dendritic cell function in the generation of influenza-specific, cytolytic T lymphocytes, *J. Exp. Med.* **176:**519–529.

Notarangelo, L. D., Duse, M., and Ugazio, A. G., 1992, Immunodeficiency with hyper-IgM (HIM), *Immunodef. Rev.* **3:**101–122.

Nussenzweig, M. C., Jiang, W., Swiggard, W. J., Mirza, A., Peng, M., and Steinman, R. M., 1995, Molecular characterization of a 205 Kd protein that is abundant on dendritic cells and identified with the NLDC-145 monoclonal antibody, *J. Cell Biochem. [Suppl.]* **21A:**20.

O'Doherty, U., Steinman, R. M., Peng, M., Cameron, P. U., Gezelter, S., Kopeloff, I., Swiggard, W. J., Pope, M., and Bhardwaj, N., 1993, Dendritic cells freshly isolated from human blood express CD4 and mature into typical immunostimulatory dendritic cells after culture in monocyte-conditioned medium, *J. Exp. Med.* **178:**1067–1078.

O'Doherty, U., Peng, M., Gezelter, S., Swiggard, W. J., Betjes, M., Bhardwaj, N., and Steinman, R. M., 1994, Human blood contains two subsets of dendritic cells, one immunologically mature and the other immature, *Immunology* **82:**487–493.

Paglia, P., Girolomoni, G., Robbiati, F., Granucci, F., and Ricciardi-Castagnoli, P., 1993, Immortalized dendritic cell line fully competent in antigen presentation initiates primary T cell responses *in vivo*, *J. Exp. Med.* **178:**1893–1901.

Pavli, P., Woodhams, C. E., Doe, W. F., and Hume, D. A., 1990, Isolation and characterization of antigen-presenting dendritic cells from the mouse intestinal lamina propria, *Immunology* **70:**40–47.

Péguet-Navarro, J., Moulon, C., Caux, C., Dalbiez-Gauthier, C., Banchereau, J., and Schmitt, D., 1994, Interleukin 10 inhibits the primary allogeneic T cell response to human epidermal Langerhans cells, *Eur. J. Immunol.* **24:**884–891.

Péguet-Navarro, J., Dalbiez-Gauthier, C., Rattis, F. M., Van Kooten, C., Banchereau, J., and Schmitt, D., 1995, Functional expression of CD40 antigen on human epidermal Langerhans cells, *J. Immunol.* **155:**4241–4247.

Pelletier, M., Perreaut, C., Landry, D., David, M., and Montplaisir, S., 1984, Ontogeny of human epidermal Langerhans cells, *Transplantation* **38:**544–546.

Porcelli, S., Morita, C. T., and Brenner, M. B., 1992, CDlb restricts the response of human CD4-8-T lymphocytes to a microbial antigen, *Nature* **360:**593–597.

Prickett, T. C. R., McKenzie, J. L., and Hart, D. N. J., 1988, Characterization of interstitial dendritic cells in human liver, *Transplantation* **46:**754–761.

Pugh, C. W., MacPherson, G. G., and Steer, H. W., 1983, Characterization of nonlymphoid cells derived from rat peripheral lymph, *J. Exp. Med.* **157**:1758–1779.

Puré, E., Inaba, K., Crowley, M. T., Tardelli, L., Witmer-Pack, M. D., Ruberti, G., Fathman, G., and Steinman, R. M., 1990, Antigen processing by epidermal Langerhans cells correlates with the level of biosynthesis of MHC class II molecules and expression of invariant chain, *J. Exp. Med.* **172**:1459–1469.

Qian, S., Demetris, A. J., Murase, N., Rao, A. S., Fung, J. J., and Starzl, T. E., 1994, Murine liver allograft transplantation: Tolerance and donor cell chimerism, *Hepatology* **19**:916–924.

Rahemtulla, A., Fung-Leung, W. P., Schilham, M. W., Kundig, T. M., Sambhara, S. R., Narendram, A., Arabian, A., Wakeham, A., Paige, C. J., and Zinkernagel, R. M., 1991, Normal development and function of $CD8^+$ cells but markedly decreased helper cell activity in mice lacking CD4, *Nature* **353**:180–184.

Ray, A., Schmitt, D., Dezutter, D. C., Fargier, M. C., and Thivolet, J., 1989, Reappearance of CDla antigenic sites after endocytosis on human Langerhans cells evidenced by immunogoldlabeling, *J. Invest. Dermatol.* **92**: 217–224.

Reid, C. D. L., Fryer, P. R., Clifford, C., Kirk, A., Tikerpae, J., and Knight, S. C., 1990, Identification of hematopoietic progenitors of macrophages and dendritic Langerhans cells (DL-CFU) in human bone marrow and peripheral blood, *Blood* **76**:1139–1149.

Reid, C. D. L., Stackpoole, A., Meager, A., and Tikerpae, J., 1992, Interactions of tumor necrosis factor with granulocyte–macrophage colony-stimulating factor and other cytokines in the regulation of dendritic cell growth *in vitro* from early bipotent $CD34^+$ progenitors in human bone marrow, *J. Immunol.* **149**:2681–2688.

Reis e Sousa, C., Stahl, P. D., and Austyn, J. M., 1993, Phagocytosis of antigens by Langerhans cells *in vitro*, *J. Exp. Med.* **178**:509–519.

Ren, Y., Silverstein, R. L., Allen, J., and Savill, J., 1995, CD36 gene transfer confers capacity for phagocytosis of cells undergoing apoptosis, *J. Exp. Med.* **181**:1857–1862.

Rhodes, J. M., and Agger, R., 1987, Comparison of membrane antigens of mouse dendritic cell types, *Immunol. Lett.* **16**:107–112.

Romani, N., Koide, S., Growley, M., Witmer-Pack, M., Livingstone, A. M., Fathman, G. G., Inaba, K., and Steinman, R. M., 1989a, Presentation of exogenous protein antigens by dendritic cells to T cell clones, *J. Exp. Med.* **169**:1169–1178.

Romani, N., Lenz, A., Glassl, H., Stossel, H., Stanzl, U., Majdic, O., Fritsch, P., and Schuler, G., 1989b, Cultured human Langerhans cells resemble lymphoid dendritic cells in phenotype and function., *J. Invest. Dermatol.* **93**:600–609.

Romani, N., Gruner, S., Brang, D., Kämpgen, E., Lenz, A., Trockenbacher, B., Konwalinka, G., Fritsch, P. O., Steinman, R. M., and Schuler, G., 1994, Proliferating dendritic cell progenitors in human blood, *J. Exp. Med.* **180**:83–93.

Ronchese, F., and Hausmann, B., 1993, B lymphocytes *in vivo* fail to prime naive T cells but can stimulate antigen-experienced T lymphocytes, *J. Exp. Med.* **177**:679–690.

Ross, E. L., Barker, J. N. W. N., Allen, M. H., Chu, A. C., Groves, R. W., and MacDonald, D. M., 1994, Langerhans' cell expression of the selectin ligand, sialyl Lewis x, *Immunology* **81**:303–308.

Rossi, G., Heveker, N., Thiele, B., Gelderblom, H., and Steinbach, F., 1992, Development of a Langerhans cell phenotype from peripheral blood monocytes, *Immunol. Lett.* **31**:189–197.

Rowden, G., 1981, The Langerhans cells, *Crit. Rev. Immunol.* **3**:94–180.

Rubin, D., Weiner, H. L., Fields, B. N., and Greene, M. I., 1981, Immunologic tolerance after oral administration of reovirus: requirement for two viral gene products for tolerance induction, *J. Immunol.* **127**:1697–1701.

Rusten, L. S., Jacobsen, F. W., Lesslauer, W., Loetscher, H., Smeland, E. B., and Jacobsen, S. E., 1994, Bifunctional effects of tumor necrosis factor alpha (TNF alpha) on the growth of mature and primitive human hematopoietic progenitor cells: Involvement of p55 and p75 TNF receptors, *Blood* **83**:3152–3159.

Saeland, S., Duvert, V., Moreau, I., and Banchereau, J., 1993, Human B cell precursors proliferate and express CD23 after CD40 ligation, *J. Exp. Med.* **178**:113–120.

Sagebiel, R. W., and Reed, T. H., 1968, Serial reconstruction of the characteristic granule of the Langerhans cell, *J. Cell. Biol.* **36**:595–608.

Salem, M., Delwel, R., Touw, I., Mahmoud, L. A., Elbasousy, E. M., and Lowenberg, B., 1990, Modulation of colony stimulating factor-(CSF) dependent growth of acute myeloid leukemia by tumor necrosis factor, *Leukemia* **4**:37–43.

Sallusto, F., and Lanzavecchia, A., 1994, Efficient presentation of soluble antigen by cultured human dendritic cells is maintained by granulocyte/macrophage colony-stimulating factor plus interleukin 4 and downregulated by tumor necrosis factor alpha, *J. Exp. Med.* **179:**1109–1118.

Sallusto, F., Cella, M., Danieli, C., and Lanzavecchia, A., 1995, Dendritic cells use macropinocytosis and the mannose receptor to concentrate macromolecules in the major histocompatibility complex class II compartment: Downregulation by cytokines and bacterial products, *J. Exp. Med.* **182:**389–400.

Santiago-Schwarz, F., Belilos, E., Diamond, B., and Carsons, S. E., 1992, TNF in combination with GM-CSF enhances the differentiation of neonatal cord blood stem cells into dendritic cells and macrophages, *J. Leukocyte Biol.* **52:**274–281.

Santiago-Schwarz, F., Divaris, N., Kay, C., and Carsons, S. E., 1993, Mechanisms of tumor necrosis factor–granulocyte-macrophage colony-stimulating factor-induced dendritic cell development, *Blood* **82:**3019–3028.

Sato, N., Caux, C., Kitamura, T., Watanabe, Y., Arai, K., Banchereau, J., and Miyajima, A., 1993, Expression and factor-dependent modulation of the Interleukin-3 receptor subunits on human hematopoietic cells, *Blood* **82:** 752–761.

Scheicher, C., Mehlig, M., Zecher, R., and Reske, K., 1992, Dendritic cells from mouse bone marrow: *In vitro* differentiation using low doses of recombinant granulocyte–macrophage colony-stimulating factor, *J. Immunol. Methods* **154:**253–264.

Schmitt, D. A., Hanau, D., Bieber, T., Dezutter-Dambuyant, C., Schmitt, D., Fabre, M., Pauly, G., and Cazenave, J.-P., 1990, Human epidermal Langerhans cells express only the 40-kilodalton Fc gamma receptor (FcRII), *J.Immunol.* **144:**4284–4290.

Schon-Hegrad, M.A., Oliver, J., McMenamin, P.G., and Holt, P.G., 1991, Studies on the density, distribution, and surface phenotype of intraepithelial class II major histocompatability complex antigen (Ia)-bearing dendritic cells (DC) in the conducting airways, *J. Exp. Med.* **173:**1345–1356.

Schrader, C. E., Geroge, A., Kerlin, R. L., and Cebra, J. J., 1990, Dendritic cells support production of IgA and other non-IgM isotypes in clonal microculture, *Int. Immunol.* **2:**563–570.

Schuler, G., and Steinman, R. M., 1985, Murine epidermal Langerhans cells mature into potent immunostimulatory dendritic cells *in vitro*, *J. Exp. Med.* **161:**526–546.

Schuler, G., Auböck, J., and Linert, J., 1983, Enrichment of epidermal Langerhans cells by immunoadsorption to *Staphylococcus aureus* cells, *J. Immunol.* **130:**2008–2010.

Shimada, S., Caughman, S. W., Sharrow, S. O., Stephany, D., and Katz, S. I., 1987, Enhanced antigen-presenting capacity of cultured Langerhans cells is associated with markedly increased expression of Ia antigen, *J. Immunol.* **139:**2551–2555.

Sornasse, T., Flamand, V., de Becker, G., Bazin, H., Tielemans, F., Thielemans, K., Urbain, J., Oberdan, L., and Moser, M., 1992, Antigen-pulse dendritic cells can efficiently induce an antibody response *in vivo*, *J. Exp. Med.* **175:**15–21.

Spalding, D. M., and Griffin, J. A., 1986, Different pathways of differentiation of pre-B cell lines are induced by dendritic cells and T cells from different lymphoid tissues, *Cell* **44:**507–515.

Spalding, D. M., Koopman, W. J., Eldridge, J. H., McGhee, J. R., and Steinman, R. M., 1983, Accessory cells in murine Peyer's patch. I. Identification and enrichment of a functional dendritic cell, *J. Exp. Med.* **157:**1646–1659.

Spencer, S. C., and Fabre, J. W., 1990, Characterization of the tissue macrophage and the interstitial dendritic cell as distinct leukocytes normally residend in the connective tissue of rat heart, *J. Exp. Med.* **171:**1841–1851.

Spry, C. J. F., Pflug, A. J., Janossy, G., and Humphrey, J. H., 1980, Large mononuclear (veiled) cells with "Ia-like" membrane antigens in human afferent lymph, *Clin. Exp. Immunol.* **39:**750–756.

Starzl, T. E., Marchioro, T. L., Porter, K. A., Taylor, P. D., Faris, T. D., Herrmann, T. J., Hlad, C. J., and Waddell, W. R., 1965, Factors determining short-and long-term survival after orthotopic liver homotransplantation in the dog, *Surgery* **58:**131–138.

Starzl, T. E., Demetris, A. J., Trucco, M., Murase, N., Ricordi, C., Ildstad, S., Ramos, H., Todo, S., Tzakis, A., and Fung, J. J., 1993, Cell migration and chimerism after whole organ transplantation: The basis of graft acceptance, *Hepatology* **17:**1153–1156.

Steiniger, B., Klempnauer, J., and Wonigeit, K., 1984, Phenotype and histological distribution of interstitial dendritic cells in the rat pancreas, liver, heart and kidney, *Transplantation* **38:**169–174.

Steinman, R. M., 1991, The dendritic cell system and its role in immunogenicity, *Annu. Rev. Immunol.* **9:**271–296.

Steinman, R. M., and Witmer, M. D., 1978, Lymphoid dendritic cells are potent stimulators of the primary mixed leukocyte reaction in mice, *Proc. Natl. Acad. Sci. U.S.A.* **75:**5132–5136.

Steinman, R. M., Lustig, D. S., and Cohn, Z. A., 1974, Identification of a novel cell type in peripheral lymphoid organs of mice. III. Functional properties *in vivo*, *J. Exp. Med.* **139:**1431–1445.

Stern, D. M., and Nawroth, P. P., 1986, Modulation of endothelial hemostatic properties by tumor necrosis factor, *J. Exp. Med.* **163:**740–745.

Stössel, H., Koch, F., Kmpgen, E., Stöger, P., Lenz, A., Heufler, C., Romani, N., and Schuler, G., 1990, Disappearance of certain acidic organelles (endosomes and Langerhans cell granules) accompanies loss of antigen processing capacity upon culture of epidermal Langerhans cells, *J. Exp. Med.* **172:**1471–1482.

Streilein, J. W., and Grammer, S. F., 1989, *In vitro* evidence that Langerhans cells can adopt two functionally distinct forms capable of antigen presentation to T lymphocytes, *J. Immunol.* **143:**3925–3933.

Streilein, J. W., Grammer, S. F., Yoshikawa, T., Demidem, A., and Vermeer, M., 1990, Functional dichotomy between Langerhans cells that present antigen to naive and memory/effector T lymphocytes, *Immunol. Rev.* **117:**159–184.

Symington, F. W., Brady, W., and Linsley, P. S., 1993, Expression and function of B7 on human epidermal Langerhans cells, *J. Immunol.* **150:**1286–1295.

Szabolcs, P., Moore, M. A. S., and Young, J. W., 1995, Expansion of immunostimulatory dendritic cells among the myeloid progeny of human $CD34^+$ bone marrow precursors cultured with c-kit-ligand, GM-CSF, and TNFα, *J. Immunol.* **154:**5851–5861.

Tabata, N., Aiba, S., Nakagawa, S., Ohtani, H., and Tagami, H., 1993, Sialyl Lewis^x expression on human Langerhans cells, *J. Invest. Dermatol.* **101:**175–179.

Takahashi, H., Nakagawa, Y., Yokomuro, K., and Berzofsky, J. A., 1993, Induction of $CD8^+$ cytotoxic T lymphocytes by immunization with syngeneic irradiated HIV-1 envelope derived peptide-pulsed dendritic cells, *Int. Immunol.* **5:**849–857.

Takahashi, K., Naito, M., Shultz, L. D., Hayashi, S., and Nishikawa, S., 1993, Differentiation of dendritic cell populations in macrophage colony-stimulating factor-deficient mice homozygous for the osteopetrosis (*op*) mutation, *J. Leukocyte Biol.* **53:**19–28.

Takahashi, S., and Hashimoto, K., 1985, Derivation of Langerhans cell granules from cytomembrane, *J. Invest. Dermatol.* **84:**469–471.

Takigawa, M., Iwatsuki, K., Yamada, M., Okamoto, H., and Imamura, S., 1985, The Langerhans cell granule is an adsorptive endocytic organelle, *J. Invest. Dermatol.* **85:**12–18.

Tang, A., Amagai, M., Granger, L. G., Stanley, J. R., and Udey, M. C., 1993, Adhesion of epidermal Langerhans cells to keratinocytes mediated by E-cadherin, *Nature* **361:**82–85.

Tazi, A., Bouchonnet, F., Grandsaigne, M., Boumsell, L., Hance, A. J., and Soler, P., 1993, Evidence that granulocyte macrophage-colony-stimulating factor regulates the distribution and differentiated state of dendritic cells/Langerhans cells in human lung cancers, *J. Clin. Invest.* **91:**566–576.

Terhorst, C., Van Agthoven, A., Le Clair, K., Stanley, J. R., and Udey, M. C., 1981, Biochemical studies of the human thymocyte cell-surface antigens T6, T9 and T10, *Cell* **23:**771–780.

Teunissen, M. B. M., Wormeester, J., Krieg, S. R., Peters, P. J., Vogels, I. M. C., Kapsenberg, M. L., and Bos, J. D., 1990, Human epidermal Langerhans cells undergo profound morphological and phenotypical changes during *in vitro* culture, *J. Invest. Dermatol.* **94:**166–173.

Thomas, R., and Lipsky, P. E., 1994, Human peripheral blood dendritic cell subsets. Isolation and characterization of precursor and mature antigen-presenting cells, *J. Immunol.* **153:**4016–4028.

Thomas, R., Davis, L. S., and Lipsky, P. E., 1993, Isolation and characterization of human peripheral blood dendritic cells, *J. Immunol.* **150:**821–834.

Turka, L. A., Linsley, P. S., Lin, H., Brady, W., Leiden, J. M., Wei, R.-Q., Gibson, M. L., Zheng, X.-G., Myrdal, S., Gordon, D., Bailey, T., Bolling, S. F., and Thompson, C. B., 1992, T-cell activation by the CD28 ligand B7 is required for cardiac allograft rejection *in vivo*, *Proc. Natl. Acad. Sci. U.S.A.* **89:**11102–11107.

van de Rijn, M., Lerch, P. G., Bronstein, B. R., Knowles, R. W., Bhan, A. K., and Terhost, C., 1984, Human cutaneous dendritic cells express two glycoproteins T6 and M241 which are biochemically identical to those found on cortical thymocytes, *Hum. Immunol.* **9:**201–210.

Van Nieuwkerk, E. B. J., Van der Baan, S., Richters, C. D., and Kamperdijk, E. W. A., 1992, Isolation and characterization of dendritic cells from adenoids of children with otitis media with effusion, *Clin. Exp. Immunol.* **88:**345–349.

Van Voorhis, W. C., Valinsky, J., Hoffman, E., Luban, J., Hair, L. S., and Steinman, R. M., 1983, Relative efficacy of human monocytes and dendritic cells as accessory cells for T cell replication, *J. Exp. Med.* **158:**171–191.

Veldman, J. E., and Kaiserling, E., 1980, Interdigitating cells, in *The Reticulo-endothelial System, Morphology* (I. Carr and W. T. Daems, eds.), pp. 381–416, Plenum Press, New York.

Volc-Platzer, B., Stingl, G., Wolff, K., Hinterberg, W., and Schnedl, W., 1984, Cytogenetic identification of allogeneic epidermal Langerhans cells in a bone-marrow-graft recipient, *N. Engl. J. Med.* **310:**1123–1124.

Vremec, D., Zorbas, M., Scollay, R., Saunders, D. J., Ardavin, C C. F., Wu, L., and Shortman, K., 1992, The surface phenotype of dendritic cells purified from mouse thymus and spleen: Investigation of the CD8 expression by a subpopulation of dendritic cells, *J. Exp. Med* **176:**47–58.

Weih, F., Carrasco, D., Durham, S. K., Barton, D. S., Rizzo, C. A., Ryseck, R.-P., Lira, S. A., and Bravo, R., 1995, Multiorgan inflammation and hematopoietic abnormalities in mice with a targeted disruption of RelB, a member of the NF-κB/Rel family, *Cell* **80:**331–340.

Weissman, D., Li, Y., Ananworanich, J., Zhou, L.-J., Adelsberger, J., Tedder, T. F., Baseler, M., and Fauci, A. S., 1995, Three populations of cells with dendritic morphology exist in peripheral blood, only one of which is infectable with human immunodeficiency virus type 1, *Proc. Natl. Acad. Sci. U.S.A.* **92:**826–830.

Wettendorff, M., Massacrier, C., Vanbervliet, B., Urbain, J., Banchereau, J., and Caux, C., 1995, Activation of primary allogeneic $CD8^+$ T cells by dendritic cells generated in vitro from $CD34^+$ cord blood progenitor cells, in *Dendritic Cells in Fundamental and Clinical Immunology* (J. Banchereau and D. Schmitt, eds.), pp. 371–374, Plenum Press, London.

Wiktor-Jedrzejczak, W., Ratajczak, M. Z., Ptasznik, A., Sell, K. W., Ahmed-Ansari, A., and Ostertag, W., 1992, CSF-1 deficiency in the op/op mouse has differential effects on macrophage populations and differentiation stages, *Exp. Hematol.* **20:**1004–1010.

Wilders, M. M., Sminia, T., Plesch, B. E., Drexhage, H. A., Weltevreden, E. F., and Meuwissen, S. G., 1983, Large mononuclear Ia-positive veiled cells in Peyer's patches. II. Localization in rat Peyer's patches, *Immunology* **48:**461–467.

Williams, L. A., Egner, W., and Hart, D. N. J., 1994, Isolation and function of human dendritic cells, *Int. Rev. Cytol.* **153:**41–103.

Witmer, M. D., and Steinman, R. M., 1984, The anatomy of peripheral lymphoid organs with emphasis on accessory cells: Light microscopic, immunocytochemical studies of mouse spleen, lymph node and Peyer's patch, *Am. J. Anat.* **170:**465–481.

Witmer-Pack, M. D., Olivier, W., Valinsky, J., Schuler, G., and Steinman, R. M., 1987, Granulocyte/macrophage colony/stimulating factor is essential for the viability and function of cultured murine epidermal Langerhans cells, *J. Exp. Med.* **166:**1484–1498.

Witmer-Pack, M. D., Valinsky, J. O., W. and Steinman, R. M., 1988, Quantitation of surface antigens on cultured murine epidermal Langerhans cells: Rapid and selective increase in the level of surface MHC products, *J. Invest. Dermatol.* **90:**387–394.

Witmer-Pack, M. D., Hughes, D. A., Schuler, G., Lawson, L., McWilliam, A., Inaba, K., Steinma, R. M., and Gordon, S., 1993, Identification of macrophages and dendritic cells in the osteopetrotic (*op/op*) mouse, *J. Cell Sci.* **104:**1021–1029.

Wolff, K., 1967, The fine structure of the Langerhans cell granule, *J. Cell. Biol.* **35:**1484–1498.

Wolff, K., and Schreiner, E., 1970, Uptake, intracellular transport and degradation of exogeneous protein by Langerhans cells, *J. Invest. Dermatol.* **48:**50–54.

Xu, S., Ariizumi, K., Edelbaum, D., Bergstresser, P. R., and Takashima, A., 1995, Cytokine-dependent regulation of growth and maturation in murine epidermal dendritic cell lines, *Eur. J. Immunol.* **25:**1018–1024.

Young, J. W., and Steinman, R. M., 1988, Accessory cell requirements for the mixed leukocyte reaction and polyclonal mitogens, as studied with a new technique for enriching blood dendritic cells, *Cell. Immunol.* **111:** 167–182.

Young, J. W., and Steinman, R. M., 1990, Dendritic cells stimulate primary human cytolytic lymphocyte responses in the absence of $CD4^+$ helper T cells, *J. Exp. Med.* **171:**1315–1332.

Zhou, L.-J., and Tedder, T. F., 1995, Human blood dendritic cells selectively express CD83, a member of the immunoglobulin superfamily, *J. Immunol.* **154:**3821–3835.

Zhou, L.-J., Schwarting, R., Smith, H. M., and Tedder, T. F., 1992, A novel cell-surface molecule expressed by human interdigitating reticulum cells, Langerhans cells, and activated lymphocytes is a new member of the Ig superfamily, *J. Immunol.* **149:**735–742.

Chapter 12

Responses of Leukemia Cells to Hematopoietic Growth Factors

Involvement of Autocrine Growth Mechanisms, Cytogenetic Abnormalities, and Defective Maturation Signaling

Ivo P. Touw and Fan Dong

1. INTRODUCTION

All blood cells produced throughout life descend from a small number of pluripotent hematopoietic stem cells, which, in adult man, reside predominantly in the bone marrow. The pluripotent stem cells generate progenitor cells committed to proliferate and mature toward the different functional end cells. The mature blood cells have a limited life span and need to be produced continuously. The hematopoietic system adapts in a highly dynamic fashion to changes in the requirements for different blood cells, for instance, during stages of infection or following sudden loss of blood. The rapid and variable needs for the various blood cell types demand a tight and complex control mechanism that finely regulates renewal, commitment, proliferation, maturation, and survival of hematopoietic cells. The hematopoietic growth factors (HGFs) play a central role in these processes (Clark and Kamen, 1987; Metcalf, 1989).

Leukemias are neoplastic conditions characterized by the accumulation of malignant myeloid or lymphoid cells, arrested at distinct stages of development, in the bone marrow

Ivo P. Touw and Fan Dong Department of Hematology, Dr. Daniel den Hoed Cancer Center and Erasmus University, Rotterdam, The Netherlands.

Blood Cell Biochemistry, Volume 7: Hematopoietic Cell Growth Factors and Their Receptors, edited by A. D. Whetton and J. Gordon. Plenum Press, New York, 1996.

and peripheral blood. Leukemic cells escape from the balanced control of proliferation, maturation, and cell survival characteristic of the normal hematopoietic system. Leukemias present either as acute or chronic diseases. The most common types of acute leukemia are B-cell precursor acute lymphoblastic leukemia (BCP-ALL) and acute myeloblastic leukemia (AML). T-cell ALL and undifferentiated or "null" ALL represent rarer forms of acute leukemia. Chronic myeloid leukemia, characterized by the chromosomal translocation (9;22)(q34;q11), known as the Philadelphia translocation, originates in primitive multilineage hematopoietic precursors (Failkow *et al.*, 1977). The chronic lymphoid leukemias arise from the B lymphocyte stages (B-cell chronic lymphocytic leukemia and hairy-cell leukemia) or the immunoblast/plasma cell stages (multiple myeloma) (Table I).

In this chapter, we first briefly summarize some general aspects of HGF and cytokine responses of human leukemia cells. Subsequently, we discuss in more depth studies concerning the proliferation and maturation properties of progenitor cells in two pathological conditions characterized by defective myeloid maturation: AML and severe congenital neutropenia (SCN), a disease that may progress toward AML. These investigations specifically dealt with autocrine mechanisms of growth, HGF responses in relation to cytogenetic abnormalities, and the involvement of HGF receptor defects in leukemogenesis. They illustrate how the analysis of HGF responses and HGF receptor function provide new insights into the heterogeneous pathobiology and clinical behavior of leukemia and related diseases.

2. LEUKEMOGENESIS

It is generally accepted that leukemogenesis is a multistep process (Hunter, 1991; Sawyers *et al.*, 1991). The leukemic mass arises from the clonal evolution and expansion of a single-cell clone, which may initially have limited genetic damage, resulting in only marginal changes in growth properties (Sawyers *et al.*, 1991). With the acquisition of additional genetic defects, the normal growth and differentiation program becomes increasingly perturbed, which may ultimately result in full leukemic transformation. Genes affected in leukemia encode proteins that are normally involved in the regulation of cell proliferation, survival, and differentiation. With few exceptions, these proteins fall into four major categories: growth factors, growth factor receptors, cytoplasmic signal transducers, and transcription factors (Cantley *et al.*, 1991). Proteins from these different categories usually act in collaboration to accomplish full leukemic transformation (Hunter, 1991).

3. GROWTH FACTORS FOR HUMAN LEUKEMIA CELLS

The availability of recombinant growth factors and cytokines has allowed detailed characterization of the proliferative responses of human leukemia to these stimuli. Primary human leukemia cells from most patients respond to growth factors *in vitro*. Thus, leukemic cell expansion is usually not associated with growth factor independence but with more subtle deviations of growth control. The proliferative responses of leukemia cells to growth factors grossly reflect those of normal cells at comparable stages of differentiation. A

Table I
Principal Growth Factors for Human Leukemia Cells

Leukemia[a]	Normal counterpart	Growth factors[b]	References
Myeloid	Myeloid progenitors	SCF, IL-3, GM-CSF, G-CSF,	Budel *et al.* (1993)
AML		M-CSF, TPO	Delwel *et al.* (1988)
CML	Multilineage progenitors	SCF, IL-3, GM-CSF, G-CSF, M-CSF, EPO	Matsumara *et al.* (1995)
Lymphoid			
BCP-ALL	B-cell precursors	IL-3, IL-7	Touw *et al.* (1990), Wörmann *et al.* (1989)
T-ALL	(Pro)thymocytes, T lymphocytes	IL-2, IL-4, IL-7	Touw *et al.* (1990)
B-CLL	CD5$^+$ B lymphocytes	IL-2, IL-4[c], TNF	Moberts *et al.* (1989) Fluckiger *et al.* (1992)
HCL	B lymphocytes	IL-4[c]	Kluin-Nelemans *et al.* (1994)
MM	Plasma cells	IL-6	Kawano *et al.* (1988)

[a]AML, acute myeloblastic leukemia; CML, chronic myeloid leukemia; BCP-ALL, B-cell precursor acute lymphoblastic leukemia; T-ALL, T-cell acute lymphoblastic leukemia; B-CLL, B-cell chronic lymphocytic leukemia; HCL, hairy cell leukemia; MM, multiple myeloma.
[b]SCF, stem cell factors; GM-CSF, granulocyte–macrophage colony-stimulating factor; G-CSF, granulocyte colony-stimulating factor; M-CSF, macrophage colony-stimulating factor; EPO, erythropoietin; IL, interleukin; TNF, tumor necrosis factor; TPO, thrombopoietin.
[c]In combination with CD40 ligand or CD40 activating antibodies.

summary of the most important growth factors for different types of leukemia is given in Table I.

Some factors exert limited or no direct proliferation-inducing effects themselves but can strongly enhance or counteract the responses of leukemic cells to other HGFs. Examples of such growth modulatory factors are tumor necrosis factor (TNF), transforming growth factor β (TGF-β) and Kit-ligand (KL, also known as stem cell factor, Steel factor, or mast cell growth factor). Both TNF and TGF-β exert dual effects on HGF-induced proliferation of AML cells: TNF strongly inhibits responses of AML cells to granulocyte colony-stimulating factor (G-CSF) (Salem *et al.*, 1990). In contrast, synergistic effects of TNF on proliferation of AML cells are seen in combination with granulocyte–macrophage CSF (GM-CSF) or interleukin 3 (IL-3) (Salem *et al.*, 1990). The inhibitory effects of TNF on G-CSF responses of AML cells are at least in part explained by the ability of TNF to rapidly decrease G-CSF binding to the cells through induction of proteolytic cleavage of the extracellular domain of G-CSF receptor (G-CSF-R) (Elbaz *et al.*, 1991a). Conversely, the TNF-induced increased in expression of GM-CSF and IL-3 receptors on AML cells may be held responsible for the synergistic effects of TNF in combination with IL-3 or GM-CSF (Elbaz *et al.*, 1991b). The dual action of TGF-β on AML cells has also been related to differential effects on expression of G-CSF-R, GM-CSF-R, and IL-3-R (Keller *et al.*, 1992). Stem cell factor (SCF) or Kit ligand (KL), can strongly synergize with IL-3, GM-CSF, and G-CSF in inducing proliferation of AML cells, but without up-regulating the receptors for these growth factors (Budel *et al.*, 1993). The mechanisms of synergistic action of SCF are still unknown.

Knowledge of the responses of leukemia cells to HGFs contributes to the design of new clinical protocols involving the application of growth factors for the treatment of leukemia patients (Davis and Morstyn, 1992). For instance, clinical trials have been initiated in which growth factors (GM-CSF or IL-3) are being administered to AML patients prior to or during treatment with cell-cycle-specific cytoreductive agents, with the objective of increasing the antileukemic action of the drugs (Estey *et al.*, 1992). (For a fuller account of these procedures, see Chapter 15).

4. STEM CELLS AND PROGENITOR CELLS IN AML

Advances in colony culture methodology have permitted the characterization of HGF responses of human progenitor cells at different stages of development. Application of these techniques for culture of leukemic cell samples has allowed detailed studies on the growth properties and the maturation features of AML colony-forming units (AML-CFU).

Before the availability of purified recombinant HGFs, it had already become clear that AML-CFU from most patients required growth factors for maximal proliferation, at that time supplied to the cultures in crude form such as culture media conditioned by lectin-activated peripheral blood leukocytes or human placenta cells or leukocyte feeder cell layers (Griffin and Löwenberg, 1986). In the absence of these stimuli, AML-CFU either completely failed to proliferate or could form only small clusters of 4 to 20 cells. Only infrequently was spontaneous AML colony formation seen. In these early studies, it was further established that only a small subset (0.1–1%) of the AML blasts was capable of forming colonies, leading to the hypothesis that, like normal colony-forming cells, AML-CFU represent the clonogenic precursors of an end-cell population with limited or no proliferative potential. In agreement with this notion, cell-sorting experiments established that the nonproliferative AML blasts expressed a somewhat more mature immunophenotype than the AML-CFU (Griffin and Löwenberg, 1986).

Studies in SCID mice lend additional support to the concept of hierarchical organization of the blast-cell population in human AML. By transplanting cell-sorted subpopulations of human AML cells in the SCID model, Dick and colleagues demonstrated that the cells responsible for the development of leukemia were among the most primitive cells (Lapidot *et al.*, 1994). Notably, these studies further showed that transplantation of a slightly more mature fraction, containing most of the AML-CFU, did not give rise to leukemia.

The cytokines GM-CSF, G-CSF, and IL-3 are most commonly active in inducing proliferative responses in AML cells in colony culture or DNA synthesis assays (Table I). Responses to these HGFs are seen in approximately 80% of AML cases. As discussed above, SCF strongly enhances the proliferation-inducing effects of IL-3, GM-CSF, or G-CSF on AML cell proliferation *in vitro*. Significantly, the combination of SCF and PIXY 321, a synthetic GM-CSF/IL-3 fusion protein, also supports the long-term growth of human AML cells in the SCID model (Lapidot *et al.*, 1994). A number of other HGFs and cytokines have been reported to stimulate AML cell growth less frequently, such as IL-1, IL-5, IL-6, IL-11, erythropoietin (EPO), macrophages-CSF (M-CSF), and tumor necrosis factor (TNF).

The AML blasts from different patients show considerable heterogeneity with respect to cytologic features. Based on morphological and cytochemical characteristics of maturation, AML cases are classified according to the criteria of the French–American–British (FAB) Cooperative Group (1985). No strict correlation has become apparent between FAB classification of AML blasts and responses to HGFs and cytokines. For instance, AML cells with a monoblastic appearance (FAB-M5) do not preferentially respond to M-CSF, and responses to G-CSF are not confined to the AML cases with relatively more mature myeloid features (FAB-M2, M3, M4) (Miyauchi *et al.*, 1987; Delwel *et al.*, 1988). Even the stimulatory effects of thrombopoietin (TPO), the major growth factor for platelet production, are not restricted to AML cases with a megakaryoblastic cytology (FAB-M7) but are seen in all cytologic subgroups of AML (Matsumara *et al.*, 1995). Thus, maturation features of the majority of AML cells are not predictive of the HGF requirements of the AML progenitor and stem cells.

5. THE HGF RESPONSES OF AML CELLS WITH SPECIFIC CYTOGENETIC ABNORMALITIES

Comparison of the HGF responses of different cytogenetic subtypes of leukemia may help to elucidate how genetic defects affect the growth characteristics of leukemic cells. Until now, a correlation between HGF responses and specific chromosomal abnormalities has become apparent in two cytogenetic subgroups of AML: AML expressing the chromosomal translocation t(8;21) and AML expressing t(15;17). The AML t(8;21) cells proliferate not only in response to IL-3, GM-CSF, or G-CSF but also in the presence of the eosinophilic growth and differentiation factor IL-5 (see Chapter 13). Furthermore, IL-5 induces eosinophilic differentiation of AML t(8;21) cells (Ema *et al.*, 1990; Touw *et al.*, 1991). Interleukin 5 responses are rarely seen in other cytogenetic subtypes of AML. Although the molecular basis for the link between t(8;21) and eosinophilic maturation features of AML cells is not yet known, identification of the gene involved in the translocation as located on chromosome 21q22 has given an important clue for further study (Miyoshi *et al.*, 1991). This gene, termed the *AML1* gene, codes for a protein identical or highly homologous to the DNA binding α subunit of the transcription factor complex known as core-binding factor (CBF), which binds to core sites in murine leukemia virus and T-cell receptor enhancers (Wang *et al.*, 1993). The non-DNA-binding β subunit of the CBF complex stabilizes DNA binding of CBFα subunits, presumably by inducing conformational changes in the CBFα protein. Significantly, the *CBF*β gene, located on chromosome 16q22, is involved in another cytogenetic abnormality associated with AML, the pericentric inversion of chromosome 16 [inv(16)(p13q22)] (Liu *et al.*, 1993): AML patients with inv(16) present as FAB-M4Eo, characterized by abnormal eosinophilia in bone marrow and peripheral blood. As a result of this inversion, a fusion protein is formed from the first 165 amino acids of CBFβ with the tail part of a smooth muscle myosin heavy chain (SMMHC) protein, encoded by the *MYH11* gene located on 16p. It is assumed that the α-helical structure of the SMMHC causes dimerization of CBFβ/SMMHC fusion proteins, thereby preventing the formation of CBFα/CBFβ heteromers (Liu *et al.*, 1993). Although the exact role of the CBF transcription factor complex in normal hematopoiesis remains to be clarified, the fact that its dysregulation is specifically associated with acute leukemias expressing eosinophilic features suggests that CBF is involved in the transcriptional control of genes important for eosinophilic cell development.

The AML with the chromosomal translocation (15;17) is characterized by the accumulation of leukemic promyelocytes and is therefore also known as acute promyelocytic leukemia (APL). The (15;17) translocation results in the formation of a fusion protein of the PML protein (gene located on chromosome 15) with the retinoic acid receptor α (RARα) (gene located on chromosome 17). The PML–RARα fusion protein functions as a RARα with altered transactivating properties (de Thé *et al.*, 1991; Kazizuka *et al.*, 1991). The APL cells show a predominant response to G-CSF in proliferation assays (Pebusque *et al.*, 1988). A unique feature of AML t(15;17) cells is that their capacity to undergo terminal myeloid differentiation is restored on exposure to all-*trans* retinoic acid, a retinoid able to activate the PML–RAR protein. Like normal mature granulocytes, the differentiated leukemic cells with die within a period of days. This forms the basis for the successful application of retinoids as a cytoreductive treatment for APL patients (Daenen *et al.*, 1986; Meng-er *et al.*, 1988; Castaigne *et al.*, 1990; Warrell *et al.*, 1991).

6. AUTOCRINE GROWTH STIMULATION IN AML

Despite the fact that AML progenitor cells of most patients respond to HGFs, the cells usually also proliferate to some extent without the addition of HGFs to the cultures (Delwel *et al.*, 1988). Most frequently, spontaneous growth results from autocrine mechanisms (Young and Griffin, 1986; Murohashi *et al.*, 1989; Reilly *et al.*, 1989; Bradbury *et al.*, 1990). On incubation in culture, AML blasts can produce variable amounts of GM-CSF, G-CSF, M-CSF, IL-1β, IL-6, and TNF (Oster *et al.*, 1989; Delwel *et al.*, 1989). Spontaneous proliferation can often be inhibited by adding antibodies that neutralize the activities of these factors on the cultures. Both direct and indirect mechanisms of autocrine stimulation can be operational in AML cells. For instance, GM-CSF, G-CSF, and M-CSF exert direct mitogenic activities through activation of their respective membrane receptors. In contrast, IL-1β is not directly mitogenic but can induce or enhance the production and release of multiple growth factors (Oster *et al.*, 1989; Delwel *et al.*, 1990; Bradbury *et al.*, 1992). Importantly, IL-1β production in AML cells is activated by IL-1 itself, which perpetuates the autocrine loop of growth stimulation in AML cells (Oster *et al.*, 1989; Delwel *et al.*, 1990); TNF can exert similar effects, although less frequently.

It has been demonstrated in several experimental leukemias, for instance, avian leukemia, that autocrine growth mechanisms can contribute to leukemogenesis (Adkins *et al.*, 1984; Graf *et al.*, 1986; Metz *et al.*, 1991). The question whether this is also true for clinical AML remains to be answered. Thus far, specific genetic defects that can be linked directly to activation of growth factor production have not been identified in human AML. Therefore, it cannot be excluded that autocrine growth stimulation in AML simply reflects a characteristic of (a subpopulation of) normal blast cells rather than a leukemia-specific feature. The fact that HGF release by AML blasts in most cases is inducible by IL-1 or TNF, similar to monocytes and certain other nonhematopoietic cell types, would support this idea. As yet, the only clear example of clinical leukemia in which autocrine growth stimulation can definitely be associated with leukemogenesis is the group of rare cases of ALL expressing the chromosomal translocation t(5;14)(q31;q32). As a result of this translocation, the IL-3 gene located on 5q31 is constitutively activated via the IgH transcription regulatory elements located on chromosome 14 (Grimaldi and Meeker, 1989; Meeker *et al.*, 1990).

Irrespective of the underlying mechanisms, it was important to investigate whether the potential of spontaneous proliferation of AML cells could influence clinical behavior of the disease. In a study involving 114 AML patients, the probability of overall disease-free survival 3 years after diagnosis was significantly greater in cases with low autonomous proliferation than in cases with high spontaneous growth (Löwenberg *et al.*, 1993). Patients whose AML cells had intermediate levels of autonomous proliferation *in vitro* also showed intermediate rates of overall disease-free survival. Similar results were reported in a study of 50 AML patients by Hunter *et al.* (1993), in which comparisons between autonomous and nonautonomous groups were made 5 years after initial diagnosis. Thus, these independent studies established that the capacity of AML blasts for autonomous proliferation is associated with a poor prognosis.

7. DEFECTIVE MATURATION ABILITIES OF AML BLASTS IN RESPONSE TO G-CSF

Although AML progenitor cells in general require HGFs for proliferation *in vitro*, their ability to undergo maturation in response to these factors is usually limited. Despite the fact that G-CSF is one of the most potent maturation-inducing factors in normal myeloid development, G-CSF failed to induce terminal maturation of AML cells in the majority of cases. This lack of maturation could not be overcome by adding G-CSF in combination with differentiation-inducing agent all-*trans* retinoic acid to the cultures (Santini *et al.*, 1991). The genetic defects responsible for the inability of AML-CFU to mature are complex and may occur at distinct levels in the maturation signaling pathway. For instance, the *fos/jun* transcription factor complex, the Evi-1 zinc finger protein, and the cytoplasmic tyrosine kinase *fes/fps* have all been associated with the development of myelodysplastic syndromes (MDS) or AML. Their inappropriate expression and function may interfere with myeloid differentiation (Greer *et al.*, 1990; Morishita *et al.*, 1992a; Lord *et al.*, 1993). Transcription of the *Evi-1* gene is activated in cases of AML and MDS with chromosomal translocations or inversions involving chromosome band 3q26, where the human *Evi-1* locus resides (Morishita *et al.*, 1992b). Exactly how these defects interfere with HGF-driven maturation is still unclear. Given the heterogeneity of genetic lesions associated with disturbed maturation in AML, it may be anticipated that in some cases of AML defects may occur at the very beginning of the signaling pathway, i.e., in the receptors of certain HGFs. Investigation of receptor dysfunction in such cases potentially could directly reveal the link between maturation defects and disturbed HGF responses. As demonstrated in the next paragraphs, study of the function of the G-CSF receptor in cases of SCN and AML has provided such an opportunity.

8. IMPAIRED G-CSF RESPONSES IN SEVERE CONGENITAL NEUTROPENIA

Severe congenital neutropenia (SCN) is characterized by profound absolute neutropenia and a maturation arrest of myeloid progenitor cells at the promyelocyte–myelocyte stage, starting in early infancy. Marrow cells from these patients frequently display a reduced responsiveness to G-CSF *in vitro*, although neutrophilic colony formation can usually be induced with elevated G-CSF concentrations. Elevated levels of biologically active G-CSF (150–910 pg/ml) are frequently found in the serum of SCN patients. Yet, the administration of G-CSF in most cases improves granulopoiesis *in vivo* and is of therapeutic benefit (Gillio and Gabrilove, 1993). Generally, GM-CSF has no therapeutic effect in SCN. In agreement with *in vitro* observations, a considerable elevation of G-CSF levels is required to overcome the neutropenia in G-CSF-responsive SCN patients (Gillio and Gabrilove, 1993). The SCN was originally described by Kostmann (1956) as an autosomal recessive disorder in several Swedish families. Subsequently, a number of other reports appeared in which a family history of SCN was not apparent (Dale *et al.*, 1994).

Importantly, SCN patients have an increased probability of developing MDS or AML (Gillio and Gabrilove, 1993). Whether G-CSF administration contributes to the progression to leukemia in these cases is uncertain; AML has also been reported in patients with SCN who had not been treated with G-CSF (Gilman *et al.*, 1970; Rosen and Kang, 1979). These clinical observations suggest that functionally closely related defects may be involved in the pathogenesis of SCN, MDS, and AML.

9. CYTOPLASMIC SUBDOMAINS OF THE G-CSF RECEPTOR INVOLVED IN PROLIFERATION AND MATURATION SIGNALING

G-CSF induces both proliferation and differentiation of myeloid precursor cells via activation of a single receptor that belongs to the superfamily of hematopoietin receptors (Bazan, 1990). Activation of these receptors by their respective ligands is mediated through the formation of dimeric or oligomeric complexes of receptor structures. Some hematopoietic growth factor receptors form heteromeric receptor complexes, e.g., IL-2R, IL-3R, IL-5R, GM-CSF-R (Miyajima *et al.*, 1993; see also Chapters 1 and 4). In contrast, G-CSF-R forms homodimeric complexes (Fukunaga *et al.*, 1991). The extracellular region of the G-CSF-R is composed of an immunoglobulin (Ig)-like domain, the cytokine receptor homology (CRH) domain, and three fibronectin type III (FNIII) domains (Figure 1) (Fukunaga *et al.*, 1990a; Larsen *et al.*, 1990).

The intracellular part of human G-CSF-R comprises 183 amino acids (Fukunaga *et al.*, 1990b; Larsen *et al.*, 1990). Like other cytokine/hematopoietin receptors, the cytoplasmic domain of G-CSF-R lacks a kinase domain. Several stretches of amino acids in the G-CSF-R cytoplasmic domain show limited sequence homology to other members of the hematopoietic growth factor receptor superfamily. Two of these stretches, referred to as "box 1" and "box 2," are positioned in the membrane-proximal region of G-CSF-R (Figure 1). Box 1 and box 2 are also present in several other hematopoietin receptors (Fukunaga *et al.*, 1991; Murakami *et al.*, 1991). The membrane-distal cytoplasmic region of the G-CSF-R contains a third segment, called "box 3," which is shared only with the IL-6 signal transducer gp130 (Fukunaga *et al.*, 1991).

The membrane-proximal cytoplasmic region is primarily involved in proliferative signaling (Fukunaga *et al.*, 1991). The carboxy-terminal (membrane distal) region, on the other hand, is essential for transduction of maturation signals (Dong *et al.*, 1993; Fukunaga *et al.*, 1993). Deletion of the carboxy terminus of G-CSF-R not only destroys the maturation signaling capacity but also removes a negative regulatory element inhibiting proliferative signaling (Dong *et al.*, 1993). Experiments by Fukunaga *et al.* (1993) with a panel of G-CSF-R deletion mutants and chimeras of growth hormone receptor and G-CSF-R

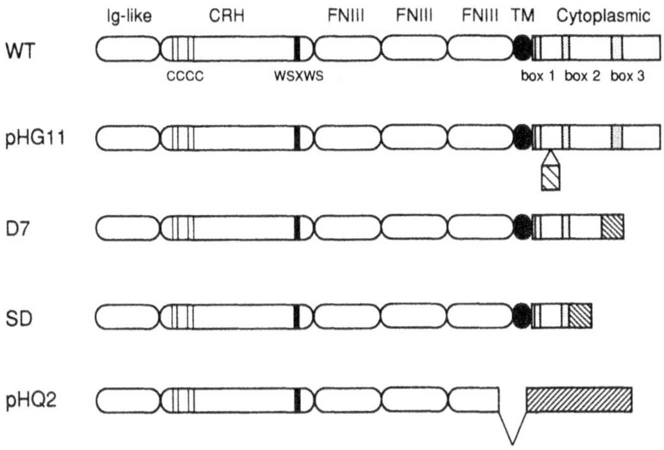

FIGURE 1. Schematic representation of structures of human G-CSF-R wild type (WT) and splice variants. The extracellular domain comprises an N-terminal immunoglobulin (Ig)-like subdomain, the cytokine receptor homology (CRH) region, which includes four conserved cysteines, a WSXWS motif, and three fibronectin III (FNIII) subdomains. The cytoplasmic regions conserved among several members of the HGF/cytokine receptor family are indicated as boxes 1, 2, and 3. Receptor isoform pHG11 contains an insertion of 27 amino acids in the cytoplasmic domain (Fukunaga *et al.*, 1990b). Isoform D7 lacks a 419-base-pair insert, which results in the coding of an altered C terminus of 34 amino acids (Larsen *et al.*, 1990). Isoform SD shares the altered C terminus with D7 but lacks the 43 amino acids directly C-terminal of box 2 (Dong *et al.*, 1994c). Isoform pHQ2 is a soluble variant lacking the entire transmembrane (TM) domain (Fukunaga *et al.*, 1990b).

indicated that mutations in the membrane-proximal cytoplasmic region could also result in the abrogation of maturation induction by G-CSF. These data demonstrate that maturation signaling, although primarily controlled by the C-terminal region of G-CSF-R, is not independent of the function of the membrane-proximal region involved in proliferation induction.

In addition to the wild-type structures, most hematopoietic growth factor receptor genes code for variant receptor proteins (including soluble receptor forms) by alternative RNA splicing. Five isoforms of human G-CSF-R have been identified (Figure 1). The physiological significance of the alternative receptor proteins is not yet clear. It is conceivable that some of them, in particular the soluble receptor forms, act as antagonists of growth factor responses. Notably, receptor forms with altered carboxy termini may be specifically hampered in transducing maturation signals (Dong *et al.*, 1993).

Some progress has been made in the identification of the intracellular substrates involved in G-CSF signaling. Recently, the involvement of the Jak (Janus kinase) family of tyrosine kinases in signaling from HGF receptors has been demonstrated (see Chapter 1). Jak proteins are rapidly phosphorylated on receptor activation and, in turn, activate molecules known as interferon-stimulated gene factors (ISGFs), alternatively referred to as STAT (signal transducer and activator of transcription) proteins. Activation of STAT molecules results in the formation of STAT complexes, which translocate to the nucleus and directly activate gene expression (Fu, 1992). This signaling pathway is activated by a number of other cytokines and hematopoietic growth factors (Larner *et al.*, 1993). Diversity within the ISGF/STAT molecules participating in complex formation induced by different growth factors has been demonstrated and may be held responsible for cell- or growth-factor-specific responses (Larner *et al.*, 1993; Ihle *et al.*, 1994). Both Jak1 and Jak2 are activated via interaction with the membrane-proximal cytoplasmic region of G-CSF-R, i.e., the region crucial for mitogenic signaling (Tian *et al.*, 1994; Dong *et al.*, 1994a). STAT complexes that are subsequently activated contain STAT3 and to a lesser extent STAT1 proteins (Tian *et al.*, 1994). Another signaling pathway directly connected to G-CSF-R involves activation of p21Ras (Bashey *et al.*, 1994). The functional contribution of Ras activation to G-CSF-induced signaling is still unknown. Ras is also activated by a number of other HGFs and cytokines, including GM-CSF, IL-2, IL-3, and EPO (Sato *et al.*, 1993; Ravichandran and Burakoff, 1994; Matsuguchi *et al.*, 1994; Damen *et al.*, 1993). Because Ras activation by some of these receptors appeared not to be crucial for mitogenic signaling, it was suggested that this pathway could be involved in other functions, such as induction of maturation and cell survival (Kinoshita *et al.*, 1995). However, because of the unique role of G-CSF in propagating neutrophil development, it is clear that specific signaling routes other than the Ras pathway must be involved in G-CSF-controlled neutrophilic maturation.

10. DEFECTIVE G-CSF-RECEPTOR FUNCTION IN CONGENITAL NEUTROPENIA AND AML: A NOVEL MECHANISM OF LEUKEMOGENESIS

Structure abnormalities of cytokine/hematopoietic growth factor receptors have been shown to contribute to different human diseases. A mutation in the EPO-R, causing a truncation of a C-terminal negative-regulatory region, has been genetically linked to familial erythrocytosis (De la Chapelle *et al.*, 1993). Truncation of the common γ chain of the receptors for IL-2, IL-4, IL-7, and IL-15 is associated with X-linked severe combined immunodeficiency (Noguchi *et al.*, 1993). Finally, mutations in the receptor for growth hormone, also a member of the cytokine receptor superfamily, have been described in patients with Laron dwarfism (Amselem *et al.*, 1989, 1991; Duquesnoy *et al.*, 1991; Berg *et al.*, 1992).

Evidence for a possible role of abnormal cytokine receptors in leukemogenesis has been obtained in mice. A mutation in the extracellular domain of the murine EPO-R, causing constitutive homodimerization and activation of the EPO-R mutant protein, could be directly associated with the development of erythroleukemia (Longmore and Lodish, 1991). Souyri *et al.* (1990) found that myeloproliferative leukemia virus (MPLV) encodes a fusion product of virus envelope protein and the C-terminal part of the hematopoietin receptor structure MPL, identified as the receptor for thrombopoietin (Lok *et al.*, 1994; Kaushansky *et al.*, 1994; Bartley *et al.*, 1994). More recently, it was shown that a mutant form of the common β chain (β_c) of the human IL-3/GM-CSF/IL-5 receptors, containing a duplication of 37 amino acids in the extracellular domain, confers growth factor independence to murine IL-3-dependent FDC-P1 cells (D'Andrea *et al.*, 1994). The FDC-P1 transfectants expressing the mutant β_c protein gave rise to leukemic outgrowth on injection into syngeneic mice, whereas FDC-P1 transfectants expressing the wild-type β_c protein, which remained factor dependent, were nontumorigenic.

Based on the observation that G-CSF responses are frequently disturbed in SCN, it had been anticipated for some time that G-CSF-R dysfunction might be involved in the pathogenesis of the disease. In agreement with this, a somatic mutation in the G-CSF-R gene was found in a case of SCN (Dong *et al.*, 1994a). This nonsense mutation resulted in the deletion of 98 amino acids from the carboxy terminus of G-CSF-R (Figure 2) (Dong *et al.*, 1993). As expected, on its ectopic expression in murine myeloid cell lines, the mutant receptor protein appeared unable to transduce maturation signals, whereas the ability to activate Jak proteins was unaffected. Because the normal allele was also expressed in the granulocytic cells of the patient, it appeared likely that the truncated receptor protein functioned in a dominant negative manner over the wild-type protein (Figure 3). These findings provided the first evidence *in vivo* that signals transduced by the G-CSF-R C terminus are indispensable for the normal development of myeloid progenitor cells. Significantly, the mutation was found in the granulocytes of the patient (who received G-CSF treatment) but not in monocytes, erythroid colony cells, T or B lymphocytes, or skin fibroblasts. These findings indicated that progenitor cells committed to the granulocyte lineage had been the target cell population for the somatic mutation.

Because committed progenitor cells are normally incapable of sustained self-renewal, the question arose how the affected progenitor cell population could persist in the patient. Because the truncated G-CSF-R protein, apart from lacking maturation signaling abilities, also misses the C-terminal element that negatively regulated proliferation, it was suggested that expression of this mutant receptor contributed to immortalization and the enhanced self-renewal capacity of the myeloid progenitor cells. On this basis, the neutropenia in this case was believed to reflect a preleukemic condition. In support of this notion, it was shown recently that the leukemic cells from two AML patients with a history of SCN also had mutations in G-CSF-R, again truncating the C-terminal maturation-inducing region (Dong *et al.*, 1995a). These were the first results demonstrating that defective G-CSF-R structures were associated with human AML. The findings also strongly support the notion that mutations in the G-CSF-R cytoplasmic domain determine the predisposition of SCN patients to develop AML.

Another G-CSF-R mutation associated with leukemia was found in the leukemic blasts of a patient with *de novo* AML (Dong *et al.*, 1995b). This mutation was discovered after

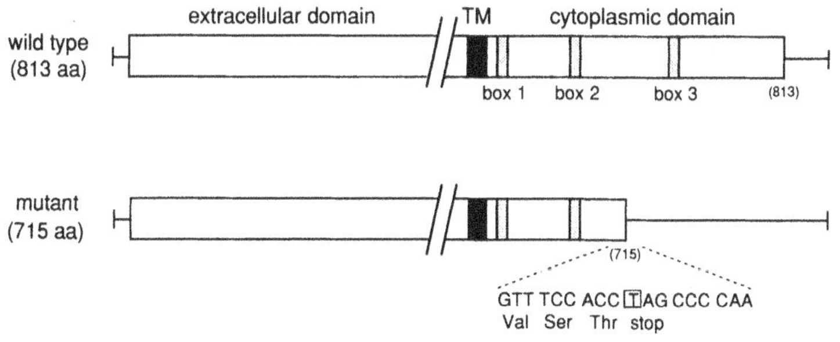

FIGURE 2 Truncation of the C-terminal maturation domain of G-CSF-R in a case of severe congenital neutropenia (Dong *et al.*, 1994a). As a result of a C-to-T transition, a TAG stop codon is introduced at amino acid position 716 (glutamine). Positions of the conserved regions and box 1 to 3 are indicated. TM, transmembrane domain.

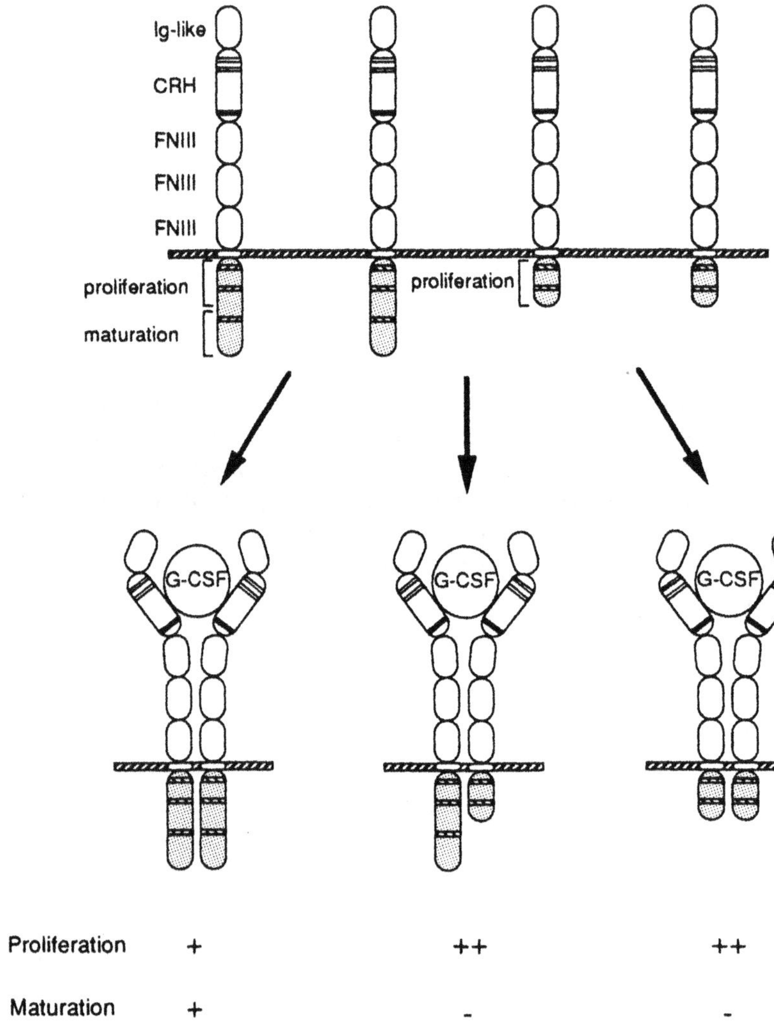

FIGURE 3. Proposed dominant negative action of truncated G-CSF-R on maturation signaling of WT G-CSF-R. For explanation of different functional domains, see Figure 1.

reverse transcriptase–polymerase chain reaction (RT-PCR) analysis with the objective of studying the expression of G-CSF-R splice variants in AML cells. The AML blasts of this patient showed high expression of the new splice variant SD (Figure 1) in which the carboxy terminus was altered by a change in the reading frame that appeared to be unable to transduce both proliferation and maturation signals on transfer to factor-dependent cell lines. In agreement with this, the AML blast cells of the patient also failed to respond to G-CSF in proliferation assays, although high responsiveness to either IL-3 or GM-CSF was noted. Analysis of genomic sequences of the G-CSF-R revealed a G-to-A transition next to the (cryptic) splice donor site involved in the alternative RNA splicing, i.e., at nucleotide position 2285 (Larsen *et al.*, 1990).

Taken together, these findings strongly suggest that genetic alterations in the G-CSF-R affecting the function of the C-terminal maturation domain, are involved in the pathogenesis of SCN and AML. They further implicate as yet unknown signaling molecules, specifically activated through the maturation domain of G-CSF-R, as additional important targets in leukemogenesis. Undoubtedly, identification of these signaling molecules and study of their function in normal and leukemic cells will shed more light on the contribution of genetic defects specifically interfering with HGF-induced maturation in the development of leukemia.

ACKNOWLEDGMENTS. This work is supported by the Dutch Cancer Society "Koningin Wilhelmina Fonds" and by the Dutch Society for Scientific Research (NWO).

11. REFERENCES

Adkins, B., Leutz, A., and Graf, T., 1984, Autocrine growth induced by *src*-related protooncogenes in transformed chicken myeloid cells, *Cell* **39**:439–445.

Amselem, S., Duquesnoy, P., Attree, O., Novelli, G., Bousnina, S., Postel-Vinay, M. C., and Goossens, M., 1989, Laron dwarfism and mutations of the growth hormone-receptor gene, *N. Engl. J. Med.* **321**:989–995.

Amselem, S., Sobrier, M. L., Duquesnoy, P., Rappaport, R., and Postel-Vinay, M. C., 1991, Recurrent nonsense mutations in the growth hormone receptor from patients with Laron dwarfism, *J. Clin. Invest.* **87**:1098–1102.

Bartley, T. D., Bogenberger, J., Hunt, P., Li, Y. S., Lu, H. S., Martin, F., Chang, M. S., Samal, B., Nichol, J. L., Swift, S., Johnson, M. J., Hsu, R. Y., Parker, V. P., Suggs, S., Skrine, J. D., Merewether, L. A., Clogston, C., Hsu, E., Hokom, M. M., Hornkohi, A., Choi, E., Pangelinan, M., Sun, Y., Mar, V., McNinch, J., Simonet, L., Jacobsen, F., Xie, C., Shutter, J., Chute, H., Basu, R., Selander, L., Trollinger, D., Sieu, L., Padilla, D., Trail, G., Elliott, G., Izumi, R., Covey, T., Crouse, J., Garcia, A., Xu, W., Castillo, J. D., Biron, J., Cole, S., Hu, M. C. T., Pacifici, R., Ponting, I., Saris, C., Wen, D., Yung, Y. P., Lin, H., and Bosselman, R. A., 1994, Identification and cloning of a megakaryocyte growth and development factor that is a ligand for the cytokine receptor Mpl, *Cell* **77**:1117–1124.

Bashey, A., Healy, L., and Marshall, C. J., 1994, Proliferative but not nonproliferative responses to granulocyte colony stimulating factor are associated with rapid activation of the $p21^{ras}$/MAP kinase signalling pathway, *Blood* **83**:949–957.

Bazan, J. F., 1990, Structural design and molecular evolution of a cytokine receptor superfamily, *Proc. Natl. Acad. Sci. U.S.A.* **87**:6934–6938.

Berg, M. A., Guevara-Aguirre, J., Rosenbloom, A. L., Rosenfeld, R. G., and Francke, U., 1992, Mutation creating a new splice site in the growth hormone receptor genes of 37 Ecuadorean patients with Laron syndrome, *Hum. Mutat.* **1**:24–32.

Bradbury, D., Rogers, S., Kozlowski, R., Bowen, G., Reilly, I. A. G., and Russell, N. H., 1990, Interleukin-1 is one factor which regulates the autocrine production of GM-CSF by the blasts of acute myeloblastic leukemia, *Br. J. Haematol.* **76**:488–493.

Bradbury, D., Rogers, S., Reilly, I. A., Kozlowski, R., Russel, N. H., 1992, Role of autocrine and paracrine production of granulocyte–macrophage colony stimulating factor and interleukin-1 beta in the autonomous growth of acute myeloblastic leukaemia cells: Studies using purified CD34-positive cells, *Leukemia* **6**:562–566.

Budel, L. M., Delwel, R., van Buitenen, C., Hoogerbrugge, H., and Löwenberg, B., 1993, Effects of Kit Ligand on acute myeloid leukemia cells *in vitro*: Effects of combinations with other cytokines, *Leukemia* **7**:426–434.

Cantley, L. C., Auger, K. R., Carpenter, C., Duckworth, B., Graziani, A., Kapeller, R., and Soltoff, S., 1991, Oncogenes and signal transduction, *Cell* **64**:281–302.

Castaigne, S., Chomienne, C., Daniel, M. T., Ballerini, P., Berger, R., Fenaux, P., and Degos, L., 1990, All *trans*-retinoic acid as a differentiation therapy for acute promyelocytic leukemia: I. Clinical results, *Blood* **76**:1704–1709.

Clark, S. C., and Kamen, R., 1987, The human colony stimulating factors, *Science* **236**:1229–1237.

Daenen, S., Vellenga, E., Van Dobbenburgh, O. A., and Halie, M. R., 1986, Retinoic acid as antileukemic therapy in a patient with acute promyelocytic leukemia and *Aspergillus* pneumonia, *Blood* **67**:559–561.

Dale, D. C., Davis, M., and Vincent, M. E., 1994, Use of filgrastim (r-metHuG-CSF) in severe chronic neutropenia, in: *Filgrastin in Clinical Practice* (G. Morstyn and T. M. Dexter, eds.), pp. 83–101, Marcel Dekker, New York.

Damen, J. E., Liu, L., Cutler, R. L., and Krystal, G., 1993, Erythropoietin stimulates the tyrosine phosphorylation of the Shc and its association with Grb2 and a 145-Kd tyrosine phosphorylated protein, *Blood* **82:**2296–2303.

D'Andrea, R., Rayner, J., Moretti, P., Lopez, A., Goodall, G. J., Gonda, T. J., and Vadas, M., 1994, A mutation of the common receptor subunit for interleukin-3 (IL-3), granulocyte macrophage colony-stimulating factor, and IL-5 that leads to ligand independence and tumorigenicity, *Blood* **83:**2802–2808.

Davis, I., and Morstyn, G., 1992, Clinical uses of growth factors, in *Balliere's Clinical Hematology*, Vol. 5, no. 3, *Growth Factors in Haemopoiesis* (B. I. Lord and T. M. Dexter, eds.), pp. 753–786, Balliere Tindal, London.

De la Chapelle, A., Träskelin, A. L., and Juvonen, E., 1993, Truncated erythropoietin receptor causes dominantly inherited benign human erythrocytosis, *Proc. Natl. Acad. Sci. U.S.A.* **90:**4495–4499.

Delwel, H. R., Salem, M., Dorssers, L., Wagemaker, G., Clark, S. C., and Löwenberg, B., 1988, Growth regulation of human acute myeloid leukemia: Effects of five recombinant hematopoietic factors in a serum-free culture system, *Blood* **72:**1944–1949.

Delwel, H. R., Van Buitenen, C., Salem, M., Bot, F., Gillis, S., Kaushansky, K., Altrock, B., and Löwenberg, B., 1989, Interleukin-1 stimulates proliferation of acute myeloid leukemia cells by induction of granulocyte-macrophage colony-stimulating factor release. *Blood* **74:**586–593.

Delwel, H. R., Schipper, P., Van Buitenen, C., Van Agthoven, T., Touw, I., and Löwenberg, B., 1990, Comparative analysis of IL-1 regulated and spontaneous growth of acute myeloid leukemia *in vitro*, *Bone Marrow Transplant.* **6**(Suppl. 1)**:**22–26.

De Thé, H. Lavau, C., Marchio, A., de Chomienne, C., Degos, L., and Dejean, A., 1991, The PML-RARα fusion mRNA generated by the t(15;17) translocation in human acute promyelocytic leukemia encodes a functionally altered RAR, *Cell* **66:**675–684.

Dong, F., Van Buitenen, C., Pouwels, K., Hoefsloot, L. H., Löwenberg, B., and Touw, I. P., 1993, Distinct cytoplasmic regions of the granulocyte colony-stimulating factor receptor involved in induction of proliferation and maturation, *Mol. Cell. Biol.* **13:**7774–7781.

Dong, F., Hoefsloot, L. H., Schelen, A. M., Broeders, L. C. A. M., Meijer, Y., Veerman, A. J. P., Touw, I. P., and Löwenberg, B., 1994, Identification of a nonsense mutation in the granulocyte–colony-stimulating factor receptor in severe congenital neutropenia, *Proc. Natl. Acad. Sci. U.S.A.* **91:**4480–4484.

Dong, F., Brynes, R. K., Tidow, N., Welte, K., Löwenberg, B., and Touw, I. P., 1995a, Mutations truncating the C-terminal maturation region of the G-CSF receptor in acute myeloid leukemia preceded by severe congenital neutropenia, *N. Engl. J. Med.* **333:**487–493.

Dong, F., Van Paassen, M., Van Buitenen, C., Hoefsloot, L. H., Löwenberg, B., and Touw, I. P., 1995b, A point mutation in the granulocyte colony-stimulating factor receptor (G-CSF-R) gene in a case of acute myeloid leukemia results in the overexpression of a novel G-CSF-R isoform, *Blood* **85:**902–911.

Duquesnoy, P., Sobrier, M. L., Amselem, S., and Goossens, M., 1991, Defective membrane expression of human growth hormone (GH) receptor causes Laron-type GH insensitivity syndrome, *Proc. Natl. Acad. Sci. U.S.A.* **88:**10272–10276.

Elbaz, O., Budel, L. M., Hoogerbrugge, H., Touw, I. P., Delwel, R., Mahmoud, L. A., and Löwenberg, B., 1991a, Tumor necrosis factor downregulates G-CSF receptor expression on human AML cells and granulocytes, *J. Clin. Invest.* **87:**838–841.

Elbaz, O., Budel, L. M., Hoogerbrugge, H., Touw, I. P., Delwel, R., Mahmoud, L. A., and Löwenberg, B., 1991b, Tumor necrosis factor regulates the expression of GM-CSF and IL-3 receptors on human acute myeloid leukemia cells, *Blood* **77:**989–995.

Ema, E., Kitano, K., Suda, T., Muroi, K., Ohta, M., Yoshida, M., Sakamoto, S., Eguchi, M., and Miura, Y., 1990, *In vitro* differentiation of leukemic cells to eosinophils in the presence of interleukin-5 in two cases of acute myeloid leukemia with the translocation (8;21)(q22;q22), *Blood* **75:**350–356.

Estey, E., Thall, P. F., Kantarjian, H., O'Brien, S., Koller, C. A., Beran, M., Gutterman, J., Deisseroth, A., and Keating, A., 1992, Treatment of newly diagnosed acute myelogenous leukemia with granulocyte–macrophage colony stimulating factor (GM-CSF) before and during continuous infusion of high-dose Ara-C and daunorubricin: Comparison to patients treated without GM-CSF, *Blood* **79:**2246–2255.

Fialkow, P. J., Jacobson, R. J., and Papayannopoulou, T., 1977, Chronic myelocytic leukemia: Clonal origin in a stem cell common to granulocyte, erythrocyte, platelet, and monocyte/macrophage, *Am. J. Med.* **63:**125–130.

Fluckiger, A. C., Rossi, J. F., Bussel, A., Bryon, P., Banchereau, J., and Defrance, T., 1992, Responsiveness of chronic lymphocytic leukemia B cells activated via surface Igs or CD40 to B-cell tropic factors, *Blood* **80:**3173–3181.

French–American–British Cooperative Group, 1985, Proposed revised criteria for the classification of acute myeloid leukemia, *Ann. Intern. Med.* **103:**620–625.

Fu, X.-Y., 1992, A transcription factor with SH2 and SH3 domains is directly activated by an interferon a-induced cytoplasmic protein tyrosine kinase(s), *Cell* **70:**323–335.

Fukunaga, R., Ishizaka-Ikeda, E., Seto, Y., and Nagata, S., 1990a, Expression cloning of a receptor for murine granulocyte colony-stimulating factor, *Cell* **61:**341–350.

Fukunaga, R., Seto, Y., Mizushima, S., and Nagata, S., 1990b, Three different mRNAs encoding human granulocyte colony-stimulating factor receptor, *Proc. Natl. Acad. Sci. U.S.A.* **87:**8702–8706.

Fukunaga, R., Ishizaka-Ikeda, E., Pan, C.-X., Seto, Y., and Nagata, S., 1991, Functional domains of the granulocyte colony-stimulating factor receptor, *EMBO J.* **10:**2855–2865.

Fukunaga, R., Ishizaka-Ikeda, E., and Nagata, S., 1993, Growth and differentiation signals mediated by different regions in the cytoplasmic domain of granulocyte colony-stimulating factor receptor, *Cell* **74:**1079–1087.

Gillio, A. P., and Gabrilove, J. L., 1993, Cytokine treatment of inherited bone marrow failure syndromes, *Blood* **81:**1669–1674.

Gilman, P. A., Jackson, D. P., and Guild, H. G., 1970, Congenital agranulocytosis: Prolonged survival and terminal acute leukemia, *Blood* **36:**576–585.

Graf, T., Weizsaecker, F. V., Grieser, S., Coll, J., Stehelin, D., Patschinsky, T., Bister, K., Bechade, C., Calothy, G., and Leutz, A., 1986, V-mil induces autocrine growth and enhanced tumorigenecity in *m-myc* transformed avian macrophages, *Cell* **45:**357–364.

Greer, P., Maltby, V., Rossant, J., Bernstein, A., and Pawson, T., 1990, Myeloid expression of the human *c-fps/fes* proto-oncogene in transgenic mice, *Mol. Cell. Biol.* **10:**2521–2527.

Griffin, J. D., and Löwenberg, B., 1986, Clonogenic cells in acute myeloblastic leukemia, *Blood* **68:**1185–1195.

Grimaldi, J. C., and Meeker, T. C., 1989, The t(5;14) chromosomal translocation in a case of acute lymphocytic leukemia joins the interleukin-3 gene to the immunoglobulin heavy chain gene, *Blood* **73:**2081–2085.

Hunter, A. E., Rogers, S. Y., Roberts, I. A., Barrett, A. J., and Russell, N., 1993, Autonomous growth of blast cells is associated with reduced survival in acute myeloblastic leukemia, *Blood* **82:**899–903.

Hunter, T., 1991, Cooperation between oncogenes, *Cell* **64:**249–270.

Ihle, J. N., Witthuhn, B. A., Quelle, F. W., Yamamoto, K., Thierfelder, W. E., Kreider, B., Silvennoinen, O., 1994, Signalling by the cytokine receptor superfamily: JAKs and STATs, *Trends Biochem. Sci.* **19:**222–227.

Kaushansky, K., Lok, S., Holly, R. D., Boudy, V. C., Lin, N., Baily, M. C., Forstrom, J. W., Buddle, M. M., Oort, P. J., Hagen, F. S., Roth, G. J., Papayannopoulou, T., and Foster, D. C., 1994, Promotion of megakaryocyte progenitor expansion and differentiation by the c-Mpl ligand thrombopoietin, *Nature* **369:**568–571.

Kawano, M., Hirano, T., Matsuda, T., Taga, T., Horii, Y., Iwato, K., Asaoku, H., Tang, B., Tanabe, O., Tanaka, H., Kuramoto, A., and Kishimoto, T., 1988, Autocrine generation and essential requirement of BSF-2/IL-6 for human multiple myelomas, *Nature* **332:**83–85.

Kazizuka, A., Miller, W. H., Umesono, K., Warrell, R. P., Frankel, S. R., Murty, V. V. V. S., Dmitrovsky, E., and Evans, R. M., 1991, Chromosomal translocation t(15;17) in human acute promyelocytic leukemia fuses RARα with a novel putative transcription factor, PML, *Cell* **66:**663–674.

Keller, J. R., Jacobsen, S. E. W., Dubois, C., Hestdal, K., and Ruscetti, F. W., 1992, Transforming growth factor-β: A bidirectional regulator of hematopoietic cell growth, *Int. J. Cell Cloning* **10:**2–11.

Kinoshita, T., Yokota, T., Arai, K., and Miyajima, A., 1995, Suppression of apoptotic death in hematopoietic cells by signalling through the IL-3/GM-CSF receptors, *EMBO J.* **14:**266–275.

Kluin-Nelemans, H. C., Beverstock, G. C., Mollevanger, P., Wessels, H. W., Hoogendoorn, E., Willemze, R., and Falkenburg, J. H. F., 1994, Proliferation and cytogenetic analysis of hairy cell leukemia upon stimulation via the CD40 antigen, *Blood* **84:**3134–3141.

Kostmann, R., 1956, Infantile genetic agranulocytosis, *Acta Paediatr. Scand. [Suppl.]* **105:**1–78.

Lapidot, T., Sirard, C., Vormoor, J., Murdoch, B., Hoang, T., Careres-Cortes, J., Minden, M., Paterson, B., Caligiuri, M. A., and Dick, J. E., 1994, A cell initiating human acute myeloid leukaemia after transplantation into SCID mice, *Nature* **367:**645–648.

Larner, A. C., David, M., Feldman, G. M., Igarashi, K.-I., Hackett, R. H., Webb, D. S. A., Sweitzer, S. M., Petricoin, E. F., III, and Finbloom, D. S., 1993, Tyrosine phosphorylation of DNA binding proteins by multiple cytokines, *Science* **261:**1730–1733.

Larsen, A., Davis, T., Curtis, B. M., Gimpel, S., Sims, J. E., Cosman, D., Park, L., Sorensen, E., March, C. J., and Smith, C. A., 1990, Expression cloning of a human granulocyte colony stimulating factor receptor: A structural mosaic of hematopoietin receptor, immunoglobulin, and fibronectin domains, *J. Exp. Med.* **172:**1559–1570.

Liu, P., Tarlé, S. A., Hajra, A., Claxton, D. F., Marlton, P., Freedman, M., Siciliano, M. J., and Collins, F. S., 1993, Fusion between transcription factor CBFβ/PEBP2β and myosin heavy chain in acute myeloid leukemia, *Science* **261:**1041–1044.

Lok, S., Kaushansky, K., Holly, R. D., Kuijper, J. L., Lofton Day, C. E., Oort, P. J., Grant, F. J., Heipel, M. D., Burkhead, S. K., Kramer, J. M., Bell, L. A., Sprecher, C. A., Blumberg, H., Johnson, R., Prunkard, D., Ching, A. F. T., Mathewes, S. L., Bailey, M. C., Forstrom, J. W., Buddle, M. M., Osborn, S. G., Evans, S. J., Sheppard, P. O., Presnell, S. R., O'Hara, P. J., Hagen, F. S., Roth, G. J., and Foster, D. C., 1994, Cloning and expression of murine thrombopoietin cDNA and stimulation of platelet production *in vivo*, *Nature* **369:**565–568.

Longmore, G. D., and Lodish, H. F., 1991, An activating mutation in the murine erythropoietin receptor induces erythroleukemia in mice: A cytokine receptor superfamily oncogene *Cell* **67:**1089–1102.

Lord, K. A., Abdollahi, A., Hoffman-Liebermann, B., and Liebermann, D. A., 1993, Proto-oncogenes of the *fos/jun* family of transcription factors are positive regulators of myeloid differentiation, *Mol. Cell. Biol.* **13:**841–851.

Löwenberg, B., Van Putten, W. L. J., Touw, I. P., Delwel, R., and Santini, V., 1993, Autonomous proliferation of leukemic cells *in vitro* as a determinant of prognosis in adult acute myeloid leukemia, *N. Engl. J. Med.* **328:**614–619.

Matsuguchi, T., Salgia, R., Hallek, M., Eder, M., Druker, B., Ernst, T. J., and Griffin, J. D., 1994, Shc phosphorylation in myeloid cells is regulated by granulocyte macrophage colony stimulating factor, interleukin-3, and Steel factor and is constitutively increased by p210BCR/ABL, *J. Biol. Chem.* **269:**5016–5021.

Matsumara, I., Kanakura, Y., Kato, T., Ikeda, H., Ishikawa, J., Horikawa, Y., Hashimoto, K., Moriyama, Y., Tsujimura, T., Nishiura, T., Miyazaki, H., and Matsuzawa, Y., 1995, Growth response of acute myeloblastic leukemia cells to recombinant human thrombopoietin, *Blood* **86:**703–709.

Meeker, T. C., Hardy, D., Willman, C., Hogan, T., and Abrams, J., 1990, Activation of the interleukin-3 gene by chromosome translocation in acute lymphocytic leukemia with eosinophilia, *Blood* **76:**285–289.

Meng-er, H., Yu-chen, Y., Shu-rong, C., Jin-ren, C., Jia Chang, L., Lin, Z., Long-jun, G., and Zhen, Y., 1988, Use of all-*trans* retinoic acid in the treatment of acute promyelocytic leukemia, *Blood* **72:**567–572.

Metcalf, D., 1989, The molecular control of cell division, differentiation, commitment and maturation in haematopoietic cells, *Nature* **339:**27–30.

Metz, T., Graf, T., and Leutz, A., 1991, Activation of cMGF expression is a critical step in avian myeloid leukemogenesis, *EMBO J.* **10:**837–844.

Miyajima, A., Mui, A. L. F., Ogorochi, T., and Sakamaki, K., 1993, Receptors for granulocyte–macrophage colony-stimulating factor, interleukin-3, and interleukin-5, *Blood* **82:**1960–1974.

Miyauchi, J., Kelleher, C. A., Yang, Y., Wong, G. G., Clark, S. C., Minden, M. D., Minkin, S., and McCulloch, E. A., 1987, The effects of three recombinant growth factors, IL-3, GM-CSF, and G-CSF, on the blast cells of acute myeloblastic leukemia maintained in short term suspension culture, *Blood* **70:**657–663.

Miyoshi, H., Shimizu, K., Kozu, T., Maseki, N., Kaneko, Y., and Ohki, M., 1991, t(8;21) breakpoints on chromosome 21 in acute myeloid leukemia are clustered within a limited region of a single gene, *AML1*, *Proc. Natl. Acad. Sci. U.S.A.* **88:**10431–10434.

Moberts, R., Hoogerbrugge, H., Van Agthoven, T., Löwenberg, B., and Touw, I., 1989, Proliferative response of highly purified B chronic lymphocytic leukemia cells in serum free culture to interleukin-2 and tumor necrosis factors alpha and beta, *Leuk. Res.* **13:**973–980.

Morishita, K., Parganas, E., Matsugi, T., and Ihle, J. N., 1992a, Expression of the *Evi-1* zinc finger gene in 32Dcl3 myeloid cells blocks granulocytic differentiation in response to granulocyte colony-stimulating factor, *Mol. Cell. Biol.* **12:**183–189.

Morishita, K., Parganas, E., Willman, C. L., Whittaker, M. H., Drabkin, H., Oval, J., Taetle, R., and Ihle, J. N., 1992b, Activation of *Evi-1* gene expression in human acute myelogenous leukemias by translocations spanning 300–400 kilobases on chromosome band 3q26, *Proc. Natl. Acad. Sci. U.S.A.* **89:**3937–3941.

Murakami, M., Narazaki, M., Hibi, M., Yawata, H., Yasukawa, K., Hamaguchi, M., Taga, T., and Kishimoto, T., 1991, Critical cytoplasmic region of the interleukin 6 signal transducer gp130 is conserved in the cytokine receptor family, *Proc. Natl. Acad. Sci. U.S.A.* **88:**11349–11353.

Murohashi, I., Tonda, S., Suzuki, T., Nazata, K., Yamashita, Y., and Nara, N., 1989, Autocrine growth mechanisms of the progenitors of blast cells in acute myeloblastic leukemia, *Blood* **74:**35–41.

Noguchi, M., Yi, H., Rosenblatt, H. M., Filipovich, A. H., Adelstein, S., Modi, W. S., McBride, O. W., and Leonard, W. J., 1993, Interleukin-2 receptor τ chain mutation results in X-linked severe combined immunodeficiency in humans, *Cell* **73:**147–157.

Oster, W., Cicco, N. A., Klein, H., Hirano, T., Kishimoto, T., Lindemann, A., Mertelsmann, R. H., and Herrmann, F., 1989, Participation of the cytokines interleukin 6, tumor necrosis factor alpha, and interleukin-1 beta secreted by acute myelogenous leukemia blasts in autocrine and paracrine growth control, *J. Clin. Invest.* **84:**451–457.

Pebusque, M. J., Lafage, M., Lopez, M., and Mannoni, P., 1988, Preferential response of acute myeloid leukemias with translocations involving chromosome 17 to human recombinant granulocyte colony-stimulating factor, *Blood* **72:**257–265.

Ravichandran, K. S., and Burakoff, S. J., 1994, The adapter protein Shc interacts with the interleukin-2 (IL-2) receptor upon IL-2 stimulation, *J. Biol. Chem.* **269:**1599–1602.

Reilly, I. A., Kozlowski, R., and Russell, N. H., 1989, Heterogeneous mechanisms of autocrine growth of AML blasts, *Br. J. Haematol.* **72:**363–369.

Rosen, R. B., and Kang, S. J., 1979, Congenital agranulocytosis terminating in acute myelomonocytic leukemia, *J. Pediatr.* **94:**406–407.

Salem, M., Delwel, R., Touw, I., Mahmoud, L. A., and Löwenberg, B., 1990, Modulation of colony stimulating factor (CSF) dependent growth of acute myeloid leukemia by tumor necrosis factor, *Leukemia* **4:**37–43.

Santini, V., Colombat, P., Delwel, R., van Gurp, R., Touw, I., and Löwenberg, B., 1991, Induction of granulocytic maturation in acute myeloid leukemia by G-CSF and retinoic acid, *Leuk. Res.* **5:**341–350.

Sato, N., Sakamaki, K., Terada, N., Arai, K., and Miyajima, A., 1993, Signal transduction by the high-affinity GM-CSF receptor: Two distinct cytoplasmic regions of the common β subunit responsible for different signaling, *EMBO J.* **12:**4181–4189.

Sawyers, C. L., Denny, C. T., and Witte, O. N., 1991, Leukemia and the disruption of normal hematopoiesis, *Cell* **64:**337–350.

Souyri, M., Vigon, I., Penciolelli, J.-F., Heard, J.-M., Tambourin, P., and Wendling, F., 1990, A putative truncated cytokine receptor gene transduced by the myeloproliferative leukemia virus immortalizes hematopoietic progenitors, *Cell* **63:**1137–1147.

Tian, S.-S., Lamb, P., Seidel, H. M., Stein, R. B., and Rosen, J., 1994, Rapid activation of the STAT3 transcription factor by granulocyte colony-stimulating factor, *Blood* **84:**1760–1764.

Touw, I., Pouwels, K., Van Agthoven, T., Van Gurp, R., Budel, L., Hoogerbrugge, H., Delwel, R., Goodwin, R., Namen, A., and Löwenberg, B., 1990, Interleukin-7 is a growth factor of precursor B and T acute lymphoblastic leukemia, *Blood* **75:**2097–2101.

Touw, I., Donath, J., Pouwels, K., Van Buitenen, C., Schipper, P., Santini, V., Hagemeijer, A., Löwenberg, B., and Delwel, R., 1991, Acute myeloid leukemias with chromosomal abnormalities involving the 21q22 region identified by their *in vitro* responsiveness to interleukin-5, *Leukemia* **5:**687–692.

Wang, S., Wang, Q., Crute, B. E., Melnikova, I. N., Keller, S. A., and Speck, N. A., 1993, Cloning and characterization of subunits of the T-cell receptor and murine leukemia virus enhancer core-binding factor, *Mol. Cell. Biol.* **13:**3324–3339.

Warrell, R. P., Frankel, S. R., Miller, W. H., Scheinberg, D. A., Itri, L. M., Hittelman, W. N., Vyas, R., Andreeff, M., Tarufi, A., Jakubowski, A., Gabrilove, J., Gordon, M. S., and Dmitrovsky, E., 1991, Differentiation therapy of acute promyelocytic leukemia with tretinoin (all-*trans*-retinoic acid), *N. Engl. J. Med.* **324:**1385–1393.

Wörmann, B., Gesner, T. G., Mufson, R. A., and LeBien, T. W., 1989, Proliferative effect of interleukin-3 on normal and leukemic human B cell precursors, *Leukemia* **3:**399–404.

Young, D. C., and Griffin, J. D., 1986, Autocrine secretion of GM-CSF in acute myeloblastic leukemia, *Blood* **68:**1178–1181.

Chapter 13

The Role of Interleukin 5 in the Production and Function of Eosinophils

Jan Tavernier, Geert Plaetinck, Yves Guisez,
Jose van der Heyden, Johan Kips, Renaat Peleman,
and Rene Devos

1. INTRODUCTION

The development of the different lineages of the hematopoietic system is regulated by a set of intercellular interactions, either by direct cellular contact or via the release of soluble factors. Proliferation, self-renewal, and lineage commitment are all strongly influenced by a complex network of external factors. Among these, cytokines play a crucial role. Both constitutive and inducible hematopoiesis are regulated by cytokines, which in the former case are secreted by stromal cells within the bone-marrow microenvironment and in the latter case reach the bone marrow via the blood circulation after being produced in inflamed tissues. Many of these cytokines influence each other's pattern of activities in a synergistic or antagonistic way. In addition, some cytokines have a (lineage-) restricted biological repertoire, and others exert a broad spectrum of activities ("pleiotropy"), resulting in functional overlaps ("redundancy") within this complex signaling network.

Interleukin 5 (IL-5) belongs to the group of cytokines [e.g., erythropoietin (EPO), granulocyte colony-stimulating factor (G-CSF)] with a restricted set of targets, as its major role (in man) is limited to the eosinophilic/basophilic lineage. Yet, the first activity-based descriptions of IL-5 were on murine B cells *in vitro*. Until the mid-1980s, descriptors including BCGF-II (B-cell growth factor; Swain *et al.*, 1988), $BCDF_{\mu}$ (B-cell differentiation

Jan Tavernier, Geert Plaetinck, Yves Guisez, Jose van der Heyden, and Rene Devos Roche Research Gent, 9000 Gent, Belgium. **Johan Kips and Renaat Peleman** Department of Respiratory Diseases, University Hospital, Depintelaan, 9000 Gent, Belgium.

Blood Cell Biochemistry, Volume 7: Hematopoietic Cell Growth Factors and Their Receptors, edited by A. D. Whetton and J. Gordon. Plenum Press, New York, 1996.

factor; Vitetta *et al.*, 1984), and TRFs (T-cell replacing factor; Ono *et al.*, 1986; Takatsu *et al.*, 1988) were used. Additional biochemical characterization together with gene cloning led to the convergence of these various B-cell activities (Kinashi *et al.*, 1986) with EDF (eosinophil differentiation factor; Sanderson *et al.*, 1986). Much in contrast to the murine system, the role of IL-5 on human B cells is still a matter of debate (e.g. Clutterbuck *et al.*, 1987).

Interleukin 5 specifically induces eosinophil (and to some extent basophil) production *in vitro* in liquid bone-marrow (Dexter *et al.*, 1977; Sanderson *et al.*, 1985) or cord blood cultures (Saito *et al.*, 1988) and in colony formation assays (Metcalf *et al.*, 1974; Enokihara *et al.*, 1988). This property is unique to IL-5 and contrasts with other cytokines such as IL-3 and GM-CSF, which, in addition to eosinophilic growth, also promote the proliferation/differentiation of other lineages. *In vivo* studies confirm this specificity (Coffman *et al.*, 1989; Dent *et al.*, 1990; Tominaga *et al.*, 1991). Because evidence for a role of eosinophils in some human diseases is accumulating, a critical role for IL-5 in these pathologies is being considered. Hence, antagonizing the activities of IL-5 is a new, important target for drug discovery programs. In this review, we restrict our attention to the effects of IL-5 on the eosinophil lineage. Effects on the basophil lineage and on murine B-cells were recently reviewed elsewhere (Denburgh, 1992; Takatsu *et al.*, 1988; Takatsu, 1992).

2. INTERLEUKIN 5

Interleukin 5 shows strong similarities with other cytokines, including IL-3, IL-4, GM-CSF, and many others, which are now often collectively referred to as hemopoietins or hemopoietic growth factors (HGFs). A common evolutionary origin is likely, as suggested by a shared three-dimensional architecture (but not primary structure), by their gene clustering and conserved gene organization, and also by common features in the regulation of their expression pattern.

2.1. The Interleukin-5 Gene

2.1.1. Gene Structure, Gene Linkage, and Chromosomal Location

Information on the structure and genomic organization of the IL-5 gene is available for both man and mouse (Kinashi *et al.*, 1986; Azuma *et al.*, 1986; Mizuta *et al.*, 1988; Campbell *et al.*, 1987, 1988). Both genes have a very similar structure containing four exons, which is a conserved feature within the HGF family (Figure 1). The human and mouse genes are linked with the IL-3, IL-4, IL-9, IL-13, and GM-CSF genes (Van Leeuwen *et al.*, 1989; Chandrasekharappa *et al.*, 1990; Warrington *et al.*, 1992; Kluck *et al.*, 1993) and form part of a growing cytokine gene cluster on chromosome 5 (region 5q31→q35, with an estimated size of 1.5 Mb of DNA) in man and on the syntenic position on chromosome 11 in the mouse, suggesting that this gene family arose from duplication events. Also located in the same region are the genes coding for interferon regulatory factor 1 (IRF-1), for the mitotic inducer CDC25C (Willman *et al.*, 1993), and for early growth response protein 1 (EGR-1). Deletions within this gene cluster have been correlated with various hematological disorders.

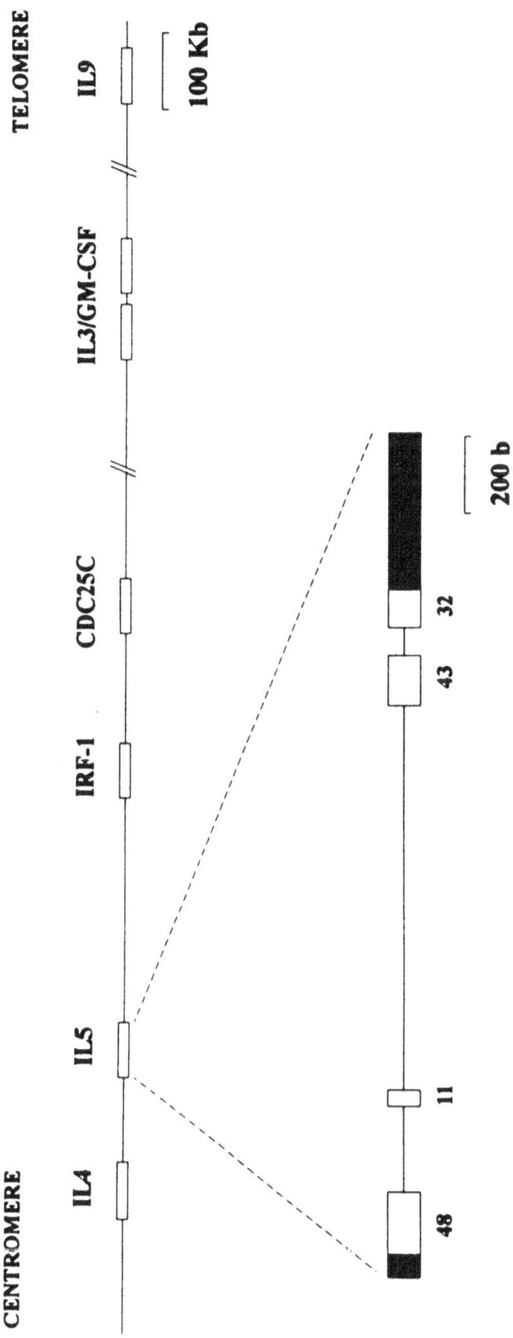

FIGURE 1. Gene linkage and structure of the human IL-5 gene. Top diagram shows the physical map of the cytokine cluster at the human 5q31 chromosome region. Open boxes represent genes (not drawn to scale). Enlarged at the bottom is the organization of the IL-5 gene. Boxes define the exons, and open segments correspond to the coding regions. Amino acid numbers present in each exon are shown below. (Willman *et al.*, 1993; Warrington *et al.*, 1992; Campbell *et al.*, 1987).

2.1.2. Gene Expression Pattern

The predominant source of IL-5 is helper T cells (Th). Indeed, eosinophilia has been recognized as a T-cell-dependent phenomenon for a long time (Basten and Beeson, 1970; Hsu *et al.*, 1976). *In vitro*, a transient pulse of IL-5 is produced by T cells on stimulation with antigen, phorbol ester, or mitogens (Tominaga *et al.*, 1988). More specifically, expression of IL-5 is often coregulated in the Th2 subset with a panel of other cytokines including IL-4, IL-6, IL-9, IL-10, and IL-13. This functional dichotomy of Th subsets (and intermediates) was first recognized in the mouse, and recent experiments suggest its existence also in man (Mosman and Coffman, 1989; Romagnani, 1994). This activity pattern of the cytokines expressed by Thl and Th2 subsets appears to correlate with delayed-type hypersensitivity and allergic disorders, respectively. Indeed, several reports mention a Th2 expression pattern by T lymphocytes (1) infiltrating conjunctiva of patients with vernal conjunctivitis (Maggi *et al.*, 1991), (2) found in allergen-induced late-phase nasal and skin lesions in atopic patients (Kay *et al.*, 1991), or (3) found in bronchoalveolar lavage from asthmatics (Robinson *et al.*, 1992).

Yet, the IL-5 expression pattern has unique characteristics, different from the Th2 cytokine cluster. Studies with certain mouse T-cell clones and human peripheral blood T cells show differential regulation of IL-5 mRNA expression compared to IL-2, IL-4, or GM-CSF mRNA expression (Bohjanen *et al.*, 1990; Naora and Young, 1994) or IL-3 mRNA expression (Van Straaten *et al.*, 1994). This observation is reminiscent of eosinophilia observed *in vivo*, which can occur without concomitant increases in other leukocyte types (Strath and Sanderson, 1986) or IgE production (Finkelman *et al.*, 1994).

Certain non-T leukocytes have also been shown to be capable of producing IL-5. This expression in mast cells (Plaut *et al.*, 1989; Bradding *et al.*, 1993) and eosinophils (Desreumaux *et al.*, 1992; Dubucquoi *et al.*, 1994; Broide *et al.*, 1992) might represent an amplification mechanism that further underscores the potentially pivotal role of IL-5 in allergic disorders.

Finally, expression of IL-5 has also been reported from malignant cells, indicating its potential for stimulating autocrine growth (Samoszuk and Nansen, 1990; Baumann and Paul, 1992; Blankenstein *et al.*, 1990).

The promoter sequences are highly homologous (77%) between human and mouse IL-5 in a promoter segment up to 150 bp upstream of the initiation site. In the region to -300 bp, several conserved promoter sequence motifs with homology to other cytokine gene promoters, including IL-2, IL-3, IL-4, and GM-CSF, have been recognized (Miyatake *et al.*, 1991), which may account for their coordinated expression pattern. For example, a conserved element (region -175 to -164) with high similarity to a T-cell-specific enhancer in the IL-2 promoter (Serfling *et al.*, 1989) is present.

In addition, independent control elements must exist that impose the differential regulation. Phorbol-ester-induced IL-5 mRNA expression is resistant to cyclosporin A (CsA) under conditions where IL-2 and GM-CSF mRNA expression is inhibited. Multiple activation pathways may exist, as suggested by the synergistic promoter stimulation by cyclic AMP and phorbol ester (Lee *et al.*, 1993). Interestingly, glucocorticoids (and also CsA) have been shown to suppress the expression of IL-5 by T lymphocytes (Corrigan *et al.*, 1993). These findings may help explain the mechanism of action of these anti-asthmatic drugs.

A full understanding of promoter regulation will await the molecular characterization of transcription factors, either shared with other promoters or unique for the IL-5 promoter, and perhaps expressed in restricted cell types.

2.2. The Interleukin-5 Protein

2.2.1. Primary Structure

Native mouse IL-5 has been isolated from T-cell supernatants (Takatsu *et al.*, 1985 ; Mackenzie *et al.*, 1987), but human IL-5 has not been purified from natural sources. The recombinant proteins have been expressed using either prokaryotic or eukaryotic expression systems: biologically active recombinant human and mouse IL-5 have been produced in *E. coli* (Proudfoot *et al.*, 1990), *S. cerevisiae* (Tavernier *et al.*, 1989; Ingley *et al.*, 1991), baculovirus-infected Sf9 insect cells (Tavernier *et al.*, 1989; Ingley *et al.*, 1991; Kunimoto *et al.*, 1991; Guisez *et al.*, 1993), and Chinese hamster ovary (CHO) cells (Minamitaki *et al.*, 1990).

A comparison of the primary structures of human, mouse, rat and sheep IL-5 is shown in Figure 2. The human IL-5 precursor is 134 residues long, including a signal peptide of 19 amino acids (Azuma *et al.*, 1986; Mita *et al.*, 1989a). The proteins show 70% sequence similarity, which is most pronounced in the C-terminal part. This high degree of similarity likely explains the cross-species reactivity and immunologic cross-reactivity observed between human and mouse IL-5.

The secreted core polypeptide of hIL-5 corresponds to a calculated molecular mass of 13,149 kDa. Accordingly, *E. coli*-derived protein migrates as a single 13-kDa band on SDS-PAA gels under reducing conditions. However, in the absence of reducing agent, a molecular mass of 27 kDa is found (Proudfoot *et al.*, 1991). Likewise, the various recombinant IL-5s expressed by eukaryotic cells have native molecular masses between 35 and 60 kDa, changing to 16 to 32 kDa after treatment with reducing agents (Tominaga *et al.*, 1990; Ingley *et al.*, 1991; Kunimoto *et al.*, 1991; Guisez *et al.*, 1993). It has been shown that IL-5 exists as a cross-linked homodimer in an antiparallel (head-to-tail) arrangement (Minamitake *et al.*, 1990; McKenzie *et al.*, 1991a) by S-S bridging between cysteine residues 44 and 86. Reduction and alkylation of the two cysteines in recombinant hIL-5 or mutation to threonine residues lead to a biologically inactive monomer (Takahashi *et al.*, 1990; Minamitake *et al.*, 1990), illustrating that dimer formation via disulfide bonds is essential for biological activity (Tavernier *et al.*, 1989; Tsujimoto *et al.*, 1989; Minamitake *et al.*, 1990; Tsuruoka *et al.*, 1990; Ingley *et al.*, 1991; McKenzie *et al.*, 1991a).

The variable high-molecular-weight range observed for secreted IL-5 is explained by the heterogeneous addition of carbohydrate (O-linked glycosylation on Thr 3 and N-linked glycosylation on Asn 28) (Tominaga *et al.*, 1990; Minamitaki *et al.*, 1990). Mouse IL-5 has an additional N-linked glycosylation at position Asn 55, which is absent in hIL-5 (Kodama *et al.*, 1992). Removal of the sugar moieties demonstrates that glycosylation is not necessary for receptor binding and biological activity *in vitro* (Tominaga *et al.*, 1990; Ingley *et al.*, 1991; Kunimoto *et al.*, 1991). Deglycosylated CHO-derived hIL-5 even shows a marked increase in activity, with O-linked glycosylation identified as the suppressor of bioactivity, while N-linked glycosylation appears to contribute to thermostabilization of the protein (Kodama *et al.*, 1993).

```
                                                  -15
Human :                            Met---ArgMetLeuLeu
Mouse :                            MetArgArgMetLeuLeu
Rat   :                            Met---ArgMetLeuLeu
Sheep :                            ---------MetHisLeu

                       -10            -5          -1
Human : HisLeuSerLeuLeuAlaLeuGlyAlaAlaTyrValTyrAla |
Mouse : HisLeuSerValLeuThrLeuSerCysValTrpAlaThrAla |
Rat   : CysLeuAsnValLeuThrLeuSerCysValTrpAlaIleAla |
Sheep : ArgLeuTrpLeuValAlaLeuGlyAlaAlaTyrValCysAla ▼

        1           5              10             15
                  *                    *     *  *  *
Human : IleProThrGluIleProThrSerAlaLeuValLysGluThrLeu
Mouse : ------MetGluIleProMetSerThrValValLysGluThrLeu
Rat   : ------MetGluIleProMetSerThrValValLysGluThrLeu
Sheep : AsnAlaValGluSerThrMetAsnArgLeuValAlaGluThrLeu

                    20             25             30
               *  *              *  *
Human : AlaLeuLeuSerThrHisArgThrLeuLeuIleAlaAsnGluThr
Mouse : ThrGlnLeuSerAlaHisArgAlaLeuLeuThrSerAsnGluThr
Rat   : IleGlnLeuSerThrHisArgAlaLeuLeuThrSerAsnGluThr
Sheep : ThrLeuLeuSerThrAsnGlnThrLeuLeuIleGlyAspGlyAsn

                    35             40          ʃ 45
                  *     *     *     *  *  *  *  *
Human : LeuArgIleProValProValHisLysAsnHisGlnLeuCysThr
Mouse : MetArgLeuProValProThrHisLysAsnHisGlnLeuCysIle
Rat   : MetArgLeuProValProThrHisLysAsnHisGlnLeuCysIle
Sheep : LeuMetIleProThrProGlnHisThrAsnHisGlnLeuCysIle

                    50             55             60
            *     *  *  *           *        *  *
Human : GluGluThrPheGlnGlyIleGlyThrLeuGluSerGlnThrVal
Mouse : GlyGluIlePheGlnGlyLeuAspIleLeuLysAsnGlnThrVal
Rat   : GlyGluIlePheGlnGlyLeuAspIleLeuLysAsnGlnThrVal
Sheep : GluGluValPheGlnGlyIleAspThrLeuLysAsnGlnThrAla

                    65             70             75
            *        *           *     *  *  *  *  *
Human : GlnGlyGlyThrValGluArgLeuPheLysAsnLeuSerLeuIle
Mouse : ArgGlyGlyThrValGluMetLeuPheGlnAsnLeuSerLeuIle
Rat   : ArgGlyGlyThrValGluIleLeuPheGlnAsnLeuSerLeuIle
Sheep : GlnGlyAspAlaValLysLysIlePheArgAsnLeuSerLeuIle

                    80             85  ʃ          90
         *  *  *  *  *     *  *     *  *  *     *  *
Human : LysLysTyrIleAspGlyGlnLysLysLysCysGlyGluGluArg
Mouse : LysLysTyrIleAspArgGlnLysGluLysCysGlyGluGluArg
Rat   : LysLysTyrIleAspGlyGlnLysGluLysCysGlyGluGluArg
Sheep : LysLysTyrIleAspLeuGlnLysArgLysCysGlyGlyGluArg
```

FIGURE 2. Comparison of the primary structures of human, mouse, rat, and sheep IL-5. Numbers above the aligned amino acid residues indicate the positions based on the human sequence. Deletions are shown by dashes. An arrowhead indicates the signal peptidase cleavage site. Conserved residues are shown by asterisks; conserved cysteines are, in addition, highlighted by a scroll sign. Predicted N-linked glycosylation sites are underlined. The amino acid sequence of human IL-5 is from Azuma *et al.* (1986); mouse IL-5 is from Kinashi *et al.* (1986); rat IL-5 is from Ueberla *et al.* (1991); the sheep IL-5 sequence is derived from a 3.5-kb genomic clone (C. E. Bryson, personal communication).

```
                    95             100            105
                        *  *  *  *  *  *     *  *  *
Human : ArgArgValAsnGlnPheLeuAspTyrLeuGlnGluPheLeuGly
Mouse : ArgArgThrArgGlnPheLeuAspTyrLeuGlnGluPheLeuGly
Rat   : ArgLysThrArgHisPheLeuAspTyrLeuGlnGluPheLeuGly
Sheep : ThrArgValLysGlnPheLeuAspTyrLeuGlnValPheLeuGly

                    110            115
         *        *  *  *        *
Human : ValMetAsnThrGluTrpIleIleGluSer
Mouse : ValMetSerThrGluTrpAlaMetGluGly
Rat   : ValMetSerThrGluTrpAlaMetGluVal
Sheep : ValIleAsnThrGluTrpThrMetGluSer
```

FIGURE 2. *(Continued)*

2.2.2. Three-Dimensional Structure

Crystals have been obtained from recombinant hIL-5 purified either from renaturated *E. coli* inclusion bodies (Graber *et al.*, 1993; Hassell *et al.*, 1993) or from the supernatant of baculovirus-infected Sf9 cells (Guisez *et al.*, 1993), and the three-dimensional structure was determined by x-ray diffraction at 2.4-Å resolution (Milburn *et al.*, 1993; Tavernier *et al.*, 1995).

The diffraction data reveal a two-domain structure in which each domain (i.e., monomer) adopts the typical cytokine fold, which has also been described for other members of the interleukin/hematopoietin family (Figure 3). Nevertheless, no significant sequence similarity exists between these proteins (Bazan, 1990). Typically, this fold consists of a four-helical bundle in an up–up, down–down array. However, IL-5 is unique among these structurally related cytokines in that one bundle is built up of three helices of one monomer, with the fourth helix (D'-helix swapping in Figure 3) being contributed by the second monomer. In addition, IL-5 also contains two short β strands located between helices A and B and C and D, respectively, forming a small antiparallel β-sheet (β_1 with β'_2 in Figure 3). Interestingly, in contrast to the noncrystallographic twofold axis found for the hIL-5 structure determined from *E. coli*-derived protein (Milburn *et al.*, 1993), hIL-5 from Sf9 cells crystallizes as an exact crystallographic dimer (Tavernier *et al.*, 1995). In the latter case, hIL-5 recovered from the crystals is fully biologically active (Guisez *et al.*, 1993). One explanation for the observed difference might be found in a differential folding behavior, as the *E. coli*-derived hIL-5 requires a denaturing–refolding procedure.

3. THE INTERLEUKIN-5 RECEPTOR

3.1. Interleukin-5 Receptor Expression and IL-5 Binding

The interleukin 5 receptor (IL-5R) is a heterodimer (or oligomer) consisting of α and β subunits (Figure 4). The α subunit is ligand-specific, which it binds with intermediate affinity in man (K_d of ~500 pM to 1 nM) and low affinity in the mouse (K_d ~ 5–10 nM). Upon association with the β subunit, a high-affinity binding of approximately 150 pM is observed for both species. The β subunit does not have any detectable affinity for IL-5 by itself (Tavernier *et al.*, 1991; Murata *et al.*, 1992; Devos *et al.*, 1990, 1991; Takaki *et al.*, 1990,

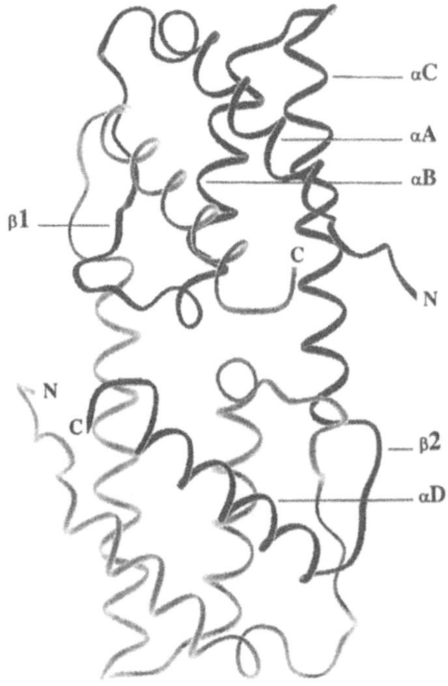

FIGURE 3. Three-dimensional structure of human IL-5. Full ribbon display of the human IL5 dimer, with each monomer shown in a different tone. The α-helices (A,B,C,D) and β-strands (1,2) are indicated, together with the N and C termini.

1991). The dissociation rate of mIL-5 from the αβ complex is considerably slower ($t_{1/2}$ > 1 hr) than from the low-affinity α binding site ($t_{1/2}$ < 2 min) (Devos *et al.*, 1991). Interleukin 5 binds to its receptor with unidirectional species specificity: mIL-5 binds with comparable affinities to both murine and human α subunits, but hIL-5 displays a 100-fold lower binding affinity for the murine α chain compared to its human counterpart.

The use of the β subunit is shared with IL-3 and GM-CSF (Tavernier *et al.*, 1991; Kitamura *et al.*, 1991), which both also have their own specific α subunits (see Chapter 1). For this reason, the β subunit is often referred to as β-common or β_c. This observation provides the molecular basis for the overlapping biological activities of these cytokines (reviewed in Lopez *et al.*, 1992a; Nicola and Metcalf, 1991). When the IL-5Rα and β_c subunits are coexpressed with the α subunits of IL-3 and GM-CSF, all three ligands will cross-compete for high-affinity binding. Interestingly, this cross-competition follows a hierarchical pattern: IL-3 = GM-CSF > IL-5. However, in the mouse, mIL-5 and mIL-3 do not cross-compete because of the presence of a distinct but very related β subunit, AIC2A, used by mIL-3 exclusively (Itoh *et al.*, 1990). The β_c subunit in the mouse is often referred to as AIC2B.

An example of such unidirectional cross-competition is shown in Figure 5. Cos 1 cells were cotransfected with hIL-5Rα, hGM-CSFRα, and β_c. After 3 days, cells were incubated on ice with either 5 nM [^{125}I]hIL-5 or 5 nM [^{125}I]GM-CSF in the absence or presence of increasing amounts of cold competitor. After cross-linking and lysis, the reaction products were analyzed by PAGE and autoradiography. As can be seen in Figure 5A, a tenfold lower concentration of unlabeled GM-CSF prevents cross-linking of [^{125}I]hIL-5 to β_c. In contrast,

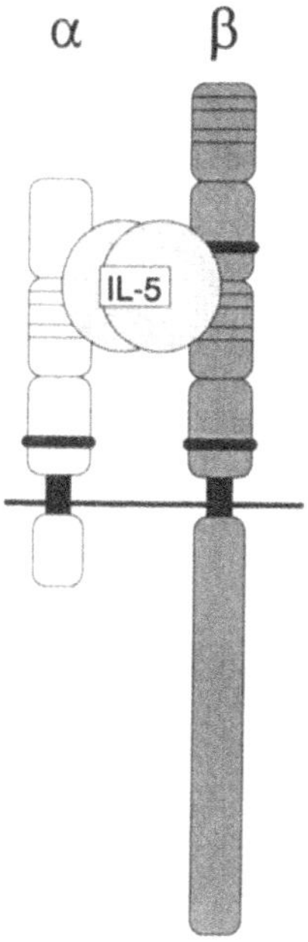

FIGURE 4. Diagrammic presentation of the IL-5/IL-5 receptor complex. The human IL5Rα and β subunits are shown, emphasizing their modular structure. Both extracellular and cytoplasmic parts are depicted. Thick and thin horizontal lines within the extracellular fibronectin type-III-like modules represent the WS-x-WS motifs and conserved cysteine residues, respectively. The IL-5 dimer is also shown.

cold IL-5 is unable to compete for binding of [^{125}I]GM-CSF to β_c even at concentrations up to 2.5 μM (Figure 5B). Competition binding could also be studied in eosinophils from a patient with eosinophilic chronic myelocytic leukemia (Eo-CML). These cells express high-affinity receptors for IL-3, IL-5, and GM-CSF. Scatchard analysis of [^{125}I]hIL-5 binding in the presence of saturating amounts of hIL-3 or hGM-CSF shows that although the receptor number is unaffected, the binding affinity for IL-5 drops two- to threefold when IL-3 or GM-CSF is present (Figure 6). This reduction corresponds to a conversion of high- to low-affinity receptors (Tavernier *et al.*, 1991). On the contrary, we were unable to block high-affinity IL-3 or GM-CSF binding with IL-5 (data not shown), again demonstrating the unidirectional nature of this cross-competition.

The biological repertoire of IL-5, and also IL-3 and GM-CSF, is dependent on the expression pattern of their receptor α subunits. In the case of human IL-5, this expression is most prominent on eosinophils and basophils. In the mouse, in addition to these cell types, IL-5-specific binding sites are also found on activated B cells and some B-cell lines (Mita *et al.*, 1988; Rolink *et al.*, 1989). No comparable expression on B cells has been described so far in man. Depending on the relative expression levels of the two receptor subunits, solely high- or both intermediate- (low- in the mouse) and high-affinity receptors can be observed. In general, on human eosinophils or eosinophilic cell lines, only one class of high-affinity receptors is found (Plaetinck *et al.*, 1990; Chihara *et al.*, 1990; Ingley and Young, 1991; Lopez *et al.*, 1991; Migita *et al.*, 1991). In contrast, on both murine eosinophils and B cells, two affinities are often detected (Mita *et al.*, 1989b; Barry *et al.*, 1991).

The exact stoichiometry of α and β_c subunits within the receptor complex remains unknown. Despite the twofold symmetrical structure of IL-5, which opens the possibility for two receptor binding sites, association with only one α subunit is seen in solution (Devos

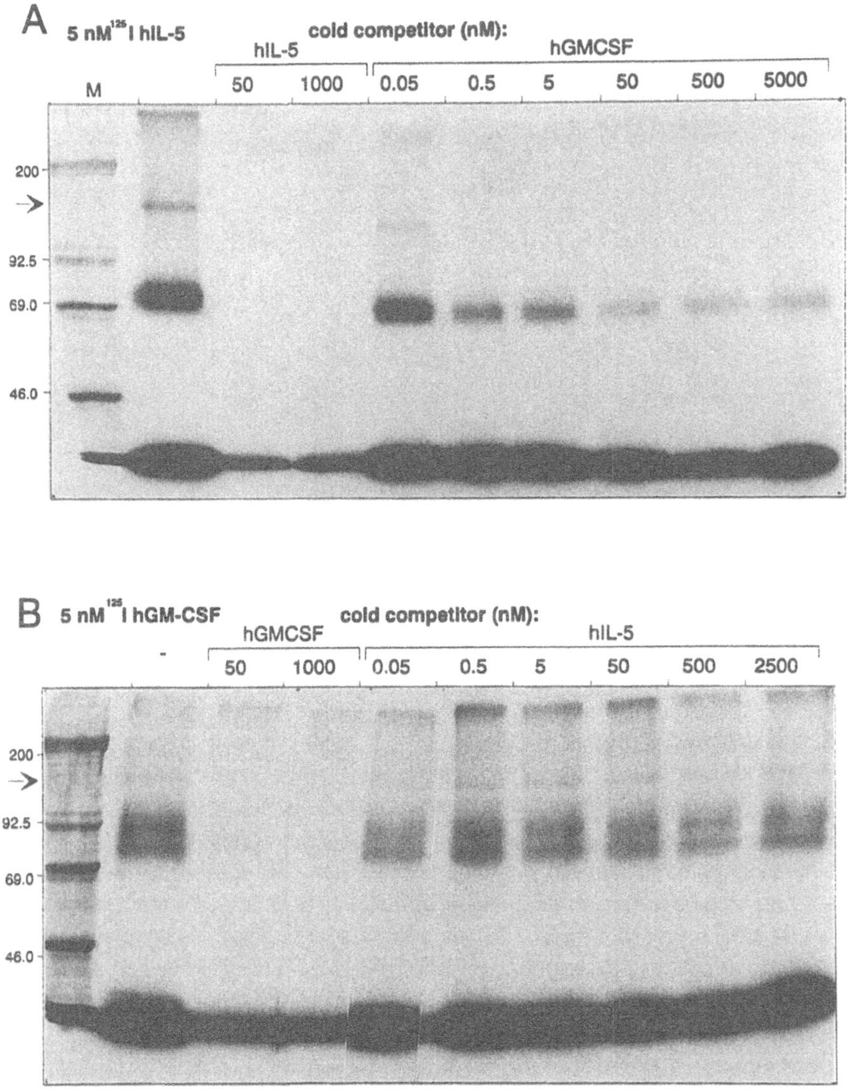

FIGURE 5. Nonreciprocal competition for the β-subunit (Cos). Chemical cross-linking of [^{125}I]hIL-5 (A) and [^{125}I]hGM-CSF (B) on Cos 1 cells expressing hIL-5Rα, hGM-CSFRα, and β_c. Transfection with the appropriate cDNAs was performed as described (Tavernier *et al.*, 1991): 10^6 cells were incubated with 5 nM radiolabeled ligand and increasing amounts of cold competitor as indicated. Cross-linking using the chemical crosslinker BS^3 (Pierce) was carried out as described (Plaetinck *et al.*, 1990). M stands for molecular weight markers. The position of the cross-linked β_c complex is indicated by arrowheads.

et al., 1993). A full high-affinity complex including α and β_c subunits cannot be reconstituted in solution (our unpublished data). In contrast to the soluble hIL-6R, where IL-6 triggers its association with the gp130 signal transducer (Yawata *et al.*, 1993), no such binding between the soluble hIL-5Rα/IL-5 complex with membrane-expressed β_c is seen. Perhaps steric hindrance or induced conformational changes may explain this observation.

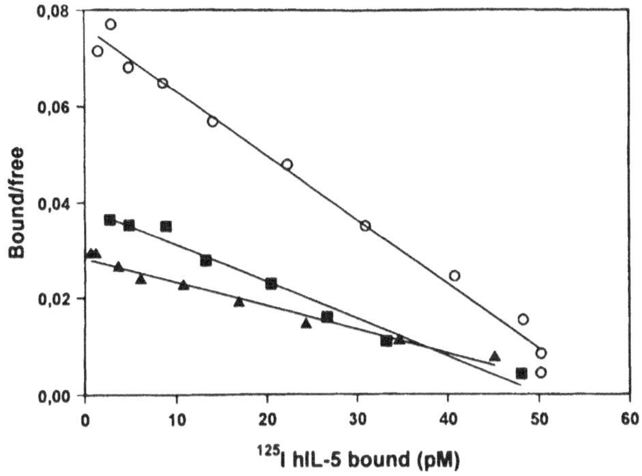

FIGURE 6. Nonreciprocal competition for the β-subunit (eoCML). catchard analysis of [^{125}I]hIL-5 binding on eosinophils from a patient with eosinophilic chronic myeloid leukemia. Eosinophilic granulocytes were isolated by density-gradient centrifugation from the peripheral blood of a single patient; 3×10^6 cells were incubated with increasing amounts of radiolabeled hIL-5. A 200-fold excess of cold IL-5 was included to determine nonspecific binding. Binding was determined in the absence (open circles, $K_d = 0.37$ nM, 1368 receptors/cell) or presence of excess cold GM-CSF (closed squares, $K_d = 0.92$ nM, 1182 receptors/cell) or IL-3 (closed triangles, $K_d = 1.13$ nM, 1350 receptors/cell). This experiment was performed in collaboration with Dr. Ivo Touw and Dr. Leo Budel (Daniel Den Hoed Kliniek, Rotterdam).

Consequently, the soluble hIL-5Rα subunit can act as an antagonist (see below). Although the data from solution binding experiments suggest the presence of only one α subunit in the complex, the stoichiometry of β_c subunits remains more elusive. It is also still unclear whether preformed $\alpha\beta_c$ complexes exist on the cell membrane. Reconstitution experiments using the GM-CSFRα subunit with β_c subunits in Cos 1 cells suggest that variability (e.g., $\alpha\beta_c$ or $\alpha\beta_c\beta_c$ complexes) can occur (Budel *et al.*, 1993).

3.2. Properties of the IL-5Rα and β-Subunit Proteins

Chemical cross-linking of hIL-5 on its receptor complex allows the detection of the α and β_c subunits as approximately 100-kDa and 160- to 170-kDa bands. Subtraction of the cross-linked IL-5 indicates sizes of 60 kDa and 120–130 kDa. The nucleotide sequence of hIL-5Rα cDNA predicts a polypeptide of 420 residues (Tavernier *et al.*, 1991, 1992; Murata *et al.*, 1992). It is characterised by a 20-residue N-terminal signal peptide, followed by a 322-amino-acid-long extracellular domain, a membrane anchor spanning 20 residues, and a 58-amino-acid-long cytoplasmic tail. The predicted molecular mass for the α chain is 45.5 kDa, suggesting N-linked glycosylation of one or more of the six potential N-glycosylation sites (and perhaps O-glycosylation). The β_c subunit totals 897 amino acids, including a predicted signal sequence of 16 residues, a 27-amino-acid-long membrane-spanning domain, and a cytoplasmic domain of 430 amino acids. This subunit is also likely glycosylated.

Both receptor chains belong to the cytokine/hematopoietin receptor superfamily (Bazan, 1990), which is characterized by a modular structure in which each module has a seven-β-sheet scaffold (deVos *et al.*, 1992). The hIL-5Rα has three such domains: a juxtamembrane module containing a canonical Trp-Ser-Xxx-Trp-Ser motif (often referred to as WS-x-WS box); a central module containing four conserved cysteines involved in two disulfide bridges; and a third N-terminal module that is related to the WS-x-WS-containing module (Tuypens *et al.*, 1992). The β_c subunit has a tandem array of the WS-x-WS- and cysteine-containing modules.

3.3. Properties of the IL-5Rα and β-Subunit Genes

The human IL-5Rα gene is located on chromosome 3 in the region 3p26, which is syntenic with the murine chromosome 6 location (Tuypens *et al.*, 1992; Isobe *et al.*, 1992). The human and mouse β_c subunit gene loci are at chromosomes 22q12.3–13.1 and at chromosome 15, respectively (Shen *et al.*, 1992; Gorman *et al.*, 1992). AIC2A and AIC2B are closely linked. The gene organization of both receptor subunits is highly conserved and reflects their relationship with the cytokine/hematopoietin receptor superfamily (Tuypens *et al.*, 1992; Imamura *et al.*, 1994). The structure of the gene for the human IL-5Rα subunit is shown in Figure 7.

The promoter of the hIL-5Rα gene was recently analyzed in detail (Sun *et al.*, 1995). By use of a luciferase reporter construct, the promoter was found to be fairly myeloid and eosinophil-lineage-specific. Furthermore, the region between -432 and -398 was implicated for the promoter activity in the eosinophilic HL-60-C115 cell line. This region does not contain consensus sequences for known transcription factors, suggesting myeloid or even eosinophil-specific regulatory elements. Further insight in the transcriptional regulation of this highly eosinophil/basophil-specific gene might help us to understand the critical steps involved in the commitment of the multipotential myeloid progenitors.

Human eosinophils express, through differential splicing, three different transcripts from the same IL5-Rα locus (Tavernier *et al.*, 1992; Tuypens *et al.*, 1992). As a result, in addition to the membrane-anchored receptor, two soluble isoforms can be produced. Intriguingly, one of these soluble variants is the predominant (>90%) transcript detected in mature, circulating eosinophils or in eosinophils obtained from cord blood cultures. This major soluble isoform arises from splicing to a soluble-specific exon, which precedes the exon encoding the transmembrane anchor (Tavernier *et al.*, 1992).

This soluble hIL-5Rα isoform can be produced in heterologous systems such as Cos 1 cells or Sf9 cells and has antagonistic properties *in vitro*. It inhibits the proliferation of IL-5-dependent cell lines and also blocks the IL-5-induced differentiation from human cord blood cultures (Tavernier *et al.*, 1991). So far. however, neither translation of the message encoding this soluble variant *in vitro* in eosinophils nor circulating soluble hIL-5Rα *in vivo* has been found. One possible explanation might be the thermolability of this soluble receptor. We have found that prolonged treatment at 37°C leads to irreversible adsorption of the protein to plastic supports (data not shown). Alternatively, this splice regulation could serve a regulatory function merely by modulating the generation of transcripts necessary for the expression of the transmembrane receptor. Some cDNAs encoding soluble receptor variants have also been detected in mouse B cells (Takaki *et al.*, 1990). They originate from a different splicing switch (Tavernier *et al.*, 1992; Imamura *et al.*, 1994). Soluble murine IL-5Rα also has antagonistic properties *in vitro*, albeit to a lesser degree than its human counterpart. This can probably be attributed to the lower affinity of the mIL-5/mIL-5Rα interaction (Tsuruoka *et al.*, 1993).

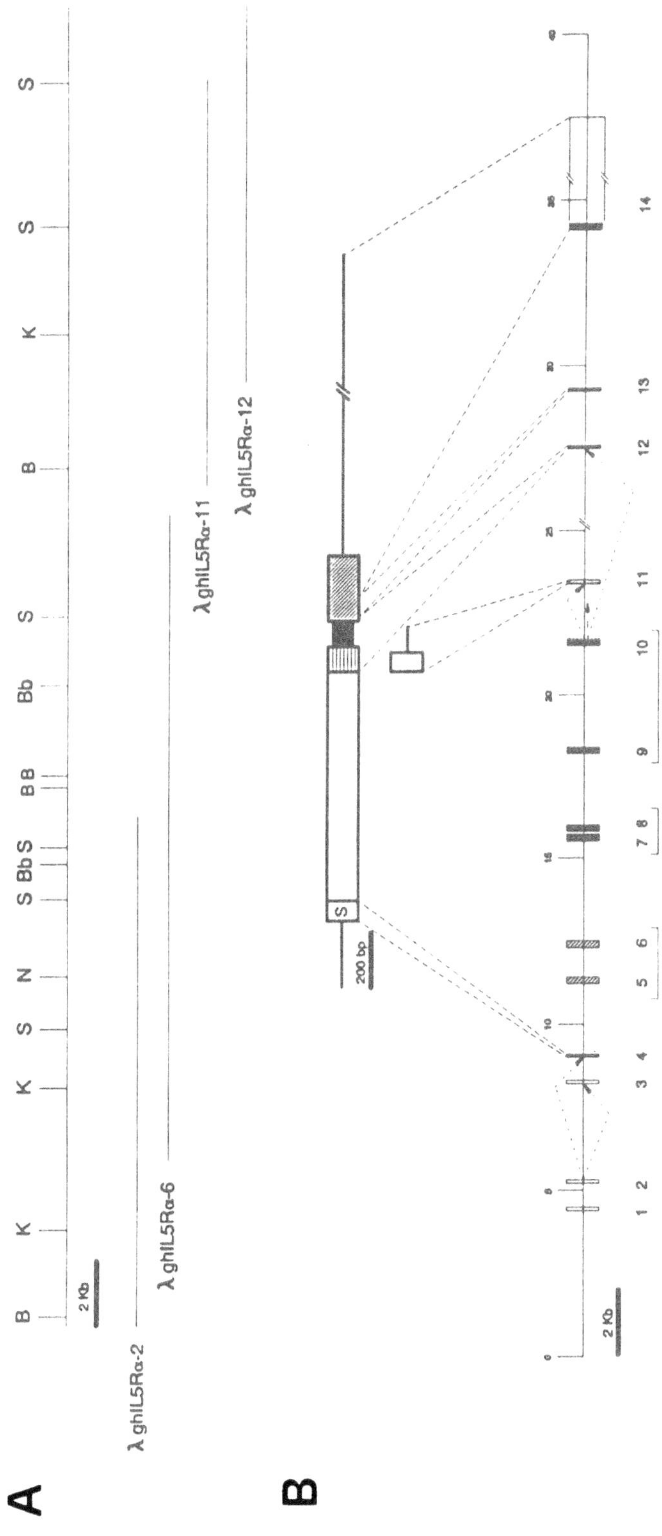

FIGURE 7. Structure of the gene for the human IL-5Rα subunit. Panel A: Restriction map and location of phage DNA inserts. S, SacI; K, KpnI; B, BamHI; Bb, Bpr PI; N, NheI. Panel B: Exon structure, relationship to functional domains, and alternative splicing pattern. Diagrams presenting two different cDNAs are shown on top. Above, membrane-anchored isoform; below, major soluble isoform; S, signal peptide. The genomic organization of the hIL-5Rα gene is shown at the bottom. Boxes represent exons and are numbered below. Paired exons with Roman numbering represent the three fibronectin type-III-like domains. Dashed lines correlate functional domains on the cDNAs with the corresponding exons. The 5′ and 3′ untranslated regions are shown by open boxes. With permission of European Cytokine Network.

4. THE INTERLEUKIN-5–INTERLEUKIN-5-RECEPTOR INTERACTION: STRUCTURE–FUNCTION RELATIONSHIPS

Insight into the interactions between IL-5 and its receptor components may ultimately lead to the development of agents that can interfere with either the α or β_c subunit. To date, structural information is limited to the ligand. Crystals of a hIL-5/shIL-5Rα complex have been obtained, but they diffract poorly (our own unpublished data). The elucidation of the three-dimensional structure of the heterotrimeric complex is still a distant goal. In contrast, much information has recently been obtained using mutagenesis studies on all three components.

Using a series of human/mouse chimeric molecules, Sanderson and colleagues have mapped the species-specific Rα interaction to the C-terminal one-third of human IL-5 (McKenzie *et al.*, 1991b). CNBr-cleavage studies confirmed this conclusion (Kodama *et al.*, 1991). We have recently performed a detailed structure–function analysis on hIL-5 (Tavernier *et al.*, 1995). Essentially three regions were identified as contributing to the interaction with the α subunit: the loop between β-strand 1 and α helix B, part of the β-strand 2, and the C-terminal region. However, as a consequence of the twofold symmetry of hIL-5, it is unclear at present whether the pharmacophore is formed by residues contributed by one or by both monomers. In both situations however, the areas identified are located close to the central axis of the hIL-5 dimer. This observation is compatible with the epitope mapping of a neutralizing (5A5) and a nonneutralizing (H30) monoclonal antibody directed against hIL-5, which bind either apically or more to the central axis of the β-strand 2, respectively (Van der Heyden *et al.*, in preparation).

Although the interaction with the α subunit is a prerequisite for receptor triggering, signaling occurs only upon association with the βc chain. A paradigm for cytokine/hematopoietin receptor activation is growth hormone (GH). Here, the ligand binds sequentially with two identical receptor subunits and induces their dimerization (Cunningham *et al.*, 1991; deVos *et al.*, 1992). Consequently, a growth hormone variant with undetectable binding to the second receptor subunit was found to have antagonistic properties (Fuh *et al.*, 1992). Mutations at position 13 of hIL-5 are deficient in their interaction with the β_c chain and likewise also exerted antagonistic activities (Tavernier *et al.*, 1994). This observation correlates well with previous reports in which homologous residues were identified in GM-CSF (residue 21; Lopez *et al.*, 1992b; Shanafelt *et al.*, 1991) and IL-3 (residue 22; Lopez *et al.*, 1992c).

To localize important hIL-5-binding regions on the hIL-5Rα subunit, we also exploited the differential cross-species activity of human and mouse IL-5. Cos 1 cells transfected with a series of chimeric receptors were tested for binding of both human IL-5, and mouse IL-5 as a control. We concluded that the first half of the N-terminal module contains most critical residues involved in species-specific ligand binding. By subsequently substituting all residues differing between human and mouse within this region, Cornelis *et al.* (1995) could demonstrate the critical role of residues D56 and E58. Using a solid-phase-based screening system, we have identified a class of compounds, isothiazolone derivatives, that are potent inhibitors of the hIL-5/hIL-5Rα interaction. These compounds act through a covalent modification of the sulfhydryl group of the free cysteine at position 66, again pointing to the important role of the N-terminal domain in ligand binding (Devos *et al.*, 1994). In addition to the N-terminal domain, both other domains need to be properly folded for ligand binding, as disruption of the disulfide bridges or of the WS-x-WS motif also abrogated ligand binding (our unpublished data).

The interaction with the β_c subunit causes a two- to threefold increase in hIL-5 binding affinity but gives a much more pronounced affinity conversion in the case of hGM-CSF (20- to 200-fold; Hayashida *et al.*, 1990) or hIL-3 (500-to 1000-fold; Kitamura *et al.*, 1991). This indicates that differences may exist in the way these three cytokines interact with the same β_c subunit and correlates with our observations on their nonreciprocal cross-competitions (see above). In a mutagenesis study, the importance of the region Y365–I368 was shown for the GM-CSF and IL-5, but not for IL-3 binding. These residues are located in a loop region and interact with the Glu 21 of hGM-CSF and, in analogy, possibly with the homologous Glu 13 of hIL-5 (Woodcock *et al.*, 1994).

5. SIGNAL TRANSDUCTION PATHWAY

The biological activities of cytokines on cells are mediated through their interaction with specific receptors that transmit the signal to the interior of the cell. Most cytokine receptors are multisubunit complexes. Some of their components are essential for intracytoplasmic signaling, whereas in some cases, others may merely be involved in ligand binding (Kishimoto *et al.*, 1994). Ligand binding can lead to the formation of homo- or heterodimers (Fukunaga *et al.*, 1991; Murakami *et al.*, 1993). This dimerization (oligomerization) of receptor molecules may be a common mechanism for receptor activation and signal transduction. Cross-linking of receptor subunits initiates a downstream signaling cascade that results in the expression of a limited set of immediate early genes. This ultimately leads to cell growth and/or differentiation. As we describe below, both IL-5 receptor subunits are required for signal transduction.

5.1. Interleukin 5 Induces Tyrosine Phosphorylation of Cellular Proteins

A number of growth factors bind to receptor protein tyrosine kinases and mediate a biological response by activating their intrinsic protein kinase activity (Ulrich and Schlessinger, 1990). In contrast, a large number of cytokines, including IL-5, GM-CSF, and IL-3, bind to receptors of the cytokine receptor superfamily. Although the cytoplasmic domains of these receptors show no homology with tyrosine kinase domains or with other known molecules or motifs involved in signaling, such as phosphatases, *src* homology domains, or nucleotide-binding proteins, cytokine binding induces rapid tyrosine phosphorylation of cellular proteins. Many tyrosine-phosphorylated proteins have been described in human and mouse cells stimulated with IL-5, GM-CSF, and IL-3 (Isfort and Ihle, 1990; Morla *et al.*, 1988; Murata *et al.*, 1990). The β_c chain, in contrast to the α chains, becomes tyrosine phosphorylated on receptor triggering (Sorensen *et al.*, 1989; Sakamaki *et al.*, 1992). The importance of tyrosine phosphorylation as a major event in signal transduction is illustrated by the observation that specific inhibitors of tyrosine kinases, such as herbimycin A, inhibit the growth response to these cytokines (Satoh *et al.*, 1992). Activation of tyrosine kinases leads to the activation of *ras* and phosphorylation of *raf*, a Ser/Thr kinase (Satoh *et al.*, 1991; Kanakura *et al.*, 1991). However, activation of *ras* or *raf* alone is not sufficient to abrogate the requirements for GM-CSF and IL-3 (Ihle *et al.*, 1995; Rein *et al.*, 1985), suggesting that these cytokines activate multiple signal transduction pathways. Although most attention has been focused on tyrosine kinases, the action of tyrosine phosphatases is also essential.

The common receptor β_c chain shared by these three cytokines provides a molecular basis for the observation that IL-5, IL-3, and GM-CSF induce similar phosphorylation patterns and have several overlapping biological activities, especially on eosinophils and basophils (Sanderson *et al.*, 1992). Both IL-3 and GM-CSF induce a growth signal in human erythroleukemic TFl cells (Kitamura *et al.*, 1989). Transfection of hIL-5Rα cDNA into these cells renders them also IL-5 responsive. As shown in Figure 8A, GM-CSF, IL-3, and IL-5 induce a similar pattern of tyrosine phosphorylation in these cells. The major phosphorylation target in these cells is a protein of around 90 kDa. Interestingly, this phosphorylation can occur at 4°C, suggesting that receptor, kinase, and substrate(s) are in close proximity to each other and may exist as a preformed complex. This and other experiments suggest that whenever receptors for these cytokines are expressed on a given cell type, the corresponding ligands elicit the same functional response. HL60 Cl15, a subclone of the promyelocytic leukemia cell line HL60, can be differentiated into eosinophilic cells. We have shown previously that these cells express IL-5 receptors (Plaetinck *et al.*, 1990). Although HL60 Cl15 cells are not growth factor dependent, IL-5 induces fast and transient tyrosine phosphorylation of a 130-kDa protein band (see Figure 8B), showing that phosphorylation can be uncoupled from cell-cycle regulation.

5.2. Identification of Receptor Domains

The β_c chain has a relatively long cytoplasmic domain (~430 amino acid residues; Itoh *et al.*, 1990; Gorman *et al.*, 1990; Hayashida *et al.*, 1990) compared to the α chains, which have rather short intracellular tails (~55 amino acid residues; Gearing *et al.*, 1989; Takaki *et al.*, 1990; Kitamura *et al.*, 1991; Tavernier *et al.*, 1991). It has been demonstrated

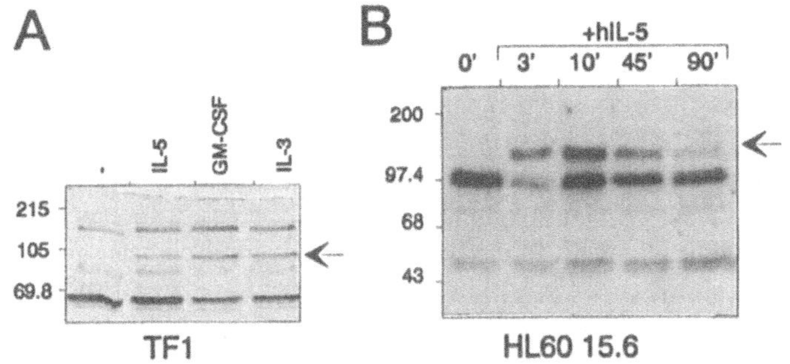

FIGURE 8. Interleukin-5-induced tyrosine phosphorylation patterns. Pattern of tyrosine phosphorylation in total cell lysates of hIL-5Rα-transfected TFl cells (A) and HL60 15.6 cells (B). About 4×10^6 TFl cells were washed, deprived of growth factors for 12 hr, and stimulated on ice with cytokines as indicated for 10 min. About 4×10^6 HL60 cells were incubated with IL-5 for different time periods at 37°C. Cells were lysed in a buffer containing 1% Triton X-100 and appropriate protease and phosphatase inhibitors. Phosphoproteins were detected by Western blotting with antiphosphotyrosine mAb 4G10 (UBI) and peroxidase-conjugated rabbit anti mouse Ab and visualized with the enhanced chemical luminescence technique (ECL, Pharmacia). Arrows indicate 90-kD (A) and 130-kD (B) phosphorylated proteins.

previously that the β_c chain is indispensable for signal transduction. Studies using a sequence of deletions in the human β_c receptor cytoplasmic domain have shown that at least two distinct regions are important for GM-CSF growth signaling (Sakamaki *et al.*, 1992; see Chapter 1). A membrane-proximal domain upstream of residue 517 containing a proline-rich element that is conserved among many cytokine receptors is essential for growth stimulation and for induction of *c-myc* and *pim-1*. A distal region of ~140 amino acid residues (between positions 626 and 763) is required for activation of *ras*, *Raf-1*, MAP kinase, and p70 S6 kinase as well as induction of *c-fos* and *c-jun* (Sato *et al.*, 1993).

On ligand binding, the β_c receptor molecule is tyrosine phosphorylated, although the significance of this event is unclear. Deletion analysis of the β_c receptor cytoplasmic domain suggests that the phosphorylation site is dispensable for mitogenic signaling (Sakamaki *et al.*, 1992). However, it might be important for transducing alternative signals involved in differentiation or activation. Because these cytokines induce tyrosine phosphorylation of SH2-containing proteins (Wang *et al.*, 1992; Corey *et al.*, 1993), phosphorylation of the receptor can generate binding sites for signaling molecules such as p85 PI_3 kinase, PLCγ, Vav, and Shc (Koch *et al.*, 1992). The latter is thought to link receptor triggering to *ras* signaling (Pelici *et al.*, 1992). Tyrosine phosphorylation may also be involved in down-regulation of the receptor, as phosphorylation of the mIL-3R increases its susceptibility to protease degradation (Mui *et al.*, 1992).

It has been reported that the cytoplasmic domains of the α subunits of GM-CSF/IL-3/IL-5 receptors are required for signal transduction (Sakamaki *et al.*, 1992; Takaki *et al.*, 1993; Weiss *et al.*, 1993). Deletion of the cytoplasmic domain of the α subunits does not affect the binding affinity but abrogates biological response, suggesting that the α subunits not only are required for cytokine binding but also are important for signal transduction. Moreover, it also opens the possibility that α receptor chains may transmit cytokine-specific signals. In order to map critical residues required for signal transduction in the cytoplasmic domain of the human IL-5Rα chain, we constructed a series of mutant receptors (Figure 9A). These were transfected in mouse BaF/3 cells expressing the resident mouse β_c chain (also known under the acronym AIC2B). All receptors were expressed and tested for IL-5 binding. As for the mouse receptor, the cytoplasmic domain of the hIL-5Rα subunit was found to be dispensable for ligand binding and association with the β_c chain (Figure 9B). BaF/3 cells transfected with wild-type receptor can proliferate with hIL-5. In contrast, when the hIL-5Rα lacking the entire cytoplasmic tail was used, transfected cells were unable to grow with IL-5. The cytoplasmic domain is encoded by two separate exons. The carboxy-terminal part, which corresponds to the second exon, shows only limited sequence conservation between man and mouse. Deletion of this segment did not affect growth signaling (Figure 9B). Within the region encoded by the first exon of the α chain of IL-5, a short fragment of homology with IL-3 and GM-CSF receptors can be observed (Kitamura *et al.*, 1991). This fragment contains a proline-rich element called "box 1" that was first identified in the cytoplasmic region of the IL-6 signal transducer gp130 and that is conserved in almost all members of the cytokine receptor family (Murakami *et al.*, 1991) including the β_c subunit. Substitutions of conserved prolines or hydrophobic amino acid residues within this element abrogated growth signaling. Mutations in conserved residues outside box 1 did not affect signal transduction.

By analogy with other cytokine receptor systems, it is likely that IL-5 induces heterodimers between α and β chains and that interaction of cytoplasmic regions of the

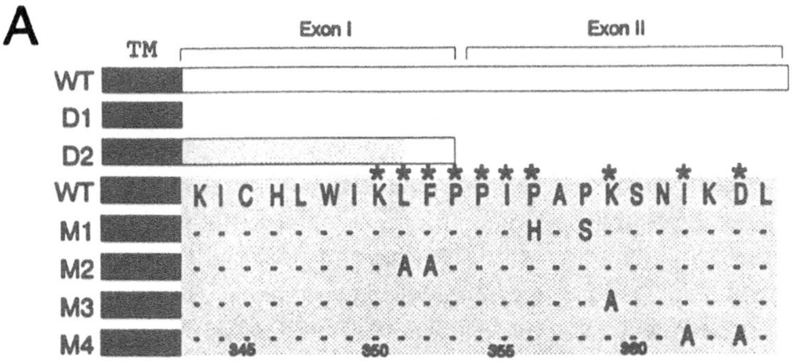

B

Ba/F3 transfection	high affinity hIL5 binding	hIL5 mβ crosslinking	proliferation		Jak2 activation		GRR EMSA	
			IL3	IL5	IL3	IL5	IL3	IL5
-	-	-	+	-	+	-	+	-
WT	+	+	+	+	+	+	+	+
D1	+	+	+	-	+	-	+	-
D2	+	nd	+	+	nd	nd	nd	nd
M1	+	+	+	-	+	-	nd	nd
M2	+	nd	+	-	+	-	nd	nd
M3	+	nd	+	+	nd	nd	nd	nd
M4	+	nd	+	+	nd	nd	nd	nd

FIGURE 9. Characteristics of mutant hIL-5Rα chains. A: Schematic representation of hIL-5 receptor α chains with mutations in the cytoplasmic domain. "WT" stands for "wild-type" receptor. D1 and D2 are truncated receptors lacking, respectively, the entire cytoplasmic domain and the region encoded by exon II. The gray shaded region spans residues 343 to 365. M1 to M4 are single or double point mutants, as indicated. The asterisks refer to residues that are conserved with IL-3Rα and GM-CSFRα. B: Properties of Ba/F3 cells transfected with the different hIL-5Rα constructs. Cells were cotransfected by electroporation with a vector coding for the receptor and a second plasmid carrying a selection marker. Stable transfectants were selected for puromycin resistance and [^{125}I]hIL-5 binding. Chemical cross-linking was carried out as described in Figure 5. Proliferation was measured by incorporation of tritiated thymidine. Jak2 activation was evaluated by immunoprecipitation with Jak2 antiserum and Western blot analysis with antiphosphotyrosine mAB as detailed in Figure 8. Induction of GRR DNA-binding activity was determined by EMSA (electrophoretic mobility shift assay) as for Figure 10. nd stands for "not done."

dimerized receptor components is important for signaling. However, receptor complexes involving more than two components cannot be excluded. In spite of its twofold symmetrical axis, human IL-5 dimer binds only one soluble receptor α subunit (Devos *et al.*, 1993). Therefore, the IL-5R complex likely contains only one α chain. The stoichiometry of binding of the β_c subunit is more elusive at present, also because association with β_c can be observed only with membrane-anchored receptors and not with soluble variants (our own personal observations). Although no physicochemical evidence for a β_c dimer has been found so far, the formation of a GM-CSF receptor complex consisting of one GM-CSF α chain and two β_c receptors has been suggested by Budel *et al.* (1993). A chimeric receptor consisting of the extracellular domain of the murine erythropoietin (EPO) receptor and the cytoplasmic region of AIC2A (a murine β_c subunit homologue) can transmit an EPO-dependent growth response, indicating that EPO-induced homodimerization of cytoplasmic β_c elements is sufficient for signaling (Zon *et al.*, 1992). On the other hand, the cytoplasmic region of IL-5Rα is absolutely required for signal transduction (Takaki *et al.*, 1993). It remains to be established whether the proline-rich sequence of IL-5Rα is essential for this dimerization (multimerization) of β_c or if it interacts directly with a downstream signaling molecule. In the latter model, cross-linking of two box-1-containing receptors, as is the case for both αβ heterodimerization or ββ homodimerization, should be sufficient to trigger signaling.

The above-described studies all evaluate the roles of various receptor components in growth stimulation. However, IL-5 induces not only proliferation but also differentiation in eosinophils. Further studies are required to determine whether IL-5-specific signal transduction pathways exist and, if so, which regions of the cytoplasmic domain of IL-5Rα are required.

5.3. Identification of Cytoplasmic Components

Despite the absence of kinase domains in their receptors, cytokines rapidly induce tyrosine phosphorylation of cellular substrate proteins as well as of the receptors. The rapid induction of tyrosine phosphorylation, the phosphorylation of receptor chains, and the detection of protein kinase activity in receptor immunoprecipitates have led to the hypothesis that a protein tyrosine kinase physically associates with the receptor and becomes activated on ligand binding. Recently, a number of studies have shown that cytokine receptors associate with and activate members of the Jak family of protein tyrosine kinases including Jak1, Jak2, Tyk2, Jak3 (previously called L-Jak) and probably more (Ihle *et al.*, 1994). Members of this family have tandem nonidentical catalytic domains and a large extracatalytic segment (Darnell *et al.*, 1994). Involvement of Jak2 tyrosine kinase in IL-3-mediated signal transduction has been reported (Silvennoinen *et al.*, 1993). In Ba/F3 cells transfected with hIL-5Rα, Jak2 also becomes phosphorylated on IL-5 stimulation (Figure 9B). Interestingly, phosphorylated β_c receptor could be coprecipitated with anti-Jak2 antibodies, suggesting that these molecules are physically associated. Mutations in the cytoplasmic domain of the hIL-5Rα that knock out biological activity also abrogate activation of Jak2 (Figure 9B). This shows that phosphorylation of Jak2 also requires an intact proline-rich box 1 sequence. Moreover, it is an indication that Jak2 is a crucial step in IL-5 signaling.

Activated Jaks phosphorylate the receptors as well as cytoplasmic proteins belonging to the family of DNA-binding proteins called STATs (signal transducers and activators of transcription) (Ihle *et al.*, 1994). These factors form a complex with homologous or heterologous subunits and migrate to the nucleus, where they bind to specific DNA regions. Importantly, tyrosine phosphorylation is required for complex formation and nuclear translocation. Interleukin 5 as well as IL-3 and GM-CSF induce phosphorylation of a 90-kDa protein in transfected TF1 cells (see Figure 8A). This size corresponds to what has been reported for STAT proteins. It has been reported that certain STAT proteins recognize the IFN-γ response region (GRR) located in the promoter of the FcγRI gene (Larner *et al.*,

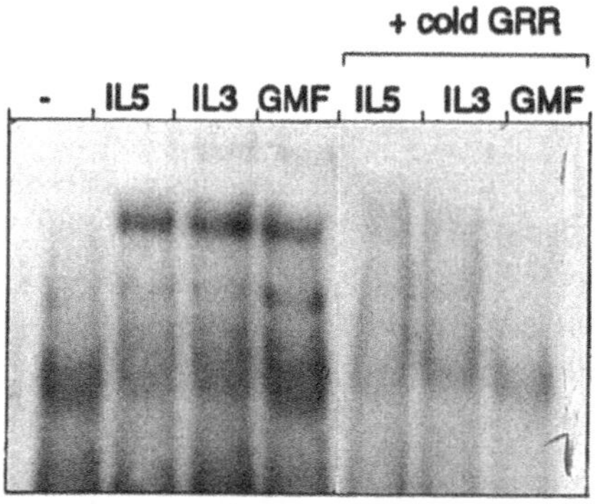

FIGURE 10. Band-shift assay. Activation of DNA-binding activity in cytoplasmic extracts of TF1 cells. Washed and growth-factor-deprived TF1 cells were stimulated on ice for 10 min with different growth factors as indicated (10^7/point). NP40 lysates were incubated with a ^{32}P labeled double-stranded oligonucleotide AGCATGTTTCAAG-GATTTGAGATGTATTTCCCAGAAAAG, corresponding to a fragment in the promotor of the human FcγRI gene. Protein–DNA complexes were analyzed by gel electrophoresis and autoradiography.

1993). Electrophoretic mobility shift assays with a ^{32}P-labeled GRR probe shows that IL-5, IL-3, and GM-CSF activate DNA-binding complexes in cytoplasmic extracts of TF1 cells (Figure 10). Remarkably, stimulation with cytokines and subsequent lysis were carried out at 4°C, suggesting that receptor, kinase, and DNA-binding proteins must be in intimate contact with each other at the moment of receptor activation.

In addition to tyrosine kinases it is evident that protein tyrosine phosphatases also play an essential role in signal transduction. Recently it has been shown that hematopoietic cell phosphatase (HCP) specifically associates with the β_c chain through its amino-terminal SH2 domain and is able to dephosphorylate the receptor chain (Yi *et al.*, 1993). Overexpression of HCP dramatically suppresses cell growth, indicating that it is essential for the down-regulation of the response. Its role in IL-5-mediated signaling is likely but remains to be shown.

6. INTERLEUKIN 5 AND EOSINOPHILS: *IN VITRO* OBSERVATIONS

6.1. Effects on Eosinophil Production

The effects of IL-5 on eosinophil production *in vitro* have mainly been studied by two methodologies: liquid culture systems and colony-formation assays in semisolid supports (starting from, e.g., bone marrow or cord blood cells). A straightforward, sensitive assay system based on liquid cultures was developed by Sanderson and collaborators (Strath *et al.*, 1985) and ultimately led to the identification and characterization of human IL-5 (Sanderson *et al.*, 1986).

The use of such assay systems to study the requirements for eosinophil proliferation and differentiation, however, did not yield a definitive picture so far. Three major conclusions could be drawn from the *in vitro* work. First, IL-5 is not the only cytokine able to stimulate eosinophil production *in vitro*. In fact, depending on the assay system used and the species analyzed, IL-3 and GM-CSF can act as equipotent or even better inducers of eosinophil production (e.g., Saito *et al.*, 1988; Clutterbuck *et al.*, 1989). Second, Sanderson and colleagues developed a two-step culturing system of human bone-marrow cells in which both methods were applied consecutively (Clutterbuck and Sanderson, 1990). They showed that IL-3 was able to increase the number of IL-5-responsive precursors. This study, together with the observation that IL-5 can induce a transient pulse of eosinophilic growth only in liquid bone-marrow cultures (indicating that IL-5 is not capable of stimulating eosinophil precursors) led to the proposition that IL-5 is a late-acting factor (see Sanderson, 1993). This view is supported by other studies using 5-fluorouracil (5-FU)-pretreated cell cultures (Yamaguchi *et al.*, 1988a,b; Warren and Moore, 1988) or using progenitor phenotyping (Ema *et al.*, 1990a). Third, in contrast to IL-3 and GM-CSF, which stimulate the production of eosinophils, neutrophils, and macrophages, IL-5 is basophil/eosinophil lineage specific (Saito *et al.*, 1988; Campbell *et al.*, 1988). The molecular basis for this difference lies in the expression pattern of the cytokine-specific receptor α subunits (see above).

However, caution must be taken when interpreting these data because (1) depending on the methodologies used, apparent discrepancies exist between the observations of different laboratories, and (2) there is a pronounced difference between the *in vitro* and *in vivo* observations (see Section 6.2). There is now little doubt that IL-5 is the central cytokine involved in eosinophil expansion *in vivo*. Furthermore, eosinophilia can occur *in vivo* without a change in the other lineages, possibly indicating the existence of an IL-5-dependent progenitor type. This observation argues against a critical role for other cytokines such as IL-3 in *in vivo* eosinophil development. It also follows from these studies that the IL-5 activity is not limited to late-stage progenitors but that IL-5 can promote the full pathway of eosinophil outgrowth. The causes underlying these discrepancies between the *in vitro* and *in vivo* data are not understood. Perhaps the *in vitro* culturing methods may not completely reflect the *in vivo* bone-marrow microenvironment.

6.2. Effects on Eosinophil Function

Interleukin 5 can affect the life span, motility, and effector functions of circulating eosinophils. Although most studies were performed using eosinophils purified from peripheral blood, more recently, useful continuous acute myelogenous leukemia (AML)-derived cell lines were established that can be induced to eosinophilic differentiation (Fischkoff, 1988; Fabian *et al.*, 1992; Paul *et al.*, 1993). Some reports also describe the use of primary cells from patients with AML responding to IL-5 (Ema, 1990b; Touw, 1991). It should be noted that all studies are performed on circulating eosinophils and thus exclude potential activating or priming effects occurring during transendothelial migration.

Interleukin 5 can promote the enhanced survival of eosinophils *ex vivo* by inhibition of the apoptotic process (Yamaguchi *et al.*, 1991). This phenomenon can be used as a very sensitive assay (Begley *et al.*, 1986; Owen *et al.*, 1989). Interestingly, transforming growth factor β (TGF-β) reverses this effect (Alam *et al.*, 1994).

Eosinophils have the ability to migrate toward sites of allergic inflammation. Three distinct steps involved in this phenomenon (adhesion, chemotaxis, and transendothelial migration) are influenced by IL-5. Interleukin 5 thus has chemotactic properties on eosinophils (Yamaguchi *et al.*, 1988b; Wang *et al.*, 1989), but is also able to prime eosinophils toward enhanced responsiveness for chemokines (IL-8 and RANTES; Schweizer *et al.*, 1994) and PAF (Numao *et al.*, 1989). It promotes eosinophil adhesion by up-regulation of adhesion markers (integrin CD11b; Walsh *et al.*, 1990). This LFA-1/ICAM-1 interaction has been shown to be a critical component for eosinophil tissue localization *in vivo* (Wegner *et al.*, 1990). In addition, eosinophil extravasation is also mediated via the VLA-4/VCAM-1 interaction (Walsh *et al.*, 1991; Pretolani *et al.*, 1994). Finally, Schleimer and collaborators have shown that eosinophil pretreatment with IL-5 leads to an enhanced transendothelial migration (Ebisawa *et al.*, 1994). The mechanisms underlying the selective accumulation of eosinophils is not yet fully understood but might be a result of a combination of the abovementioned factors. In addition to IL-5, an eosinophil-specific chemokine (eotaxin; Jose *et al.*, 1994) might also contribute to this eosinophil-specific migration.

Interleukin 5 modulates the functional status of eosinophils; both direct effects and priming effects have been described. Importantly, eosinophils are not (or very poorly) capable of responding in many *in vitro* assays unless they become activated by exogeneous stimuli such as IL-5, suggesting the role *in vivo* of such stimuli. Activation enhancement was observed in antibody-dependent tumor-cell killing- , phagocytosis-, and respiratory burst assays (Lopez *et al.*, 1986, 1988). Another important aspect of IL-5-mediated eosinophil activation is the synthesis and release of granule proteins (Fujisawa *et al.*, 1990; Carlson *et al.*, 1993; Paul *et al.*, 1994) and proteoglycan biosynthesis (Rothenberg *et al.*, 1988b). Again, whenever tested, IL-5 appears as an eosinophil-specific stimulus, whereas IL-3 and GM-CSF, which are equally potent on eosinophils, also display activities on other lineages.

Another marker for eosinophil activation that is influenced by IL-5 is the density status of the eosinophils. A shift from the normodense to the hypodense state correlates with activation of eosinophils, as seen by several criteria including leukotriene synthesis and enhanced parasite killing (DeSimone, 1982; Winqvist *et al.*, 1982; Prin *et al.*, 1984; Kajita, 1985). This lower gradient-sedimentation density is observed for eosinophils purified from the peripheral blood of patients suffering from asthma, helminthic infections, or hypereosinophilic syndromes. *In vitro* culturing of normodense eosinophils with IL-5 leads to a shift toward the hypodense state and a concomitant activation (Owen *et al.*, 1987; Rothenberg *et al.*, 1989). By use of antibodies against IL-5, its role in this activation *in vivo* could be demonstrated (Owen *et al.*, 1989, 1990).

7. INTERLEUKIN 5, EOSINOPHILS, AND DISEASE

In this section, we review some of the data available to date on the importance of IL-5 and eosinophils in the pathogenesis of human disease. Eosinophils can occur in association with numerous disease entities (Table I). We limit this overview to a few striking examples, namely, the hypereosinophilic syndrome, helminth infections, and allergic disorders, bronchial asthma in particular.

7.1. Interleukin 5 and Eosinophils: *In Vivo* Observations

Although a number of cytokines including IL-3 and GM-CSF can affect eosinophil production and activation *in vitro* (*vide supra*), IL-5 appears to be the most important cytokine in the development of eosinophil-associated diseases. In man, a substantial amount of data is present associating eosinophils with increased production of IL-5 or expression of its mRNA. Serum from patients with the hypereosinophilic syndrome contains increased levels of IL-5, but not IL-3 or GM-CSF (Owen *et al.*, 1989). Similarly, in patients with onchocerciasis, the posttreatment eosinophilia is accompanied by a rise in serum IL-5, but not IL-3 or GM-CSF (Limaye *et al.*, 1991). Increased amounts of IL-5 and mRNA encoding

Table I
Eosinophil-Derived Mediators

Granule constituents
Lysophospholipase (Charcot–Leyden crystal protein)
Cationic proteins: MBP, ECP, EPO, EDN
Arylsulfatase B, β-glucuronidase, collagenase, ...
Inflammatory mediators
PAF
LTC_4
PGE_1/PGE_2
Oxygen radicals
Substance P
Cytokines
IL-3, IL-5, GM-CSF
IL-1α, IL-6, IL-8, MIP-1α
TGF-α, TGF-β_1

for IL-5 are also detected in bronchoalveolar lavage (BAL) fluid and endobronchial biopsies from asthmatic subjects (in close correlation with eosinophil counts) (Walker *et al.*, 1992; Hamid *et al.*, 1991).

This descriptive analysis, however, does not provide proof of the functional importance of IL-5 in the induction of eosinophilia *in vivo*. Because of obvious ethical and practical limitations, evaluation of the causal link between increased IL-5 production and eosinophilia is, at present, possible only in animal models. Here, three different types of approaches provide strong evidence for IL-5 being the principal cytokine involved in eosinophilia *in vivo*.

1. Administration of IL-5 has been shown to cause eosinophilia *in vivo* in a variety of models (Van Oosterhout *et al.*, 1993; Iwama *et al.*, 1992).

2. Two groups have reported on the generation of IL-5 transgenic mice. Dent *et al.* (1990) used the human CD2 DCR (dominant control region) ligated to the mouse IL-5 gene in order to obtain T-cell-specific expression of IL-5, and Tominaga *et al.* (1991) used the mouse metallothionin (MT) promoter to drive the production of IL-5. The expression of IL-5 in both of these mice is sufficient to induce a life-long eosinophilia and suggests that IL-5 *in vivo* does induce the full pathway of eosinophil differentiation and production. As expected, the eosinophil distribution in the CD2-IL-5 mice was consistent with the predicted T-cell expression (bone marrow, spleen, and peritoneum). In addition to these sites, significant numbers of eosinophils were found in the liver and skeletal muscle of the MT-IL-5 transgenic mice. As pointed out by Sanderson (1993), this important difference between the two strains of mice demonstrates that IL-5 does not act solely systemically but might also be involved in the localization of eosinophils to the tissues where it is being produced. The IL-5-transgenic mice remained healthy, suggesting that expansion of this cell population had no pathological consequences. This shows that additional signals are required for triggering the degranulation of eosinophils, resulting in the release of tissue-damaging substances. An interesting observation was made by Strath *et al.* (1992), when the CD2-IL-5-transgenic mice were infected with *Mesocestoides corti*. Although the level of

IL-5 in these mice further increased during this infection, the number of eosinophils decreased during the period corresponding to high-level serum IL-5. This was shown to be caused by a decrease in eosinophil progenitor level and suggested that a mechanism is active that prevents the overproduction of eosinophils.

3. Converse to the two approaches described above, pretreatment with anti-IL-5 monoclonal antibodies completely abolishes parasite-induced eosinophilia both in blood and in tissues of mice (Coffman *et al.*, 1989). Experiments in mice and guinea pigs similarly indicate that allergen-induced airway eosinophilia can also be inhibited by pretreatment with anti-IL-5 monoclonal antibodies (Chand *et al.*, 1992; Gulbenkian *et al.*, 1992; Nakajima *et al.*, 1992; Van Oosterhout *et al.*, 1993; Mauser *et al.*, 1993).

Taken together, all these data strongly argue against the redundancy that could be anticipated from *in vitro* studies with IL-3 and GM-CSF and suggest that, *in vivo*, IL-5 is by far the most important cytokine involved in the regulation of eosinophilia. Apparently, IL-5 *in vivo* can support the full differentiation pathway of eosinophils, and its unique status may suggest the existence of a so far unidentified IL-5-dependent eosinophilic progenitor.

7.2. Eosinophils and Disease

Eosinophils are known to express a wide range of membrane receptors, including receptors for complement, immunoglobulins (IgG, IgE, IgA), and cytokines, in addition to adhesion molecules and HLA-DR molecules (Gleich *et al.*, 1991).

Upon stimulation, eosinophils have the capacity to release a variety of mediators, including inflammatory lipid mediators such as PAF or LTC_4, oxygen metabolites and the typical granule-derived cationic proteins (Table II). The release of these various substances may account for much of the pathological changes observed in a variety of diseases.

Table II
Eosinophil-Associated Diseases[a]

Parasitic infections
Allergic diseases
asthma, allergic rhinitis, urticaria, allergic bronchopulmonary aspergillosis (ABPA), drug hypersensitivity reactions
Toxic reactions
Eosinophilia myalgia syndrome (L-tryptophan), toxic oil syndrome
Neoplastic and myeloproliferative disorders
Idiopathic hypereosinophilic syndrome, eosinophilic leukemia, lymphoma, angioimmunoblastic lymphadenopathy, immunotherapy with interleukin-2
Connective tissue diseases
Hypersensitivity vasculitis, Churg–Strauss syndrome, hypereosinophilic fasciitis
Respiratory tract disorders
Eosinophilic pneumonia, Loeffler's syndrome, hypersensitivity pneumonitis
Skin disorders
Atopic dermatitis, scabies, eosinophilic cellulitis
Immunodeficiency syndromes
Wiskott–Aldrich, hyperimmunoglobulin E syndrome
Gastrointestinal diseases
Eosinophilic gastroenteritis, inflammatory bowel diseases

[a]Adapted from Mahanty and Nutman (1993).

It has become increasingly clear that in addition, eosinophils also produce a range of cytokines through which they can interact with other cells. Human eosinophils have been reported to contain mRNA coding for cytokines enhancing their own growth such as GM-CSF, IL-3, and IL-5 (Kita *et al.*, 1991; Moqbel *et al.*, 1991; Broide *et al.*, 1992; Dubucquoi *et al.*, 1994). Proinflammatory cytokines that can be formed by human eosinophils include IL-1α, IL-6, IL-8, TNF-α, and MIP-1α (Weller *et al.*, 1993; Hamid *et al.*, 1992; Braun *et al.*, 1993; Costa *et al.*, 1993). In addition, eosinophils also express mRNA for TGF-α and TGF-β_1, two cytokines believed to play an important role in the induction of fibrosis in a variety of disorders (Wong *et al.*, 1990; Ohno *et al.*, 1992; Elovic *et al.*, 1994).

Moreover, it has to be pointed out that in diseases such as bronchial asthma, helminth infections, or hypereosinophil syndrome, an increased number of circulating eosinophils display a hypodense phenotype, which is associated with eosinophil activation. On stimulation, these hypodense cells release larger quantities of mediators, such as LTC_4, and have an increased helminthotoxic effect. It is thought that the activation or priming of eosinophils is largely caused by growth factors such as IL-3, IL-5, and GM-CSF (Lopez *et al.*, 1988; Silberstein *et al.*, 1986; Rothenberg *et al.*, 1988b).

7.2.1. The Idiopathic Hypereosinophilic Syndrome

This is a syndrome of unknown etiology characterized by multiple organ dysfunction in the presence of a sustained increase of circulating eosinophils (>1500/mm^3 for at least 6 months) in the absence of known causes of eosinophilia (Fauci *et al.*, 1982). The multiple-organ dysfunction includes as one of its main features cardiac involvement. The heart shows signs of fibrosis, which are thought to result from the release of toxic metabolites and possibly profibrogenic cytokines such as TGF-β_1 from infiltrating eosinophils.

The hypereosinophilic syndrome (HES) is associated with increased levels of circulating IL-5 (Owen *et al.*, 1989). A major source of IL-5 is CD4$^+$ T-helper 2 (Th2)-like cells (Mosman and Coffman, 1989). The association of HES with clonal proliferation of type 2 helper T cells, potentially representing a premalignant condition, has recently been reported (Cogan *et al.*, 1994).

7.2.2. Parasitic Infections

The immunologic response to helminth infections commonly consists of Th2-like cytokine production, resulting in production of specific IgE and eosinophilia (Urban *et al.*, 1992).

Several lines of evidence implicate the eosinophil as an effector cell in the protection against these parasites. *In vitro* studies have demonstrated that eosinophil granule-derived cationic proteins can kill helminths (Butterworth *et al.*, 1979). Furthermore, *in vitro* studies have indicated that eosinophils have the ability to kill larval parasites through an antibody-dependent cell-mediated cytotoxicity reaction (ADCC) (Butterworth, 1984; Capron *et al.*, 1987). The role of eosinophils in the immunity to parasites was further substantiated by animal studies showing that polyclonal anti-eosinophil serum reduced *in vivo* immunity to *Schistosoma mansoni* (Mahmoud *et al.*, 1975).

The observation in man that a peak in the serum level of IL-5 precedes the increase in blood eosinophilia observed after treatment in patients infected with *Onchocerca* further supported the hypothesis of T helper 2 cell activation with IL-5 production and eosinophilia as the crucial element in protection against helminths (Limaye *et al.*, 1991). In line with this hypothesis, it was reported that in mice, infection with the *Trichuris muris* parasite induces a Th1-type response in strains unable to expel the parasite, whereas in resistant strains, a Th2-type response was observed (Else *et al.*, 1992).

Despite various observations demonstrating its helminthotoxic potential, the importance of the eosinophil as an effector cell in the Th2-type driven response to helminths remains a matter of debate. Animal studies, mainly in mice, have shown that pretreatment with monoclonal antibodies to IL-5 completely abolishes the blood eosinophilia and infiltration of eosinophils into tissues of animals infected with helminths such as *S. mansoni*. However, the eradication of the eosinophilia has no influence on the immunity of the animals to the parasites (Sher *et al.*, 1990). These and other studies (Herndon and Kayes, 1992; Urban *et al.*, 1992) would therefore argue against a functional role of eosinophils *in vivo* in the defense against helminths. Whether these findings can be extrapolated to human disease awaits further studies, allowing for direct interference with eosinophil function in man.

7.2.3. Bronchial Asthma

The disease process that probably has received most interest with regard to eosinophils as potential effector cells is bronchial asthma. Chronic mucosal airway inflammation is a cardinal feature of asthma and is characterized, among other elements, by the presence of activated, $EG2^+$ eosinophils in the bronchial mucosa (Djukanovic *et al.*, 1990). Airway inflammation has been attributed a central role in the pathogenesis of bronchial hyperresponsiveness (BHR), another key feature of asthma, although the precise functional importance of each different cell type and mediator remains to be fully established.

Eosinophils clearly have the potential to cause bronchoconstriction and alter airway responsiveness through the release of a variety of inflammatory mediators. These include the very potent bronchoconstrictor and proinflammatory sulfidopeptide leukotrienes, vasodilating prostaglandins, toxic oxygen radicals, and also the granule-derived cationic proteins, whose release is considered to be one of the main causes of epithelial damage (Table I). Major basic protein (MBP) and eosinophil cationic protein (ECP) are both known to be cytotoxic to bronchial epithelium (Gleich *et al.*, 1991). Both MBP and ECP have been immunolocalized to sites of epithelial damage in airway tissue from patients with severe asthma (Filly *et al.*, 1982; Dahl *et al.*, 1988). Increased levels of MBP have been reported in sputum collected during an acute exacerbation, reaching concentrations that are cytotoxic to bronchial epithelium *in vitro* (10 μg/ml)(Frigas *et al.*, 1981). Furthermore, MBP has been reported to increase responsiveness of guinea pig tracheal tissue *in vitro* (Flavahan *et al.*, 1988), whereas intraepithelial application of MBP in canine trachea augments the responsiveness of underlying smooth muscle to acetylcholine *in vivo* (Brofman *et al.*, 1989). It has subsequently also been shown on human tissue that MBP can interfere with binding of acetylcholine to the muscarinic M_2 receptor, thereby potentially enhancing vagally induced bronchoconstriction in asthma (Jacoby *et al.*, 1993).

Various descriptive studies have linked the presence of eosinophils and their cationic proteins to the pathogenesis of asthma: BAL fluid from patients with chronic asthma has been found to contain not only increased numbers of eosinophils but also increased amounts of MBP, and MBP levels and eosinophil counts correlated with the degree of bronchial hyperresponsiveness (Wardlaw *et al.*, 1988). Others have reported increased ECP levels in

BAL or serum from asthmatics, again in correlation with parameters of disease severity (Venge, 1994). A rise in serum levels was found to correlate with the increase in airway responsiveness during the pollen season in patients with seasonal allergic symptoms (Rak *et al.*, 1988). Conversely, BAL ECP levels decline in parallel with clinical improvement in steroid-treated patients (Adelroth *et al.*, 1990). Studies on bronchial biopsies taken from subjects with mild asthma further confirm the presence of activated eosinophils both in patients with intrinsic and extrinsic asthma (Jeffery *et al.*, 1989; Beasley *et al.*, 1989; Azzawi *et al.*, 1990; Bousquet *et al.*, 1990; Bentley *et al.*, 1992). Their presence has been linked to the local production of IL-5. The BAL fluid from asthmatics contains both increased levels of IL-5 (Walker *et al.*, 1992) and higher numbers of cells expressing mRNA coding for IL-5 (Robinson *et al.*, 1992).

Similarly, cells containing IL-5 mRNA were detected in bronchial biopsies from six of ten patients with mild asthma. The presence of IL-5 mRNA was correlated with an increased infiltration of the bronchial mucosa by eosinophils. Moreover, these patients tended to have more severe disease than those not expressing IL-5 mRNA (Hamid *et al.*, 1991). Finally, following an allergen inhalation challenge in atopic asthmatics, BAL and biopsies were shown to contain an increased number of cells expressing mRNA for IL-5, in association with increased numbers of $EG2^+$ eosinophils (Robinson *et al.*, 1993a; Bentley *et al.*, 1993). Conversely, a course of oral steroids diminished both eosinophil numbers and the number of cells expressing mRNA for IL-5 in bronchial biopsies (Robinson *et al.*, 1993b). In these various studies, the main source of IL-5 is thought to be the $CD4^+$ T helper cell, although other inflammatory cells such as mast cells and eosinophils themselves might be additional important sources of IL-5 (Broide *et al.*, 1992; Bradding *et al.*, 1993).

A substantial amount of evidence therefore supports an important role for IL-5-dependent eosinophil activation in the pathogenesis of asthma and the accompanying bronchial hyperresponsiveness. However, correlations among IL-5, eosinophils, and bronchial hyperresponsiveness do not prove a causal link *per se*. That IL-5 and eosinophils do indeed cause bronchial hyperresponsiveness remains to be proven by direct interference with their action. Again, to date, this is only possible in animal studies. In guinea pigs, it has been confirmed that administration of IL-5 causes an influx of eosinophils into the airways and an increase in airway responsiveness (Van Oosterhout *et al.*, 1993; Iwama *et al.*, 1992, 1993). Moreover, in the same model, pretreatment with anti-IL-5 monoclonal antibodies inhibits the antigen-induced eosinophil recruitment and airway hyperreactivity (Van Oosterhout *et al.*, 1993). In *Ascaris*-sensitive primates from the wild, pretreatment with monoclonal antibodies to ICAM-1, one of the adhesion molecules that is thought to play an important role in the recruitment of eosinophils from the bloodstream into the tissues (Resnick and Weller, 1993), again inhibits eosinophil recruitment to the airways and the increase in airway responsiveness following exposure to aerosolized *Ascaris* ova (Wegner *et al.*, 1990).

Finally, in mice, it was reported that not only anti-IL-5 but also depletion of $CD4^+$ T lymphocytes prevents antigen-induced airway hyperresponsiveness and pulmonary eosinophilia (Nakajima *et al.*, 1992; Gavett *et al.*, 1994). Taken together, all these data seem to confirm the hypothesis that eosinophil influx into the airways, driven by T helper 2 cell-derived IL-5 production, is the main cause of airway hyperresponsiveness. However, the pathogenesis of bronchial hyperresponsiveness might be more complicated, as demonstrated by several animal models. Although depletion of $CD4^+$ T helper cells in A/J mice inhibits antigen induced airway hyperreactivity, pretreatment of BALB/c mice with anti-IL-5 only inhibits allergen-induced airway eosinophilia but not the increase in airway responsiveness (Nagai *et al.*, 1993). Similarly, in a rat model, cyclosporin A was reported to inhibit the eosinophil influx following allergen challenge but not the increase in airway responsiveness (Elwood *et al.*, 1991).

These data therefore indicate that although IL-5-dependent eosinophil infiltration in the airways might contribute to the pathogenesis of bronchial hyperresponsiveness, additional factors seem to be involved, reflecting the complex etiology of this disease.

8. CONCLUSIONS

In the past decade, IL-5 has emerged as the principal cytokine acting on the eosinophil lineage. It acts as an eosinophil proliferation factor and as such is often referred to as an eosinophil-specific colony-stimulating factor (eo-CSF). It is also active on mature eosinophils, enhancing their effector capabilities and directing them to sites of allergic inflammation. During the same period studies in several animal model systems have highlighted the putative role of eosinophils in the tissue damage that accompanies allergic inflammation. Corroborating this IL-5–eosinophil–disease axis, recent reports have indeed shown a correlation between IL-5 and eosinophil-associated diseases in man.

Hence, taking into account the lineage specificity of IL-5, an IL-5 antagonist might have considerable clinical applications. The availability of such antagonist will also help to answer some remaining important questions. Can the role *in vivo* of IL-5 be compensated by the related cytokines IL-3 or GM-CSF in its absence? Indeed, in a case of acute lymphocytic leukemia with accompanying eosinophilia, activation of the IL-3 gene was reported (Meeker *et al.*, 1990). On the contrary, studies using transgenic mice or administration of IL-3 or GM-CSF indicated only marginal effects on the eosinophilic lineage. The complete inhibition of eosinophilia *in vivo* by antibodies directed against IL-5 supports this view. The availability of mice with a disruption of the IL-5 gene will further help to clarify this issue. Such mice will also be very useful to learn whether IL-5 has unexpected roles besides its effects on the eosinophil/basophil lineage (and B cells in the mouse). A related question is the role of eosinophils in phenomena other that allergic inflammation. What might be the effect of eliminating the eosinophil lineage for a prolonged time on other functions *in vivo*? The association of eosinophils with tumors has been described, in some cases accompanied by a positive correlation with prolonged survival (reviewed by Sanderson, 1992).

What are the possibilities of generating an IL-5 antagonist? Perhaps the most straightforward approach is to try to interfere with the interaction of IL-5 with its receptor components. Three different protein-based antagonists can be considered: IL-5 mutants (*vide supra*), the soluble IL-5Rα subunit, or Fab fragments of a humanized anti-hIL-5 monoclonal antibody. Alternatively, small-molecular-weight derivatives might be selected using high-flux screening assays. So far, with the exception of the highly toxic isothiazolone derivatives, such screening has been unsuccessful. It remains to be seen whether the extremely large interaction surface between IL-5 and its α subunit is amenable for blocking with small compounds. Perhaps the interaction with the β subunit could present a more manageable target. One can also envisage blocking the downstream signaling pathway, but no IL-5-specific signaling mechanism has been described so far (*vide supra*).

It should also be stressed that important gaps exist in our understanding of eosinophil proliferation and activation. The interconnections between the cytokine network and the different stages of eosinophil development are certainly not fully understood, and insight into this will require further refinement of both cellular and molecular approaches. For example, a remarkable observation of the past few years has been that eosinophils can synthesize both agonists (IL-5, GM-CSF) and antagonists (shIL-5Rα, TGF-β) of their own function. Are dysregulations of this agonism–antagonism balance involved in pathologies? Perhaps, over the longer term, understanding the production and release of these endogenous factors might lead to new targets for drug development. Examples of such approaches would be to influence the splicing pattern involved in the control of the receptor α-subunit isoforms or to modulate the expression pattern of the Th2 cytokine gene cluster. Evidently, such approaches will await advances in understanding gene regulation, but once they are within reach, applications can be expected in various areas.

ACKNOWLEDGMENTS. We wish to thank Dr. S. Ackerman, Dr. C. Bryson, and Dr. I. Touw for making unpublished data available to us; Prof. W. Fiers, Prof. R. Pauwels, and Dr. M. Steinmetz for their support and suggestions throughout our work; and Prof. C. Sanderson and S. Cornelis for many helpful discussions. The technical assistance of Tania Tuypens, Annick Verhee, Ina Faché, and Freya Van Houtte and the layout work of Johan Bostoen are also greatly appreciated.

9. REFERENCES

Adelroth, E., Rosenhall, L., Johansson, S-Å., Linden, M., and Venge, P., 1990, Inflammatory cells and eosinophilic activity in asthmatics investigated by bronchoalveolar lavage, *Am. Rev. Respir. Dis.* **142:**91–99.

Alam, R., Forsythe, P., Stafford, S., and Fukuda, Y., 1994, Transforming growth factor β abrogates the effects of hematopoietins on eosinophils and induces their apoptosis, *J. Exp. Med.* **179:**1041–1045.

Azuma, C., Tanabe, T., Konishi, M., Kinashi, T., Noma, T., Matsuda, F., Yaoita, Y., Takatsu, K., Hammarström, L., Smith, C. I. E., Severinson, E., and Honjo, T., 1986, Cloning of cDNA for human T-cell replacing factor (interleukin-5) and comparison with the murine homologue, *Nucleic Acids Res.* **14:**9149–9158.

Azzawi, M., Bradley, B., Jeffery, P. K., Frew, A. J., Wardlaw, A. J., Knowles, G., Assoufi, B., Collins, J. V., Durham, S., and Kay, A. B., 1990, Identification of T lymphocytes, macrophages and activated eosinophils in the bronchial mucosa in intrinsic asthma, *Am. Rev. Respir. Dis.* **147:**1407–1413.

Barry, S. C., McKenzie, A. N., Strath, M., and Sanderson, C. J., 1991, Analysis of IL5 receptors on murine eosinophils: A comparison with receptors on B13 cells, *Cytokine* **3:**339–344.

Basten, A., and Beeson, P. B., 1970, Mechanisms of eosinophilia. II. Role of the lymphocyte, *J. Exp. Med.* **131:**1288–1305.

Baumann, M. A., and Paul, C. C., 1992, Interleukin-5 is an autocrine growth factor for Epstein–Barr virus-transformed B lymphocytes, *Blood* **79:**1763–1767.

Bazan, J. F., 1990, Structural design and molecular evolution of a cytokine receptor superfamily, *Proc. Natl. Acad. Sci. U.S.A.* **87:**6934–6938.

Beasley, R., Roche, W. R., Roberts, J. A., and Holgate, S. T., 1989, Cellular events in the bronchi in mild asthma and after bronchial provocation, *Am. Rev. Respir. Dis.* **139:**806–817.

Begley C. G., Lopez, A. F., Nicola, N. A., Wang, E. A., Warren, D. J., Sanderson, C. J., Vadas, M. A., and Metcalf, D., 1986, Purified colony-stimulating factors enhance the survival of human neutrophils and eosinophils *in vitro*: A rapid and sensitive microassay for colony-stimulating factors, *Blood* **68:**162–166.

Bentley, A. M., Menz, G., Storz, C., Robinson, D. S., Bradley, S., Jeffery, P. K., Durham, S. T., and Kay, A. B., 1992, Identification of T lymphocytes, macrophages and activated eosinophils in the bronchial mucosa in intrinsic asthma, *Am. Rev. Respir. Dis.* **146:**500–506.

Bentley, A. M., Meng, Q., Robinson, D. S., Hamid, Q., Kay, A. B., and Durham, S. R., 1993, Increases in activated T lymphocytes, eosinophils, and cytokine mRNA expression for interleukin-5 and granulocyte/macrophage colony-stimulating factor in bronchial biopsies after allergen inhalation challenge in atopic asthmatics, *Am. J. Respir. Cell Mol. Biol.* **8:**35–42.

Blankenstein, T., Li, W., Ueberla, K., Qin, Z., Tominaga, A., Takatsu, K., Yamaguchi, N., and Diamantstein, T., 1990, Retroviral interleukin 5 gene transfer into interleukin 5-dependent growing cell lines results in autocrine growth and tumorigenicity, *Eur. J. Immunol.* **20:**2699–2705.

Bohjanen, P. R., Okajima, M., and Hodes, R. J., 1990, Differential regulation of interleukin 4 and interleukin 5 gene expression. A comparison of T-cell induction by anti-CD3 antibody or by exogenous cytokines, *Proc. Natl. Acad. Sci. U.S.A.* **87:**5283–5287.

Bousquet, J., Chanez, P., Lacoste, J. Y., Barnéon, G., Ghavanian, N., Enander, I., Venge, P., Ahlstedt, S., Simony-Lafontaine, J., Godard, P., and Michel, F.-G., 1990, Eosinophilic inflammation in asthma, *N. Engl. J. Med.* **323:**1033–1039.

Bradding, P., Feather, I. H., Wilson, S., Bardin, P. G., Heusser, C. H., Holgate, S. T., and Howarth, P. H., 1993, Immunolocalization of cytokines in the nasal mucosa of normal and perennial rhinitic subjects. The mast cell as a source of IL-4, IL-5 and IL-6 in human allergic mucosal inflammation, *J. Immunol.* **151:**3853–3865.

Braun, R. K., Franchini, M., Erard, R., Rihs, S., De Vries, I. J. M., Blaser, K., Hansel, T. T., and Walker, C., 1993, Human peripheral blood eosinophils produce and release interleukin-8 on stimulation with calcium ionophore, *Eur. J. Immunol.* **23:**956–960.

Brofman, J. D., White, S. R., Blake, J. S., Munoz, N. M., Gleich, G. J., and Leff, A. R., 1989, Epithelial augmentation of trachealis contraction caused by major basic protein of eosinophils, *J. Appl. Physiol.* **66:** 1867–1873.

Broide D. H., Paine, M. M., and Firestein, G. S., 1992, Eosinophils express interleukin 5 and granulocyte-macrophage colony stimulating factor mRNA at sites of allergic inflammation in asthmatics, *J. Clin. Invest.* **90:**1414–1424.

Budel, L. M., Hoogerbrugge, H., Pouwels, K., van Buitenen, C., Delwel, R., Löwenberg, B., and Touw, I. P., 1993, Granulocyte-macrophage colony-stimulating factor receptors alter their binding characteristics during myeloid maturation through up-regulation of the affinity converting β-subunit (KH97), *J. Biol. Chem.* **268:** 10154–10159.

Butterworth, A. E., 1984, Cell-mediated damage to helminths, *Adv. Parasitol.* **23:**143–235.

Butterworth, A. E., Wassom, D. L., Gleich, G. J., Loegering, D. A., and David J. R., 1979, Damage to schistosomula of *Schistosoma mansoni* induced directly by eosinophil major basic protein, *J. Immunol.* **122:**221–229.

Campbell, H. D., Tucker, W., Hort, Y., Martinson, M. E., Mayo, G., Clutterbuck, E., Sanderson, C. J., and Young, I., 1987, Molecular cloning, nucleotide sequence, and expression of the gene encoding human eosinophil differentiation factor (interleukin 5), *Proc. Natl. Acad. Sci. U.S.A.* **84:**6629–6633.

Campbell, H. D., Sanderson, C. J., Wang, Y., Hort, Y., Martinson, M. E., Tucker, W. Q. J., Stellwagen, A., Strath, M., and Young, I. G., 1988, Molecular cloning, nucleotide sequence and expression of the gene encoding murine eosinophil differentiation factor. A comparison with other eosinophilopoietic lymphokines and identity with BCGFII (interleukin-5), *Eur. J. Biochem.* **174:**345–352.

Capron, A., Dessaint, J. P., Capron, M., Ouma, J. H., and Butterworth, A. E., 1987, Immunity to schistosoma; progress towards vaccine, *Science* **238:**1065–1072.

Carlson, M., Peterson, C., and Venge, P., 1993, The influence of IL-3, IL-5, and GM-CSF on normal human eosinophil and neutrophil C3b-induced degranulation, *Allergy* **48:**437–442.

Chand, N., Harrison, J. E., Rooney, S., Pillar, J., Jakubicki, R., Nolan, K., Diamantis, W., and Sofia, R. D., 1992, Anti-IL-5 monoclonal antibody inhibits allergic late phase bronchial eosinophilia in guinea pigs: A therapeutic approach, *Eur. J. Pharmacol.* **211:**121–123.

Chandrasekharappa, S. C., Rebelsky, M. S., Firak, T. A., Le Beau, M. M., and Westbrook, C. A., 1990, A long-range restriction map of the interleukin-4 and interelukin-5 group on chromosome 5, *Genomics* **6:**94–99.

Chihara, J., Plumas, J., Gruart, V., Tavernier, J., Prin, L., Capron, A., and Capron, M., 1990, Characterization of a receptor for interleukin 5 on human eosinophils: Variable expression and induction by granulocyte/macrophage colony-stimulating factor, *J. Exp. Med.* **172:**1347–1351.

Clutterbuck, E. J., and Sanderson, C. J., 1990, Regulation of human eosinophil precursor production by cytokines: A comparison of recombinant human interleukin-1 (IL-1), rhIL-3, rhIL-5, rhIL-6 and rh granulocyte–macrophage colony-stimulating factor, *Blood* **75:**1774–1779.

Clutterbuck, E. J., Shields, J. G., Gordon, J., Smith, S. H., Boyd, A., Callard, R. E., Campbell, H. D., Young, I. G., and Sanderson, C. J., 1987, Recombinant human interleukin-5 is an eosinophil differentiation factor but has no activity in standard human B cell growth factor assays, *Eur. J. Immunol.* **17:**1743–1750.

Clutterbuck, E. J., Hirst, E. M. A., and Sanderson, C. J., 1989, Human interleukin-5 (IL-5) regulates the production of eosinophils in human bone marrow cultures: Comparison and interaction with IL-1, IL-3, IL-6 and GMCSF, *Blood* **73:**1504–1512.

Coffman, R. L., Seymour, B. W., Hudak, S., Lackson, J., and Renninck, D., 1989, Antibody to interleukin-5 inhibits helminth-induced eosinophilia in mice, *Science* **245:**308–310.

Cogan, E., Schandené, L., Crusiaux, A., Cochaux, P., Velu, T., and Goldman M., 1994, Clonal proliferation of type 2 helper T cells in a man with the hypereosinophilic syndrome, *N. Engl. J. Med.* **330:**535–538.

Corey, S., Eguinoa, A., Puyana-Theall, K., Bolen, J. B., Cantley, L., Mollinedo, F., Jackson, T. R., Hawkins, P. T., and Stephens, L. R., 1993, Granulocyte macrophage colony-stimulating factor stimulates both association and activation of phosphoinositide 3OH-kinase and *src*-related tyrosine kinase(s) in human myeloid derived cells, *EMBO J.* **12:**2681–2387.

Cornelis, S., Plaetinck, G., Devos, R., Van der Heyden, J., Tavernier, J., Sanderson, C. J., Guisez, Y., and Fiers, W., 1995, Detailed analysis of the IL5-IL5Rα interaction: Characterization of crucial residues on the legand and the receptor, *EMBO J.* **14:**3395–3402.

Corrigan, C. J., Haczku, A., Gemou-Engesaeth, V., Doi, S., Kikuchi, Y., Takatsu, K., Durham, S. R., and Kay, A. B., 1993, CD4 T-lymphocyte activation in asthma is accompanied by increased serum concentrations of interleukin-5. Effect of glucocorticoid therapy, *Am. Rev. Respir. Dis.* **147:**540–547.

Costa, J. J., Matossian, K., Beil, W. J., Wong, D. T. W., Gordon, J. R., Dvorak, A. M., Weller, P. F., and Galli, S. J., 1993, Human eosinophils can express the cytokines TNFα and MIP1-α, *J. Clin. Invest.* **91:**2673–2684.

Cunningham, B. C., Ultsch, M., deVos, A. M., Mulkerrin, M. G., Clauser, K. R., and Wells, J. A., 1991, Dimerization of the extracellular domain of the human growth hormone receptor by a single hormone molecule, *Science* **254:**821–825.

Dahl, R., Venge, P., and Fredens, K., 1988, Eosinophils, in *Asthma: Basic Mechanisms and Clinical Management* (P. J. Barnes, I. W. Rodger, and N. C. Thomson, eds.), pp. 115–129, Academic Press, London.

Darnell J. E., Kerr, M., and Stark, G. R., 1994, Jak–STAT pathways and transcriptional activation in response to IFNs and other extracellular signaling proteins, *Science* **264:**1415–1421.

Denburgh, J. A., 1992, Basophil and mast cell lineages *in vitro* and *in vivo*, *Blood* **79:**846–860.

Dent, L. A., Strath, M., Mellor, A. L., and Sanderson, C. J., 1990, Eosinophilia in transgenic mice expressing interleukin 5, *J. Exp. Med.* **172:**1425–1431.

DeSimone, C., Donell, G., Meli, D., Rosati, F., and Sorice, F., 1982, Human eosinophils and parasitic diseases. II. Charactization of two cell fractions isolated at different densities, *Clin. Exp. Immunol.* **48:**249–255.

Desreumaux, P., Janin, A., Colombel, J. F., Prin, L., Plumas, J., Emilie, D., Torpier, G., Capron, A., and Capron, M., 1992, Interleukin 5 mRNA expression by eosinophils in the intestinal mucosa of patients with coeliac disease, *J. Exp. Med.* **175:**293–296.

deVos, A. M., Ultsch, M., and Kossiakoff, A. A., 1992, Human growth hormone and extracellular domain of its receptor: Crystal structure of the complex, *Science* **255:**306–312.

Devos, R., Vandekerckhove, J., Rolink, A., Plaetinck, G., Van der Heyden, J., Fiers, W., and Tavernier, J., 1990, Amino acid sequence of a mouse interleukin 5 receptor protein reveals homology with a mouse interleukin 3 receptor protein, *Eur. J. Immunol.* **21:**1315–1317.

Devos, R., Plaetinck, G., Van der Heyden, J., Cornelis, S., Vandekerckhove, J., Fiers, W., and Tavernier, J., 1991, Molecular basis of a high affinity murine interleukin 5 receptor, *EMBO J.* **10:**2133–2137.

Devos, R., Guisez, Y., Cornelis, S., Verhee, A., Van der Heyden, J., Manneberg, M., Lahm, H.-W., Fiers, W., Tavernier, J., and Plaetinck, G., 1993, Recombinant soluble human interleukin-5 (hIL-5) receptor molecules. Cross-linking and stoichiometry of binding to IL-5, *J. Biol. Chem.* **268:**6581–6587.

Devos, R., Guisez, Y., Plaetinck, G., Cornelis, S., Tavernier, J., Van der Heyden, J., Foley, J., and Scheffler, J. E., 1994, Covalent modification of the interlmeukin-5 receptor by isothiazolones leads to inhibition of the binding on interleukin-5, *Eur. J. Biochem.* **225:**635–640.

Dexter, T. M., Allen, T. D., and Lajtha, L. G., 1977, Conditions controlling the proliferation of haemopoietic stem cells in culture, *J. Cell. Physiol.* **91:**335–344.

Djukanovic, R., Roche, W. R., Wilson, J. W., Beasley, C. R. W., Twentyman, O. P., Howarth, P. H., and Holgate, S. T., 1990, Mucosal inflammation in asthma, *Am. Rev. Respir. Dis.* **142:**434–457.

Dubucquoi, S., Desreumaux, P., Janin, A., Klein, O., Goldman, M., Tavernier, J., Capron, A., and Capron, M., 1994, Interleukin 5 synthesis by eosinophils: Association with granules and immunoglobulin-dependent secretion, *J. Exp. Med.* **179:**703–708.

Ebisawa, M., Liu, M. C., Yamada, T., Kato, M., Lichtenstein, L. M., Bochner, B. S., and Schleimer, R. P., 1994, Eosinophil transendothelial migration induced by cytokines. II. Potentiation of eosinophil transendothelial migration by eosinophil-active cytokines, *J. Immunol.* **152:**4590–4596.

Elovic, A., Wong, D. T. W., Weller, P. F., Mattosian, K., and Galli, S. J., 1994, Expression of TGF-α and TGF-β1 mRNA and product by eosinophils in nasal polyps, *J. Allergy Clin. Immunol.* **93:**864–869.

Else, K. J., Hultner, L., and Grencis, R. K., 1992, Cellular immune responses to the murine nematode parasite Trichuris muris. II: Differential induction of Th-cell subsets in resistant versus susceptible mice, *Immunology* **75:**232–237.

Elwood, W., Lotvall, J. O., Barnes, P. J., and Chung K. F., 1992, Effect of dexamethasone and cyclosporin A on allergen-induced airway hyperresponsiveness and inflammatory cell responses in sensitized brown-Norway rats, *Am. Rev. Respir. Dis.* **145:**1289–1294.

Ema, H., Suda, T., Nagayoshi, K., Miura, Y., Civin, C. I., and Nakauchi, H., 1990a, Target cells for granulocyte colony-stimulating factor, interleukin-3 and interleukin-5 in differentiation pathways of neutrophils and eosinophils, *Blood* **76:**1956–1961.

Ema, H., Kitano, K., Suda, T., Sato, Y., Muroi, K., Ohta, M., Yoshida, M., Sakamoto, S., Eguchi, M., and Miura, Y., 1990b, *In vitro* differentiation of leukemic cells to eosinophils in the presence of interleukin-5 in two cases of acute myeloid leukemia with the translocation (8;21)(q22;q22), *Blood* **75:**350–356.

Enokihara, H., Nagashima, S., Noma, T., Kajitani, H., Hamaguchi, H., Saito, K., Furusawa, S., Shishido, H., and Honjo, T., 1988, Effect of human recombinant interleukin 5 and G-CSF on eosinophil colony formation, *Immunol. Letters* **18:**73–76.

Fabian, I., Lass, M., Kletter, Y., and Golde, D. W., 1992, Differentiation and functional activity of human eosinophilic cells from an eosinophil HL-60 subline: Response to recombinant hematopoietic growth factors, *Blood* **80:**788–794.

Fauci, A. S., Harley J. B., Roberts W. C., Ferrans, V. J., Gralnick H. R., and Bjornson, B. H., 1982, The idiopathic hypereosinophilic syndrome, *Ann. Intern. Med.* **97:**78–92.

Filley, W. V., Holley, K. E., Kephart, G. M., and Gleich, G. J., 1982, Identification by immunofluorescence of eosinophil granule major basic protein in lung tissues of patients with bronchial asthma, *Lancet* **2:**11–16.

Finkelman, F. D., Madden, K. B., Cheever, A. W., Katona, I. M., Morris, S. C., Gately, M. K., Hubbard, B. R., Gause, W. C., and Urban, J. F., 1994, Effects of interleukin 12 on immune responses and host protection in mice infected with intestinal nematode parasites, *J. Exp. Med.* **179:**1563–1572.

Fischkoff, S. A., 1988, Graded increase in probability of eosinophilic differentiation of HL-60 promyelocytic leukemia cells induced by culture under alkaline conditions, *Leuk. Res.* **12:**679–688.

Flavahan, N. A., Slifman, N. R., Gleich, G. J., and Vanhoutte, P. M., 1988, Human eosinophil major basic protein causes hyperreactivity of respiratory smooth muscle: Role of the epithelium, *Am. Rev. Respir. Dis.* **138:**685–688.

Frigas, E., Loegering, D. A., Solley, G. O., Farrow, G. M., and Gleich, G. J., 1981, Elevated levels of the eosinophil granule major basic protein in the sputum of patients with bronchial asthma, *Mayo Clin. Proc.* **56:**345–353.

Fuh, G., Cunningham, B. C., Fukunaga, R., Nagata, S., Goeddel, D. V., and Wells, J. A., 1992, Rational design of potent antagonists to the human growth hormone receptor, *Science* **256:**1677–1680.

Fujisawa, T., Abu-Ghazaleh, R., Kita, H., Sanderson, C. J., and Gleich, G. J., 1990, Regulatory effect of cytokines on eosinophil degranulation, *J. Immunol.* **144:**642–646.

Fukunaga, R., Ishizada-Ikeda, E., Pan, C.-X., Seto, Y., and Nagata, S., 1991, Functional domains of the granulocyte colony-stimulating factor, *EMBO J.* **10:**2855–2865.

Gavett, S. H., Chen, X., Finkelman, F., and Wills-Karp, M., 1994, Depletion of murine CD4$^+$ T lymphocytes prevents antigen-induced airway hyperreactivity and pulmonary eosinophilia, *Am. J. Respir. Cell Mol. Biol.* **10:**587–593.

Gearing, D. P., King, J. A., Gough, N. M., and Nicola, N. A., 1989, Expression cloning of a receptor for human granulocyte–macrophage colony stimulating factor, *EMBO J.* **8:**3667–3676.

Gleich, G. J., Adolphson, C. R., and Leiferman, K. M., 1991, The biology of the eosinophilic leukocyte, *Annu. Rev. Med.* **44:**85–101.

Gorman, D. M., Itoh, N., Kitamura, T., Schreurs, J., Yonehara, S., Arai, K., and Miyajima, A., 1990, Cloning and expression of a gene encoding an interleukin 3 receptor-like protein: Identification of another member of the cytokine family, *Proc. Natl. Acad. Sci. U.S.A.* **87:**5459–5465.

Gorman, D. M., Itoh, N., Jenkins, N. A., Gilbert, D. J., Copeland, N. G., and Miyajima, A., 1992, Chromosomal localization and organization of the murine genes encoding the β subunits (AIC2A and AIC2B) of the interleukin 3, granulocyte/macrophage colony-stimulating factor, and interleukin 5 receptors, *J. Biol. Chem.* **267:**15842–15848.

Graber, P., Bernard, A. R., Hassell, A. M., Milburn, M. V., Jordan, S. R., Proudfoot, A. E. I., and Fattah, D., and Wells, T. N. C., 1993, Purification, characterisation and crystallisation of selenomethionyl recombinant human interleukin-5 from *Escherichia coli*, *Eur. J. Biochem.* **212:**751–755.

Guisez, Y., Oefner, C., Winkler, F. K., Schlaeger, E.-J., Zulauf, M., Van der Heyden, J., Plaetinck, G., Cornelis, S., Tavernier, J., Fiers, W., Devos, R., and D'Arcy, A., 1993, Expression, purification and crystallization of fully active, glycosylated human interleukin-5, *FEBS Lett.* **331:**49–52.

Gulbenkian, A. R., Egan, R. W., Fernandez, X., Jones, H., Kreutner, W., Kung, T., Payvandi, F., Sullivan, L., Zurcher, J. A., and Watnik, A. S., 1992, Interleukin-5 modulates eosinophil accumulation in allergic guinea pig lung, *Am. Rev. Respir. Dis.* **146:**263–265.

Hamid, Q., Azzawi, M., Ying, S., Moqbel, R., Wardlaw, A. J., Corrigan, C. J., Bradley, B., Durham, S. R., Collins, J. V., Jeffery, P. K., Quint, D. J., and Kay, A. B., 1991. Expression of mRNA for interleukin-5 in mucosal biopsies from asthma, *J. Clin. Invest.* **87:**1561–1566.

Hamid, Q., Barkans, J., Meng, Q., Ying, S., Abrams, J. S., Kay, A. B., and Moqbel, R., 1992, Human eosinophils synthesize and secrete interleukin-6 *in vitro*, *Blood* **80:**1496–1501.

Hassell, A. M., Wells, T. N. C., Graber, P., Proudfoot, A. E. I., Anderegg, R. J., Burkhart, W., Jordan, S. R., and Milburn, M. V., 1993, Crystallization and preliminary X-ray diffraction studies of recombinant interleukin-5, *J. Mol. Biol.* **229:**1150–1152.

Hayashida, K., Kitamura, T., Gorman, D. M., Arai, K., Yokota, T., and Miyajima, A., 1990, Molecular cloning of a second subunit of the GM-CSF receptor. Reconstitution of a high affinity GM-CSF receptor, *Proc. Natl. Acad. Sci. U.S.A.* **87:**9655–9659.

Herndon, F. J., and Kayes, S. G., 1992, Depletion of eosinophils by anti-IL5 monoclonal antibody treatment of mice infected with *Trichinella spiralis* does not alter parasite burden or immunologic resistance to reinfection, *J. Immunol.* **149:**3642–3647.

Hsu, C. K., Hsu, S. H., Whitney, R. A., and Hansen, C. T., 1976, Immunopathology of schistomsomiasis in athymic mice, *Nature* **262:**397–399.

Ihle, J. N., Keller, J., Rein, A., Cleveland, J., and Rapp, U., 1995, in *Cancer Cells 3–Growth Factors and Transformation* (J. Feramiso, ed.), pp. 211–219, Cold Spring Harbor, New York.

Ihle, J. N., Witthuhn, B. A., Quelle, F. W., Yamamoto, K., Thierfelder, W. E., Kreider, B., and Silvennoinen, O., 1994, Signaling by the cytokine receptor superfamily: Jaks and STATs, *Trends Biochem. Sci.* **19:**222–227.

Imamura, F., Takaki, S., Akagi, K., Ando, M., Yamamura, K. I., Takatsu, K., and Tominaga, A., 1994, The murine interleukin-5 receptor α-subunit gene: Characterization of the gene structure and chromosome mapping, *DNA Cell. Biol.* **13:**283–292.

Ingley, E., and Young, I. G., 1991, Characterization of a receptor for interleukin-5 on human eosinophils and the myeloid leukemia line HL-60, *Blood* **78:**339–344.

Ingley, E., Cutler, R. L., Fung, M.-C., Sanderson, C. J., and Young, I.G., 1991, Production and purification of recombinant human interleukin-5 from yeast and baculovirus expression systems, *Eur. J. Biochem.* **196:** 623–629.

Isfort, R. J., and Ihle, J. N., 1990, Multiple hematopoietic growth factors signal through tyrosine phosphorylation, *Growth Fact.* **2:**213–220.

Isobe, M., Kumura, Y., Murata, Y., Takaki, S., Tominaga, A., Takatsu, K., and Ogita, Z., 1992, Localization of the gene encoding the α subunit of human interleukin-5 receptor (IL5Rα) to chromosome region 3p24–3p26, *Genomics* **14:**755–757.

Itoh, N., Yonehara, S., Schreurs, J., Gorman, D. M., Maruyama K., Ishii, A., Yahara, I., Arai, K., and Miyajima, A., 1990, Cloning of an interleukin-3 receptor gene: A member of a distinct receptor gene family, *Science* **247:** 324–327.

Iwama, T., Nagai, H., Suda, H., Tsuruoka, N., and Koda, A., 1992, Effect of murine recombinant interleukin-5 on the cell population in guinea-pig airways, *Br. J. Pharmacol.* **105:**19–22.

Iwama, T., Nagai, H., Tsuruoka, N., and Koda, A., 1993, Effect of murine rcombinant interleukin-5 on bronchial reactivity in guinea-pigs, *Clin. Exp. Allergy* **23:**32–38.

Jacoby, D. G., Gleich, G. J., and Fryer, A. D., 1993, Human eosinophil major basic protein is an endogenous allosteric antagonist at the inhibitory muscarinic M_2 receptor, *J. Clin. Invest.* **91:**1314–1318.

Jeffery, P. K., Wardlaw, A. J., Nelson, F. C., Collins, J. V., and Kay, A. B., 1989, Bronchial biopsies in asthma, *Am. Rev. Respir. Dis.* **140:**1745–1753.

Jose, P. J., Griffiths-Johnson, D. A., Collins, P. D., Walsh, D. T., Moqbel, R., Totty, N. F., Truong, O., Hsuan, J. J., and Williams, T. J., 1994, Eotaxin: A potent eosinophil chemoattractant cytokine detected in a guinea pig model of allergic airway inflammation, *J. Exp. Med.* **179:**881–887.

Kajita, T., Yui, Y., Mita, H., Taniguchi, N., Saito, H., Mishima, T., and Shida, T., 1985, Relese of leukotriene C_4 from eosinophils and its relation to the cell density, *Int. Arch. Allergy Appl. Immunol.* **78:**406–410.

Kanakura, Y., Druker, B., Wood, K. W., Mamon, H. J., Okuda, K., Roberts, T. M., and Griffin, J. D., 1991, Granulocyte–macrophage stimulating factor and interleukin 3 induce rapid phosphorylation and activation of the proto-oncogene *Raf-1* in a human factor-dependent myeloid cell line, *Blood* **77:**243–250.

Kay, B., Ying, S., Varney, V., Gaga, M., Durham, S. R., Moqbel, R., Wardlaw, A. J., and Hamid, Q., 1991, Messenger RNA expression of the cytokine gene cluster, interleukin 3 (IL-3), IL-4, IL-5, and granulocyte–macrophage colony stimulating factor, in allergen-induced late-phase cutaneous reactions in atopic subjects, *J. Exp. Med.* **173:**775–778.

Kinashi, T., Harada, N., Severinson, E., Tanabe, T., Sideras, P., Konishi, M., Azuma, C., Tominaga, A., Bergstedt-Lindqvist, S., Takahashi, M., Matsuda, F., Yaoita, Y., Takatsu, K., and Honjo, T., 1986, Cloning of cDNA encoding T-cell replacing factor and identity with B-cell growth factor II, *Nature* **324:**70–73.

Kishimoto, T., Taga, T., and Akira, S., 1994, Cytokine signal transduction, *Cell* **76:**253–262.

Kita, H., Ohnishi, T., Okubo, Y., Weiler, D., Abrams, J. S., and Gleich, G. J., 1991, Granulocyte/macrophage colony-stimulating factor and interleukin 3 release from human peripheral blood eosinophils and neutrophils, *J. Exp. Med.* **174:**745–748.

Kitamura, T., Tange, T., Terasawa, T., Chiba, S., Kuwaki, T., Miyagawa, K., Piao, Y.-F., Miyazono, K., Urabe, A., and Takaku, F., 1989, Establishment and characterization of a unique human cell line that proliferates dependently on GM-CSF, IL-3, or erythropoietin, *J. Cell. Physiol.* **140:**323–334.

Kitamura, T., Sato, N., Arai, K., and Miyajima, A., 1991, Expression cloning of the human IL-3 receptor cDNA reveals a shared β subunit for human IL-3 and GM-CSF receptors, *Cell* **66:**1165–1174.

Kluck, P. M. C., Wiegant, J., Raap, A. K., Vrolijk, H., Tanke, H. J., Willemze, R., and Landegent, J. E., 1993, Order of human hematopoietic growth factor and receptor genes on the long arm of chromosome 5, as determined by fluorescence *in situ* hybridization, *Ann. Hematol.* **66:**15–20.

Koch, C. A., Anderson, D., Moran, M. F., Ellis, C., and Pawson, T., 1992, SH2 and SH3 domains: Elements that control interactions of cytoplasmic signaling proteins, *Science* **252:**668–671.

Kodama, S., Tsuruoka, N., and Tsujimoto, M., 1991, Role of the C-terminus in the biological activity of hIL5, *Biochem. Biophys. Res. Commun.* **178:**514–519.

Kodama, S., Endo, T., Tsujimoto, M., and Kobata, A., 1992, Characterization of recombinant murine interleukin 5 expressed in Chinese hamster ovary cells, *Glycobiology* **2:**419–427.

Kodama, S., Tsujimoto, M., Tsuruoka, N., Sugo, T., Endo, T., and Kobata, A., 1993, Role of sugar chains in the *in vitro* activity of recombinant human interleukin 5, *Eur. J. Biochem.* **211:**903–908.

Kunimoto, D. Y., Allison, K. C., Watson, C., Fuerst, T., Armstrong, G. D., Paul, W., Strober, W., 1991, High-level production of murine interleukin-5 (IL-5) utilizing recombinant baculovirus expression. Purification of the rIL-5 and its use in assessing the biological role of IL-5 glycosylation, *Cytokine* **3:**224–230.

Larner, A. C., David, M., Feldman, G. M., Igarachi, K.-I., Hackett, R. H., Webb, D. S. A., Sweitzer, S. M., Petricoin, E. F. III, and Finbloom, D. S., 1993, Tyrosine phosphorylation of DNA binding proteins by multiple cytokines, *Science* **261:**1730–1733.

Lee, H., Koyano-Nakagawa, N., Naito, Y., Nishida, J., Arai, N., Arai, K. I., and Yokota, T., 1993, cAMP activates the IL-5 promoter synergistically with phorbol ester through the signaling pathway involving protein kinase A in mouse thymoma line EL-4, *J. Immunol.* **151:**6135–6142.

Limaye, A. P., Abrams, J. S., Silver, J. E., Awadzi, K., Francis, H. F., Ottesen, E. A., and Nutman, T. B., 1991, Interleukin-5 and the posttreatment eosinophilia in patients with onchocerciasis, *J. Clin. Invest.* **88:**1418–1421.

Lopez, A. F., Begley, C. G., Williamson, D. J., Warren, D. J., Vadas, M. A., and Sanderson, C. J., 1986, Murine eosinophil differentiation factor. An eosinophil-specific colony-stimulating factor with activity on human cells, *J. Exp. Med.* **163:**1085–1099.

Lopez, A. F., Sanderson, C. J., Gamble, J. R., Campbell, H. D., Young, I. G., and Vadas, M. A., 1988, Recombinant human interleukin-5 is a selective activator of human eosinophil function, *J. Exp. Med.* **167:**219–224.

Lopez, A. F., Vadas, M. A., Woodcock, J. M., Milton, S. E., Lewis, A., Elliott, M. J., Gillis, D., Ireland, R., Olwell,

E., and Park, L. S., 1991, Interleukin-5, interleukin-3, and granulocyte–macrophage colony-stimulating factor cross-compete for binding to cell surface receptors on human eosinophils, *J. Biol. Chem.* **266:**24741–24747.

Lopez, A. F., Elliott, M. J., Woodcock, J., and Vadas, M. A., 1992a, GM-CSF, IL-3 and IL-5: Cross-competition on human haemopoietic cells, *Immunol. Today* **13:**495–500.

Lopez, A. F., Shannon, M. F., Hercus, T., Nicola, N. A., Cambareri, B., Dottore, M., Layton, M. J., Eglinton, L., and Vadas, M. A., 1992b, Residue 21 of human granulocyte–macrophage colony-stimulating factor is critical for biological activity and for high but not low affinity binding, *EMBO J.* **11:**909–916.

Lopez, A. F., Shannon, M. F., Barry, S., Phillips, J. A., Cambareri, B., Dottore, M., Simmons, P., and Vadas, M. A., 1992c, A human interleukin 3 analog with increased biological and binding activities, *Proc. Natl. Acad. Sci. U.S.A.* **89:**11842–11846.

Mackenzie, D. T., Filutowitz, H. I., Swain, S. L., and Dutton, R. W., 1987, Purification and partial sequence analysis of murine B cell growth factor II (interleukin 5), *J. Immunol.* **139:**2661–2668.

Maggi, E., Biswas, P., Del Prete, G., Parronchi, P., Macchia, D., Simonelli, C., Emmi, L., De Carli, M., Tiri, A., Ricci, M., and Romagnani, S., 1991, Accumulation of Th-2-like helper T-cells in the conjunctiva of patients with vernal conjunctivitis, *J. Immunol.* **146:**1169–1174.

Mahanty, S., and Nutman, T. B., 1993, Eosinophilia and eosinophil related disorders, in *Allergy, Principles and Practice*, 4th ed. (E. Middleton, C. E. Reed, E. F. Ellis, N. F. Adkinson, J. W. Yunginnger, and W. W. Busse, eds.), pp. 1077–1108, Mosby Year Book, St. Louis.

Mahmoud, A. A. F., Warren, K. S., and Peters, P., 1975, A role for the eosinophil in acquired resistance to *Schistosoma mansoni* infection as determined by anti-eosinophil serum, *J. Exp. Med.* **142:**805–813.

Mauser, P. J., Pitman, A., Witt, A., Fernandez, X., Zurcher, J., Kung, T., Jones, H., Watnick, A. S., Egan, R. W., Kreutner, W., and Adams, G. K., 1993, Inhibitory effect of the TRFK-5 anti-IL-5 antibody in a guinea pig model of asthma, *Am. Rev. Respir. Dis.* **148:**1623–1627.

McKenzie, A. N. J., Ely, B., and Sanderson, C. J., 1991a, Mutated interleukin-5 monomers are biologically inactive, *Mol. Immunol.* **28:**155–158.

McKenzie, A. N. J., Barry, S. C., Strath, M., and Sanderson, C. J., 1991b, Structure–function analysis of interleukin-5 utilizing mouse/human chimeric molecules, *EMBO J.* **10:**1193–1199.

Meeker, T. C., Hardy, D., Willman, C., Hogan, T., and Abrams, J., 1990, Activation of the interleukin-3 gene by chromosomal translocation in acute lymphocytic leukemia with eosinophilia, *Blood* **76:**285–289.

Metcalf, D., Parker, J. W., Chester, H. M., and Kincade, P. W., 1974, Formation of eosinophilic-like granulocytic colonies by mouse bone marrow cells *in vitro*, *J. Cell. Physiol.* **84:**275–290.

Migita, M., Yamaguchi, N., Mita, S., Higuchi, S., Hitoshi, Y., Yoshida, Y., Tomonaga, M., Matsuda, I., Tominaga, A., and Takatsu, K., 1991, Characterization of the human IL-5 receptors on eosinophils, *Cell. Immunol.* **133:** 484–497.

Milburn, M. V., Hassell, A. M., Lambert, M. H., Jordan, S. R., Proudfoot, A. E. I., Graber, P., and Wells, T. N. C., 1993, A novel dimer configuration revealed by the crystal structure at 2.4 Å resolution of human interleukin-5, *Nature* **363:**172–176.

Minamitake, Y., Kodama, S., Katayama, T., Adachi, H., Tanaka, S., and Tsujimoto, M., 1990, Structure of recombinant human interleukin 5 produced by Chinese hamster ovary cells, *J. Biochem.* **107:**292–297.

Mita, S., Harada, N., Naomi, S., Hitoshi, Y., Sakamoto, K., Akagi, M., Tominaga, A., and Takatsu, K., 1988, Receptors for T cell-replacing factor (TRF)/interleukin 5 (IL-5): Specificity, quantitation and its implication, *J. Exp. Med.* **167:**863–878.

Mita, S., Hosoya, Y., Kubota, I., Nishihara, T., Honjo, T., Takahashi, T., and Takatsu, K., 1989a, Rapid methods for purification of human recombinant interleukin-5 (IL-5) using the anti-murine IL-5 antibody-coupled immuno-affinity column, *J. Immunol. Methods* **125:**233–241.

Mita, S., Tominaga, A., Hitoshi, Y., Honjo, T., Sakamoto, K., Akagi, M., Kikuchi, Y., and Takatsu, K., 1989b, Characterization of high-affinity receptors for interleukin 5 (IL-5) on IL-5 dependent cell lines, *Proc. Natl. Acad. Sci. U.S.A.* **86:**2311–2315.

Miyatake, S., Shlomai, J., Arai, K., and Arai, N., 1991, Characterization of the mouse GM-CSF promoter: Nuclear factors that interact with an element shared by three lymphokine genes: GM-CSF, IL-4 and IL-5, *Mol. Cell. Biol.* **11:**5894–5901.

Mizuta, T., Tanabe, T., Nakakubo, H., Noma, T., and Honjo, T., 1988, Molecular cloning and structure of the mouse interleukin-5 gene, *Growth Factors* **1:**51–57.

Moqbel, R., Hamid, Q., Ying, S., Barkans, J., Hartnell, A., Tsicopoulos, A., Wardlaw, A. J., and Kay, A. B., 1991,

Expression of mRNA and immunoreactivity for the granulocyte/macrophage colony-stimulating factor in activated human eosinophils, *J. Exp. Med.* **174:**749–752.

Morla, A. O., Schreurs, J., Miyajima, A., and Wang, Y. J., 1988, Hematopoietic growth factors activate the tyrosine phosphorylation of distinct sets of proteins in interleukin-3-dependent murine cell lines, *Mol. Cell. Biol.* **8:**2241–2218.

Mosman, T. R., and Coffman, R. L., 1989, TH_1 and TH_2 cells: Different patterns of lymphokine secretion lead to different functional properties, *Annu. Rev. Immunol.* **7:**145–173.

Mui, A. L., Kay, R. J., Humphries, R. K., and Krystal, G., 1992, Ligand-induced phosphorylation of the murine interleukin 3 receptor signals its cleavage, *Proc. Natl. Acad. Sci. U.S.A.,* **89:**10812–10815.

Murakami, M., Narazaki, M., Hibi, M., Yawata, H., Yasukawa, K., Hamaguchi, M., Taga, T., and Kishimoto, T., 1991, Critical cytoplasmic region of the IL-6 signal transducer, gp130, is conserved in the cytokine receptor family, *Proc. Natl. Acad. Sci. U.S.A.* **88:**11349–11353.

Murakami, M., Hibi, M., Nakagawa, N., Nakagawa, T., Yasukawa, K., Yamanishi, K., Taga, T., and Kishimoto, T., 1993, IL-6 induced homodimerization of gp 130 and associated activation of a tyrosine kinase, *Science* **260:**1808–1810.

Murata, Y., Yamaguchi, N., Hitoshi, Y., Tominaga, A., and Takatsu, K., 1990, IL5 and IL3 induce serine and tyrosine phosphorylations of several cellular proteins in an IL5-dependent cell line, *Biochem. Biophys. Res. Commun.* **173:**1102–1108.

Murata, Y., Takaki, S., Migita, M., Kikuchi, Y., Tominaga, A., and Takatsu, K., 1992, Molecular cloning and expression of the human interleukin 5 receptor, *J. Exp. Med.* **175:**341–351.

Nagai, H., Yamaguchi, S., Inagaki, N., Tsuruoka, N., Hitoshi, Y., and Takatsu, K., 1993, Effect of anti-IL5 monoclonal antibody on allergic bronchial eosinophilia and airway hyperresponsiveness in mice, *Life Sci.* **53:**243–247.

Nakajima, H., Iwamoto, I., Tomoe, S., Matsumura, R., Tomioka, H., Takatsu, K., and Yoshida, S., 1992, $CD4^+$ T-lymphocytes and interleukin-5 mediate antigen-induced eosinophil infiltration into the mouse trachea, *Am. Rev. Respir. Dis.* **146:**374–377.

Naora, H., and Young, I. G., 1994, Mechanisms regulating the mRNA levels of interleukin-5 and two other coordinately expressed lymphokines in the murine T lymphoma EL4.23, *Blood* **83:**3620–3628.

Nicola, N. A., and Metcalf, D., 1991, Subunit promiscuity among hemopoietic growth factor receptors, *Cell* **67:** 1–4.

Numao, T., Fukuda, T., Akutsu, I., Makino, S., Enokihara, H., and Honjo, T., 1989, Selective enhancement of eosinophil chemotaxis by recombinant human interleukin 5, *J. Allergy Clin. Immunol.* **83:**298.

Ohno, I., Lea, R. G., Flanders, K. C., Clark, D. A., Banwattt, D., Dolovich, J., Denburg, J., Harley, C. B., Gauldie, J., and Jordana, M., 1992, Eosinophils in chronically inflamed human upper airway tissues express transforming growth factor β1 gene (TGFβ1), *J. Clin. Invest.* **89:**1662–1668.

Ono, S., Hayashi, S., Takahama, Y., Dobashi, K., Katoh, Y., Nakanishi, K., Paul, W. E., and Hamaoka, T., 1986, Identification of two distinct factors, B151-TRF1 and B151-TRF2, inducing differentiation of activated B cells and small resting B cells into antibody-producing cells, *J. Immunol.* **137:**187–196.

Owen, W. F., Rothenberg, M. E., Silberstein, D. S., Gasson, J. C., Stevens, R. L., Austen, K. F., and Soberman, R. J., 1987, Regulation of human eosinophil viability, density, and function by granulocyte/macrophage colony-stimulating factor in the presence of 3T3 fibroblasts, *J. Exp. Med.* **166:**129–141.

Owen, W. F., Rothenberg, M. E., Petersen, J., Stevens, R. L., Silberstein, D. S., McKenzie, D. T., Austen, K. F., and Owen W. F., 1989, Interleukin 5 and phenotypically altered eosinophils in the blood of patients with the idiopathic hypereosinophilic syndrome, *J. Exp. Med.* **170:**343–348.

Owen, W. F., Petersen, J., Sheff, D. M., Folkerth, R. D., Anderson, R. J., Corson, J. M., Sheffer, A. L., and Austen, K. F., 1990, Hypodense eosinophils and interleukin 5 activity in the blood of patients with the eosinophilia–myalgia syndrome, *Proc. Natl. Acad. Sci. U.S.A.* **87:**8647–8651.

Paul, C. C., Tolbert, M., Mahrer, S., Singh, A., Grace, M. J., and Baumann, M. A., 1993, Cooperative effects of interleukin-3 (IL-3), IL-5 and granulocyte–macrophage colony-stimulating factor: A new myeloid cell line inducible to eosinophils, *Blood* **81:**1193–1199.

Paul, C. C., Ackerman, S. J., Mahrer, S., Tolbert, M., Dvorak, A., and Baumann, M. A., 1994, Cytokine induction of granule protein synthesis in an eosinophil-inducible human myeloid cell line, AML14, *J. Leukocyte Biol.* **56:**74–79.

Pelicci, G., Lanfrancone, L., Grignani, F., McGlade, J., Cavallo, F., Forni, G., Nicoletti, I., Grignani, F., Pawson, T., and Pelicci, P. G., 1992, A novel transforming protein (SHC) with an SH2 domain is implicated in mitogenic signal transduction, *Cell* **70:**93–99.

Plaetinck, G., Van der Heyden, J., Tavernier, J., Faché, I., Tuypens, T., Fischkoff, S., Fiers, W., and Devos, R., 1990, Characterization of interleukin 5 receptors on eosinophilic sublines from human promyelocytic leukemia (HL60) cells, *J. Exp. Med.* **172:**683–691.

Plaut, M., Pierie, J. H., Watson, C. J., Hanley-Hyde, J., Nordan, R. P., and Paul, W. E., 1989, Mast cell lines produce lymphokines in response to cross-linkage of $Fc_{\varepsilon}RI$ or to calcium ionophores, *Nature* **339:**64–67.

Pretolani, M., Ruffié, C., Lapa e Silva, J.-R., Joseph, D., Lobb, R. R., and Vargaftig, B. B., 1994, Antibody to very late activation antigen 4 prevents antigen-induced bronchial hyperreactivity and cellular infiltration in the guinea pig airways, *J. Exp. Med.* **180:**795–805.

Prin, L., Charon, J., Capron, M., Gosset, P., Taelman, H., Tonnel, A.-B., and Capron, A., 1984, Heterogeneity of human eosinophils. II. Variability of respiratory burst activity related to cell density, *Clin. Exp. Immunol.* **57:** 735–742.

Proudfoot, A. E. I., Fattah, D., Kawashima, E. H., Bernard, A., and Wingfield, P. T., 1990, Preparation and characterization of human interleukin-5 expressed in recombinant *Escherichia coli*, *Biochem. J.* **270:** 357–361.

Proudfoot, A. E. I., Davies, J. G., Turcatti, G., and Wingfield, P. T., 1991, Human interleukin-5 expression in *Escherichia coli*: Assignment of the disulfide bridges of the purified unglycosylated protein, *FEBS Lett.* **283:**61–64.

Rak, S., Lowhagen, D., and Venge, P., 1988, The effect of immunotherapy on bronchial hyperresponsiveness and eosinophilic cationic protein in pollen-allergic patients, *J. Allergy Clin. Immunol.* **82:**470–480.

Rein, A., Keller, J., Schultz, A. M., Holmes, K. L., Medicus, R., and Ihle, J. N., 1985, Infection of immune mast cells by Harvey sarcoma virus: Immortalization without loss of requirement for interleukin-3, *Mol. Cell. Biol.* **5:**2257–2264.

Resnick, M. B., and Weller, P. F., 1993, Mechanisms of eosinophil recruitment, *Am. J. Respir. Cell Mol. Biol.* **8:**349–355.

Robinson, D. S., Hamid, Q., Ying, S., Tsicopoulos, A., Barkans, J., Bentley, A., Corrigan, C., Durham, S., and Kay, B., 1992, Predominant TH2-like bronchoalveolar T-lymphocyte population in atopic asthma, *N. Engl. J. Med.* **326:**298–304.

Robinson, D., Hamid, Q., Bently, A., Ying, S., Kay, B., and Durham, S., 1993a, Activation of $CD4^{+}$ T cells, increased T_{H2}-type cytokine mRNA expression, and eosinophil recruitment in bronchoalveolar lavage after allergen inhalation challenge in patients with atopic asthma, *J. Allergy Clin. Immunol.* **92:**313–324.

Robinson, D., Hamid, Q., Ying, S., Bentley, A., Assoufi, B., Durham, S., and Kay, A. B., 1993b, Prednisolone treatment in asthma is associated with modulation of bronchoalveolar lavage cell interleukin-4, interleukin-5 and interferon gamma cytokine gene expression, *Am. Rev. Respir. Dis.* **148:**401–406.

Rolink, A., Melchers, F., and Palacios, R., 1989, Monoclonal antibodies reactive with the mouse interleukin 5 receptor, *J. Exp. Med.* **169:**1693–1705.

Romagnani, S., 1994, Human TH1 and TH2 subsets: "Eppur si muove," *Eur. Cytokine Netw.* **5:**7–12.

Rothenberg, M. E., Pomerantz, J. L., Owen, W. F., Avraham, S., Soberman, R. J., Austen, K. F., and Stevens, R. L., 1988, Characterization of a human eosinophil proteoglycan, and augmentation of its biosynthesis and size by interleukin 3, interleukin 5 and granulocyte/macrophage colony-stimulating factor, *J. Biol. Chem.* **263:** 13901–13908.

Rothenberg, M. E., Petersen, J., Stevens, R. L., Silberstein, D. S., McKenzie, D. T., Austen, K. F., and Owen, W. F., 1989, IL-5-dependent conversion of normodense human eosinophils to the hypodense phenotype uses 3T3 fibroblasts for enhanced viability, accelerated hypodensity, and sustained antibody-dependent cytotoxicity, *J. Immunol.* **143:**2311–2316.

Saito, H., Hatake, K., Dvorak, A. M., Leiferman, K. M., Donnenberg, A. D., Arai, N., Ishizaka, K., and Ishizaka, T., 1988, Selective differentiation and proliferation of hematopoietic cells induced by recombinant human interleukins, *Proc. Natl. Acad. Sci. U.S.A.* **85:**2288–2292.

Sakamaki, K., Miyajima, I., Kitamura, T., and Miyajima, A., 1992, Critical cytoplasmic domains of the common beta subunit of the human GM-CSF, IL-3 and IL-5 receptors for growth signal transduction and tyrosine phosphorylation, *EMBO J.* **11:**3541–3545.

Samoszuk, M., and Nansen, L., 1990, Detection of interleukin-5 mRNA in Reed–Sternberg cells of Hodgkin's disease with eosinophilia, *Blood* **75:**13–16.

Sanderson, C. J., 1992, Interleukin-5, eosinophils, and disease, *Blood* **79:**3101–3109.

Sanderson, C. J., 1993, Interleukin-5 and the regulation of eosinophil production, in *Immunopharmacology of Eosinophils*, Academic Press, Orlando, pp. 11–24.

Sanderson, C. J., O'Garra, A., Warren, D. J., and Klaus, G. G. B., 1986, Eosinophil differentiation factor also has B-cell growth factor activity: Proposed name interleukin 4, *Proc. Natl. Acad. Sci. U.S.A.* **83:**437–440.

Sanderson, C. J., Warren, D. J., and Strath, M., 1985, Identification of a lymphokine that stimualtes eosinophil differentiation in vitro: Its relationship to IL3, and functional properties of eosinopils produced in cultures, *J. Exp. Med.* **162:**60–74.

Sato N., Sakamaki, K., Terada, N., Arai, K.-I., and Miyajiman, A., 1993, Signal transduction by the high affinity GM-CSF receptor: Two distinct cytoplasmic regions of the common β subunit responsible for different signaling, *EMBO J.* **12:**4181–4189.

Satoh, T., Nakafuku, M., Miyajima, A., and Kaziro, Y., 1991, Involvement of Ras p21 protein in signal-transduction pathways from interleukin-2, interleukin-3, and granulocyte–macrophage colony stimulating factor, but not from interleukin-4, *Proc. Natl. Acad. Sci. U.S.A.* **88:**3314–3318.

Satoh, T., Uehara, Y., and Kaziro, Y., 1992, Inhibition of interleukin 3 and granulocyte–macrophage colony-stimulating factor stimulated increase of active *ras*-GTP by herbimycin A, a specific inhibitor of tyrosine kinase, *J. Biol. Chem.* **267:**2537–2541.

Schweizer, R. C., Welmers, B., Raaijmakers, J., Zanen, P., Lammers, J.-W., and Koenderman, L., 1994, RANTES- and interleukin-8-induced responses in normal human eosinophils: Effects of priming with interleukin 5, *Blood* **83:**3697–3704.

Serfling, E., Barthelmas, R., Pfeuffer, I., Schenk, B., Zarius, S., Swoboda, R., Mercurio, F., and Karin, M., 1989, Ubiquitous and lymphocyte-specific factors are involved in the induction of the mouse interleukin 2 gene in T lymphocytes, *EMBO J.* **8:**465–475.

Shanafelt, A. B., Miyajima, A., Kitamura, T., and Kastelein, R. A., 1991, The amino-terminal helix of GM-CSF and IL-5 governs high affinity binding to their receptors, *EMBO J.* **10:**4105–4112.

Shen, Y., Baker, E., Callen, D. F., Sutherland, G. R., Willson, T. A., Rakar, S., and Gough, N. M., 1992, *Cytogenet. Cell. Genet.* **61:**175–177.

Sher, A., Coffman, R. L., Hieny, S., and Cheever, A. W., 1990, Ablation of eosinophil and IgE responses with anti-IL-5 or anti-IL-4 antibodies fails to affect immunity against *Schistosoma mansoni* in the mouse, *J. Immunol.* **145:**3911–3916.

Silberstein, D. S., Owen, W. F., Gasson, J. C., DiPersio, J. F., Golde, D. W., Bina, J. C., Soberman, R., Austen, K. F., and David, J. R., 1986, Enhancement of human eosinophil cytotoxicity and leukotriene synthesis by biosynthetic (recombinant) granulocyte–macrophage colony-stimulating factor, *J. Immunol.* **137:**3290–3294.

Silvennoinen O., Witthühn, B., Quelle , F. W., Cleveland, J. L., Yi, T., and Ihle, J. N., 1993, Structure of the Jak2 protein kinase and its role in IL-3 signal transduction, *Proc. Natl. Acad. Sci. U.S.A.* **90:**8429–8433.

Sorensen, P., Mui, A. L.-F., and Krystal G., 1989, Interleukin-3 stimulates the tyrosine phosphorylation of the 140-kilodalton interleukin-3 receptor, *J. Biol. Chem.* **264:**19253–19259.

Strath, M., and Sanderson, C. J., 1986, Detection of eosinophil differentiation factor and its relationship to eosinophilia in Mesocestoides corti-infected mice, *Exp. Hematol.* **14:**16–20.

Strath, M., Warren, D. J., and Sanderson, C. J., 1985, Detection of eosinophils using an eosinophil peroxidase assay. Its use as an assay for eosinophil differentiation factors, *J. Immunol. Methods* **83:**209–215.

Strath, M., Dent, L. A., and Sanderson, C. J., 1992, Infection of IL5 transgenic mice with *Mesocestoides corti* induces very high levels of IL5 but depressed the production of eosinophils, *Exp. Hematol.* **20:**229–234.

Sun, Z., Yergeau, D. A., Tuypens, T., Tavernier, J., Paul, C. C., Baumann, M. A., Tenen, D. G., and Ackerman, S. J., 1995, Identification and characterization of a functional promoter region in the human eosinophil IL-5 receptor α-subunit gene, *J. Biol. Chem.* **270:**1–10.

Swain, S. L., McKenzie, D. T., Dutton, R. W., Tonkonogy, S. L., and English, M., 1988, The role of IL4 and IL5: Characterization of a distinct helper T cell subset that makes IL4 and IL5 (TH2) and requires priming before induction of lymphokine secretion, *Immunol. Rev.* **102:**77–105.

Takahashi, T., Yamaguchi, N., Mita, S., Yamaguchi, Y., Suda, T., Tominaga, A., Kikuchi, Y., Miura, Y., and Takatsu, K., 1990, Structural comparison of murine derived T-cell-replacing factor with T-cell (B151K12)-rIL-5: Dimer formation is essential for the expression of biological activity, *Mol. Immunol.* **9:**911–920.

Takaki, S., Tominaga, A., Hitoshi, Y., Mita, S., Sonoda, E., Yamaguchi N., and Takatsu, K., 1990, Molecular cloning and expression of the murine interleukin-5 receptor, *EMBO J.* **9:**4367–4374.

Takaki, S., Mita, S., Kitamura, T., Yonehara, S., Yamaguchi, N., Tominaga, A., Miyajima, A., and Takatsu, K.,

1991, Identification of the second subunit of the murine interleukin-5 receptor: Interleukin-3 receptor-like protein, AIC2B is a component of the high affinity interleukin-5 receptor, *EMBO J.* **10:**2833–2838.

Takaki, S, Murata, Y., Kitamura, T., Miyajima, A., Tominaga, A., and Takatsu, K., 1993, Reconstitution of the functinal receptors for murine and human interleukin 5, *J. Exp. Med.* **177:**1523–1529.

Takatsu, K., 1992, Interleukin 5, *Curr. Opin. Immunol.* **4:**299–306.

Takatsu, K., Harada, N., Hara, Y., Takahama, Y., Yamada, G., Dobashi, K., and Hamoaka, T., 1985, Purification and physicochemical characterization of murine T cell replacing factor (TRF), *J. Immunol.* **134:**382–389.

Takatsu, K., Tominaga, A., Harada, N., Mita, S., Matsumoto, M., Takashi, T., Kikuchi, Y., and Yamaguchi, N., 1988, T-cell replacing factor (TRF)/interleukin-5 (IL-5): Molecular and functional properties, *Immunol. Rev.* **102:**107–135.

Tavernier, J., Devos, R., Van der Heyden, J., Hauquier, G., Bauden, R., Faché, I., Kawashima, E., Vandekerckhove, J., Contreras, R., and Fiers, W., 1989, Expression of human and murine interleukin-5 in eukaryotic systems, *DNA* **8:**491–501.

Tavernier, J., Devos, R., Cornelis, S., Tuypens, T., Van der Heyden, J., Fiers, W., and Plaetinck, G., 1991, A human high affinity interleukin-5 receptor (IL5R) is composed of an IL-5-specific alpha chain and a beta chain shared with the receptor for GM-CSF, *Cell* **66:**1175–1184.

Tavernier, J., Tuypens, T., Plaetinck, G., Verhee, A., Fiers, W., and Devos, R., 1992, Molecular basis of the membrane-anchored and two soluble isoforms of the human interleukin 5 receptor α subunit, *Proc. Natl. Acad. Sci. U.S.A.* **89:**7041–7045.

Tavernier, J., Tuypens, T., Verhee, A., Plaetinck, G., Devos, R., Van der Heyden, J., Guisez, Y., and Oefner, C., 1995, Identification of receptor binding domains on human interleukin 5 (IL5) and design of an IL5-derived receptor antagonist, *Proc. Natl. Acad. Sci. U.S.A.* **92:**5194–5198.

Tominaga, Matsumoto, M., Harada, N., Takahashi, T., Kikuchi, Y., and Takatsu, K., 1988, Molecular properties and regulation of mRNA expression for murine T cell-replacing factor, *J. Immunol.* **140:**1175–1181.

Tominaga, A., Takahashi, T., Kikuchi, Y., Mita, S., Naomi, S., Harada, N., Yamaguchi, N., and Takatsu, K., 1990, Role of carbohydrate moiety of IL-5. Effect of tunicamycin on the glycosylation of IL-5 and the biological activity of deglycosylated IL-5, *J. Immunol.* **144:**1345–1352.

Tominaga, A., Takaki, S., Koyama, N., Katoh, S., Matsumoto, R., Migata, M., Higoshi, Y., Hosoya, Y., Yamauchi, S., Kanai, Y., Miyazaki, J. I., Usuku, G., Yamamura, K. I., and Takatsu, K., 1991, Transgenic mice expressing a B cell growth and differentiation factor gene (interleukin 5) develop eosinophilia and autoantibody production, *J. Exp. Med.* **173:**429–437.

Touw, I., Donath, J., Pouwels, K., van Buitenen, C., Schipper, P., Santini, V., Hagemeijer, A., Löwenberg, B., and Delwel, R., 1991, Acute myeloid leukemias with chromosomal abnormalities involving the 21q22 region identified by their *in vitro* responsiveness to interleukin-5, *Leukemia* **5:**687–692.

Tsujimoto, M., Adachi, H., Kodama, S., Tsuruoka, N., Yamada, Y., Tanaka, S., Mita, S., and Takatsu, K., 1989, Purification and characterization of recombinant human interleukin 5 expressed in Chinese hamster ovary cells, *J. Biochem. (Tokyo)* **106:**23–28.

Tsuruoka, N., Funakoshi, K., Kodama, S., and Tsujimoto, M., 1990, Interaction of interleukin-5 with its receptors on murine leukemic BCL1 cells and its implication in biological activity, *Cell. Immun.* **125:**354–362.

Tsuruoka, N., Yamashiro, K., and Tsujimoto, M., 1993, Purification of soluble murine interleukin 5 (IL-5) receptor α expressed in Chinese hamster ovary cells and its action as an IL-5 antagonist, *Arch. Biochem. Biophys.* **307:**133–137.

Tuypens, T., Plaetinck, G., Baker, E., Sutherland, G., Brusselle, G., Fiers, W., Devos, R., and Tavernier, J., 1992, Organization and chromosomal localization of the human interleukin 5 receptor α-chain gene, *Eur. Cytokin. Netw.* **3:**451–459.

Ueberla, K., Li, W., Qin, Z., Richter, G., Raabe, T., Diamantstein, T., and Blankenstein, T., 1991, The rat interleukin-5 gene: Characterization and expression by retroviral gene transfer and polymerase chain reaction, *Cytokine* **3:**72–81.

Ulrich, A., and Schlessinger, J., 1990, Signal transduction by receptors with tyrosine kinase activity, *Cell* **61:** 203–212.

Urban, J. F., Madden, K. B., Svetic, A., Cheever, A. L, Trotta, P. P., Gause, W. C., Katona, I. M., and Finkelman F. D., 1992, The importance of TH_2 cytokines in protective immunity to nematodes, *Immunol. Rev.* **127:** 205–220.

Van der Heyden, Tuypens, T., Devos, R., Plaetinck, G., Van Ostade, X., Guisez, Y., and Tavernier, J., 1996, Binding characteristics of monoclonal antibodies directed against human interleukin 5 (in press).

Van Leeuwen, B. H., Martinson, M. E., Webb, G. C., and Young, I. G., 1989, Molecular organization of the cytokine gene cluster, involving the human IL-3, IL-4, IL-5 and GM-CSF genes, on human chromosome 5, *Blood* **73:**1142–1148.

Van Oosterhout, A. J. M., Ladenius, R. C., Salvelkoul, H. F. J., Van Ark, I., Delsman, K. C., and Nijkamp, F. P., 1993, Effect of anti-IL5 on airway hyperreactivity and eosinophils in guinea pigs, *Am. Rev. Respir. Dis.* **147:**548–552.

Van Straaten, J. F. M., Dokter, W. H. A., Stulp, B. J., and Vellenga, E., 1994, The regulation of interleukin 5 and interleukin 3 expression in human T cells, *Cytokine* **6:**229–234.

Venge, P., 1994, Soluble markers of allergic inflammation, *Allergy* **49:**1–8.

Vitetta, E. S., Brooks, K., Chen, Y., Isakson, P., Jones, S., Layton, J., Mishra, G. C., Pure, E., Weiss, E., Word, C., Yuan, D., Tucker, P., Uhr, J. W., and Krammer, P. H., 1984, T-cell derived lymphokines that induce IgM and IgG secretion in activated murine B cells, *Immunol. Rev.* **78:**137–157.

Walker, C., Bode, E., Boer, L., Hansel, T. T., Blaser, K., and Virchow, J. C., 1992, Allergic and nonallergic asthmatics have distinct patterns of T-cell activation and cytokine production in peripheral blood and bronchoalveolar lavage, *Am. Rev. Respir. Dis.* **146:**109–115.

Walsh, G. M., Hartnell, A., Wardlaw, A. J., Kurihara, K., Sanderson, C. J., and Kay, A. B., 1990, IL-5 enhances the *in vitro* adhesion of human eosinophils, but not neutrophils, in a leucocyte integrin (CD11/18)-dependent manner, *Immunology* **71:**258–267.

Walsh, G. M., Mermod, J.-J., Hartnell, A., Kay, A. B., and Wardlaw, A. J., 1991, Human eosinophil, but not neutrophil, adherence to IL-1-stimulated human umbilical vascular endothelial cells is $\alpha_4\beta_1$ (very late antigen-4) dependent, *J. Immunol.* **146:**3419–3423.

Wang, J. M., Rambaldi, A., Biondi, A., Chen, Z. G., Sanderson, C. J., and Mantovani, A., 1989, Recombinant human interleukin 5 is a selective eosinophil chemoattractant, *Eur. J. Immunol.* **19:**701–704.

Wang, L. M., Keegan, A. D., Paul, W. E., Heidaran, M. A., Gutkind, J. S., and Pierce, J. H., 1992, IL-4 activates a distinct signal transduction cascade from IL-3 in factor dependent myeloid cells, *EMBO J.* **11:**4899–4906.

Wardlaw, A. J., Dunnette S., Gleich, G. J., Collins, J. U., and Kay, A. B., 1988, Eosinophils and mast cells in bronchoalveolar lavage in subjects with mild asthma, *Am. Rev. Respir. Dis.* **137:**62–69.

Warren, D. J., and Moore, M. A., 1988, Synergism among interleukin 1, interleukin 3, and interleukin 5 in the production of eosinophils from primitive hemopoietic stem cells, *J. Immunol.* **140:**94–99.

Warrington, J., Bailey, S., Armstrong, E., Aprelikova, O., Alitalo, K., Dolganov, G., Wilcox, A., Sikela, J., Wolfe, S., Lovett, M., and Vasmuth J., 1992, A radiation hybrid map of 18 growth factor, growth factor receptor, hormone receptor, or neurotransmitter receptor genes on the distal region of the long arm of chromosome 5, *Genomics* **13:**803–808.

Wegner, C. D., Gundel, R. H., Reilly, P., Haynes, N., Letts, L. G., and Rothlein, R., 1990, Intercellular adhesion molecule (ICAM-1) in the pathogenesis of asthma, *Science* **247:**456–459.

Weiss, M., Yokoyama, C., Shikama, Y., Naugle, C., Druker, B., and Sieff, C. A., 1993, Human granulocyte–macrophage colony-stimulating factor receptor signal transduction requires the proximal cytoplasmic domains of the α and β subunits, *Blood* **82:**3298–3303.

Weller, P. F., Rand, T. H., Barrett, T., Elovic, A., Wong, D. T., and Finberg, R. W., 1993, Accessory cell function of human eosinophils: HLA-DR dependent, MHC-restricted antigen-presentation and interleukin-Ia formation, *J. Immunol.* **150:**2554–2562.

Willman, C. L., Sever, C. E., Pallavicini, M. G., Harada, H., Tanaka, N., Slovak, M. L., Yamamoto, H., Harada, K., Meeker, T. C., List, A. F., and Taniguchi, T., 1993, Deletion of IRF-1, mapping to chromosome 5q31.1, in human leukemia and preleukemic myelodysplasia, *Science* **259:**968–971.

Winqvist, I., Olofsson, T., Olsson, I., Persson, A.-M., and Hallberg, T., 1982, Altered density, metabolism and surface receptors of eosinophils in eosinophilia, *Immunology* **47:**531–539.

Wong, D. T. W., Weller, P. F., Galli, S. J., Elovic, A., Rand, T. H., Gallagher, G. T., Chiang, T., Chou, M. Y., Matossian, K., McBride, J., and Todd, R., 1990, Human eosinophils express transforming growth factor α, *J. Exp. Med.* **172:**673–681.

Woodcock, J. M., Zacharakis, B., Plaetinck, G., Bagley, C. J., Qiyu, S., Hercus, T. R., Tavernier, J., and Lopez, A. F., 1994, Three residues in the common b chain of the human GM-CSF, IL-3 and IL-5 receptors are essential for GM-CSF and IL-5 but not IL-3 high affinity binding and interact with Glu^{21} of GM-CSF, *EMBO J.* **13:**5176–5185.

Yamaguchi, Y., Suda, T., Suda, J., Eguchi, M., Miura, Y., Harada, N., Tominaga, A., and Takatsu, K., 1988a, Purified interleukin 5 supports the terminal differentiation and proliferation of murine eosinophilic precursors, *J. Exp. Med.* **167:**43–56.

Yamaguchi, Y., Hayashi, Y., Sugama, Y., Miura, Y., Kasahara, T., Kitamura, S., Torisu, M., Mita, S., Tominaga, A., and Takatsu, K., 1988b, Highly purified murine interleukin 5 (IL-5) stimulates eosinophil function and prolongs *in vitro* survival, *J. Exp. Med.* **167:**1737–1742.

Yamaguchi, Y., Suda, T., Ohta, S., Tominaga, K., Miura, Y., and Kasahara, T., 1991, Analysis of the survival of mature human eosinophils: Interleukin-5 prevents apoptosis in mature human eosinophils, *Blood* **78:**2542–2547.

Yawata, H., Yasukawa, K., Natsuka, S., Murakami, M., Yamasaki, K., Hibi, M., Taga, T., and Kishimoto, T., 1993, Structure–function analysis of human IL-6 receptor: Dissociation of amino acid residues required for IL-6 binding and for IL-6 signal transduction through gp130, *EMBO J.* **12:**1705–1712.

Yi, T., Mui, A. L.-F., Krystal, G., and Ihle, J. N., 1993, Hematopoietic cell phosphatase associates with the Interleukin-3 (IL-3) receptor b chain and down-regulates IL-3 induced tyrosine phosphorylation and mitogenesis, *Mol. Cell. Biol.* **13:**7577–7586.

Zon, L. I., Moreau, J.-F., Koo, J.-W., Mathey-Prevot, B., and D'Andrea, A. D., 1992, The erythropoietin receptor transmembrane region is necessary for activation by the Friend virus gp55 glycoprotein, *Mol. Cell. Biol.* **12:**2949–2957.

Chapter 14

Stem Cell Factor

Ian K. McNiece and Robert A. Briddell

1. INTRODUCTION

The production of mature hematopoietic cells is a tightly regulated process involving many intermediate cell types and regulatory proteins. A specific group of glycoproteins called colony-stimulating factors (CSFs) have been identified, and these CSFs interact with their respective target cells in each of the hematopoietic lineages to give rise to mature cells.

For many years investigators searched for regulatory proteins that control the proliferation and differentiation of stem cells. The existence of such a growth factor had been suggested by mutant strains of mice. Mutations at loci known as the *Sl*, or Steel locus (Sarvella and Russell, 1956), and the *W*, or white spotting locus (Russell, 1979a), resulted in defects in the hematopoietic, gonadal, and melanocyte lineages (Russell, 1970, 1979a). Elegant transplantation studies demonstrated that the defect in the *Sl* mutations derived from an ineffective stromal microenvironment, and the defect in the *W* mutations was intrinsic to stem cells (Bernstein, 1970). Several groups demonstrated that the *W* locus encoded the protooncogene *c-kit*, a transmembrane tyrosine kinase receptor (Geissler *et al.*, 1988a,b; Chabot *et al.*, 1988; Yarden *et al.*, 1987). Further analysis of the *Sl* and *W* mutations led Investigators to propose that the product of the *Sl* locus was the ligand for *c-kit* (Yarden *et al.*, 1987; Geissler *et al.*, 1988a,b). Several groups subsequently identified the ligand for *c-kit* and demonstrated that it was the product of the *Sl* locus. This protein was isolated independently as stem cell factor (SCF), mast cell growth factor (MGF) and *c-kit* ligand (KL) (Zsebo *et al.*, 1990a,b; Anderson *et al.*, 1990; Huang *et al.*, 1990; Williams *et al.*, 1990).

The SCF is produced by stromal cells, including fibroblasts, as both a soluble protein and a membrane-bound protein (Anderson *et al.*, 1990; Toksoz *et al.*, 1992). The two forms

Ian K. McNiece and Robert A. Briddell Department of Developmental Hematology, Amgen Inc., Thousand Oaks, California 91320.

Blood Cell Biochemistry, Volume 7: Hematopoietic Cell Growth Factors and Their Receptors, edited by A. D. Whetton and J. Gordon. Plenum Press, New York, 1996.

of SCF result from alternate splicing of the DNA, resulting in two proteins consisting of 248 amino acids (SCF^{248}) and 220 amino acids (SCF^{220}), respectively. SCF^{248} is cleaved in the extracellular domain to form the soluble form of SCF, which consists of 165 amino acids (Toksoz *et al.*, 1992). The native soluble SCF exists as a dimeric glycoprotein with a molecular mass of approximately 45 kDa (Zsebo *et al.*, 1990b). SCF^{220} exists as a membrane-bound protein because of the lack of exon 6, which is believed to encode for a proteolytic cleavage site. The full biological function of the membrane-bound form (SCF^{220}) is not well understood, and the majority of studies to date have evaluated soluble SCF. (Details of the signaling events associated with SCF receptor occupation can be found in Chapter 5.)

2. *IN VITRO* BIOLOGY

Stem cell factor alone has little stimulatory effect on bone marrow or peripheral blood cells. In agar culture of mouse bone marrow, SCF stimulates the formation of neutrophil colonies. Long-term liquid culture of mouse bone marrow in the presence of SCF generates pure populations of mast cells. The SCF has little if any stimulatory effect alone on human bone marrow (McNiece *et al.*, 1991a,b; Williams *et al.*, 1992; Bernstein *et al.*, 1991; Ulich *et al.*, 1991a,b).

The major stimulatory effect of SCF is in combination with other growth factors. A synergistic increase in colony numbers is obtained when SCF is added with CSFs or interleukins. For example, SCF synergizes with granulocyte–macrophage colony-stimulating factor (GM-CSF), granulocyte colony-stimulating factor (G-CSF), interleukin 3 (IL-3), and colony-stimulating factor 1 (CSF-1) in cultures of both mouse and human hematopoietic cells (McNiece *et al.*, 1991a,b; Williams *et al.*, 1992; Bernstein *et al.*, 1991; Ulich *et al.*, 1991a,b). The stimulatory effect of SCF is observed on a range of cells from primitive progenitor cells through to mature progenitor cells (Williams *et al.*, 1992; Bernstein *et al.*, 1991). On all types of progenitor cells, SCF has little if any effect on the lineage of differentiation. For example, in the presence of GM-CSF, all colonies stimulated by the combination of SCF plus GM-CSF are mixed granulocyte/macrophage colonies. In addition to stimulating increased numbers of colonies, SCF, in combination with other growth factors, stimulates larger colonies containing more cells.

All lineages of hematopoietic development are stimulated by SCF. In combination with erythropoietin (Epo), SCF stimulates the formation of burst-forming unit erythroid (BFU-E)-derived colonies. Recent data have demonstrated a synergistic effect of SCF with thrombopoietin (TPO) to stimulate megakaryocyte colony formation (Hunt *et al.*, 1994). Stem cell factor acts in combination with GM-CSF, G-CSF, IL-3, and CSF-1 to stimulate myeloid progenitors. Both B- and T-lymphoid development are stimulated by SCF. In combination with interleukin 7 (IL-7), SCF acts on primitive mouse B-cell and T-cell precursors (McNiece *et al.*, 1991b; Namen *et al.*, 1991).

The mechanism of action of SCF is not clearly understood. It is known to bind to its receptor, *c-kit*, which is a tyrosine kinase receptor. However, as SCF alone has little stimulatory effect, alternative second message signaling pathways must be activated. Some investigators have proposed a sequential model for the action of combinations of growth factors in which the first factor expands a population of cells that respond to the second growth factor. Alternatively, the role of SCF may be to induce increased receptor expression for the second growth factor. This model is not supported by sequential addition experiments, which demonstrate the requirement of both growth factors at the initiation of the culture. Another model is that the interaction of the receptors on signaling pathways may induce a unique cascade of signaling not initiated by either factor alone.

3. POTENTIAL *IN VITRO* CLINICAL USES OF SCF

3.1. *Ex Vivo* Expansion

Several studies have described the ability of SCF to synergize with other growth factors to expand the number of mature cells, mature progenitor cells, or primitive progenitor cells. In liquid culture of purified mouse bone marrow cells ($Lin^-/Sca\text{-}1^+$), SCF plus interleukin 6 (IL-6) was shown to stimulate the generation of primitive progenitor cells (high proliferative potential colony-forming cells; HPP-CFC), while SCF plus IL-3 generated mature progenitor cells (granulocyte–macrophage colony-forming cells; GM-CFC) (Williams *et al.*, 1992). Similar studies using human bone marrow cells or peripheral blood progenitor cells (PBPC) have demonstrated the ability of growth factor cocktails to expand mature cells approximately 100-fold and GM-CFC 20-fold or higher (McNiece *et al.*, 1991a). In all combinations of growth factors evaluated, the presence of SCF is essential for optimal expansion.

The potential clinical uses of *ex vivo* expanded cells include the elimination of thrombocytopenia and neutropenia, tumor purging, and the improvement of suboptimal bone marrow or leukapheresis harvests. The clinical development of *ex vivo* expansion involves a number of regulatory issues. These issues include a procedure that is reproducible while maintaining sterility. Because of these issues, a procedure has been identified that can result in at least 20-fold expansion of GM-CFC in large-scale cultures (Shieh *et al.*, 1994). CD34-expressing ($CD34^+$) cells from a leukapheresis harvest are isolated and cultured in three 1-liter Teflon culture bags at 1.0×10^4 $CD34^+$ cells/ml in defined medium (DM, Amgen Inc.). The DM contains no fetal calf serum or other animal products. Comparison of this medium to a commercially available medium found greater expansion with DM than with Ex-Vivo 15 medium (Biowhittaker Laboratories Inc.). The addition of autologous plasma to either DM or Ex Vivo 15 results in increased expansion; however, the DM in the absence of autologous plasma results in equivalent expansion to the Ex Vivo 15 plus autologous plasma. Different lots of autologous plasma have been found to give highly variable expansion and in some instances to inhibit expansion. Therefore the development of the DM provides a highly reproducible culture medium and eliminates the variability introduced with media requiring the addition of autologous plasma (Shieh *et al.*, 1994).

Maximal expansion in the DM has been obtained with the combination of SCF/IL-3/IL-6/G-CSF, all added at 100 ng/ml (Shieh *et al.*, 1994). The use of culture bags enables the transfer of cells for washing in a closed system with sterile transfer sets. Under these conditions, maximal expansion of GM-CFC is obtained at day 10 of culture without refeeding or further addition of factors. A comprehensive list of quality assurance and quality control (QA/QC) will be necessary for validation of the final cell product. As the parameters for optimal *ex vivo* expansion have been defined, what remains to be accomplished is to perform the clinical studies to determine the potential of this technology.

3.2. Gene Therapy

The current technology employed for the transfer of genes into hematopoietic cells is based on retroviral transduction into stem cells. Retroviruses will not transfect quiescent cells, so growth factors are used to stimulate quiescent and slowly cycling cells into an active cell cycle. Several studies in animal models have demonstrated efficient transduction.

It has been shown that primitive murine multipotential colony-forming unit spleen cells (CFU-S) can be expanded up to threefold by culturing for 6 days with SCF, IL-3, and IL-6 (Bodine *et al.*, 1992). A more remarkable expansion of four- to eightfold was seen when murine bone marrow cells were enriched for CFU-S via elutriation and lineage subtraction before culture. Combinations of SCF and IL-3 or IL-3 and IL-6 maintain CFU-S number but are not as effective at expansion (Bodine *et al.*, 1989, 1992). The long-term repopulating ability of murine bone marrow cells measured in competitive repopulation experiments has also been shown to be enhanced severalfold by *in vitro* culture in SCF, IL-3, and IL-6 (Bodine *et al.*, 1989, 1992). The IL-3 and IL-6 alone maintain but do not expand cells with repopulating ability (Bodine *et al.*, 1991). $CD34^+$ cells isolated from rhesus bone marrow expand severalfold in 4 days of culture with SCF, IL-3, and IL-6 (Donahue *et al.*, 1991). Autologous transplantation with these cultured cells after total body irradiation (TBI) has produced rapid trilineage reconstitution.

There is extensive evidence in murine models that retroviral vectors can introduce foreign genes at high efficiency into hematopoietic progenitor cells and stem cells (Bodine *et al.*, 1989; Nienhuis *et al.*, 1991; Karlson *et al.*, 1991; Lim *et al.*, 1987). A number of variables appear to be critical to effect efficient gene transfer into primitive progenitor cells and stem cells; these include 5-fluorouracil (5-FU) pretreatment of the donor animal prior to marrow harvest, use of very high-titer viruses, and the inclusion of hematopoietic growth factors in the medium during transduction (Bodine *et al.*, 1989, 1990, 1991; Wieder *et al.*, 1991). Culture in the presence of growth factors for at least 48 hr appears to be necessary for efficient retroviral gene transfer into murine stem cells (Bodine *et al.*, 1989; Lim *et al.*, 1987). This is presumably related to the increased likelihood of proviral integration in actively cycling cells. Pretreatment of the donor with 5-FU and *in vitro* culture in IL-3 and IL-6 produce long-term engraftment with marked cells in up to 90% of recipient mice with up to 20% of circulating cells positive for the provirus (Bodine *et al.*, 1989, 1991).

Experience in larger animals is less extensive. Two groups working with dogs have shown gene transfer in 0.1–10% of circulating cells or progenitor cells of multiple lineages up to 2 years posttransplantation (Schuening *et al.*, 1991; Carter *et al.*, 1992). Early primate studies using short exposure to low-titer viral supernatant without addition of growth factors could not demonstrate transduction of long-term repopulating cells despite transient circulation of marked cells presumably derived from successfully transduced committed progenitor cells (Kantoff *et al.*, 1987). Incorporation of the modifications found to increase gene transfer efficiency in the murine models has also improved the results in primates. Six of six primates transplanted with 5-FU-primed marrow transduced in coculture with a high-titer but helper-contaminated producer line in the presence of IL-3 and IL-6 were positive for proviral DNA in hematopoietic tissue at 6–12 months (Bodine *et al.*, 1990). Three of seven animals transplanted with CD34-enriched transduced marrow had the marker gene detectable in marrow, peripheral blood, purified granulocytes, and T cells at levels of 1–10% for more than 100 days posttransplant. However, the presence of high-level helper virus contamination of the producer cell line used in these experiments and chronic viremia in these animals make interpretation of these results difficult (Bodine *et al.*, 1990). Three of these animals subsequently developed T-cell lymphomas related to the helper virus contamination.

Preclinical studies designed to assess directly the feasibility and efficiency of gene transfer to human stem cells or long-term repopulating cells are not possible. Various surrogates have been proposed, including the long-term bone marrow culture-initiating cell assay (LTBMCIC) (Sutherland *et al.*, 1990); however, it is not known if this assay really predicts the behavior *in vivo* of human stem cells. Human LTBMCIC have been successfully marked by retroviruses (Hughes *et al.*, 1989; Fraser *et al.*, 1990; Nolta and Kohn, 1990). Several different laboratories have shown that up to 40% of colony-forming unit cells (GM-CFC) cultured out of 5- to 8-week long-term bone marrow cultures initiated with retrovirally transduced marrow are marked with the proviral genome. All of these studies have found that culture for at least 3 days with either hematopoietic growth factors or autologous stroma is necessary for efficient transduction. The transduction process had no effect on the number of total LTBMCIC present or on clonogenic progenitor cell production by these cultures. These published studies used higher-titer viruses. Of note in all of these studies is that the transduction efficiency of GM-CFC was nearly identical to the transduction efficiency of LTBMCIC.

4. *IN VIVO* BIOLOGY

The biological effects of SCF have been studied in rodents, dogs, nonhuman primates, and, most recently, in humans. In all species studied, SCF stimulates a broad range of biological responses within hematopoietic tissues. The pharmacokinetics of SCF have been studied in mice and more recently in humans. In mice, the soluble molecule demonstrated striking localization to the lungs and was associated with mast cell degranulation (Lynch *et al.*, 1992). Doses of more than 100 μg/kg per day of SCF are required to elicit significant biological responses in healthy animals, although lower doses have elicited significant responses in mice with the *Steel*^{Dicke} (*Sl*^{d}) mutation. In rodent models, polyethylene glycol-modified SCF (SCF-PEG) has a prolonged serum half-life and has been shown to have greater biological activity than unmodified SCF.

5. TREATMENT OF ANIMALS WITH SCF

5.1. Changes in Peripheral Blood of Animals Stimulated by SCF

Zsebo *et al.* (1990a,b) described *in vivo* biological effects of recombinant rat SCF (rrSCF) given to *Sl/Sl*^{d} mice. The rrSCF stimulated erythrocyte and mast cell production, partially correcting the macrocytic anemia and mast cell deficiency, and also stimulated a thrombocytosis and a leukocytosis with increases in circulating neutrophils, lymphocytes, and monocytes (Zsebo *et al.*, 1990a,b). In normal mice and rats, rrSCF stimulates a dose-dependent leukocytosis (Ulich *et al.*, 1991a,b; Molineux *et al.*, 1991). Similar observations were made for recombinant human SCF (rhSCF) given to baboons and for recombinant canine SCF (rcSCF) given to dogs (Rosen *et al.*, 1990; Dale *et al.*, 1991; Schuening *et al.*, 1993). Splenectomy of normal mice was essential for optimal effects of rrSCF, as the spleen acts as a "trap" for leukocytes and progenitor cells. This effect was not obtained with baboons or dogs, suggesting that splenectomized mice are a better model for hematopoiesis.

Neutrophilia was a common response in all species studied. In rodents and nonhuman primates, a lymphocytosis was observed, and in nonhuman primates increases in circulating monocytes, eosinophils, and basophils were seen (Andrews *et al.*, 1991). In gray collies that have cyclic hematopoiesis, rcSCF increases neutrophil production and induces basophilia (Dale *et al.*, 1991). In baboons, both B and T lymphocytes, including phenotypically immature ($CD10^+$) B cells, were increased in the circulation (Andrews *et al.*, 1991), consistent with the observations that SCF can enhance the proliferation *in vitro* of immature B cells (McNiece *et al.*, 1991a,b). Mast cells are increased in tissues throughout the body of mice and primates treated with SCF (Tsai *et al.*, 1991; Wershil *et al.*, 1992; Galli *et al.*, 1993), and mast cell activation is also stimulated by SCF administration (Dale *et al.*, 1991; Schuening *et al.*, 1993).

Although SCF stimulates neutrophil and mast cell production, there are currently little data on the function of leukocytes in animals given SCF. Steinshamm *et al.* (1993) reported that the combination of G-CSF plus SCF protects mice against lethal doses of *Candida albicans* better than comparable doses of G-CSF or SCF alone. The SCF may also play an important role in stimulating resistance to certain parasitic infections (Grencis *et al.*, 1993).

Sl/Sl^d mice given rrSCF have increased erythrocyte production (Zsebo *et al.*, 1990a,b). In baboons, rhSCF stimulates reticulocytosis and a transient increase in erythrocytes (Andrews *et al.*, 1991) as well as more rapid recovery of erythrocyte mass following acute blood loss than occurs without rhSCF (Rosen *et al.*, 1990). Baboons given rhSCF for more than 2 weeks develop iron-deficient erythropoiesis (Andrews *et al.*, 1991). Normal mice and rats given rrSCF can have a transient increase in reticulocytes, although little change in erythrocyte mass has been reported. In contrast, dogs given rcSCF have not shown any change in either reticulocyte count or erythrocyte mass (Dale *et al.*, 1991; Schuening *et al.*, 1993).

Thrombocytosis was observed in *Sl/Sl^d* mice given rrSCF (Zsebo *et al.*, 1990a,b), whereas in normal mice and rats, rrSCF did not produce any change in their platelet counts (Ulich *et al.*, 1991; Molineux *et al.*, 1991). Baboons and dogs given rhSCF and rcSCF, respectively, also had no increase in platelet counts but rather a slight decrease in platelets during the period of SCF administration (Andrews *et al.*, 1991; Dale *et al.*, 1991; Schuening *et al.*, 1993). This is of interest, as megakaryocytes are increased in the marrows of dogs, baboons, and mice given SCF (Andrews *et al.*, 1991; Schuening *et al.*, 1993; Hunt *et al.*, 1992), and studies *in vitro* have clearly demonstrated the presence of functional *c-kit* receptors on megakaryocyte progenitor cells as well as on megakaryocytes (Briddell *et al.*, 1991; Avraham *et al.*, 1992; Tanaka *et al.*, 1992). It is possible that the major effect of SCF is to promote the proliferation of immature megakaryocytes and their precursors but that other factors play a more important role in the final stages of platelet generation, release, and survival (Tanaka *et al.*, 1992).

5.2. Changes in Bone Marrow in Animals Stimulated by SCF

Mice and rats given rrSCF show no change in marrow cellularity, although spleen size and cellularity may be increased (Molineux *et al.*, 1991; Ulich *et al.*, 1991a,b). In dogs and nonhuman primates, marrow cellularity increases with administration of high-dose SCF, and megakaryocytes are increased in the marrow (Andrews *et al.*, 1991; Rosen *et al.*, 1990; Schuening *et al.*, 1993). In both rodents and nonhuman primates, a transient left shift or increase in morphologically immature myeloid and erythroid elements in marrow occurs early in the course of SCF administration. In baboons, this left shift is associated with the expansion of $CD34^+$ cells and an increase in marrow cellularity during the first week of treatment with SCF (Ulich *et al.*, 1991a; Andrews *et al.*, 1991). With continued SCF administration, the morphology of the marrow returns to normal, although it remains hypercellular in dogs and primates. Spleen size also increases in nonhuman primates treated with SCF. In rodents and primates, there is an associated increase in mast cells in the marrow (Ulich *et al.*, 1991a,b; Galli *et al.*, 1993).

5.3. Changes in Progenitor Cells in Animals Stimulated by SCF

In all animal model studies, administration of SCF has produced increases in the number of detectable progenitor cells of multiple types in hematopoietic tissues, including marrow, blood, and spleen.

In rodents, progenitor and marrow-repopulating cells are increased in the spleen and circulation of treated animals but may be reduced in marrow (Molineux *et al.*, 1991; Briddell *et al.*, 1993; Bodine *et al.*, 1993; Fleming *et al.*, 1993). In nonhuman primates and dogs, progenitor cells in marrow increase in association with the increase in marrow cellularity (Andrews *et al.*, 1992a,b; Schuening *et al.*, 1993). In blood, progenitor cells can be increased up to 1000-fold in animals given high doses of SCF compared with pretreatment values or untreated control animals (Andrews *et al.*, 1992a,b; Schuening *et al.*, 1993). The SCF stimulates a redistribution of progenitor and marrow-repopulating cells from marrow to blood, and, in rodents, to the spleen as well (Andrews *et al.*, 1991; Bodine *et al.*, 1993; Fleming *et al.*, 1993). The mechanism by which this occurs is not currently known. Whether stem cells as opposed to more mature progenitor cells are increased in number in the body as a result of SCF treatment remains uncertain (Bodine *et al.*, 1993; Fleming *et al.*, 1993). Nevertheless, in all species studied, administration of SCF clearly increases the circulation of cells that can stably repopulate hematopoiesis in lethally irradiated recipients (Andrews *et al.*, 1992a,b; Briddel *et al.*, 1993; Bodine *et al.*, 1993, Fleming *et al.*, 1993; de Revel *et al.*, 1994), a matter of some clinical significance (see Chapter 15).

5.4. *In Vivo* Synergy of SCF with Other Growth Factors

Studies have been conducted on the ability of rrSCF-PEG and rhG-CSF alone and in combination to mobilize peripheral blood progenitor cells (PBPC) with marrow-repopulating ability in mice (Briddell *et al.*, 1993; Yan *et al.*, 1994). Splenectomized BDF1 mice were injected intravenously with rrSCF at 200 μg/kg per day for 7 days, and the peripheral blood was harvested by cardiac puncture. A fivefold increase in peripheral WBC numbers resulted (compared with control mice). Although a cell dose of 10×10^6 low-density mononuclear cells (LDMNC) from normal mice was required to rescue lethally irradiated mice, as few as 0.25×10^6 LDMNC from rrSCF mice resulted in 100% survival. Marked increases in circulating mature (GM-CFC) and primitive (HPP-CFC) progenitor cells were obtained in rrSCF-treated mice compared with control mice (Yan *et al.*, 1994). Because rhG-CSF has previously been shown to mobilize PBPC, the effect of combined rrSCF plus rhG-CSF treatment on splenectomized mice was examined. Greater than additive increases in circulating LDMNC, GM-CFC, and HPP-CFC were obtained compared with the increases in mice treated with rrSCF or rhG-CSF alone. In three seperate experiments, the mobilization potentials of optimal doses of rhG-CSF (200 μg/kg per day for 7 days), low doses of rrSCF (25 μg/kg for 7 days), and combined administration of the two factors were compared. These studies demonstrated that mice treated with the combination of rrSCF plus rhG-CSF had approximately 1.5-fold higher numbers of circulating WBC, fivefold higher GM-CFC, twofold higher HPP-CFC, and greater than fivefold higher number of cells with *in vivo* repopulating ability compared with mice treated with rhG-CSF alone (Briddell *et al.*, 1993).

5.5. Engraftment of Animals Transplanted with PBPC

In canine models, treatment with the combination of rcSCF (25 μg/kg per day) plus rhG-CSF (10 μg/kg per day) for 7 days dramatically increased the level of PBPC (de Revel *et al.*, 1994). Peripheral blood mononuclear cells (10^8/kg) mobilized by the cytokines, alone or in combination, were capable of rescuing lethally irradiated dogs, whereas an equal number of peripheral blood mononuclear cells from control animals (no cytokine pretreatment) were not. Additionally, time to engraftment, as defined by 0.5×10^9 granulocytes/liter in the peripheral blood after transplant, was shown to be reduced when the combination group was compared to either the rhG-CSF-alone group or the high-dose (200 μg/kg per day) rcSCF-alone group (de Revel *et al.*, 1994). There was a trend for more rapid platelet engraftment (to 20×10^9/liter) in the combination group, but it was not significant.

In baboons, low doses of rhSCF (causing no change in peripheral WBCs when administered alone) were found to significantly increase the numbers of both peripheral WBCs and multiple-lineage hematopoietic progenitors when administered in combination with rhG-CSF (Andrews *et al.*, 1994). Administration of rhSCF at 25 μg/kg per day plus rhG-CSF at 10 or 100 μg/kg per day for 14 days resulted in maximal WBC counts (day 14) that were 2.0- and 1.5-fold higher, respectively, than the maximal WBC counts for baboons given rhG-CSF alone at these doses. In the 25 μg/kg per day rhSCF plus 100 μg/kg per day rhG-CSF group, the mean maximal WBC count was approximately 150×10^9/liter on day 14. Baboons administered rhSCF at 25 μg/kg per day plus rhG-CSF at 250 μg/kg per day did not develop WBC counts higher than those of baboons administered 250 μg/kg per day rhG-CSF alone.

Additionally, the kinetics of PBPC mobilization were studied (Andrews *et al.*, 1994). When baboons were treated with rhG-CSF alone, a wave of progenitor cells was released into the circulation, but the combination of rhSCF plus rhG-CSF resulted in a more continual increase of PBPC that reached a higher level than that observed with rhG-CSF alone, with peripheral PBPC reaching the maximum level at 12 days and continuing at this level through the end of treatment. This is consistent with a cytokine-combination-induced expansion of an earlier population with subsequent maturation and release into the circulation. The increased progenitor cell population included megakaryocyte progenitor cells (MK-CFC), suggesting that the use of rhSCF plus rhG-CSF-mobilized PBPC may lead to improved engraftment of platelets following transplantation.

Sequential cytokine administration was also evaluated (Andrews *et al.*, 1994). Baboons that were treated with rhSCF for 7 days followed by rhG-CSF alone for 7 days showed no significant changes in leukocytes or circulating PBPC during the first 7 days. During the second 7 days, there were increases in leukocytes and PBPC similar to those observed for rhG-CSF alone. Thus, sequential cytokine administration did not enhance the biological response to rhG-CSF when compared with administration of the two factors together.

5.6. Radiation Protection by SCF

Administration of high doses of SCF can protect mice from otherwise lethal doses of total-body irradiation (TBI) by increasing the rate of neutrophil and platelet recovery (Zsebo *et al.*, 1993). Administration of neutralizing anti-SCF antibodies can decrease the irradiation survival of normal mice (Neta *et al.*, 1993). In mice, the timing of SCF administration appears to be critical for radioprotection in that SCF was effective if given as a single dose shortly before but not following irradiation, although for optimal protection SCF was given both before and after irradiation. In dogs, rcSCF was effective when given after total-body irradiation (Schuening *et al.*, 1993). The mechanism by which SCF produces its irradiation-protective effects is not known, although it has been hypothesized that SCF might bring more stem cells into S phase, thereby increasing their resistance to irradiation.

5.7. *In Vivo* Toxicity of SCF

In animal models, toxicities related to administration of SCF by either continuous intravenous infusion or subcutaneous injection have been limited. In mice, rrSCF enhanced mast cell activation (Lynch *et al.*, 1992; Wershil *et al.*, 1992), and rcSCF given to dogs induced transient muzzle edema and rhinorrhea (Dale *et al.*, 1991; Schuening *et al.*, 1993). Rapid bolus intravenous infusion of rcSCF led to hypotensive shock and death in one dog, whereas in mice rapid bolus infusion of rrSCF produced hypotension and pulmonary edema (Lynch *et al.*, 1992). Until now, all studies of SCF *in vivo* have been limited to relatively short periods of factor administration, and whether additional toxicities would occur with prolonged administration is not known.

5.8. Pharmacokinetics in Animals

In rats, plasma clearance of intravenously administered [^{125}I]r-metHuSCF is characterized by a rapid distribution-phase half-life of 9.1 $\pm$ 0.8 minutes (mean $\pm$ SD) and an elimination-phase half-life of 117 $\pm$ 54 min. Initially, [^{125}I]r-metHuSCF was concentrated in plasma and kidneys. Radioactive material was rapidly removed from all tissues, resulting in a decrease of approximately 90% in radioactivity between 1 and 24 hr after injection. This rapid loss appears to be mediated primarily by excretion into the urine as lower-molecular-weight degradation metabolites (Amgen, unpublished data).

In a crossover study comparing routes of administration, cynomolgus monkeys were given 50 or 500 μg/kg r-metHuSCF, both intravenously and subcutaneously, separated by a 2-week washout period. Serum curves from intravenous doses were biphasic with initial half-lives of 26 and 16 min for the 50- and 500-μg/kg doses and terminal half-lives of 2.9 and 3.7 hr, respectively. Serum curves from subcutaneous doses increased to peak concentrations of 58 and 633 ng/ml at 3.7 and 3.0 hr for the low and high doses, respectively. The terminal half-lives for subcutaneous doses were 4.7 and 6.6 hr for the low and high doses, respectively, and these half-lives were considerably longer than the corresponding half-lives after intravenous dosing (Amgen, unpublished data).

6. CLINICAL TRIALS WITH SCF

6.1. Concentrations of Endogenous SCF

In plasma from 257 healthy volunteers, the mean concentrations of circulating endogenous SCF (± SD) was 3.3 ± 1.1 ng/ml (Langley *et al.*, 1993). Concentrations of circulating endogenous SCF also have been evaluated in patients with evidence of hematopoietic dysfunction (Holmberg *et al.*, 1992). Sera from patients with aplastic anemia or myelodysplasia or following allogeneic bone-marrow transplantation (alloBMT) were evaluated by immunoassay for the presence of soluble endogenous SCF. In 16 individuals with aplastic anemia, eight individuals with myelodysplasia, and normal control patients, the mean concentrations of SCF (±SD) were 2.99 ± 1.27 ng/ml, 2.34 ± 1.12 ng/ml, and 3.64 ± 0.496 ng/ml, respectively.

Following alloBMT, serum concentrations of endogenous SCF decreased below normal at day 2 and remained subnormal beyond the period of initial engraftment (defined as ANC $\geq 0.5 \times 10^9$/liter and an unsupported platelet count $\geq 20 \times 10^9$/liter). Thus, myelosuppression in these situations was not accompanied by increased levels of circulating soluble endogenous SCF (Holmberg *et al.*, 1992). Cairo *et al.* (1993) found that endogenous concentrations of SCF may be decreased during neutropenia after BMT, but the concentrations did not correlate significantly with myeloid engraftment.

6.2. Clinical Pharmacokinetics

In phase I studies, rhSCF was administered by daily subcutaneous injection for 14 days prior to myelosuppressive chemotherapy at doses of 5, 10, 25, or 50 μg/kg per day (Young *et al.*, 1993). Blood samples were obtained at 12 time points up to 24 hr following doses on days 1 and 14, and serum concentrations were measured by enzyme immunoassay. Baseline concentrations of SCF measured by this assay were approximately 1 ng/ml. Model-independent techniques were used to determine the area under the serum concentration curve (AUC) and the maximal serum concentration (C_{max}) and time (T_{max}). The mean AUC on day 1 increased in proportion to the dose, and at each dose level the AUC increased about twofold from day 1 to day 14. The C_{max} increased in the same manner as AUC on day 1 and from days 1 to 14. The T_{max} occurred at 12 to 17 hr on day 1; T_{max} decreased to 2 to 11 hr on day 14. Trough concentrations before and 24 hr after dose on day 14 were nearly equal, suggesting that the steady-state concentration had been reached. These steady-state trough values were about twofold larger than dose 1 trough values.

Additional blood samples were obtained at 15 time points up to 72 hr following a single dose of rhSCF (25 μg/kg per day) in six patients to further evaluate the pharmacokinetics. Based on these additional data points, the half-life of rhSCF has been calculated to be 35 hr.

These data suggest that the pharmacokinetic model for rhSCF is linear and that slow absorption from repeated injections leads to a cumulative twofold increase in serum concentration at steady state.

6.3. Phase I Studies

Patients with advanced non-small-cell lung cancer were enrolled in a phase I study (Crawford *et al.*, 1993). Prior to any chemotherapy, a 14-day course of rhSCF was given, followed by 14 days of observation. Subsequently, patients were treated with etoposide and carboplatin followed by an additional 14 doses of rhSCF. The rhSCF was given subcutaneously at doses of 5, 10, 25, or 50 μg/kg per day. Premedication with a combination of H_1 and H_2 antihistamines, with or without ephedrine, was tried in subsets of patients to assess its effectiveness as prophylaxis against adverse reactions.

The study has been completed with a total enrollment of 35 patients. Ten patients were removed from the study prematurely: three because of adverse events, and seven because of progression of the underlying disease.

Patients with advanced carcinoma of the breast were enrolled in a second phase I study (Demetri *et al.*, 1993). A 14-day course of rhSCF was given as part of an initial 21-day period to evaluate the effects of rhSCF alone. Subsequently, patients received one cycle of cyclophosphamide and doxorubicin without any adjuvant growth factor, followed by up to five cycles of the same chemotherapy regimen with rhSCF. The comparison group received rhG-CSF. The doses of rhSCF and the premedication regimen were identical to those used in the previously described study.

The study has been completed with a total enrollment of 26 patients (21 received rhSCF, and five received rhG-CSF). Twenty-three patients were prematurely removed from the study: ten because of adverse reactions, four for disease progression on study, seven because of administrative or investigator's decisions, and two for other reasons.

The most frequently observed adverse events were mild to moderate dermatologic reactions at the injection sites including edema, erythema, pruritus, skin hyperpigmentation, and urticaria (Crawford *et al.*, 1993; Demetri *et al.*, 1993). Essentially all patients who received rhSCF by subcutaneous injection developed a raised pruritic wheal with surrounding erythema at each rhSCF injection site. These reactions, which were most prominent clinically 90 to 120 min after administration, occurred with each dose, and did not appear to change in intensity over the 14-day treatment period. Most other related adverse events occurred as multisymptom systemic anaphylactoid reactions. These reactions included mild to severe dermatologic reactions at distant sites with or without respiratory symptoms (including cough, sore throat, throat tightness, and dyspnea), all of which were transient and reversible.

These clinical observations are consistent with the findings from studies demonstrating that r-metHuSCF can induce mast-cell hyperplasia as well as induce mast-cell activation and mediator release (Costa *et al.*, 1993). Consequently, a prophylactic antihistamine regimen was incorporated into the protocols of the phase I and phase I/II trials. The efficacy of the premedication regimen could not be established in these studies.

Hematologic data show no consistent changes during rhSCF administration before chemotherapy other than a modest dose-related increase in WBC and platelet counts in patients treated with rhSCF after chemotherapy. There were no consistent changes in patients' bone marrow measured by cellularity, M/E ratio, and differential count. At the highest dose (50 μg/kg per day), increases in circulating BFU-E and GM-CFC suggested mobilization of progenitor cells. Serum samples showed no evidence of antibodies to rhSCF in any patient. Blood chemistries showed no trends overall.

6.4. Phase I/II Studies

A phase I/II study has been evaluated in breast cancer patients [high-risk (>10 nodes positive) stage II breast cancer, stage III breast cancer with locally advanced disease, or stage IV disease] (McNiece *et al.*, 1993). These patients were appropriate candidates for high-dose chemotherapy requiring cellular support. The study compared PBPC harvests obtained after mobilization with rhSCF at different doses in combination with a standard dose of rhG-CSF (10 μg/kg per day). A cohort given rhG-CSF alone during the collection phase serves as the comparison group. Enrollment in a second group, 5 μg/kg per day r-metHuSCF alone for mobilization (cohort B), was suspended after five of five patients failed to have stable engraftment after mobilized PBPC were transplanted following high-dose chemotherapy. For all patients, cytokine administration and PBPC harvest preceded a high-dose (myeloablative) chemotherapy regimen; PBPC transplantation was followed by rhG-CSF support during the recovery period.

Duration of dosing with the combination cytokines was 7, 10, or 13 days with leukapheresis performed on days 5 to 7, 8 to 10, or 11 to 13, respectively. The rhG-CSF-alone comparison group received 10 μg/kg per day for 7 days with leukapheresis performed on days 5 to 7.

The median WBCs increased from normal baseline values to 58.7×10^9/liter for the rhG-CSF-alone group at 7 days and to $>50 \times 10^9$/liter for all the groups that received rhSCF plus rhG-CSF (McNiece *et al.*, 1993). This leukocytosis was maintained over the course of dosing. Most of this leukocytosis was caused by the increased ANC. Patients who received rhSCF alone showed no increase in WBC or ANC over baseline. Median mononuclear cell (MNC) counts in peripheral blood were also increased in all groups that received rhSCF plus rhG-CSF compared to rhG-CSF alone and rhSCF alone. Median hemoglobin concentrations remained largely unchanged through the dosing period. Median platelet counts were affected by leukaphersis, as expected.

The number of $CD34^+$ cells/ml of blood was increased for all cohorts receiving rhSCF plus rhG-CSF compared with rhG-CSF alone. Peak median total $CD34^+$ cells for rhG-CSF alone was 3.16×10^4/ml on day 5. For the rhSCF-plus-rhG-CSF cohorts, the peak was two- to threefold higher. $CD34^+$ cell counts in the rhSCF-alone cohort were never greater than 0.3×10^4/ml, tenfold less than for rhG-CSF alone (McNiece *et al.*, 1993).

For each day's leukapheresis product, MNC, $CD34^+$ cells, GM-CFC, and BFU-E were measured, and the total yield was calculated by summing the result from the three leukapheresis procedures for each variable. Patients included in this analysis had results available from all 3 days of leukapheresis. Median MNC yield per kilogram generally increased in cohorts combining doses of rhSCF of at least 10 μg/kg per day with rhG-CSF over the yield obtained with rhG-CSF alone (9.7×10^8 MNC/kg).

Median $CD34^+$ total leukapheresis yield was 3.2×10^6/kg for rhG-CSF-alone cohort, and rhSCF plus rhG-CSF resulted in median yields $\geq 7.2 \times 10^6$/kg, two to three times that seen with rhG-CSF alone. For doses of rhSCF greater than 10 μg/kg in combination with rhG-CSF, a plateau effect was obtained with an increase in $CD34^+$ total cell yield. Median total GM-CFC yield and median total BFU-E yield also were increased over yields with rhG-CSF alone at doses of rhSCF $\geq$10 μg/kg per day in combination with rhG-CSF.

During the treatment phase, patients were followed for hematological recovery after high-dose chemotherapy and PBPC transplantation (Glaspy *et al.*, 1994).

Patients treated with rhSCF alone had significantly delayed ANC and platelet recovery. All patients in this cohort required infusion of cryopreserved bone marrow to reestablish hematopoiesis. Despite these patients reaching the leukapheresis product MNC target (4.0×10^8 MNC/kg) required for transplant, the progenitor-cell analysis of the leukapheresis product in combination with the recovery data from this patient cohort demonstrated that rhSCF alone was not adequate for mobilization.

The ANC recoveries were excellent, as expected in cohorts of rhG-CSF alone or in combination with rhSCF, and platelet recovery was rapid. There was a suggestion of improved platelet recoveries in cohorts using higher doses of rhSCF in combination with rhG-CSF.

7. REFERENCES

Anderson, D. M., Lyman, S. D., Baird, A., Wignall, J. M., Eisenman, J., Rauch, C., March, C. J., Boswell, H. S., Gimpel, S. D., Cosman, D., and Williams, D. E., 1990, Molecular cloning of mast cell growth factor, a hematopoietin that is active in both membrane bound and soluble forms, *Cell* **63**:225–233.

Andrews, R., G., Knitter, G. H., Bartelmez, S. H., Langely, K. E., Farrar, D., Hendren, R. W., Appelbaum, F. R., Bernstein, I. D., and Zsebo, K. M., 1991, Recombinant human stem cell factor, a *c-kit* ligand, stimulates hematopoiesis in nonhuman primates, *Blood* **78**:1975–1980.

Andrews, R. G., Bartelmez, S. H., Knitter, G. H., Myerson, D., Bernstein, I. D., Appelbaum, F. R., and Zsebo, K. M., 1992a, A *c-kit* ligand, recombinant human stem cell factor, mediates reversible expansion of multiple $CD34^+$ colony-forming cell types in blood and marrow of baboons, *Blood* **80**:920–927.

Andrews, R. G., Bensinger, W. I., Knitter, G. H., Bartelmez, S. H., Longin, K., Bernstein, I. D., Appelbaum, F. R., and Zsebo, K. M., 1992b. The ligand for *c-kit*, stem cell factor, stimulates the circulation of cells that engraft lethally irradiated baboons, *Blood* **80**:2715–2720.

Andrews, R. G., Briddell, R. A., Knitter, G. H., Opie, T., Bronsden, M., Myerson, D., Appelbaum, F. R., and McNiece, I. K., 1994, *In vivo* synergy between recombinant human stem cell factor and recombinant human granulocyte colony-stimulating factor in baboons: Enhanced circulation of progenitor cells, *Blood* **84**(3): 800–810.

Avraham, H., Vannier, E., Cowley, S., Jiang, S. X., Chi, S., Dinarello, C. A., Zsebo, K. M., and Groopman, J. E., 1992, Effects of the stem cell factor, *c-kit* ligand, on human megakaryocytic cells, *Blood* **79**:365–371.

Bernstein, I. D., Andrews, R. G., and Zsebo, K. M., 1991, Recombinant human stem cell factor enhances the formation of colonies by $CD34^+$ and $CD34^+$ lin^- cells, and the generation of colony-forming cell progeny from $CD34^+$ lin^- cells cultured with interleukin-3, granulocyte–colony stimulating factor, or granulocyte-macrophage colony-stimulating factor, *Blood* **77**:2316–2321.

Bernstein, S. E., 1970, Tissue transplantations as an analytical and therapeutic tool in hereditary anemias, *Am. J. Surg.* **119**:448–452.

Bodine, D. M., Karlsson, S., and Nienhuis, A. W., 1989, Combination of interleukins 3 and 6 preserves stem cell function in culture and enhances retrovirus-mediated gene transfer into hematopoietic stem cells, *Proc. Natl. Acad. Sci. U.S.A.* **86**:8897–8901.

Bodine, D.M., McDonough, K. T., Brandt, S. J., Ney, P. A., Aricola, B., Byrne, E., and Nienhuis, A. W., 1990, Development of a high-titer retrovirus producer cell line capable of gene transfer into rhesus monkey hematopoietic stem cells, *Proc. Natl. Acad. Sci. U.S.A.* **87**:3738–3744.

Bodine, D.M., Mcdonough, K. T., Seidel, N. E., and Nienhuis, A. W., 1991, Survival and retrovirus infection of murine hematopoietic stem cells *in vitro*: Effects of 5-FU and method of infection, *Exp. Hematol.* **19:**206–212.

Bodine, D. M., Orlic, D., Birkett, N. C., Seidel, N. E., and Zsebo, K. M., 1992, Stem cell factor increases colony-forming unit-spleen number *in vitro* in synergy with interleukin-6, and *in vivo* in *Sl/Sl*-d mice as a single factor, *Blood* **79:**913–919.

Bodine, D. M., Seidel, N. E., Zsebo, K. M., and Orlic, D., 1993, *In vivo* administration of stem cell factor to mice increases the absolute number of pluripotent hematopoietic stem cells, *Blood* **82:**445–455.

Briddell, R. A., Bruno, E., Cooper, R. J., Brandt, J. E., and Hoffman, R., 1991, Effect of *c-kit* ligand on *in vitro* human megakaryocytopoiesis, *Blood* **78:**2854–2859.

Briddell, R. A., Hartley, C. A., Smith, K. A., and McNiece, I. K., 1993, Recombinant rat stem cell factor synergizes with recombinant human granulocyte colony-stimulating factor *in vivo* in mice to mobilize peripheral blood progenitor cells that have enhanced repopulating potential, *Blood* **82:**1720–1723.

Cairo, M. S., Gillan, E. R., Weinthal, J., Yancik, S., Van de Ven, C., Mo, W., Shen, V., Buzby, J. S., and Suen, Y., 1993, Decreased endogenous circulating steel factor (SLF) levels following allogenic and autologous BMT: Lack of an inverse correlation with post-BMT myeloid engraftment, *Bone Marrow Transplant* **11:**155–161.

Carter, R. F., Abrams-Ogg, A. C. G., Dick, J. E., Kruth, S. A., Valli, V. E., Kamel-Reid, S., and Dube, I. D., 1992, Autologous transplantation of canine long-term marrow culture cells genetically marked by retroviral vectors, *Blood* **79:**356.

Chabot, B., Stephenson, D. A., Chapman, V. M., Besmer, P., and Bernstein, A., 1988, The proto-oncogene *c-kit* encoding a transmembrane tyrosine kinase receptor maps to the *W* locus, *Nature* **335:**88.

Costa, J. J., Demetri, G. D., Hayes, D. F., Merica, E. A., Menchaca D. M., and Galli, S. J., 1993, Increased skin mast cells and urine methyl histamine in patients receiving recombinant methionyl human stem cell factor, *Proc. Am. Assoc. Cancer. Res.* **34:**211 (Abstract).

Crawford, J., Lau, D., Erwin, R., Rich, W., McGuire, B., and Meyers, F., 1993, A phase I trial of recombinant methionyl human stell cell factor (SCF) in patients (pts) with advanced non-small cell lung carcinoma (NSCLC), *Proc. Am. Soc. Clin. Oncol.* **12:**135 (Abstract 338).

Dale, D. C., Hammond, W. P., and Zsebo, K, M., 1991, Stem cell factor therapy for cyclic hematopoiesis in grey collie dogs, *Blood* **78**(Suppl):375 (Abstract).

Demetri, G., Costa, J., Hayes, D., Sledge, G., Galli, S., Hoffman, R., Merica, E., Rich, W., Harkins, B., McGuire, B., and Gordon, M., 1993, A phase I trial of recombinant methionyl human stem cell factor (SCF) in patients with advanced breast carcinoma pre- and post chemotherapy (chemo) with cyclophosphamide (C) and doxorubicin (A), *Proc. Am. Soc. Clin. Oncol.* **12:**142 (Abstract).

de Revel, T., Appelbaum, F. R., Storb, R., Schuening, F., Nash, R., Deeg, J., McNiece, I., Andrews, R., and Graham, T., 1994, Effects of granulocyte colony stimulating factor and stem cell factor, alone and in combination on the mobilization of peripheral blood cells that engraft lethally irradiated animals, *Blood* **83**(12):3795–3799.

Donahue, R. E., Kessler, S. W., Bodine, D. B., Agricola, B. A., Byrne, E. R., Metzger, M. E., McDonough, K. T., Bacher, J. D., Zsebo, K. M., and Nienhuis, A. W., 1991, Gene transfer into primate immunoselected $CD34^+$ cells: *In vivo* evidence that CD34 cells are responsible for both short- and long-term reconstitution of lymphoid and myeloid populations, *Blood* **78:**79a.

Fleming, W. H., Alpern, E. J., Uchida, N., Ikuta, K., and Weissman, I. L., 1993, Steel factor influences the distribution and activity of murine hematopoietic stem cells *in vivo*, *Proc. Natl. Acad. Sci. U.S.A.* **90:**3760–3764.

Fraser, C., Eaves, C. J., Szilvassy, S. J., and Humphries, R. K., 1990, Expansion *in vitro* of retrovirally marked totipotent hematopoietic stem cells, *Blood* **76:**1071–1079.

Galli, S. J., Iemura, A., Garlick, D. S., Gamba-Vitalo, C., Zsebo, K. M., and Andrews, R. G., 1993, Reversible expansion of primate mast cell populations *in vivo* by stem cell factor, *J. Clin. Invest.* **91:**148–152.

Geissler, E. N., Cheng, S. V., Gusella, J. F., and Housman, D., 1988a, Genetic analysis of the dominant white-spotting locus (*W*) region on mouse chromosome 5: Identification of cloned DNA markers near *W*, *Proc. Natl. Acad. Sci. U.S.A.* **85:**9635–9642.

Geissler, E. N., Ryan, M. A., and Housman, D. E., 1988b, The dominant-white spotting (*W*) locus of the mouse encodes the *c-kit* protooncogene, *Cell* **55:**185–196.

Glaspy, J., McNiece, I. K., LeMaistre, F., Menchaca, D., Briddell, R., Lill, M., Jones, R., Tami, J., Morstyn, G., Brown, S., and Shpall, E. J., 1994, Effects of stem cell factor (rhSCF) and filgrastim (rhG-CSF) on mobilization of peripheral blood progenitor cells (PBPC) and on hematological recovery post transplant: Preliminary phase I/II study results, *Br. J. Hematol.* **87**(Suppl. 1):156.

Grencis, R. K., Else, K. J., Huntley, J. F., and Nishikawa, S. I., 1993, The *in vivo* role of stem cell factor (*c-kit* ligand) on mastocytosis and host protective immunity to the intestinal nematode *Trichinella spiralis* in mice, *Parasite Immunol.* **15:**55–59.

Holmberg, L. A., Yancik, S., Zsebo, K., and Torok-Storb, B., 1992, Circulating levels of kit ligand (KL) in serum of patients with aplastic anemia, myelodysplasia or following allogeneic bone marrow transplantation, *Exp. Hematol.* **20:**775 (Abstract 269).

Huang, E., Nocka, K., Beier, D. R., Chu, T-Y., Buck, J., Lahm, H-W., Wellner, D., Ledner, P., and Besmer, P., 1990, The hematopoietic growth factor KL is encoded by the *Sl* locus and is the ligand for the *c-kit* receptor, the product of the *W* locus, *Cell* **63:**226–234.

Hughes, P. F. D., Eaves, C. J., Hogge, D. E., and Humphries, R. K., 1989, High efficiency gene transfer to human hematopoietic cells maintained in long-term marrow culture, *Blood* **74:**1915–1925.

Hunt, P., Zsebo, K. M., Hokom, M. M., Hornkohl, A., Birkett, N. C., del Castillo, J. C., and Martin, F., 1992, Evidence that stem cell factor is involved in the rebound thrombocytosis that follows 5-fluorouracil treatment, *Blood* **80:**904–911.

Hunt, P., Bartley, T., Li, Y.-S., Bogenberger, J., Lu, H., Samal, B., Martin, F., Chang, M. S., Parker, V., and Bosselman, B., 1994, Purification and cloning of a megakaryocyte growth and development factor: A novel cytokine found in aplastic plasma, *Exp. Hematol.* **22**(8)**:**838 (Abstract).

Kantoff, P. W., Gillio, A. P., McLachlin, J. R., Bordignnon, C., Eglitis, M. A., Kernan, N. A., Moen, R. C., Kohn, D. B., Yu, S. F., Karson, E., Karlsson, S., Zweibel, J., Gilboa, E., Blaese, R. M., Nienhuis, A., O'Reilly, R. J., and Anderson, W. F., 1987, Expression of human deaminase in nonhuman primates after retrovirus-mediated gene transfer, *J. Exp. Med.* **166:**219–227.

Karlson, S., 1991, Treatment of genetic defects in hematopoietic cell function by gene transfer, *Blood* **78:**2481–2492.

Langley, K. E., Bennett, L. G., Wypych, J., Yancik, S. A., Liu, X-D., Westcott, K. R., Chang, D. G., Smith, K. A., and Zsebo, K. M., 1993, Soluble stem cell factor in human serum, *Blood* **81**(3)**:**656–660.

Lim, B., Williams, D. A., and Orkin, S. H., 1987, Retrovirus-mediated gene transfer of human adenosine deaminase: Expression of functional enzyme in murine hematopoietic stem cells *in vivo*, *Mol. Cell. Biol.* **7:**3459–3470.

Lynch, D. H., Jacobs, C., DuPont, D., Eisenman, J., Foxworthe, D., Martin, U., Miller, R. E., Roux, E., Liggit, D., and Williams, D. E., 1992, Pharmacokinetic parameters of recombinant mast cell growth factor (rMGF), *Lymphokine Cytokine Res.* **11:**233–236.

McNiece, I. K., Langley, K. E., and Zsebo, K. M., 1991a, Recombinant human stem cell factor synergises with GM-CSF, G-CSF, IL-3 and Epo to stimulate human progenitor cells of the myeloid and erythroid lineages, *Exp. Hematol.* **19:**226–231.

McNiece, I. K., Langley, K. E., and Zsebo, K. M., 1991b, The role of recombinant stem cell factor in early B cell development: Synergistic interaction with IL-7, *J. Immunol.* **146:**3785–3790.

McNiece, I., Glaspy, J., LeMaistre, F., Briddell, R., Menchaca, D., and Shpall, E. J., 1993, Effects of recombinant methionyl human stem cell factor (rhSCF) and filgrastim (rhG-CSF) on mobilization of peripheral blood progenitor cells: Preliminary laboratory results from a phase I/II study, *Blood* **82**(Suppl.)**:**84 (Abstract).

Molineux, G., Migdalska, A., Szmitkowski, M., Zsebo, K., and Dexter, T. M., 1991, The effects on hematopoiesis of recombinant stem cell factor (ligand for *c-kit*) administered *in vivo* to mice either alone or in combination with granulocyte-colony stimulating factor, *Blood* **78:**961–966.

Namen, A. E., Widmer, M. B., Voice, R., Christensen, S., Braddy, S., Lyman, S. D., and Williams, D. E., 1991, A ligand for the *c-kit* proto-oncogene (MGF) stimulates lymphoid progenitor cells *in vitro*, *Exp. Hematol.* **19:**497 (Abstract).

Neta, R., Williams, D., Selzer, F., and Abrams, J., 1993, Inhibition of *c-kit* ligand/steel factor by antibodies reduces survival of lethally irradiated mice, *Blood* **81:**324–327.

Nienhuis, A. W., McDonough, K. T., and Bodine, D. M., 1991, Gene transfer into hematopoietic stem cells, *Cancer* **67:**2700–2712.

Nolta, J. A., and Kohn, D. B., 1990, Comparison of the effects of growth factors on retroviral vector-mediated gene transfer and the proliferative stauts of human hematopoietic progenitor cells, *Hum. Gene Ther.* **1:**257.

Rosen, B., Catchatourian, R., Egrie, J., Gould, S. A., Greenberg, R. A., Langley, K. E., Sehgal, L. R., and Zsebo, K. M., 1990, The *in vivo* effects of recombinant human stem cell factor (RhSCF) on hematopoiesis in nonhuman primates, *Blood* **76**(Suppl.)**:**163 (Abstract).

Russell, E. S., 1970, Abnormalities of erythropoiesis associated with mutant genes in mice, in *Regulation of*

Hematopoiesis, (A. L. Gordon, ed.), pp. 649–656, Appelton-Century-Crofts, NewYork.

Russell, E. S., 1979a, Analysis of pleiotropism at the *W* locus in the mouse. Relationship between the effects of *W* and W^v substitution on hair pigmentation and erythrocytes, *Genetics* **34:**708–712.

Russell, E. S., 1979b, Hereditary anemias of the mouse: A review for geneticists, *Adv. Genet.* **20:**357–364.

Sarvella, P. A., and Russell, L. B., 1956, *Steel*, a new dominant gene in the mouse, *J. Hered.* **47:**123–127.

Schuening, F. G., Kawahara, K., Miller, A. D., To, R., Goehle, S., Stewart, D., Mullally, K., Fisher, L., Graham, T. C., Appelbaum, F. R., Hackman, R., Osborne, W. R. A., and Storb, R., 1991, Retrovirus-mediated gene transduction into long-term repopulating marrow cells of dogs, *Blood* **78:**2568–2576.

Schuening, F. G., Appelbaum, F. R., Deeg, H. J., Sullivan-Pepe, M., Graham, T. C., Hackman, R., Zsebo, K. M., and Storb, R., 1993, Effects of recombinant canine stem cell factor, a *c-kit* ligand, and recombinant granulocyte colony-stimulating factor on hematopoietic recovery after otherwise lethal total body irradiation, *Blood* **81:**20–26.

Shieh, J.-H., Chen, Y.-F., Briddell, R., Stoney, G., and McNiece, I. K., 1994, High purity of blast cells in CD34 selected populations are essential for optimal *ex vivo* expansion of human GM-CFC, *Exp. Hematol.* **22**(8)**:**756 (Abstract).

Steinshamm, S., Bergh, K., and Waage, A., 1993, Effects of stem cell factor and granulocyte colony-stimulating factor on granulocyte recovery and *Candida albicans* infection in granulocytopenic mice, *J. Infect. Dis.* **168:** 1444–1448.

Sutherland, H. J., Lansdorp, P. M., Henkelman, D. H., Eaves, A. C., and Eaves, C. J., 1990, Functional characterization of individual human hematopoietic stem cells cultured at limiting dilution on supportive marrow stromal layers, *Proc. Natl. Acad. Sci. U.S.A.* **87:**3584–3588.

Tanaka, R., Koike, K., Imai, T., Shiohara, M., Kubo, T., Amano, Y., Komiyama, A., and Nakahata, T., 1992, Stem cell factor enhances proliferation, but not maturation of murine megakaryocytic progenitors in serum-free culture, *Blood* **80:**1743–1749.

Toksoz, D., Zsebo, K. M., Smith, K. A., Mu, S., Brankow, D., Suggs, S., Martin, F. M., and Williams, D., 1992, Support of human hematopoiesis in long-term bone marrow cultures by murine stromal cells selectively expressing the membrane bound and secreted forms of the human homolog of the steel gene product, stem cell factor, *Proc. Natl. Acad. Sci. U.S.A.* **89:**7350–7354.

Tsai, M., Shih, L. S., Newlands, G. F. J., Takeishi, T., Langley, K. E., Zsebo, K. M., Miller, H. R. P., Geissler, E. N., and Galli, S. J., 1991, The rat *c-kit* ligand, stem cell factor, induces the development of connective tissue type and mucosal mast cells *in vivo*: Analysis by anatomical distribution, histochemistry, and protease phenotype, *J. Exp. Med.* **174:**125–131.

Ulich, T. R., Delcastillo, J., McNiece, I. K., Yi, E. S., Alzona. C. P., Yin, S. M., and Zsebo, K. M., 1991a, Stem cell factor in combination with granulocyte colony-stimulating factor (CSF) or granulocyte-macrophage CSF synergistically increases granulopoiesis *in vivo*, *Blood* **78:**1954–1962.

Ulich, T. R., Delcastillo, J., Yi, E. S., Yin, S. M., McNiece, I., Yung, Y. P., and Zsebo, K. M., 1991b, Hematologic effects of stem cell factor *in vivo* and *in vitro* in rodents, *Blood* **78:**645–648.

Wershil, B. K., Tsai, M., Geissler, E. N., Zsebo, K. M., and Galli, S. J., 1992, The rat *c-kit* ligand stem cell factor, induces *c-kit* receptor-dependent mouse mast cell activation *in vivo*. Evidence that signaling through the *c-kit* receptor can induce expression of cellular function, *J. Exp. Med.* **175:**245–255.

Wieder, R., Cornetta, K., Kessler, S. W., and Anderson, W. F., 1991, Increased efficiency of retroviral-mediated gene transfer and expression in primate bone marrow after 5-fluorouracil-induced hematopoietic suppression and recovery, *Blood* **77:**448–457.

Williams, D. E, Eisenman, J., Baird, A., Rauch, C., van Ness, K., March, C. J., Park, L. S., Martin, U., Mochizuki, D. Y., Boswell, H. S., Burgess, G. S., Cosman, D., and Lyman, S. D., 1990, Identification of a ligand for the *c-kit* proto-oncogene, *Cell* **63:**213–224.

Williams, N., Bertoncello, I., Kavnoudias, H., Zsebo, K., and McNiece, I., 1992, Recombinant rat stem cell factor stimulates the amplification and differentiation of fractionated mouse stem cell populations, *Blood* **79:**58–64.

Yan, X. O., Briddell, R., Hartley, C., Stoney, G., Samal, B., and McNiece, I. K., 1994, Mobilization of long-term hematopoietic reconstituting cells in mice by the combination of SCF plus G-CSF, *Blood* **84**(3)**:**795–799.

Yarden, Y., Kuang, W., and Yang-Feng, T., 1987, Human proto-oncogene *c-kit*: A new cell surface receptor tyrosine kinase for an unidentified ligand, *EMBO J.* **6:**3341.

Young, J. D., Crawford, J., Gordon, M., Hsieh, A., Yyres, M., Meyres, F., Fare, J., and Demetri, G., 1993, Pharmacokinetics (pk) of recombinant methionyl human stem cell factor (SCF) in patients (pts) with lung and breast cancer in phase I trials, *Proc. Am. Assoc. Cancer Res.* **34:**217 (Abstract).

Zsebo, K. M., Williams, D. A., Geissler, E. N., Broudy, V. C., Martin, F. H., Atkins, H. L., Hsu, R.-Y., Birkett, N. C., Okino, K. H., Murdock, D. C., Jacobsen, F. W., Langley, K. E., Smith, K. A., Takeishi, T., Cattanach, B. M., Galli, S. J., and Suggs, S. J., 1990a, Stem cell factor is encoded at the *Sl* locus of the mouse and is the ligand for the *c-kit* tyrosine kinase receptor, *Cell* **63:**213.

Zsebo, K. M., Wypych, J., McNiece, I. K., Lu, H., Smith, K., Karkare, S., Sachdev, R., Yuschenkoff, V., Birkett, N., Williams, L., Satyagal, V., Tung, W., Bosselman, B., Mendiaz, E., and Langley, K., 1990b, Identification, purification and biological characterization of hematopoietic stem cell factor from buffalo rat liver-conditioned medium, *Cell* **63:**195–199.

Zsebo, K. M., Smith, K. A., Hartley, C. A., Greenblatt, M., Cooke, K., Rich, W., and McNiece, I. K., 1993, Radioprotection of mice by recombinant rat stem cell factor, *Proc. Natl. Acad. Sci. U.S.A.* **89:**9464–9468.

Chapter 15

Clinical Use of Myeloid Growth Factors

Andrew Weaver and Nydia G. Testa

1. INTRODUCTION

With the development of DNA technology, hematopoietic growth factors have been produced in sufficient quantities to enable them to enter the clinic, and there has been an explosion of clinical trials over the last 5 years attempting to define their role. Some of the more commonly used ones are listed in Table I. There are an ever-increasing number of conditions in which hematopoietic growth factors are used, and for any one condition an increasing number of growth factors or combination of factors are being tested in the form of clinical trials. Many would argue that our access to these powerful molecules has, in many ways, outstripped our knowledge of how to use them effectively.

Information on the basic physiological actions of the hematopoietic growth factors can be found elsewhere in this book. We aim in this chapter to review the major clinical areas where hematopoietic growth factors are currently being used, the evidence to support their use, as well as any evidence to the contrary. We also discuss what the future might hold for the clinical use of these factors. The major areas discussed below are their use in delivery of standard and high-dose chemotherapy, bone marrow transplantation, peripheral blood progenitor cell (PBC) mobilization and PBC transplantation, treatment of neutropenic sepsis, and, last but not least, treatment of leukemia.

2. STANDARD-DOSE CHEMOTHERAPY

The development of effective chemotherapy agents and combination regimens has been a milestone in oncology and, as a consequence, has resulted in previously fatal

Andrew Weaver and Nydia G. Testa Cancer Research Campaign Departments of Experimental Haematology and Medical Oncology, Christie Hospital NHS Trust, Manchester M20 9BX, United Kingdom.

Blood Cell Biochemistry, Volume 7: Hematopoietic Cell Growth Factors and Their Receptors, edited by A. D. Whetton and J. Gordon. Plenum Press, New York, 1996.

Table I
Hematopoietic Growth Factors and Their Primary Effector Cells

Growth factor		Form	Effector cell
Granulocyte colony-stimulating factor	G-CSF	Glycosylated and nonglycosylated	Neutrophil
Granulocyte–macrophage colony-stimulating factor	GM-CSF	Glycosylated and nonglycosylated	Eosinophil, neutrophil, monocyte, macrophage
Macrophage colony-stimulating factor	M-CSF		Monocyte, macrophage
Interleukin 3	IL-3 or multiCSF		Mast cell, basophil, macrophage, megakaryocyte, red blood cell
GM-CSF/IL-3 fusion protein	PIXY-321		Eosinophil, neutrophil, monocyte, macrophage, mast cell, basophil, megakaryocyte, red blood cell
Erythropoietin	EPO	α and β	Red blood cell

conditions becoming potentially curable. Failure of chemotherapy treatment may ultimately follow the development of resistant clones of malignant cells within the tumor. This resistance by the malignant cells may be present at the beginning of a tumor's life or may develop following exposure to chemotherapy. In the former case, clearly the prognosis is extremely poor, because without novel therapeutic strategies or alternative treatments such as radiotherapy, the resistance cannot be overcome. In the latter case, the aim of treatment, where possible, is to eradicate the tumor before the emergence of resistant clones that are immune to the chemotherapy.

The relationship between the dose of chemotherapy and tumor response was first demonstrated by Skipper (1967) in animal experiments using tumor models such as L1210 leukemia. In these models the dose response is curvilinear, with a steep response to the antitumor agent in the linear portion of the graph (Frei and Canellos, 1980). Dose reduction in the linear phase of the graph results first in a loss of capacity to cure the tumor before there is a loss in response rate. In other words, complete remissions will continue to be observed in animals, but the last few residual cells will not be ablated, and the animal will ultimately relapse.

Successful chemotherapy regimens, with very few exceptions, consist of a combination of active, non-cross-resistant drugs that are administered at an effective dose. Dose intensity is of fundamental importance in the successful outcome of chemotherapy against a patient's cancer. Both the response rate and, ultimately, the cure rate may be sacrificed if the dose intensity is decreased. Unfortunately in clinical practice dose intensity is often reduced because of toxic side effects of the treatment, and although the dose reduction may alleviate the side effects to some extent, it may also allow the tumor to escape control (DeVita, 1989). Thus, decreased dose intensity has a dramatic effect on the prognosis of the patient.

As with many therapeutic manipulations in medicine, there is a price to pay. As the dose of chemotherapy is increased, very often the degree of neutropenia is more marked as a result. The myeloid growth factors were first used in supporting patients undergoing conventional-dose chemotherapy. The first published nonrandomized study in small-cell lung cancer by Bronchud and colleagues (1987) tested the effect of G-CSF given to patients for 14 days on alternate cycles only of doxorubicin, ifosphamide, and etoposide (AIE), so that each patient acted as his or her own control. Cycles supported by G-CSF had shorter and less severe episodes of neutropenia and also fewer febrile episodes.

The results of the first randomized, placebo-controlled trial in patients with small-cell lung cancer were published in 1991 by Crawford and colleagues. In this study, 199 patients were evaluable, and they received up to six cycles of combination chemotherapy (cyclophosphamide, doxorubicin, and etoposide). The incidence of severe (grade IV) neutropenia was reduced in the G-CSF group, relative to the placebo group, as was the median duration of episodes (6 days versus 1 day across all cycles). A 50% reduction in the frequency of infection, as manifested by fever with neutropenia, was produced; decreases in the rate of infections confirmed by culture, mean number of days of hospitalization, and intravenous antibiotic use were of similar magnitude. This positive effect persisted throughout the six cycles of chemotherapy. The percentage of patients who qualified to receive their next planned course of chemotherapy was significantly increased. However, the number of complete responses and overall survival were the same with or without G-CSF.

Granulocyte–macrophage CSF was evaluated by Antman and colleagues (1988) for patients receiving chemotherapy for sarcoma. A first cycle containing GM-CSF had fewer days of neutropenia when compared to a non-GM-CSF-containing second cycle (3.5 days compared to 7.4 days). The results in this study may have been influenced by the cumulative effects of cytotoxic drug administration. In a further, nonrandomized study investigating the efficacy of GM-CSF by Herrmann and colleagues (1990), patients with a variety of tumors received GM-CSF-containing second cycles and were noted to have less neutropenia, which resulted in fewer days of fever and reduced antibiotic requirement. In a small randomized study in patients with International Federation of Gynecology and Obstetrics (FIGO) stage III or IV ovarian cancer, patients were allocated into one of three dose levels of GM-CSF or placebo. Neutropenia was significantly reduced in those receiving GM-CSF, particularly at the highest dose level (6 μg/kg per day). In addition, platelet counts were increased in GM-CSF-treated patients (de Vries *et al.*, 1991).

Two further randomized studies involving patients with small-cell lung cancer deserve describing. In the first, patients were randomized to receive either cisplatin, vincristine, doxorubicin, and etoposide (COAE) alone or COAE plus G-CSF. Febrile episodes were significantly reduced and dose intensity increased, but the overall total response rate did not improve in those patients receiving G-CSF. Although these early results suggest that COAE plus G-CSF improved the clinical outcome, the follow-up period was too short, and the number of patients too few, to reach a meaningful conclusion on the benefit of adding G-CSF to standard-dose treatment (Fukuoka *et al.*, 1992).

The second study is a randomized trial performed by the European G-CSF Lung Cancer Study Group, comparing cisplatin, doxorubicin, and etoposide (CAE) plus G-CSF with CAE plus placebo in 130 patients with small-cell lung cancer (Trillet-Lenoir *et al.*, 1993). The results were very consistent with many other studies involving similar patients: febrile neutropenia and antibiotic use were reduced in the G-CSF-treated group, treatment delays occurred more frequently in the placebo group, but response rates and survival were similar in both groups (Table II).

Finally two randomized studies (Pettengell *et al.*, 1992; Engelhard *et al.*, 1993) in patients with non-Hodgkin lymphoma treated with standard-dose chemotherapy and randomized to growth factor support showed a significant benefit in terms of hematological

Table II
Randomized Studies Investigating the Role of G-CSF in Small-Cell Lung Cancer: Results for Chemotherapy Plus G-CSF as Compared to the Control Group

Study	Number of patients	Dose intensity	Febrile episodes	Response (CR + PR)[a]	Overall survival
Crawford *et al.* (1991)	207	Higher	Fewer	NS[b]	NS
Fukuoka *et al.* (1992)	64	Higher	Fewer	NS	$*p < 0.05$
Trillet-Lenoir *et al.* (1993)	130	Higher	Fewer	NS	NS

[a]CR, complete remission; PR, partial remission.
[b]NS, not significant.

outcome. The Christie Hospital group randomized 80 patients with high-grade disease to receive a weekly chemotherapy regimen (VAPEC-B) alone or chemotherapy with G-CSF. There were fewer and shorter treatment delays in the G-CSF group ($p = 0.02$). In addition, more of the control patients experienced at least one dose reduction in their myelosuppressive drugs ($p = 0.01$). This study did not show increased survival in those receiving G-CSF (median follow-up 15 months).

Engelhard and colleagues (1993) treated similar patients with COP-BLAM (cyclophosphamide, vincristine, prednisolone, doxorubicin, procarbazine, bleomycin) chemotherapy and randomized patients to GM-CSF. This trial confirmed a higher response rate (69% versus 48%) in those patients receiving the growth factor support, but this again has not translated into improved survival.

In conclusion, myeloid growth factors used in conjunction with standard-dose chemotherapy are well tolerated and produce relatively mild side effects, such as bone pain occurring in approximately 20%, which can be relieved by acetaminophen. A rise in serum lactate dehydrogenase (LDH) and uric acid may occur during treatment, particularly with higher doses of G-CSF. Rashes and influenza-type side effects may occur with GM-CSF, especially as the clinician escalates the dose of growth factor.

Most randomized studies to date confirm that the addition of G-CSF or GM-CSF growth factors to chemotherapy results in fewer and shorter episodes of neutropenic fever and hence permits delivery of increased dose intensity of treatment. As yet, in the setting of conventional-dose chemotherapy this has rarely translated into increased response rates or survival for the patient, though the majority of studies, whether randomized or not, did not have response or survival rates as a primary study objective. The exception to this finding is the study of Fukuoka and colleagues (1992) showing an overall survival advantage in favor of those small-cell lung cancer patients receiving growth factor in addition to chemotherapy ($p < 0.05$).

Despite many studies in bladder carcinoma (Gabrilove *et al.*, 1988; Kotake *et al.*, 1991; Aso and Akaza, 1992), breast cancer (Chevallier *et al.*, 1993; Venturini *et al.*, 1992), and sarcoma (Chevallier *et al.*, 1992) in addition to the abovementioned trials, we feel there is no clear evidence supporting the use of G-CSF or GM-CSF in patients receiving conventional, standard-dose chemotherapy except to alleviate a few days of fever, and whether this is clinically justified if one considers the cost–benefit ratio is debatable. Clearly, more randomized clinical trials with this objective in mind are needed.

3. HIGH-DOSE CHEMOTHERAPY

In chemosensitive malignancies in which combination chemotherapy has improved response rates and overall survival, dose intensity is of fundamental importance. The South West Oncology Group (SWOG) found that half-dose cyclophosphamide, doxorubicin, vincristine, and prednisolone (CHOP) was inferior to full-dose CHOP in a nonrandomized retrospective analysis (12% versus 41% cure in favor of those patients with high-grade lymphoma receiving full-dose CHOP) (Coltman *et al.*, 1986).

In Hodgkin disease, a comparison of studies of nitrogen mustard, vincristine, procarbazine, and prednisolone (MOPP) chemotherapy at different centers shows variability in treatment outcome. Response rates were very similar, but the more important parameter of relapse-free survival varied markedly, being 55% in the NCI series (DeVita *et al.*, 1980) but only 36% at 8 years in the Bonadonna (1987) series and 31% at 5 years in the SWOG series (Coltman *et al.*, 1978). These differences may be explained in terms of varying amounts of chemotherapy actually administered to the patients rather than variability in prognostic factors between the groups. DeVita (1989) calculated the relative dose intensity of chemotherapy from the different centers and correlated it with complete remission rates and disease-free survival. Table III shows that there was a trend for more favorable outcome for patients receiving MOPP at a greater dose intensity (Hellman *et al.*, 1989). The correlation was even more striking when comparisons were made with the actual chemotherapy that was given rather than with the planned doses. Reductions of 29% and 38% in the Eastern Cooperative Oncology Group (ECOG) and Milan series, respectively, resulted in 33% and 35% decreases in overall disease-free survival.

In advanced ovarian cancer the effect of dose intensity of first-line chemotherapy on clinical outcome has been studied retrospectively by Levin and Hryniuk (1987). They calculated the average dose intensity of drugs relative to a standard cyclophosphamide, hexamethylmelamine, doxorubicin, and cisplatin (CHAP) regimen and found that the average relative dose intensity, especially of cisplatin, correlated significantly with clinical response and also with the median survival time. Furthermore, there was a distinct advantage for combination regimens containing cisplatin compared with single-agent alkylating

Table III
Dose Intensity and Outcome in Patients with Hodgkin Disease Treated with MOPP Regimens[a]

Study	Relative dose intensity versus NCI MOPP	Actual relative dose intensity	Complete remission (%)	Patients free of disease (%)	Follow-up (Years)
NCI	1.0	0.85	84	55	15
Stanford	0.95	0.64	72	30	5
BNLI	0.82		52	30	5
SEG	0.82	0.64	46	16	2
CALGB	0.81		74	37	5
ECOG	0.77	0.60	73	37	5
Milan	0.76	0.53–0.66	74	36	8
SWOG	0.70		78	31	5

[a]Adapted from Hellman *et al.* (1989).

agents. Additional evidence in support of this retrospective analysis comes from experimental studies that show a steep dose–response relationship for cisplatin, the most active single agent in ovarian cancer (Behrens *et al.*, 1985).

Although there is a substantial amount of experimental evidence (Skipper, 1967; Frei and Canellos, 1980) in support of the concept of dose-intensive treatment resulting in higher response rates because of increased tumor kill, it has by no means been proven in the clinical setting that delivery of higher-than-standard-dose chemotherapy is necessarily translated into improved survival. Data like those summarized above led logically to trials studying delivery of high-dose chemotherapy with the use of growth factors to alleviate the most frequent dose-limiting toxicity, myelosuppression. Dose intensification may be achieved by increasing the dose of one or more of the drugs being administered in a combination regimen or, alternatively, by decreasing the interval between treatment cycles.

Prior to formal phase I and II studies of intensive chemotherapy, several groups made attempts at redefining the toxicities of individual chemotherapeutic agents and their combinations when used in conjunction with hematopoietic growth factors (Edmonson *et al.*, 1990; Lichtman *et al.*, 1989). For some of these treatments, the dose-limiting toxicity has remained hematological; often growth factors support the neutrophil count adequately, but the platelet counts are critically lowered as a result of the increased dose intensity. For other treatments the growth factors are sufficient to overcome the hematological toxicities, only to be replaced by a nonhematological dose-limiting toxicity.

In order to accelerate marrow recovery following high-dose treatment, several investigators have found the timing and the dose of the growth factor administration important (Havermann *et al.*, 1991). Gianni *et al.* (1990a) and Vadhan-Raj *et al.* (1992) both found it important to commence GM-CSF as soon as possible following chemotherapy, although the cytotoxic drugs should probably be allowed to "wash out" before the growth factors simulate myeloid proliferation.

Vadhan-Raj *et al.* (1992) has also demonstrated that patients with sarcoma who received GM-CSF prior to chemotherapy showed a paradoxical increase in myeloprotection, which may permit dose intensification. Several mechanisms for this are postulated: either the GM-CSF increases the myeloid cell mass or, following the withdrawal of GM-CSF and prior to the chemotherapy, a quiescent state in the bone marrow may be induced, hence protecting the stem cells and early progenitor cells. However, because of the risks inherent in such an approach, including the uncertainty about the timing of the protection versus possible deleterious effects (Molineux *et al.*, 1994), most trials administer the growth factors following chemotherapy.

Granulocyte CSF was used to increase the intensity of doxorubicin treatment in patients with breast and ovarian cancer by both increasing the dose and reducing the interval between cycles. With this approach it was possible to escalate the doxorubicin dose intensity 4.5-fold before nonhematological dose-limiting toxicities were reached. A high response rate was reported, but the duration of the remission was disappointingly short (Bronchud *et al.*, 1989). However, when Hoekman and colleagues (1991) attempted to increase dose intensity in breast cancer patients by escalating both doxorubicin and cyclophosphamide with GM-CSF support, cumulative myelotoxicity (see below) and prolonged thrombocytopenia occurred. This forced dose reductions of both drugs from cycle 2 onward in all patients treated with the highest dose. Also, fever associated with GM-CSF led to a high rate of hospitalization and antibiotic use.

An EORTC study has shown that the treatment intervals can be reduced in patients with metastatic breast cancer when supported with G-CSF. Patients received epirubicin (100 mg/m^2) and cyclophosphamide (830 mg/m^2), and cycle intervals were reduced from 22 to 13 days, with G-CSF given on days 2–11. The clinical response rate was 94%, with a complete remission rate of 25% (Piccart *et al.*, 1991).

Neidhart initially reported in 1989 the use of G-CSF, followed in 1990 with the use of GM-CSF, to enable delivery of high-dose cyclophosphamide, etoposide, and cisplatin, without bone marrow transplantation, to patients with chemotherapy-resistant malignancies. The use of growth factor support resulted in accelerated neutrophil recovery, and with G-CSF at the highest dose (60 μg/kg per day), the median duration of antibiotic therapy was reduced, although the duration of hospitalization was not reduced (Table IV). It was noted that the two highest dose levels of GM-CSF resulted in marked side effects such as fever, headaches, and edema. The authors recommended that GM-CSF is optimal at 500 μg/m^2 and starting on day 4 after chemotherapy.

The results from various studies have shown conflicting results in regard to the effects of chemotherapy and growth factors on platelets. Several studies have described prolonged thrombocytopenia (Hoekman *et al.*, 1991; Shea *et al.*, 1990; O'Dwyer *et al.*, 1992). In some nonrandomized comparisons the opposite effect was noted, and GM-CSF appeared to support the platelet count (Edmonson *et al.*, 1990; Gianni *et al.*, 1990a; Steward *et al.*, 1990). The effect seen depended more on the chemotherapy regimen than anything else and is variable and inconsistent.

As mentioned above, dosage and scheduling of growth factors may be important. Increasing the frequency at which GM-CSF is administered may well have a beneficial effect in terms of dose delivery, as shown by Edmonson and colleagues (1989). They investigated increasing the dose of carboplatin in combination with cyclophosphamide (1 g/m^2). When a daily schedule of 20 μg/day GM-CSF is used, the carboplatin dose could not be increased beyond 300 mg/m^2. But if the GM-CSF administration was changed to 10 μg/kg twice daily, then the carboplatin dose could be escalated to 700 mg/m^2.

It may be important not only to deliver high-dose chemotherapy at the initiation of

Table IV
G-CSF and GM-CSF following Dose-Intensive Chemotherapy in Patients with Resistant Malignancies[a]

	Day of starting cytokine	ANC × 10^9/liter (days)[b]		Duration of antibiotics (days)
		<0.1	<0.5	
Historical controls		8.5	12	9
G-CSF				
40 μg/day		7.0	6	5
60 μg/day		5.5	6	5
GM-CSF				
500 μg/m^2	Day 6	8.0	—	
	Day 4	6.5*	—	
750 μg/m^2	Day 6	8.0	—	
1000 μg/m^2	Day 6	7.0	—	

[a]Adapted from Neidhart *et al.* (1989, 1990).
[b]ANC, absolute neutrophil count. *Significant decrease versus control.

treatment but also to sustain dose intensity over a number of treatment cycles. Several studies have reported that GM-CSF does not abrogate cumulative myelotoxicity following multiple cycles of chemotherapy. Cebon and colleagues showed the importance of the schedule of GM-CSF administration relative to chemotherapy administration. Even though an optimal schedule was identified and the frequency of GM-CSF injections was increased, dose-intensive chemotherapy could not be supported beyond three cycles for patients with small-cell lung cancer receiving double-dose carboplatin (200 mg/m^2 for 3 days) and etoposide (240 mg/m^2 for 3 days) supported with twice-daily injections of subcutaneous GM-CSF (Cebon *et al.*, 1994). A similar result was seen by Rusthoven and colleagues (1992) when they were unable to increase and sustain carboplatin dose beyond 600 mg/m^2. After two or three cycles at this dose, delays or dose reductions were required because of the prolonged cytopenias.

In conclusion, early trials of dose-intensive chemotherapy with growth factor support confer a modest advantage in terms of infection rates. However, there is little evidence that any beneficial effects on neutrophil counts can be sustained as the patient progresses through successive cycles of chemotherapy. Likewise, these early trials have already revealed that when dose intensity can be achieved in terms of avoiding severe neutropenia, the myeloid growth factors fail to improve the prolonged thrombocytopenia that commonly occurs. However, a recently described growth factor, c-Mpl ligand or thrombopoietin (TPO), shows a marked and apparently specific stimulating effect on the megakaryocyte lineage, which results in increased platelet production, and has just been introduced into phase I clinical trials in America and Australia (Metcalf, 1994). If the hopes invested in TPO are realized, the last lineage that determines hematopoietic limiting toxicity could be manipulated in the clinic.

The other major problem the clinician may well experience is that, even if the dose-intensive treatment is relatively sparing in terms of hematological toxicity, a non-hematological toxicity may supervene, e.g., neuropathy or bowel toxicity.

Although treatment has been intensified in terms of dose, and response rates have been correspondingly increased, there is very little evidence yet available to show that this translates into improved survival for the patient. Until there is a clear demonstration by well-designed clinical trials that growth factor administration as supportive treatment to chemotherapy leads to improved survival, it is too early to recommend the routine use of growth factors for dose intensification outside the setting of a clinical trial. Until such trials are forthcoming, myeloid growth factors may find their most useful role in high-dose therapies such as myeloablative treatments followed by transplantation, or as agents for mobilization of peripheral blood progenitor cells.

4. BONE MARROW TRANSPLANTATION

Hematopoietic growth factors have been used in conjunction with bone marrow transplantation for several years, and their clinical benefits have been reported in numerous phase II and III clinical trials. The majority of investigators have used the growth factors to stimulate the recovery of neutrophils and hence reduce infectious sequelae. Most experience has been with G-CSF and GM-CSF, and little has been published on the clinical use and effectiveness of M-CSF and IL-3 in this setting.

4.1. Autologous Bone Marrow Transplantation

The administration of G-CSF following autologous bone marrow transplantation significantly reduces the time to neutrophil recovery (Sheridan *et al.*, 1989; Taylor *et al.*, 1989; Peters *et al.*, 1989; Peters, 1989). In these early studies, G-CSF (filgrastim) was used at relatively high doses (20–75 μg/kg per day for up to 28 days) and given by intravenous or subcutaneous infusion. A variety of conditioning chemotherapy regimens were used for several different malignancies including Hodgkin disease, breast cancer, and malignant myeloma. The median time to absolute neutrophil count recovery (>0.05 × 10^9/liter) was 14 days or less in patients who received G-CSF. These data compare with the median time to recovery of approximately 20–30 days in historical controls treated without G-CSF.

In studies of high-dose chemotherapy and autologous bone marrow transplantation, more than 90% of patients may experience febrile neutropenia. There is little evidence that G-CSF has reduced this frequency. Sheridan and colleagues did show, however, that the reduced duration of neutropenia in autologous transplants supported by G-CSF was associated with a reduced number of days with fever (7 versus 9.4 days). They also noted a significantly reduced duration of parenteral antibiotic use (11 versus 18 days) as well as a significant reduction in the length of time for which total parenteral nutrition was required when compared to matched historical controls (10 versus 16 days). Parenteral nutrition is a reasonably good measure of severe mucositis associated with prolonged periods of neutropenia in the bone marrow transplantation setting. As a result of these reduced recovery times, there was a significant reduction in the time a patient spent in isolation (10 versus 18 days). Likewise, there was a trend for the patient to spend fewer days hospitalized (23 versus 30 days) (Sheridan *et al.*, 1989).

The efficacy of GM-CSF following autologous bone marrow transplant has been demonstrated in several nonrandomized studies and subsequently confirmed in randomized placebo-controlled trials (Nemunaitis *et al.*, 1988; Blazar *et al.*, 1989; Brant *et al.*, 1988; Devereux *et al.*, 1989; Lazarus *et al.*, 1990; Link, 1990; Bennett *et al.*, 1990; Link *et al.*, 1990a; Rabinowe *et al.*, 1990; Advani *et al.*, 1990; Gorin *et al.*, 1990; Nemunaitis *et al.*, 1991b; Michon *et al.*, 1990). In these studies, patients with a wide variety of malignancies have been treated with many different conditioning regimens consisting of chemotherapy with or without irradiation. However, different forms of GM-CSF (derived from yeast, *E. coli*, or mammalian cells) have been extensively investigated using different doses as well as schedules, which may call into doubt the validity of some of the comparisons made by different investigators.

The majority of nonrandomized studies have shown that GM-CSF is associated with shortened time to neutrophil recovery. Nemunaitis and colleagues (1988) studied patients with lymphoid malignancies and found that the time to absolute neutrophil recovery (>0.5 × 10^9/liter), was significantly shorter in patients receiving >60 μg/kg per day GM-CSF than in historical controls (17 versus 25 days). However, Brandt (1988) showed no difference in this parameter in breast cancer or melanoma patients receiving GM-CSF following their transplant compared with historical controls.

Nemunaitis *et al.* (1991a) also reported a randomized, double-blind study conducted at three centers in which 128 patients were evaluable. It showed that GM-CSF accelerated neutrophil recovery to >0.5 × 10^9/liter by 7 days. In another large series reported by Advani and colleagues (1990), 231 patients were evaluable, and neutrophil recovery to >0.05 × 10^9/

liter was reached by 16 days in those receiving GM-CSF as opposed to 19 days in those receiving placebo. This difference was highly significant.

A multicenter, randomized, double-blind European study in non-Hodgkin lymphoma and acute lymphoblastic leukemia found a significantly shortened time to neutrophil recovery in those treated with GM-CSF compared to those patients receiving placebo (15 days compared to 28 days, $p = 0.0001$) (Link *et al.*, 1990a).

The effect of GM-CSF on platelet recovery is not well established. Nemunaitis' randomized, placebo-controlled trial (1991a) showed no significant difference in the time to platelet independence between the two groups, yet the earlier nonrandomized study by the same group had reported that patients receiving GM-CSF had reduced times to platelet transfusion independence from 38 days to 29 days.

In view of these findings, some investigators have attempted to improve platelet recovery by using IL-3 after transplantation. Interleukin 3 may encourage proliferation of the megakaryocyte progenitors, resulting in expansion of the platelet population. Nemunaitis and colleagues (1992) conducted a phase I study in Seattle: 17 patients with lymphoid malignancies were treated with IL-3 followed autologous bone marrow transplantation. The IL-3 had little effect on platelet recovery, though it did appear to hasten neutrophil recovery. Two further studies have shown similar results, with some effect on neutrophil recovery but little evidence for an effect on platelet recovery (Peters *et al.*, 1991; Fibbe *et al.*, 1991).

Although GM-CSF stimulates macrophage production (unlike G-CSF at clinical doses), there is little evidence to show that GM-CSF reduces the incidence of fungal infections following bone marrow transplant.

Despite the studies mentioned above, there are still important questions that remain unanswered. Neither the optimal dose nor schedule of GM-CSF is known, and several pilot studies suggest that a dose administered over a 2-hr period is, perhaps, less effective than if it were administered by continuous infusion.

Most of the published trials have used GM-CSF produced in yeast, and the comparative effectiveness and toxicity of GM-CSF produced in *E. coli* and Chinese hamster ovary cells is uncertain. But clearly, both GM-CSF and G-CSF are capable of reducing the duration of severe neutropenia following autologous bone marrow transplantation by approximately 7 days (Table V). This may lead to a reduction of life-threatening infectious complications and also to earlier discharge from hospital.

4.2. Allogeneic Bone Marrow Transplantation

There have been several studies reported on the use of GM-CSF following allogeneic transplantation. Four of those listed in Table VI are nonrandomized, and two are randomized studies (McKenzie *et al.*, 1990; Singer *et al.*, 1990a,b; De Witte *et al.*, 1992; Powles *et al.*, 1990). Singer and colleagues (1990a,b) reported a nonrandomized study in leukemic patients and demonstrated that GM-CSF significantly accelerated neutrophil recovery to $>0.1 \times 10^9$/liter compared to historical controls (4 versus 12 days). Recovery to $>1.0 \times 10^9$/liter took 13 days in GM-CSF-treated patients compared to 21 days in controls, though this difference was not significant.

Nemunaitis and colleagues (1991a,b) reported a nonrandomized dose-escalation study showing that those patients treated with GM-CSF had reduced neutropenia, days of fever, and hospitalization, though benefit was seen only in those patients who were not receiving

Table V
Comparative Studies of Growth Factor Effects following Autologous Bone Marrow Transplantation

Study	Tumor type[b]	Number of patients receiving growth factor	Growth factor	Control group	Duration of ANC <0.5 × 10^9/liter (days) GF[a]	Control	Documented infection
Sheridan *et al.* (1989)	Mixed	16	G-CSF	Historical	12*	20	NS
Taylor *et al.* (1989)	Hodgkin's	25	G-CSF	Historical	14*	22	NS
Peters *et al.* (1989)	Breast and myeloma	45	G-CSF	Historical	14*	19	NS
Nemunaitis *et al.* (1988)	Lymphoid	15	GM-CSF	Historical	17*	25	+
Brandt *et al.* (1988)	Breast and melanoma	19	GM-CSF	Historical	NS		+
Advani *et al.* (1990)	Solid tumor and NHL	69	GM-CSF	Placebo/ randomized	16*	19	+
Link *et al.* (1990a)	ALL and NHL	79	GM-CSF	Placebo/ randomized	15*	28	*

[a]GF, growth factor group. *Significant difference in favor of growth factor. [+]Difference in favor of growth factor but not reaching significance.
[b]ALL, acute lymphoblastic leukemia; NHL, non-Hodgkin lymphoma.

methotrexate as prophylaxis against graft-versus-host disease. There did not appear to be an increased incidence of graft-versus-host disease.

In a multicenter randomized, double-blind trial of 57 patients with diagnoses comprising leukemia, aplastic anemia, or multiple myeloma, GM-CSF significantly accelerated engraftment of neutrophils and monocytes compared with placebo. Mortality from both infective and noninfective causes was reduced in the GM-CSF group compared to the control patients, but the difference was not statistically significant. However, the incidence of bronchopneumonia was significantly lower in the GM-CSF-treated patients (two versus eight patients). Likewise, the number of days on systemic antibiotics was 14 in the GM-CSF group compared with 20 days in the control group with 30 evaluable patients ($p = 0.03$) (De Witte *et al.*, 1992).

Powles and colleagues (1990) reported a double-blind, placebo-controlled trial of 40 patients with leukemia. Again, GM-CSF accelerated neutrophil recovery after allogeneic bone marrow transplant, but they were unable to report significant clinical benefits. The median neutrophil count at 14 days was 1.9×10^9/liter in the GM-CSF group versus 0.46×10^9/liter in the control groups (significant difference). Overall, neutrophil recovery to $>0.5 \times 10^9$/liter was not significantly different in the two groups. However, in those patients receiving total-body irradiation as part of their conditioning regimen, the median time to recovery was significantly shorter if they, in addition, received GM-CSF (13 versus 18 days). There was no difference in the time to discharge between the two groups. However,

Table VI
Comparative Studies of GM-CSF following Allogeneic Bone Marrow Transplantation

Study	Tumor type[b]	Control group	No. of patients treated/control	Neutropenia[a] (10^9/liter) <0.5	<1.0	Documented infection	Use of antibiotic
Singer *et al.* (1990a)	Lymphoid, AML, CML	Historical	14/185	na	+	na	na
McKenzie *et al.* (1990)	Pediatrics	Historical	13/na	++	na	na	na
Singer *et al.* (1990b)	Leukemia	Placebo/ randomized	29/28	na	na	+	+
Powles *et al.* (1990)	Leukemia	Placebo/ randomized	20/20	+	na	−	−

[a]na, Not available; ++, significant difference in favor of GM-CSF; +, difference in favor of GM-CSF, but not reaching significance; −, no significant difference.
[b]AML, acute myeloid leukaemia; CML, chronic myeloid leukaemia.

fever caused by the study drug appeared to increase the use of empirical antibiotics. Importantly, there was no effect on the leukemic relapse rate in the GM-CSF group.

Granulocyte-CSF has been less well studied after allogeneic transplantation than GM-CSF. In a phase I/II study, with G-CSF being administered to patients with leukemia, seven patients received an allogeneic bone marrow transplant; G-CSF was started 5 days after the transplant and continued until absolute neutrophil count (ANC) recovery reached 10×10^9/liter. The mean duration of recovery to an ANC of 0.5×10^9/liter was 11.3 days with G-CSF compared with 26.8 days in historical controls. This study also showed that severe infections that could not be controlled with antibiotics alone responded to G-CSF plus antibiotics (Teshima *et al.*, 1989).

The results of a randomized, placebo-controlled, phase II trial of G-CSF in both adults and children receiving autologous or allogeneic transplants were published in 1992 (Harousseau, 1992; Linch, 1992). Patients from 52 centers in eight European countries were randomized to receive either G-CSF (5 μg/kg per day) or placebo. The G-CSF was given daily until the neutrophil count was $>1.0 \times 10^9$/liter and maintained for 3 consecutive days or until day 28. A total of 298 patients were evaluable for efficacy; 152 received G-CSF, and 146 received placebo. The G-CSF significantly accelerated neutrophil recovery (Table VI).

Subgroup analysis showed similar figures for autologous transplant (n = 224), allogeneic (n = 66) and children (n = 46). The accelerated neutrophil recovery with G-CSF translated into a shorter hospitalization (25 versus 29 days), a reduced period of total parenteral nutrition (6 versus 24 days), reduced intravenous antibiotic use (15 versus 19 days), and a decrease in the period of neutropenic fever (3 versus 5 days). Microbiologically proven infections were also reduced (8 versus 12 days). All these results were statistically significant.

Macrophage CSF and IL-3 have much weaker effects on hematopoietic reconstitution than either GM-CSF or G-CSF (Nemunaitis *et al.*, 1992). An early study, reported in 1988, randomized 115 patients with a variety of malignancies to receive purified human urinary

M-CSF or placebo following autologous and allogeneic transplantation. Recovery to 0.5×10^9/liter neutrophils took 23 days in the patients receiving M-CSF compared with 27 days in the control group (Masaoka *et al.*, 1988). However, this reduction in duration of neutropenia is probably a result of the release of G-CSF and GM-CSF from monocytes and not a direct action of M-CSF on neutrophils.

Little data are available on the use of IL-3 alone in the bone marrow transplant setting, apart from the few studies mentioned earlier when IL-3 was used in an attempt to enhance platelet engraftment. Interleukin 3 can produce a multilineage hematopoietic response, although the onset of the IL-3 effect on recovery is observed later than that seen with more lineage-specific factors such as G-CSF (Mertelsmann, 1991).

Erythropoietin (EPO) has been used in addition to GM-CSF by Pene and colleagues (1993) in attempts to reduce the time for red-cell recovery. They found that the combination of the two growth factors did not improve red-cell or platelet recovery.

Link and colleagues (1990b) demonstrated an increased reticulocyte count and decreased period of red cell transfusion in 12 patients treated with EPO after allogeneic bone marrow transplantation compared with 44 historical controls. Clearly, if a significant effect on red cell production results from the use of EPO following transplantation, additional (myeloid) growth factors will still be required to enhance neutrophil recovery. The use of two growth factors in this situation will certainly increase the cost of the treatment, and as yet there is little substantive evidence that EPO offers any additional clinical benefit in patients treated by bone marrow transplantation.

In summary, if the results of the use of myeloid growth factors after bone marrow transplantation are taken together, then a good case can be made for the use of GM-CSF and G-CSF. These factors lead to reduced periods of neutropenia and hence to less of a requirement for antibiotics and to a shorter hospitalization as a result. This may well be translated into better quality of life for the transplant patient. The period of neutropenia can be reduced on average by 7 days following the use of either factor, and there may be some clinical advantages of using G-CSF instead of GM-CSF in that GM-CSF may produce fever during the time of administration, which may lead to an increased empirical use of antibiotics and ultimately longer hospitalization.

At low doses, GM-CSF is well tolerated, but as the dose is escalated, significant toxicity may result, such as bone pain, fever, thrombophlebitis, and capillary leak syndromes in addition to central venous thrombosis. These more serious side effects are seen at doses >30 μg/kg. Although G-CSF has been less extensively reported in clinical trials involving bone marrow transplantation, it appears to be better tolerated than GM-CSF in this setting. Graft-versus-host disease is not enhanced by the use of myeloid growth factors. To date there is little evidence to support the use of combinations of growth factors such as IL-3 or EPO plus G-CSF or GM-CSF.

5. THE USE OF PERIPHERAL BLOOD CELLS IN TRANSPLANTATION

An immense amount of interest and effort has been spent over the past decade trying to optimize mobilization of primitive hematopoietic cells into the circulation. Several clinical advantages have become apparent by their use: autologous PBC transplants are associated with shorter recovery times of both neutrophils and platelets compared with bone marrow transplants. This recovery time is reduced when PBC are mobilized with growth factors and chemotherapy compared with chemotherapy alone (Sheridan *et al.*, 1992). Transplants of PBC also lead to shorter periods of antibiotic use and hospital stay (Sheridan *et al.*, 1992; Advani *et al.*, 1992; Pettengell *et al.*, 1993). Other advantages are that the apheresis can be performed as an outpatient procedure, avoiding the need for anesthesia, and it is feasible in patients with marrow hypoplasia or malignant infiltration as well as in those who have received pelvic irradiation (Korbling *et al.*, 1992). Growth factors alone may be used to mobilize progenitor cells into the blood, and an apheresis performed in order to harvest these cells. Alternatively, an aliquot of whole blood may be removed following mobilization and subsequently reinfused after further medical intervention (usually the administration of a cycle of chemotherapy (Ossenkoppele *et al.*, 1993).

Although there is evidence that certain regimens may mobilize malignant cells into the circulation as well as progenitor cells, the amount of contamination of the apheresis product appears to depend on the tumor burden (i.e., the clinical stage of the patient) at the time the patient presents, prior to mobilization. However, this contamination appears less compared with the amount of tumor cells present in bone marrow infiltrated by malignancy (Brugger *et al.*, 1994). A further advantage of PBCs is the ability to divide the apheresis product into a defined number of aliquots, and then, following high-dose cyclic chemotherapy, an aliquot of the product can be reinfused to support the patient. Often G-CSF or GM-CSF is administered following the reinfusion of the progenitor cells, although the need to do this has not been established. This high-dose sequential regimen, described initially by Gianni *et al.* (1994), allows clinicians, perhaps for the first time, to deliver high-dose cyclic chemotherapy and hence truly investigate whether high-dose chemotherapy can offer the patient a clear clinical advantage.

5.1. Normal Blood

It is now over 30 years since Goodman and colleagues first noticed that the mononuclear cell component of peripheral blood contained small numbers of cells that were capable of hematopoietic reconstitution in lethally irradiated mice (Goodman and Hodgson, 1962; McCredie *et al.*, 1971). Initial experimental and later human studies suggested that these mononuclear cells included a population of progenitor cells that were characterized by their capacity to form colonies when plated in agar (McCredie *et al.*, 1971; Barr *et al.*, 1975). This led investigators to consider transplanting peripheral blood mononuclear cells in humans.

Because the apheresis contains progenitor cells, the harvests are commonly referred to as PBPC (peripheral blood progenitor cells). However, as the apheresis products contain several cell types, and also because the role of progenitor cells in hematopoietic reconstitution after transplantation has not been defined in patients, we have used the term PBC here.

Initial studies suggested that PBC were capable of short-term engraftment (i.e., weeks to months) but that long-term reconstitution (i.e., months to years) was much less effective (Micklem *et al.*, 1975; Hershko *et al.*, 1979). In fact, the initial enthusiasm over the application of PBC in a transplant setting was tempered by these early reports of failure. But by 1986, several authors reported success with autologous PBC transplants in disorders such as acute myeloid leukemia and Burkitt lymphoma (Reiffers *et al.*, 1986; McCarthy and Goldman, 1984; Stiff *et al.*, 1987).

5.2. After Chemotherapy

Initial attempts at PBC transplantation utilized mononuclear cells obtained by apheresis of peripheral blood from unstimulated patients. Kessinger, as early as 1988, reported reconstitution following myeloablative therapy using PBC collected in the steady state. Typically, patients required a median of eight aphereses to produce a final product containing a median of 8×10^4 GM-CFC (granulocyte–macrophage progenitor cells)/kg.

Richman (1976) observed that myelosuppressive chemotherapy induced a marked increase in progenitor cell numbers in the peripheral blood, typically during the second and third weeks following the cytotoxic treatment. The increase is over and above the initial baseline levels; i.e., there is mobilization of progenitor cells into the blood. Apheresis products collected during the period when the white cell count is rapidly recovering are found to be markedly enriched in progenitor cells (To *et al.*, 1987). By careful timing of the apheresis in relation to the chemotherapy administration, there may be a ten- to 30-fold increase in GM-CFC numbers over baselines values. Gianni *et al.* (1990b) and To *et al.* (1987) have both shown that these apheresis products are capable of long-term hematopoietic reconstitution.

Although many chemotherapy drugs mobilize progenitor cells, not all produce increases in progenitor cell numbers in the blood. Even among the agents that are effective at mobilizing progenitor cells, there is great variability between the different drugs as to the extent of mobilization produced. Some drugs such as busulfan are known to be toxic to both bone marrow stem cells and their stromal microenvironment. These agents should be avoided if at all possible when one is considering mobilization of primitive cells in patients. Not only is there variation in mobilization evoked by different drugs, but there is up to a 100-fold interpatient variation in the numbers of progenitor cells mobilized. Cyclophosphamide, an alkylating agent commonly used in the treatment of a variety of malignancies, is the best-documented and the most widely used drug to date for mobilizing cells into the circulation. Although myelosuppressive, it is relatively nontoxic to early hematopoietic cells, and, in addition to mobilizing colony-forming cells, Sutherland and colleagues (1994) have demonstrated that putative stem cells, the long-term culture-initiating cells, are released simultaneously with or slightly before the clonogenic cells.

Etoposide is a topoisomerase II inhibitor and also an effective agent for mobilizing progenitor cells. Gianni *et al.* (1992) found a 7.4-fold increase in peak colony-forming cell numbers with etoposide alone. Etoposide, like cyclophosphamide appears to have low extramedullary toxicity and in preclinical studies appears to spare hematopoietic primitive cells (Kushner *et al.*, 1987). This contrasts with results obtained using single-agent doxorubicin (100 mg/m^2) in previously untreated patients with breast cancer. It mobilizes progenitors after the first dose and not thereafter, suggesting that doxorubicin may be toxic to the marrow progenitors or stroma (unpublished data).

Single-agent chemotherapeutic regimens are rarely employed in patients receiving dose-intensive treatments, and patients are more likely to be treated with combination regimens. A number of cytotoxic combinations have been used to mobilize primitive hematopoietic cells, but comparison between them are difficult because mobilization varies greatly depending on previous treatment, disease extent, and pathology in addition to the mobilizing chemotherapy used. The mechanism by which cells are mobilized from the bone marrow into the peripheral blood is unknown at present and is currently being investigated. Research into adhesion molecules, receptor configuration changes, and receptor biology in general are all areas currently being explored to try to elucidate the mobilization mechanism.

5.3. The Use of Growth Factors Alone

Growth factors may be administered alone in order to mobilize primitive cells into the circulation and are an attractive proposition for cancer patients in remission, for candidates for gene therapy, and for normal donors for allogeneic transplantation. The amount of cell mobilization observed with each type of growth factor differs greatly. G-CSF has consistently been shown to mobilize progenitor cells extremely well in a number of studies using different dosages and schedules. In contrast, erythropoietin, also a relatively lineage-specific growth factor at clinical doses, mobilizes progenitor cells poorly (Table VII).

Growth factors acting on early hematopoietic cells, such as IL-3 and stem cell factor (SCF), are weak mobilizers of progenitor cells when administered alone but are potent in combination with other factors. Molineux and colleagues (1991) demonstrated the synergism between SCF and G-CSF in mice, and de Revel (1994) showed similar findings in dogs. Glaspy and colleagues (1994) have reported early results in breast cancer patients showing enhanced mobilization with the combination of G-CSF and SCF compared with that seen for each factor alone.

A limited number of studies have been published concerning IL-3 in combination with GM-CSF. Interleukin 3 alone did not increase progenitor numbers significantly, but the combination of the two factors did (Table VIII). The G-CSF treatment may increase the number of circulating progenitor cells about 60-fold (Table VIII). Schedules for administering G-CSF to result in optimal yields of progenitor cells have not been established. It appears that there is a dose response in terms of progenitor cell numbers released into the circulation with increasing doses of G-CSF (Zeller *et al.*, 1994). Investigators such as Bensinger (1993) and Stoppa (1994) have used doses as high as 32 μg/kg per day in an attempt to maximize yields, but at doses greater than 10 μg/kg per day, there is a significant increase in side effects, most notably bone pain. Spitzer (1994b) failed to show enhancement of the apheresis product when patients were mobilized with both G-CSF and GM-CSF compared to mobilizing with G-CSF alone. Also, Ahmed (1994) failed to demonstrate a dose response in terms of mobilization of progenitors when GM-CSF was used at increasing dosages.

Table VII
Median Times to Neutrophil Recovery following Bone Marrow Transplantation with and without G-CSF Support[a]

	Median time to neutrophil recovery (days)		
Treatment arm	$>0.2 \times 10^9$/liter	$>0.5 \times 10^9$/liter	$>1.0 \times 10^9$/liter
G-CSF	11*	14*	16*
Placebo	15	20	27

[a]Adapted from Harousseau (1992). *Significant difference in favor of G-CSF.

Table VIII
Mobilization of Progenitor Cells by Growth Factors Alone

Growth factor	Dose	Prior treatment	Number of patients	Peak CFU-GM/ml	Fold increase	Reference
G-CSF	12 μg/kg per day	Yes	21	3,300 (600–6,000)	58	Sheridan *et al.* (1992)
	10 μg/kg per day	No	11	5,504 (1,849–12,869)	66	Weaver *et al.* (1995)
	10 μg/kg per day	Yes	5	845 (320–3,472)		Ossenkoppele (1994)
	15 μg/kg per day	No	10	3,200 ± 700	9.4	Schwinger *et al.* (1993)
	5 μg/kg per day	No	2	891, 1,787	13, 9	Dreger *et al.* (1994)
	1–40 μg/kg per day	Yes	30	2,258 ± 3,624	11.7	Stoppa *et al.* (1994)
GM-CSF	4–64 μg/kg per day	No	12	300 (0–1,815)	18	Socinski *et al.* (1988)
	8 μg/kg per day	Yes	7	638 ± 497	4.4	Aglietta *et al.* (1989)
	20 μg/kg per day	Yes	6	880 ± 1,280	8.4	Villeval *et al.* (1990)
EPO	0.4–0.12 IU/kg 3×/week	No	11	70 ± 50	3.0	Ganser *et al.* (1989b)
IL-3	60–500 μg/m^2	Yes	13	184 (18–409)	1.8	Ottomann *et al.* (1990)
	5–10 μg/kg per day	No	13	1,805 ± 756	3.0	D'Hondt *et al.* (1993)

The efficacy of purified human M-CSF in terms of progenitor cell mobilization is very poor when it is administered at a conventional dose of 8 × 10^6 U/day (Eto *et al.*, 1994).

5.4. The Use of Growth Factors in Allogeneic Transplantation

Experimental and preclinical studies demonstrated that not only progenitor cells but (stem) cells capable of repopulating the hematopoietic tissue permanently are mobilized by G-CSF treatment (Molineux *et al.*, 1990). This evidence ultimately led to the use of PBC for allogeneic transplants.

To date there have been approximately 100 allogeneic transplants worldwide, with the majority of investigators using G-CSF alone before apheresis (Russell *et al.*, 1993). Early published results indicate that reconstitution is successful without any evidence of graft failure to date. Likewise, graft-versus-host disease appears no worse when compared with bone marrow transplantation.

Overall, this method of PBC mobilization appears encouraging from both the donors' and the recipients' point of view. Not only has it expanded the panel of available donors, but it has put them at less risk, as they no longer require a general anesthetic to obtain the cells for transplant. There appear few, and usually only mild, adverse effects on the donors following G-CSF injections. Current studies may show that allogeneic transplants using PBC engraft more rapidly than bone marrow cells and ultimately may result in lower overall costs for transplantation.

5.5. The Use of Growth Factors Combined with Chemotherapy

Chemotherapy followed by either G- or GM-CSF results in an additive or synergistic increase in the numbers of progenitor cells mobilized (Table IX). Socinski and colleagues observed a 62-fold increase in progenitor cells when GM-CSF was administered in addition to cyclophosphamide (Socinski *et al.*, 1988). Gianni *et al.* (1989) similarly demonstrated an 80-fold increase in the number of progenitor cells following cyclophosphamide alone but a 550-fold increase following GM-CSF and cyclophosphamide. Demuynck *et al.* (1992) demonstrated a 59-fold increase in progenitor cell numbers with a combination of chemotherapy and G-CSF compared with a median 13-fold increase above baseline with chemotherapy alone.

Several investigators have found that the addition of growth factors to standard chemotherapy has reduced the median number of aphereses required to collect adequate cell numbers. Schwartzberg (1992) showed that the addition of G-CSF to standard chemotherapy resulted in a reduction from five to three aphereses, with a median 3.8-fold increase in CFC-GM in the final apheresis product. Subsequent transplantation of these cells resulted in a reduced duration of neutropenia and thrombocytopenia compared with non-growth-factor mobilized cells.

Several groups have reported that, with careful timing of the apheresis procedure, it is often possible to collect sufficient progenitor cells from a single apheresis (Jones *et al.*, 1994; Pettengell *et al.*, 1993). Apheresis is normally commenced when the leukocyte count is starting to rise exponentially during the recovery phase following myelosuppressive chemotherapy. The optimal method of mobilizing progenitor cells would provide maximal

Table IX
Comparative Studies of Chemotherapy and Growth Factor Mobilization of Progenitor Cells

Tumor	Number of patients	Regimen	Peak CFU-GM/ml	Fold	Reference
Breast	37	Cyclophosphamide + GM-CSF	8,000	55	Gianni *et al.*
	5	Cyclophosphamide + G-CSF	33,000	200	(1993)
	22	Cyclophosphamide + IL-3	3,400	25	
NHL	50	Baseline	16		Pettengell *et al.*
		Doxorubicin/cyclophosphamide	142	9	(1993)
		Doxorubicin/etoposide	734	46	
		Doxorubicin/cyclophosphamide/G-CSF	1,927	120	
		Doxorubicin/etoposide/G-CSF	5,018	314	
Ovary	15	Baseline	83	—	Weaver *et al.*
	15	Cyclophosphamide/G-CSF	4,515	54	(1995)
Myeloma	5	Cyclophosphamide	1,351		Boiron *et al.*
	10	Cyclophosphamide/GM-CSF	1,764		(1993)

enrichment of the peripheral blood, with the least toxicity in achieving this aim, and associated with the least risk of tumor contamination. To date this method has still to be defined.

In summary, peripheral blood is an attractive source of hematopoietic progenitor and stem cells both for transplantation and as support for repeated cycles of high-dose chemotherapy. A wide variety of growth factors and chemotherapeutic agents used alone or in combination can mobilize blood progenitor cells. Many different regimens are currently in use, and the optimal regimen remains to be defined. Because apheresis is an efficient method of collecting mononuclear cells, performing multiple aphereses compensates for suboptimal mobilizing regimens. Conversely, with an effective mobilizing regimen, a single apheresis may collect sufficient progenitor and stem cells.

The use of PBC to support each cycle of high-dose chemotherapy treatment has for the first time enabled clinicians to investigate the potential advantages of this treatment over conventional-dose treatment. At least in some protocols, the cell mobilization is of such magnitude that 1 unit of blood is sufficient to provide adequate support, with the consequent simplification of clinical protocols and a reduction in costs (Pettengell *et al.*, 1995).

6. USE OF GROWTH FACTORS IN LEUKEMIA AND MYELODYSPLASTIC SYNDROMES

The potential and reported interactions of the hematopoietic growth factors with malignant myeloid cells are extensive and are of great concern when one considers the use of growth factors in the treatment of the leukemias. The initial question to be asked was whether the myeloid growth factors would stimulate the proliferation of leukemic cells *in vivo* as they often do *in vitro*, thus resulting in more frequent and/or earlier relapse.

To evaluate the safety and efficacy of GM-CSF following induction therapy for acute myeloid leukemia, GM-CSF was given to patients with poor prognosis (>65 years of age or with relapsed disease). Patients were first demonstrated to have established bone marrow aplasia with fewer than 5% blasts; then they were given intravenous GM-CSF at 250 $\mu g/m^2$. The neutrophil recovery time following the GM-CSF was reduced to a median of 6 or 9 days, depending on which chemotherapy regimen was used, when compared with a historical series of patients (Buchner *et al.*, 1991).

The accelerated myeloid recovery was associated with fewer early deaths and an improved complete response rate (50% versus 32%). The GM-CSF did not affect platelet recovery, frequency of infections, or the durations of remissions. In two patients, blast cells appeared following the initiation of GM-CSF. In one of these patients, withdrawal of the growth factor resulted in regression of the blast population. It is pertinent to comment, however, that early normal bone marrow regeneration may also be accompanied by a transient excess of normal blasts. Insufficient patients have been studied to assess the true risk of leukemic proliferation associated with GM-CSF treatment, but anecdotal evidence suggests that in certain circumstances, leukemic cell growth may be stimulated by GM-CSF so that when the growth factor is withdrawn, the proliferation ceases (Buchner *et al.*, 1991). As mentioned above, the Powles *et al.* (1990) study using GM-CSF in the setting of allogeneic bone marrow transplantation showed no adverse effect on the leukemic relapse rate.

In order to clarify whether or not G-CSF alters the course of leukemia, a randomized multicenter study (Ohno *et al.*, 1990) was first performed with patients with acute leukemia (myeloid or lymphoid) and a poor prognosis. Chemotherapy was given until bone marrow examination showed severe hypoplasia; then the patients were randomized to receive G-CSF or to act as a control. Patients allocated to G-CSF received 200 $\mu g/m^2$ starting 2 days after completion of induction therapy and continued until their neutrophil counts were >1.5 $\times 10^9$/liter. Caution was taken to avoid giving G-CSF to patients when large numbers of blast cells remained in the marrow. Forty-eight patients were treated with G-CSF, and 50 patients acted as controls. The regrowth of leukemic blast cells was compared by examining the bone marrow 21 to 40 days after the end of induction. There was no difference between the two groups, even when the regrowth was compared in the subgroups excluding ALL. Fifty percent of the patients treated with G-CSF obtained a complete remission, compared with 36% in the control arm, although the difference was not significant. There was no difference in the duration of remission between the two groups. Because the G-CSF-treated group had a higher remission rate, this group tended to have a better event-free survival ($p = 0.12$).

Other groups have found evidence to the contrary. A report by Usuki and colleagues (1994) showed that patients with AML who developed neutropenic sepsis and were treated with G-CSF and antibiotics had stimulation of their leukemic blast cells during the time G-CSF was administered. The amount of proliferation appeared to depend on the leukemic cell mass at the start of G-CSF administration.

Further randomized studies are at present being performed to determine whether or not G-CSF can be used safely in patients with AML. Until these results are available, caution is advisable: clinical judgment must be exercised to decide whether G-CSF should be used in life-threatening infections or in patients with uncontrollable infections in severe neutropenic states.

Early studies performed in patients with myelodysplastic syndromes (MDS) have shown that approximately 15% of reported patients progressed to acute myeloid leukemia. These tended to be the patients with more than 10–15% blast cells in the pretreatment marrow (Thompson *et al.*, 1989; Vadhan-Raj *et al.*, 1987; Antin *et al.*, 1990; Ganser *et al.*, 1989b; Herrmann *et al.*, 1989). Studies have also shown that G-CSF can improve the white blood count and the absolute neutrophil count, usually at a dose of 3 $\mu g/kg$, though in resistant cases doses of 5 $\mu g/kg$ may be required. On stopping the G-CSF, blood counts fall to pretreatment levels after a 2- to 4-week period (Negrin *et al.*, 1989). Negrin and colleagues (1990) performed a second, longer-term study. Eleven patients were treated for 6 months with G-CSF, which produced a rise in neutrophil counts, and on stopping the treatment, blood levels fell to pretreatment values. On rechallenging the patients with G-CSF, blood counts were shown to improve. There was no evidence from either of these studies that G-CSF altered the natural history of the disease, i.e., either slowed or accelerated the rate of progression to acute myeloid leukemia. Two other studies have reported similar findings (Kobayashi *et al.*, 1989; Yoshida *et al.*, 1991).

Because differentiation induction is one of the aims of G-CSF treatment in MDS, an important question is whether the responding neutrophils are derived from the abnormal clone or from residual normal cells. Several approaches have been used to address this question. Marrow cytogenetics have been evaluated before and after treatment. In most patients, the cytogenetic pattern found prior to treatment persisted following G-CSF therapy. In one patient with trisomy 8, this was particularly instructive because this individual had all abnormal metaphases prior to treatment. This responsive patient had an elevated neutrophil count that persisted for 18 months. Bone marrow cytogenetics performed after 2, 6, 12, and 18 months of treatment all showed trisomy 8 in virtually all metaphases (Negrin *et al.*, 1990). These findings suggest that the neutrophils present were derived from the abnormal clone.

A more direct approach is to study restriction fragment length polymorphisms (RFLP) on women who had heterozygous expression of certain X-linked genes such as pyruvate glycerol kinase (PGK) or hypoxanthinephosphoribosyltransferase (HPRT). Four women were analyzed in this manner, and one patient was found to be polymorphic at the PGK gene (Negrin *et al.*, 1990). Homozygosity was found in bone marrow prior to treatment, indicative of clonal hematopoiesis. After treatment with G-CSF, this patient experienced a fivefold increase in her absolute neutrophil count. Neutrophils isolated at the time of maximal response had the same homozygous pattern on RFLP analysis, indicating that they were derived from the abnormal clone. This result is in contrast to a single patient analyzed by this approach following treatment with GM-CSF in whom polyclonal hematopoiesis resulted (Vadhan-Raj *et al.*, 1989).

Another approach in the treatment of leukemia using growth factors is to combine them with chemotherapy, with the aim of increasing cytotoxicity. Certain chemotherapeutic agents, such as cytosine arabinoside, are cell-cycle specific. That is, cytotoxicity of the drug is variable depending on the cycle status of the target cell. Clarkson and colleagues (1970) demonstrated that up to 10% of leukemic blast cells may not be cycling in any 8- to 10-day period.

In vitro studies have shown that sensitivity to cytosine arabinoside can be increased by the addition of hematopoietic growth factors (Cannistra *et al.*, 1989; Bhalla *et al.*, 1988). Both GM-CSF and IL-3 have been shown to increase the cytotoxicity of cytosine arabinoside, and several clinical studies have been reported that exploit this effect to some extent (Bettelheim *et al.*, 1991; Wiley *et al.*, 1991; Estey *et al.*, 1992; Cannistra *et al.*, 1991). In Bettelheim's study, reported in 1991, 18 previously untreated patients received standard induction chemotherapy concomitantly with intravenous GM-CSF. The GM-CSF was commenced 24–48 hr prior to the chemotherapy, with the intention of maximizing the proportion of cycling cells. A complete response rate of 83% was reported, with a 3-day reduction in the duration of neutropenia when compared with historical controls. Similar findings have been reported in patients with relapsed or refractory acute myeloid leukemia (Estey *et al.*, 1992; Cannistra *et al.*, 1991). However, taking the studies overall, the true value of GM-CSF in combination with cytotoxic chemotherapy for patients with AML remains unproven. Randomized studies will be required to establish if there is a useful role for GM-CSF or other myeloid growth factors to encourage cell cycling and hence improve response rates.

7. TREATMENT OF NEUTROPENIC SEPSIS

Neutropenic infections commonly occur in oncology in two clinical situations: following chemotherapy and in patients with myelodysplastic syndromes. Over the years there have been many anecdotal or uncontrolled studies with G-CSF, GM-CSF, and IL-3, showing transient rises in neutrophil counts following their administration (Kurzrock *et al.*,

1991). In myelodysplastic syndromes, doses as low as 0.3 μg/kg of either G-CSF or GM-CSF produce a rise in neutrophil count. On stopping the administration of growth factor, the count rapidly returns to its original level.

Schuster (1990) has reported a randomized study in myelodysplastic patients that supports the contention that GM-CSF decreases the rate of infectious complications. The combination of erythropoietin and either G-CSF or GM-CSF, in low dosage, decreases or even eliminates the need for red cell transfusion in approximately 20% of patients with myelodysplastic syndrome. This erythropoietin effect is most marked in patients with initially low circulating baseline erythropoietin levels (Hellstrom-Lindberg *et al.*, 1992; Hansen *et al.*, 1992). A case can be made for the use of such combinations in patients prone to infection, as a prophylactic measure, or in those with low erythropoietin levels and transfusion dependency. Much less information is available, and very little has been published, on the use of the myeloid growth factors in patients who have actually developed fever and infection while neutropenic. The reason for this may be that chemotherapy-induced neutropenic sepsis is transient and commonly responsive to antibiotics, and although it is likely that growth factors may decrease the time that the patient is neutropenic, it does not automatically follow that outcomes such as duration of hospitalization or antibiotic use will be beneficially affected.

Despite modern management, mortality remains approximately 10% among patients with documented infection (Pizzo *et al.*, 1986). Myeloid growth factors may have greater impact in those patients failing to respond to the introduction of broad-spectrum antibiotics or in those with infections that carry a poor prognosis (e.g., systemic fungal infections and *Pseudomonas aeruginosa* septicemia, and M-CSF may be the factor of choice). Current investigations into the clinical role of M-CSF include its use in immunocompromised patients with chronic intracellular infections, in conjunction with antibiotics (Nemunaitis *et al.*, 1992).

The roles of hematopoietic growth factors not only in stimulating cell production but also on inducing function of the mature neutrophils and monocytes/macrophages may also indicate a role in the treatment of patients at risk of infection, such as those with extensive burns, who are not cytopenic.

8. INHIBITORS

8.1. Macrophage Inflammatory Protein

In the late 1960s, interest was developing in inhibitory molecules that act by a negative feedback principle to regulate cellular proliferation. Lord and his colleagues (1976, 1977) separated, from normal bone marrow, an activity that could take proliferating primitive hematopoietic cells out of cycle and prevent the recruitment of quiescent cells into a proliferation cycle. The specificity of this extract for hematopoietic primitive cell proliferation made it an ideal candidate for a physiologically active regulator (Lord *et al.*, 1976; Tejero *et al.*, 1984). Furthermore, its mode of assay, by way of protection against cycle-specific cytotoxic agents together with *in vivo* effectivity, pointed to its potential for protection of hematopoietic stem cells during cytotoxic chemotherapy, particularly in cancer treatments (Lord *et al.*, 1976; Tejero *et al.*, 1984). The active principle has subsequently been shown to be a macrophage inflammatory protein (MIP), specifically the 1α form.

The fact that MIP-1α activity induces proliferative quiescence means that it should be potentially useful for protecting normal stem cells during chemotherapy, particularly if it does turn out to be a pan-stem-cell inhibitor as suggested. The nature of protracted chemotherapy is such that although initial treatments designed to kill S-phase cells will not kill many pluripotent stem cells, killing of the more mature cells will result in recruitment of quiescent cells into cycle, thereby rendering them susceptible to subsequent cycles of chemotherapy. Therefore, administering MIP-1α before the second cycle of treatment should have a protective effect on the stem cells, thus enabling chemotherapy to be delivered as planned, with consequently fewer delays in treatment. Such clinical studies are currently being performed, and the results are awaited with interest.

However, MIP also has effects on proliferation and development of hematopoietic progenitor cells. Broxmeyer (1989) demonstrated that committed progenitor cells displayed enhanced levels of colony formation, especially when suboptimal doses of hematopoietic growth factors were present. This was shown for M-CSF and GM-CSF but not G-CSF in murine experimental studies, suggesting that only certain growth factors can synergize with MIP in stimulating proliferation. Likewise, no growth effect of MIP on erythroid progenitors has been demonstrated.

8.2. Seraspenide

In the late 1970s, two groups reported data demonstrating the existence of inhibitors that act on hematopoietic stem cells. Frindel and Guigon (1977) showed that a dialyzable extract of low molecular weight from fetal calf bone marrow is able to inhibit CFU-S recruitment in mice. The active molecule was finally purified and sequenced after a decade of biochemical research (Lenfant *et al.*, 1989; Wdzieczak-Bakala *et al.*, 1983). It is a tetrapeptide *N*-acetyl-seryl-aspartyl-lysyl-proline (AcSDKP). The synthetic product is known as seraspenide.

The naturally occurring compound is synthesized in mice both *in vitro* and *in vivo* (Wdzieczak-Bakala *et al.*, 1990). The tetrapeptide appears to derive from thymosine-β4, which contains this amino acid sequence. Its biosynthesis involves an endopeptidase. AcSDKP appears to play a role in maintaining cells in G_0 or early G_1 phase, probably by blocking their entry into S phase. However, it is ineffective in late G_1 and during S phase (Monpezat and Frindel, 1989).

AcSDKP levels were determined in healthy individuals by enzyme immunoassay, with highest recorded levels in purified lymphocytes, while significant levels were not detectable in red blood cells or platelets. *In vitro* experiments on human cells have shown that seraspenide lowers the proportion of cycling progenitor cells (Guigon *et al.*, 1990). Incubation for 24 hr of normal human mononuclear cells with seraspenide leads to significant growth inhibition of CFU-GM, BFU-E, and in some patients CFU-E, with a marked dose–response effect (Bonnet *et al.*, 1992). The mechanism of action of the inhibitory effects on hematopoietic stem cells is at present unclear, although several mechanisms have been postulated, including the idea that seraspenide enhances hematopoietic cell binding through activation of the stromal cells, therefore altering the microenvironment of the stem cells. Alternatively, the compound may have a direct action on the hematopoietic stem cells, preventing their recruitment into a proliferative phase.

A phase I–II double-blind crossover study was undertaken in 1991 in cancer patients receiving two consecutive cycles of chemotherapy with cytosine arabinoside or ifosfamide. A protective effect of seraspenide was demonstrated, as judged by leukocyte and neutrophil numbers, the depth of their nadir, and the duration of hypoplasia (Carde *et al.*, 1992). Seraspenide was reported to cause virtually no toxic side effects. Clearly, these early results are promising, but further clinical investigation is needed.

9. MISCELLANEOUS USES OF GROWTH FACTORS

9.1. Treatment of Nonhealing Ulcers with GM-CSF and Skin Grafts

Pojda and Struzyna (1994) have reported on the use GM-CSF to enhance split-skin engraftment in a small series of patients who previously all had at least one unsuccessful autogenic skin graft for long-standing ulceration (pressure wounds, varicose ulceration or residual ulceration following burns).

Before transplantation, the split-thickness skin grafts were incubated for 4–5 hr at 37°C in culture medium supplemented with 100 ng/ml of GM-CSF. In three patients the only successful grafts were the ones pretreated with GM-CSF, whereas the control skin grafts were rejected. In two patients both the GM-CSF-treated and control grafts were successful, and in one patient all the grafts were rejected (Pojda and Struzyna, 1994).

9.2. Improvement in Wound Healing

Several investigators have demonstrated the use of growth factors, most notably GM-CSF, to encourage wound healing. Single intradermal injections of GM-CSF have been administered locally into healing biopsy wounds of patients with leprosy, thereby encouraging repair by inducing Langerhans-cell recruitment and keratinocyte growth (Kaplan and Cohn, 1993). Boente and colleagues (1993) have shown improvement in Kaposi sarcoma lesions in AIDS patients by the local perilesional injections of GM-CSF.

10. CONCLUSIONS

When the myeloid growth factors were first introduced into the clinic over 5 years ago, there was great hope that they could minimize the major dose-limiting side effect of neutropenia following the delivery of standard-dose chemotherapy. In this setting, the growth factors have had only a modest impact in terms of clinical benefit for the patient. Despite the reduced duration of fever and shortened hospital stays observed in those patients treated with chemotherapy and growth factor support, this has not translated into improved response rates or survival for the vast majority of patients. The most likely reason for this finding is that growth factors used in this supportive role only allow for a modest increase in dose intensity. A very similar situation also holds true for many of the higher-dose-intensive studies carried out to date. Even if the growth factors may help overcome the hematological toxicity of a chemotherapy regimen, this side effect may merely be replaced by a non-hematological dose-limiting toxicity.

There have been some very encouraging results in certain tumor types, namely, lymphomas and leukemias, by the use of bone marrow transplantation and growth factor support. Growth factors used in the posttransplant setting certainly reduce the duration of neutropenia as well as the duration of fever, which results in shortened hospital stays. They may well have the additional benefit of improving the patient's quality of life, though this area of investigation is still in its early days. More published results are available for the use of GM-CSF than for G-CSF in this context, though a case for the use of G-CSF in favor of GM-CSF can be made in view of the fact that in some of the GM-CSF studies there was an increased use of empiric antibiotics, due to GM-CSF causing fever (one of its known side-effects). Also in some of the studies GM-CSF was delivered at relatively high doses, which again resulted in increased side-effects, many of which do not occur with G-CSF.

Over the last 2 to 3 years, bone marrow transplants have been increasingly replaced by peripheral blood cell transplants because of distinct advantages with the latter form of therapy. The use of PBC instead of bone marrow in allogeneic transplants appears full of promise, and the good early clinical results will progress to controlled clinical trials.

It appears likely that, as time progresses, PBC treatments will become even more popular as more and more centers develop experience in mobilization regimens and harvesting techniques. It is in the role as agents for cell mobilization that growth factors are having a major impact and obvious clinical benefit. The use of high-dose sequential chemotherapy with PBC support allows the clinician, perhaps for the first time, effectively to increase the dose intensity of chemotherapy, and as a result investigate whether dose-intensive treatment offers a clear and beneficial clinical advantage.

11. REFERENCES

Advani, R., Greenberg, P., and Gulati, S., 1990, Randomized placebo controlled trial of granulocyte–macrophage colony-stimulating factor in patients undergoing autologous bone marrow transplantation, *Blood* **76** (Suppl. 1)**:**525.

Advani, R., Chao, N. J., Horning, S. J., 1992, Granulocyte–macrophage colony-stimulating factor (GM-CSF) as an adjunct to autologous hemopoietic stem-cell transplantation for lymphoma, *Ann. Intern. Med.* **116:**183–189.

Aglietta, M., Piacibello, W., and Sanavio, F., 1989, Kinetics of human hemopoietic cells after *in vivo* administration of granulocyte–macrophage colony-stimulating factor, *J. Clin. Invest.* **83:**551–557.

Ahmed, T., Preti, R. A., Lake, D. E., Farley, T., Cook, P., Hohwart, C., Razis, E., Beer, M., Syed, N., and Ciavarella, D., 1994, Comparison of 2 different doses of sargramostim to mobilize peripheral blood progenitors, *Blood* **84**(Suppl. 1)**:**414.

Antin, J. H., Weinberg, D. S., and Rosenthal, D. S., 1990, Variable effect of recombinant human granulocyte–macrophage colony-stimulating factor on bone marrow fibrosis in patients with myelodysplasia, *Exp. Hematol.* **18:**266–270.

Antman, K., Griffin, J., Elias, A., Socinski, M. A., Ryan, L., Cannistra, S. A., Oette, D., Whitley, M., Frei, E. III, and Schnipper, L. E., 1988, Effect of recombinant human granulocyte–macrophage colony-stimulating factor on chemotherapy-induced myelosuppression, *N. Engl. J. Med.* **319:**593–598.

Aso, Y., and Akaza, H., 1992, Effect of recombinant human granulocyte colony-stimulating factor in patients receiving chemotherapy for urogenital cancer, *J. Urol.* **147:**1060–1064.

Barr, R. D., Whang-Penn, J., and Perry, S., 1975, Hematopoietic stem cells in human peripheral blood, *Science* **190:**284–285.

Behrens, B. C., Grotzingen, K. R., and Hamilton, T. C., 1985, Cytotoxicity of 3 cisplatin analogues in drug sensitive and a new cisplatin resistant human ovarian cancer cell line, *Proc. Am. Assoc. Cancer Res.* **26:**262.

Bennett, C. L., Greenberg, P., and Gulati, S. C., 1990, GM-CSF decreases duration of cytopenia and hospitalization, and in-hospital costs in patients with Hodgkin's disease treated with high-dose chemotherapy and autologous bone marrow transfusion (ABMT), *Blood* **76**(Suppl. 1):152.

Bensinger, W., Singer, J., Appelbaum, F., and Lilleby, K., 1993, Autologous transplantation with peripheral blood mononuclear cells collected after administration of recombinant human granulocyte colony-stimulating factor, *Blood* **81:**3158–3163.

Bettelheim, P., Valent, P., Andreeff, M., Tafuri, A., Haimi, J., Gorischek, C., Muhm, M., Sillaber, C. H., Haas, O., Vieder, L., Mauer, D., Schulz, G., Speiser, W., Geissler, K., Kier, P., Hinterberger, W., and Lechner, K., 1991, Recombinant human granulocyte–macrophage colony-stimulating factor in combination with standard induction chemotherapy in *de novo* acute myeloid leukemia, *Blood* **77:**700–711.

Bhalla, K., Birkhofer, M., Arlin, Z., Grant, S., Lutzky, J., and Graham, G., 1988, Effect of recombinant GM-CSF on the metabolism of cytosine arabinoside in normal and leukemic human bone marrow cells, *Leukemia* **2:** 810–813.

Blazar, B. R., Kershey, J. U., and McGlave, P. B., 1989, *In vivo* administration of recombinant human granulocyte–macrophage colony-stimulating factor in acute lymphoblastic leukemia patients receiving purged autografts, *Blood* **73:**849–857.

Boente, P., Sampaio, C., Brandao, M., Moreira, E. D., Badaro, R., and Jones, T. C., 1993, Local peri-lesional therapy with rhGM-CSF for Kaposi's sarcoma, *Lancet* **341:**1154.

Boiron, J. M., Marit, G., and Faberes, C., 1993, Collection of peripheral blood stem cells in multiple myeloma following single high-dose cyclophosphamide with and without recombinant human granulocyte–macrophage colony-stimulating factor (rhGM-CSF), *Bone Marrow Transplant* **12:**49–55.

Bonnadonna, G., 1987, Hodgkin's disease: The Milan Cancer Institute experience with MOPP and ABVD, in *Third International Conference on Malignant Lymphoma*, Conference Proceedings.

Bonnet, D., Lemoine, F., Baillou, C., Najman, A., and Guigon, M., 1992, The tetrapeptide acetyl-N-Ser-Asp-Lys-Pro inhibits the response of human purified marrow $CD34^+$ cells to growth factors, *Exp. Hematol.* **20:**759.

Brandt, S. J., Peters, W. P., and Atwater, S. K., 1988, Effect of recombinant human granulocyte–macrophage colony-stimulating factor on hematopoietic reconstitution after high dose chemotherapy and autologous bone marrow transplantation, *N. Engl. J. Med.* **318:**869–876.

Bronchud, M. H., Scarffe, J. H., Thatcher, N., and Crowther, D., 1987, Phase I/II study of recombinant human granulocyte colony-stimulating factor in patients receiving intensive chemotherapy for small cell lung cancer, *Br. J. Cancer* **56:**809–813.

Bronchud, M. H., Howell, A., and Crowther, D., 1989, The use of granulocyte colony-stimulating factor to increase the intensity of treatment with doxorubicin in patients with advanced breast and ovarian cancer, *Br. J. Cancer* **60:**121–125.

Broxmeyer, H. E., Sherry, B., and Lilu, P., 1989, Myelopoietic enhancing effects of murine macrophage marrow granulocyte/macrophage progenitor cells, *J. Exp. Hematol.* **170:**1583–1594.

Brugger, W., Bross, K., Glatt, M., Weber, F., Mertelsmann, R., and Kanz, L., 1994, Mobilization of tumor cells and hemopoietic progenitor cells into peripheral blood of patients with solid tumours, *Blood* **83**(3)**:**636–640.

Buchner, T., Hiddemann, W., Koenigsmann, M., Zuhlsdorf, M., Wormann, B., Boeckmann, A., Freire, E. A., Innig, C., Maschmeyer, G., Ludwig, W. D., Sauerland, M. C., Heinecke, A., and Schulz, G., 1991, Recombinant human granulocyte–macrophage colony-stimulating factor after chemotherapy in patients with acute myeloid leukemia at higher age or after relapse, *Blood* **78:**1190–1197.

Cannistra, S. A., Groshek, P., and Griffin, J. D., 1989, Granulocyte–macrophage colony-stimulating factor enhances the cytotoxic effects of cytosine arabinoside in acute myeloblastic leukemia and in the myeloid blast crisis phase of chronic myeloid leukemia, *Leukemia* **3:**328–334.

Cannistra, S. A., DiCarlo, J., Groshek, P., Kanakura, Y., Berg, D., Mayer, R. J., and Griffin, J. D., 1991, Simultaneous administration of granulocyte–macrophage colony-stimulating factor and cytosine arabinoside for the treatment of relapsed acute myeloid leukemia, *Leukemia* **5:**230–238.

Carde, P., Chastang, C., Goncalves, E., Mathieu-Tubiana, N., Vuillemin, E., Delwail, V., and Corbion, O., 1992, Seraspenide (AcSDKP): Etude en phase I-II d'un inhibiteur de l'hematopoiese protegeant de la toxicite de monochimiotherapies aracytine et ifosfamide, *C.R. Acad. Sci. (Paris)* **315**(Serie III)**:**545–550.

Cebon, J., and Lieschke, G. J., 1994, Granulocyte–macrophage colony-stimulating factor for cancer treatment, *Oncology* **51:**177–188.

Cebon, J., Layton, J., Pavlovic, R., Iaria, J., Dyer, W., and Morstyn, G., 1991, Endogenous cytokine production in response to sepsis in neutropaenic and non-neutropaenic patients, *Blood* **78**(Suppl.):8 (Abstract 23).

Chevallier, B., Bui, N. B., and Bonichon, F., 1992, Efficacy of rG-CSF on haematologic tolerance to MAID chemotherapy in sarcoma patients and impact on dose intensity, *Proc. Am. Soc. Clin. Oncol.* **11**:1446.

Chevallier, B., Challet, P. H., and Merrouche, Y., 1993, Glycosylated rHuG-CSF (lenograstim) prevents morbidity from FEC HD chemotherapy in inflammatory breast cancer, *Proc. Am. Soc. Clin. Oncol.* **12**:137.

Clarkson, B., Strife, A., Fried, J., Sakai, Y., Ota, K., Ohkita, T., and Masuda, R., 1970, Studies of cellular proliferation in human leukemia IV. Behaviour of leukemia cells in 3 adults with acute leukemia given continuous infusions of ^{3}H-thymidine for 8–10 days, *Cancer* **26**:1–19.

Coltman, C. A., Jr., Jones, S. E., and Grozea, P. N., 1978, Bleomycin in combination with MOPP for the management of Hodgkin's disease: Southwest Oncology Group experience, in *Bleomycin: Current Status and New Developments* (S. K. Carter and S. T. Crooke, eds.), pp. 227–242, Academic Press, Orlando.

Coltman, C. A., Dahlberg, S., and Jones, S. E., 1986, CHOP is curative in thirty percent of patients with large-cell lymphoma: A twelve year Southwest Oncology Group follow-up, in *Advances in Cancer Chemotherapy: Update on Treatment for Diffuse Large-Cell Lymphoma* (A. T. Skarin, ed.), pp. 71–82, John Wiley, New York.

Crawford, J., Ozer, H., and Stoller, R., 1991, Reduction by granulocyte colony-stimulating factor of fever and neutropenia induced by chemotherapy in patients with small cell lung cancer, *N. Engl. J. Med.* **325**:164–170.

Demuynck, H., Pettengell, R., Testa, N. G., and Dexter, T. M., 1992, The engraftment capacity of peripheral blood progenitor cells (PBPC) mobilized with chemotherapy +/− G-CSF, *Int. J. Cell Cloning* **10**(Suppl. 1):59–61.

de Revel, T., Appelbaum, F. R., Storb, R., Scheuning, F., Nash, R., Deeg, J., McNiece, I., Andrews, R., and Graham, T., 1994, Effects of granulocyte colony-stimulating factor and stem cell factor, alone and in combination, on the mobilisation of peripheral blood cells that engraft lethally irradiated dogs, *Blood* **83**:3795–3799.

Devereux, S., Linch, D. C., and Gribben, J. G., 1989, GM-CSF accelerates neutrophil recovery after autologous bone marrow transplantation for Hodgkin's disease, *Bone Marrow Transplant* **4**:49–54.

DeVita, V. T., 1989, Principles of chemotherapy, in *Cancer Principles and Practice of Oncology* (V. T. DeVita, S. Hellman, and S. A. Rosenberg, eds.), pp. 276–300, J. B. Lippincott, Philadelphia.

DeVita, V. T., Simon, R. M., and Hubbard, S. M., 1980, Curability of advanced Hodgkin's disease with chemotherapy. Long term follow-up of MOPP treated patients at the NCI, *Ann. Intern. Med.* **92**:587–595.

DeVita, V. T., Jr., Hubbard, S. M., and Longo, D. L., 1987, The chemotherapy of lymphomas: Looking back, moving forward, *Cancer Res.* **47**:5810.

de Vries, E. G. E., Biesma, B., Willemse, P. H. B., Mulder, N. H., Stern, A. C., Aalders, J. G., and Vellenga, E., 1991, A double blind placebo controlled study with granulocyte–macrophage colony-stimulating factor during chemotherapy for ovarian carcinoma, *Cancer Res.* **51**:116–122.

De Witte, T., Gratwohl, A., Van Der Lely, N., Bacigalupo, A., Stern, A. C., Speck, B., Schattenberg, A., Nissen, C., Gluckman, E., and Fibbe, W. E., 1992, Recombinant human granulocyte–macrophage colony-stimulating factor accelerates neutrophil and monocyte recovery after allogeneic T-cell depleted bone marrow transplantation, *Blood* **79**:1359–1365.

D'Hondt, V., Weynants, P., and Humblet, Y., 1993, Dose dependent interleukin-3 stimulation of thrombopoiesis and neutropoiesis in patients with small cell lung carcinoma before and following chemotherapy: A placebo controlled randomised phase Ib study, *J. Clin. Oncol.* **11**:2063–2071.

Dreger, P., Haferlach, T., and Eckstein, V., 1994, G-CSF mobilized peripheral blood progenitor cells for allogeneic transplantation: Safety, kinetics of mobilization and composition of the graft, *Br. J. Haematol.* **87**:609–613.

Edmonson, J. H., Long, H. J., and Jeffries, J. A., 1989, Amelioration of chemotherapy-induced thrombocytopaenia by GM-CSF: Apparent dose and schedule dependency, *J. Natl. Cancer Inst.* **81**:1510–1512.

Edmonson, J. H., Colon-Otero, G., Long, H. J., Fitch, T. R., Jeffries, J. A., Buckner, J. C., Rubin, J., and Braich, T. A., 1990, Granulocyte–macrophage colony-stimulating factor (GM-CSF) permits escalation of carboplatin (CBDCA) doses, *Proc. Am. Soc. Clin. Oncol.* **9**:85.

Engelhard, M., Gerhartz, H., and Brittinger, G., 1993, Cytokine efficiency in the treatment of high grade malignant non-Hodgkin's lymphomas: Results of a randomised double-blind placebo-controlled study with intensified COP-BLAM +/− rhGM-CSF, in *Proceedings of the 5th International Conference on Malignant Lymphoma, Lugano*, p. 29.

Estey, E., Thall, P. F., Kantarjian, H., O'Brien, S., Koller, C. A., Beran, M., Gutterman, J., Deisseroth, A., and Keating, M., 1992, Treatment of newly diagnosed acute myelogenous leukemia with granulocyte–macrophage colony-stimulating factor (GM-CSF) before and during continuous-infusion high-dose ara-C + daunorubicin: Comparison to patients treated without GM-CSF, *Blood* **79**:2246–2255.

Eto, T., Takamatsu, Y., Harada, M., Harada, N., Akashi, K., Teshima, T., Inaba, S., Okamura, T., and Niho, Y., 1994, Effects of macrophage colony-stimulating factor (M-CSF) on the mobilization of peripheral blood stem cells, *Bone Marrow Transplant.* **13**(2):125–129.

Fibbe, W. E., Raemaekers, J., Verdonck, L. F., Schouten, H. C., van Imhof, G., and Tromp, B., 1991, Recombinant human interleukin-3 after autologous bone marrow transplantation for malignant lymphoma: Aphase I/II multicentre study, *Blood* **78:**163.

Frei, E. III, and Canellos, G. P., 1980, Dose: A critical factor in cancer chemotherapy, *Am. J. Med.* **69:**585–594.

Frindel, E., and Guigon, M., 1977, Inhibition of CFU entry into cycle by a bone marrow extract, *Exp. Hematol.* **5:**74–76.

Fukuoka, M., Takada, M., and Masuda, N., 1992, Dose intensive weekly chemotherapy with or without recombinant human granulocyte colony-stimulating factor (G-CSF) in extensive stage small cell lung cancer, *Proc. Am. Soc. Clin. Oncol.* **11:**967.

Gabrilove, J. L., Jakubowski, A., and Scher, H., 1988, Effect of granulocyte colony-stimulating factor on neutropaenia and associated morbidity due to chemotherapy for transitional-cell carcinoma of the urothelium, *N. Engl. J. Med.* **318:**1414–1422.

Ganser, A., Volkers, B., Greher, J., Ottmann, O. G., Walther, F., Becher, R., Bergmann, L., Schulz, G., and Hoelzer, D., 1989b, Recombinant human granulocyte–macrophage colony-stimulating factor in patients with myelodysplastic syndromes—a phase I/II trial, *Blood* **73:**31–37.

Gianni, A. M., Siena, S., and Bregni, M., 1989, Granulocyte–macrophage colony-stimulating factor to harvest circulating haemopoietic stem cells for autotransplantation, *Lancet* **2:**580–585.

Gianni, A., Bregni, M., Siena, S., Orazi, A., Stern, A. C., Gandola, L., and Bonnadonna, G., 1990a, Recombinant granulocyte–macrophage colony-stimulating factor reduces hematological toxicity and widens clinical applicability of high dose cyclophosphamide treatment in breast cancer and non-Hodgkin's lymphoma, *J. Clin. Oncol.* **8:**768–778.

Gianni, A. M., Siena, S., and Bregni, M., 1990b, Factors influencing the yield of circulating hematopoietic progenitor cells harvested from cancer patients after high-dose cyclophosphamide treatment, with or without RHGM-CSF infusion, in *Biology and Clinical Application of Granulocyte–Macrophage Colony-Stimulating Factor* (W. E. Fibbe, ed.), pp. 20–28, Sandoz/Schering-Plough, Uden.

Gianni, A. M., Bregni, M., Siena, S., Magni, M., Di Nicola, M., Lombardi, F., Tarella, C., Pileri, A., and Bonnadonna, G., 1992, Granulocyte–macrophage colony-stimulating factor or granulocyte colony-stimulating factor makes high-dose etoposide a safe out-patient regimen that is effective in lymphoma and myeloma patients, *J. Clin. Oncol.* **10:**1955–1962.

Gianni, A. M., Bregni, M., and Siena, S., 1994, 5-year update of the Milan Cancer Institute randomized trial of high-dose sequential (HDS) vs. MACOP-B therapy for diffuse large-cell lymphomas, *Proc. Am. Soc. Clin. Oncol.* **13:**1263.

Glaspy, J., McNiece, I., LeMaistre, F., Menchaca, D., Briddell, R., Lill, M., Jones, R., Morstyn, G., and Shpall, E. J., 1994, Effects of stem cell factor and filgrastim on mobilization of peripheral blood progenitor cells and on hematological recovery posttransplant: Early results from a phase I/II study, *Proc. Am. Soc. Clin. Oncol.* **13:**76.

Goodman, J. W., and Hodgson, G. S., 1962, Evidence for stem cells in the peripheral blood of mice, *Blood* **19:** 702–714.

Gorin, N. C., Coiffier, B., and Pico, J., 1990, Granulocyte–macrophage colony-stimulating factor (GM-CSF) shortens aplasia duration after autologous bone marrow transplantation (ABMT) in non-Hodgkin's lymphoma: A randomized placebo controlled double-blind study, *Blood* **76**(Suppl. 1):542.

Guigon, M., Bonnet, D., Lemoine, F., Kobari, L., Parmentier, C., Mary, J. Y., and Najman, A., 1990, Inhibition of human bone marrow progenitors by the synthetic tetrapeptide AcSDKP, *Exp. Hematol.* **18:**1112–1115.

Hansen, P. B., Johnsen, H. E., Hippe, E., and Ralfkiaer, E., 1992, Recombinant human granulocyte–macrophage colony-stimulating factor plus human erythropoietin may improve anaemia in selected patients with myelodysplasia, *Blood* **80**(Suppl. 1):84b.

Harrousseau, J. L., 1992, Lenograstim after bone marrow transplantation; results of a European multicenter randomised study in 315 patients, in *Satellite Symposium to the 24th Congress*, International Society of Haematology, London.

Havemann, K., Klausmann, M., Wolf, M., Fischer, J. R., Drings, P., and Oster, W., 1991, Effect of rhGM-CSF on haemopoietic reconstitution after chemotherapy in small cell lung cancer, *J. Cancer Res. Clin. Oncol.* **117**(Suppl. 4):S203–207.

Hellman, S., Jaffe, E. S., and DeVita, V. T., Jr., 1989, Hodgkin's disease, in *Cancer Principles and Practice of Oncology* (V. T. DeVita, S. Hellman, and S. A. Rosenberg, eds.), pp. 1696–1740, J. B. Lippincott, Philadelphia.

Hellstrom-Lindberg, E., Birgegard, G., Carlsson, M., Carneskog, J., and Dahl, I. M., 1992, Treatment of myelodysplastic syndromes (MDS) with a combination of G-CSF and erythropoietin—A synergistic effect on the erythropoietic response, *Blood* **80**(Suppl. 1)**:**84a.

Herrmann, F., Lindemann, A., Klein, H., Lubbert, M., Schulz, G., and Mertelsmann, R., 1989, Effect of recombinant human granulocyte–macrophage colony-stimulating factor in patients with myelodysplastic syndrome with excess blasts, *Leukemia* **3:**335–338.

Herrmann, F., Schulz, G., Wiesier, M., Kolbe, K., Nicolay, U., Noack, M., Lindemann, A., and Mertelsmann, R., 1990, Effect of granulocyte–macrophage colony-stimulating factor on neutropenia and related morbidity induced by myelotoxic chemotherapy, *Am. J. Med.* **88:**619–624.

Hershko, C., Ho, W. G., and Gale, R. P., 1979, Cure of aplastic anaemia in paroxysmal nocturnal haemoglobinuria by marrow transfusion from identical twin: Failure of peripheral leukocyte transfusion to correct marrow aplasia, *Lancet* **1:**945–947.

Hoekman, K., Wagstaff, J., van-Groeningen, C. J., Vermorken, J. B., Boven, E., and Pinedo, H. M., 1991, Effects of recombinant human granulocyte–macrophage colony-stimulating factor on myelosuppression induced by multiple cycles of high-dose chemotherapy in patients with advanced breast cancer, *J. Natl. Cancer Inst.* **83:**1546–1553.

Jones, H. M., Jones, S. A., Watts, M. J., Khwaja, A., Mills, W., Fielding, A., Goldstone, A. H., and Linch, D. C., 1994, Development of a simplified single-apheresis approach for peripheral-blood progenitor-cell transplantation in previously treated patients with lymphoma, *J. Clin. Oncol.* **12**(8)**:**1693–1702.

Kaplan, G., and Cohn, Z. A., 1993, Effects of intradermal recombinant human granulocyte–macrophage colony-stimulating factor on Langerhans cell recruitment and keratinocyte growth, in *Hemopoietic Growth Factors and Mononuclear Phagocytes* (R. Van Furth, ed.), pp. 205–210, S. Karger, Basel.

Kessinger, A., Armitage, J. O., Landmark, J. D., Smith, D. M., and Weisenburger, D. D., 1988, Autologous peripheral haematopoietic stem cell transplantation restores function following marrow ablative therapy, *Blood* **71:**723–727.

Kobayashi, Y., Okabe, T., and Ozawa, K., 1989, Treatment of myelodysplastic syndromes with recombinant human granulocyte colony-stimulating factor: A preliminary report, *Am. J. Med.* **86:**178–182.

Korbling, M., Juttner, C., Henon, P., and Kessinger, A., 1992, Autologous blood stem cell versus bone marrow transplantation, *Bone Marrow Transplant.* **10**(Suppl. 1)**:**144–148.

Kotake, T., Miki, T., and Akaza, H., 1991, Effect of recombinant granulocyte colony-stimulating factor (G-CSF) on chemotherapy-induced neutropenia in patients with urogenital cancer, *Cancer Chemother. Pharmacol.* **27:**253–257.

Kurzrock, R., Talpez, M., Estey, E. H., and Gutterman, J. U., 1991, Haemopoietic growth factors in bone marrow failure states, *Cancer Bull.* **43:**215–23.

Kushner, B. H., Kwon, J.-H., Gulati, S. C., and Castro-Malaspina, H., 1987, Preclinical assessment of purging with VP-16-213: Key role for long-term marrow cultures, *Blood* **69:**65–71.

Lazarus, H., Andersen, J., and Oette, D., 1990, Phase II ECOG trial of recombinant human granulocyte–macrophage colony-stimulating factor (rhGM-CSF) after autologous bone marrow transplant (AuBMT) for relapsed non-Hodgkin's lymphoma (NHL), *Proc. Am. Soc. Clin. Oncol.* **9:**15.

Lenfant, M., Wdzieczak-Bakala, J., Guittet, E., Prome, J. C., Sotty, D., and Frindel, E., 1989, Inhibitor of hematopoietic pluripotent stem cell proliferation: Purification and determination of its structure, *Proc. Natl. Acad. Sci. U.S.A.* **36:**779–782.

Levin, L., and Hryniuk, W. M., 1987, Dose intensity analysis of chemotherapy regimens in ovarian carcinoma, *J. Clin. Oncol.* **5:**756–767.

Lichtman, S., Ratain, M., Budman, D. R., Norton, L., Egorin, M., Propert, K., Modeas, C., Paterson, D., and Van Echo, D. A., 1989, Phase I trial of recombinant granulocyte–macrophage colony-stimulating factor (GM-CSF) plus cyclophosphamide in solid tumours, *Blood* **74:**328.

Linch, D. C., 1992, Lenograstim. A new glycosylated rHuG-CSF; pharmacology and clinical profile in bone marrow transplants. Chugai–Rhone-Poulenc Symposium, *Br. J. Haematol.* **82:**274–275.

Link, H., 1990, Recombinant human granulocyte–macrophage colony-stimulating factor (rhGM-CSF) for hematological reconstitution after autologous bone marrow transplantation, in: *Hematopoietic Growth Factors in Clinical Applications* (R. Mertelsmann and F. Herrmann, eds.), pp. 313–335, Marcel Dekker, New York.

Link, H., Boogaerts, M., and Carella, A., 1990a, Recombinant human granulocyte–macrophage colony-stimulating factor (rhGM-CSF) after autologous bone marrow transplantation for acute lymphoblastic leukemia and non-Hodgkin's lymphoma: A randomized double-blind multicentre trial in Europe, *Blood* **76** (Suppl. 1)**:**152.

Link, H., Diedrich, H., and Ebell, W., 1990b, Recombinant human erythropoietin after allogeneic bone marrow transplantation, *Bone Marrow Transplant* **5**(Suppl. 2)**:**219a.

Lord, B. I., Mori, K. J., Wright, E. G., and Lajtha, L. G., 1976, An inhibitor of stem cell proliferation in normal bone marrow, *Br. J. Haematol.* **34:**441–445.

Lord, B. I., Mori, K. J., and Wright, E. G., 1977, A stimulator of stem cell proliferation in regenerating bone marrow, *Biomed. Exp.* **27:**223–226.

Masaoka, T., Motoyoshi, K., and Takaku, I., 1988, Administration of human urinary colony-stimulating factor after bone marrow transplantation, *Bone Marrow Transplant* **3:**121–127.

McCarthy, D. M., and Goldman, J. M., 1984, Transfusion of circulating stem cells, *CRC Crit. Rev. Clin. Lab. Sci.* **20:**1–24.

McCredie, K., Hersh, E., and Freireich, E., 1971, Cells capable of colony formation in the peripheral blood of man, *Science* **171:**293–294.

McKenzie, S., New, M., and Hirschfeld, N., 1990, GM-CSF in pediatric bone marrow transplantation: A phase I/II study, *Blood* **76**(Suppl. 1)**:**553.

Mertelsmann, R., 1991, Hematopoietins: Biology, pathophysiology, and potential as therapeutic agents, *Ann. Oncol.* **2**(4)**:**251–263.

Metcalf, D., 1994, Thrombopoietin—at last, *Nature* **369:**519.

Michon, J., Bouffet, E., and Bernard, J. L., 1990, Administration of recombinant human GM-CSF after autologous bone marrow transplantation (ABMT). A study of 21 stage IV neuroblastoma patients undergoing a double intensification regimen, *Proc. Am. Soc. Clin. Oncol.* **9:**184.

Micklem, H. S., Anderson, N., and Ross, R., 1975, Limited potential of circulating hemopoietic stem cells, *Nature* **256:**41–43.

Molineux, G., Pojda, Z., Hampson, I. N., Lord, B. I., and Dexter, T. M., 1990, Transplantation potential of peripheral blood stem cells induced by granulocyte colony-stimulating factor, *Blood* **76**(10)**:**2153–2158.

Molineux, G., Migdalska, A., Szmitkowski, M., Zsebo, K., and Dexter, T. M., 1991, The effects on hematopoiesis of recombinant stem cell factor (ligand for *c-kit*) administered *in vivo* to mice either alone or in combination with granulocyte colony-stimulating factor, *Blood* **78:**961–966.

Molineux, G., Migdalska, A., Haley, J., Evans, G. S., and Dexter, T. M., 1994, Total marrow failure induced by pegylated stem cell factor administered before 5-fluorouracil, *Blood* **83:**3491–3499.

Monpezat, J. P., and Frindel, E., 1989, Further studies on the biological activities of the CFU-S inhibitory tetrapeptide AcSDKP on GM-CFC and on the possible involvement of T lymphocytes in AcSDKP response, *Exp. Hematol.* **17:**1077–1080.

Negrin, R. S., Haeuber, D. H., and Nagler, A., 1989, Treatment of myelodysplastic syndromes with recombinant human granulocyte colony-stimulating factor. A phase I–II trial, *Ann. Inter. Med.* **110:**976–984.

Negrin, R. S., Haeuber, D. H., and Nagler, A., 1990, Maintenance treatment of patients with myelodysplastic syndromes using recombinant human granulocyte colony-stimulating factor, *Blood* **76:**36–43.

Neidhart, J., Mangalik, A., and Kohler, W., 1989, Granulocyte colony-stimulating factor stimulates recovery of granulocytes in patients receiving dose-intensive chemotherapy without bone marrow transplantation, *J. Clin. Oncol.* **7:**1685–1692.

Neidhart, J., Stidley, G., and Ferguson, J., 1990, GM-CSF decreases duration of cytopenia in patients receiving dose intensive therapy with cyclophosphamide, etoposide and cisplatin, *Proc. Am. Soc. Clin. Oncol.* **9:**753.

Nemunaitis, J., Singer, J. W., and Buckner, C. D., 1988, Use of recombinant human granulocyte–macrophage colony-stimulating factor in autologous marrow transplantation for lymphoid malignancies, *Blood* **72**(2)**:** 834–836.

Nemunaitis, J., Buckner, C. J., Appelbaum, F. R., Higano, C. S., Mor, M., Bianco, J., Epstein, C., Lipani, J., Hansen, J., Storb, R., Thomas, E. D., and Singer, J. W., 1991a, Phase I/II trial of recombinant human granulocyte–macrophage colony-stimulating factor following allogeneic bone marrow transplantation, *Blood* **77:**2065–2071.

Nemunaitis, J., Rabinowe, S. N., and Singer, J. W., 1991b, Recombinant human granulocyte–macrophage colony-stimulating factor after autologous bone marrow transplantation for lymphoid cancer, *N. Engl. J. Med.* **324**(24)**:**1773–1778.

Nemunaitis, J., Buckner, C. D., Appelbaum, F. R., Lilleby, K., Wolff, S., and Bierman, P., 1992, Phase I trial with recombinant interleukin-3 (IL-3) in patients with lymphoid cancer undergoing autologous bone marrow transplantation (ABMT), *Blood* **80:**85.

O'Dwyer, P. J., LaCreta, F. P., Schilder, R., Nash, S., McAleer, C., Miller, L. L., Hudes, G. R., and Ozols, R. F., 1992, Phase I trial of thiotepa in combination with recombinant human granulocyte–macrophage colony-stimulating factor, *J. Clin. Oncol.* **10:**1352–1358.

Ohno, R., Tomonaga, M., and Kobayashi, T., 1990, Effect of granulocyte colony-stimulating factor after intensive induction therapy in relapsed or refractory acute leukemia: A randomized controlled study, *N. Engl. J. Med.* **323:**871–877.

Ossenkoppele, G. J., Jonhoff, A. R., and Huijgens, P. C., 1993, Peripheral blood progenitors mobilised by G-CSF (filgrastim) and reinfused as unprocessed autologous whole blood shorten the pancytopenic period following high-dose melphalan in multiple myeloma, *Bone Marrow Transplant* **11:**103–108.

Ottmann, O. G., Ganser, A., Seipelt, G., Eder, M., Schulz, G., and Hoelzer, D., 1990, Effects of human recombinant human interleukin-3 on human hematopoietic progenitor and precursor cells *in vivo*. *Blood* **76:**1494–1502.

Pene, R., Appelbaum, F. R., Fisher, L., Lilleby, K., Nemunaitis, J., Storb, R., and Buckner, C. D., 1993, Use of granulocyte–macrophage colony-stimulating factor and erythropoietin in combination after autologous marrow transplantation, *Bone Marrow Transplant* **11:**219–222.

Peters, W. P., 1989, The effect of recombinant human colony-stimulating factors hemopoietic reconstitution following autologous bone marrow transplantation, *Semin. Hematol.* **26**(Suppl. 2)**:**18–23.

Peters, W. P., Kurtzberg, J., and Atwater, S., 1989, Comparative effects of r-HuG-CSF and r-HuGM-CSF on haemopoietic reconstitution and granulocyte function following high dose chemotherapy and autologous bone marrow transplantation(ABMT), *Proc. Am. Soc. Clin. Oncol.* **8:**702.

Peters, W. P., Hussein, A. M., Kurtzberg, J., Ross, M., Vredenburgh, J., Gilbert, C., and Coniglio, D., 1991, Use of recombinant human interleukin-3 (IL-3) in patients (PTS) with metastatic breast cancer receiving high dose chemotherapy (HDC) and chemo-immunologically purged autologous bone marrow transplantation, *Blood* **78:**162.

Pettengell, R., Gurney, H., Radford, J. A., Deakin, D. P., James, R., Wilkinson, P. M., Testa, N. G., and Crowther, D., 1992, Granulocyte colony-stimulating factor to prevent dose limiting neutropenia in non-Hodgkin's lymphoma; a randomised controlled trial, *Blood* **80:**1430–1436.

Pettengell, R., Morgenstern, G. R., Woll, P. J., Chang, J., Rowlands, M., Young, R., Radford, J. A., Scarffe, J. H., Testa, N. G., and Crowther, D., 1993, Peripheral blood progenitor cell transplantation in lymphoma and leukemia using a single apheresis, *Blood* **82**(12)**:**3770–3777.

Pettengell, R., Woll, P. J., Thatcher, N., Dexter, T. M., and Testa, N. G., 1995, Multicyclic, dose-intensive chemotherapy supported by sequential reinfusion of haemopoietic progenitors in whole blood, *J. Clin. Oncol.* **13:**148–156.

Piccart, M., van der Schueren, E., and Bruning, P., 1991, High dose-intensity (DI) chemotherapy (CT) with epi-adriamycin (E), cyclophosphamide (C) and r-metHuG-CSF (AMGEN) in breast cancer (BC) patients, *Eur. J. Cancer* **27**(Suppl. 2)**:**S56 (Abstract 308).

Pizzo, P. A., Hathorn, J. W., and Hiemenz, J., 1986, A randomized trial comparing ceftazidime alone with combination antibiotic therapy in cancer patients with fever and neutropaenia, *N. Engl. J. Med.* **315:**552–558.

Pojda, Z., and Struzyna, J., 1994, Treatment of non-healing ulcers with rhGM-CSF and skin grafts, *Lancet* **343:**1100.

Powles, R., Smith, C., Milan, S., Treleaven, J., Millar, J., McElwain, T., Gordon-Smith, E., Milliken, S., and Tiley, C., 1990, Human recombinant GM-CSF in allogeneic bone marrow transplantation for leukaemia: Double-blind placebo controlled trial, *Lancet* **336:**1417–1420.

Rusthoven, J., Sheperd, F. A., and Goss, P. E., 1992, Phase 1 trial of GM-CSF with high dose carboplatin and etoposide for the treatment of small cell lung cancer: A study of the National Cancer Institute of Canada Clinical Trials Group (letter), *J. Natl. Cancer Inst.* **84:**59–60.

Rabinowe, S., Freedman, A., and Demetri, G., 1990, Randomized double blind trial of rhGM-CSF in patients with B-cell non-Hodgkin's lymphoma (B-NHL) undergoing high dose chemoradiotherapy and monoclonal anti-body purged autologous bone marrow transplantation (ABMT), *Blood* **76**(Suppl. 1)**:**161.

Reiffers, J., Bernard, P., and David, B., 1986, Successful autologous transplantation with peripheral blood hemopoietic cells in a patient with acute leukemia, *Exp. Hematol.* **14:**312–315.

Richman, C. M., Winer, R. S., and Yankee, R. A., 1976, Increase in circulating stem cells following chemotherapy in man, *Blood* **47:**1031–1039.

Russell, N. H., Hunter, A., Rogers, S., Hanley, J., and Anderson, D., 1993, Peripheral blood stem cells as an alternative to marrow for allogeneic transplantation, *Lancet* **341:**1482.

Schuster, N. U., Larson, R. A., Thompson, J. A., Coiffier, B., Bennett, J. M., and Israel, R. J., 1990, Granulocyte–macrophage colony-stimulating factor (GM-CSF) for myelodysplastic syndrome (MDS); results of a multicentre randomised trial, *Blood* **76**(Suppl.)**:**1263a.

Schwartzberg, L., West, W., Birch, R., and Blanco, R., 1992, G-CSF improves peripheral blood stem cells (PBSC) mobilization and hematologic reconstitution after high-dose chemotherapy. *Int. J. Cell Cloning* **10**(Suppl. 1)**:**193.

Schwinger, W., Mache, C., Urban, C., Beaufort, F., and Toglhofer, W., 1993, Single dose of filgrastim (rhG-CSF) increases the number of hematopoietic progenitors in the peripheral blood of adult volunteers, *Bone Marrow Transplant.* **11:**489–492.

Shea, T. C., Mason, J. R., Newton, B., Storniolo, A. M., Taetle, R., and Green, M. R., 1990, High-dose carboplatin (CBDCA) with recombinant human granulocyte–macrophage colony stimulating factor (rHu-GM-CSF): A phase 1 study, *Proc. Am. Soc. Clin. Oncol.* **9:**286 (Abstract 1109).

Sheridan, W. P., Morstyn, G., and Wolf, M., 1989, Granulocyte colony-stimulating factor and neutrophil recovery after high dose chemotherapy and autologous bone marrow transplantation, *Lancet* **2:**890–895.

Sheridan, W. P., Begley, C. G., and Juttner, C. A., 1992, Effect of peripheral-blood progenitor cells mobilised by filgrastim (G-CSF) on platelet recovery after high-dose chemotherapy, *Lancet* **339:**640–644.

Singer, J. W., Nemunaitis, J., and Bianco, J. C., 1990a, RhGM-CSF following allogeneic bone marrow transplantation from unrelated marrow donors: A phase II study, *Blood* **76**(Suppl. 1):566.

Singer, J. W., Nemunaitis, J., and Buckner, C. D., 1990b, Phase I/II trial of rhGM-CSF after allogeneic bone marrow transplantation (BMT), *J. Cell Biochem.* (Suppl.) **14a:**333.

Skipper, H. E., 1967, Criteria associated with destruction of leukaemia and solid tumour cells in animals, *Cancer Res.* **27**(1)**:**2636–2645.

Socinski, M. A., Cannistra, S. A., Elias, A., Antman, K. H., Schnipper, L., and Griffin, J. D., 1988, Granulocyte–macrophage colony-stimulating factor expands the circulating hemopoietic progenitor cell compartment in man, *Lancet* **1:**1194–1198.

Spitzer, G., Mathews, M., Dunphy, F., Bowers, C., Petruska, P., Velasquez, W., Broun, G., Kronmueller, N., Niemeyer, R., McIntyre, W., and Adkins, D., 1994b, Are growth factor (GF) combinations superior to G-CSF alone for mobilization of peripheral blood stem cells (PBSC): A randomized study comparing G-CSF with GM-CSF for PBSC mobilization, *Blood* **84**(Suppl. 1)**:**415.

Steward, W. P., Scarffe, J. H., Dirix, L. Y., Chang, J., Radford, J. A., Bonnem, E., and Crowther, D., 1990, Granulocyte–macrophage colony-stimulating factor (GM-CSF) after high-dose melphalan in patients with advanced colon cancer, *Br. J. Cancer* **61:**749–954.

Stiff, P. J., Koester, A. R., and Weidner, M. K., 1987, Autologous bone marrow transplantation using unfractionated cells cryopreserved in dimethylsulfoxide and hydroxyethyl starch without controlled rate freezing, *Blood* **70:** 974–978.

Stoppa, A. M., Blaise, D., and Viens, P., 1984, Phase I study of *in vivo* lenograstim (rHuG-CSF) for stem cell collection demonstrates improved neutrophil recovery after autologous bone marrow transplantation, *Bone Marrow Transplant.* **13:**541–547.

Sutherland, H. J., Eaves, C. J., Lansdorp, P. M., Phillips, G. L., and Hogge, D. E., 1994, Kinetics of committed and primitive blood progenitor mobilization after chemotherapy and growth factor treatment and their use in autotransplants, *Blood* **83:**3808–3814.

Taylor, K. M., Jagannath, S., and Spitzer, G., 1989, Recombinant human granulocyte colony-stimulating factor hastens granulocytic recovery after high-dose chemotherapy and autologous bone marrow transplantation, *J. Clin. Oncol.* **7**(12)**:**1791–1799.

Tejero, C., Testa, N. G., and Lord, B. I., 1984, The cellular specificity of haemopoietic stem cell proliferation regulators, *Leuk. Res.* **8:**1073–1083.

Teshima, H., Ishikawa, J., and Kitawama, H., 1989, Clinical effects of recombinant human granulocyte colony-stimulating factor in leukemia patients: A phase I/II study, *Exp. Hematol.* **17:**853–858.

Thompson, J. A., Lee, D. J., Kidd, P., Rubin, E., Kaufmann, J., Bonnem, E. M., and Fefer, A., 1989, Subcutaneous granulocyte–macrophage colony-stimulating factor in patients with myelodysplastic syndrome: Toxicity, pharmacokinetics and hematological effects, *J. Clin. Oncol.* **7:**629–637.

To, L. B., Dyson, P. G., and Branford, A., 1987, Peripheral blood stem cells collected in very early remission produce rapid and sustained autologous hematopoietic reconstitution in acute non-lymphoblastic leukaemia, *Bone Marrow Transplant.* **2:**103–108.

Trillet-Lenoir, V., Green, J., and Manegold, C., 1993, Recombinant granulocyte colony-stimulating factor reduces the infectious complications of cytotoxic chemotherapy, *Eur. J. Cancer* **29A:**319–324.

Usuki, K., Iki, S., Endo, M., Ito, K., Kitasume, K., Watanabe, M., and Urabe, A., 1994, Granulocyte colony-stimulating factor in acute myeloid leukemia, *Blood* **84**(Suppl. 1)**:**86.

Vadhan-Raj, S., Keating, M., LeMaistre, A., Hittelman, W. N., McCredie, K., Trujillo, J. M., Broxmeyer, H. E., Henney, C., and Gutterman, J. U., 1987, effects of recombinant human granulocyte–macrophage colony-stimulating factor in patients with myelodysplastic syndromes, *N. Engl. J. Med.* **317:**1545–1552.

Vadhan-Raj, S., Broxmeyer, H. E., and Spitzer, G., 1989, Stimulation of non-clonal hematopoiesis and suppression of the neoplastic clone after treatment with recombinant human granulocyte–macrophage colony-stimulating factor in a patient with therapy-related myelodysplastic syndrome, *Blood* **74:**1491.

Vadhan-Raj, S., Broxmeyer, H. E., Hittleman, W. N., Papadopoulos, N. E., Chawla, S. P., Fenoglio, C., and Cooper, S., 1992, Abrogating chemotherapy induced myelosuppression by recombinant granulocyte–macrophage colony-stimulating factor in patients with sarcoma: Protection at the progenitor cell level, *J. Clin. Oncol.* **10:** 1266–1277.

Venturini, M., Sertoli, M. R., and Ardizzoni, A., 1992, Prospective randomized trial of accelerated FEC chemotherapy with or without GM-CSF in advanced breast carcinoma, *Proc. Am. Soc. Clin. Oncol.* **11:**36.

Villeval, J.-L., Duhrsen, U., Morstyn, G., and Metcalf, D., 1990, Effect of recombinant human granulocyte–macrophage colony-stimulating factor on progenitor cells in patients with advanced malignancies, *Br. J. Haematol.* **74:**36–44.

Wdzieczak-Bakala, J., Guigon, M., Lenfant, M., and Frindel, E., 1983, Further purification of a CFU-S inhibitor: *In vivo* effects after cytosine arabinoside treatment, *Biomed. Pharmacother.* **37:**467–471.

Wdzieczak-Bakala, J., Fache, M. P., Lenfant, M., Frindel, E., and Sainteny, F., 1990, AcDSKP, an inhibitor of CFU-S proliferation, is synthesized in mice under steady state conditions and secreted by bone marrow in long term culture, *Leukemia* **4:**235–237.

Weaver, A., Wrigley, E., Jenkins, B., Matcham, J., Ewen, C., Dexter, T. M., Testa, N. G., and Crowther, D., 1995, Effects of filgrastim (G-CSF) alone or in combination with chemotherapy on the mobilisation of peripheral blood progenitor cells: Early results from a pilot study of high dose chemotherapy in ovarian cancer, in *Proceedings of International Conference on Recent Advances in Hematopoietic Stem Cell Transplantation—Clinical Progress, New Technologies and Gene Therapy*, San Diego, Abstract 16, p. 379.

Wiley, J., Cebon, J., Woodruff, R., Jamieson, G., Snook, M., McKendrick, J., Szer, J., Dennington, P., Sheridan, W., Begley, G., van der Weyden, M., and Fox, R., 1991, Granulocyte–macrophage colony-stimulating factor followed by chemotherapy in the treatment of poor prognosis acute myeloid leukemia, *Blood* **78**(suppl)**:**42a.

Yoshida, Y., Hirashima, K., Asano, S., and Takaku, F., 1991, A phase II trial of recombinant human granulocyte colony-stimulating factor in the myelodysplastic syndromes, *Br. J. Haematol.* **78:**378–384.

Zeller, W., Cassens, U., Stockschlader, M., Kruger, W., Kabisch, H., Kuhnl, P., Hossfeld, D. K., and Zander, A. R., 1994, Higher dose of G-CSF increases yield of mobilized $CD34^+$ cells, *Blood* **84**(Suppl. 1)**:**413.

Index

GPSR Compliance

The European Union's (EU) General Product Safety Regulation (GPSR) is a set of rules that requires consumer products to be safe and our obligations to ensure this.

If you have any concerns about our products, you can contact us on ProductSafety@springernature.com

In case Publisher is established outside the EU, the EU authorized representative is:

Springer Nature Customer Service Center GmbH
Europaplatz 3
69115 Heidelberg, Germany

Batch number: 10762209

Printed by Printforce, the Netherlands